A. V. Narlikar

High Temperature Superconductivity 1

Springer

Berlin
Heidelberg
New York
Hong Kong
London
Milan
Paris
Tokyo

http://www.springer.de/engine/

A. V. Narlikar (Ed.)

High Temperature Superconductivity 1

With 398 Figures and 37 Tables

 Springer

Prof. Dr. Anant V. Narlikar

Inter-University Consortium
for DAE Facilities
University Campus
Khandwa Road
Indore 452 017 (MP), India

ISBN 3-540-40631-X Springer-Verlag Berlin Heidelberg New York

Cataloging-in-Publication Data applied for.
Bibliographic information published by Die Deutsche Bibliothek. Die Deutsche Bibliothek lists this publication in the Deutsche Nationalbibliografie; detailed bibliographic data is available in the Internet at <http://dnb.ddb.de>.

Springer-Verlag Berlin Heidelberg New York
a member of BertelsmannSpringer Science+Business Media GmbH

http://www.springer.de

Typesetting: Dataconversion by author
Cover-design: medio, Berlin
Printed on acid-free paper   62 / 3020 hu – 5 4 3 2 1 0

CONTRIBUTORS

ADACHI, S.
Superconductivity Research Laboratory
International Superconductivity Technology Center
1-10-13 Shinonome, Koto-ku
Tokyo 135-0062, JAPAN

BEAUQUIS, Sandrine
Laboratoire des Matériaux et du Génie Physique
LMGP-ENSPG-INPG, CNRS-UMR5628, BP 46
F-38402 Saint Martin d'Hères, FRANCE

DOUINE, Bruno
GREEN Faculté des Sciences
BP 239 F-54506 Vandoeuvre Cedex, FRANCE

FENG, Y.
Northwest Institute for Nonferrous Metal Research,
P.O.Box 51, Xi'an,710016 CHINA, and
Superconductivity Research Laboratory
International Superconductivity Technology Center
1-10-13 Shinonome, 1-chome
Koto-ku, Tokyo 135, JAPAN

FLUKIGER, R.
Group of Applied Physics,
University of Geneva
CH-1211 Geneva, SWITZERLAND

FRAGALA, Ignazio L.
Dipartimento di Scienze Chimiche

Universita' di Catania, and I.N.S.T.M.
UdR di Catania, Viale Andrea Doria 6
I-95125 Catania, ITALY

GALEZ, Philippe
Laboratoire d'Instrumentation et des Materiaux d'Annecy (LAIMAN),
ESIA, Université de Savoie, Domaine Universitaire
B.P.808, F-74016- Annecy Cedex, FRANCE

GLOWACKI, Bartek A.
Department of Materials Science and Metallurgy
University of Cambridge, Pembroke Street
Cambridge CB2 3QZ, ENGLAND, and
IRC in Superconductivity, Cavendish Laboratory
University of Cambridge, Madingley Road, West Cambridg Site
Cambridge CB3 0HE, ENGLAND

GOYAL, A.
Metals & Ceramics Division,
Oak Ridge National Laboratory
Oak Ridge, TN 37831, USA

HAYAKAWA, Naoki
Department of Electrical Engineering,
Nagoya University
Furo-cho, Chikusa-ku
Nagoya 464-8603, JAPAN

HOPFINGER, Thomas
Laboratoire d'Instrumentation et des Materiaux d'Annecy (LAIMAN)
ESIA Université de Savoie,
Domaine Universitaire
B.P.808, F-74016- Annecy Cedex, FRANCE

HOTT, Roland
Forschungszentrum Karlsruhe,
Institut für Festkörperphysik,
P.O. Box 3640, D-76021 Karlsruhe, GERMANY

IKUTA, Hiroshi
Center for Integrated Research in Science and Engineering
Nagoya University,
Furo-cho, Chikusa-ku
Nagoya 464-8603, JAPAN

INOUE, N.
Superconductivity Research Laboratory,
International Superconductivity Technology Center
1-10-13 Shinonome
Koto-ku, Tokyo 135-0062, JAPAN

JIMENEZ, Carmen
Laboratoire des Matériaux et du Génie Physique
LMGP-ENSPG-INPG, CNRS-UMR5628, BP 46
F-38402 Saint Martin d'Hères, FRANCE

JONDO, Timothé Koffi

Laboratoire d'Instrumentation et des Materiaux d'Annecy (LAIMAN)
ESIA, Université de Savoie,
Domaine Universitaire
B.P.808, F-74016- Annecy Cedex, FRANCE, and
Université du Benin, Lomé, Togo, FRANCE

JORDA, Jean-Louis
Laboratoire d'Instrumentation et des Materiaux d'Annecy (LAIMAN)
ESIA, Université de Savoie,
Domaine Universitaire
B.P.808, F-74016- Annecy Cedex, FRANCE, and

VIII

Institut National Polytechnique de Grenoble
LMGP, F-38402 – Saint Martin d'Hères Cedex, FRANCE

KOSHIZUKA, N.
Superconductivity Research Laboratory
International Superconductivity Technology Center
1-10-13 Shinonome, 1-chome
Koto-ku, Tokyo 135 JAPAN

KWASNITZA, K.
Group of Applied Physics,
University of Geneva
CH-1211 Geneva, SWITZERLAND

LEVEQUE, Jean
GREEN Faculté des Sciences
BP 239 F-54506 Vandoeuvre Cedex, FRANCE

MALANDRINO, Graziella
Dipartimento di Scienze Chimiche,
Universita' di Catania, and I.N.S.T.M. UdR di Catania, Viale Andrea
Doria 6
I-95125 Catania, ITALY

MORIWAKI, Y.
Superconductivity Research Laboratory,
International Superconductivity Technology Center
1-10-13 Shinonome
Koto-ku, Tokyo 135-0062, JAPAN

MURAKAMI, M.
Superconductivity Research Laboratory,
International Superconductivity Technology Center
1-10-13 Shinonome, 1-chome

Koto-ku, Tokyo 135 JAPAN

NETTER, Denis
GREEN Faculté des Sciences
BP 239 F-54506 Vandoeuvre Cedex, FRANCE

OGAWA, A.
Superconductivity Research Laboratory,
International Superconductivity Technology Center
1-10-13 Shinonome
Koto-ku, Tokyo 135-0062, JAPAN

OKUBO, Hitoshi
Department of Electrical Engineering,
Nagoya University
Furo-cho, Chikusa-ku,
Nagoya 464-8603, JAPAN

PHOK, Sovannary
Institut National Polytechnique de Grenoble, LMGP,
F-38402 – Saint Martin d'Hères Cedex, FRANCE

PRADHAN, A. K.
Superconductivity Research Laboratory
International Superconductivity Technology Center
1-10-13 Shinonome, 1-chome
Koto-ku, Tokyo 135, JAPAN

RUTTER, N. A.
Metals & Ceramics Division,
Oak Ridge National Laboratory
Oak Ridge, TN 37831, USA

X

SHIOHARA, Yuh

Superconductivity Research Laboratory
International Superconductivity Technology Center
2-4-1 Mutsuno, Atsuta-ku
Nagoya 456-8587, JAPAN

_SU, X.D._

Group of Applied Physics,
University of Geneva
CH-1211 Geneva, SWITZERLAND

SUGANO, T.

Superconductivity Research Laboratory,
International Superconductivity Technology Center
1-10-13 Shinonome
Koto-ku, Tokyo 135-0062, JAPAN

TAKAGI, K.

Advanced Research Laboratory, Hitachi, Ltd.
Kokubunji, Tokyo 185-8601, JAPAN

TANABE, K.

Superconductivity Research Laboratory
International Superconductivity Technology Center
1-10-13 Shinonome,
Koto-ku, Tokyo 135-0062, JAPAN

TSUKAMOTO, A.

Advanced Research Laboratory, Hitachi, Ltd.
Kokubunji, Tokyo 185-8601, JAPAN

WEISS, François

Laboratoire des Matériaux et du Génie Physique,
LMGP-ENSPG-INPG, CNRS-UMR5628, BP 46
F-38402 Saint Martin d'Hères, FRANCE

WITZ, G.
Group of Applied Physics,
University of Geneva,
CH-1211 Geneva, SWITZERLAND

WU, X. -J.
Superconductivity Research Laboratory
International Superconductivity Technology Center
1-10-13 Shinonome,
Koto-ku, Tokyo 135-0062, JAPAN

YAMADA, Yutaka,
Superconductivity Research Laboratory
International Superconductivity Technology Center
2-4-1 Mutsuno, Atsuta-ku
Nagoya 456-8587, JAPAN

YAN, G.
Northwest Institute for Nonferrous Metal Research, P.O.Box 51
Xi'an, 710016 CHINA

ZHAO, Y.
School of Materials Sciences and Engineering,
University of New South Wales, Sydney, 2052, NSW, AUSTRALIA, and
Superconductivity Research Laboratory
International Superconductivity Technology Center
1-10-13 Shinonome, 1-chome
Koto-ku, Tokyo 135 JAPAN

ZHOU, L.
Northwest Institute for Nonferrous Metal Research
P.O.Box 51, Xi'an, 710016 CHINA

PREFACE

The discovery of high temperature superconductors (HTS) in 1986 brought in its wings a new set of puzzles, but also exciting promises and problems of a multidisciplinary nature. Popular imagination was enraptured with the novel possibilities and expectations from new materials that could potentially revolutionise the very basis of modern existence. The central challenge was how to exploit the new materials in the form of bulk samples, wires, tapes and thin films in varied electrical installations, appliances and electronic devices, all functioning in a superconducting way at liquid nitrogen temperature. The underlying problem was that these new materials were hard and brittle ceramics, and moreover many of their transport properties too were generally unfriendly for their conductor fabrication or making device structures. As a result of these new opportunities and challenges, a whole new inter-disciplinary agenda swung into action.

Today, even after over fifteen years of the discovery, the mechanism driving the superconductivity at high temperature, remains shrouded in mystery and controversy. But at the more practical level, recent years have seen a dramatic development in the fabrication of these materials and their applications. There is a need for an extended source of information on applied superconductivity focusing on HTS Materials and Engineering Applications. The information in research journals is generally not accessible in a consolidated form to bring out the actual overall status of the field. Here we have two books, the Book-1 *on* Materials *and the other,* the Book-2, *on* Engineering Applications, *both carrying the common title of* Advances in High Temperature Superconducting Materials and Engineering Applications. *The books take stock of the recent advancements in the state of art materials technology and engineering applications of HTS at the global level. The books carry extended original research papers and technical review articles written by physicists, chemists, materials scientists and engineers, all noted experts in the field from different countries. The interdisciplinary coverage of the volumes should interest and benefit graduate students, researchers and specialists from aforesaid academic faculties and also from the electrical and electronic industries. At the same time, the topical reviews, written by experienced researchers, should interest the undergraduate and the non-specialist reader. Wherever possible, the necessary attention has also been paid to actual future applications in a digestive way and tasks for future work have been elaborated. Every care has been taken to ensure that the prime focus of the two volumes remains firmly anchored to the theme of interrelation between materials fabrication and engineering applications of HTS and there is no digression to various other superconductivity areas not directly relevant to applications issues. A major part of literature presently available on applied superconductivity, mostly as conference proceedings and a few edited books, seems to significantly lack this focus. These two books together are perhaps the first attempt, to bring out a comprehensive status of simultaneous advancements in materials engineering and engineering applications of HTS in a rapidly changing scenario of applied superconductivity.*

Book-1 comprises 13 chapters focusing on major technical advancement in HTS materials processing for applications. At the start there is an interesting overview of HTS for applications. Advances in bulk materials, wires, tapes, conductors in varied forms and thin films of all the technically important HTS systems, e.g., YBCO, BSCCO, Hg- and Tl-based cuprates, MgB_2, are presented in relation to their various applications. The relevant areas like microstructures, flux *pinning, A.C. losses etc., having a direct bearing on applications, have received an adequate coverage in various chapters. A successful exploitation of HTS in superconducting power engineering installations often demands compatible high quality insulating materials which have to withstand very large magneto-mechanical stresses at cryogenic temperatures. There is an extended paper in this book, covering this important area, interfacing HTS with practical engineering applications.*

The book on applications, i.e., Book-2, carries 16 chapters covering numerous application areas where HTS is making a frontal impact. As with the other book, the two initial chapters highlight and consolidate some prominent happenings in HTS applications. Advances in SQUIDS, SQUID applications, SQUID microscopy, microwave filters, thin film devices, bolometers, HTS magnets – both electromagnets as well as permanent magnets, magnetic energy storage, levitation applications, power cables, power transformers, a.c. machines, motors, fault current limiters etc., form the mainstay of this book.

Contents of the two books together, I feel, make an unbelievable success story for applied superconductivity. It is astonishing to see such a spectacular progress occurring within a short span of 15 years. Formidable challenges originally perceived by HTS materials have been effectively met and the work has gone a long way in substantially fulfilling the initial promises and expectations for their exciting engineering applications. Although the choice of materials systems available for applications seems restricted, there does not seem any real need of a fresh discovery of a new HTS material, unless that happens to be the room temperature superconductor ! With remarkable progress achieved in understanding the complex phase relationships in existing cuprates, ingenious ways of their processing and dramatic improvements in their properties through dopings and additives, it has now been possible to produce these materials in wires, tapes, thin films or bulk form to suit the desired application requirements. Extraordinary gains of HTS technology in respect of several applications listed above are presented in chapters of Book-2. HTS microwave filters and HTS SQUIDS seem to emerge as niche market areas for HTS electronics. Similarly, the fault current limiters, power cables, a.c. machines, motors etc. have attended a competitive marketable status. It is important for the industries to take stock of these developments for their possible large scale commercialisation. Most electronic industries today, however, seem to look askance at cooling to cryogenic temperatures and want to confine cooling at best to the level of a fan unless the performance advantages are immense. The latter kind of situation is expected to hold for HTS technology where the liquid nitrogen temperatures can be realised with remarkable ease. A changeover to any new technology invariably carries higher initial costs that are generally outweighed by long term gains. This should happen also with the HTS technology. The present costs of HTS cables and conductors and also of HTS SQUID elements, for example, are high, but they should go down with time as the materials further improve and their demand rises.

I am thankful to more than 70 contributors from 12 countries for their most commendable efforts in preparing and timely sending their contributions. I am grateful to Roland Hott for making several interesting suggestions and sending various useful information that helped me to organize them. Thanks are further extended to IUC-DAEF, Indore for providing facilities and the required infrastructure for doing this work.

May, 2003   *Anant V. Narlikar*

CONTENTS

**MATERIALS ASPECTS OF HIGH TEMPERATURE SUPERCONDUCTORS
FOR APPLICATIONS**

Roland HOTT

FASCINATING THALLIUM-BASED SUPERCONDUCTING CUPRATES AND SUBSYSTEMS : FORMATION AND STABILITY

Jean-Louis JORDA, Philippe GALEZ, Sovannary PHOK, Thomas HOPFINGER, and Timothé Koffi JONDO

MELT PROCESSED RE-Ba-Cu-O BULK SUPERCONDUCTORS

Hiroshi IKUTA

COATED CONDUCTORS AND HTS MATERIALS BY CHEMICAL DEPOSITION PROCESSES

Sandrine BEAUQUIS, Carmen JIMENEZ and François WEISS

TlBaCaCuO SUPERCONDUCTING THIN FILMS BY AN EX-SITU MOCVD BASED APPROACH: FROM BaCaCuO(F) MATRICES TO DEVICES

Graziella MALANDRINO and Ignazio L. FRAGALA'.

Hg-BASED SUPERCONDUCTING THIN FILMS FOR DC SQUIDS BY IMPROVED FABRICATION PROCESS

N. INOUE, A. TSUKAMOTO, Y. MORIWAKI, T. SUGANO, X. –J. WU, A. OGAWA, S. ADACHI, K. TAKAGI and K. TANABE

ELECTRICAL INSULATION FOR SUPERCONDUCTING POWER APPARATUS

Naoki HAYAKAWA and Hitoshi OKUBO

MODELING CURRENT FLOW IN GRANULAR SUPERCONDUCTORS AND IMPLICATIONS FOR POTENTIAL APPLICATIONS

N. A. RUTTER and A. GOYAL

CRITICAL CURRENT DENSITY, FLUX PINNING AND MICROSTRUCTURE IN MgB_2 SUPERCONDUCTORS

Y. FENG , Y. ZHAO, G. YAN, A.K. PRADHAN, L. ZHOU, N. KOSHIZUKA and M. MURAKAMI

AC LOSSES UNDER SELF-FIELD IN A SUPERCONDUCTING TUBE

Jean LEVEQUE, Bruno DOUINE, Denis NETTER

MATERIALS ASPECTS OF
HIGH-TEMPERATURE SUPERCONDUCTORS
FOR APPLICATIONS

Roland HOTT

Forschungszentrum Karlsruhe, Institut für Festkörperphysik,
P. O. Box 3640, 76021 Karlsruhe, GERMANY

Email: roland.hott@ifp.fzk.de

1. Introduction

Sixteen years after their discovery [1], cuprate high-temperature superconductors (HTS) are still one of the topical centers of interest in physics. The mechanism of superconductivity is still under lively debate [2,3,4]. The description of their normal state properties [5,6] has turned out to represent an even bigger challenge to Solid State Physics theory. A major part of the progress in clarifying these questions is due to the improvement of sample quality [7,8,9]. This may be attributed in part to the efforts of Applied Superconductivity to create a materials basis for the multitude of applications, which were envisioned in the unprecedented scientific euphoria raised by the HTS discovery[1].

The present situation of HTS materials science resembles in many aspects the history of semiconductor materials development half a century ago [11]. However, in sharp contrast in particular to the case of silicon, HTS are *multi-element* compounds based on complicated sequences of *oxide layers* [12,13,14]. In addition to the impurity problem due to undesired additional elements which gave early semiconductor research a hard time in establishing reproducible materials properties, intrinsic local stochiometry defects arise in HTS from the

---

[1] It has been estimated that in the hottest days of this "Woodstock of physics" in 1987 [10] about one third of *all* physicists all over the world tried to contribute to the HTS topic. Meanwhile, the number of publications on HTS has increased to > 100 000.

insertion of cations in the wrong layer and defects of the oxygen sublattice. As additional requirement for the optimization of the superconducting properties, the oxygen content has to be adjusted in a compound-specific off-stochiometric ratio [15], but nevertheless with a spatially homogeneous microscopic distribution of the resulting oxygen vacancies or interstitials [16].

The discovery of High-Temperature Superconductivity has thus challenged research in a class of complicated compounds which otherwise would have been encountered on the classical research route of systematic investigation with gradual increase of materials complexity only in a far future. In the quest for better HTS materials quality, the very short superconducting coherence length of the order of the dimensions of the crystallographic unit cell did not allow easy progress. On the other hand, this experimental restriction leads to a close connection of macroscopically measurable superconducting properties with micro-structural details which permits a kind of atomic scale monitoring of the HTS materials quality. The huge efforts in HTS materials science have thus paved the way for the improved preparation of oxide materials in general, e. g. ferroelectric oxides [17] or manganites [18,19]. The broad spectrum of unusual physical properties of this new generation of "sophisticated oxides" has opened new perspectives for fundamental as well as for applied research [20,21].

Today, reproducible preparation techniques for a number of HTS material species are available which provide a first materials basis for applications. As a stroke of good fortune, the optimization of these materials with respect to their superconducting properties seems to be in accord with the efforts to improve their stability in technical environments in spite of the only metastable chemical nature of these substances under such conditions [22,23,24].

2. Atomic Structure and Classification

The structural element of HTS compounds related to the location of mobile charge carriers are stacks of a certain number $n = 1, 2, 3, ...$ of CuO_2 layers which are "glued" on top of each other by means of intermediate Ca layers (See Fig.1&2) [12,13,25,26]. Counterpart of these *active blocks* of $(CuO_2/Ca/)_{n-1}CuO_2$ stacks are *charge reservoir blocks* $EO/(AO_x)_m/EO$ with $m = 1, 2$ monolayers of a quite arbitrary oxide AO_x (A = Bi [13], Pb [25], Tl [13], Hg [13], Au [27], Cu [13], Ca [28], B [25], Al [25], Ga [25]; see Table 1)[2] "wrapped" on each side by a monolayer of alkaline earth oxide EO with E = Ba, Sr (see Fig. 1&2). The HTS structure results from alternating stacking of these two block units. The choice of BaO or SrO as "wrapping" layer is not arbitrary but depends on the involved AO_x since it has to provide a good spatial adjustment of the CuO_2 to the AO_x layers.

The general chemical formula[3] $A_mE_2Ca_{n-1}Cu_nO_{2n+m+2+y}$ (see Fig. 2) is conveniently abbreviated as **A-m2(n-1)n** [26] (e. g. $Bi_2Sr_2Ca_2Cu_3O_{10}$: Bi-2223) neglecting the indication of the alkaline earth element[4] (see Tab.1). The family of all $n = 1, 2, 3,...$ representatives with common AO_x will be referred here as "A-HTS", e. g. Bi-HTS.

The most prominent compound $YBa_2Cu_3O_7$ (see Fig. 1), the first HTS discovered with a critical temperature T_c for the onset of superconductivity above the boiling point of liquid

---

[2] HTS compounds based on "oxide mixtures" $A^{(1)}_s A^{(2)}_{1-s} O_x$ are neglected in this consideration.

[3] $y = m (x - 1)$ in the oxygen stoichiometry factor.

[4] A more precise terminology $A-m^{(E)}2(n-1)n$ has been suggested [26].

nitrogen [29], is traditionally abbreviated as **"YBCO"** or **"Y-123"** ($Y_1Ba_2Cu_3O_{7-\delta}$). It also fits into the general HTS classification scheme as a modification of Cu-1212 where Ca is completely substituted by Y. This substitution introduces extra charge next to the CuO_2 layers due to the higher valence of Y (+3) compared to Ca (+2). The HTS compounds $RBa_2Cu_3O_{7-\delta}$ ("RBCO", "R-123") where R can be La [13] or any rare earth element [30] except for Ce or Tb [31] can be regarded as a generalization of this substitution scheme. The lanthanide contraction of the R ions provides an experimental handle on the distance between the two CuO_2 layers of the active block of the doped Cu-1212 compound [32]. $Y_1Ba_2Cu_4O_8$ (**"Y-124"**) is the $m = 2$ counterpart Cu-2212 of YBCO. The **"214"** HTS compounds $E_2Cu_1O_4$ (s. Tab1), e. g. $La_{2-x}Sr_xCuO_4$ ("LSCO") or $Nd_{2-x}Ce_xCuO_4$ ("NCCO") are a bit exotic in this ordering scheme but may also be represented here as "0210" with $m = 0$, $n = 1$ and $E_2 = La_{2-x}Sr_x$ and $E_2 = Nd_{2-x}Ce_x$, respectively[5].

Further interesting chemical modifications of the basic HTS compositions are the introduction of fluorine [33,34] or chlorine [35] as more electronegative substituents of oxygen. This often results in a slight improvement of T_c: For Hg-1223, $T_c = 135$ K can thus be raised to 138 K [36], the highest T_c reported by now under normal pressure conditions (164 K at 30 GPa [37]). Another interesting modification which is realized for many HTS compounds are oxycarbonate HTS [38,39,40]. The inclusion of carbon into the HTS structure[6] with neutral or even slightly beneficial effect on the superconducting properties [40] is amazing since carbon impurities as secondary phases in HTS samples are usually very detrimental in this respect.

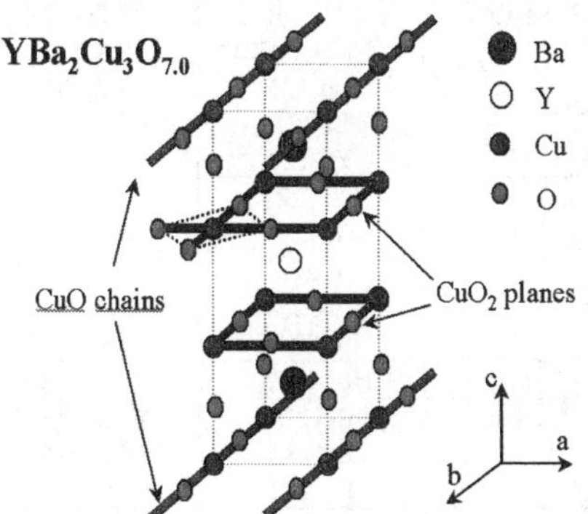

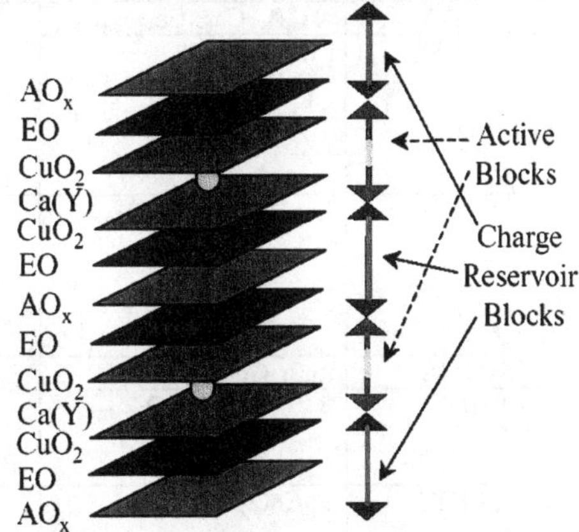

Fig. 1 Crystal structure of $YBa_2Cu_3O_7$ ("YBCO"). The presence of the CuO chains introduces an orthorhombic distortion of the unit cell (a = 0.382 nm, b = 0.389 nm, c = 1.167 nm [65]).

Fig. 2 General structure of a cuprate HTS A-m2(n-1)n ($A_mE_2Ca_{n-1}Cu_nO_{2n+m+2+y}$) for $m = 1$. For $m = 0$ (or 2) the missing (additional) AO_x layer per unit cell leads to a (a/2, b/2, 0)"side step" of the unit cells adjoining in c-axis direction.

---

[5] The missing BaO / SrO wrapping layer apparently leads to a much more sensitive behaviour of the superconducting properties with respect to structural disorder next to the CuO_2 layers.

[6] The carbon atoms are integrated as CO_3 units into the lattice structure of the AO_x layer.

HTS Family	Stochiometry	Notation	Compounds	Highest T_c	
Bi-HTS	$Bi_mSr_2Ca_{n-1}Cu_nO_{2n+m+2}$		Bi-1212	102 K	[41]
		Bi-m2(n-1)n,	Bi-2201	34 K	[42]
	$m = 1, 2$	BSCCO	Bi-2212	90 K	[43]
	$n = 1, 2, 3 \ldots$		Bi-2223	110 K	[13]
			Bi-2234	110 K	[44]
Pb-HTS	$Pb_mSr_2Ca_{n-1}Cu_nO_{2n+m+2}$	Pb-m2(n-1)n	Pb-1212	70 K	[45]
			Pb-1223	122 K	[46]
Tl-HTS	$Tl_mBa_2Ca_{n-1}Cu_nO_{2n+m+2}$		Tl-1201	50 K	[13]
	$m = 1, 2$	Tl-m2(n-1)n,	Tl-1212	82 K	[13]
	$n = 1, 2, 3 \ldots$	TBCCO	Tl-1223	133 K	[47]
			Tl-1234	127 K	[48]
			Tl-2201	90 K	[13]
			Tl-2212	110 K	[13]
			Tl-2223	128 K	[49]
			Tl-2234	119 K	[50]
Hg-HTS	$Hg_mBa_2Ca_{n-1}Cu_nO_{2n+m+2}$		Hg-1201	97 K	[13]
		Hg-m2(n-1)n,	Hg-1212	128 K	[13]
		HBCCO	Hg-1223	135 K	[51]
	$m = 1, 2$		Hg-1234	127 K	[51]
	$n = 1, 2, 3 \ldots$		Hg-1245	110 K	[51]
			Hg-1256	107 K	[51]
			Hg-2212	44 K	[52]
			Hg-2223	45 K	[53]
			Hg-2234	114 K	[53]
Au-HTS	$Au_mBa_2Ca_{n-1}Cu_nO_{2n+m+2}$	Au-m2(n-1)n	Au-1212	82 K	[27]
123-HTS	$RBa_2Cu_3O_{7-\delta}$	R-123, RBCO	Y-123, YBCO	92 K	[32]
	$R = $ Y, La, Pr, Nd, Sm,		Nd-123, NBCO	96 K	[32]
	Eu, Gd, Tb, Dy, Ho,		Gd-123	94 K	[54]
	Er, Tm, Yb, Lu		Er-123	92 K	[7]
			Yb-123	89 K	[30]
Cu-HTS	$Cu_mBa_2Ca_{n-1}Cu_nO_{2n+m+2}$	Cu-m2(n-1)n	Cu-1223	60 K	[13]
			Cu-1234	117 K	[55]
	$m = 1, 2$		Cu-2223	67 K	[13]
	$n = 1, 2, 3 \ldots$		Cu-2234	113 K	[13]
			Cu-2245	< 110 K	[13]
Ru-HTS	$RuSr_2GdCu_2O_8$	Ru-1212	Ru-1212	72 K	[56]
B-HTS	$B_mSr_2Ca_{n-1}Cu_nO_{2n+m+2}$	B-m2(n-1)n	B-1223	75 K	[57]
			B-1234	110 K	[57]
			B-1245	85 K	[57]
214-HTS	E_2CuO_4	LSCO	$La_{2-x}Sr_xCuO_4$	51 K	[58]
		"0201"	Sr_2CuO_4	25 (75)K	[59]
		Electron-Doped HTS	$La_{2-x}Ce_xCuO_4$	28 K	[60]
		PCCO	$Pr_{2-x}Ce_xCuO_4$	24 K	[61]
		NCCO	$Nd_{2-x}Ce_xCuO_4$	24 K	[61]
			$Sm_{2-x}Ce_xCuO_4$	22 K	[62]
			$Eu_{2-x}Ce_xCuO_4$	23 K	[62]
	$Ba_2Ca_{n-1}Cu_nO_{2n+2}$	"02(n-1)n"	"0212"	90K	[63]
			"0223"	120K	[63]
			"0234"	105K	[63]
			"0245"	90K	[63]
Infinite-Layer HTS	$ECuO_2$	*Electron-Doped I. L.*	$Sr_{1-x}La_xCuO_2$	43 K	[64]

Table 1 Classification and reported T_c values of HTS compounds.

3. Structural Obstacles for Supercurrents

The High-T_c euphoria in 1987 arrived soon at a first hangover after the first measurements of critical current densities: The J_c values of only several 100 A/cm^2 obtained in the best polycrystalline material available at that time were below the current density level of copper wires at room-temperature at negligible energy dissipation. The situation was even worse since even weak magnetic fields turned out to be sufficient to kill these low supercurrents. This did not look like a continuation of the success story of LTS materials where magnet applications represent nowadays by far the biggest market and provide the basis of a substantial wire industry. Meanwhile, this "weak-link behavior" of HTS has been analyzed in detail and can now be overcome by special processing techniques.

Electrical currents create magnetic fields which tend to suppress superconductivity. This prevented for a long time high-current applications of superconductors. A first step towards this goal was the discovery of type-II superconductors where the magnetic penetration depth λ is substantially longer than the superconducting coherence length ξ and allows therefore the penetration of magnetic fields into the superconducting bulk. This coexistence of magnetic fields and superconductivity leads to a substantial reduction of the loss of superconductive condensation energy that has to be paid for magnetic field penetration and enables the survival of superconductivity even in strong magnetic fields. Fortunately, HTS are extreme type-II superconductors with $\lambda > 100$ nm and $\xi \sim 1$ nm.

Superconductivity in HTS is believed to have its origin in the physics of the CuO$_2$ layers where the mobile charges are located[7] [2,3]. The superconductive coupling between these CuO$_2$ layers within a given (CuO$_2$/Ca/)$_{n-1}$CuO$_2$ stack ("*interlayer coupling*") is much weaker than the *intralayer coupling* within the CuO$_2$ layers, but still much stronger than the coupling between the (CuO$_2$/Ca/)$_{n-1}$CuO$_2$ stacks which can be described as Josephson coupling (see Fig. 3). This quasi-2-dimensional nature of superconductivity in HTS leads to a pronounced anisotropy of the superconducting properties with much higher supercurrents along the CuO$_2$ planes than in the perpendicular direction, a property which is not at all appreciated with respect to technical applications but can be compensated by additional engineering efforts.

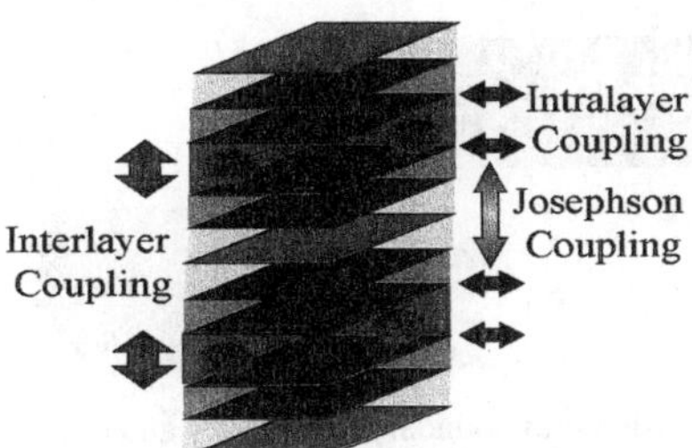

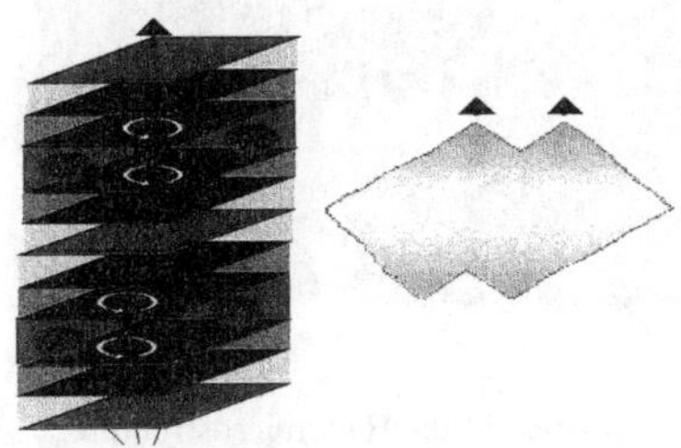

Fig. 3 Hierarchy of the superconducting coupling between the different structural elements of cuprate HTS.

Fig. 4 Quasi-disintegration of magnetic vortex lines into pancake vortices due to weak superconducting interlayer coupling and schematic overlap of neighboring vortices.

---

[7] In the present theoretical picture, the charge reservoir blocks play only a passive role in providing the doping charge as well as the storage space for extra oxygen ions and cations introduced on doping.

However, being type-II superconductor is not enough for a flow of strong currents without dissipation: The magnetic vortices that are introduced into the superconducting material by the magnetic field need to be fixed by *pinning centers,* or else dissipation by the *flux flow* of the vortices is induced. In "technical" superconductors, material imperfections of the dimension of the coherence length do this job by blocking superconductivity from these regions which provide then a natural "parking area" for vortex cores.

Material imperfections of the dimension of the coherence length are easily encountered in HTS due to their small coherence lengths, e. g. for YBCO values of ξ_{ab} = 1.6 nm, ξ_c = 0.3 nm for $T \to 0$ K [66] which are already comparable to the lattice parameters (YBCO: a = 0.382 nm, b = 0.389 nm, c = 1.167 nm [65]). However, the low ξ_c, i.e. the weak superconductive coupling between the $(CuO_2/Ca/)_{n-1}CuO_2$ stacks causes new problems. The thickness of the charge reservoir blocks $EO/(AO_x)_m/EO$ in-between these stacks is larger than ξ_c with the result that due to the low Cooper pair density vortices are here no longer well-defined (see Fig. 4).

This leads to a quasi-disintegration of the vortices into stacks of *"pancake vortices"* which are much more flexible entities than the continuous quasi-rigid vortex lines in conventional superconductors and therefore require individual pinning centers. The extent of this quasi-disintegration is different for the various HTS compounds since ξ_c is on the order of the thickness of a single oxide layers: Hence the number of layers in the charge reservoir blocks $EO/(AO_x)_m/EO$ makes a significant difference with respect to the pinning properties and thus to their supercurrents in magnetic fields. This is one of the reasons why YBCO ("Cu-**1**212") has a higher supercurrent capability in magnetic fields than the Bi-HTS Bi-**2**212 and Bi-**2**223 which for manufacturing reasons are still the most prominent HTS conductor materials[8].

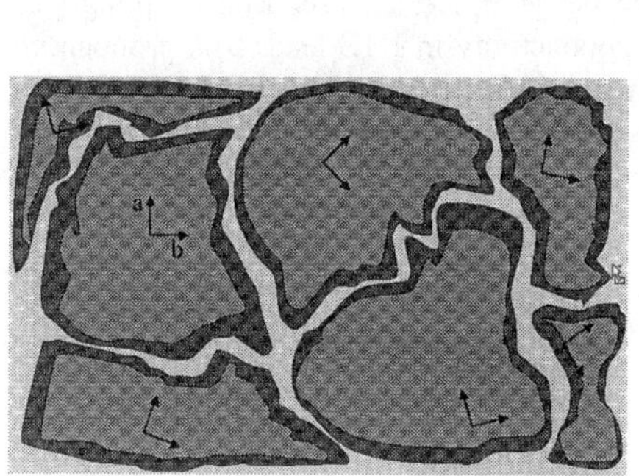

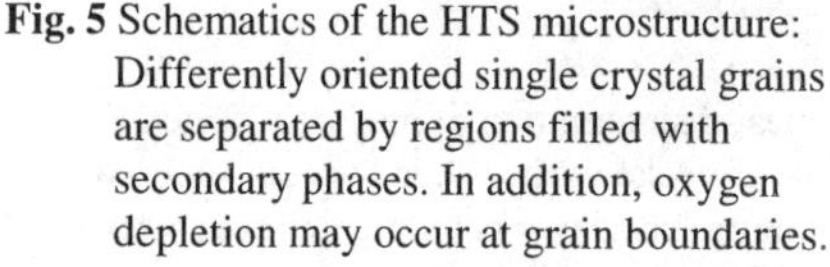

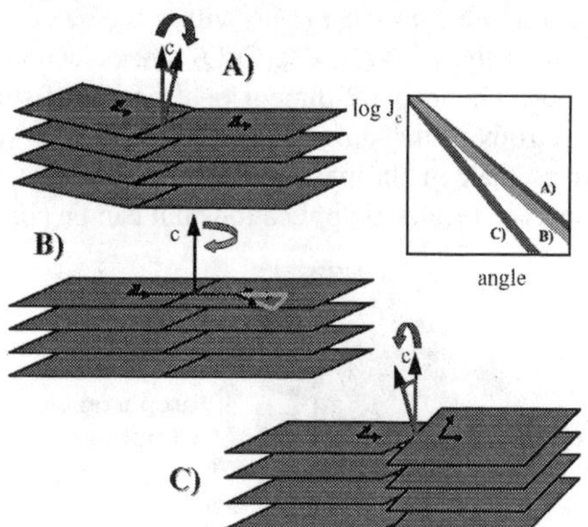

Fig. 5 Schematics of the HTS microstructure: Differently oriented single crystal grains are separated by regions filled with secondary phases. In addition, oxygen depletion may occur at grain boundaries.	**Fig. 6** Basic grain boundary geometries and experimentally observed J_c reduction $J_c \sim e^{-\alpha/\alpha_0}$ as function of the misalignment angle α: $\alpha_0 \approx 5°$ for A) and B), $\alpha_0 \approx 3°$ for C) independent of temperature [71,72].

---

[8] In addition, in the Cu-HTS family the AO_x layer is formed by CuO chain structures (see Fig. 1). There are indications that the CuO chains become superconducting via proximity effect. This leads to stronger Josephson coupling in c-axis direction and thus to the smallest superconductive anisotropy among all HTS families [67].

Beside these intrinsic obstacles for the transport of supercurrent in single-crystalline HTS materials there are additional hurdles since HTS materials are in general not a homogeneous continuum but rather a network of linked grains (see Fig. 5). The mechanism of crystal growth is such that all material that can not be fitted into the lattice structure of the growing grains is pushed forward into the growth front with the consequence that in the end all remnants of secondary phases and impurities are concentrated at the boundaries in-between the grains. Such barriers impede the current transport and have to be avoided by careful control of the growth process, in particular of the composition of the offered material.

Another obstacle for supercurrents in HTS is misalignment of the grains: Exponential degradation of the supercurrent transport is observed as a function of the misalignment angle (see Fig. 6). One of the reasons for this behavior is the d-symmetry of the superconducting order parameter (see Fig. 7) which meanwhile seems to be established for most of the HTS compounds [68]. The J_c reduction as a function of the misalignment angle α turns out to be much larger than what is expected from d-wave symmetry only [69,70,71]. This extra J_c degradation as well as the change of the current-voltage characteristics of the transport behavior[9] [75] are believed to arise from structural defects such as dislocations [76] and deviations from stoichiometry. In particular, the loss of oxygen at the grain surfaces [77] leads to a decrease of doping with respect to the grain bulk value and thus to a local degradation of the superconducting properties according to the temperature-doping phase diagram of HTS (see Fig. 8) [78]. Recently it has been demonstrated that for YBCO this effect can be partially eliminated by means of Ca-doping [79] which widens the range of acceptable grain misalignment in technical HTS material. In conclusion, manufacturing of technically applicable HTS materials requires well-defined preparation techniques to obtain chemically clean grains with small misorientation of their crystal axes.

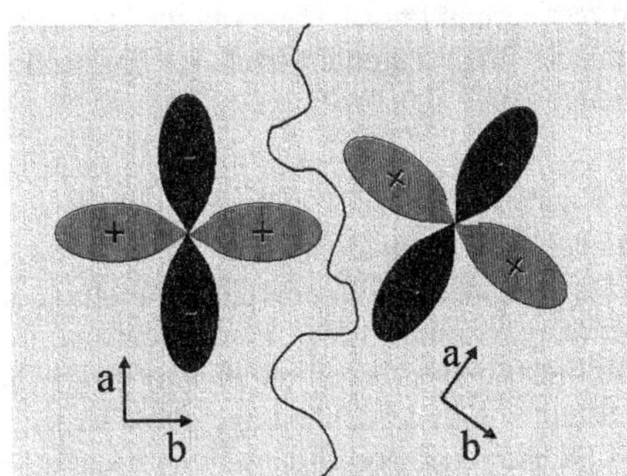

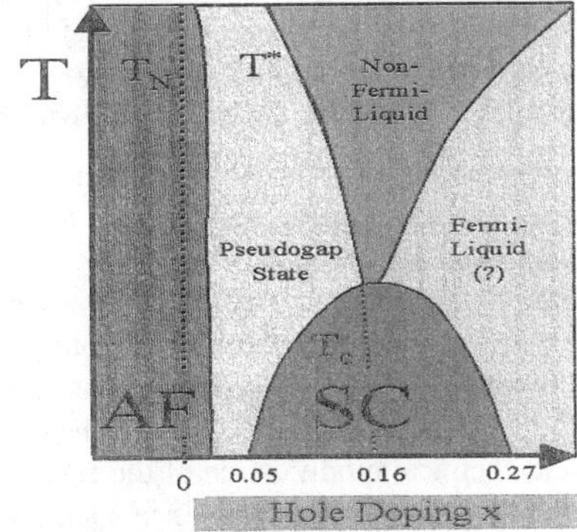

Fig. 7 Schematics of a boundary grain between HTS grain. Misorientation of the superconducting d-wave order parameter leads to partial cancellation of the supercurrents modified by the faceting of the grain boundaries.

Fig. 8 Schematic HTS temperature-doping phase diagram dominated by the interplay of antiferromagnetism ("AF") and superconductivity ("SC")[2,80,81].

---

[9] In addition to the quantitative effect on increasing misalignment angles of a reduction of the supercurrents, the current-voltage characteristics shows in addition a qualitative change from a flux-flow behavior for smaller misalignment angles [71,73] to overdamped Josephson junction behavior for larger misalignment angles [74].

4. Technically Applicable HTS Materials

The development of technically applicable HTS materials has progressed on several routes. Epitaxial HTS *thin films* achieve excellent superconducting properties ($T_c > 90$ K; critical current density J_c (77 K, 0 T) $> 10^6$ A/cm^2; microwave surface resistance R_s(77 K, 10 GHz) < 500 $\mu\Omega$, R_s(T,f) $\propto f^2$) that are well-suited for superconductive electronics. They are already in use in commercial and military microwave filter systems. HTS *Josephson junctions* based either on Josephson tunneling through ultra-thin artificial barriers or on the Josephson junction behavior of HTS grain boundaries have become available that can be used for the construction of highly sensitive magnetic field sensors ("Superconducting QUantum Interference Devices", "SQUIDs"). They are also tested for active electronic devices that can broaden the range of HTS thin-film applications. HTS single crystals are not suitable for applications due to their small size and their low J_c values as a consequence of a low density of pinning centers. However, the "quick and dirty" version *melt-textured* HTS *bulk material* shows superb magnetic pinning properties and are already applied as high-field permanent magnets e.g. in magnetic bearings and motors. In spite of the ceramic nature of the cuprate oxides, flexible HTS *wire* or *tape conductor* material is obtained either by embedding HTS as thin filaments in a silver matrix or by HTS coating of metal carrier tapes.

Among these technical HTS materials, the Bi-HTS conductor material represents the only exception to the rule that strong biaxial texture is necessary to achieve technically relevant currents[10]. One reason is the exceptional softness of the Bi-22(n-1)n HTS as a consequence of the weak Van-der-Waals bonding between the two neighboring BiO_x layers in the charge reservoir layers which allows alignment via mechanical processing steps like rolling or pressing. Another reason is the good mechanical and electrical contact with the Ag matrix due to similar melting temperatures which allows high current flow under the relaxed condition that a little detour via Ag is possible in case the direct current transfer between the HTS grains is blocked. This applies as well to "Ag-impregnated" Bi-2212 bulk which is commercially available in sizes of several 10 cm with homogeneous J_c (77 K,0 T) of several kA/cm^2 [82,83].

Much higher critical current densities have been achieved in polycrystalline HTS materials, but their use implies a higher risk for applications. In all these cases, substantial J_c reduction compared to the J_c (77 K,0 T) ~ 1 MA/cm^2 encountered in well-textured material with a sufficiently high amount of pinning centers always indicates a high degree of structural and electrical inhomogeneity with potentially devastating consequences: Forced current flow through defective materials regions may lead to a local quenching of superconductivity and the creation of "hot spots" [84]. The deposited quench energy leads then again to a structural and chemical modification of the HTS material in the neighborhood of these "hot spots". This enlarges the quench zone since such modifications are liable to spoil the conditions for good supercurrent transfer between neighboring HTS grains as discussed above. This spreading of the quench zone may finally end up in the destruction of the superconductor. This quench mechanism is different from the situation in classical superconductors where the quenching is caused by insufficient heat transfer due to the freezing of the phonon mechanism at LHe temperature. At the much higher operating temperature of HTS this is not such a critical issue since the heat distribution by phonons is here still very effective which brings about a comfortably high specific heat [85].

---

[10] Even though biaxial texture could be helpful in improving J_c by up to one order of magnitude.

5. Thin Films

YBCO [86,87] and Tl-HTS (Tl-2212) [88] thin films are nowadays produced in quantities of several thousand wafers per year for commercial and military microwave filters. RE-123 films (RE = Nd, Sm, Gd, Dy,....) are investigated as YBCO alternatives with slightly higher T_c values [89,90,91]. However, the required higher processing temperature increases the risk of cracks in the films [89]. Bi-HTS thin films are too soft for applications: High quality Bi-HTS films can be wiped away by cotton wool! Hg-HTS films are still restricted to small substrate sizes since fabrication in sealed quartz tubes is required in order to obtain good and stable films [92].

As for *YBCO*, *thermal coevaporation* has established as commercial *deposition method* [93, 94]. Substrates with sizes up to $\varnothing$ = 9" and 20 x 20 cm^2 can be coated on one side at a time at typical rates of 20 - 30 nm/min. Double-sided coating can be done in two successive deposition steps. Meanwhile, *pulsed laser deposition (PLD)* comes close to these possibilities [95]. *Magnetron sputtering* at deposition rates of up to 10 nm/min [91,96] is widely used in research laboratories. An extension to larger wafer sizes is still in preparation. *Molecular beam epitaxy (MBE)* has now also successfully demonstrated large area (DyBCO) deposition [97], but the much higher complexity of the deposition machines and lower deposition rates are intrinsic disadvantages. *Metalorganic chemical vapor deposition (MOCVD)* has the potential of deposition rates up to 1 μm/min [98], but the deposition of high quality YBCO films has only been reported for much lower rates. A special problem is the stability of the Ba precursor [99]. Recent quality improvement of YBCO thin films produced by *sol-gel* methods [100] is promising with respect to low cost production since the films can be prepared by fast non-vacuum deposition of precursor films with subsequent annealing to form the epitaxial films.

The higher T_c > 100 K of *Tl-2212* allows a higher operation temperature than YBCO. Precursor deposition followed by an annealing step in Tl$_2$O$_3$ vapor is the regular fabrication method due to the high volatility of Tl [88]. Recently, improvement of the microwave properties has been achieved, however, at the expense of similar tendencies towards crack formation for increasing film thickness as in YBCO [101].

Substrates have to provide a suitably lattice-matched crystal matrix to align the HTS grains in uniform orientation [94]. Most suitable for applications is the c-axis orientation of the HTS films where the CuO$_2$ planes are parallel to the substrate which preserves the isotropy of the electrical properties in the film plane. Fortunately, this film orientation is promoted by the faster growth of HTS films along the CuO$_2$ plane directions compared to the c-axis direction. Matching of the thermal expansion coefficients of the substrates and the HTS films is an additional requirement since the HTS deposition takes place at 650 – 900 °C. Different contraction of HTS film and substrate on cooling to room or even cryogenic temperature leads to mechanical stress that can be tolerated by the film only up to a certain maximum thickness without crack formation [102]. This limits the thickness of YBCO films on the technically most interesting substrates silicon and sapphire to ~ 50 nm [103] and ~ 250-300 nm [104], respectively. Furthermore, these substrates have to be buffered e.g. by thin layers of CeO$_2$ and Y-stabilized ZrO$_2$ (YSZ) to prevent poisoning of superconductivity by diffusion of Si or Al into the HTS films. Buffer layers are also beneficial with respect to improved lattice match.

Wet etching [105] can be used for *patterning* of device structures with sizes down to the µm range. However, undercutting of the photoresist stencils causes problems in providing smooth sidewalls [106]. *Ar ion beam etching* is therefore the preferred method for the fabrication of µm size structures [106,107]. In combination with e-beam lithography and additional preparation techniques even sub-µm size structures can be realized [108,109].

While HTS single layer circuits have already reached a commercial stage of reproducibility, *multilayer circuits* can still be produced only at low yield. Due to the high volatility of Tl, multilayer technology with its need of high temperature deposition steps is only available for YBCO [110,111,112]. Recent improvement with respect to electrical insulation and high current capability in the different YBCO-layers as well as to the integrability of Josephson junctions gives rise to hope for an improved realization of HTS circuits with higher complexity [113,114,115,116].

6. Josephson Junctions

In HTS junctions, the d-wave symmetry of the superconducting wave function [117] and the vicinity of the metal-insulator transition leads to an intrinsically much more sensitive electrical transport behavior compared to classical superconductor junctions [118]: Bound Andreev states and the suppression of the order parameter at interfaces (depending on their orientation with respect to the crystal axes) introduce additional physical problems concerning spatial J_c homogeneity. YBCO is the prevailing materials choice for HTS Josephson junctions due to its multilayer compatibility.

Ramp-type junctions [119] are still the most promising junction type for HTS digital circuits. The reproducibility of the contact parameters has reached a level which allows the production of digital circuits with a complexity of 10 – 100 junctions [121;115,116]. For previous barrier concepts based on thin insulating layers of (Ga-doped) PrBCO [109,119] or Co-doped YBCO [120,121] the J_c homogeneity was limited by the growth control of this barrier layer on the ramp [119]. In *"interface-engineered"* junctions [115,116,123,124,125], instead of depositing a barrier layer a "natural barrier" (probably "pseudo-cubic" YBCO [126,127]) is formed by Ar ion etch treatment of the ramp base electrode [123]. Improved reproducibility and stability in subsequent high temperature treatments for multi-layer preparation have been reported [115,116]. Unfortunately, the barrier properties have turned out to show substantial dependence on the preparation conditions [124,125]. Another limiting factor is the geometrical uncertainty with respect to the ramp angle. Recent success in the fabrication of YBCO-Au-Nb ramp-type junctions [128] enables now the use of the d-wave symmetry related π phase shift as a new functional principle for the construction of novel electronic devices.

c-axis microbridge (CAM) junctions are c-axis interconnects between two superconducting layers separated by an insulator [129]. This is an attractive geometry, with low parasitic inductance suitable for multilayer circuits. Such junctions can be made by the planarization of a mesa or growth into a window in the insulator. The junction behavior is based on an Y rich interface prepared by the ion milling of the YBCO surface due to different milling rates of the cations. This places the mesa junctions in the class of interface engineered junctions with similar limitations. Geometrical uncertainties arise here from the preparation of submicron-size contacts.

Step-edge junctions [130,131,132] are still in use for 1- or 2-junction circuits such as SQUIDs [133]. Earlier hopes to control the growth behavior of the HTS films in the region of the substrate step in a sufficient way to achieve better reproducibility [130; 131] have not been fulfilled [134]. Stability of the junctions is an issue.

SNS step-edge junctions [135] with Ag-Au alloy as normal metal are also in use for SQUID magnetometers [137]. The point-contact-like nature of the transport from the normal metal to the YBCO electrodes and vice versa [135] indicates a stochastic flow of the supercurrent which gives little hope for a systematic improvement of the reproducibility of these contacts.

Bicrystal junctions [108,138,139,140,141,142] are widely used for demonstrator circuits [143,144,145]. They can easily be produced by single-film deposition followed by a simple photolithographic step at a reasonable degree of reproducibility. In multilayer technology even stacked junctions have been realized [113,143]. As for commercial circuits, the cost of the bicrystal substrates and the design restrictions imposed by the positioning of the contacts along the grain boundary appear to be prohibitive. The microscopic limit of J_c homogeneity is given by the meandering of the grain boundary of the deposited HTS films which is connected intrinsically to the HTS film growth [146], and by the implications of d-wave symmetry for such a grain boundary geometry [117,147]. The reproducibility of junction parameters seems to be limited by defects in the substrate bicrystal grain boundary [148] which leads to an irregular course of its meandering.

Biepitaxial contacts [149,150,151,152] have slipped out of technological interest since only 45° [001] tilt grain-boundary contacts with low J_c and $I_c R_N$ values have been developed to a more or less reproducible stage. Unfortunately, this geometry is a pathological case due to the d-wave symmetry of the superconducting wave function of HTS [153,154].

Junctions based on weak links introduced by *focused electron beam irradiation* ("FEBI") [155,156,157,158] have reached a high level of microscopic contact uniformity and have been used for the realization of single-layer digital circuits [159,160]. The contacts can be placed at an arbitrary position on the chip and allow in combination with focused ion-beam etching an absolute and relative positioning of the contacts with an accuracy of ~ 50 nm and ~ 1 nm, respectively. However, the strong temperature dependence of the I_c values requires precise temperature stabilization. Due to the slow writing of the electron beam (~ 1 µm/min), each contact needs a "manufacturing time" of ~ 1 min. As for a similar approach based on ion implantation [161], compatibility of the junction fabrication with multilayer processing is unclear since irradiation of a YBCO layer will also affect buried layers and subsequent high-temperature process steps will affect already fabricated junctions by annealing.

Microbridges over substrate regions damaged by a *focused (Ga-) ion beam* ("FIB") [162,163] show Josephson contact like behavior that is based on the irregular growth behavior of the YBCO films on these substrate regions. Junctions with the desired RSJ-like IV-characteristic are therefore a rare exception [163].

There is still hope for classical *SIS sandwich contacts* once the oxide deposition will be controlled on an atomic scale [164,165,166].

Among non-YBCO contacts, *intrinsic Josephson junctions* in Bi-HTS [167,168,169, 170,171] or Tl-HTS [171] mesa structures are of interest with regard to rf oscillators and other active rf devices. The problem is not so much the preparation of the junctions that has already been realized in thin film technique [172] but the suitable microwave coupling arrangement that overcomes the impedance mismatch to conventional microwave circuits [173].

7. Wire & Tape Conductors

Bi-HTS /Ag tapes were the first practical HTS conductors [175] and have already reached a commercial stage: Tapes with cross sectional areas of ~ 1 mm^2 are available that can transport critical currents I_c (77 K,0 T) up to 130 A [176] over km lengths. In the "Powder-in-Tube" fabrication, Ag tubes are filled with Bi-HTS precursor powder and are subject to various thermomechanical processing steps where the tubes are flattened out to tapes that include the Bi-HTS material as thin filaments. The fragility of the Bi-22(n-1)n grains along neighboring Bi-oxide planes[11] is the physical starting-point for the successful mechanical alignment of the Bi-HTS grains by means of rolling or pressing which orients the grains preferably with the ab-planes parallel to the tape surface. Silver is the only matrix material that is chemically not reacting with HTS compounds and allows for sufficient oxygen diffusion. The common melting of Ag and Bi-HTS around ~ 850° C gives rise to a close mechanical and electrical contact that helps to bridge non-superconducting regions by means of low-resistivity shorts. However, the cost and the softness of Ag are critical issues for technical applications.

Among the two "practical" Bi-HTS systems Bi-2212 and Bi-2223, Bi-2212 has the advantage of a much better control of the chemical phase development [177,178]: Bi-2212 forms from a single phase precursor whereas Bi-2223 requires a multi-phase starting mixture. Today, a quite homogeneous superconductive connection of the HTS grains can be achieved in Bi-2212 filaments [179] with *engineering critical current densities* (calculated with respect to the total conductor cross section) J_{eng} (4.2 K, 0 T) >70 kA/cm^2 over several 100 m conductor length [180]. By clever arrangement of Bi-2212 tapes, J_{eng} (4.2 K, 24 T) > 20 kA/cm^2 has been achieved in a practical wire conductor independent of the direction of the applied magnetic field. By new processing techniques Bi-2212 tapes have been fabricated with *superconducting critical current densities* J_c (4.2 K, 10 T) > 500 kA/cm^2 (calculated with respect to the HTS part of the conductor cross section) already close to the limit of epitaxial thin films [181]. This demonstrates good perspectives for Bi-2212 conductor with respect to applications in high magnetic fields at low temperature [182].

Bi-2223/Ag tapes were for a long time in the center of interest of HTS conductor development since their T_c > 100 K allows operation at LN$_2$ temperature [183,184,185]. However, since the phase evolution of Bi-2223 is much more complicated compared to Bi-2212, the pronounced percolative superconductive current flow in the Bi-2223 filaments could not yet be overcome [186]. Moreover, the upper field limit of ~ 1 T at 77 K [187] restricts the LN$_2$ operation of Bi-2223 tapes to low-field applications such as cables [188,189] or transformers. For operation at technically interesting magnetic field levels of several Tesla, Bi-2223 windings have to be cooled to T_{op} < 30 K. However, in this temperature region Bi-2212 conductors may be competitive due to simper processing and thus lower fabrication cost [181].

Cost arguments have always been in favor of HTS coating of simple robust metal carrier tapes. However, the transfer of YBCO thin film techniques to such a conductor fabrication has taken a lot of efforts since biaxial texture without any interruption is required over the whole

---

[11] The two BiO$_x$ layers in the charge reservoir layers SrO/(BiO$_x$)$_2$/SrO of the Bi-HTS are attached to each other only by weak van-der-Waals-like bonding.

conductor length. Meanwhile, J_c (77 K, 0 T) ~ 1 MA/cm^2 is achieved over a length of ~ 10 meters which allows first practical demonstrations. This status of conductor length development is comparable to the situation of Bi-HTS conductors 10 years ago.

Tl-1223 [190] and Hg-1212 [191] are under investigation with respect to HTS coatings with T_c > 100 K. However, the difficulties of the required thin film techniques suggest that these HTS are at present no practical alternatives to YBCO. Earlier efforts with respect to a fabrication of Tl-1223 conductors by means of processing techniques developed for Bi-HTS conductors have also not been successful [192].

7. A. Bi-2212

Bi-2212 tape conductors with a HTS fill factor of up to 60 % can be produced by simple surface coating of Ag bands. By improved processing ("Pre-Annealing and Intermediate Rolling", "PAIR") tapes with J_c (4.2 K, 10 T) > 500 kA/cm^2 have been fabricated [181]. Trade-off between J_c and the thickness of the Bi-2212 layers [193] limits the expectations for J_{eng} (4.2 K, 20 T) of optimized long tapes to 50 - 60 kA/cm^2 [194].

As practical demonstration, such tape conductor was assembled in a cable similar to a "Rutherford cable", the standard conductor for accelerator magnets: It consisted of 427 Bi-2212 filaments in a AgMgSb alloy outer sheath to allow for higher stress tolerance compared to pure Ag [195]. This wire exhibited excellent I_c (4.2 K, self-field) = 285 A and J_c (4.2 K, self-field) = 340 kA/cm^2 and a room temperature tensile strength of 180 MPa. For a full cable I_c (4.2 K, self-field) = 3500 A was obtained in a short sample, I_c (4.2 K, self-field) > 2700 A in 17-m-long sections cut from a continuous 80-m-long cable. Beside a possible application for next generation accelerator magnets, the cable is also investigated with respect to its use in Superconducting Magnetic Energy Storage devices ("SMES") [196].

By clever arrangement of differently oriented Bi-2212 tapes[12], a wire conductor with high J_{eng} (4.2 K, 24 T) > 20 kA/cm^2 independent of the direction of the applied magnetic field has been constructed which is suitable for the fabrication of conventional solenoid magnets ("ROtation-Symmetric Arranged Tape-in tube wire", "ROSATwire") [180,197]. The Bi-HTS fill factor could be increased to ~ 40 % [198].

This development aims at insertion magnets for 30 T class magnets where the Bi-HTS coils have to generate an additional field of several Tesla in a background field of ~ 20 T [199]. The most ambitious goal is the use of such magnet systems for high-field NMR: The extreme requirements of temporal stability can probably only be fulfilled in a superconducting magnet system operated in persistent mode which relies on an extremely slow decay of the supercurrents [200]. However, the demonstrated decay times are still not yet sufficient to achieve this goal.

---

[12] The ROSATwire conductor is an assembly of an equal number of tapes rotated around the tape axis by 0, 120 and 240 degrees, respectively. In an external magnetic field, the field component perpendicular to the ab-planes and thus to the tape surface has the strongest effect on J_c. This results for the ROSATwire tape arrangement in a J_c suppression which is nearly independent of the conductor orientation.

7. B. Bi-2223

Bi-2223 Powder-in-Tube tapes have seen for more than ten years a continuous performance increase. The critical current densities in the HTS filaments were raised up to J_c (77 K, 0 T) ~ 70 kA/cm^2 in short samples [201]. Better understanding of the mechanical deformation process [202] and of the Bi-2223 phase development during thermal annealing [203] helped to establish a production of several 10 km of such tape conductors with J_{eng}(77 K, 0 T) ~ 15 kA/cm^2 at a total conductor cross section of ~ 1 mm^2 [183,184,185]. Locally, J_c (77 K, 0 T) up to 180 kA/cm^2 has been observed [186], but this seems to be the upper limit for Bi-2223/Ag tapes [204]. The reason are non-superconducting inclusions and pores in the HTS filaments leading to a percolative supercurrent flow along the Bi-2223 grains which are well connected but make up at most only ~ 2/3 of the filament cross-section [205]. With an HTS fill factor of high-J_c Bi-2223 conductors of at most 35 % an intrinsic limit J_{eng} (77 K, 0 T) ~ 60 kA/cm^2 has to be expected which is still comparable to the limit of ~ 100 kA/cm^2 estimated for YBCO coated conductors (assuming a future development of J_c (77 K, 0 T) ~ 1 MA/cm^2 in 5 μm thick YBCO coatings on 50 μm thick metal carrier tapes).

However, for applications in higher magnetic field a substantially lower operating temperature is required with the additional benefit of a J_c increase of up to a factor of 7 compared to 77 K (J_c(T, 0 T) $\approx$ J_c(77 K, 0 T) [7 − T/13 K] [206]). Another economical disadvantage with respect to YBCO coated conductors is the mandatory use of silver as matrix material. Previous cost estimations of 10 \$/kA m for full-scale production levels [201] (at present: 200 - 300 \$/kA m) have turned out to be too optimistic [207] and are meanwhile adjusted to 25 -50 \$/kA m in 2006. Moreover, while the good electrical contact of the HTS filaments with the Ag matrix is helpful in bridging disconnected HTS regions by means of low resistance shorts [208], frequent use of this current rerouting prevents persistent-mode operation of Bi-2223 coils.

Bi-2223 conductors have overcome many problems on the way to a technical HTS material. The problem of the softness of the Ag matrix has been solved by applying dispersed MgO in the Ag matrix [209] or by adding a thin layer of stainless steel reinforcement to both sides of the tapes which enables the tapes to withstand a tensile stress of 300 MPa and a tensile strain of 0.45% at 77 K [176]. Promising approaches have been developed to tackle the problem of ac losses which arise from electromagnetic coupling of the HTS filaments [210,211] and are of concern for all power applications since they determine the necessary cooling power [212,213,214,215,216]: Insulating $BaZrO_3$ [212] or $SrCO_3$ barriers [213] around the HTS filaments and higher resistive Ag alloys such as AgAu or AgPd as matrix material as well as twisting of the filaments (down to twist pitches l_p ~ 1 cm [213]) reduces these coupling currents. A novel wire arrangement of horizontal and vertical stacks of Bi-2223 tapes reduces the I_c-anisotropy with respect to the orientation of an external magnetic field [217].

In conclusion, Bi-2223 conductors have arrived at a practical level of technical applicability, however, still at a quite high cost level. At present, the biggest psychological handicap for Bi-HTS conductors are the great expectations for a soon arrival of a cheaper and better YBCO coated conductor.

7. C. YBCO

YBCO-coated metal tapes are promising with respect to lower-cost HTS conductor operating even at LN_2 temperature in magnetic fields up to several Tesla. $J_c(77\text{ K, }0\text{ T})$ = 1 - 2 MA/cm^2 has been achieved in short samples by a number of YBCO coating methods choosing different routes to biaxial YBCO texture. One route is the deposition of a textured buffer layer on (untextured) metal with the assistance of an *ion beam* that *introduces orientation-selective growth.* With stainless steel as attractive choice for the metal, 10 m long high-J_c tapes have been demonstrated [218,219]. Another route is the use of *cube-textured metal bands.* Buffer layers (CeO_2,Y-stabilized ZrO_2 (YSZ),$Gd_2Zr_2O_7$ (GZO), In-Sn oxide (ITO), ...) have to be applied here as well in order to compensate the lattice mismatch and to prevent poisoning of superconductivity in the YBCO coating by in-diffusion of metal atoms. Ni and Ni alloys are used here as metal substrates that show the required cube-texture after appropriate metallurgical and heat treatment.

Most of the tested buffer materials are insulators with the consequence that a disruption of the superconducting current path can not be bridged by a low-resistance short via the metal tape as in the case of Bi-HTS/Ag tapes. Practical solutions of this problem are the deposition of a Au layer on top of the YBCO coating [218] or use of metallic oxide buffer layers [220].

$J_c(77\text{ K, }0\text{ T}) \sim 1\ MA/cm^2$ is obtained in YBCO coatings with a thickness of up to $\sim 3\ \mu m$ [218,221]. For $\sim 100\ \mu m$ thick metal carrier tapes this results in a HTS fill factor of only $\sim 3\ \%$. The engineering current density J_{eng} calculated with respect to the total conductor cross-section of $J_{eng}(77\text{ K, }0\text{ T}) \sim 30\ kA/cm^2$ is therefore still close to what is achieved in present commercial Bi-2223/Ag tapes. 50 μm thick metal carrier tapes are available. Ca doping and optimization of the grain architecture may help to transfer the high intra-grain $J_c(77\text{ K, }0\text{ T}) \sim 5\ MA/cm^2$ [222] into macroscopic current densities even for the case of a grain alignment of only 10° [223].

By *Ion Beam Assisted Deposition* ("*IBAD*") [221,224,225] tapes with $J_c(77\text{K, }0\text{ T})$ = 2.2 MA/cm^2 ($I_c(77\text{K, }0\text{ T})$ = 78 A) measured over a length of 10 m [218] and $J_c(77\text{ K, }0\text{ T})$ = 0.5 MA/cm^2 over a length of 46 m have been fabricated. The deposition of the required $\sim 1\ \mu m$ thick YSZ buffer under assistance of an additional ion beam selecting the required biaxial texture is painfully time-consuming. This problem can be overcome by MgO-IBAD-buffering where a buffer thickness of ~ 10 nm is already sufficient for biaxial texture [226]. Meanwhile, $J_c(77\text{K, }0\text{ T})$ = 0.9 MA/cm^2 ($I_c(77\text{K, }0\text{ T})$ = 14.4 A) has been achieved with this technique over a length of 0.8 m [230]. This renders again the YBCO laser deposition as the time-limiting fabrication step at present coating rates of up to 60-70 nm $\times$ m^2/h [218]. The cost of the laser deposition is an economical issue: IBAD-YBCO tapes are still more expensive than Bi-2223/Ag tapes [218].

Inclined Substrate Deposition ("*ISD*") achieves biaxial alignment of the YSZ buffer without assistance of an additional ion-beam using a high rate laser deposition and appropriate inclination of the metal substrate with respect to the laser plume [227]. This makes the YSZ buffer deposition by a factor 10 faster than IBAD. However, the degree of biaxial alignment is not yet sufficient: $J_c(77\text{ K, }0\text{ T}) \sim 1\ MA/cm^2$ has been achieved up to now only in short samples [228].

In *Rolling-Assisted-Biaxially-Textured-Substrates* ("*RABiTS*"), cube-texture of a Ni (alloy) substrate tape is generated by conventional rolling with heavy deformation (> 95 %) to a roll textured tape, followed an annealing step which results in a recrystallization into the desired biaxially textured cubic phase [229,231]. Sophisticated buffering techniques have been tested in order to avoid or remove the oxide layer from the tape surface since it destroys the unique biaxial orientation of the substrate and may blast the YBCO coating due to the volume expansion accompanying the oxidation. With a $CeO_2/YSZ/CeO_2$ buffer system J_c (77 K, 0 T) > 1 MA/cm^2 has been achieved in small samples [229], but only J_c (77 K, 0 T) ~ 0.6 MA/cm^2 over 0.8 m tape length [230].

"Surface-Oxidation Epitaxy" ("*SOE*") is a similar approach where cube-textured pure Ni tape is used [232] which is oxidized under controlled condition to form cube-textured NiO. The actual texture "seen" by the YBCO coating is here not the Ni texture but the texture of the NiO which is suitably lattice matched to YBCO. J_c (77 K, 0 T) ~ 0.1 MA/cm^2 demonstrated in short samples with additional MgO buffering is not yet satisfactory.

Biaxially textured Ag-substrates are investigated for YBCO coating as well [233]. However, the use of expensive and soft Ag carrier tapes in combination with the low HTS fill factor of the tape coating approach is not very promising. Moreover, the achieved degree of textured grains is still well below the level of nearly 100 % which is already obtained for Ni and Ni alloy tapes.

Substantial cost-reduction can be expected if less costly deposition methods than the technically successful but expensive vacuum physical vapor deposition approaches can be employed. Recent progress in *solution-based* YBCO *coating* (dip coating, spray pyrolysis, sol-gel) with first J_c (77 K, 0 T) ~ 1 MA/cm^2 samples is most promising [234,235]. Recently, MOD-YBCO coated tapes with I_c (77 K, 0 T) = 140A over 7.5 m tape length have been reported. *Liquid Phase Epitaxy* is another promising option [236].

8. Bulk Material

Melt-textured YBCO pellets [237] with good superconducting properties (J_c (77 K, 1 T) > 10 kA/cm^2, J_c (50 K, 10 T) > 100 kA/cm^2) can be grown reproducibly in sizes up to a diameter d = 6 cm even in complex shapes [238] comparable to what has been achieved in NdFeB permanent magnets. The present production cost of > 1000 EUR / kg (including quality control) is still a factor of ~ 10 higher than for NdFeB [239] but it can be expected to come closer this cost level once a production of several tons per year can be established here as well. Sizes up to d = 10 cm have been realized [240] but joining of smaller size pellets with good superconducting properties of the connections [241] seems to be a more practical solution[13]: For larger pellets full oxidation becomes increasingly difficult even though it is not based on oxygen bulk diffusion [243] but on oxygen transport along microcracks [244,245].

---

[13] The product of J_c and sample size r = d/2 is probably the most appropriate quantitative criterion for the superconducting quality of bulk samples since it is proportional to the magnetic field trapping capability [242]. Even if the J_c of joints is at present still substantially below the J_c of high-quality bulk samples a higher $J_c \times$ r product can be obtained by joining such samples.

The strong pinning allows the "freezing" of high magnetic fields. In single domain cylinders with a diameter of 30 mm an induction of 1.3 T can thus be fixed at 77 K [238], 16 T have been achieved at 24 K in-between two d = 2.2 cm samples [246], very recently even 17 T at 29 K in-between two d = 2.65 cm / h = 1.5 cm samples [247]. This exceeds the potential of conventional permanent magnets based on ferromagnetic spin polarization by an order of magnitude and can be used for high-field "cryomagnets". The present field limit of YBCO cryomagnets does not stem from the pinning, but from the involved mechanical forces which exceed at high fields the fracture toughness of the YBCO ceramic [248] (20 - 30 MPa in YBCO without additions [249]) and leads without appropriate mechanical stabilization to cracking of the pellets. Ag addition [246,250] were shown to be helpful in providing higher intrinsic stability. Passivation by resin impregnation [248,251] protects the samples against humidity and thus improves the long-term stability.

The combination of melt-textured YBCO with permanent magnets leads to levitation properties that are attractive for bearing applications. In contrast to classic magnetic bearings, superconducting magnetic bearings are self-stabilized due to flux pinning. The strong levitation forces (e.g. > 80 N for zero field cooled d = 3 cm YBCO pellets at 77 K at a distance of 0.5 mm from a 0.4 T / d = 2.5 cm SmCo magnet [238]) with force densities of > 10 N/cm^2 and the stiffness are already limited by the magnetic field strength of the permanent magnet [252,253,254]. YBCO pellets are even under consideration as substitute for LHe-cooled NbTi-coils in the Japanese Maglev project [255].

In contrast to YBCO, high-quality rare-earth 123 samples can not be fabricated in air but require oxygen-reduced processing atmosphere. For melt-textured Nd-123 [256,257] and Sm-123 [258] improved superconducting properties have been reported but only for sample sizes up to d ~ 3 cm. Nevertheless, the freezing of 2.0 T maximum induction in d ~ 5 cm sized Gd-123 samples at 77 K and trapping of 3.3 T in between two such samples at 77 K [259] is beyond what has been achieved for YBCO. Ternary mixtures of Rare-Earth-123 [260] have demonstrated even better pinning of stronger magnetic field at 77 K [261].

Bi-2212 rods and tubes are commercially available in sizes up to d = 30 cm and 50 cm length with electrical properties sufficient for many applications such as current leads or fault current limiters [82].

9. Conclusion

In sixteen years since their discovery, High-Temperature Superconductors proceeded quite far towards the goal of a materials basis from which engineers can choose well-specified products to build technical systems. However, the HTS spectrum of such industrial feedstock is still limited. Nevertheless, from the experience gained in the development of these materials it has become clear that certain promised benefits can be expected from a technical use of HTS within the next couple of years.

Acknowledgement

I would like to thank W. Goldacker, T. Habisreuther, M. Lakner, G. Linker, T. Scherer, C. W. Schneider, P. Seidel, M. Siegel, and T. Wolf for critically reading the manuscript and for many helpful comments.

References

[T] J. G. Bednorz, K. A. Müller, Z. Phys. B **64** (1986) 189

[2] J. Orenstein, A. J. Millis, Science **288** (2000) 468

[3] P. W. Anderson, Science **235** (1987) 1196

[4] P. W. Anderson, *The Theory of Superconductivity in High-T_c Cuprates Superconductors* (Princeton University Press, Princeton, 1997)

[5] B. Batlogg, H. Y. Hwang, H. Takagi, R. J. Cava, H. L. Kao, J. Kwo, Physica C **235-240** (1994) 130

[6] B. Batlogg, Solid State Comm. **107** (1998) 639

[7] A. Erb, E. Walker, J.-Y. Genoud, R. Flükiger, Physica C **282-287** (1997) 89

[8] G. Yang, J. S. Abell, C. E. Gough, Appl. Phys. Lett. **75** (1999) 1955

[9] A. Yamamoto, W. Z. Hu, F. Izumi, S. Tajima, Physica C **351** (2001) 329

[10] R. F. Service, Science **271** (1996) 1806

[11] A. B. Fowler, Physics Today, October 1993, 59

[12] C. W. Chu, J. Supercond. **12** (1999) 85

[13] C. W. Chu, IEEE Trans. Appl. Supercond. **7** (1997) 80

[14] B. Raveau, in *High-T_c Superconductivity 1996: Ten Years after the Discovery*, NATO ASI Series E Vol. **343** (Kluwer Academic Publishers, 1997), p.109

[15] J. L. Tallon, G. V. M. Williams, J. W. Loram, Physica C **338** (2000) 9

[16] H. Lütgemeier, S. Schmenn, P. Meuffels, O. Storz, R. Schöllhorn, Ch. Niedermayer, I. Heimaa, Yu. Baikov, Physica C **267** (1996) 191

[17] C. H. Ahn, J. M. Triscone, N. Archibald, M. Decroux, R. H. Hammond, T. H. Geballe, O. Fischer, M. R. Beasley, Science **269** (1995) 373

[18] M. B. Salomon, M. Jaime, Rev. Mod. Phys.**73** (2001) 583

[19] J. Z. Sun, Physica C **350** (2001) 215

[20] A. Moreo, S. Yunoki, E. Dagotto, Science **283** (1999) 2034

[21] Ch. Renner, G. Aeppli, B.-G. Kim, Y.-A. Soh, S. W. Cheong, Science **416** (2002) 518

[22] R. Bormann, J. Nölting, Appl. Phys. Lett. **54** (1989) 2148

[23] A. Mawdsley, J. L. Tallon, M. R. Presland, Physica C **190** (1992) 437

[24] E. L. Brosha, P. K. Davis, F. H. Garzon, I. D. Raistrick, Science **260** (1993) 196

[25] H. Yamauchi, M. Karppinen, S. Tanaka, Physica C **263** (1996) 146

[26] H. Yamauchi, M. Karppinen, Supercond. Sci. Technol. **13** (2000) R33

[27] P. Bordet, S. LeFloch, C. Chaillout, F. Duc, M. F. Gorius, M. Perroux, J. J. Capponi, P. Toulemonde, J. L. Tholence, Physica C **276** (1997) 237

[28] N. L. Wu, Z. L. Du, Y. Y. Xue, I. Rusakova, D. K. Ross, L. Gao, Y. Cao, Y. Y. Sun, C. W. Chu, M. Hervieu, B. Raveau, Physica C **315** (1999) 227

[29] M. K. Wu, J. R. Ashburn, C. J. Torng , P. H. Hor, R. L. Meng, L. Gao, Z. J. Huang, Y. Q. Wang, C. W. Chu, Phys. Rev. Lett. **58** (1987) 908

[30] J. G. Lin, C. Y. Huang, Y. Y. Xue, C. W. Chu, X. W. Cao, J. C. Ho, Phys. Rev. B **51** (1995) 12900

[31] J. W. Chu, H. H. Feng, Y. Y. Sun, K. Matsuishi, Q. Xiong, C. W. Chu, *HTS Materials, Bulk Processing and Bulk Applications* - Proceedings of the 1992 TCSUH Workshop, ed. by C. W. Chu, W. K. Chu, P. H. Hor and K. Salama (Singapore: World Scientific, 1992), p. 53; TCSUH preprint 92:043

[32] G. V. M. Williams, J. L. Tallon, Physica C **258** (1996) 41

[33] M. Al-Mamouri, P. P. Edwards, C. Greaves, M. Slaski, Nature **369** (1994) 382

[34] T. Kawashima, Y. Matsui, E. Takayama-Muromachi, Physica C **257** (1996) 313

[35] Z. Hiroi, N. Kobayashi, M. Takano, Nature **371** (1994) 139

[36] S. N. Putilin, E. V. Antipov, A. M. Abakumov, M. G. Rozova, K. A. Lokshin, D. A. Pavlov, A. M. Balagurov, D. V. Sheptyakov, M. Marezio, Physica C **338** (2001) 52

[37] L. Gao, Y. Y. Xue, F. Chen, Q. Xiong, R. L. Meng, D. Ramirez, C. W. Chu, J. H. Eggert, H. K. Mao, Phys. Rev. B **50** (1994) 4260

[38] K. Kinoshita and T. Yamada, Nature **357** (1992) 313

[39] M. Sato, J. Akimitsu, H. Takahashi, N. Mori, Physica C **271** (1996) 79

[40] H. Kumakura, H. Kitaguchi, K. Togano, T. Kawashima, E. Takayama-Muromachi, S. Okayasu, Y. Kazumata, IEEE Trans. Appl. Supercond. **5** (1995) 1399

[41] P. Zoller, J. Glaser, A. Ehmann, C. Schultz, W. Wischert, S. Kemmler-Sack, T. Nissel, R. P. Huebener, Z. Phys. B **96** (1995) 505

[42] D. L. Feng, A. Damascelli, K. M. Shen, N. Motoyama, D. H. Lu, H. Eisaki, K. Shimizu, J.-i. Shimoyama, K. Kishio, N. Kaneko, M. Greven, G. D. Gu, X. J. Zhou, C. Kim, F. Ronning, N. P. Armitage, Z.-X. Shen, Phys. Rev. Lett. **88** (2002) 107001

[43] P. Wagner, F. Hillmer, U. Frey, H. Adrian, T. Steinborn, L. Ranno, A. Elschner, I. Heyvaert, Y. Bruynseraede, Physica C **215** (1993) 123

[44] S. Lösch, H. Budin, O. Eibl, M. Hartmann, T. Rentschler, M. Rygula, S. Kemmler-Sack, R. P. Huebener, Physica C **177** (1991) 271

[45] H. Yamauchi, T. Tamura, X.-J. Wu, S. Adachi, S. Tanaka, Jpn. J. Appl. Phys. **34** (1995) L349

[46] T. Tamura, S. Adachi, X.-J. Wu, T. Tatsuki, K. Tanabe, Physica C **277** (1997) 1

[47] A. Iyo, Y. Tanaka, Y. Ishiura, M. Tokumoto, K. Tokiwa, T. Watanabe, H. Ihara, Supercond. Sci. Technol. **14** (2001) 504

[48] A. Iyo, Y. Aizawa, Y. Tanaka, M. Tokumoto, K. Tokiwa, T. Watanabe, H. Ihara, Physica C 357-360, 324 (2001)

[49] D. Tristan Jover, R. J. Wijngaarden, R. Griessen, E. M. Haines, J. L. Tallon, R. S. Liu, Phys. Rev. B **54** (1996) 10175

[50] Z. Y. Chen, Z. Z. Sheng, Y. Q. Tang, Y. F. Li, L. M. Wang, D. O. Pederson, Supercond. Sci. Technol. **6** (1993) 261

[51] E. V. Antipov, A. M. Abakumov, S. N. Putilin, Supercond. Sci. Technol. **15** (2002) R31

[52] C. Acha, S. M. Loureiro, C. Chaillout, J. L. Tholence, J. J. Capponi, M. Marezio, M. Nunez-Regueiro, Solid State Comm. **102** (1997) 1

[53] T. Tatsuki, A. Tokiwa-Yamamoto, A. Fukuoka, T. Tamura, X.-J. Wu, Y. Moriwaki, R. Usami, S. Adachi, K. Tanabe, S. Tanaka, Jpn. J. Appl. Phys. **35** (1996) L205

[54] E. Stangl, S. Proyer, M. Borz, B. Hellebrand, D. Bäuerle, Physica C **256** (1996) 245

[55] H. Ihara, Physica C **364-365** (2001) 289

[56] P. W. Klamut, B. Dabrowski, S. M. Mini, M. Maxwell, S. Kolesnik, J. Mais, A. Shengelaya, R. Khasanov I. Savic, H. Keller, T. Graber, J. Gebhardt, P. J. Viccaro, Y. Xiao, Physica C **364-365** (2001) 313;

B. Lorenz, R. L. Meng, J. Cmaidalka, Y. S. Wang, J. Lenzi, Y. Y. Xue, C. W. Chu, Physica C **363** (2001) 251

[57] T. Kawashima, Y. Matsui, E. Takayama-Muromachi, Physica C **254** (1995) 131

[58] I. Bozovic, G. Logvenov, I. Belca, B. Narimbetov, I. Sveklo, Phys. Rev. Lett. **89** (2002) 107001

[59] S. Karimoto, H. Yamamoto, T. Greibe, M. Naito, Jpn. J. Appl. Phys. **40** (2001) L127

[60] M. Naito, M. Hepp, Jpn. J. Appl. Phys. **39** (2000) L485

[61] L. Alff, S. Meyer, S. Kleefisch, U. Schoop, A. Marx, H. Sato, M. Naito, R. Gross,
Phys. Rev. Lett. **83** (1999) 2644

[62] C. Q. Jin, Y. S. Yao, S. C. Liu, W. L. Zhou, W. K. Wang, Appl. Phys. Lett. **62** (1993) 3037

[63] A. Iyo, Y. Tanaka, M. Tokumoto, H. Ihara, Physica C **366** (2001) 43

[64] C. U. Jung, J. Y. Kim, S. M. Lee, M.-S. Kim, Y. Yao, S. Y. Lee, S.-I. Lee, D. H. Ha
Physica C **364-365** (2001) 225

[65] D. R. Harshman, A. P. Mills, Jr., Phys. Rev. B **45** (1992) 10684

[66] N. P. Plakida, *High-Temperature Superconductivity,* Springer-Verlag Berlin-Heidelberg 1995

[67] K. Tanaka, A. Iyo, Y. Tanaka, K. Tokiwa, M. Tokumoto, M. Ariyama, T. Tsukamoto,
T. Watanabe, H. Ihara, Physica B **284-288** (2000) 1081;

T. Watanabe, S. Miyashita, N. Ichioka, K. Tokiwa, K. Tanaka, A. Iyo, Y. Tanaka, H. Ihara,
Physica B **284-288** (2000) 1075

[68] C. C. Tsuei, J. R. Kirtley, Rev. Mod. Phys. **72** (2000) 969;

C. C. Tsuei, J. R. Kirtley, Physica C **367** (2002) 1

[69] H. Hilgenkamp and J. Mannhart, B. Mayer, Phys. Rev. B **53** (1996) 14586

[70] P. A. Nilsson, Z. G. Ivanov, H. K. Olsson, D. Winkler, T. Claeson, E. A. Stepantsov,
A. Ya. Tzalenchuk, J. Appl. Phys. **75** (1994) 7972

[71] B. Götz, "Untersuchung der Transporteigenschaften von Korngrenzen in Hochtemperatur-
Supraleitern", PhD-thesis, University of Augsburg, 2000

[72] H. Hilgenkamp, J. Mannhart, Rev. Mod. Phys. **74** (2002) 485

[73] G. Blatter, M. V. Feigel'man, V. B. Geshkenbein, A. I. Larkin, V. M. Vinokur,
Rev. Mod. Phys. **66** (1994) 1125

[74] K. K. Likharev, *Dynamics of Josephson Junctions and Circuits,*
Gordon & Breach Science Publishers, 1986, p.30

[75] J. Betouras, R. Joynt, Physica C **250** (1995) 256

[76] J. Alarco, E. Olsson, Phys. Rev. B **52** (1995) 13625

[77] D. Agassi, D. K. Christen, S. J. Pennycook, Appl. Phys. Lett. **81** (2002) 2803

[78] H. Hilgenkamp, J. Mannhart, Appl. Phys. Lett. **73** (1998) 265

[79] G. Hammerl, A. Schmehl, R. R. Schulz, B. Goetz, H. Bielefeldt, C. W. Schneider,
H. Hilgenkamp, J. Mannhart, Nature **162** (2000) 162;

C. W. Schneider, R. R. Schulz, B. Goetz, A. Schmehl, H. Bielefeldt, H. Hilgenkamp,
J. Mannhart, Appl. Phys. Lett. **75** (1999) 850;

H. Hilgenkamp, C. W. Schneider, R. R. Schulz, B. Goetz, A. Schmehl, H. Bielefeldt,
J. Mannhart, Physica C **326-327** (1999) 7

[80] J. L. Tallon, J. W. Loram, Physica C **349** (2001) 53

[81] C. M. Varma, Phys. Rev. B **55** (1997) 14554

[82] P. F. Herrmann, E. Beghin, J. Bock, C. Cottevieille, A. Leriche, T. Verhaege,
Inst. Phys. Conf. Ser. No. **158** (1997) 825

[83] M. Noe, K. -P. Juengst, F. N. Werfel, S. Elschner, J. Bock, A. Wolf, F. Breuer,
Physica C **372-376** (2002) 1626

[84] A. Usoskin, A. Issaev, H. C. Freyhardt, M. Leghissa, M. P. Oomen, H. -W. Neumueller,
Physica C **378-381** (2002) 857

[85] T. Kiss, S. Noda, S. Nishimura, K. Ohya, D. Utsunomiya, Yu. A. Ilyn, H. Okamoto,
Physica C **357-360** (2001) 1165

[86] W. Prusseit, S. Furtner, R. Nemetschek, Supercond. Sci. Technol. **13** (2000) 519

[87] V. Matijasevic, Z. Lu, T. Kaplan, C. Huang, Inst. Phys. Conf. Ser. No. **158** (1997) 189

[88] D. W. Face, R. J. Small, M. S. Warrington, F. M. Pellicone, P. J. Martin,
Physica C **357-360** (2001) 1488;
D. W. Face, C. Wilker, J. J. Kingston, Z.-Y. Shen, F. M. Pellicone, R. J. Small, S. P. McKenna,
S. Sun, P. J. Martin, IEEE Trans. Appl. Supercond. **7** (1997) 1283

[89] R. Semerad, J. Knauf, K. Irgmaier, W. Prusseit, Physica C **378-381** (2002) 1414

[90] K. Sudoh, Y. Ichino, M. Itoh, Y. Yoshida, Y. Takai, I. Hirabayashi,
Jpn. J. Appl. Phys. **41** (2002) L983

[91] J. Geerk, P. Ratzel, H. Rietschel, G. Linker, R. Heidinger, R. Schwab,
IEEE Trans. Appl. Supercond. **9** (1999) 1543;
R. Krupke, G. Ulmer, R. Schneider, M. Kurzmeier, G. Linker, J. Geerk,
Physica C **279** (1997) 153

[92] R. S. Aga, Jr., Y.-Y. Xie, S. L. Yan, J. Z. Wu, S. S. Han, Appl. Phys. Lett. **79** (2001) 2417

[93] H. Kinder, P. Berberich, W. Prusseit, S. Rieder-Zecha, R. Semerad, B. Utz,
Physica C **282-287** (1997) 107

[94] R. Wördenweber, Supercond. Sci. Technol. **12** (1999) 86

[95] M. Lorenz, H. Hochmuth, D. Natusch, M. Kusunoki, V. L. Svetchnikov, V. Riede, I. Stanca,
G. Kästner, D. Hesse, IEEE Trans. Appl. Supercond. **11** (2001) 3209

[96] J. Schneider, J. Einfeld, P. Lahl, Th. Königs, R. Kutzner, R. Wördenweber,
Inst. Phys. Conf. Ser. No. **158** (1997) 221

[97] M. Naito, S. Karimoto, H. Yamamoto, H. Nakada, K. Suzuki,
IEEE Trans. Appl. Supercond. **11** (2001) 3848

[98] A. Abrutis, J. P. Sénateur, F. Weiss, V. Kubilius, V. Bigelyte, Z. Saltyte, B. Vengalis,
A. Jukna, Supercond. Sci. Technol. **10** (1997) 959

[99] H. Nagai, Y. Yoshida, Y. Ito, S. Taniguchi, I. Hirabayashi, N. Matsunami, Y. Takai,
Supercond. Sci. Technol. **10** (1997) 213;
S. Yamamoto, K. Nagata, S. Sugai, A. Sengoku, Y. Matsukawa, T. Hattori, S. Oda,
Jpn.. J. Appl. Phys. **38** (1999) 4727

[100] T. Araki, T. Niwa, Y. Yamada, I. Hirabayashi, J. Shibata, Y. Ikuhara, K. Kato, T. Kato,
T. Hirayama, J. Appl. Phys. 92 (2002) 3318

[101] H. Schneidewind, M. Manzel, G. Bruchlos, K. Kirsch, Supercond. Sci. Technol. **14** (2001) 200

[102] Y. Yamada, J. Kawashima, J. G. Wen, Y. Niiori, I. Hirabayashi,
Jpn. J. Appl. Phys. **39** (2000) 1111

[103] P. Seidel, S. Linzen G. Kaiser, F. Schmidl, Y. Tian, A. Matthes, S. Wunderlich,
H. Schneidewind, Proc. SPIE **3481** (1998) 366

[104] A. G. Zaitsev, G. Ockenfuss, R. Wördenweber, Inst. Phys. Conf. Ser. No. **158** (1997) 25

[105] R. P. Vasquez, B. D. Hunt, M. C. Foote, Appl. Phys. Lett. **53** (1988) 2692;
W. Eidelloth, R. L. Sandstrom, Appl. Phys. Lett. **59** (1991) 1632

[106] H. Schneidewind, F. Schmidl, S. Linzen, P. Seidel, Physica C **250** (1995) 191

[107] L. Alff, G. M. Fischer, R. Gross, F. Kober, A. Beck, K. D. Husemann, T. Nissel, F. Schmidl,
C. Burckhardt, Physica C **200** (1992) 277

[108] F. Herbstritt, T. Kemen, A. Marx, R. Gross, J. Appl. Phys. **91** (2002) 5411

[109] P. V. Komissinski, B. Högberg, A. Ya. Tzalenchuk, Z. Ivanov, Appl. Phys. Lett. **80** (2002) 1022

[110] W. H. Mallison, S. J. Berkowitz, A. S. Hirahara, M. J. Neal, K. Char,
Appl. Phys. Lett. **68** (1996) 3808

[111] B. D. Hunt, M. G. Forrester, J. Talvacchio, J. D. McCambridge, R. M. Young,
Appl. Phys. Lett. **68** (1996) 3805

[112] R. Scharnweber, N. Dieckmann, M. Schilling, Appl. Phys. Lett. **70** (1997) 2189

[113] H. Takashima, N. Kasai, A. Shoji, Jpn. J. Appl. Phys. **41** (2002) L1062

[114] D. Cassel, Th. Ortlepp, R. Dittmann, B. Kuhlmann, H. Toepfer, M. Siegel, F. H. Uhlmann, Physica C **372-376** (2002) 139

[115] Y. Soutome, T. Fukazawa, A. Tsukamoto, K. Saitoh, K. Takagi, Physica C **372-376** (2002) 143

[116] T. Satoh, M. Hidaka, S. Tahara, J. G. Wen, N. Koshizuka, S. Tanaka, IEEE Trans. Appl. Supercond. **11** (2001) 770

[117] J. Mannhart, H. Hilgenkamp, Supercond. Sci. Technol. **10** (1997) 880

[118] R. Gross, L. Alff, A. Beck, O. M. Froehlich, D. Koelle, A. Marx, IEEE Trans. Appl. Supercond. **7** (1997) 2929

[119] D. H. A. Blank, H. Rogalla, J. Mater. Res. **12** (1997) 2952;

K. Verbist, O. I. Lebedev, M. A. J. Verhoeven, R. Wichern, A. J. H. M. Rijnders, D. H. A. Blank, F. Tafuri, H. Bender, G. Van Tendeloo, Supercond. Sci. Technol. **11** (1998) 13;

K. Verbist, O. I. Lebedev, G. Van Tendeloo, M. A. J. Verhoeven, A. J. H. M. Rijnders, D. H. A. Blank, H. Rogalla, Appl. Phys. Lett. **70** (1997) 1167

[120] L. Antognazza, K. Char, T. H. Geballe, Appl. Phys. Lett. **70** (1997) 3152

[121] J. Talvacchio, M. G. Forrester, B. D. Hunt, J. D. McCambridge, R. M. Young, X. F. Zhang, D. J. Miller, IEEE Trans. Appl. Supercond. **7** (1997) 2051

[122] J. R. LaGraff, H. Chan, J. M. Murduck, S. H. Hong, Q. Y. Ma, Appl. Phys. Lett. **71** (1997) 2199

[123] B. H. Moeckly, K. Char, Appl. Phys. Lett. **71** (1997) 2526

[124] J.-K. Heinsohn, R. Dittmann, J. Rodriguez Contrera, J. Scherbel, A. Klushin, M. Siegel, C. L Jia, A. Golubov, M. Yu. Kupryanov, J. Appl. Phys. **89** (2001) 3852

[125] Y. Ishimaru, Y. Wu, O. Horibe, H. Tano, T. Suzuki, Y. Tarutani, U. Kawabe, K. Tanabe, Jpn. J. Appl. Phys. **41** (2002) 1998;

Y. Ishimaru, Y. Wu, O. Horibe, H. Tano, T. Suzuki, Y. Yoshida, M. Horibe, H. Wakana, S. Adachi, Y. Takahashi, Y. Oshikubo, H. Sugiyama, M. Iiyama,Y. Tarutani, K. Tanabe, Physica C **378-381** (2002) 1327

[126] S. Linzen, J. Kräußlich, A. Köhler, P. Seidel, B. Freitag, W. Mader, Physica C **290** (1997) 323

[127] Y. Huang, K. L. Merkle, B. H. Moeckly, K. Char, Physica C **314** (1999) 36

[128] H. J. H. Smilde, H. Hilgenkamp, G. Rijnders, H. Rogalla, D. H. A. Blank, Appl. Phys. Lett. **80** (2002) 4579;

H. J. H. Smilde, H. Hilgenkamp, G. J. Gerritsma, D. H. A. Blank, H. Rogalla, Physica C **350** (2001) 269;

H. J. H. Smilde, Ariando, D. H. A. Blank, G. J. Gerritsma, H. Hilgenkamp, H. Rogalla, Phys. Rev. Lett. **88** (2002) 057004

[129] P. J. Hirst, R. G. Humphreys, J. S. Satchell, M. J. Wooliscoft, C. L. Reeves, G. William, A. J. Pidduck, H. Willis, IEEE Trans. Appl. Supercond. **11** (2001) 143

[130] K. Herrmann, G. Kunkel, M. Siegel, J. Schubert, W. Zander, A. I. Braginski, C. L. Jia, B. Kabius, K. Urban, J. Appl. Phys. **78** (1995) 1131

[131] M. Gustafsson, E. Olsson, H. R. Yi, D. Winkler, T. Claeson, Appl. Phys. Lett. **70** (1997) 2903;

F. Lombardi, Z. G. Ivanov, G. M. Fischer, E. Olsson, T. Claeson, Appl. Phys. Lett. **72** (1998) 249

[132] E. Il'ichev, V. Zakosarenko, V. Schultze, H.-G. Meyer, H. E. Hoenig,V. N. Glyantsev, A. Golubov, Appl. Phys. Lett. **72** (1998) 731

[133] Y. Zhang, W. Zander, J. Schubert, F. Rüders, H. Soltner, M. Banzet, N. Wolters, X. H. Zeng, A. I. Braginski, Appl. Phys. Lett. **71** (1997) 704

[134] P. Larsson, A. Ya. Tzalenchuk, Z. Ivanov, J. Appl. Phys. **90** (2001) 3450

[135] M. Bode, M. Grove, M. Siegel, A. I. Braginski, J. Appl. Phys. **80** (1996) 6378

[137] M. S. DiIorio, S. Yoshizumi, K.-Y. Yang, Appl. Phys. Lett. **67** (1995) 1926

[138] J. Mannhart, P. Chaudhari, Physics Today, November 2001, p.48

[139] Z. G. Ivanov, E. A. Stepantsov, T. Claeson, F. Wenger, S. Y. Lin, N. Khare, P. Chaudhari,
Phys. Rev. B **57** (1998) 602

[140] T. Minotani, S. Kawakami, T. Kiss, Y. Kuroki, K. Enpuku,
Jpn. J. Appl. Phys. **36** (1997) L 1092

[141] H. Q. Li, R. H. Ono, L. R. Vale, D. A. Rudman, S. H. Liou, Appl. Phys. Lett. **71** (1997) 1121

[142] F. Schmidl, S. Linzen, S. Wunderlich, P. Seidel, Appl. Phys. Lett. **72** (1998) 602

[143] B. Oelze, B. Ruck, E. Sodtke, A. F. Kirichenko, M. Yu. Kupriyanov, W. Prusseit,
Appl. Phys. Lett. **70** (1997) 658

[144] Y. Chong, B. Ruck, R. Dittmann, C. Horstmann, A. Engelhardt, G. Wahl, B. Oelze,
E. Sodtke, Appl. Phys. Lett. **72** (1998) 1513

[145] Y. H. Kim, J. H. Kang, J. M. Lee, T. S. Hahn, S. S. Choi, S. J. Park, Physica C **280** (1997) 304

[146] X.-F. Zhang, V. R. Todt, D. J. Miller, J. Mater. Res. **12** (1997) 3029

[147] M. Carmody, L. D. Marks, K. L. Merkle, Physica C **370** (2002) 228

[148] E. B. McDaniel, S. C. Gausepohl, C.-T. Li, M. Lee, J. W. P. Hsu, R. A. Rao, C. B. Eom,
Appl. Phys. Lett. **70** (1997) 1882

[149] B. Vuchic, K. L. Merkle, K. Char, D. B. Buchholz, R. P. H. Chang, L. D. Marks,
J. Mater. Res. **11** (1996) 2429

[150] K. Y. Constantinian, G. A. Ovsyannikov, A. D. Mashtakov, J. Ramos, Z. G. Ivanov, J. Mygind,
N. F. Pedersen, Physica C **273** (1996) 21

[151] S. Nicoletti, H. Moriceau, J. C. Villegier, D. Chateigner, B. Bourgeaux, C. Cabanel,
J. Y. Laval, Physica C **269** (1996) 255

[152] F. Tafuri, S. Shokhor, B. Nadgorny, M. Gurvitch, F. Lombardi, A. Di Chiara,
Appl. Phys. Lett. **71** (1997) 125;

A. Di Chiara, F. Lombardi, F. M. Granozio, U. Scotti di Uccio, M. Valentino, F. Tafuri,
A. Del Vecchio, M. F. De Riccardis, L. Tapfer, Physica C **273** (1996) 30

[153] J. Mannhart, H. Hilgenkamp, B. Mayer, Ch. Gerber, J. R. Kirtley, K. A. Moler, M. Sigrist,
Phys. Rev. Lett. **77** (1996) 2782

[154] E. Il'ichev, Physica C **367** (2002) 79

[155] A. J. Pauza, W. E. Booij, K. Herrmann, D. F. Moore, M. G. Blamire, D. A. Rudman,
L. R. Vale, J. Appl. Phys. **82** (1997) 5612;

K. Herrmann, A. J. Pauza, F. Baudenbacher, J. S. Santiso, D. F. Moore,
Physica C **274** (1997) 309;

W. E. Booij, A. J. Pauza, E. J. Tarte, D. F. Moore, M. G. Blamire,
Phys. Rev. B **55** (1997) 14600

[156] S. K. Tolpygo, M. Gurvitch, Appl. Phys. Lett. **69** (1996) 3914

[157] B. M. Hinaus, M. S. Rzchowski, B. A. Davidson, J. E. Nordman, K. Siangchaew, M. Libera,
Phys. Rev. B **56** (1997) 10828

[158] S.-J. Kim, H. Myoren, J. Chen, K. Nakajima, T. Yamashita, M. Esashi,
Jpn. J. Appl. Phys. **36** (1997) L 1096

[159] B. Ruck, B. Oelze, E. Sodtke, Supercond. Sci. Technol. **10** (1997) 991;

B. Ruck, B. Oelze, R. Dittmann, A. Engelhardt, E. Sodtke, W. E. Booij, M. G. Blamire,
Appl. Phys. Lett. **72** (1998) 2328

[160] S. Shokhor, B. Nadgorny, M. Gurvitch, V. Semenov, Yu. Polyakov, K. Likharev, S. Y. Hou, J. M. Phillips, Appl. Phys. Lett. **67** (1995) 2869

[161] A. S. Katz, A. G. Sun, S. I. Woods, R. C. Dynes, Appl. Phys. Lett. **72** (1998) 2032

[162] S. Morohashi, J. Wen, Y. Enomoto, N. Koshizuka, Jpn. J. Appl. Phys. **36** (1997) 5086

[163] K. Saitoh, T. Utagawa, Y. Enomoto, Appl. Phys. Lett. **72** (1998) 2754

[164] E. Fujimoto, H. Sato, T. Yamada, H. Akoh, Appl. Phys. Lett. **80** (2002) 3985

[165] T. Kito, Y. Yoshinaga, S. Izawa, M. Inoue, A. Fujimaki, H. Hayakawa, Physica C **378-381** (2002) 1322

[166] M. Sato, G. A. Alvarez, T. Utagawa, K. Tanabe, T. Morishta, Jpn. J. Appl. Phys. **41** (2002) 5572

[167] G. Hechtfischer, R. Kleiner, A. V. Ustinov, P. Müller, Phys. Rev. Lett. **79** (1997) 1365; G. Hechtfischer, R. Kleiner, K. Schlenga, W. Walkenhorst, P. Müller, H. L. Johnson, Phys. Rev. B **55** (1997) 14638

[168] A. Yurgens, D. Winkler, T. Claeson, N. V. Zavaritsky, Appl. Phys. Lett. **70** (1997) 1760; N. Mros, V. M. Krasnov, A. Yurgens, D. Winkler, T. Claeson, Phys. Rev. B **57** (1998) 8135

[169] A. Odagawa, M. Sakai, H. Adachi, K. Setsune, T. Hirao, K. Yoshida, Jpn. J. Appl. Phys. **36** (1997) L 21; A. Odagawa, M. Sakai, H. Adachi, K. Setsune, Jpn. J. Appl. Phys. **37** (1998) 486

[170] A. Irie, Y. Hirai, G. Oya, Appl. Phys. Lett. **72** (1998) 2159

[171] K. Schlenga, R. Kleiner, G. Hechtfischer, M. Mößle, S. Schmitt, P. Müller, Ch. Helm, Ch. Preis, F. Forsthofer, J. Keller, H. L. Johnson, M. Veith, E. Steinbeiß, Phys. Rev. B **57** (1998) 14518

[172] F. Schmidl, A. Pfuch, H. Schneidewind, E. Heinz, L. Dörrer, A. Matthes, P. Seidel, U. Hübner, M. Veith, E. Steinbeiss, Supercond. Sci. Technol. **8** (1995) 740

[173] H. B. Wang, P. H. Wu, J. Chen, K. Maeda, T. Yamashita, Appl. Phys. Lett. **80** (2002) 1604

[174] P. Vase, R. Flükiger, M. Leghissa, B. Glowacki, Supercond. Sci. Technol. **13** (2000) 71

[175] K. Heine, J. Tenbrink, M. Thöner, Appl. Phys. Lett. **55** (1989) 2441

[176] L. Masur, D. Parker, M. Tanner, E. Podtburg, D. Buczek, J. Scudiere, P. Caracino, S. Spreafico, P. Corsaro, M. Nassi, IEEE Trans. Appl. Supercond. **11** (2001) 3256

[177] Th. Lang, D. Buhl, S. Al-Wakeel, D. Schneider, L. J. Gauckler, Physica C **281** (1997) 283; Th. Lang, D. Buhl, L. J. Gauckler, Physica C **275** (1997) 284; D. Buhl, T. Lang, L. J. Gauckler, Supercond. Sci. Technol. **10** (1997) 32

[178] R. Funahashi, I. Matsubara, T. Ogura, K. Ueno, H. Ishikawa, Physica C **273** (1997) 337

[179] M. O. Rikel, D. Wesolowskii, A. A. Polyanskii, X. Y. Cai, K. Marken, H. Miao, E. E. Hellstrom, Physica C **372-376** (2002) 1839

[180] J. Sato, K. Ohata, M. Okada, K. Tanaka, H. Kitaguchi, H. Kumakura, T. Kiyoshi, H. Wada, K. Togano, Physica C **357-360** (2001) 1111

[181] H. Kitaguchi, K. Itoh, T. Takeuchi, H. Kumakura, H. Miao, H. Wada, K. Togano, T. Hasegawa, T. Kizumi, Physica C **320** (1999) 253; T. Hasegawa, T. Koizumi, N. Ohtani, H. Kitaguchi, H. Kumakura, K . Togano, H. Miao, Supercond. Sci. Technol. **13** (2000) 23

[182] H. Kumakura, Supercond. Sci. Technol. **13** (2000) 34

[183] J. Kellers, L. J. Masur, Physica C **372-376** (2002) 1040

[184] T. J. Arndt, B. Fischer, J. Gierl, H. Krauth, M. Munz, A. Szulczyk, M. Leghissa, H.-W. Neumueller, IEEE Trans. Appl. Supercond. **11** (2001) 3261

[185] J. Fujikami, T. Kaneko, N. Ayai, S. Kobayashi, K. Hayashi, H. Takei, K. Sato,
Physica C **378-381** (2002) 1061

[186] A. Polyanskii, D. M. Feldmann, S. Patnaik, J. Jiang, X. Cai, D. Larbalestier, K. DeMoranville,
D. Yu, R. Parella, IEEE Trans. Appl. Supercond. **11** (2001) 3269

[187] Q. Li, M. Suenaga, T. Kaneko, K. Sato, Ch. Simmon, Appl. Phys. Lett. **71** (1997) 1561

[188] P. Corsaro, M. Bechis, P. Caracino, W. Castiglioni, G. Cavalleri, G. Coletta, G. Colombo,
P. Ladiè, A. Mansoldo, R. Mele, S. Montagnero, C. Moro, M. Nassi, S. Spreafico, N Kelley,
C. Wakefield, Physica C **378-381** (2002) 1168

[189] T. Masuda, T. Kato, H. Yumura, M. Watanabe, Y. Ashibe, K. Ohkura, C. Suzawa,
M. Hirose, S. Isojima, K. Matsuo, S. Honjo, T. Mimura, T. Kuramochi, Y. Takahashi,
H. Suzuki, T. Okamoto, Physica C **378-381** (2002) 1174

[190] R. N. Bhattacharya, D. Banerjee, J. G. Wen, R. Padmanabhan, Y. T. Wang, J. Chen, Z. F. Ren,
A. M. Hermann, R D Blaugher, Supercond. Sci. Technol. **15** (2002) 1288

[191] J. Z. Wu, Y. Y. Xie, Z. W. Xing, R. S. Aga, Physica C **382** (2002) 62

[192] E. Bellingeri, R. E. Gladyshevskii, F. Marti, M. Dhallé, R. Flükiger,
Supercond. Sci. Technol. **11** (1998) 810

[193] H. Kitaguchi, H. Miao, H. Kumakura, K. Togano, Physica C **320** (1999) 71

[194] H. Kitaguchi, Physica C **335** (2000) 26

[195] Y. Aoki, T. Koizumi, N. Ohtani, T. Hasegawa,
L. Motowidlo, R. S. Sokolowski, R. M. Scanlan, S. Nagaya, Physica C **335** (2000) 1

[196] M. Minami, T. Nakano, S. Akita, H. Kasahara, H. Tada, M. Takahashi, Y. Nara, T. Yamanaka,
H. Sakaguchi, Physica C **357-360** (2001) 1323

[197] M. Okada, Supercond. Sci. Technol. **13** (2000) 29

[198] K. Tanaka, M. Okada, K. Ohata, J. Sato, H. Kitaguchi, H. Kumakura, K. Togano,
Physica C **357-360** (2001) 1102;
K. Ohata, J. Sato, M. Okada, K. Tanaka, H. Kitaguchi, H. Kumakura, K. Togano, T. Kiyoshi,
H. Wada, Physica C **357-360** (2001) 1107

[199] T. Wakuda, M. Okada, S. Awaji, K. Watanabe, Physica C **357-360** (2001) 1293

[200] T. Hase, Y. Murakami, S. Hayashi, Y. Kawate, T. Kiyoshi, H. Wada, S. Sairote, R. Ogawa,
Physica C **335** (2000) 6;
K. Fukushima, K. Tanaka, T. Wakuda, M. Okada, K. Ohata, J. Sato, T. Kiyoshi, H. Wada,
Physica C **357-360** (2001) 1297

[201] A. P. Malozemoff, W. Carter, S. Fleshler, L. Fritzemeier, Q. Li, L. Masur, P. Miles, D. Parker,
R. Parrel, E. Podtburg, G. N. Riley, Jr., M. Rupich, J. Scudiere, W. Zhang,
IEEE Trans. Appl. Supercond. **9** (1999) 2469

[202] Z. Han, P. Skov-Hansen, T. Freltoft, Supercond. Sci. Technol. **10** (1997) 371;
M. Polak, J. A. Parrell, A. A. Polyanskii, A. E. Pashitski, D. C. Larbalestier,
Appl. Phys. Lett. **70** (1997) 1034;
T. Sakai, H. Utsunomiya, Y. Saito, T. Hanamachi, M. Shinkawa, Physica C **277** (1997) 189

[203] T. R. Thurston, P. Haldar, Y. L. Wang, M. Suenaga, N. M. Jisrawi, U. Wildgruber,
J. Mater. Res. **12** (1997) 891;
M. Quilitz, W. Goldacker, Supercond. Sci. Technol. **11** (1998) 577

[204] G. Grasso, F. Marti, Y. Huang, A. Perin, R. Flükiger, Physica C **281** (1997) 271

[205] M. Dhallé, M. Cuthbert, M. D. Johnston, J. Everett, R. Flükiger, S. X. Dou, W. Goldacker,
T. Beales, A. D. Caplin, Supercond. Sci. Technol. **10** (1997) 21

[206] T. Kaneko, T. Hikata, M. Ueyama, A. Mikumo, N. Ayai, S. Kobayashi, N. Saga, K. Hayashi,
K. Ohmatsu, K. Sato, IEEE Trans. Appl. Supercond. **9** (1999) 2465

[207] P. M. Grant, T. P. Sheahen, cond-mat/0202386

[208] T. B. Peterson, U. Welp, G. W. Crabtree, N. Vasanthamohan, J. P. Singh, M. T. Lanagan,
V. K. Vlasko-Vlasov, V. I. Nikitenko, Appl. Phys. Lett. **71** (1997) 134;

M. Polak, W. Zhang, J. Parrell, X. Y. Cai, A. Polyanskii, E. E. Hellstrom, D. C. Larbalestier,
M. Majoros, Supercond. Sci. Technol. **10** (1997) 769

[209] J. Tenbrink, M. Wilhelm, K. Heine, H. Krauth, IEEE Trans. Appl. Supercond. **3** (1993) 1123

[210] M. Däumling, Supercond. Sci. Technol. **11** (1998) 590

[211] F. Gömöry, L. Gherardi, R. Mele, D. Morin, G. Crotti, Physica C **279** (1997) 39

[212] Y. B. Huang, R. Flükiger, Physica C **294** (1998) 71;

K. Kwasnitza, St. Clerc, R. Flükiger, Y. B. Huang, Physica C **299** (1998) 113;

K. Kwasnitza, S. Clerc, R. Flükiger, Y. Huang, Cryogenics **39** (1999) 829

[213] H. Eckelmann, M. Däumling, M. Quilitz, W. Goldacker, Physica C **295** (1998) 198;

W. Goldacker, J. Krelaus, R. Nast, H. Eckelmann,
IEEE Trans. Appl. Supercond. **11** (2001) 2947;

H. Eckelmann, R. Nast, C. Schmidt, W. Goldacker,
IEEE Trans. Appl. Supercond. **11** (2001) 2955

[214] M. P. Oomen, J. Rieger, M. Leghissa, J. J. Rabbers, B. Ten Haken, Physica C **340** (2000) 87;

M. P. Oomen, J. Rieger, M. Leghissa, B. Ten Haken, Supercond. Sci. Technol. **13** (2000) 1101

[215] M. Sugimoto, A. Kimura, M. Mimura, Y. Tanaka, H. Ishii, S. Honjo, Y. Iwata,
Physica C **279** (1997) 225

[216] K. Yamashita, F. Sumiyoshi, H. Hayashi, Physica C **357-360** (2001) 1218

[217] G. Grasso, R. Flükiger, Supercond. Sci. Technol. **10** (1997) 223;

X. -D. Su, G. Witz, K. Kwasnitza, R. Flükiger, Supercond. Sci. Technol. **15** (2002) 1184

[218] A. Usoskin, J. Dzick, A. Issaev, J. Knoke, F. García-Moreno, K. Sturm, H. C. Freyhardt,
Supercond. Sci. Technol. **14** (2001) 676 ;

Superconductor Week, October 14, 2002

[219] R. F. Service, Science **295** (2002) 787

[220] T. Aytug, J. Z. Wu, C. Cantoni, D. T. Verebelyi, E. D. Specht, M. Paranthaman, D. P. Norton,
D. K. Christen, R. E. Ericson, C. L. Thomas, Appl. Phys. Lett. **76** (2000) 760

[221] Q. X. Jia, S. R. Foltyn, P. N. Arendt, J. F. Smith, Appl. Phys. Lett. **80** (2002) 1601;

S. R. Foltyn, Q. X. Jia, P. N. Arendt, L. Kinder, Y. Fan, J. F. Smith,
Appl. Phys. Lett. **75** (1999) 3692

[222] D. M. Feldmann, D. C. Larbalestier, D. T. Verebelyi, W. Zhang, Q. Li, G. N. Riley,
R. Feenstra, A. Goyal, D. F. Lee, M. Paranthaman, D. M. Kroeger, D. K. Christen,
Appl. Phys. Lett. **79** (2001) 3998

[223] G. Hammerl, A. Herrnberger, A. Schmehl, A. Weber, K. Wiedenmann, C. W. Schneider,
J. Mannhart, Appl. Phys. Lett. **81** (2002) 3209

[224] K. Kakimoto, Y. Iijima, T. Saitoh, Physica C **378-381** (2002) 937;

Y. Iijima, K. Kakimoto, K. Takeda, Physica C **357-360** (2001) 952

[225] A. Knierim, R. Auer, J. Geerk, G. Linker, O. Meyer, H. Reiner, R. Schneider,
Appl. Phys. Lett. **70** (1997) 661

[226] J. R. Groves, P. N. Arendt, S. R. Foltyn, Q. Jia, T. G. Holesinger, H. Kung, R. F. DePaula,
P. C. Dowden, E. J. Peterson, L. Stan, L. A. Emmert, Physica C **382** (2002) 47

[227] Y. Sato, S. Honjo, Y. Takahashi, K. Muranaka, K. Fujino, T. Taneda, K. Ohmatsu, H. Takei,
Physica C **378-381** (2002) 1118;

T. Taneda, K. Muranaka, K. Fujino, K. Ohmatsu, H. Takei, Y. Sato, S. Honjo, Y. Takahashi, Physica C **378-381** (2002) 944

[228] R. Metzger, M. Bauere, R. Semerad, P. Berberich, H. Kinder, IEEE Trans. Appl. Supercond. **11** (2001) 2826

[229] A. Goyal, D. F. Lee, F. A. List, E. D. Specht, R. Feenstra, M. Paranthaman, X. Cui, S. W. Lu, P. M. Martin, D. M. Kroeger, et al. , Physica C **357-360** (2001) 903;

E. D. Specht, F. A. List, D. F. Lee, K. L. More, A. Goyal, W. B. Robbins, D. O'Neill, Physica C **382** (2002) 342

[230] J. Badin, Viewgraphs "2002 DOE Wire Workshop", January 22, 2002, St. Petersburg, FL, http://www.eren.doe.gov/superconductivity/docs/06badin_presentation.ppt

[231] J. Eickemeyer, D. Selbmann, R. Opitz, B. de Boer, B. Holzapfel, L. Schultz, U. Miller, Supercond. Sci. Technol. **14** (2001) 152

[232] T. Watanabe, K. Wada, Y. Ohashi, M. Ozaki, K. Yamamoto, T. Maeda, I. Hirabayashi, Physica C **378-381** (2002) 911;

T. Watanabe, K. Matsumoto, T. Maeda, T. Tanigawa, I. Hirabayashi, Physica C **357-360** (2001) 914

[233] T. Doi, M. Mori, H. Shimohigashi, Y. Hakuraku, K. Onabe, M. Okada, N. Kashima, S. Nagaya, Physica C **378-381** (2002) 927

[234] Q. Li, W. Zhang, U. Schoop, M. W. Rupich, S. Annavarapu, D. T. Verebelyi, C. L. H. Thieme, V. Prunier, X. Cui, M. D. Teplitsky et al. , Physica C **357-360** (2001) 987;

M. Paranthaman, T. Aytug, S. Sathyamurthy, D. B. Beach, A. Goyal, D. F. Lee, B. W. Kang, L. Heatherly, E. D. Specht, K. J. Leonard et al. , Physica C **378-381** (2002) 1009

[235] H. Hiei, K. Yamagiwa, Y. Takahashi, S. B. Kim, Yasuji Yamada, J. Shibata, T. Hirayama, H. Ikuta, I. Hirabayashi and U. Mizutani, Physica C **357-360** (2001) 942;

H. Fuji, T. Honjo, Y. Nakamura, T. Izumi, Y. Shiohara, R. Teranishi, M. Yoshimura, Y. Iijima, T. Saitoh, Physica C **378-381** (2002) 1013

[236] T. Izumi, X. Yao, K. Hasegawa, M. Kai, Y. Tokunaga, S. Asada, Y. Nakamura, T. Izumi, T. Watanabe, Y. Shiohara, Physica C **378-381** (2002) 989;

Y. Yamada, T. Suga, H. Kurosaki, S. B. Kim, T. Maeda, Y. Yamada, I. Hirabayashi, Y. Iijima, K. Kakimoto, T. Saitoh, Physica C **378-381** (2002) 971

[237] G. Desgardin, I. Monot. B. Raveau, Supercond. Sci. Technol. **12** (1999) R115

[238] D. Litzkendorf, T. Habisreuther, R. Müller, S. Kracunovska, O. Surzhenko, M. Zeisberger, J. Riches, W. Gawalek, Physica C **372-376** (2002) 1163

[239] A. W. Kaiser, H. J. Bornemann, R. Koch, Inst. Phys. Conf. Ser. No. **158** (1997) 837

[240] T. Fujimoto, M. Mitsuru, N. Masahashi, T. Kaneko, Inst. Phys. Conf. Ser. No. **167** (1999) 79

[241] T. Prikhna, W. Gawalek, V. Moshchil, A. Surzhenko, A. Kordyuk, D. Litzkendorf, S. Dub, V. Melnikov, A. Plyushchay, N. Sergienko, A. Koval', S. Bokoch, T. Habisreuther, Physica C **354** (2001) 333;

A. A. Kordyuk, V. V. Nemoshkalenko, A. Plyushchay, T. Prikhna, W. Gawalek, Supercond. Sci. Technol. **14** (2001) 41

[242] M. Zeisberger, T. Habisreuther, D. Litzkendorf, R. Müller, O. Surzhenko, W. Gawalek, Physica C **372-376** (2002) 1890;

A. A. Kordyuk, V. V. Nemoshkalenko, R. V. Viznichenko, W. Gawalek, T. Habisreuther, Appl. Phys. Lett. **75** (1999) 1595

[243] A. Erb, J.-Y. Genoud, M. Dhalle, F. Marti, E. Walker, R. Flükiger, Inst. Phys. Conf. Ser. No. **158** (1997) 1109

[244] M. Murakami, Inst. Phys. Conf. Ser. No. **167** (1999) 7

[245] T. Yamada, H. Ikuta, M. Yoshikawa, Y. Yanagi, Y. Itoh, T. Oka, U. Mizutani,
Physica C **378-381** (2002) 713

[246] G. Krabbes, G. Fuchs, P. Verges, P. Diko, G. Stöver, S. Gruss, Physica C **378-381** (2002) 636;
S. Gruss, G. Fuchs, G. Krabbes, P. Verges, G. Stöver, K.-H. Müller, J. Fink, L. Schultz,
Appl. Phys. Lett. **79** (2001) 3131;

[247] M. Tomita, M. Murakami, Nature **421** (2003) 517

[248] M. Tsuchimoto, H. Takashima, M. Tomita, M. Murakami, Physica C **378-381** (2002) 718

[249] K. Katagiri, A. Murakami, T. Sato, T. Okudera, N. Sakai, M. Muralidhar, M. Murakami,
Physica C **378-381** (2002) 722

[250] G. Krabbes, Th. Hopfinger, C. Wende, P. Diko, G. Fuchs,
Supercond. Sci. Technol. **15** (2002) 665;
P. Diko, G. Fuchs, G. Krabbes, Physica C **378-381** (2002) 636

[251] M. Tomita, M. Murakami, Supercond. Sci. Technol. **13** (2000) 722;
M. Tomita, M. Murakami, Physica C **378-381** (2002) 842

[252] H. Teshima, M. Sawamura, M. Morito, M. Tsuchimoto, Cryogenics **37** (1997) 505

[253] P. Stoye, W. Gawalek, P. Görnert, A. Gladun, G. Fuchs,
Inst. Phys. Conf. Ser. No. **158** (1997) 1535

[254] W. Hennig, D. Parks, R. Weinstein, R.-P. Sawh, Appl. Phys. Lett. **72** (1998) 3059

[255] H. Fujimoto, H. Kamijo, Physica C **335** (2000) 83

[256] T. Wolf, A.-C. Bornarel, H. Küpfer, R. Meier-Hirmer, B. Obst, Phys. Rev. B **56** (1997) 6308

[257] M. Matsui, N. Sakai, M. Murakami, Physica C **378-381** (2002) 732

[258] H. Ikuta, H. Ishihara, Y. Yanagi, Y. Itoh, U. Mizutani, Supercond. Sci. Technol. **15** (2002) 606

[259] S. Nariki, N. Sakai, M. Murakami, Supercond. Sci. Technol. **15** (2002) 648;
S. Nariki, N. Sakai, M. Murakami, Physica C **378-381** (2002) 631

[260] M. Muralidhar, M. Jirsa, N. Sakai, M. Murakami, Supercond. Sci. Technol. **16** (2003) R1;
M. Muralidhar, M. Jirsa, N. Sakai, M. Murakami, Appl. Phys. Lett. **79** (2001) 3107;
M. Muralidhar, M. Jirsa, N. Sakai, M. Murakami, Supercond. Sci. Technol. **15** (2002) 688;
M. Muralidhar, M. Murakami, Supercond. Sci. Technol. **15** (2002) 683;
M. Muralidhar, M. Murakami, Physica C **378-381** (2002) 627

[261] M. Muralidhar, N. Sakai, N. Chikumoto, M. Jirsa, T. Machi, M. Nishiyama, Y. Wu,
M. Murakami, Phys. Rev. Lett. **89** (2002) 237001

FASCINATING THALLIUM-BASED SUPERCONDUCTING CUPRATES AND SUBSYSTEMS : FORMATION AND STABILITY

Jean-Louis JORDA[a,b], Philippe GALEZ[a], Sovannary PHOK[b], Thomas HOPFINGER[a], Timothé Koffi JONDO[a,c]

[a] Laboratoire d'Instrumentation et des Materiaux d'Annecy (LAIMAN), ESIA, Université de Savoie, Domaine Universitaire, B.P.808, F-74016- Annecy Cedex, France
[b] Institut National Polytechnique de Grenoble, LMGP, F-38402 – Saint Martin d'Hères Cedex, France
[c] Permanent address: Université du Benin, Lomé, Togo

1. INTRODUCTION

After the discovery by Sheng and Hermann [1,2] of superconductivity above the liquid nitrogen temperature in thallium-based cuprates, an intensive work was undertaken in order to understand the relations between the crystallographic structure and the superconductivity in the two identified series $Tl_mBa_2Ca_{n-1}Cu_nO_x$ (m=1,2; n=0-4) [3-5], hereafter referred to as Tl-m2(n-1)n, differing by the number m of TlO layers per formula unit. Reviews on the systems have been soon writen by Hervieu et al.[6], Greenblatt et al.[7], Siegal et al. [8] , and more recently by Gladyshevskii and Galez [9]. A special attention was first devoted to Tl-2201, which may be prepared with two different crystal-type structure, orthorhombic and tetragonal, and which is the only example in the high T_c's cuprates showing a superconducting transition temperature up to 90 K with the two crystal symmetries [4,10-14]. The interest however rapidly turned over to the third member of the single TlO layer series, Tl-1223 with an optimum critical temperature T_c= 122-125 K [2,15]. Intragrain critical current densities J_c in the range of 10^4-10^5 A.cm^{-2} at 77 K in a magnetic field of 1 Tesla [16,17], have been measured for the $(Tl,Pb,Bi)(Ba,Sr)_2Ca_2Cu_3O_x$ substituted systems and fluorine addition has been recognized to decrease the degradation of J_c under magnetic field [18,19]. Despite the suspicious approach to thallium compounds, due to their toxicity, such intrinsic superconducting properties resulted in an impressive number of investigations on crystallographic structure [see ref.9 for a review], preparation methods, chemistry and related superconducting state of these substituted (Tl,M)-1223 phases, with M= Pb, Bi, Hg, Cu, Ru, Li or rare-earths [20-31]. When the production of high quality wires and tapes for power applications became the challenge for high temperature superconductors, technological constraints favoured searches on the yttrium or bismuth- based cuprates but the thallium systems continued to be seriously considered as potential candidates [32-35]. With the so-called second generation of superconductors, especially the coated conductors, critical currents densities of about 10^6 A.cm^{-2} have been measured on substituted (Tl,Pb)-1223 films prepared by screen printing [36], electrodeposition [37] or spray pyrolysis [38]. Similar performances have been reached very recently with unsubstituted Tl-1223 thick films prepared by a two-step process implying the spray pyrolysis of a $Ba_2Ca_2Cu_3O_x$ precursor [39-41]. This is a clear indication that thallium substitution is not the only key to improve superconducting transport properties. In addition, it is evident but not unprofitable to note that partial substitution of Tl and/or Ba in thallium cuprates increases the number of constituents of the thermodynamic system and then, due to the Gibbs rule, increases the difficulty to describe the equilibrium state. In reality the knowledge of the path for the formation of the phases of interest as well as the phase relations in the multicomponent Tl-Ba-Ca-Cu-O system is of major importance because they give the possibility to control the nature and the quantity of impurity phases and moreover, indicates the solid-liquid equilibrium fields which is of primordial interest for high temperature grain growth. These two aspects have been partially explored in the past years. E.Ruckenstein and C.T Cheung [42,43] first reported the reaction pathways for the formation of the Tl-2223, Tl-1223, Tl-1234 and Tl-1212 phases at 867°C. They discovered that the single TlO layer compounds are formed following the sequence Tl-1201 $\rightarrow$ Tl-2201$\rightarrow$ Tl-2212$\rightarrow$ Tl-2223 $\rightarrow$ Tl-1223 or Tl$\rightarrow$1234. Using BaO_2 as barium source, they also stressed the role of a "glassy" phase prior to the crystallisation of the superconducting compounds and noted the effect of the initial grain size of the CaO and CuO starting powder on the final state of the reaction. On the other hand, Aselage et al.[44,45] reported on the stability of Tl-based superconductors with Ba:Ca:Cu = 2:1:2, 2:2:3 and 2:3:4, as a function of the Tl_2O partial pressure at a temperature of 880°C at constant oxygen pressure. They found that depending on the Tl_2O partial pressure, intergrowth mixtures of Tl-2223 and Tl-1223 may be observed and, using a two-zone furnace they proposed stability phase diagrams useful for the preparation of bulk and film materials.

However, both the reaction pathways for the formation of the phases and the thermodynamic equilibrium cannot be well understood if the lower order systems are doubtful and also if the mechanisms for phase formation are not known. For these reasons and considering the lack of thermodynamic informations on the binary and ternary constituting systems of the quaternary $TlO_{1.5}$-BaO-CaO-CuO, we have undertaken systematic studies of phase equilibrium, combined with reaction kinetics. This continuous effort reflects our thought process on the study of materials, applied here to superconducting thallium cuprates [46-55].

In this chapter we critically review our results on the formation of selected phases, reaction kinetics and phase equilibrium diagrams related to the thallium cuprates. In the reaction pathway, the nature of the starting compounds (metal oxide, carbonates...) is evaluated and the role of binary and ternary phases is evidenced. The relations between the formation of the phases and the superconducting properties namely the critical temperatures T_c, the irreversibility lines (IRL) and the critical current densities J_c are also discussed, essentially for Tl-1223.

2. BASICS FOR SOLID STATE REACTION KINETICS

When reactions kinetics are discussed for the formation of high temperature superconducting cuprates, they are almost restricted to oxygen diffusion [56,57] and isothermal grain growth process, usually analysed with an Avrami formalism [58-60]. Because more details concerning the mechanisms which are rate limiting in a solid state reaction may be found by kinetics studies, we summarize here basics for such studies.

If we note α the conversion extend of the reaction, that is the proportion of the produced material B in a reaction A $\rightarrow$ B, the reaction rate $d\alpha/dt$, may be expressed as $d\alpha/dt = f(\alpha).k(T)$ (see for instance [61-65]) which applies for both isothermal and non isothermal experiments. In this equation $f(\alpha)$ describes the kinetic model for the decomposition and $k(T)$ reflects the change of the rate coefficient with the temperature. The identification of $f(\alpha)$ is generally time consuming and indeed, for heterogeneous reactions, several functions may satisfactorily fit the data with an honest accuracy. Sestak and Berggren [66] proposed that for a large majority of processes $f(\alpha) = \alpha^n.(1-\alpha)^m.[-Ln(1-\alpha)]^p$ with at least one of the exponents (n, m or p) being zero. Note that this expression is valid as well for reactions rate-limited by 1-dimension diffusion (parabolic law) with m=p=0 and n = -1, as for generalized Johnson-Mehl and Avrami nucleation and growth (n=0) mechanism (see [67] for a review). Note also that another approach to reaction mechanisms consists in using the integral equation for the reaction rate $\int \frac{d\alpha}{f(\alpha)} = g(\alpha) = k(T).t$. In almost all cases, $k(T)$ is well represented by an Arrhenius-type equation $k(T) = k_0.exp(- E_A/RT)$, k_0 is the frequency factor, E_A the apparent activation energy and R the gas constant. Generally (and successfully) applied for both homogeneous and heterogeneous reactions, the use of the Arrhenius function for $k(T)$ received a recent theoretical justification at least for solid state reactions involving ionic crystals [68].

A great number of methods, based on DTA and TG data on small samples, has been developed to calculate the kinetic parameters of solid state reactions. An example of rapid analysis of a DTA/TG run is due to Phadnis and Deshpande [69]. The method is based on the use of a power law for $f(\alpha)$, $f(\alpha) = (1-\alpha)^m$. Noting that, at the maximum of the transformation rate $(d\alpha/dt)_{max}$, the second derivative $d^2\alpha/dt^2$ is zero, elementary mathematics allow to found the pseudo- reaction order m and the apparent activation energy E_A from a plot of $Ln[(d\alpha/dt)/(1-\alpha)^m]$ vs 1/T. Much more accurate are the isoconversion methods at different heating rates β in non isothermal experiments or at different temperatures for isothermal treatment of kinetics [70-74]. Indeed, for a constant α, the reaction rate depends only, in an Arrhenius way, on the temperature. It is then easy to find E_A and k_0 even for unknown $f(\alpha)$. When a reaction involves a weight change of the reactants and/or products, for example when the overall oxygen content varies, the conversion α can be obtained from TG experiments: $\alpha = [w(t)-w_0]/[w_f-w_0]$ where w_0 and w_f are the initial and final mass of the sample, respectively and $w(t)$ the mass at the time t.

3. THE SINGLE CATION SYSTEMS FOR THE SUPERCONDUCTING CUPRATES

The preparation of the thallium-based superconducting cuprates often includes a first reaction treatment having for objectives to decompose the alkaline-earth salts or peroxides, nitrates, carbonates... and to form, with CuO, a multiphase mixed-oxide precursor with chemical formula "$Ba_2Ca_{n-1}Cu_nO_z$". This first step raises the first problem due to the barium and calcium sources which are highly sensitive to water and carbon. Thallium is usually added in the form of thallium sesqui-oxide Tl_2O_3 and, in the case of fluorine doping, TlF is often preferred to more stable fluorides, TlF_3 or CuF_2. In the stability range of these components, the thermodynamic systems may be considered as single component systems, but when the temperature is increased decomposition may occur and the result depends on the oxygen partial pressure. The variance is then increased and the systems must be described as multicomponent.

3.1. The barium sources

The choice of the barium source is directed by the necessity to avoid the formation of barium carbonate which has a high decomposition temperature and which affects, under particular conditions, the phase equilibrium of the superconducting systems. As a matter of fact, the carbon impurities are known either to form new cuprate series at high pressure [75,76] or to favour the formation of Tl-2223 detrimental to Tl-1223 [77,78] at ambient pressure, for carbon content $y > 0.11$ in $TlBa_2Ca_2Cu_3O_{8.5+\delta}C_y$. The use of barium carbonate as a Ba source precursor must then be restricted to systems for which high temperature treatments (T>980°C) are possible. In figure 1 we have reported the differential thermal analysis (DTA) coupled with gravimetric measurements (TG) showing the decomposition in an argon flow of $BaCO_3$ for a heating rate of 6°C/min. From the DTA curve we can see the structural transitions from orthorhombic Pmcn space group, a= 5.27(2) Å, b= 9.05(5) Å, c= 6.39(1) Å (measured at 650°C) to trigonal SPCG R-3m, a= 6.21(1) Å, c= 10.55(5) Å (measured at 850°C) at 795°C and to cubic Fm-3m space group, a = 6.960 Å at 978°C. These temperature values well agree with previous works [79,80] performed in oxygen. The decomposition of the carbonate, deduced from the deviation of the baseline of the TG signal, starts at 955°C, just below the formation temperature of the cubic-type structure, and ends at 1300°C. From the weight change it is possible to have access to the kinetics for the decomposition and particularly to the activation energy.

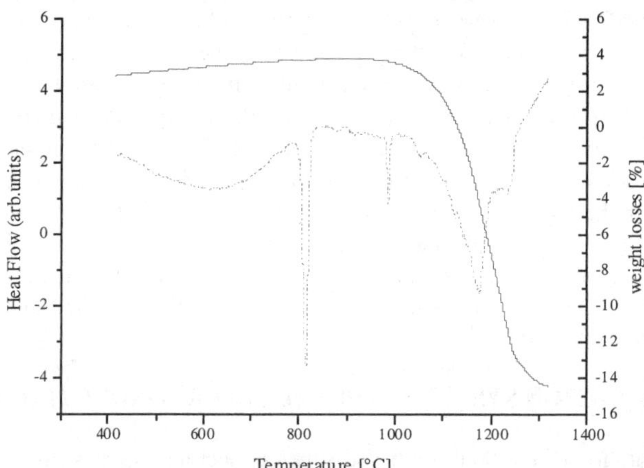

Fig.1. DTA/TG of barium carbonate in argon flow. Heating rate $\beta = 6$°C/min

We first observe on the heat flow curve a small endothermic signal at 1035°C, due to the eutectic reaction L = BaCO$_3$ + BaO. The α value at this temperature is 2.5%, then, almost all the BaO formation from the carbonate occurs in presence of a liquid phase. At 1245°C a liquidus line is crossed. For the accelerating part ($d^2\alpha/dt^2 > 0$) of the carbonate decomposition, and for temperatures lower than 1200°C, we found, using the formalism of Phadnis and Deshpande [69], m=1.6 and an activation energy E_A = 302 kJ.mol^{-1} which well compares with previous experimental studies [81,82].

In order to avoid the formation of carbo-oxide superconducting phases, the use of the barium peroxide BaO$_2$ as barium source is frequent. We have recently shown that three mechanisms are implied in the decomposition of the barium peroxide, in argon or oxygen flows, in the temperature range 773 K< T< 1273 K [53]. They are evidenced by the change of the activation energy with α in isoconversion analysis (fig.2) using a differential or an integral formalism [83-86]. The activation energy of the oxygen removal in the BaO$_{2-x}$ solid solution, up to α = 0.25 is 168 $\pm$ 4 kJ.mol^{-1} may be associated with a 3D diffusion mechanism as suggested by Triblehorn and Brown [87]. The disturbed region for 0.25<α<0.7 is due to the

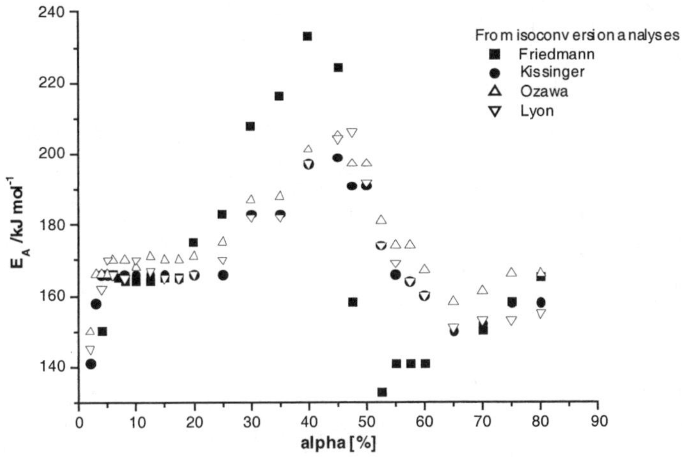

Fig.2: Activation energies for the decomposition of BaO$_2$ in argon

crossing of a monovariant (BaO$_{2-x}$)-(BaO$_{1+y}$) equilibrium line or an invariant equilibrium with a liquid phase, both implying monoxide nucleation and growth. Finally, for α >0.7, a slow increase of E_A from 140 kJ.mol^{-1} to 160 kJ.mol^{-1} is assumed to be due to a monovariant (BaO$_{1+y}$)- liquid equilibrium. Similar analysis in oxygen flowing with a heating rate as low as β= 1 K.min^{-1}, for which the equilibrium state is expected to be reached, allowed us to confirm the peritectic formation of the BaO$_{2-x}$ solid solution (fig.3), previously suggested by the assessment of Zimmerman et al. [88] in contradiction with a congruent melting behaviour [89]. It is interesting to note here that both the barium peroxide and monoxide have been found, from high temperature X-ray diffraction to coexist up to the peritectic temperature T= 1080 K as shown in figure 4 . The solid solutions in oxygen, at this temperature extend up to x

34

= 0.27 and y = 0.14 for (BaO_{2-x}) and (BaO_{1+y}) respectively. It has to be mentioned that the formation of metastable BaO_z phases suggested by Mayorova et al. [90] was not confirmed here.

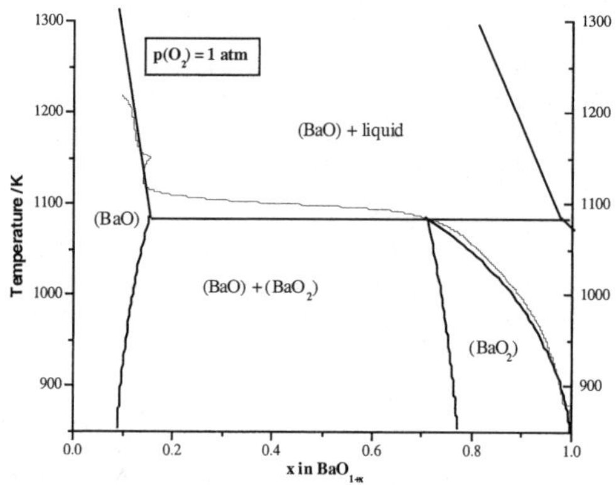

Fig.3: The phase equilibrium diagram of the BaO_2- BaO system in oxygen

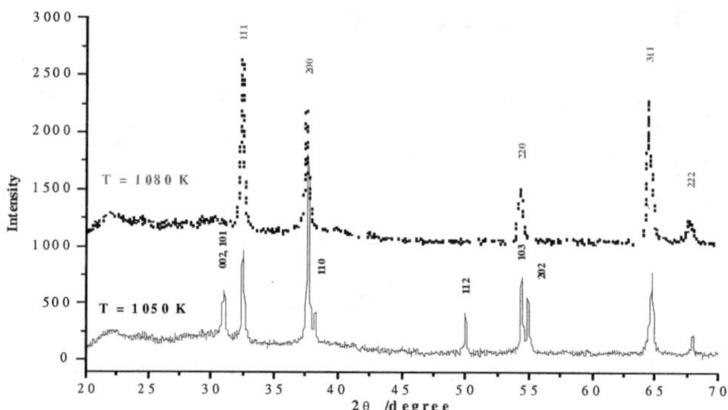

Fig.4 : High temperature X-ray diffraction of the barium oxide in oxygen. At 1050 K the peroxide and the monoxide coexist. BaO is the stable phase at 1080 K

The reversibility of the reaction $BaO_{2-x} = BaO_{1+y} + [1-(x+y)]/2\ O_2$ must be taken into account in the study of higher order systems because it may influence the equilibrium states and/or the local oxygen content of the phases under investigation.

3.2. The thallium sources

The toxicity of thallium compounds is the reason for a general use of reactions in closed systems. For instance, in a two step process for the preparation of the bulk or coated superconducting phases, a sintered (respectively coated) precursor $Ba_2Ca_{n-1}Cu_nO_x$ is thallinated in sealed tubes. Following the IPCS recommendation [91] dust should be controlled at $0.1mg/m^3$. It is important to realise, simultaneously with the high toxicity of the element , that thallium compounds are, in small quantity, of industrial use: the sulfide for photocells, the bromide and iodide as infrared materials and special glasses with high refraction index are produced using the oxide Tl_2O_3.

The decomposition of Tl_2O_3 has been studied by Wahlbeck et al.[92] using Knudsen effusion mass spectroscopy and thermogravimetry. From vapour pressure measurements the authors deduced that Tl_2O_3 is stabilized when placed in superconducting Tl-2223 with the change in the chemical potential given by $\mu(Tl_2O_3,SC)$ - $\mu(Tl_2O_3, pure)$ = -33.84 ±3 $kJ.mol^{-1}$. In their experiments, Walbeck et al. [92] concluded to a congruent vaporization of Tl_2O_3. Holstein [93], in a detailed thermodynamic study of the volatilisation of thallium oxides showed that Tl_4O_3 is the stable condensed phase at oxygen partial pressures less than 10^{-1} atm. up to 990 K and deduced the TlO phase diagram in which, for 1 atm. of oxygen, solid Tl_2O_3 is stable at least up to 1113 K. We have analysed by DTA/TG with a heating rate β= 1°C/min (fig.5), the decomposition in oxygen flowing of a Tl_2O_3 specimen weighting 36 mg, placed in a Al_2O_3 crucible.

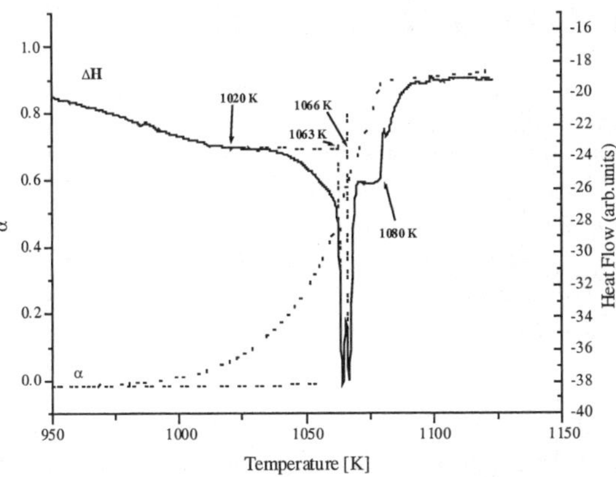

Fig.5: Decomposition of Tl_2O_3 in oxygen flow- Heating rate 1°C/min

In figure 5 where the TG has been converted to the extend of the reaction α, we see that Tl_2O_3 starts to decompose at 975 K and is completely evaporated at 1080K. In this temperature range two endothermic reactions are observed which are believed to be due to the remaining solid material in the crucible, not yet decomposed.

For temperatures lower than 950 K we did not found any significant weight losses which could correspond to the formation of Tl_2O so that we have to conclude that sublimation of Tl_2O_3 occurs at 975 K before melting. In our experimental conditions, for approximately 50% of the sample evaporated, the remaining solid Tl_2O_3 melts at 1063 K supporting the result of Lamoreaux et al. [94] for which solid Tl_2O_3 is the stable phase for T<1060 K. The conversion of liquid Tl_2O_3 to liquid Tl_2O is assumed to correspond to the second DTA endotherm at 1066 K for which a change in the decomposition rate is observed.

Fluorine doping has been reported to improve the superconducting properties of the thallium cuprates. The group of Tachikawa and Kikuchi [95,96] studied the effect of fluorine addition on different compounds of the Tl-series, and especially of Tl-1223 for which they found that TlF substitution for Tl_2O_3 acts like an effective flux enhancing the formation of the phase; Sun et al. [97] on the basis of Mössbauer experiments discussed on fluorine substitution on the oxygen sites in the Tl-2201 cell; Bellingeri et al. [19] and Hamdan et al. [98] found that fluorine additions favour the formation of substituted (Tl,Pb)-1223 and improve the transport properties. Recently Phok et al. [41] reported similar results on unsubstituted Tl-1223 coated conductors using TlF as fluorine source. It is argued that such effects are due to the low melting temperature of TlF. The presence of a liquid phase during the formation of the superconducting phases could be beneficial to the formation kinetics and the grain growth. As a matter of fact, TlF was found to melt at 266°C without decomposition , in flowing oxygen and the liquid evaporates at 578°C. In closed system, for the formation of superconducting phases at temperatures near 900°C, the TlF vapor is thus expected to react with the precursor system similarly Tl_2O_3. Then, improved microstructures and superconducting properties are not due to liquid TlF, but more likely, to the lowering of the melting temperature of the fluorine doped cuprates.

4. THE BINARY AND TERNARY CATION SYSTEMS

If we consider the equilibrium between condensed phases in the Tl-Ba-Ca-Cu-O system, the use of the metal-oxides as components reduces the number of independent variables allowing then a quaternary $TlO_{1.5}$-BaO-CaO-CuO phase diagram representation. The knowledge of this system implies the knowledge of 6 binaries and 4 ternaries. However, at high temperature the volatility of the thallium oxide, the possible formation of the barium peroxide and of Cu_2O have to be taken into account and increase the difficulty of the study. In this section we shall focuss our attention on the binary or pseudo-binary systems which have been found to have some importance in the formation of the superconducting phases.

4.1. The $TlO_{1.5}$-BaO system

The binary $TlO_{1.5}$-BaO system, studied by Jondo et al.[46] under an oxygen pressure of 1 atm. is represented in figure 6. The known crystallographic data are reported in table 1.

Table1. Summary on the crystal structure of the compounds identified in the $TlO_{1.5}BaO$

Compound	Crystal system	Cell parameters	references
Tl_2BaO_4	Orthorhombic Pnma	a= 10.46(3) Å, b=11.93(3)Å, c= 3.476(4) Å	[99, 100]
$Tl_6Ba_4O_{13}$	Orthorhombic	a= 5.748 Å, b= 7.164 Å, c= 9.572 Å	[101]
$Tl_2Ba_2O_5$	Orthorhombic Pcmn	a= 6.264 Å, b= 17.258 Å, c= 6.05 Å	[102]
$Tl_2Ba_4O_7$	Tetragonal	a= 5.33 Å, c= 23.04 Å	[46]

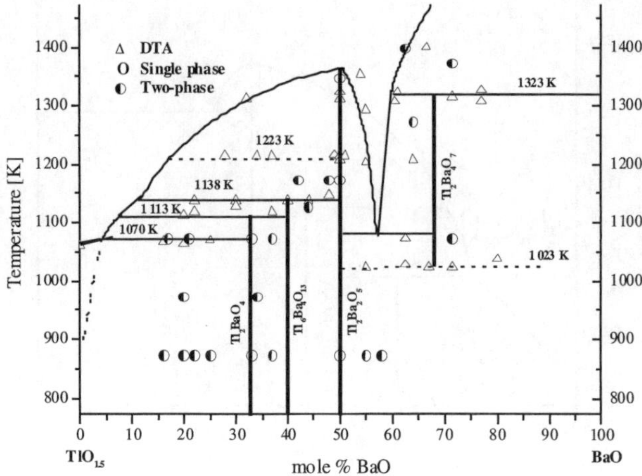

Fig.6: Phase diagram of the $TlO_{1.5}$-BaO system.

The compound $Tl_4Ba_3O_9$ and the rich-thallium phases $Tl_{10}Ba_2O_{17}$ and $Tl_{14}Ba_6O_{27}$ have been also reported by Lopato et al.[99] and Zhou et al.[100], respectively. They were not confirmed by the investigation of Jondo et al. [46] who, on the other hand observed the formation $Tl_2Ba_4O_7$ above 1023 K which was assumed for the calculation of the lattice parameters reported in table 1, to be isotypic with tetragonal $Tl_2Sr_4O_7$ described by Von Scheng and Muller-Buschbaum [103]. Amongst these four compounds, $Tl_2Ba_2O_5$ which melts congruently at 1373 K appeared to be an interesting precursor for the preparation of the two TlO-layer superconducting phases especially, as we will see in the following for $Tl_2Ba_2CuO_6$.

4.2. The $TlO_{1.5}$-CaO system

The system has not been yet reported in the literature. Goutenoire et al. [104-106] discovered the new calcium thallate series $Tl_2Ca_nO_{n+3}$ and solved the crystallographic structure of the members with n = 1-3, including an intergrowth structure with n= 1.5. A first thermodynamic exploration of the system due to T.K.Jondo [107] is presented here. It is based on X-ray diffraction characterization and DTA/TG measurements taking into account the mass losses attributed to Tl_2O_3 departure for the determination of the invariant reactions in the system. The phase equilibrium diagram is tentatively reproduced in figure 7.

The DTA signals reported in this figure may be interpreted on the basis of the existence of the four compounds first identified by the Raveau group [104-106], formed by peritectic reaction except $Tl_2Ca_3O_6$ which seems to be congruently melting. With this assumption, some endothermic signals observed at temperatures above 1100 K, mostly in the rich-thallium oxide side of the phase diagram cannot be explained. They could be due a liquid-vapour equilibrium or to the reaction of the liquid with the alumina crucibles used for the DTA experiments.

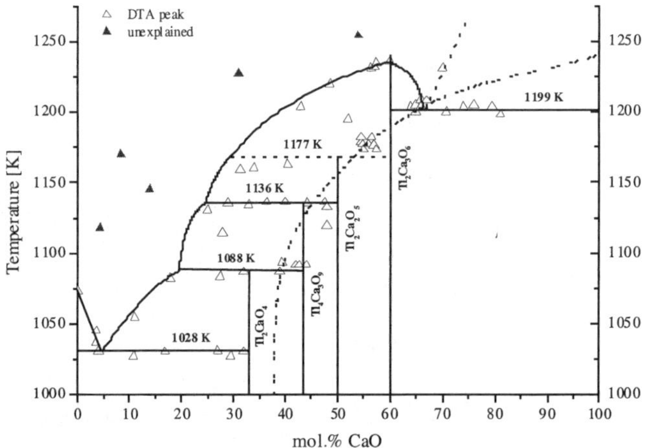

Figure 7. The $TlO_{1.5}$-CaO system

We have also reported in figure 7 the change in composition with the temperature for a sample containing initially 38 mole % CaO. It has to be noticed that contradictory to the $TlO_{1.5}$-BaO system, the thallium oxide is not stabilised in any of the compounds in the $TlO_{1.5}$-CaO system. All the samples prepared for this study showed a decomposition temperature only 50°C higher than for pure thallium oxide. The eutectic reaction at 1028 K between pure $TlO_{1.5}$, Tl_2CaO_4 and the liquid may explain this behaviour.

For the completion of this section devoted to the thallium-based binary systems, we have to mentioned that the $TlO_{1.5}$-CuO system has been studied by T.K.Jondo et al [47] in oxygen flowing and appears to be a single eutectic system at 1033 K, with a possible 0.12 at.% copper solubility in thallium oxide [108].

4.3. The Ba-Cu-O puzzle

In 1988, early after the discovery of high T_c's superconductors, Abbattista et al.[109] noticed that the knowledge of the ternary Ba-Cu-O system is a constrained path for correlating the superconducting properties of $YBa_2Cu_3O_7$ with the composition. Fifteen years later, this remark still remains valid for an increased number of superconducting systems, including the thallium phases. The system was yet investigated from thermodynamic and structural viewpoints. Related to the study of the Y-Ba-Cu-O system, Nevriva et al. [110] confirmed the previous work of Roth et al. [111] concerning the congruent melting of $BaCuO_2$ and specified the eutectic temperature in the $BaCuO_2$-CuO tie-line. De Leeuw et al. [112] investigated essentially the compound $Ba_2CuO_{3+\delta}$, prepared in air and measured, by iodometric titration, the oxygen excess, $\delta= 0.3$, in slowly cooled $Ba_2CuO_{3+\delta}$. They also prepared $BaCu_2O_2$ by rapid quenching from the melt at about 1100°C. We note that these two compounds do not belong to the BaO-CuO line. The authors did not find any evidence for the formation of Ba_3CuO_4 suggested by Frase et al. [113] and Abbattista et al. [109] but prepared $Ba_3Cu_5O_{8+z}$ in oxygen atmosphere up to 800°C. The first phase diagram of the BaO-CuO system is due to Zhang et al. [114]. It is based on the analysis of samples prepared in a controlled atmosphere (Ar + 0.21 atm.O_2) or in air at 750°C or 800°C. The equilibrium phase diagram shows the presence of two intermediate phases, Ba_2CuO_3 and $BaCuO_2$. The former is reported to be formed by a peritectic reaction from the liquid and BaO and to exist with two allotropic varieties, tetragonal at high temperature and orthorhombic below 810°C. $BaCuO_2$ is proposed to melt following a synthetic reaction implying two immiscible liquids: $L_1 + L_2 \rightarrow BaCuO_2$. A new investigation which includes measurements at low oxygen partial pressure was reported by T.B.Lindemer and E.D.Specht [115] in their study of the solid-liquid equilibrium of the BaO-Cu-CuO system. The ternary regions $BaCuO_2$-Cu-CuO, Ba_2CuO_3-$BaCu_2O_2$-$BaCuO_2$ and BaO-Cu-$BaCu_2O_2$ are thermodynamically characterized by DTA/TGA. The pseudo-binary BaO-CuO section at 0.21 atm.O_2 confirms the structural change for Ba_2CuO_3 at 810°C but shows that the phase is not stable for temperatures lower than 305°C. $BaCuO_2$ is shown to be formed by a peritec reaction from the liquid and Ba_2CuO_3 and to decompose at 813 K. The $Ba_2Cu_3O_5$ compound, extensively studied by Thompson et al. [116,117] and by Brosha et al. [118], stable at room temperature, undergoes a peritectoid transformation giving $BaCuO_2$ + CuO at 789°C. Finally thermodynamic evaluations of the Ba-Cu-O system have been performed by Voronin and Degterov [119] for solid state equilibrium, by A.Pisch [120] and V.A.Lysenko [121] for the equilibrium implying a liquid phase. Reference papers for the measurement of the specific heats and heats of formation of $MeCuO_2$ and Me_2CuO_3 (Me= Ca, Sr, Ba) may be found in an assessment of thermochemical functions by Mrovec et al. [122].

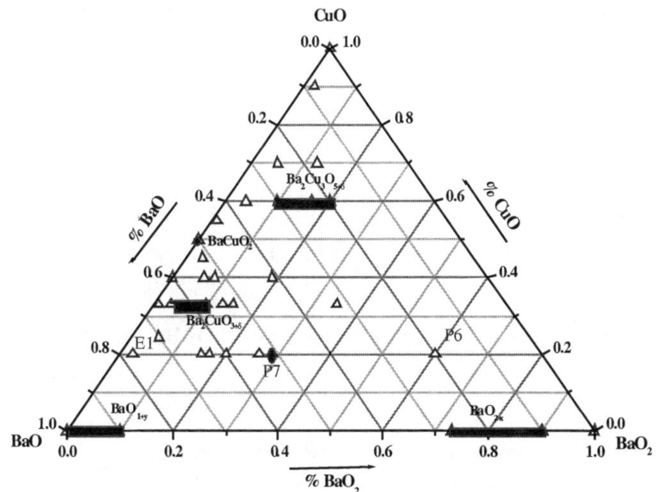

Fig.8: Gibbs compositional triangle for the BaO_2-BaO-CuO ternary system

On the basis of our studies on the BaO_2-BaO system, we recently re-visited the Ba-Cu-O system and demonstrated that, in oxygen flowing, the equilibrium phase diagram must consider the barium peroxide, and then, the system has to be described in a ternary BaO_2-BaO-CuO section [39,123,124] in which the homogeneity fields of the compounds with oxygen excess may be clearly represented as shown in figure 8.

We confirmed the existence of $Ba_2CuO_{3+\delta}$, $BaCuO_2$ and $Ba_2Cu_3O_{5+\delta}$ and were able to precise the extension of the oxygen solubility by DTA/TG. The triangular symbols in figure 8 are for endothermic reactions observed from these analysis; P6, P7 are peritectic equilibrium in the ternary system, for which we have retained the convention of V.A.Lysenko [121], while E1 is a eutectic reaction in the binary field $Ba_2CuO_{3+\delta}$ - BaO_{1+y}. This description supports the presence of 6 invariant equilibrium with four coexisting phases (fig.9). Examples are given in figure 10 for the formation of orthorhombic $Ba_2CuO_{3.39}$ at 720°C, and the class II transformation at 820°C corresponding to the equilibrium $Ba_2CuO_{3.39} + BaO_{2-x} = BaO_{1+y} + L$ which is an extension in the ternary field of the peritectic formation of the barium peroxide

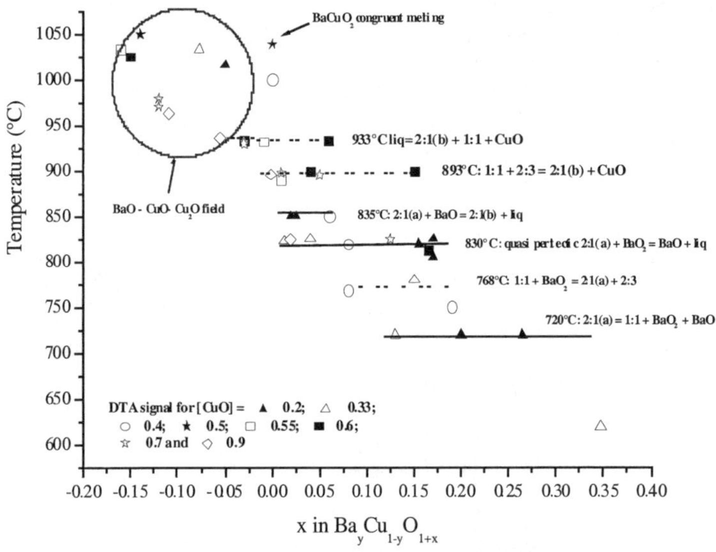

Fig.9: Invariant 4-phase equilibrium in the BaO_2-BaO-CuO system

The low temperature formation of $Ba_2CuO_{3+\delta}$ could be observed by DTA/TG coupled with high temperature X-Ray diffraction for a single phase sample prepared at 850°C and cooled by switching off the furnace power. All the diffraction lines were indexed with an orthorhombic lattice, Immm space group [112] with cell parameters a= 4.093(4) Å, b= 3.898(2) Å and c= 12.924(6) Å. The phase is metastable at room temperature.

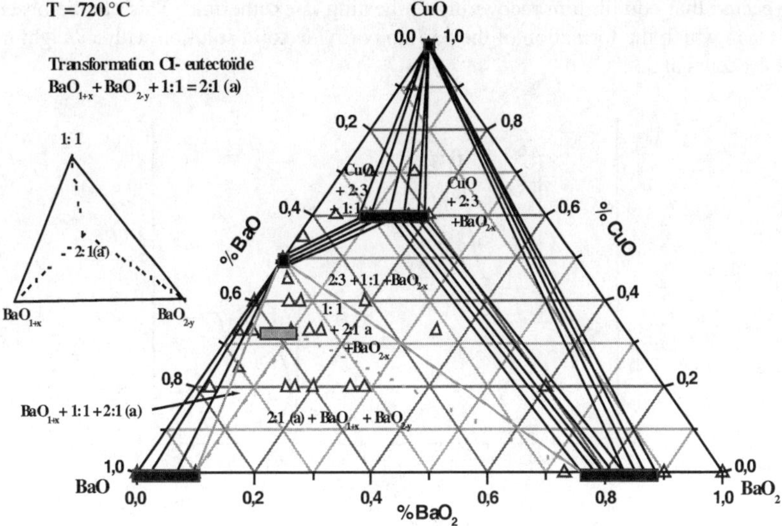

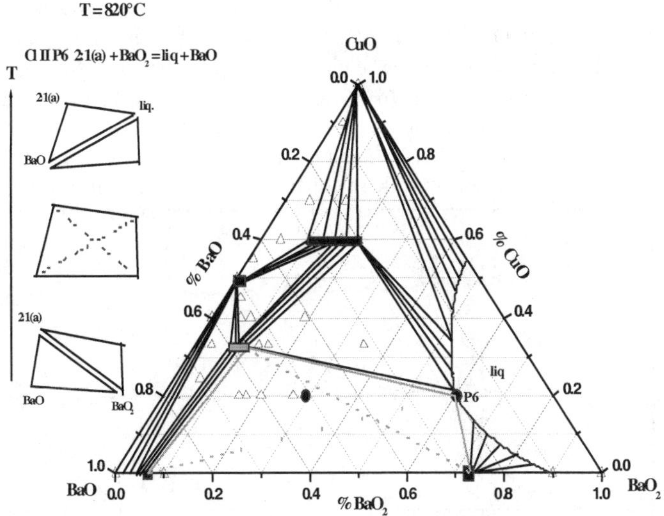

Fig.10 Invariant reactions in the BaO_2-BaO-CuO system. Top: formation of $Ba_2CuO_{3.39}$; Bottom: class II pseudo peritectic reaction (see text)

It is expected that equilibrium recovering on heating is exothermic. This can be observed in figure 11, in which the formation of the barium peroxide solid solution with a weight increase of 3.64% occurs at 334°C.

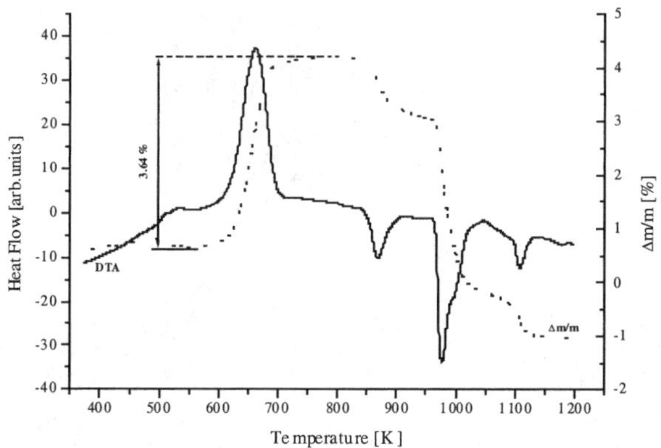

Fig.11: DTA/TG of pure $Ba_2CuO_{3.39}$ in oxygen. Heating rate β= 10 K/min

The corresponding composition, $Ba_2CuO_{4.28}$, calculated from the oxygen content in the high temperature compound $\delta = 0.1$ [125], lies in the ternary field $BaCuO_2$- $Ba_2Cu_3O_{5+\delta}$-BaO_{2-x}.
The endothermic peak at 572°C is attributed to the crossing of a monovariant line in the pseudo binary $BaCuO_2$-BaO_{2-x} system. The associated oxygen losses are due to the desappearance of $Ba_2Cu_3O_{5+\delta}$ in this new equilibrium field. The formation of orthorhombic $Ba_2CuO_{3+\delta}$ with high oxygen content (δ= 0.39) occurs at 720°C and the transition from orthorhombic to tetragonal-type structure, I4/mm space group [112] at 835°C. At this temperature, the cell size is a= 4.032(2) Å and c= 13.03(1) Å. This transformation sequence has been confirmed by high temperature X-ray diffraction reproduced in figure 12

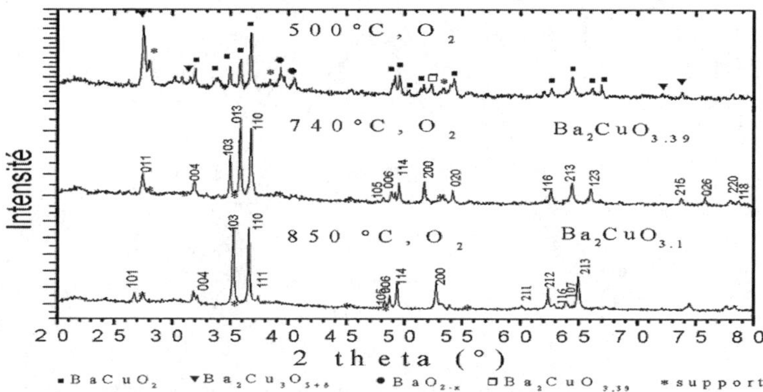

Fig.12: High temperature X-ray diffraction showing the evolution of the equilibrium state as a function of the temperature from ref.39

The presence of a small impurity diffraction line, attributed to $Ba_2CuO_{3.39}$ in the data at 500°C is due to the kinetics for the equilibrium recovery. The indexed diffraction peaks at 740°C and 850°C corresponds to the orthorhombic and tetragonal phases, respectively. The melting temperature of $Ba_2CuO_{3.1}$ in oxygen is 1016°C, in excellent agreement with the result of Lindemer and Specht [115]. It has to be noticed that in flowing argon the tetragonal phase, formed at 600°C, undergoes a new structural transformation at 700°C, leading to stoichiometric Ba_2CuO_3. According to Abbattista et al. [109] the X-ray diffraction pattern of this phase could be indexed with an orthorhombic structure, isotypic with Me_2CuO_3 (Me= Ca, Sr) and cell parameters valuated at 700°C: a= 4.137(4) Å, b= 3.799(4) Å and c= 13.156(7) Å. The transition sequence orthorhombic (LT) – (600°C) $\rightarrow$ tetragonal – (700°C) $\rightarrow$ orhorhombic (HT) was found to be "inversible", the oxygen content of the phases on cooling being not necessary the same as that on heating.

An important compound frequently encountered during the preparation of the high T_c phases

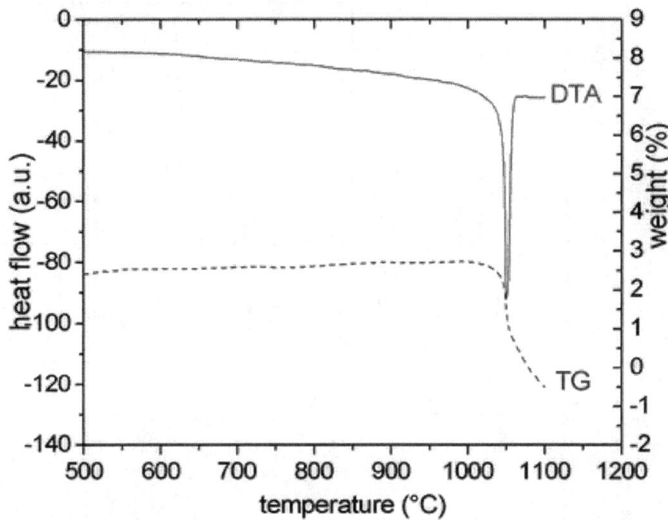

Fig.13: DTA/TG for $BaCuO_2$ in oxygen. Heating rate β = 10 K/min

is the equimolar $BaCuO_2$ compound which was found to melt congruently at 1047°C (fig.13). It is interesting to note that the weight losses starts at the melting temperature which is a clear indication that the solid phase is stoichiometric in oxygen. At 1100°C, the mass reduction is 3.4% , which is consistent with a total $Cu^{2+} \rightarrow Cu^+$ conversion. The crystal structure of $BaCuO_2$ is cubic, space group Im-3m [126,127] with a= 18.354(5) Å.

$Ba_2Cu_3O_{5+\delta}$ decomposes at 898°C following the invariant class II reaction : $Ba_2Cu_3O_{5+\delta}$ + $BaCuO_2$ = CuO + $Ba_2CuO_{3.1}$. In the Ba:Cu = 2:3 tie line, the liquidus surface is crossed at 1020°C with the composition $Ba_{0.33}Cu_{0.67}O_{0.86}$, in the $BaCuO_2$-$BaCu_2O_2$-CuO ternary field [119]

44

4.4. The CaO-CuO and TlO$_{1.5}$-CaO-CuO systems

Three intermediate phases have been identified in the binary CaO-CuO system : Ca$_2$CuO$_3$, Ca$_{0.8}$CuO$_{1.93}$ and CaCu$_2$O$_3$ [128-133]. Phase equilibrium and thermodynamic assessments have been reported [130, 134-136] in which it appeared that the main problem in this system is related to the low kinetics for the formation and decomposition of Ca$_2$CuO$_3$ and Ca$_{0.8}$CuO$_{1.93}$ and to the reduced stability domains of all the phases in the system represented in figure 14.

The invariant equilibrium corresponding to the decomposition of Ca$_2$CuO$_3$ and to the formation of Ca$_{0.8}$CuO$_{1.93}$ have been established by long stepwise annealing and by the analysis of the phases prepared using a sol-gel process [137].

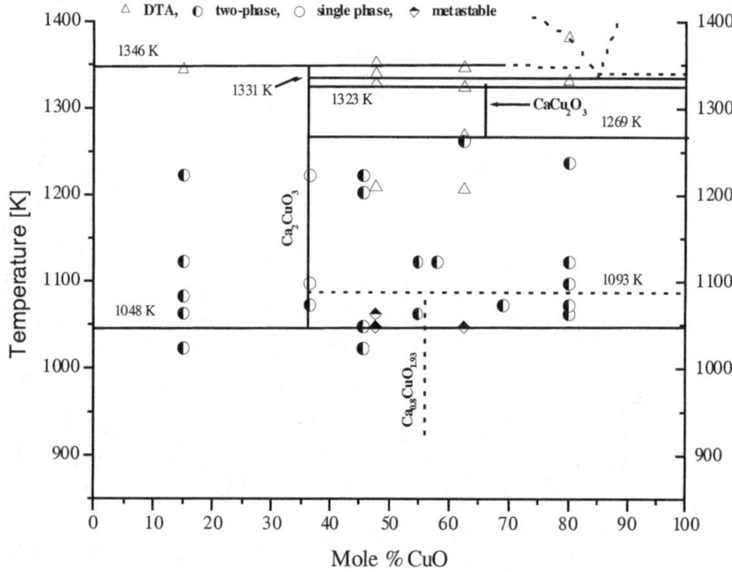

Fig.14: Phase relations in the CaO-CuO system at 1 atm of oxygen

In figure 14, the DTA signals at 1200 K are associated to the decomposition on heating, of Ca$_{0.8}$CuO$_{1.93}$ prepared from commercial CaO and CuO powders; they give an idea of the importance of the kinetics in this system. The sol-gel process provides, after a first firing at 450°C submicronic powder with high reactivity which allowed us to prepare, for instance a nearly single phase Ca$_{0.8}$CuO$_{1.93}$ after a heat treatment for 2 h at 800°C (fig.15).

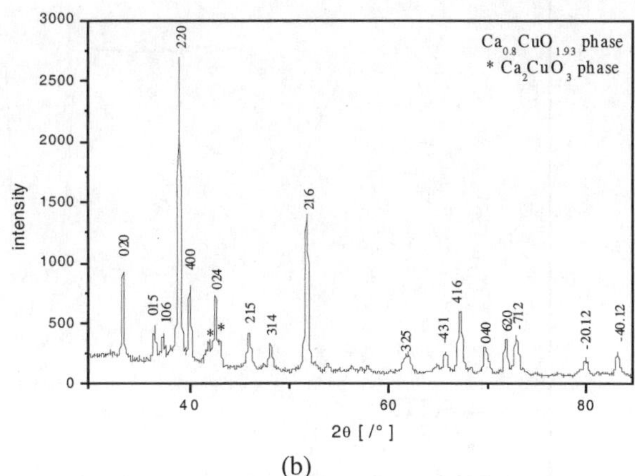

(a)   (b)

Fig.15: Ca$_{0.8}$CuO$_{1.93}$ prepared by a sol gel process (a) after a first firing at 450°C in oxygen. (b) X-ray diffraction pattern after annealing in oxygen at 800°C for 2 h.

For comparison, the same result is obtained after a sintering time of 100 h at the same temperature for sample prepared by classic powder metallurgy. The Ca$_{0.8}$CuO$_{1.93}$ phase has been represented in dashed lines in the phase diagram because, with 0.13 oxygen atom in excess per mole, it does not belong to the section CaO-CuO. This experimental diagram is in fairly good agreement with the modelisation of Risold et al. [135], excepted for CaCu$_2$O$_3$ which was calculated to be formed by peritectic reaction but which shows here a peritectoïd decomposition at 1323 K.

The grain growth and the eutectoïd decomposition of Ca$_2$CuO$_3$, has to be taken into account during the preparation of the superconducting Tl-1223 phase, using Ba:Ca:Cu = 2:2:3 as precursor because the equilibrium state of this mixture is in reality two-phase BaCuO$_2$ + Ca$_2$CuO$_3$. The latter phase has an orthorhombic lattice, isotypic with Sr$_2$CuO$_3$ (Immm Space Group) with a= 12.208 Å, b= 3.768 Å and c= 3.249 Å. The crystal structure of Ca$_{0.8}$CuO$_{1.93}$ has been recently solved and refined by Galez et al. [133] using X-ray and neutron powder diffraction. It was found monoclinic, P2/c space group, with a= 10.9456(4) Å, b= 6.3192(2) Å c= 6.3192(2) Å and β= 104.952(2)°. The oxygen content, deduced from the refinement is exactly the same as that observed by Mathews et al. [138] from reduction and emf measurements. The phase composition and the displacement of atoms with respect to their position in the NaCuO$_2$-type substructure are due to the minimisation of Ca-Ca repulsion and by a relaxation towards a more regular octahedral environment for Ca atoms. In vacuum, the decomposition of Ca$_{0.8}$CuO$_{1.93}$ is shown in figure 16 to start at 723°C [39,137]. The associated weight losses of 1.64% exactly correspond to the departure of 0.13 oxygen atoms in excess per formula unit and confirm both crystal structure and thermochemical analyses.

In addition, by the use of the thermogravimetric measurement and a kinetics treatment by Phadnis and Deshpande [69], we found that the transformation rate is well represented by dα/dt = (1-α)$^{1/2}$.k(T), indicating a 2D diffusion process compatible with the structural data.

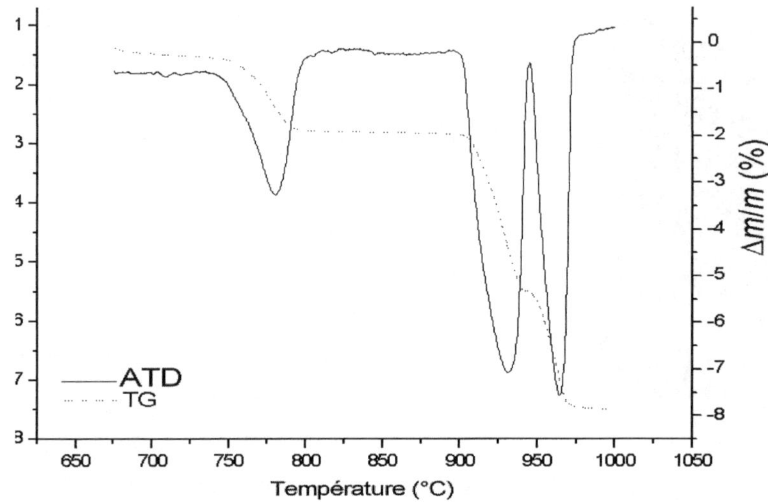

Fig.16: Decomposition of $Ca_{0.8}CuO_{1.93}$ in vacuum. Heating rate β= 5 K/min

As a matter of fact, the crystal structures of decomposing $Ca_{0.8}CuO_{1.93}$ and formed Ca_2CuO_3 contain linear chains of edge-shared CuO_4 squares directed along a unique axis [129,133]. The diffusion of copper and oxygen in the planes formed by these chains could be the mechanism for the transformation.

The $Ca_{0.8}CuO_{1.93}$ phase supports thallium substitution for calcium atoms and then is supposed to have an extension in the ternary $TlO_{1.5}$-CaO-CuO phase diagram. Galez at al. [133] demonstrated that thallium substitution increases the thermal stability of the compound but results in incommensurate structures with modulation vectors depending on the thallium content. In their investigation on the phase relationships in $TlO_{1.5}$-CaO-CuO, Phok et al. [40] found evidence for a possible thallium substitution for calcium up to 23 %. The Tl-rich compound decomposes at 1000°C with total release of thallium so that possible oxygen excess in the structure cannot be measured. Nevertheless, it is obvious that the complete solid solution $(Ca,Tl)_{1-x}CuO_z$ has to be represented in a quaternary Tl-Ca-Cu-O system rather than in a ternary $TlO_{1.5}$-CaO-CuO section for which no ternary compound was detected. At 750°C, in the ternary section, the binary equilibrium join CuO to the 4 compounds in the $TlO_{1.5}$-CaO system.

Amongst the precursor systems for thallium cuprates, not yet investigated or still doubtful, we have to mention BaO-CaO-CuO and $TlO_{1.5}$-CaO-BaO. In the BaO-CaO-CuO system, a number of ternary compounds have been identified, depending on the starting components used and on the conditions for the synthesis. Wu and Ruckenstein [139] reported the existence of a Ba_2CaCuO_4 which was then used for the preparation of Tl-2212 but a single phase sample with this composition was never confirmed. In a preliminary investigation of the ternary system at 920-950°C, Kubat-Martin et al. [140,141] isolated few purple single crystals of $Ba_4CaCu_3O_x$, with a cubic cell, Im3m space group, also identified by Abbattista et al. [142] and Lin and Wu [143]. The latter authors, in their study of the subsolidus relations at 900°C and 1000°C in oxygen, found that $Ba_7CaCu_3O_x$ is the correct composition for the previously $Ba_6CaCu_3O_x$ phase reported by Abbattista et al. [142] and that a $(Ba_{1-y}Ca_y)CuO_2$

solid solution exist up to y= 0.08 in the ternary field. This phase sequence is drastically changed when operating in air containing a few hundreds ppm of CO_2. In contrast, Li et al. [144] in their investigation of the system in air at 850°C, did not find any ternary phases nor extended solid solution. As evidence, more work is needed for a complete understanding of the phase relations and formation.

To our knowledge, there are no works reported for the $TlO_{1.5}$-BaO-CaO system.

5. PHASE EQUILIBRIUM FOR SUPERCONDUCTING THALLIUM CUPRATES

5.1. The $TlO_{1.5}$ – BaO – CuO system and the $Tl_2Ba_2CuO_{6+\delta}$ phases

We have already mentioned that, due to congruent melting which stabilizes the thallium, $Tl_2Ba_2O_5$ is an appropriate precursor for the preparation of superconducting Tl-2201. As reported by T.K.Jondo et al. [47], the $Tl_2Ba_2O_5$ – CuO isopletic line is really a quasi-binary section in the $TlO_{1.5}$-BaO-CuO ternary system. This prevents the formation of $BaCuO_2$ which has an important Curie-Weiss magnetic contribution at low temperatures, interfering with the superconducting signal, in contrast with the conventional route using single metallic oxide components. In this binary section, Tl-2201 is formed by peritectic reaction from $Tl_2Ba_2O_5$ and the liquid at 930°C. In the Cu-rich corner, a eutectic reaction occurs at 896°C for an estimated copper content of 83 mole %. $Tl_2Ba_2CuO_{6+\delta}$ then prepared in flowing oxygen at 650°C using $Tl_2Ba_2O_5$ as a precursor, is not superconducting. Its structure is orthorhombic, Fmmm space group [145], and the thallium content, measured by plasma emission spectrometry, is higher than 1.9 atoms per formula unit. Based on cell size measurements, a= 5.432(5) Å, b= 5.493(5) and c= 23.12(1) Å, from part to part of the $Tl_2Ba_2CuO_{6+\delta}$ compound, it was not possible to deduce a Tl/Cu substitution but High Resolution Electron Microscopy on samples prepared under high oxygen pressure (10 MPa) [12], susceptibility measurements [11] and resonant synchrotron X-ray diffraction studies [146,147] do not exclude such behaviour.

In open system, the tetragonal form of $Tl_2Ba_2CuO_{6+\delta}$, I4/mmm space group [4], is thallium deficient and was found to become superconducting at temperatures depending on the further heat treatments. In table 2, we have reported some results given in the literature concerning the conditions for the formation of tetragonal Tl-2201. Obviously from this table, superconductivity in tetragonal Tl-2201 is not related to the thallium content of the samples.

Table 2. Survey of the literature summarizing the formation of tetragonal Tl-2201

Formation	Tl content	cell size	$T_c(K)$	ref.
815°C/8h + 840°C/4h - Q	1.604	a= 3.866 Å; c= 23.24 Å	70	[148]
880°C/3h – P_{O2} = 1 atm.	1.8	a= 3.858 Å; c= 23.152 Å	20	[149]
880°C/ 3h	1.8	a= 3.874 Å; c= 23.14 Å	NS	[150]
630°C/10h + 850°C/5 min + 630°C/10h + 900°C, p_{O2} = 3 atm	1.83	a= 3.851 Å; c= 23.12 Å	<4.2	[151]
Item + 500°C/2h in argon	1.83	a= 3.855 Å; c= 23.03 Å	50	[151]

In our experiments based on weight measurements, the composition of tetragonal Tl-2201 prepared in open system is $Tl_{1.8}Ba_2CuO_{6+\delta}$. A small thallium homogeneity range, down to $Tl_{1.7}Ba_2CuO_6$, is also found and recovery of the orthorhombic cell was observed after addition to tetragonal $Tl_{1.7}Ba_2CuO_{6+\delta}$ of the appropriate amount of Tl_2O_3 followed by a new sintering treatment at 650°C in oxygen flow. In this atmosphere, the tetragonal phase melts at 915°C, a temperature 15°C lower than the orthorhombic cell.

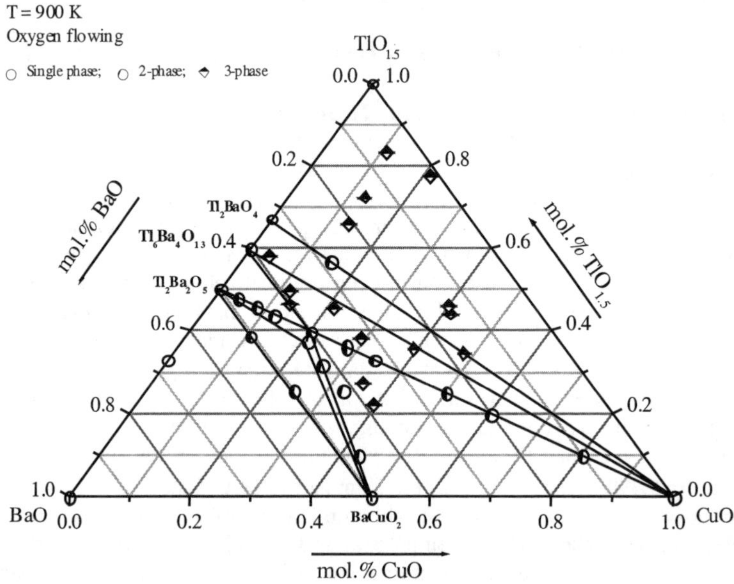

Fig.17 Isothermal section (900 K) of TlO$_{1.5}$-BaO-CuO in oxygen

The equilibrium states between the two structural varieties is included in the isothermal (900 K) ternary section $TlO_{1.5}$-BaO-CuO (fig.17) [47,151]. In reality, the BaO-rich corner of this section is not completely solved due to the BaO_2-BaO equilibrium which has not been considered here. Nevertheless, it is noteworthy that, excepted the two structural forms of $Tl_2Ba_2CuO_{6+\delta}$, there is no other stable ternary phase in the system. In addition, the phase field ranging between orthorhombic $Tl_2Ba_2CuO_{6+\delta}$ and tetragonal $Tl_{1.8}Ba_2CuO_{6+\delta}$, is not a thallium solid solution but a two-phase field. In an inert atmosphere (Ar or He) or at low oxygen partial pressures (10^{-2} atm), departure of thallium from orthorhombic Tl-2201 occurs at 500°C. High temperature X-ray diffraction in helium showed that the orthorhombic phase is completely converted to tetragonal at 670°C and simultaneously, $BaCuO_2$ appears as a minor impurity phase. The two compounds were found to coexist in equilibrium for temperatures higher than 500°C, after thallium losses from the orthorhombic phase. The temperature dependence of the a and b lattice parameters measured from the (200), (020) and (115) diffraction lines of the orthorhombic cell is reported in figure 18, comparatively to a "virtual" tetragonal cell $a^* = \sqrt{2}.$ a(tetrag.) .

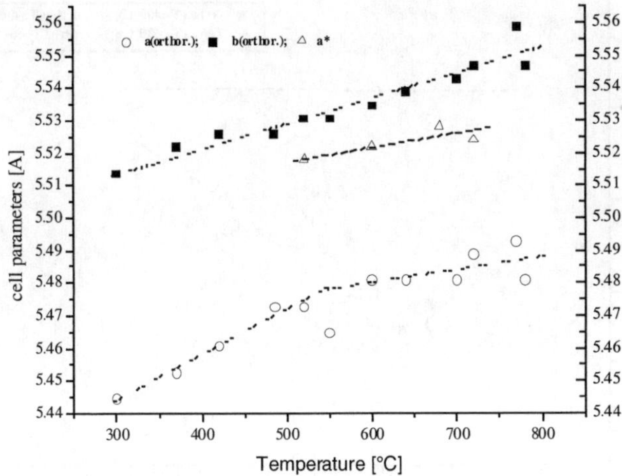

Fig.18: Temperature dependence of the lattice parameters for the orthorhombic and tetragonal cells for Tl-2201 in helium gas.

As the tetragonal phase growth thallium oxide is removed from the orthorhombic cell which vanishes above 800°C in the experiment reported in figure 18 because the thallium content per formulae unit has reached 1.8. This behaviour is different the continuous transformation observed in the $YBa_2Cu_3O_{7-\delta}$ superconducting phase [152] and expected from the models given for the orthorhombic-tetragonal transition, based either on thallium vacancies order-disorder transition [149,153] or on thallium splitting into two pieces in the orthorhombic phase, along the a-axis, implying static displacements of both thallium and copper atoms [11]. However, the 20% difference between the thermal expansion coefficients for a ($\alpha_a = 15.7 \ 10^{-6}$ /°C) and b ($\alpha_b = 12.5 \ 10^{-6}$ /°C) in the orthorhombic cell supports the assumption of a directional defect suggested by the two models.

$Tl_2Ba_2O_5$ as precursor material was also used to prepare $Tl_2Ba_2CuO_{6+\delta}$ under high isostatic pressures at high temperature by Opagiste et al. [12,48,154] and then to study the superconducting properties. The orthorhombic non superconducting phase was obtained after heating at 900-930°C in 100 atm of oxygen a precursor pellet for 30 min, whereas the tetragonal phase was formed at the same pressure in argon or helium at 850°C for 12 min. All samples with the tetragonal structure were superconducting with critical temperatures around 90 K. Using a high gas pressure for the preparation reduces the weight losses for both the orthorhombic and tetragonal phases to 0.8% and 1.5 % , respectively, leading, if thallium oxide vaporization is assumed, to $Tl_{1.97}Ba_2CuO_{6+\delta}$ and $Tl_{1.93}Ba_2CuO_{6+\delta}$ consistent with single crystal X-ray refinement of a tetragonal sample. Further heat treatments at temperatures below 600°C are expected to modify the oxygen stoichiometry leaving unchanged the thallium content. The superconducting orthorhombic phase was then prepared and relations between the superconducting properties, the oxygen pressure during annealing and the crystal structures were established as may be seen in figure 19 from data of Opagiste et al. [48]

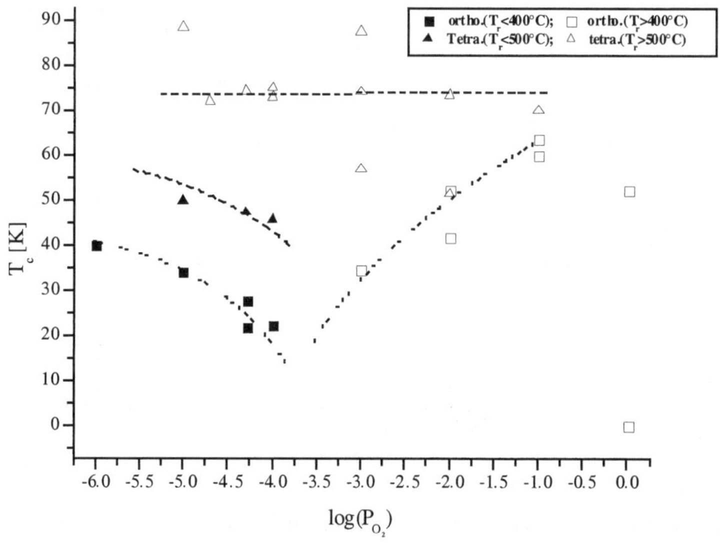

Fig. 19: Critical temperature dependence on oxygen pressure used
during the annealing of orthorhombic and tetragonal T-2201

This figure summarises the general trends found in the Tl-2201 compounds. All samples used in this study were quenched, so that the oxygen content, even if not available, is fixed for a given temperature. The phase prepared in oxygen atmosphere is orthorhombic and not superconducting. Annealing at high temperature in oxygen pressures in the range 10^{-3} atm – 1 atm. has for major effect to create thallium vacancies, and perhaps copper substitution on the thallium sites. Simultaneously, the oxygen content in the sample, fixed by the couple T-PO$_2$ of the annealing conditions, allows the appearance of superconductivity. Low temperature annealing (T<400°C) in low oxygen partial pressure (10^{-6}atm.<PO$_2$< 10^{-4}atm.) only modify the oxygen stoichiometry and a clear relation between the critical temperature and the oxygen content is observed, as well as the reversible orthorhombic-tetragonal transition. When an inert atmosphere is used for the preparation, the tetragonal superconducting phase is formed. High temperature heat treatments in low oxygen partial pressure do not affect too much the critical temperature but appearance of impurities and vanishing Meissner effect attest for the degradation of the samples which become multiphase with a tetragonal Tl-2201 phase of fixed thallium deficient composition. The dependence T_c-oxygen content for the tetragonal phase is observed in samples annealed at T< 500°C. Evidence that the superconductivity is related to the oxygen composition in Tl-2201, independent on the structure-type is given by the relation between T_c and the c-cell parameter reported in figure 20 from the data of Opagiste et al.[48], Y.Shimakawa [11] and Wagner et al.[13].

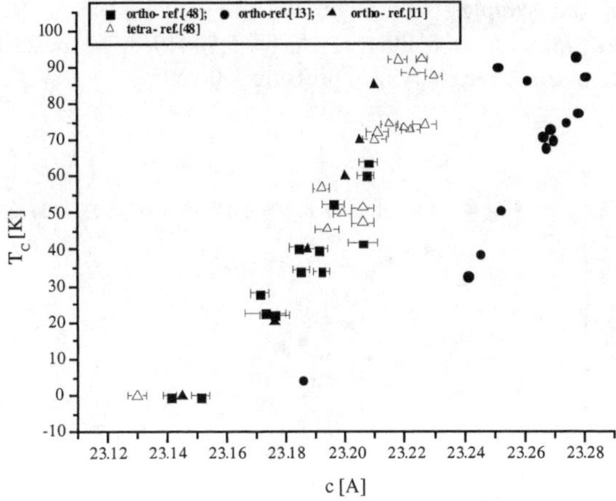

Fig.20: Correlation between T_c and the c-cell parameter in Tl-2201

For all data used for this figure, the linear dependence of T_c with c with $dT_c/dc \cong 10^3$ K/ Å is remarkable. Wagner et al.[13] found that this behaviour is related to the distances copper – apical oxygen O2 and adjacent thallium-O3 bonds in the TlO layers. They also explain the deviation observed in their samples with $T_c > 80$ K, to the linear T_c-c relation by singularities in the distance thallium-apical oxygen O2 and a second oxygen defect not identified. In highly oxygenated orthorhombic samples not superconducting, or with a low T_c, Shimakawa [11] pointed out the presence of extra X-ray diffraction peaks due to structural incommensurate modulations.

The possibility to prepare samples with a fixed thallium stoichiometry gave us the opportunity to study the kinetics for the possible structural transitions from the orthorhombic to the tetragonal phase and inversely [14]. Orthorhombic samples prepared under high oxygen pressure (HPO) were submitted to DTA/TG investigations in flowing oxygen or argon. We found that weight losses occurred around 450°C and are associated with an endothermic peak in the DTA signal, indicating a first order phase transition, but the samples still remained orthorhombic (LPO). As thallium is not expected to diffuse out of the cell at this temperature, we assumed that only oxygen was concerned by the transformation, namely, 0.04 and 0.08 oxygen atom losses for experiments in oxygen and argon, respectively. The transformation rate $d\alpha/dt$ of the reaction (HPO) $\rightarrow$ (LPO) is described by a first order kinetic model $f(\alpha)= (1-\alpha)$ indicating that nucleation of small particles and 2D growth is the limiting mechanism, with an activation energy $E_A = 150(5)$ kJ.mol^{-1} (fig.21). This value has to be compared with activation energies for the oxygen diffusion in $YBa_2Cu_3O_{7-\delta}$ which have been found to range from 48 kJ.mol^{-1} to 125 kJ.mol^{-1} [155-157].

The samples used for the kinetic study were too small for a complementary detailed structural transition analysis but consistency with the result of Shimakawa [11] has to be noticed. In addition, in a recent investigation by high-resolution time-of flight neutron and synchrotron X-ray powder diffraction experiments, Pederzolli et al.[158] evidenced a structural phase transition for high oxygenated $Tl_2Ba_2CuO_{6+\delta}$ at 400°C in argon. Neutron refinements of the

52

crystal structure of the samples with high oxygen content (δ= 0.192) concluded to a monoclinic cell, F112/m, with a= 5.43846(7) Å, b= 5.50419(8) Å, c= 23.1563(3) Å and γ = 90.189° for a sample treated in high oxygen pressure (600 atm.).

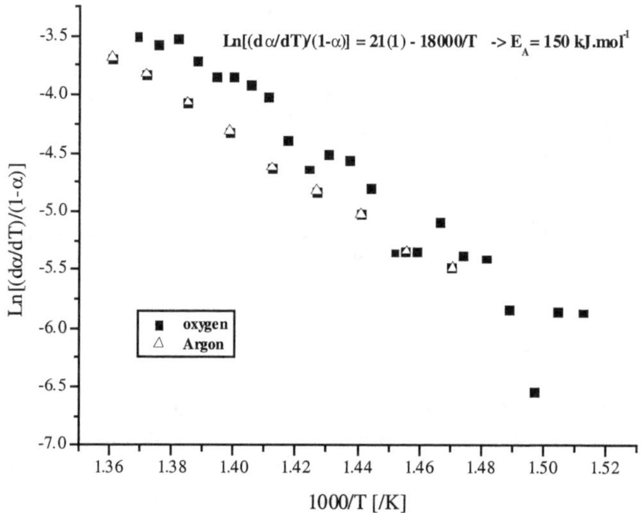

Fig.21: Arrhenius analysis for the HPO $\rightarrow$ LPO transition for Tl-2201, in oxygen and argon flows, described by a first order reaction mechanism

The transition to the orthorhombic form with T_c > 50 K then occurs through the departure of interstitial oxygen atoms in the O(4) site wedged between the two TlO sheets. It is also accompanied by a sudden increase of the c- parameter, correlated with the distances Cu-O(2) and Tl-O(2) but also, according to Wagner et al [13] with Tl-O(3). From these authors the oxygen excess in the O(4) site does not exceed δ= 0.13 for non superconducting samples, in agreement with our weight measurements showing that the orthorhombic phase extends from δ = 0.15 for samples annealed in oxygen to δ= 0.11 for samples annealed in argon.

The tetragonal to orthorhombic conversion has been studied starting from tetragonal Tl-2201 (HPT) prepared under high helium or argon pressure (100 atm.). The compound is superconducting at 91.8 K. The transition in flowing oxygen occurs at 200-250°C with a weight increase of 0.3 %, i.e. 0.16 oxygen atoms per formula unit but no enthalpy change could be associated suggesting, in that case where the thallium composition remains constant, a continuous transformation. After this treatment, the sample was found orthorhombic and is expected to have the same oxygen content than that resulting of the HPO $\rightarrow$ LPO transformation in oxygen, namely 6.15 per formula unit. We can then deduce that $\delta \cong$ -0.01 for the starting HPT phase, then underdoped, superconductivity being due either to mixed thallium valence Tl^{3+}- Tl^{1+}, or to spontaneous charge transfer to the CuO_2 layers. A similar kinetic analysis to that undertaken for the HPO-LPO transition provided a reaction order m= 1.5 and an activation energy E_A = 78.3 kJ.mol⁻¹. Obviously two different limiting mechanisms governs the two transitions. The lack of enthalpy change for the HPT-LPO transformation, suggest an order –disorder transition which could be due to a rearrangement of the oxygen atoms in the O(3) and O(4) sites. Indeed, the neutron diffraction studies of references [148,160] yielded large isotropic thermal coefficient for oxygen atoms in the O(3) site, indicating a great mobility for these atoms, consistent with the ability to be vacant. This assumption has to be related to the conclusion of Wagner et al. [13] which states that O(4) is not the only oxygen defect site in Tl-2201 influencing both phase transition and superconductivity.

5.2. The quaternary $TlO_{1.5}$-BaO-CaO-CuO system

The formation of the superconducting phases $Tl_{(1;2)}Ba_2Ca_{n+1}Cu_nO_x$ implies the study of the quaternary $TlO_{1.5}$-BaO-CaO-CuO system. The phase relations in the system need a 3-D representation as in figure 22 which is the compositional diagram for the most frequently reported compounds in the system without taking into account neither the possibility of cationic solid solutions nor the oxygen stoichiometry deviation. The compounds have been named following the cationic ratios Tl:Ba:Ca:Cu.

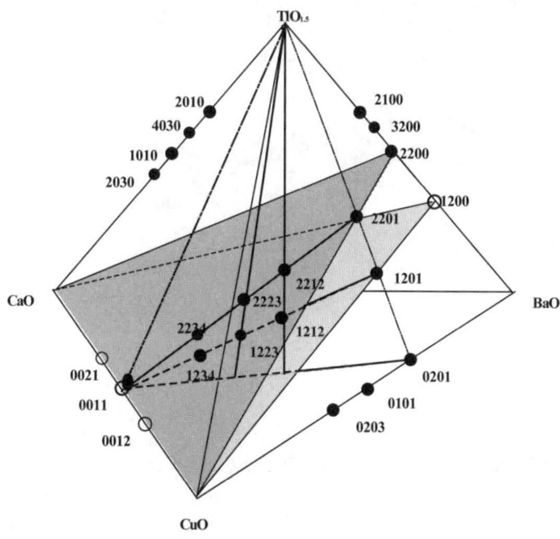

Fig.22: Schematic representation of the $TlO_{1.5}$-BaO-CaO-CuO system

For nominal compositions, we can see that all the superconducting phases lie in the $TlO_{1.5}$-(Ba_2CuO_3)-$(CaCuO_2)$ ternary section which can be used for decreasing the number of degree of freedom of the system. Unfortunately, most of the phases, mainly in the double TlO layer series have homogeneity ranges which cannot be visualised in such ternary section. A derived Janecke's coordinates have been used hereafter to give a better view of the quaternary phase equilibrium. Considering three main components constituting a Gibbs triangle, here BaO,CaO and CuO, the fourth component ($TlO_{1.5}$) is represented on an axis perpendicular to the plane of this triangle. The Janecke's compositions refer to one mole of the main components. Then if n_{BaO}, n_{CaO}, n_{CuO} and $n_{TlO1.5}$ are the number mole of BaO, CaO, CuO and $TlO_{1.5}$, respectively, the Janecke's coordinates for BaO, for instance will be $x_j(BaO) = n_{BaO}/(n_{BaO}+n_{CaO}+n_{CuO})$. In this system, pure $TlO_{1.5}$ is rejected to infinite. Usually the composition of this fourth element is projected in the plane of the main components but we found here that a 3-D representation is more adequate. The diagram is displayed in figure 23.

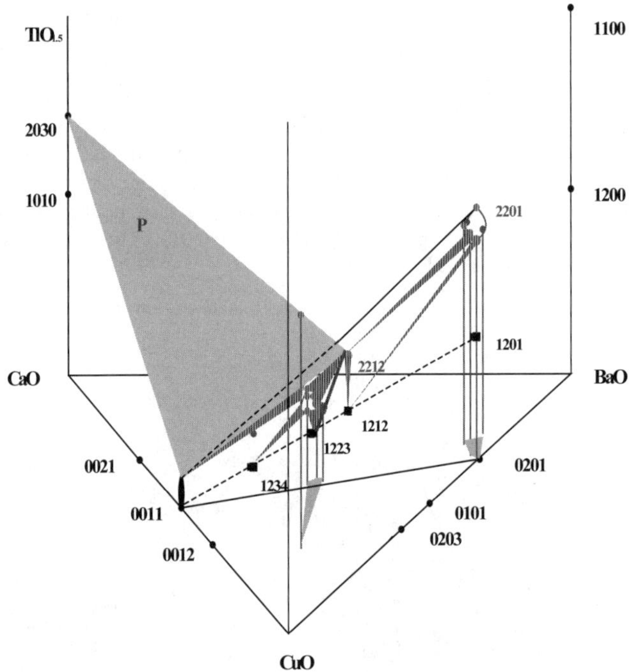

Fig.23: Solid-state equilibrium in the $TlO_{1.5}$-BaO-CaO-CuO system, using Janeckes's coordinates

The compositions in the diagram correspond to single phase samples determined by either structure refinements or by chemical analysis (EDX, wet chemistry, plasma spectroscopy). They are reported in table 3 with some boundary values measured in multiphase samples. The dashed fields are the two phase fields determined by Hopfinger [160], Galez et al. [161] and Aselage et al. [44]. Projections of the solid solutions for the Tl-2201 and Tl-2223 phases on the BaO-CaO-CuO plane reflect possible substitution Tl/Ca. The ternary field $Tl_2Ca_3O_6$ - $(Ca_{1-x}Tl_x)CuO_2$- $Tl_2Ba_2CaCu_2O_z$ is of great importance for the formation of the Tl-2223 phase, as we shall see later in this chapter. It is interesting to note that all the double TlO-layer phases are thallium deficient, partly due to vacancies, partly due to Ca or Cu substitution on the Tl site. Such defects have not been reported in a so large extend for the single TlO-layer series.

Table 3: Summary of single phase superconducting samples in $TlO_{1.5}$-BaO-CaO-CuO

Phases	Janecke Coordinates				method	Cell (Å)	T_c(K)	ref.
	$TlO_{1.5}$	BaO	CaO	CuO				
Tl-2201	0.634	0.654	0	0.346	SC[a]	a=3.8714(5), c=23.269(7)	90	[154]
Tl-2201	0.6	0.628	0	0.372	SC-EDX[b]	a=3.8608(4), c=23.1332(9)	12.4	[162]
Tl-2201	0.577	0.67	0	0.33	SC	a=3.8686(3), c=23.259(3)	110	[163]
Tl-2201	0.591	0.67	0	0.33	ND[c]	a=5.4511(3),b=5.5107(2), c=23.1988(11)	< 4	[148]
Tl-2212	0.313	0.375	0.250	0.375	ND	a=3.85269(5),c=29.3290(5)	-	[160]
Tl-2223	0.250	0.278	0.305	0.417	WC[d]-EDX UVPES[e]	a=3.861(4), c=35.76(3)	112	[164]
Tl-2223	0.216	0.27	0.324	0.405	"	a=3.856(2), c=35.76(2)	118	[164]
Tl-2223	0.2	0.286	0.286	0.428	WC	a=3.844(2), c=35.586(15)	117	[165]
Tl-2223	0.203	0.267	0.332	0.411	ND	a=3.84842(4),c=35.6837(5)	116	[160]
Tl-2223	0.415	0.385	0.154	0.461	EDX*	-	-	[166]
Tl-2234	0.182	0.222	0.343	0.444	XRD[f]	a=3.84877(5),c=42.0494(9)	-	[167]
Tl-1201	0.444	0.741	0	0.259	TEM	a=3.869(2), c=9.694(9)	<4.2	[149]
Tl-1201	0.333	0.67	0	0.33	-	a=3.859(4), c=9.229(9)	25	[168]
Tl-1212	0.245	0.415	0.170	0.415	SC	a=3.8566(4), c=12.754(2)	120	[169]
Tl-1212	0.255	0.383	0.199	0.418	EDX*	-	-	[170]
Tl-1223	0.114	0.286	0.286	0.428	XRD-EDX	a=3.84597(6),c=15.8681(2)	114	[160]
Tl-1223	0.218	0.305	0.238	0.457	EDX*	-	97	[170]
Tl-1234	0.111	0.223	0.33	0.446	XRD	a=3.84809(5),c=19.0005	114	[171]
Tl-1234	0.109	0.208	0.323	0.469	SC	-	118	[172]

[a]: Single crystal, [b]: Scanning electron microscopy with quantitative energy system analysis, [c]: neutron diffraction, [d]: Wet chemistry, [e]: Ultraviolet photoemission spectroscopy, [f]: X-ray powder diffraction and Rietveld analysis, * EDX of grains in multiphase samples.

The phase properties of $Tl_2Ba_2CuO_{6+\delta}$ have been already discussed. A great number of studies appeared in the literature because this member of the thallium cuprates is a unique exemple in the field of high temperature superconductivity: it may be prepared with orthorhombic and tetragonal structures, but contradictory to $YBa_2Cu_3O_{7-\delta}$, T_c may be changed from 0 K to 90 K in the two crystal symmetries, depending on the oxygen and thallium concentrations; the structural phase transition, due to disorder in the TlO layers, offers the possibility to correlate T_c with the cation content on the Tl-site, including Tl itself and vacancies, the oxygen composition, the bond distances and the lattice parameters. Here are met the good conditions carry out fundamental studies on the superconducting properties. For this purpose, heavily overdoped single crystal have been grown using a self-flux method in order to measure the upper critical field [173] and the crystal structure of thin films have been determined for studying the pairing symmetry [174,175]. The non congruent melting at 930°C in flowing oxygen of orthorhombic $Tl_{1.9}Ba_2CuO_{6+\delta}$ has been noticed. We have also mentioned that the tetragonal phase $Tl_{1.77}Ba_2CuO_{6+\delta}$ melts at 915°C but the invariant equilibrium implying 4 phases at T and P constant is not established. Due to the proximity of the two phases it is however reasonable to think that the peritectic reaction in the quasi-binary $Tl_2Ba_2O_5$-CuO system extends in the ternary section, so that the formation of tetragonal Tl-2201 should also imply $Tl_2Ba_2O_5$ and a liquid.

Except for the crystal structure determination and the characterization of the superconducting state, $Tl_2Ba_2CaCu_2O_x$ received less attention than Tl-2201. For this reason the phase field is not seriously known. However, with Tl-2212 single crystals extracted from a self-flux growth with a Tl:Ba:Ca:Cu= 2:2:2:3 composition Morosin et al. [176] found that 21 % of the calcium sites were occupied by thallium. This result has been confirmed by Galez et al. [161] in a recent neutron powder diffraction investigation in which the structure refinement offers the possibility for the thallium-rich limit of the Tl-2212 phase to have the composition $Tl_2Ba_2Ca_{0.74}Cu_2O_x$ due to 26% of thallium on the Ca sites and 13% of thallium vacancies on the double TlO layer. This refinement also excludes, in addition, thallium substitution by copper. From samples prepared at high gas pressure, Opagiste [177,178] studied the reversibility of the mixed state and showed that Tl-2212, similarly to Tl-2201, is in the overdoped field. The thin film technology reveals that Tl-2212 with T_c = 110 K, has low surface resistances ($R_s \cong 130$ $\mu\Omega$) suitable for microwave applications [179,180]. The DTA/TG investigation of $Tl_2Ba_2CaCu_2O_z$ in oxygen gas shows 3 endothermic peaks in the temperature range 910°C-940°C indicating that the melting is not congruent. The composition of the liquid at 940°C is $Tl_{1.4}Ba_2CaCu_2O_z$. This equilibrium behaviour differs completely with the non equilibrium reaction mechanism observed by Cheung and Ruckenstein [43] indicating that the Tl-2212 phase is formed from a "$Tl_2Ba_2Cu_3O_z$" liquid which directly reacts with large CaO grains at 867°C in air. In reality, this composition is located on the $Tl_2Ba_2CuO_5$-CuO tie-line, in which a eutectic invariant in oxygen atmosphere was observed at 896°C (see section V.1).

With T_c= 127 K [181], $Tl_2Ba_2Ca_2Cu_3O_x$ is the thallium cuprate with the highest superconducting temperature. Similarly to the preceding members, the double TlO layer of Tl-2223 is highly thallium deficient and is expected to accept Ca substitution. Manthiram [164,165] prepared single phase Tl-2223 samples from Tl= 1.4 to Tl = 2 with calcium excess up to 0.4 per formula unit and most of the reports, even in the early time of its discovery, agree with a convenient thallium deficiency and calcium off stoichiometry, $Tl_{1.85}Ba_2Ca_{2\pm\delta}Cu_3O_z$ for the preparation of the "pure" compound [5,181,182]. It is difficult to precisely define the phase limits for the same reasons as those encountered for the single and double CuO_2-layers phases: if Tl/Ca substitution is generally found in the structure refinements, there are still discussions about the copper behaviour. In fact, Hopfinger [52, 160] showed that the difficulty to prepare the phase with a high degree of purity is related to the ternary equilibrium between $Tl_2Ca_3O_6$, Tl-2212 and the higher limit of the solid solution $(Ca,Tl)_{1-x}CuO_z$ represented by the darkened plan P in figure 23. More precisely, the complex equilibrium in this range of the phase diagram have been partially established (figure 24), implying 4 new four-phase equilibrium fields:

- $Tl_2Ca_3O_6 + (Ca,Tl)_{1-x}CuO_z + $ Tl-2212 $+ CuO$
- Tl-1223 $+ BaCuO_2 + (Ca,Tl)_{1-x}CuO_z + CaO$
- $Tl_2Ca_3O_6 + (Ca,Tl)_{1-x}CuO_z + $ Tl-2212 $+ $ Tl-2223
- Tl-2212 $+ $ Tl-2223 $+ $ Tl-1223 $+ $ Tl-1212

The existence of the two last, very narrow, regions is due to the off stoichiometry of Tl-2212 and Tl-2223. They are the key for understanding the formation of Tl-2223. These additional fields complete and question some of the nine four-phase regions previously identified by Majewski et al.[170] on the basis of EDX analysis.

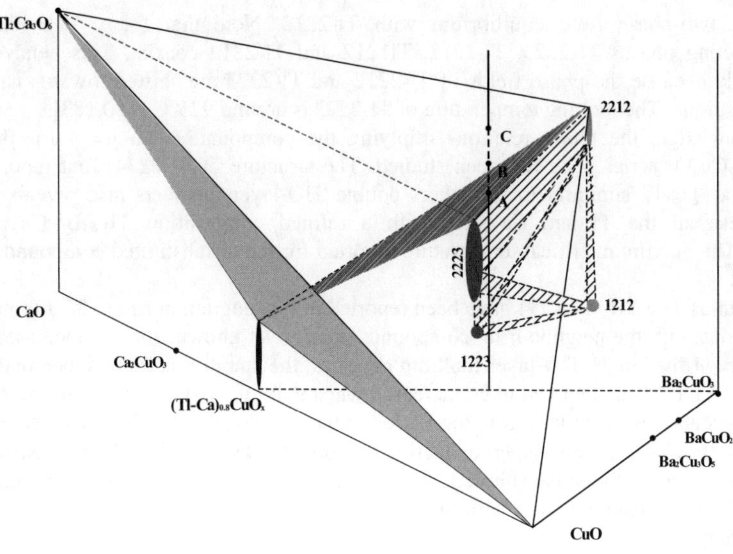

Fig.24: Phase relations in the region of the superconducting compounds

The line Tl_x-223 in figure 24 may be used to understand the formation of Tl-2223 as a function of the initial thallium quantity x. For x > 2, above the point C, Tl-2223 cannot be formed because the representative composition of the samples lies in a quaternary phase field having $Tl_2Ca_3O_6$, Tl-2212 and CuO as one of the limiting ternary section. This view is similar to that of Aselage et al. [44,45] in the stability diagrams obtained by controlling both the oxygen and the Tl_2O partial pressures by the use of a two-zone furnace. Indeed Tl-2212 is the only superconducting phase found to be formed in all the tie-lines investigated by these authors [44,45] when the thallium source temperature is greater than 775°C for which $p(Tl_2O)$ = 1.2 10^{-2} atm and $p(O_2)$= 0.08 atm. Lowering the thallium oxide partial pressure down to $p(Tl_2O)$= 7.4 10^{-3} atm results in the formation of Tl-2223 and the transformation is found reversible in the Tl-223 and Tl-234 tie lines. The non superconducting equilibrium phases are CuO and Tl-rich oxides, one of them found to be $Tl_2Ca_3O_6$. When the thallium content is decreased to values near x= 2, between the points C and B in figure 24 the representative composition lies between the $Tl_2Ca_3O_6$- Tl2212- CuO and $Tl_2Ca_3O_6$- Tl-2212-$(Ca,Tl)_{1-x}CuO_2$ planes, and again, Tl-2212 is the only superconducting phase in this field. For 2>x>1.7 from point B to point A, the composition of the samples lies inside the equilibrium tetraedron $Tl_2Ca_3O_6$- Tl-2212- Tl-2223-$(Ca,Tl)_{1-x}CuO_2$. Finally, with 1.7>x>1.4 Tl-2223 is the major phase in a two-phase field equilibrium with Tl-2212. Note that for x= 1.3 the four superconducting phases Tl-2223, Tl-2212, Tl1212 and Tl-2212 coexist. This behaviour is possible only because the phase fields of Tl-2212 and Tl-2223 are shifted towards calcium-rich compositions. The melting temperature of Tl-2223 is around 925°C [160,183].

To our knowledge, the phase relations implying the compounds with n= 4 for the two $Tl_mBa_2Ca_{n-1}Cu_nO_x$ series have not been studied. The structure of Tl-2234, first reported by Hervieu et al. [184], similarly to the others double TlO-layer members also reveals partial disorder between the Tl and Ca sites with a refined composition $Tl_{1.64}Ba_2Ca_{3-y}Cu_4O_x$ [167,185]. The maximum critical temperature reported for the unsubstituted compound is 119 K.

The compounds Tl-1201, Tl-1234 have been reported in the diagram in figure 23 although the phase relations with the neighbouring compounds are not yet known. As it is the case for all the members of the single TlO-layer thallium cuprates, the stability and the superconducting properties of Tl-1201 are claimed to be increased when thallium and/or barium are substituted by other elements. This is at least true for T_c: for instance $Tl_{0.5}Pb_{0.5}Sr_2CuO_5$, $TlBaSrCuO_5$ and $Tl_{0.8}(CrO_4)_{0.2}Ba_2CuO_{4.2-\delta}$ are superconducting at 60 K, 48 K and 42 K, respectively [186,187,188], while the unsubstituted compound is reported to be nonsuperconducting [149,189,190] or superconducting with only T_c= 9.5 K [187]. Similar results are obtained by fluorine doping [191].

The first crystals of Tl-1212 with T_c = 103 K have been extracted from a melt mixture containing also Tl-2212 and Tl-2223 [169,192]. The structure refinement indicates a composition $Tl_{1.169}Ba_2Ca_{0.831}Cu_2O_{6.747}$ which results of thallium sites fully occupied by thallium and a partial thallium substitution on the Ca sites with Ca/Tl = 0.831/0.169(8). The thallium site is found to contain two types of atoms, one off the four fold site (x= 0.1105,0,1/2) supposed to be Tl^+, the second (Tl^{3+}) on the four fold site with a ratio off/on =0.772/0.228 [169,193]. A great number of compounds were prepared with the Tl-1212-type structure by thallium and barium substitutions. Their cell size and T_c are listed in ref.[9].

Due to its potential use in power applications, $TlBa_2Ca_2Cu_3O_x$ with T_c = 115-120 K is surely the phase which has been, and still is, the most studied thallium cuprate. In fact, since the measurement by Liu et al.[17] of critical current densities J_c= 1.24 10^5 A/cm^2 in $(Tl_{0.5}Pb_{0.5})Sr_2Ca_2Cu_3O_9$, the structure, the superconducting properties, the phase formation and the equilibrium as well as the processing methods of substituted systems $(Tl,Pb,Bi,Hg...)(Ba,Sr)_2Ca_2Cu_3O_x$ have been extensively scanned. The unsubstituted phase was excluded from these works ought to the admitted ideas that it was more difficult to prepare and that the enhanced superconducting properties, particularly the high critical current densities and the high temperature irreversibility lines, were due to defects related with substitutions. Establishing the phase relations, we were able to prepare the unsubstituted Tl-1223 phase, either in open system or in sealed tubes, with a high degree of purity and with superconducting properties similar, if not enhanced, to those of the substituted phases. This last point is reflected in the irreversibility fields vs. the temperature reported in figure 25 for $(Tl_{0.5}Pb_{0.5})(Sr_{1.6}Ba_{04})Ca_2Cu_3O_x$ and unsubstitued Tl-1223 [52,160]. The behaviour of $YBa_2CuO_{7-\delta}$ is given for comparison. The hight temperature values of B_{irr} (the HT 1223 line in fig.25) were measured on some unsubstituted Tl-1223 samples which are thought to be in a non equilibrium state.

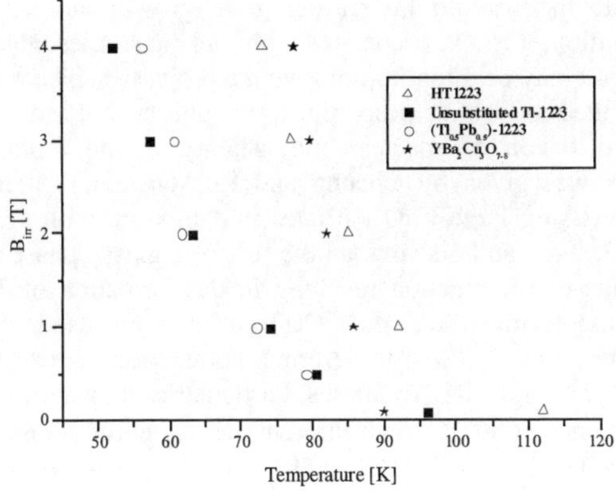

Fig.25: Irreversibility field for (Tl,Pb)- and Tl-1223 phases.

In addition, the formation of the unsubstituted phase favours the growth of platelet-like grains which is beneficial for the processing of tapes and coated conductors for power applications. A small homogeneity field in the thallium deficient direction (Tl= 0.9), the presence of a ternary equilibrium Tl-1223-BaCuO$_2$-(Ca,Tl)$_{1-x}$CuO$_z$ for Tl $\cong$ 0.8 and the existence of a new quaternary phase field Tl-1223-BaCuO$_2$-(Ca,Tl)$_{01-x}$CuO$_z$-CaO complete the phase relationships at the vicinity of unsubstituted Tl-1223. The formation temperature of the phase is 880°C in sealed tubes [160] whereas the substituted phase may be obtained with about 90% purity at 830°C after a flash reaction at 940°C for 1 hour [194]. The decomposition and melting temperatures of Tl-1223 at 938°C and 943°C, respectively are about 50°C lower than for (Tl,Pb)-1223. If we compare the formation and the decomposition temperatures of substituted and unsubstituted Tl-1223, the argument of decreased stability in the unsubstituted system is obvious. However this disadvantage is balanced by the decrease of the variancy of the system and by a reaction pathway which favours the platelet-like grain morphology.
Tl-1234, first identified by Ihara et al.[195] in a mixture containing small amounts of Tl-2212 and Tl-2223 phases, was recently found by EDX analysis on multiphase samples to show a homogeneity range due to calcium and barium substitutions on the Tl sites, limited, if copper is normalised with 4 atoms per formula unit by the compositions Tl$_{0.87-1.3}$Ba$_{1.9-2.05}$Ca$_{2.55-3.1}$Cu$_4$O$_x$ [172]. Ogborne and Weller [171] refined the structure of a slightly calcium deficient (Ca=2.96 /formula unit) compound with T$_c$=114 K. The critical current density at 50 K was estimated by Zhang et al. [196] to be 1.6 10^5 A.cm^{-2}. It is noteworthy that copper substituted (Tl,Cu)-1234 thin films were prepared by Khan et al. [197] with T$_c$= 110K and a critical current density J$_c$= 10^6 A.cm^{-2} at 77 K.

6. REACTION PATHWAY AND KINETICS FOR THE FORMATION OF THE THALLIUM CUPRATES

The knowledge of the reaction pathway for the formation of the compounds cannot be dissociated of phase equilibrium studies because it gives information about the expected impurity phases, provides a procedure for the preparation and sometimes, as it is the case for

Tl-1223, may indicate the way to favour the grain growth and consequently the grain connectivity. By definition, it is also associated with kinetic studies which estimate a limiting reaction mechanism and may be a tool to improve the properties. The reaction pathway is not only governed by the thermodynamic properties of the phases, but it depends also on physical parameters such as diffusion of species, homogeneity of the mixture or grain size. A demonstration of this was given by Cheung and Ruckenstein [43] in their study on the formation of Tl-2223. Using large CaO particles in a mixture with the stoichiometric ratio Tl:Ba:Ca:Cu = 2:2:2:3, these authors stressed the role of a glassy phase surrounding the CuO grains at the first stage of the reaction resulting in the formation of Tl-2212, whereas the expected Tl-2223 phase forms when small CaO particles are used. In the same series of publication [42,43], they were the first to describe the reaction pathway for the formation of Tl-2223, Tl-1223, Tl-1234 and Tl-1212 phases. Particularly, they showed that the single TlO layered phases are obtained at 867°C in air through the sequential transformations Tl-1201 $\rightarrow$ Tl-2201 $\rightarrow$ Tl-2212 $\rightarrow$ Tl-2223 $\rightarrow$ Tl-1223 $\rightarrow$Tl-1234.

6.1. Experiments in open system

We reproduced and specified these results for the formation of Tl-1223 and Tl-2223 in open and closed systems. In a first series of experiments, in open system, the samples were prepared by Tl_2O_3 additions to the multiphase precursors $Ba_2Ca_2Cu_3O_x$ and $Ba_2Ca_{1.65}Cu_3O_x$ [160,197]. In table 4 we have reported the observed phases after the different heat treatments. The two first heat treatments (I) and (II) were performed in air, the last one in a sealed quartz tube with an oxygen pressure of 0.3 atm. at room temperature. For the two different thallium content, we see that the first compounds which are formed are the double TlO-layer phases . With a calcium deficiency when the initial Tl content is less than 2, , the superconducting phases Tl-2201 and Tl-2212 are observed after the first treatment at 800°C.

Table 4: Pathway for the formation of the superconducting thallium cuprates

Initial Tl:Ba:Ca:Cu	Heat treatment	Final Tl	Obs. S.C. phases	Characteristics
1.6:2:1.65:3	I	1.1	2201	a= 3.872(3) Å, c= 23.15(4) Å
			2212	a= 3.868(2) Å, c= 29.35(2) Å
	I + II	1.3	2212	a= 3.862(3) Å, c= 29.40(2) Å
			2223	a= 3.854(2) Å, c= 35.68(2) Å
2.7:2:2:3	I	1.7	2212	a= 3.882(2) Å, c= 29.43(2)Å
			tr.2223	a= 3..88(1)Å, c= 35.3 Å
	I + II	1.7	2212	a= 3.871(2) Å, c= 29.42(2) Å
			2223	a= 3.852(5) Å, c= 35.84(5) Å
	I + II + III	1.1	2212	a= 3.862(4) Å, c= 29.53(4) Å
			2223	a=3.865(5)Å,c= 35.32(8) Å
			tr.1223	

I: 800°C/10h in air, II: 875°C/15 min+800°C/12h in air, III: 905 °C/6h insealed quartz tube

They are identified by XRD and by the measurement of the a.c.susceptibility. The latter shows a T_c onset at 108 K characteristic of Tl-2212 and a second transition around 90 K, better revealed by the derivative χ" of the real part χ' (figure 26) attributed to tetragonal Tl-

2201. A flash annealing at 875°C is necessary to improve the formation of Tl-2223 but once initialised, the phase proportion increases during further annealing at 800°C. It is noteworthy that even with a final thallium content as low as 1.1 per formula unit, Tl-1223 is not formed. The main unreacted phases are $BaCuO_2$, $Tl_2Ca_3O_6$ and CaO after the first heat treatment at 800°C. Tl-2201 and $Tl_2Ca_3O_6$ vanishes after the second heating. Note also that in these experiments in open system, Tl_2O_3 was added after each treatment in order to compensate for losses.

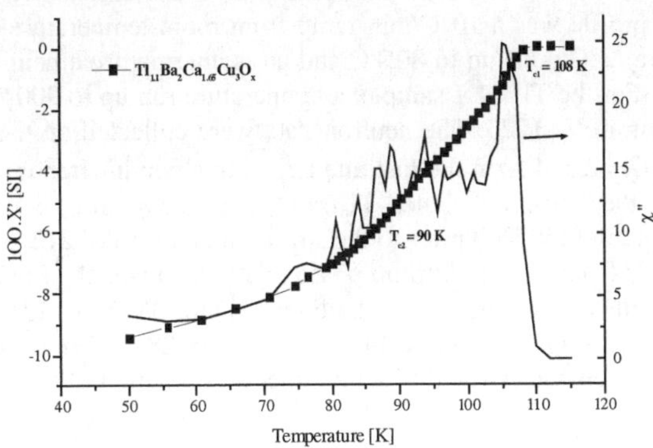

Fig.26: A.C.Susceptibility of $Tl_{1.1}Ba_2Ca_{1.65}Cu_3O_x$ at the first stage of the formation of Tl-1223

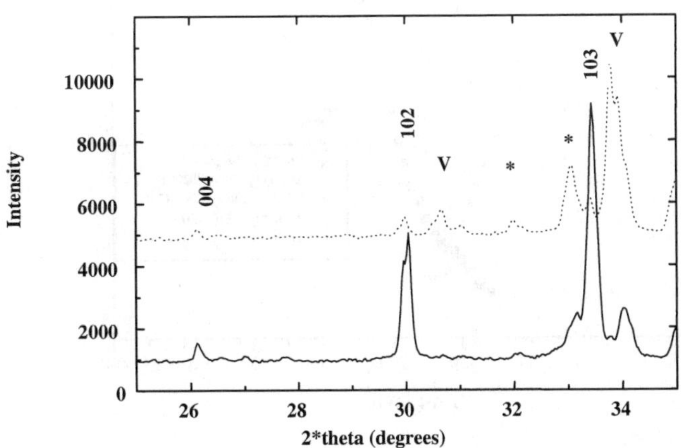

Fig.27: Low angle X-ray diffraction showing the growth of Tl-1223 in a $Tl_{0.8}Ba_2Ca_2Cu_3O_x$ sample after annealing at 905°C (dashed) and 915°C in sealed tube. v is for Tl-2223 and * is $BaCuO_2$

For high initial thallium content, the first stage shows essentially Tl-2212 as a superconducting phase. Only the last heat treatment in sealed quartz tubes at temperatures between 900°C-925°C, improves the formation of Tl-1223 (figure 27)

6.2. In-situ neutron diffraction in closed system

After these preliminary results, a more precise investigation was undertaken by *in situ* neutron diffraction [52,161] on three samples with nominal compositions $Tl_{1.1}Ba_2Ca_2Cu_3O_x$ $Tl_{1.7}Ba_2Ca_2Cu_3O_x$ and $Tl_{2.3}Ba_2Ca_2Cu_3O_x$ prepared from a precursor mixture $BaCuO_2$ + Ca_2CuO_3 and placed in evacuated quartz tubes. For the thallium-rich compositions, the typical temperature profile was a 10°C/min ramp from room temperature to 425°C followed by a low heating rate (2°C/min) up to 805°C and an isothermal treatment at this temperature for at least 200 min. For the Tl = 1.1 sample, a temperature run up to 900°C was necessary to observe the formation of Tl-1223. The neutron data were collected on the high flux powder diffractometer D1B (λ= 2.524 Å) at the Institute Laue Langevin in Grenoble. It was found that in such conditions, the barium thallates $Tl_6Ba_4O_{13}$, Tl_2BaO_4 and an unknown phase are formed at low temperature (507°C) prior to the superconducting Tl-2212 phase which appears at 587°C. As expected from the equilibrium phase relations, this is the final state observed for Tl=2.3 for which, with an expected final thallium content Tl≅2, Tl-2212 can coexist with $(Ca,Tl)_{1-x}CuO_y$ and $Tl_2Ca_3O_x$, clearly identified in figure 28 which shows the normalized integrated intensities of characteristic diffraction lines as a function of time.

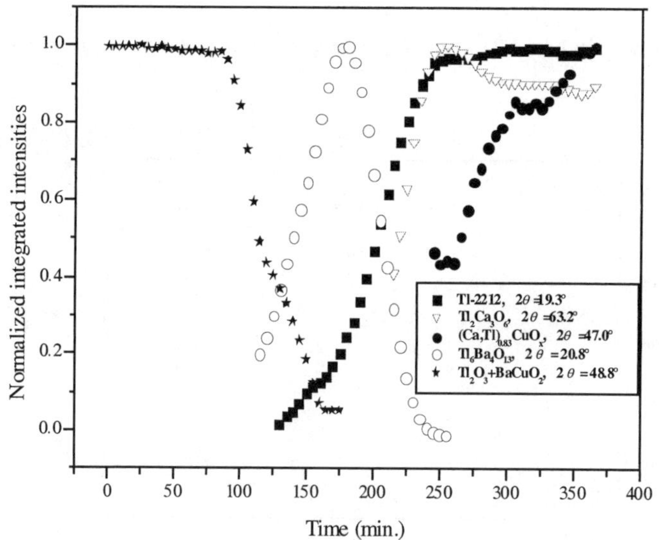

a: high thallium content sample: $Tl_{2.3}Ba_2Ca_2Cu_3O_x$

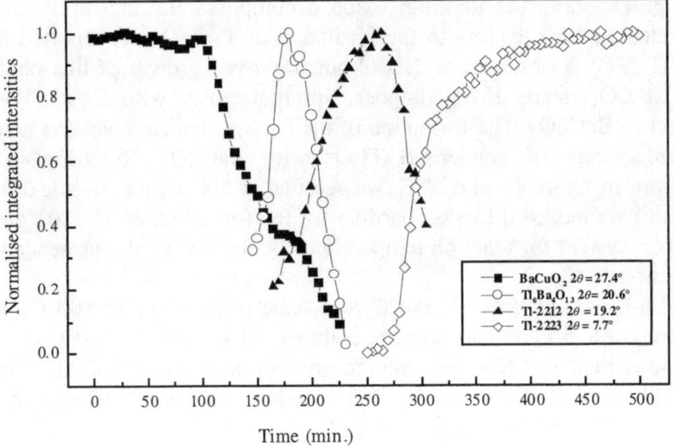

b: low thallium content sample $Tl_{1.7}Ba_2Ca_2Cu_3O_x$

Fig.28: Pathway for the formation of Tl-2212 and Tl-2223 from normalized integrated intensities of characteristic reflections of some parent phases.

For the sample with Tl= 1.7, conversion from Tl-2212 to Tl-2223 is observed at about 800°C, just below the temperature plateau fixed by the experiment and reaches completion at t = 300 min. After cooling, the major phase in the sample is Tl-2223 with well shaped grains (fig.29) indicating that all the reaction pathway occurred in the solid state.

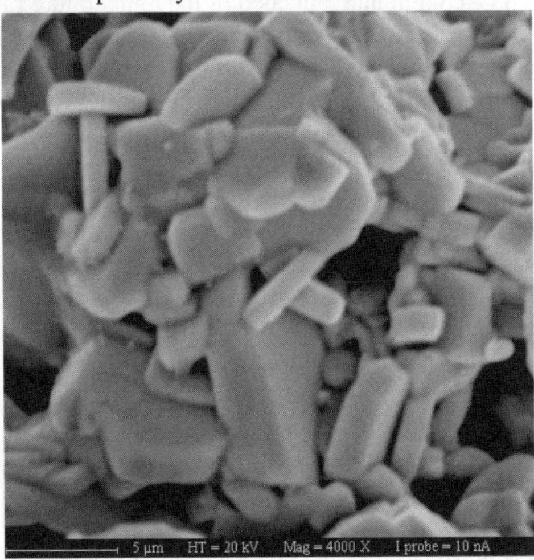

Fig.29: Secondary electron micrograph of $Tl_{1.7}Ba_2Ca_2Cu_3O_x$ after *in situ* neutron diffraction study at high temperature

Therefore, in closed system, the thallium oxide decomposes $BaCuO_2$ at about 500°C and reacts with the released BaO leading to the formation of Tl_2BaO_4 not reported in figure 28, and $Tl_6Ba_4O_{13}$. Tl-2212 is observed at 525°C but improved growth of this phase occurs at 587°C, when $Tl_6Ba_4O_{13}$ starts to decompose, simultaneously with Ca_2CuO_3, remarkably stable with respect to $BaCuO_2$. The formation of the barium thallate phases is here due to the low reaction temperature of condensed Tl_2O_3 with $BaCuO_2$. Referring to the phase equilibrium diagram in figure 17 at 627°C, we see that at 500°C, the tie-lines Tl_2BaO_4-CuO and $Tl_6Ba_4O_{13}$-CuO are indeed diffusion barriers for the formation of Tl-2201 preceding that of Tl-2212. In open system for reaction temperatures about 800°C, the presence of these tie-lines may be questioned.

The sample $Tl_{1.1}Ba_2Ca_2Cu_3O_x$ follows exactly the same pathway up to 800°C [198]. In fact Tl-1223 is observed at 875°C. For lower temperatures, for instance at 870°C during 2 hours, $(Ca,Tl)_{1-x}CuO_y$ rather than Tl-1223 was found to grow in presence of Tl-2223. The formation rate is however enhanced at 890°C and the final product was found to lie in the quaternary section $Tl_2Ca_3O_x$-$(Ca,Tl)_{0.83}CuO_x$-Tl2223-Tl-1223.

6.3. Coated conductors

The *in situ* neutron diffraction study confirmed our preliminary results in open system. They were reinforced by isothermal reaction kinetics study on thick films [39,41,199]. For this purpose, the coated conductors were prepared in a two step process, implying first the spray pyrolysis of the precursor with the cationic ratio Ba:Ca:Cu = 2:2:3 and then an *ex-situ* thallination. The thallium source was a pellet with composition $Tl_{0.9}Ba_2Ca_2Cu_3O_x$, placed in contact with the precursor film. Both the pellet and the film were wrapped in gold foil, sealed in a quartz tube with $p(O_2)$ = 0.5 atm and annealed in the temperature range 880°C<T<910°C for 30 min. to 420 min. The weight fractions of the superconducting phases were measured from the integrated intensities of selected X-ray diffraction lines and the transformation rate α was computed from these weight fractions. The microstructure evolution of the samples was observed by scanning electron microscopy. An example is given in figure 30 for the isothermal treatment at 880°C. As for ceramics, at controlled pressure, the first superconducting phase which was found to be formed after 30 min. at 880°C is Tl-2212 (fig.30.a) with small elongated and not textured grains. The texture comes with the formation of Tl-2223, the main superconducting phase after 90 min. at the same temperature (fig.30.b). It is improved with the emergence of Tl-1223 which coexists with Tl-2223 after 420 min (c). For a longer heating time, precipitates of the $(Ca,Tl)_{1-x}CuO_y$ appear, due to the reaction of the superconducting film with the material substrate, here $LaAlO_3$. A quantitative analysis of this behaviour at 880°C is given in figure 31. The reaction rate is drastically increased at higher temperatures. For example at 910°C, Tl-1223 is the main observed phase after 30 min and it is largely decomposed after 180 min due to the reaction with the substrate. From a kinetics point of view, the reaction order mechanism $f(\alpha) = (1-\alpha)^n$ with $n \cong 2$ was found the best fit for our data. Note that this model , introduced by Friedman [83], is generally well adapted for a large variety of solid state reactions. It reveals however the difficulty to separate the major contribution to the kinetics.

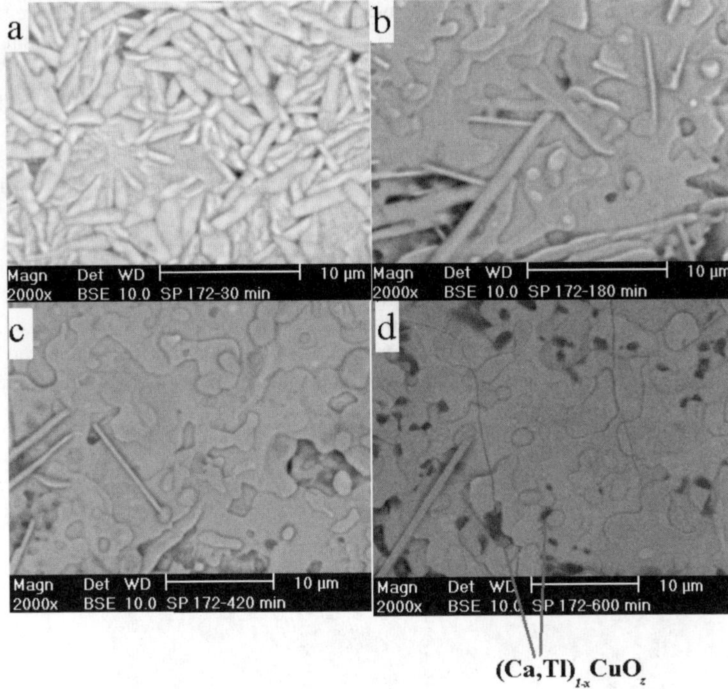

$(Ca,Tl)_{1-x}CuO_z$

Fig.30: Backscattered electron images of thallinated films at 880°C after 30(a), 90(b), 420(c) and 600 min.(d)

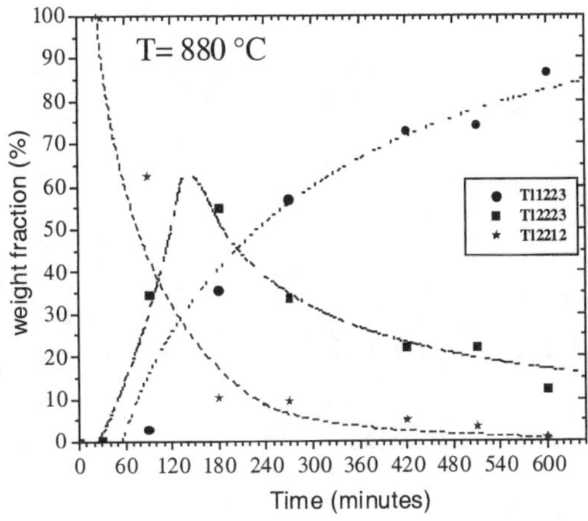

Fig.31: Weight fractions of the superconducting phases as a function of time at 880°C

Nevertheless, with the usual restrictions regarding kinetics analysis, it is reasonable to assume a complex limiting mechanism implying a 2-D growth, linked to the amount of the decomposing phases. The activation energy for such mechanism is $E_a = 225$ kJ.mol^{-1}, half the minimum value found for the formation, from $Bi_2Sr_2CaCu_2O_x$, of the bismuth cuprate Bi-2223 for which the limiting mechanism is nucleation [200].

We have already mentioned, in the paragraph III.1, that the superconducting properties of the thallium cuprates may be drastically improved by fluorine doping. The reason for this is still doubtful but possible substitution of fluorine on the oxygen sites [19], the low melting temperature of TlF used as fluorine source and acting as a flux in the formation of Tl-1223 [95,96], the lowering of the solidus temperature in the fluorine doped phase… are valuable arguments. As a matter of fact, Phok et al. [39-41] succeeded in the preparation of fluorine doped Tl-1223 coated conductor with the same process than undoped film but at 885°C, and with an optimal treatment time of 200 min. A backscattered electron image is given in figure 32 where the resulting layers are very dense with a limited number of steps with respect to the F-free films.

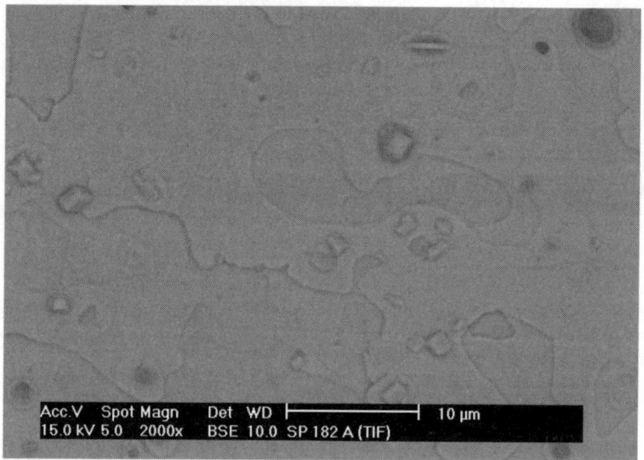

Fig.32: Backscattered electron micrograph of a fluorine doped Tl-1223 coated conductor

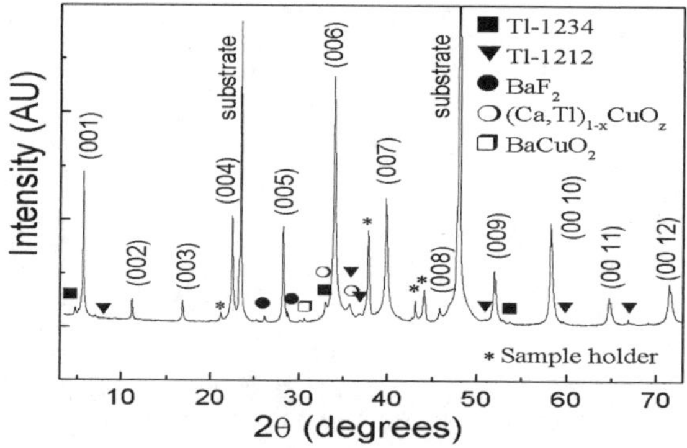

Fig.33: The X-ray diffraction pattern for a F-containing Tl-1223 film

The presence of Tl-1234 and Tl-1212 as new superconducting impurity phases in the X-ray diffraction analysis (Fig. 33) indicates a different equilibrium state due to the formation of BaF_2. The superconducting critical temperatures of the fluorine doped films were found in the range $T_c= 103$ K-116 K. This increased scatter of T_c's is believed to reflect effective fluorine substitution. Transport critical current densities of $7x10^5$ A.cm^{-2} were measured on 1.6x0.1 mm^2 microbridges.

The results of these studies evidenced that, whatever the superconducting phase to be prepared, the reaction pathway implies the double TlO layer members, in a general sequence (M) $\rightarrow$ Tl-2212 $\rightarrow$ Tl-2223$\rightarrow$ Tl1223 where (M) contains Tl-2201 or not, depending on the conditions used for preparing the phases (open or closed system). In addition to the observations reported by Cheung and Ruckenstein [42,43],the formation temperatures of the superconducting compounds have been specified. It is noteworthy that similarly to the case of the bismuth Bi-2223 phase, the formation of Tl-1223 from Tl-2223 is restricted to a very limited temperature range, from 880°C to be efficient up to 910°C to avoid phase decomposition. The role of the equilibrium ternary sections $Tl_2Ca_3O_6$-Tl-2212- CuO and $Tl_2Ca_3O_6$-$(Ca,Tl)_{1-x}CuO_y$ -Tl-2212 on the formation of Tl-2223 and therefore Tl-1223 has been pointed out. It has been also shown that all reactions leading to Tl-1223 occurred in the solid state. Moreover, the changes in the grain morphology at the formation of Tl-2223 from Tl-2212 is of great interest for the growth of textured Tl-1223. A special attention was devoted to the transformation Tl-2223 $\rightarrow$ Tl-1223

6.4. Intergrowth of structures in the Tl-Ba-Ca-Cu-O system

The high thallium deficiency frequently observed in the $Tl_2Ba_2Ca_{n-1}Cu_nO_x$ series offers the possibility of Tl-disorders and order-disorder transitions as suggested in $Tl_2Ba_2CuO_{6+\delta}$ but also of the formation of intergrowth structures along the c-axis in the pathway of formation of the superconducting phases. Shortly after the discovery of the high-T_c thallium cuprates, Iijima et al. [201,202] reported on the presence of intergrowth structures in the homologous Tl2-series, between Tl-2212 and Tl-2223. Another example for such defect structure may be found in the single TlO layer series, between Tl-1223 and Tl-1234, in a publication of Ogborne et al. [203]. It was also found that thallium deficiency may be accommodated by random occurrence of double TlO by single TlO layers as reported by Holstein et al. [204,205] and Malandrino et al. [206] for $Tl_{1+x}Ba_2Ca_2Cu_3O_x$ thin films. Depending on the preparation conditions, ordered intergrowths have been observed in thin film and bulk materials constituting then intermediate phases. For instance Aselage et al. [45] in their work on phase stability mentioned (Tl-1223)$_{0.9}$(Tl-2223)$_{0.1}$, Zhang et al. [207] identified two types of intermediate structures. The first one is described as a transient state corresponding to the early stage of the transformation from Tl-2212 to Tl-1212 and consisting in an ordered arrangement of high defective (30%) and fully occupied (70% Tl^{3+} , 20%Ca^{2+}, 10% Cu^{2+}) Tl-layers. The space group of this state is I4mm and the phase is still superconducting at 103 K. The second intermediate structure is based on ordered intergrowths between the Tl-12(n-1)n and Tl-2(n-1)n phases, especially between Tl-1223 and Tl-2223 for which the space groups have been tentatively proposed and intermediate critical temperatures attributed. More recently Hopfinger et al. [160,198] identified 3 new ordered intergrowth phases between Tl-2223 and Tl-1223, in a non-equilibrium sample with composition $Tl_{1.05}Ba_2Ca_2Cu_3O_x$. The stacking sequences of these phases are 1223/2223, 1223/2223/2223 and 1223/ 2223/2223/2223, for which complete structure models have been derived and refined. An illustration of the existence in the same grain of domains with different thallium content is given in figure 34.

Fig.34: Back scattered electron microphotograph of a 50-μm long Tl-1223 grain showing two intergrowth phases.

In this figure, the dark regions (top and bottom of the grain) correspond to Tl-1223. The central part of the grain with light-dark and light aspects are richer in thallium, with compositions intermediate between Tl-1223 and Tl-2223 and are associated with two intergrowth phases. In this study, it was confirmed that, as suggested by Zhang et al. [207] there is a characteristic superconducting transition temperature for each intermediate phase as shown in figure 35 which is the real part χ' of the a.c.susceptibility of the powdered $Tl_{1.05}Ba_2Ca_2Cu_3O_x$ sample as a function temperature. Four critical temperatures may be clearly separated in the first derivative of χ' (inset of fig.35)

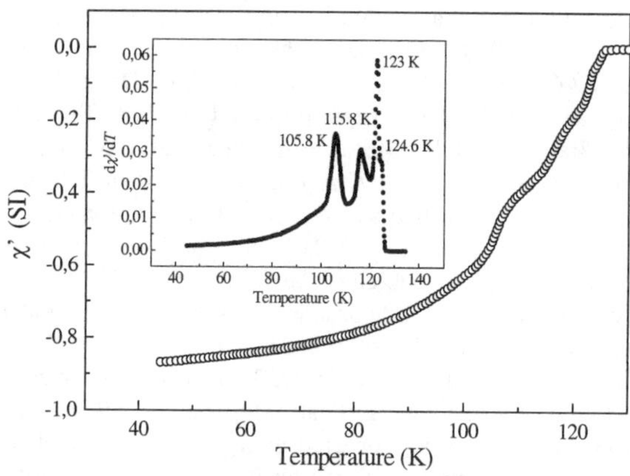

Fig.35: AC susceptibility of non equilibrium powdered $Tl_{1.05}Ba_2Ca_2Cu_3O_x$

Although it is difficult to assign unambiguously the transition temperature to the different intermediate phases, it is expected that T_c increases with increasing number of TlO bilayers. In this case, consistency has been found between relative weight fractions estimated from the diamagnetic signal and the quantitative analysis from X-ray diffraction data.

From a superconducting point of view, the existence of these ordered intermediate phases, could be the reason for improved transport properties as suggested by the HT irreversibility line reported in figure 25 for bulk samples. From a non equilibrium viewpoint, they suggest that the transition from a double TlO layer compound to the corresponding monolayered phase occurs by a progressive and ordered de-intercalation of the TlO layers. Note that such model does not require nucleation for the formation of the monolayer from the parent bilayer, but only rearrangement of the layers following the diffusion of the Tl and O species, which is consistent with the results of the kinetics analysis for Tl-1223. In this special case of Tl-1223, the promising consequence of the proposed mechanism is the formation of plate-like grains retained from the initial Tl-2223 phase. This behaviour has been discussed in the preceding paragraph for the Tl-1223 thick films formation

7. CONCLUSION

The studies which have been developed in this report illustrate our conception on the work needed for the determination of intrinsic properties of materials, here unsubstituted thallium based superconductors. The thermodynamic and kinetic approach were facilitated by the reduction of the independent constituents of the system, from 6 for the substituted phases to 4 if we restrict the analysis to a quaternary oxide system. Nevertheless, the complete understanding of phase equilibrium requires the knowledge of the 4 components, 6 binary and 4 ternary phase equilibrium diagram. We contributed for some of them but more experiments have still to be done concerning, for instance, the evolution of the single phase fields with the change of the temperature, the determination of many polyvariant domains, the study of the ternary BaO-CaO-CuO system.. If now we turn towards the substituted phases, the most simple being the (Tl,Pb)-(Ba,Sr)-Ca-Cu-O system, the difficulty increases rapidly. We have to consider 2 supplementary constituents, 15 binary, 20 ternary, 15 quaternary and 6 quinary systems ! Add to this the difficulty with TG experiments to differentiate the thallium from the lead during the high temperature investigations. It is evident that, in the literature flow concerning the substituted phases, only partial information may be obtained on the phase equilibrium, and to our knowledge, they are mostly related to the Tl-1223 phase [26, 209,210], the most interesting compound for power applications. As already mentioned in the introduction, improved superconducting properties have been observed for $(Tl,Pb)(Ba,Sr)_2Ca_2Cu_3O_x$ [18,19,36-38] and particularly for the "Hitachi group composition" $(Tl_{0.5}Pb_{0.5})(Sr_{0.8}Ba_{0.2})_2Ca_2Cu_3O_z$ [16]. It seems interesting to briefly summarise and compare both unsubstituted and substituted phases (table 4).

From a thermodynamic viewpoint, the temperature range between the formation and the decomposition (or melting) of the compound gives some advantage for preparing the substituted phase which was found, in addition, to have a large compositional homogeneity range [26,209] due to the existence of solid solutions towards Pb and Sr directions (fig.36)

Table 4: Comparison of some basic properties of superconducting Tl-1223 and (Tl,Pb)-1223 phases

	Unsubstituted Tl-1223	**Substituted (Tl,Pb)(Ba,Sr)-1223**
Formation	Implies the 2 TlO layer series $.. \rightarrow$ Tl-2212$\rightarrow$Tl2223$\rightarrow$ Tl1223	... (Tl,Pb)1212$\rightarrow$(Tl,Pb)-1223
	875°C<T<938°C	850°C<T<980°C
Kinetics	from Tl-2223: de-intercalation of the TlO layers + 2-D growth E_A=225 kJ.mol^{-1}	from Tl-1212 grain-growth model (Avrami) 115<E_A<164 kJ.mol^{-1}
crystallog.	metastable intergrowth structures (Tl-2223)$_n$-(Tl-1223)$_m$	
supercond.	T_c = 116-117 K B_{irr}(77K) = 0.5 T High Temp.Line B_{irr}(77K) = 2 T J_c(77 K) coated cond. e=0.1μm, 7.5x10^5 A.cm^2	T_c =116-124 K B_{irr}(77 K) 0.5 T J_c(77 K) = 1.1x10^6 A.cm^2

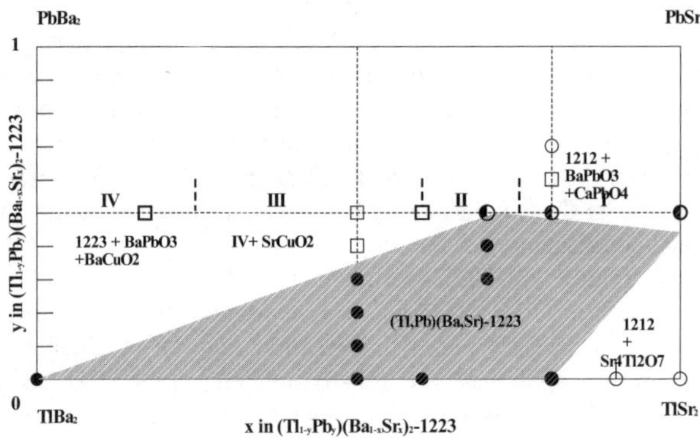

Fig.36: Schematic diagram showing the homogeneity field of (Tl,Pb)-1223 at 950°C

Tl-1212 in the rich-Sr corner with $Sr_4Tl_2O_7$ are the major phases for Pb< 0.4-0.5. Note that $TlSr_2Ca_2Cu_3O_z$ was obtained by Adashi et al. [211] using high pressure synthesis and high temperature (1100°C). On the $(Tl_{0.5}Pb_{0.5})(Ba_{1-x}Sr_x)$-1223 four fields may be differentiate with the main phases being (Tl,Pb)-1212 and $CaPbO_4$ for x>0.8. For lower x values, (Tl,Pb)-1223 coexists mainly with the doubly substituted perovskite $(Tl,Pb)(Ba,Sr)PbO_3$ [212,213]. The possibility to form so extended (Tl,Pb)(Ba,Sr)-1223 solid solution may be translated by "increased stability" of the phase. The reaction pathway for the formation is completely different to that of the unsubstituted phase. It implies substituted (Tl,Pb)-1212 which have to be further decomposed prior nucleation and growth of (Tl,Pb)-1223. The mechanism is detrimental to the platelet like grain morphology coming from the Tl-2223 phase for the unsubstituted phase. Both (Tl,Pb)-1223 and Tl-1223 phases have similar critical transition temperatures and for coated conductors, the critical current at 77 K is nearly the same. A high temperature irreversibility line was observed for unsubstituted Tl-1223 thought to be due to possible ordered intergrowths.

From these remarks we can conclude that the knowledge of the phase relations, complemented with the characterization of the pathway for the formation of the compounds are the conditions for the preparation of high quality superconducting unsubstituted thallium cuprates, ceramics or tapes, with properties similar if not enhanced compared to the substituted phases.

Acknowledgements

The authors are grateful to the French Ministry of Research and to the Region Rhône-Alpes for the financial support of this work. One of us (T.K.Jondo) is pleased to acknowledge the CIES for supporting stages at the LAIMAN

REFERENCES

[1] Z.Z.Sheng, A.M.Hermann, Nature 332[1988]55

[2] Z.Z.Sheng, A.M.Hermann, Nature 332[1988]138

[3] R.M.Hazen, L.W.Finger, R.J.Angel, C.T.Prewitt, N.I.Ross, C.G.Hadidiacos, P.J.Heaney, D.R.Veblen, Z.Sheng, A.El Ali, A.M.Hermann, Phys.Rev.Letters 60[1988]1657

[4] C.C.Torardi, M.A.Subramanian, J.C.Calabrese, J.Gopalakrishnan, E.M.McCarron, K.J.Morissey, T.R.Askey, R.B.Flippen, U.Chowdhry, A.W.Sleight, Phys.Rev.B38[1988]225

[5] S.S.P.Parkin, V.Y.Lee, E.M.Engler, A.I.Nazzal, T.C.Huang, G.Gorman, R.Savoy, R.Beyers, Phys.Rev.Letters 60[1988]2539

[6] M.Hervieu, C.Martin, C.Michel, J.Provost and B.Raveau, J.Solid State Chem.75[1988]212

[7] M.Greenblatt, S.S.Li, E.H.McMills, K.V.Ramanujachary, Studies of High Temperature Supercond., Ed.A.Narlikar, Nova, New York, Vol.5[1990]5

[8] M.P.Siegal, E.L.Venturini, M.Morosin, T.L.Aselage, J.Mat.Research 12[1997]2825

[9] R.E.Gladyshevskii, P.Galez, Handbook of Superconductivity, Ed. C.P.Poole Jr, Academic Press, S.Diego [2000]

[10] J.B.Parise, C.C.Torardi, M.A.Subramanian, J.Gopalakrishnan, A.W.Sleight, E.Prince, Physica C 159[1989]239

[11] Y.Shimakawa, Physica C 204[1993]247

[12] C.Opagiste, M.Couach, A.F.Khoder, T.Graf, A.Junod, G.Triscone, J.Muller, T.K.Jondo, J.L.Jorda, R.Abraham, M.Th.Cohen-Adad, L.A.Bursill, O.Leckel, M.G.Blanchin Physica C 205[1993]247

[13] J.L.Wagner, O.Chmaissem, J.D.Jorgensen, D.G.Hinks, P.G.Radaelli, B.A.Hunter and W.R.Jensen, Physica C 277[1997]170

[14] J.L.Jorda, T.K.Jondo, R.Abraham, M.T.Cohen-Adad, C.Opagiste, M.Couach, A.F.Khoder and G.Triscone J.Alloys and Compounds 215[1994]135

[15] M.A.Subramanian, C.C.Torardi, J.Gopalakrishnan, P.L.Gai, J.C.Calabresse, T.R.Askew, R.B.Flippen and A.W.Sleight, Science 242[1988]249

[16] T.Doi, M.Okada, A.Soeta, T.Yuasa, K.Aihara, T.Kamo, S.P.Matsuda, Physica C 183 [1991]67

[17] R.S.Liu, D.N.Zheng, J.W.Loram, K.A.Mirza, A.M.Campbell, P.P.Edwards, Appl.Phys.Letters 60[1992]1019

[18] A.Kikuchi, T.Kinoshita, N.Nishikawa, S.Komiya and K.Tachikawa, Jpn.J.Appl.Phys.34[1995]L167

[19] E.Bellingeri, R.Gladyshevskii, F.Marti, M.Dhalle, R.Flükiger, Supercond.Sci.Technol.11[1998]810

[20] R.E.Gladyshevskii, Ph.Galez, K.Lebbou, J.Allemand, R.Abraham, M.Couach, R.Flükiger, J.L.Jorda, M.T.Cohen-Adad, Physica C 267[1996]93

[21] J.L.Jorda, R.Abraham, M.T.Cohen-Adad, M.Couach and A.F.Khoder, Material Letters 11[1991]326

[22] M.Couach, A.F.Khoder, F.Monnier, J-L.Jorda, M-T.Cohen-Addad in "High Temperature Superconductors", P.Vincenzini Ed., Elsevier Sci.Publishers B.V.[1991]353

[23] B.A.Glowacki, S.P.Ashworth, Physica C 200[1992]140

[24] P.E.D.Morgan, T.Doi, R.M.Houley, J.R.Porter, ISS'92, Kobe, Japan, Advances in Superconductivity V, Springer Verlag, 1993

[25] J.C. Moore, D.M.C.Hyland, C.J.Salter, C.J.Eastell, S.Fox, C.R.M.Grovenor, M.J.Goringe, Physica C 276[1997]202

[26] K.Lebbou, S.Trosset, R.Abraham, M.T.Cohen-Adad, M.Ciszek, W.Y.Liang, Physica C 304[1998]21

[27] M.Eder, G.Gritzner, Supercond.Sci.Technol.13[2000]1302

[28] G.Gritzner, M.Eder, A.Cigàn, J.Manka, G.Plesch, V.Zrubec, Physica C366[2002]169

[29] P.J.Kung, M.P.Maley, P.G.Wahlbeck and D.E.Peterson, J.Mater.Res.8[1993]713

[30] R.Awad, Supercond.Sci.Technol.15[2002]933

[31] T.S.Kayed , H.Ozkan, N.M.Gasanly, Supercond.Sci.Technol.14[2001]738

[32] M.Jergel, A.Conde Gallardo, C.Falcony-Guajardo, V.Strbik Supercond.Sci.Technol.9[1996]1

[33] R.E.Gladyshevskii, A.Perrin, B.Henzel, R.Flükiger, R.Abraham, K.Lebbou, M-T.Cohen-Addad, J-L.Jorda, Physica C 255[1995]113

[34] R.Flükiger, R.E.Gladyshevskii, E.Bellingeri, J.Supercond.11[1998]

[35] D.Y. Jeong, S.M.Baek, H.K.Kim, B.J.Kim, E.Y.Lee, Y.C.Kim, Physica C 377[2002]445

[36] O.Heiml, G.Gritzner, Supercond.Sci.Technol.15[2002]956

[37] R.N.Bhattacharya, H.L.Wu, Y-T.Wang, R.D.Blaugher, S.X.Yang, D.Z.Wang, Z.F.Ren, Y.Tu, D.T.Verebelyi, D.K.Christen, Physica C 333[2000]59

[38] W.Li, D.Z.Wang, J.Y.Lao, Z.F.Ren, J.W.Wang, M.Paranthaman, D.T.Verebelyi and D.K.Christen, Supercond.Sci.Technol.12[1999]L1

[39] S.Phok, PhD thesis, University of Savoie, Dec.2002

[40] S.Phok, Ph.Galez, J-L.Jorda, Z.Supardi, D.de Barros, P.Odier, A.Sin, F.Weiss, Physica C 372-376[2002]876

[41] S.Phok, Ph.Galez, J-L.Jorda, C.Peroz, C.Villard, D.de Barros, F.Weiss, Appl.Supercond.Confer., Houston (2002). To appear in IEEE Transactions on Applied Superconductivity

[42] E.Ruckenstein and C.T.Cheung, J.Mater.Res.4[1989]116

[43] C.T.Cheung and E.Ruckenstein, J.Mater.Res.5[1990]245

[44] T.L.Aselage, E.L.Venturini, S.B.Van Deusen, T.J.Headley, M.O.Eatough and J.A.Voigt Physica C 203[1992]25

[45] T.L.Aselage, E.L.Venturini and S.B.Van Deusen, J.Appl.Phys.75[1994]2

[46] T.K.Jondo, R.Abraham, M.Th.Cohen-Adad, J-L.Jorda, J.Alloys and Compounds 186[1992]347

[47] T.K.Jondo, C.Opagiste, J-L.Jorda, M.Th.Cohen-Adad, F.Sibieude, M.Couach, A.F.Khoder, J.Alloys and Compounds 195[1993]53

[48] C.Opagiste, G.Triscone, M.Couach, T.K.Jondo, J-L.Jorda, A.Junod, A.F.Khoder, J.Muller, Physica C 213[1993]17

[49] J-L.Jorda, K.Lebbou, Ph.Galez, R.Abraham, J.Alloys and Compounds 256[1997]34

[50] J-L.Jorda, J.Thermal Analysis 48[1997]585

[51] K.Lebbou, R.Abraham, S.Trosset, M.Th.Cohen-Adad, Mater.Research Bull.32[1997]1017

[52] Th.Hopfinger, M.Lomello-Tafin, J-L.Jorda, Ph.Galez, M.Couach, R.Gladyshevskii, J-L.Soubeyroux, Physica C 351[2001]53

[53] J-L.Jorda, T.K.Jondo, J.Alloys and Compounds 327[2001]167

[54] Ph.Galez, Th.Hopfinger, J-L.Soubeyroux, M.Lomello-Tafin, C.Opagiste, Ch.Bertrand and J-L.Jorda, Physica C 372-376[2002]1137

[55] S.Phok, T.K.Jondo, J-L.Jorda Physica C 372-376[2002]1191

[56] J.L.Routbort, S.J.Rothman, J.Appl.Phys.76[1994]5615

[57] W.Goldacker, R.Nast, J.Krelaus IEEE Trans. on Appl.Supercond. 11[2001]3407

[58] Y.S.Sung, E.E.Hellstrom, J.Amer.Ceram.Soc.78[1995]2004

[59] E.Giannini, E.Bellingeri, F.Marti, M.Dhallé, V.Honkimäki and R.Flükiger, Advances in Supercond.12[2000]637

[60] E.Bellingeri, PhD.Thesis, Univ.Geneva, July 2000

[61] E.J.Mittemeijer, J.Mat.Sci.27[1992]3977

[62] S.Vyazovkin, C.A.Wight, Ann.Rev.Phys.Chem.48[1997]125

[63] J.P.Elder, Thermochim.Acta 318[1998]229

[64] F.W.Wilburn, D.Dollimore, Thermochim.Acta 357-358[2000]141

[65] A.K.Galwey, M.E.Brown, J.Thermal Analysis and Calorimetry, 60[2000]863

[66] J.Sestak, G.Berggren, Thermochim.Acta 3(1971)1

[67] J.W.Christian, " The Theory of Transformations in Metals and Alloys" Pergamon Press, New York, (1975)

[68] A.K.Galwey, M.E.Brown, Proc.R.Soc.Lond.A 450[1995]501

[69] A.B.Phadnis, V.V.Deshpande, Thermochim.Acta 42(1980)109

[70] S.Vyasovkin, C.A.Wight, Annu.Rev.Phys.Chem. 48[1997]125

[71] M.J.Starink, J.Mat.Sci.32[1997]6505

[72] J.Sempere, R.Nomen and R.Serra, J.Thermal Analysis and Calorimetry 56[1999]843

[73] J.Malek, T.Mitsuhashi, J.M.Criado, J.Mat.Res.16[2001]1862

[74] J.D.Sewry, M.E.Brown, Thermochim Acta 390[2002]217

[75] M.Huvé, C.Michel, A.Maignan, M.Hervieu, C.Martin and B.Raveau, Physica C 205[1993]219

[76] M.A.Alario-Franco, C.Chaillout, J.J.Capponi, J-L.Tholence, B.Souletie, Physica C 222[1994]52

[77] Y.S.Sung, X.F.Zhang, P.J.Kostic and D.J.Miller, Appl.Phys.Letters 69[1996]3420

[78] A.Iyo, Y.Ishiura, Y.Tanaka, P.Badica, K.Tokiwa, T.Watanabe, H.Ihara, Physica C370[2002]205

[79] J.J.Lander J.Amer.Chem.Soc.73(1951)5893

[80] J.A.Pask, L.K.Templeton, "Kinetics of High Temperature Processes" MIT Press, Cambridge, 1959

[81] T.K.Basu, A.W.Searcy, J.Chem.Soc., Fraday Trans.I 72(1976)1889

[82] M.D.Judd, M.I.Pope, J.Thermal Analysis 4(1972)31

[83] H.L.Friedman, J.Polym.Sci.66(1965)137

[84] H.E.Kissinger, Anal.Chem.29(1957)1702

[85] T.Ozawa, Bull.Chim.Soc.Jpn. 38(1965)1881

[86] R.E.Lyon, Thermochim.Acta 297[1997]117

[87] M.J.Tribelhorn, M.E.Brown, Thermochim.Acta 255[1995]143

[88] E.Zimmermann, K.Hack, D.Neuschütz, Brite Euram Report, BREU0203C [1993]

[89] R.S.Roth, C.J.Rawn, M.D.Hill, NIST Special Publ.n°804[1991]225

[90] A.F.Mayorava, S.N.Mudretsova, M.N.Mamontov, P.A.Levashov and A.D.Rusin Thermochimica Acta, 217[1993]241

[91] IPCS International Programme on Chemical Safety, Health and Safety Guide n°102 [1996]

[92] P.G.Wahlbeck, R.R.Richards and D.L.Myers, J.Chem.Phys.95[1991]9122

[93] W.L.Holstein, J.Phys.Chem.97[1993]4224

[94] R.H.Lamoreaux, D.L.Hildelbrand, L.Brewer, J.Phys.Chem.Ref.Data 16[1987]419

[95] K.Tachikawa, A.Kikuchi, T.Kinoshita, S.Komiya, IEEE Trans.Appl.Superconductivity 5[1995]2019

[96] A.Kikuchi, K.Inoue, K.Tachikawa, Physica C 337[2000]180

[97] G.F.Sun, Y.Xin, D.F.Lu, K.W.Wong, Y.Zhang, J.G.Steven, Solid State Commun. 101[1997]849

[98] N.M.Hamdan, Kh.A.Ziq, A.S.Al-Harthi, Physica C 314[1999]125

[99] L.M.Lopato, Z.A.Yaremenko, S.G.Tresvyatskii, Ukr.Khim.Zh. 32(1966)437

[100] W.Zhou, R.S.Liu and P.P.Edwards, J.Solid State Chem.91[1991]32

[101] W.Zhou, R.S.Liu, P.P.Edwards, J.Solid State Chem.87[1990]472

[102] R.V.Von Scheng, H.K.Muller-Buschbaum, Z.Anorg.Allg.Chem. 400(1974)197

[103] R.V.Von Scheng, H.K.Muller-Buschbaum, Z.Anorg.Allg.Chem. 396(1973)113

[104] F.Goutenoire, V.Caignaert, M.Hervieu, C.Michel, and B.Raveau J.Solid State Chem.115[1995]508

[105] F.Goutenoire, V.Cagnaert, M.Hervieu, and B.Raveau, J.Solid State Chem.119[1995]134

[106] F.Goutenoire, V.Caignaert, M.Hervieu, C.Michel, and B.Raveau, J.Solid State Chem.114[1995]428

[107] T.K.Jondo, Private Communication [2001]

[108] W.Zhou, R.S.Liu, R.Janes, P.P.Edwards, Materials Letters 9[1990]169

[109] F.Abbattista, C.Brisi, M.Lucco-Borlera and M.Vallino, Nuvo Cimento Ital.Fis.D 10[1988]611

[110] M.Nevriva, E.Pollert, L.Matejkova, A.Triska, J.of Crystal Growth 91[1988]434

[111] R.S.Roth, K.L.Davis, J.R.Dennis, Advan.Ceram.Mat.2[1987]303

[112] D.M.De Leeuw, C.A.H.A.Mutsaers, C.Langereis, H.C.A.Smoorenburg and P.J.Rommers, Physica C 152[1988]39

[113] K.G.Frase, E.G.Liniger and D.R.Clarke, J.Amer.Ceram.Soc.70[1987]C-204

[114] W.Zhang, K.Osamura and S.Ochiai, J.Amer.Cer.Soc.73[1990]1958

[115] T.B.Lindemer, E.D.Specht, Physica C 255[1995]81

[116] J.G.Thompson, J.D.Fitz Gerald, R.L.Withers, P.J.Barlow, J.S.Anderson, Mat.Res.Bull.24[1989]505

[117] J.G.Thompson, T.J.White, R.L.Withers, J.D.Fitz Gerald, P.J.Barlow, S.J.Collocott, Materials Forum 14[1991]27

[118] E.L.Brosha, F.H.Garzon, I.D.Raistrick, J.Solid State Chem. 122[1996]176

[119] G.F.Voronin and S.A.Degterov, J.Solid State Chem.110[1994]50

[120] A.Pisch, PhD. Thesis, INP Grenoble [1994]

[121] V.A.Lysenko, Inorganic Materials 34[1998]995

[122] M.Mrovec, J.Leitner, M.Nevriva, D.Sedmidubsky, J.Stejskal, Thermochim.Acta 318[1998]63

[123] T.K.Jondo, S.Phok, Ph.Galez, J.L.Jorda, EMRS Strasbourg, 17-21.06/2002. Accepted for publication in Cryogenics [2003]

[124] S.Phok, T.K.Jondo, Ph.Galez, J.L.Jorda to be published

[125] W.Zhang, K.Osamura, Jpn.J.Appl.Phys.29[1990]L1090

[126] R.Kipka, H.Müller-Buschbaum, Z.Naturforsch.32B(1977)121

[127] E.F.Paulus, G.Miehe, H.Fuess, I.Yehia, and U.Löchner, J.Solid State Chem.90[1991]17

[128] C.Teske, H.Müller-Buschbaum, Z.Anorg.Allg.Chem.370(1969)134

[129] C.Teske, H.Müller-Buschbaum, Z.Anorg.Allg.Chem.379(1970)234

[130] R.S.Roth, C.J.Rawn, J.J.Ritter, B.P.Butron, J.Amer.Ceram.Soc.72[1989]1545

[131] O.Milat, G.Van Tendeloo, S.Amerlinkx, T.G.N.Babu, C.Greaves, Solid State Comm. 79[1991]1059

[132] O.Milat, G.Van Tendeloo, S.Amerlinkx, T.G.N.Babu, C.Greaves, J.Solid State Chem. 101[1992]92

[133] Ph.Galez, M.Lomello-Tafin, Th.Hopfinger, Ch.Opagiste, Ch.Bertrand, J.Solid State Chem. 151[2000]170

[134] R.O.Suzuki, P.Bohac, L.J.Gauckler, J.Amer.Ceram.Soc.77[1994]41

[135] D.Risold, B.Hallstedt, L.J.Gauckler, J.Amer.Ceram.Soc. 78[1995]2655

[136] T.K.Jondo, M.Lomello-Tafin, J.C.Marty, Ch.Bertrand, Ph.Galez, J.L.Jorda, 26èmes Journées d'Etude des Equilibres entre Phases, March 23-24/2000, Marseille (France). ISBN-2-9511531-0-4, p72

[137] S.Phok, T.K.Jondo, Ph.Galez, J.L.Jorda, J.Phys.IV France 11[2001] Pr10-99

[138] T. Mathews, J.P. Hajra and K.T. Jacob, Chem. Mater. 5 n° 11 [1993] 1669.

[139] N.L.Wu, E.Ruckenstein, Mater.Lett.7[1988]169

[140] K.A.Kubat-Martin, E.Garcia and D.E.Peterson, Physica C 172[1990]75

[141] K.A.Kubat-Martin, G.H.Kwei, A.C.Lawson and D.E.Peterson, J.Solid State Chem.100[1992]130

[142] F.Abbattista, A.Delmastro, D.Mazza, M.Vallino, Mater.Chem.Phys.28[1991]33

[143] S.H.Lin, N.L.Wu, Physica C 262[1996]33

[144] J.Q.Lin, C.C.Lam, J.Feng, K.C.Hung and E.C.L.Fu, Z.Metallkd.88[1997]9

[145] T.C.Huang, V.Y.Lee, R.Karimi, R.Beyers and S.S.P.Parkin, Mat.Res.Bull. 23[1988]1307

[146] M.A.G.Aranda, D.C.Sinclair, and J.P.Attfield, Phys.Rev.B 51[1995]12747

[147] M.A.G.Aranda, D.C.Sinclair, and J.P.Attfield, Physica C 235-240[1994]965

[148] A.W.Hewat, P.Bordet, J.J.Capponi, J.Chenavas, M.Ghodino, E.A.Hewat, J.L.Hodeau, and M.Marezio, Physica C 156[1988]369

[149] S.S.P.Parkin, V.Y.Lee, A.I.Nazal, R.Savoy, T.H.Huang, G.Gorman, P.Beyers, Phys.Rev.B 38[1988]6351

[150] Y.Shimakawa, Y.Kubo, T.Manako, T.Saroh, J.Ichibashi and H.Igarashi, Physica C 154[1989]279

[151] J.L.Jorda, T.K.Jondo, R.Abraham, M.T.Cohen-Adad, C.Opagiste, M.Couach, A.Khoder and F.Sibieude, Physica C 205[1993]177

[152] J.D.Jorgensen, M.A.Beno, D.G.Hinks, L.Soderholm, K.J.Volin, R.L.Hitterman, J.D.Grace, I.K.Schuller, C.U.Segre, K.Zhang and M.S.Kleefish, Phys.Rev.B 36[1987]3608

[153] P.Bordet, J.J.Capponi, J.Chenavas, M.Godinho, A.W.Hewat, E.A.Hewat, J.L.Hodeau, A.M.Spieser, J.L.Tholence and M.Marezio, J.Alloy and Compounds 150[1989]109

[154] C.Opagiste, M.Couach, A.F.Khoder, R.Abraham, T.K.Jondo, J.L.Jorda, M.Th.Cohen-Adad, A.Junod, G.Triscone, J.Muller, J.Alloys and Compounds 195[1993]47

[155] K.N.Tu, N.C.Yeh, S.I.Park and C.C.Tsuei, Phys.Rev.B 39(304)1989

[156] Z.G.Fan, Y.X.Zhuang, G.Yang, R.Shao and G.F.Zhang, J.Alloys and Compounds 200[1993]33

[157] J.L.Routbort, S.J.Rothman, J.Appl.Phys.76[1994]5615

[158] D.R.Pederzolli, G.M.Wltschek and J.P.Attfield, Chem.Comm.[1997]435

[159] Y.Shimakawa, Y.Kubo, T.Manako, H.Igarashi, F.Izumi, H.Asano, Phys.Rev.B 42[1990]10165

[160] Th.Hopfinger, PhD. thesis, Université de Savoie, France [1999]

[161] Ph.Galez, J.L.Soubeyroux, Th.Hopfinger, Ch.Opagiste, M.Lomello-Tafin, Ch.Bertrand and J-L.Jorda, Supercond. Sci.Technol.14[2001]583

[162] R.S.Liu, S.D.Hughes, R.J.Angel, T.P.Hackwell, A.P.Mackenzie and P.P.Edwards, Physica C198[1992]203

[163] N.N.Kolesnikov, V.E.Korotkov, M.P.Kulakov, R.P.Shibaeva, V.N.Molchanov, R.A.Tamazyan and V.I.Simonov, Physica C 195[1992]219

[164] C.S.Gopinath, S.Subramanian, M.Paranthaman, A.M.Hermann, Physica C 214[1993]153

[165] K.S.Nanjundaswami, A.Manthiram and J.B.Goodenough, Physica C 207[1993]339

[166] D.E.Morris, M.R.Chandrachood, A.P.B.Sinha, Physica C 175[1991]156

[167] D.M.Ogborne and M.T.Weller, Physica C 201[1992]53

[168] H.C.Ku, M.F.Tai, J.B.Shi, M.J.Shieh, S.W.Hsu, G.H.Hwang, D.C.Ling, T.J.Watson-Yang and T.Y.Lin, Jap.J.Appl.Phys.6[1989]L923

[169] B.Morosin, D.S.Ginley, P.F.Hlava, M.J.Carr, R.J.Baighman, J.E.Schirber, E.L.Venturini and J.F.Kwak, Physica C 152[1988]413

[170] P.Majewski, A.Jalowicki, F.Aldinger, Physica C 341-348[2000]403

[171] D.M.Ogborne and M.T.Weller, Physica C 230[1994]153

[172] P.Badica, A.Iyo, A.Crisan, H.Ihara, Supercond.Sci.Technol.15[2002]975

[173] M.Hasegawa, K.Izawa, A.Shibata, Y.Matsuda, Physica C 377[2002]459

[174] H.Q.Chen, L.G.Johanson, V.Langer, Z.G.Ivanov, Physica C371[2002]83

[175] J.G.Wen, D.Z.Wang, S.X.Yang, Z.F.Ren, Physica C 366[2002]203

[176] B.Morosin, D.S.Ginley, R.J.Baughman and C.P.Tigges, Physica C172[1991]413

[177] C.Opagiste, PhD Thesis [1994], Université J.Fourier, Grenoble, France

[178] C.Opagiste, M.Couach, A.Khoder, G.Triscone, J.Ljorda, M.Th.Cohen-Adad, Physica B 194-196[1994]1809

[179] S.L.Yan, L.Fang, M.S.Li, H.L.Cao, Q.X.Song, J.Yan, X.D.Zhou, J.M.Hao, Superconduc.Sci.Technol.7[1994]681

[180] H.Schneidewind, M.Manzel, G.Bruchlos, K.Kirsch, Supercond.Sci.Technol. 14[2001]200

[181] T.Kaneko, H.Yamauchi, S.Tanaka, Physica C178[1991]377

[182] R.S.Liu and P.P.Edwards, Physica C 179[1991]353

[183] S.X.Dou, H.K.Liu, A.J.Bourdillon, N.X.Tan, N.Savvides, C.Andrikidis, R.B.Roberts and C.C.Sorrell, Supercond.Sci.Technol.1[1988]83

[184] M.Hervieu, A.Maignan, C.Martin, C.Michel, J.Provost and B.Raveau, Mod. Phys.Letters B2[1988]1103

[185] P.L.Gai,M.A.Subramanian, J.Gopalakrishnan, E.D.Boyes, Physica C159[1989]801

[186] M.H.Pan and M.Greenblatt, Physica C176[1991]80

[187] I.K.Gopalakrishnan, J.V.Yakhmi and R.M.Iyer, Physica C 175[1991]183

[188] F.Letouzé, C.Martin, M.Hervieu, A.Maignan, M.Daturi, C.Michel, B.Raveau, Physica C 292[1997]32

[189] J.C.Barry, Z.Iqbal, B.L.Ramakrishna, H.Eckhardt, F.Reidinger and R.Sharma, J.Appl.65[1989]5207

[190] T.Manako, Y.Shimakawa, Y.Kubo, T.Satoh and H.Igarashi, Physica C 158[1989]143

[191] M.A.Subramanian, Mater.Res.Bull.29[1994]119

[192] D.S.Ginley, B.Morosin, R.J.Baughman, E.L.Venturini, J.E.Schirber, J.F.Kwak, J.Cryst.Growth 91[1988]456

[193] B.Morosin, E.L.Venturini, R.G.Dunn, P.P.Newcomer, Physica C 288[1997]255

[194] K.Lebbou, PhD thesis, University C.Bernard, Lyon 1, France 1996

[195] H.Ihara, R.Sugise, M.Hirabayashi, N.Terada, M.Jo, K.Hayashi, A.Negishi, M.Tokumoto, Y.Kimura and T.Shimomura, Nature 334[1988]510

[196] L.Zhang, J.Z.Liu and R.N.Shelton, Supercond.Sci.Technol.11[1998]1045

[197] N.A.Khan, Y.Sekita and H.Ihara, Supercond.Sci.Technol. 15[2002]613

[198] J.L.Jorda, Th.Hopfinger, M.Couach, P.Pugnat, C.Bertrand and Ph.Galez J.Supercond.11[1997]87

[199] Th.Hopfinger, O.O.Shcherban, Ph.Galez, R.E.Gladyshevskii, M.Lomello-Tafin, J.L.Jorda, M.Couach, J.Alloys and Compounds 333[2002]237

[200] S.Phok, Ph.Galez, J.L.Jorda, D.De Barros, C.Villard, F.Weiss, EMRS, Strasbourg, France, June 17-21, 2002

[201] J.C.Grivel, R.Flükiger, J.Alloys and Compounds 241[1996]127

[202] S.Iijima, T.Ichihashi, Y.Kubo, Jpn.J.Appl.Phys.27[1988]L817

[203] S.Iijima, T.Ichihashi, S.Shimakawa, T.Manako, Y.Kubo, Jpn.J.Appl.Phys.27[1988]L1054

[204] D.M.Ogborne, M.T.Weller, P.C.Lancaster, C.Grovenor, L.Romano, Supercond.Sci. Technol.5[1992]78

[205] W.L.Holstein, L.A.Parisi, C.R.Fincher, P.L.Gai, Physica C 212[1993]110

[206] W.L.Holstein, J.Appl.Phys.74[1993]4963

[207] G.Malandrino, G.G.Condorelli, I.L.Fragalla, F.Miletto Granozio, U.Scotti di Uccio, M.Valentino, Supercond.Sci.Technol.9[1996]570

[208] X.F.Zhang, Y.S.Sung, D.J.Miller, Physica C 301[1996]221

[209] S.Trosset-Jarnieux, PhD thesis, University C.Bernard, Lyon 1, France [1997]

[210] L.P.Cook, W.Wong-Ng, P.Paranthaman, J.Res.Natl.Inst.Stand.Technol. 101[1996]675

[211] S.Adachi, T.Shibata, T.Takutsi, T.Tamura, K.Tanabe, S.Fujihara and T.Kimura, Physica C [1999?2000]

[212] J.L.Jorda, K.Lebbou, M.Couach, Ph.Galez, S.Trosset, M.T.Cohen-Adad, LT21 Proceedings, Prague, August 1996, Czech.J.Phys.46[1996]1471

[213] J.L.Jorda, K.Lebbou, Ph.Galez, R.Abraham, J.Alloys and Compounds, 256[1997]34

MELT PROCESSED RE-Ba-Cu-O BULK SUPERCONDUCTORS

Hiroshi IKUTA

Center for Integrated Research in Science and Engineering
Nagoya University
Furo-cho, Chikusa-ku, Nagoya 464-8603, Japan

1 INTRODUCTION

Melt processing is one of the most important and successful techniques for the fabrication of a $REBa_2Cu_3O_y$ (RE123, RE=rare earth) high temperature superconductor with a high critical current density (J_c). This technique, first proposed by Jin *et al.*[1], proved to be very effective in overcoming the weak-coupling problem of grain boundaries in high temperature superconductors. The technique was soon improved by Salama *et al.*[2] and Murakami *et al.*[3], and owing to the continuous worldwide efforts since then, J_c had been increased beyond the level that is assumed to be necessary for applications.

A high J_c is, however, not the only requirement to be fulfilled for applications of melt-grown bulk superconductors. Large grain diameters are often very important and significant progress had been accomplished in this respect as well, and it is now possible to grow grains to diameters of several centimeters with the c-axis aligned, which is very important due to the large anisotropy of J_c in high temperature superconductors. Nevertheless, the performance of a large grain sample is usually not as large as expected from the high J_c values of smaller samples. This is because the magnitude and spatial variation of J_c within a bulk superconductor are highly dependent on the microstructural and compositional characteristics of the sample. The J_c values reported in the literature are usually determined from magnetic hysteresis measurements on a small sample cut from a large grain. However, the microstructure e.g., the density of the secondary phase that is important for flux pinning and/or various defects like micro-cracks or impurity phases, is not homogeneous throughout a large grain sample. As a result, the superconducting properties show a certain inhomogeneity, and the macroscopic performance of the sample is not simply related to the J_c value determined from a small sample. Rather than a sample that only exhibits an extraordinarily high J_c in a small part, a large grain with reasonably high J_c throughout the entire sample will show a better performance as a whole. For practical applications, therefore, it is very important to evaluate the total performance of the superconductor, and to establish a technique to grow large grain samples with homogeneous superconducting properties.

In the first part of this review, therefore, we will mostly concentrate on the bulk properties of the melt-processed superconductors. We are more interested in the performance of the sample as a whole rather than the microscopic or structural properties. This will distinguish the present article from many other excellent reviews covering details of the fabrication issues and J_c properties [4–7]. To evaluate the performance of a large sample, the trapped field is one of the most important parameters, because it directly reveals the macroscopic properties of the sample. The trapped field corresponds to the magnetization in the remnant state, and is determined by the pinning effect of the sample. If defects or cracks exist inside the sample, that part cannot pin magnetic flux lines and the density of trapped magnetic flux decreases. If the distribution of the secondary phase, the pinning sites, is not homogeneous, the magnetic field that the sample actually trap is not as high as what would be estimated from the J_c value that might be extraordinarily high only in a small part of the sample.

In the next section, we will give an overview of the most general methods used for the preparation of the samples, and discuss on the characterization of the performance of a bulk sample through the measurement of the trapped field. Following to that, we will review the reported performances of large grain samples, mostly the trapped field, of several materials. It will be shown that the trapped field of bulk superconductors had been greatly increased in the past 5-6 years, and we will see that the trapped field has reached a very high level, well above the requirement for many applications.

There are two types of applications for melt-processed bulk superconductors, "passive" and "active" applications. Levitation is an important passive application, for which a detailed review is published recently [8]. In active applications, a bulk sample that had trapped magnetic fluxes is used as a source of magnetic field. For this kind of applications, the study of magnetization technique is very important. This is because the quasi-static magnetization method that is usually employed in evaluating the performance of the sample would not be feasible for many actual applications, as will be discussed later. An important technique to overcome this problem is pulsed-field magnetization, and we will devote a section to discuss this technique in detail.

2 PROCESSING AND CHARACTERIZATION

2.1 Melt processing

The basic of the melt processing technique is simple: it makes full use of the fact that RE123 compounds decompose into solid RE_2BaCuO_5 (RE211) (or $Nd_4Ba_2Cu_2O_{10}$ (Nd422) if RE=Nd) and liquid Ba-Cu-O phases by a peritectic melting, and crystallizes through the reverse process. For $YBa_2Cu_3O_y$ (Y123), the decomposition is described by the following equation:

$$Y123 \rightarrow Y211 + \text{liq. (Ba-Cu-O)} + x\,O_2$$

Because this is a peritectic decomposition, the melt contains a significant amount of solid 211 particles. The solidification follows the same equation to the opposite direction, and the 211 phase particles react with the liquid, increasing the size of the 123 phase while the size of 211 particles decreases. If the process is conducted with an infinitely long time, the whole sample will turn into the 123 phase. However, the process usually proceeds before the reaction between the 211 phase and liquid completes, and small unreacted 211 phase particles are entrapped in the 123 phase due to the incomplete peritectic interaction.

Y211 particles entrapped in the Y123 phase matrix, or more precisely the interface between the 123 and 211 phases, have the ability to pin magnetic flux [9]. By using a starting composition that is in excess of the 211 phase, the amount of unreacted 211 particles can be increased. Consequently, an enhanced pinning effect can be expected. The addition of an excess of 211 phase has also a practical benefit, *i.e.*, it increases the viscosity of the melt. Therefore, the sample can stand without to be sustained in a crucible even above the peritectic decomposition temperature, which minimizes the contamination of the sample from the surrounding. For these two reasons, it is common to use starting composition that is in excess of the 211 phase. Alternatively, melt-processed Y-Ba-Cu-O (YBCO) [1] can be prepared from Y_2O_3 admixed starting precursors instead of using a composition that is excess in Y211, which was reported to possess an increased process window for a stable growth [10].

For a given composition, the decrease of the 211 particle size means the increase in the numbers of the non-superconducting inclusions. Because the interface between the 123 and 211 phases acts as a pinning center, small 211 particles is very effective in increasing J_c, and several techniques have been adopted to inhibit coarsening of the 211 phase during the melt processing. Adding Pt or PtO_2 is commonly used for this purpose [11–14]. It is also known that CeO_2 has a similar effect [14, 15]. Especially for the Nd-Ba-Cu-O system, the effect of Pt on refining Nd422 particles seems not to be as large as in the other compounds, while CeO_2 addition was effective for decreasing the size of the non-superconducting particles [16]. Rapid heating to the peritectic decomposition temperature, lowering the maximum temperature and minimizing the holding time at the maximum temperature are the general points of notice in order to avoid coarsening of the 211 phase particles by the Ostwald ripening mechanism. Refining the size of the 211 particles in the starting composition is also effective to decrease the size of the non-superconducting inclusions in the finished sample.

The precursor for the melt-process is usually a pellet that is made from premixed powders. Some groups use mixed powders of the pre-sintered 123 and 211 phases, and some groups use mixed powders of rare-earth-, Ba-, and Cu-oxides. In either case, the starting powders are mixed with a ratio so the nominal composition is excess in the 211 phase. Pt or CeO_2 is added as already mentioned. The pellets are made by compressing the starting powders, and sometimes, but not necessary, a cold-isostatic-pressing (CIP) is adopted. After the melt-process, the sample is generally shrunken because of the increase in the density, but the shape is approximately preserved.

It is necessary to fabricate samples free from grain boundaries, because they act as weak links in high temperature superconductors. Furthermore, it is important to align the *c*-axis direction perpendicular to the sample surface because of the anisotropy of high temperature

---

[1]The material composing a melt-processed superconductor is referred to as REBCO in this article because it consists of several different phases. RE123 and RE211 are used to refer to the pure superconducting and secondary phases, respectively.

superconductors. For these purposes, a top seeded melt growth (TSMG) technique has been developed which involves seeding a small single crystal on the top of the sample with similar lattice parameters to provide a heterogeneous grain nucleation. MgO is one of the commonly used material for the seed crystal, but with this material, the c-axis of the prepared bulk sample is usually aligned parallel to the top surface of the sample. In order to align the c-axis perpendicular to the sample surface, a 123 material is used as a seed crystal, and is put on the sample so the c-axis of the seed crystal is directing perpendicular to the sample surface [17, 18]. The seed crystal should not melt during the process. Therefore, a 123 material with a higher peritectic temperature, usually Sm123 or Nd123, is used as a seed crystal.

If the melting or decomposition temperature of the seed material is sufficiently higher than the sample, the seed crystal may be placed on the top of the sample before starting the melt-growth. This cold-seeding has an advantage that it enables an accurate control in placing the seed crystals on the sample. However, if the melting point of the seed material is not sufficiently high, cold-seeding may cause a partial decomposition and an inter-diffusion between the sample material because the seed crystal is exposed to a temperature higher than the peritectic temperature of the material to be grown. To avoid this problem, the seeding is often performed after the temperature is decreased to or below the peritectic temperature. Even with this hot-seeding technique, it is known that the seed material may inter-diffuse with the sample material, and it is important to use a seed crystal with a peritectic temperature as high as possible. This is particularly problematic for the processing of a material with a high peritectic temperature like NdBCO and SmBCO. A practical way to reduce the peritectic temperature in order to increase the difference in the decomposition temperature of the seed crystal and that of the material to be grown is adding Ag into the starting composition [19]. It was also reported that a combined addition of Ag_2O and Au further decreases the peritectic decomposition temperature, at least for NdBCO [20].

2.2 Characterization

Among the various parameters characterizing a superconducting material, the critical temperature (T_c) and the critical current density (J_c) is with no doubt the most important parameters when applications are concerned. These parameters are usually determined from a small piece of sample, and reflect only the local properties of the bulk sample. However, for applications, it is the total performance of the bulk sample that is important, which is sometimes not directly related to the local J_c values. Besides the fact that the superconducting properties can strongly depend on the position in the sample, the performance of the whole sample is determined by structural defects like microcracks, voids, grain-boundaries, etc. Geometrical effects also influence the performance of a bulk sample. Hence, it is important to characterize the performance of the bulk superconductor as a whole when applications are concerned. The most direct and practical way for this purpose is to measure the trapped field, which is the remnant field in the sample after an external field is applied and then removed.

For an ideal sample, the trapped field is expected to distribute like the schematic drawing shown in Fig. 1(a). Here we assume a cylindrical sample and plot the magnetic flux density along a diametrical direction, neglecting the demagnetization effect. As shown in Fig. 1(a), the field profile should have a single peak located at the center. The gradient of the field distribution is determined by J_c, and with the assumption of a constant J_c as in Bean's critical state model [21], the magnetic flux density at the center is proportional to both J_c and the radius of

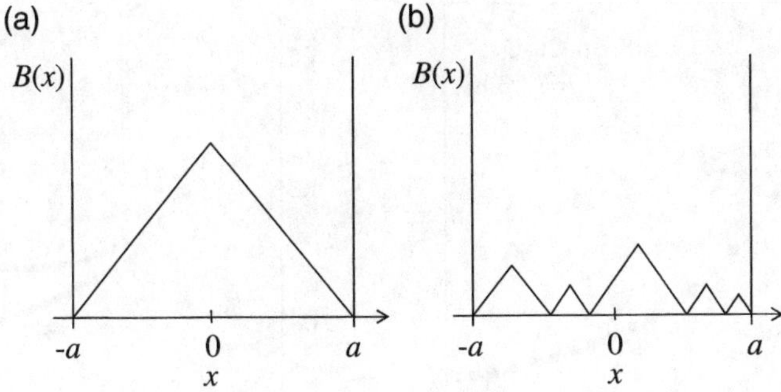

Figure 1: Schematic drawing of the trapped magnetic flux density of (a) an ideal and (b) a cracked cylindrical bulk sample. These pictures depict the magnetic flux density along a diameter of the top surface of the sample, neglecting the demagnetization effect.

the sample. However, if there are cracks inside the sample, those regions cannot hold magnetic fluxes, and the field distribution would have several peaks, as shown in Fig. 1(b). The trapped field is significantly lower than that of Fig. 1(a) because each domain is independently magnetized, which reduces considerably the effective size of the sample. Even with no cracks, the shape of the field distribution would be distorted and the trapped field be reduced if there were weak-links that causes deterioration of the local J_c. Hence, the shape of the distribution and the peak value of the trapped magnetic flux density are direct measures that reflect the macroscopic properties of a bulk sample.

Usually, the sample is magnetized by a field cooling (FC) method, *i.e.*, an external field is applied at a temperature higher than T_c, and the sample is cooled to the target temperature in the presence of the magnetic field. The trapped field will be measured after removing the external field. Alternative methods are zero-field-cooling (ZFC) magnetization and pulsed-field magnetization (PFM), which will be discussed in detail in Section 4. Note that the local field within the sample cannot be higher than the external field, and if the magnet that supplies the external field cannot produce a high enough field, the resulted trapped field is lower than what the sample could otherwise accomplish. For this reason, some of the values in the literature may give only the lower bound of the potential of the reported material.

The trapped field is usually measured using a Hall sensor. To measure the trapped field at a specific location, the Hall sensor might be mounted on the sample surface. To map the field distribution, however, the Hall sensor must be scanned over the sample surface. Note, that the trapped field measured by scanning a Hall sensor is smaller than the magnetic flux density at the surface. This is because the magnetic flux density strongly depends on the distance from the source of magnetic field. Assuming that the current density in a magnetized bulk superconductor is uniform and equal to J_c, the trapped field at the center can be easily calculated using the Biot-Savart law [22]. Figure 2 shows the magnetic flux density as a function of the distance (z) from the sample measured along the central axis of a cylindrical sample of a SmBCO sample that is 36 mm in diameter. The solid curve in the figure is the result of fitting the above mentioned theoretical curve to the data, which shows a quite good agreement. The result shown in Fig. 2 clearly indicates that the trapped field rapidly decreases with increasing

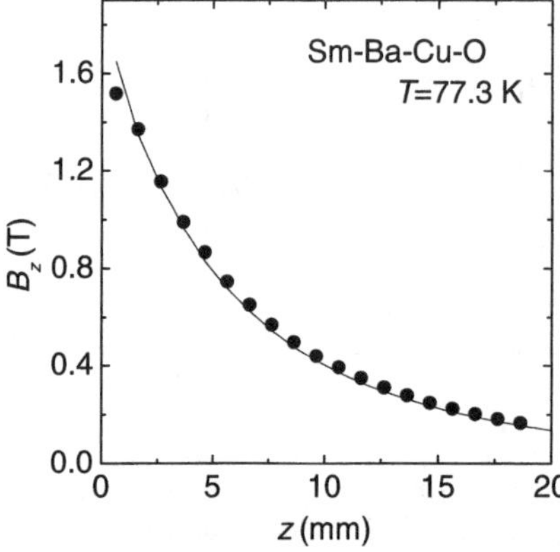

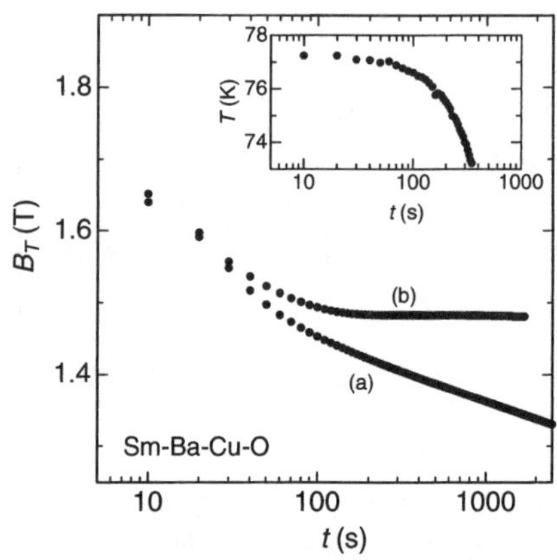

Figure 2: Trapped field as a function of the distance from the surface of a SmBCO bulk superconductor measured by moving a Hall sensor along the central axis of the cylindrical sample. The solid curve shows the result of fitting the theoretical expression described in the text to the data.

Figure 3: The trapped field as a function of time after the removal of the external field when (a) the temperature was kept at 77 K and (b) the heater that kept the sample at 77 K was switched off immediately after the external field was turned off. The temperature of the sample during the experiment of (b) is plotted in the inset.

the distance from the surface. Therefore, comparison of the trapped field values reported by different groups have to be made with care, as the distance between the sample surface and the magnetic sensor may be different. It should also be accounted for that the location where the sensor actually detects the field strength, the active area, may locate away from the bottom surface of the sensor.

It is also necessary to account for the flux creep effect when comparison of the reported performance of bulk superconductors from different groups is to be made. The flux creep effect causes a decrease in the trapped field as a logarithmic function of time [23, 24]. Because of this time dependence, the creep rate is very large during the first minutes after the external field is removed, and it is important to know when the trapped field was recorded. Practically, the decrease in trapped field will be significantly suppressed after about 30 minutes from the removal of the external field. The creep is much slower when the superconductor is not fully magnetized, *i.e.*, if the trapped field is less than the capability of the material [22] because the field gradient inside the superconductor is less steep. Similarly, the creep can be suppressed by post-cooling a fully magnetized sample [25]. This is demonstrated by the experiment shown in Fig. 3, which is a plot of the trapped field as a function of the time from the removal of the external field. The sample used here is a bulk SmBCO that is 36 mm in diameter, which was cooled using a Gifford-McMahan refrigerator. The field was measured using a Hall sensor that was mounted on the top surface of the sample. When the sample was kept at the same temperature, the trapped field followed well the $\log t$ law of the flux creep theory (curve (a)). For curve (b), on the other hand, the heater was turned off after finishing the magnetization, and the temperature started to decrease as plotted in the inset. Obviously, the decrease in the trapped field because of the creep effect was significantly suppressed, and was negligibly small after the temperature decreased by about 2 K in 200 s for the particular experiment of Fig. 3. This is because J_c is larger at lower temperature, and the field gradient produced by the magnetization at higher temperature is less steep than what the sample can support at lower temperature.

The trapped field depends also on the geometry of the sample because of the demagnetizing effect [26, 27]. Fukai *et al.* measured the trapped field of melt-processed YBCO superconductors, the thickness of which was varied by successively slicing the samples, and it was shown that the trapped field increased with the thickness. Roughly, their results show that the trapped field saturated when the thickness was comparable to the diameter for a cylindrical sample or to the length of the edge of a square sample, and increased by more than a factor of 2.5 with increasing the thickness from 1 mm to 15 mm. A theoretical calculation based on the Biot-Savart law and assuming a constant J_c value reproduced the experimental results quite well, both for the cylindrical and the square samples.

Due to the demagnetization effect, the magnetic flux densities above a bulk superconductor and inside a gap between two samples are different. The magnetic flux density in the gap region is approximately twice that achievable with a single sample. However, applications that can utilize the magnetic field in a small gap between two or more samples would be limited, and we are more interested on the magnetic field that can be used in a free space in this review. Therefore, the trapped fields discussed in the following section are those which were measured above the surface of a single sample, if not specified otherwise.

3 MATERIALS

In this section, we will take an overview of the performances reported for melt-processed bulk samples of several materials. Although, YBCO had been the most extensively studied material, we will start with SmBCO, because the successful preparation of large SmBCO samples with no cracks or weak-links in 1997-98 was one of the recent major breakthroughs in this field. Many further extensive studies on a variety of materials had been triggered by the success of SmBCO, as will be shown in the following.

3.1 Sm-Ba-Cu-O

A stoichiometric RE123 exhibits a T_c higher than Y123 if the replaced element RE is a light rare earth and it is therefore especially interesting to adopt the melt-processing technique. However, it was actually not as simple as just to apply the experiences of YBCO to a light rare earth element based superconductor, and there were several problems to overcome for the successful fabrication of a bulk form sample with high performance. The first difficulty confronting the melt-processing of REBCO materials with a light rare earth (LRE) element was the fact that a LRE element easily substitutes part of the Ba site, forming a solid solution (ss) of $LRE_{1+x}Ba_{2-x}Cu_3O_y$ [28, 29]. This leads to the decrease of the carrier density because Ba forms a divalent ion while a rare earth element a trivalent ion, and as a result, depresses T_c seriously with increasing x. However, Yoo and co-workers showed in 1994 that when the melt-process is conducted in a reduced oxygen atmosphere, the substitution is suppressed and the superconducting properties represented by T_c and J_c are greatly improved [5, 30, 31]. This is because the peritectic temperature of LRE123ss decreases with the concentration of LRE on the Ba site under an atmosphere with a reduced oxygen partial pressure [32]. In addition, J_c was found to exhibit a peak with increasing the applied magnetic field. It is generally believed that fluctuation in the superconducting properties due to a chemical modulation by local substitution of the LRE element for the Ba site is the source of a field induced pinning effect, and causes the peak of J_c in the applied field dependence. For practical applications, this so-called peak effect was taken as a great advantage over YBCO, the J_c of which rapidly decreases with the application of a magnetic field.

The successful fabrication of a bulk-sized superconductor needed to overcome yet another practical problem, *i.e.*, a LRE element based bulk superconductor, especially NdBCO and Sm-BCO, easily cracks during the growth or oxygen annealing. The group of the present author reported that they frequently observed *macro*-cracks for SmBCO, which are much severe and larger than the *micro*-cracks that are commonly observed in a melt-grown YBCO [33]. These macro-cracks of SmBCO were clearly visible from outside the sample, and often ran across the whole sample perpendicular to the top surface. While this problem must have been causing a serious trouble for many researchers, it was not explicitly discussed until it was shown that the addition of Ag_2O to the starting composition of SmBCO is effective to prevent the formation of such large cracks [33, 34]. Silver was chosen because it was already known to decrease the density of small cracks in YBCO [35, 36]. We then added 10 wt.% of Ag_2O to the starting composition, and succeeded in eliminating the large cracks of SmBCO. The effect of adding Ag_2O is also recognizable in the studies by Haseyawa and co-workers that was submitted shortly before the above mentioned work, although their aim of adding Ag was not to eliminate the macro-cracks but to reduce the peritectic temperature of the material to avoid the possible dissolving of the seed crystal [37, 38].

The performances of Ag-added SmBCO samples processed in a reduced oxygen atmosphere well exceeded that of an ordinary YBCO [34]. Figure 4 shows the trapped field of a 10 wt.% Ag_2O added SmBCO sample that is 36 mm in diameter measured at liquid nitrogen temperature. The gap between the active area of the Hall sensor and the sample surface during the mapping of the field distribution was about 1.1 mm. As shown in Fig. 4, there is only one peak at the center of the field profile, which well corresponds to the schematic drawing of Fig. 1(a). This means that the sample is free from macroscopic defects such as cracks or weak links which seriously affect the capability of trapping a magnetic field of a bulk superconductor. After completing the measurement of the field distribution, the Hall sensor was moved to the peak position, and was lowered until it touched to the sample surface. In this way, the distance between the sample surface and the active area of the Hall sensor was reduced to 0.64 mm, and the trapped magnetic flux density thus measured was 1.7 T, which was far larger than any other trapped field reported at that time including YBCO.

An extensive study of the effect of changing the amount of Ag_2O showed that 10 wt.% is the amount that is at least necessary to prevent the formation of the undesirable cracks or weak links in 30 mm diameter SmBCO samples [39]. Further increase in the amount of Ag_2O, however, resulted in a decrease in the trapped field. This was attributed to the relative decrease in the volume fraction of the superconducting phase that was brought about in exchange of increasing the amount of Ag_2O [39].

Figure 5 shows the temperature dependence of trapped field of several SmBCO/Ag bulk superconductors with different amount of Ag_2O and/or sizes measured by cooling the samples using a Gifford-McMahon refrigerator [34, 40, 41]. The trapped fields plotted in the figure

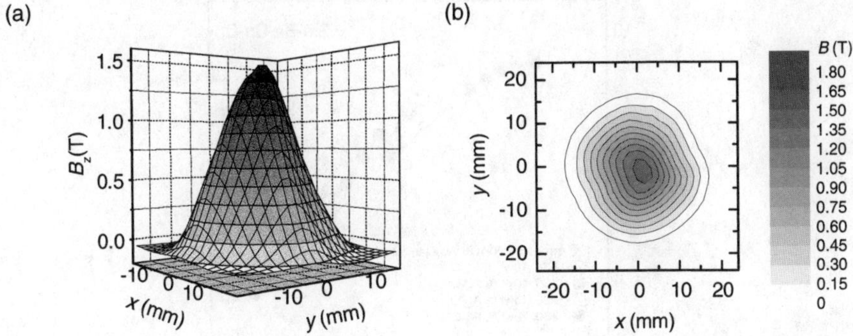

Figure 4: (a) A three-dimensional plot and (b) a counter plot of the distribution of the trapped magnetic flux density of a SmBCO bulk superconductor measured at liquid nitrogen temperature [34]. The sample is 36 mm in diameter and 10 wt.% of Ag_2O was added to the starting composition.

were measured immediately after the external field had been removed. A Hall sensor that was mounted on the center of the sample surface was used for the measurement, and the active area of this sensor located 0.55 mm above the sample surface. The figure shows that the trapped field increased with lowering the temperature because of the increase in J_c. However, the trapped field vs. temperature curve is terminated at a sample dependent trapped field with decreasing the temperature. This is because the samples were cracked when it was attempted to magnetize them at a temperature lower than the plotted data.

When a magnetic flux line is pinned by a superconductor, a force that is equally large as the pinning force is transferred to the crystal lattice where the pinning centers are embedded [42]. The trapped field and the field gradient are enormously large in a melt-processed superconductor, and hence the pinning induced magnetic pressure may lead to a cracking of the sample [43–46]. This is why the samples were cracked during the attempt of magnetizing them at lower temperatures. Because the addition of Ag strengthens the sample and the superconductor can sustain a larger magnetic pressure, the largest magnetic field that the sample trapped without being broken increased with the content of Ag_2O. On the other hand, the trapped field of the sample with less amount of Ag is generally larger when compared at the same temperature as shown in Fig. 5, because of the decrease in the volume fraction of the superconducting phase, as already pointed out. Therefore, there is a tradeoff between the highest trapped field and the trapped field at a fixed temperature. The largest trapped field of the samples shown in Fig. 5 was 2.1 T at 77 K and 9.0 T at 25 K [34, 40, 41], which is far exceeding the magnetic field produced by an ordinary Nd-Fe-B permanent magnet.

The cross in Fig. 5 indicates the magnetic flux density measured in a 2 mm gap between two 10 wt.% Ag_2O added samples that are 36 mm in diameter. The trapped field measured in this configuration is larger than that at the surface of a single sample as mentioned in Section 2.2, and was 3.55 T when measured immediately after turning off the external field at liquid nitrogen temperature.

As already mentioned, melt-processing SmBCO in air leads to a serious depression of the superconducting properties because of the formation of a solid solution of $Sm_{1+x}Ba_{2-x}Cu_3O_y$.

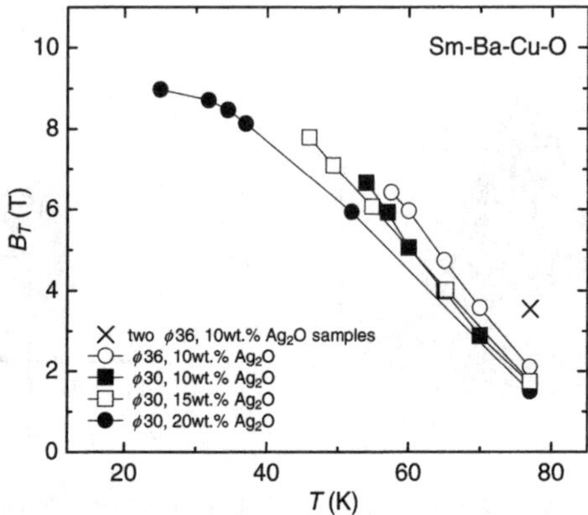

Figure 5: The temperature dependence of the trapped field of several SmBCO samples that are 30 or 36 mm in diameter and prepared from starting powders that contain various amounts of Ag_2O [34, 40, 41].

Therefore, the performance of an in-air prepared SmBCO bulk superconductor is expected to be significantly lower than that prepared under a reduced oxygen atmosphere. Surprisingly, however, Kohayashi *et al.* have reported a trapped field of 1.3 T at 77 K for a sample that is 45 mm in diameter [47]. While this value is lower than that of the samples prepared in a reduced oxygen atmosphere (Fig. 4), it is certainly much higher than what one would have expected for an in-air prepared sample. Because in-air preparation has obviously technical merits over a process that needs to control the oxygen content, this is an important result worth for further investigations.

To clarify the reason of the unexpectedly high performance of an in-air prepared SmBCO, the variation of the superconducting properties within the sample was recently studied [48]. It was found that when SmBCO is melt-grown in air, T_c is significantly depressed in the region beneath the seed crystal, but increases with the distance from the seed crystal, and near the edge, an onset T_c of 95 K was observed as shown in Fig. 6. This result indicates that Sm/Ba substitution, which takes place at the beginning, is suppressed during the sample growth. The suppression of substitution was attributed to the shift of the composition of the melt during the crystallization. Because of the depressed T_c in the region beneath the seed crystal where the crystallization starts, it is obvious that Sm/Ba substitution indeed took place at the early stage of crystallization. This means that more Sm and less Ba were consumed for the formation of the superconducting phase compared to crystallizing in a stoichiometric Sm123. Consequently, an enrichment of Ba in the melt is expected. It is known that the substitution of a light rare earth element for Ba can be suppressed when the starting composition of the precursor is shifted from the RE123-RE211 tie-line to the Ba-rich side [49, 50]. Therefore, the increase in the Ba content with the growth of the sample was concluded to be responsible for the observed suppression of Sm/Ba substitution in the region near the edge.

A variation of the local composition with the distance from the seed crystal had been indeed observed by Kambara *et al.* in an NdBCO bulk superconductor that was melt-processed in air with an isothermal condition [51]. The composition of the superconducting phase was Nd-

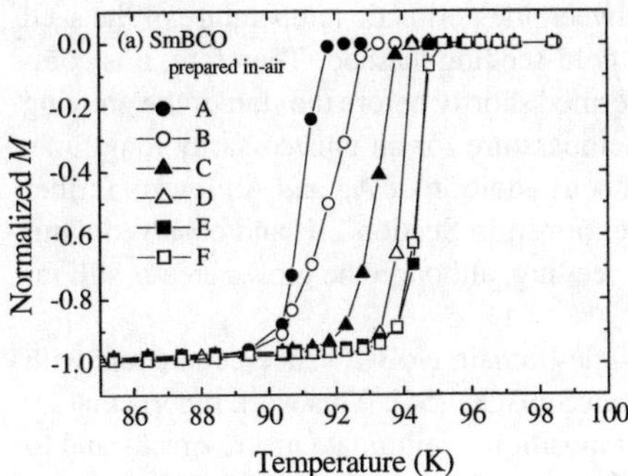
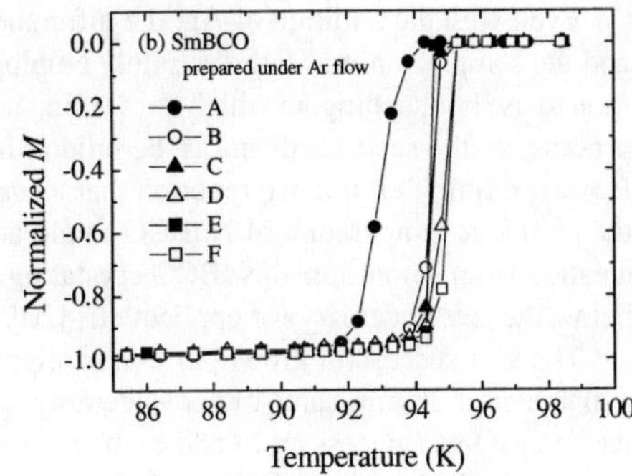

Figure 6: The temperature dependence of magnetization of several small specimens cut from SmBCO bulk samples which were prepared (a) in air and (b) under Ar flowing [48]. Both bulk samples were 30 mm in diameter, and specimens A to F were cut from inside out with an interval of about 2.5 mm.

rich near the seed crystal, corresponding to about 5% of the Ba site substituted by Nd, but the concentration of Nd decreased with the distance from the seed crystal. This observation was explained by the compositional change toward the Ba-enriched side due to the formation of solid solution, similarly as explained above. Although the samples of Fig. 6 were prepared by slow-cooling and thus the situation might be more complicated, the same mechanism probably explains the increase of T_c with the distance from the seed crystal of the in-air prepared sample in Fig. 6(a).

3.2   Nd-Ba-Cu-O

Nd-Ba-Cu-O exhibits the highest J_c among the melt-processed RE-Ba-Cu-O superconductors [5], and this material was the most enchanting one when Yoo *et al.* first reported about the oxygen-controlled melt-growth (OCMG) process [30, 31]. Many efforts have been devoted to grow a bulk form sample of this material, but it seems to be even more difficult than SmBCO to prevent NdBCO from cracking during the fabrication. Here again, the addition of Ag seems so far the only solution to overcome this problem. In addition, the difficulties of preparing a c-axis oriented single-domain NdBCO superconductor in comparison with other systems lies probably in the possession of a higher peritectic temperature, the more susceptible nature of the RE/Ba substitution to the partial pressure of oxygen, and the high growth rate from the melt [52–54], which present considerably greater challenges for the preparation of a large NdBCO sample.

The seed crystal used for aligning the c-axis perpendicular to the sample surface should not be heated to a temperature that is too close to its own peritectic temperature, in order to avoid the dissolving of the seed. This is a very hard condition to be fulfilled for NdBCO. As already mentioned in Section 2.1, however, the addition of Ag greatly reduces the peritectic temperature of the 123 compounds. For instance, it was reported that the peritectic temperature decreased almost linearly by about 20°C with increasing the Ag_2O content up to 2.5 wt.%, and remained almost constant with a further increase [54]. Therefore, it is possible to use a non-Ag-added NdBCO as a seed crystal for the preparation of a bulk Ag-added NdBCO sample. Hence, the effect of Ag-addition is twofold; makes possible to use NdBCO as a seed crystal by lowering the peritectic temperature and prevents cracking.

Even with the addition of Ag, the difference between the peritectic temperature of the seed and the sample is not as large as safely employ the cold-seeding method. Therefore, it is common to use hot-seeding, in which the seeding is performed shortly before the start of the growing process, so the seed needs not to be hold at high temperature for an unnecessarily long time. However, Smith *et al.* have reported that a simultaneous addition of Ag and Au lowers further the peritectic temperature of NdBCO as already mentioned in Section 2.1, and observed a nucleation from a non-doped NdBCO crystal by cold-seeding, although the grown area is still far below the size necessary for applications [20].

The first successful growth of *c*-axis aligned single-domain NdBCO superconductors with a diameter of 30 mm came out considerably later than SmBCO [54]. This work reports that 20 wt.% Ag_2O was necessary to add to the starting composition to eliminate macro-cracks and to prepare single domain NdBCO bulk superconductors with a high reproducibility. This is larger than the amount necessary for the preparation of SmBCO, for which the addition of 10 wt.% Ag_2O was sufficient as already mentioned. The largest trapped field at liquid nitrogen temperature of NdBCO was slightly smaller than that of a 10 wt.% Ag_2O added SmBCO sample [54]. This result was attributed to the difference in the volume fraction of the superconducting phase brought about by the unavoidably increase in the Ag content to prevent the crack formation. However, if the comparison were made between samples with the same Ag_2O content, the trapped field at 77 K of the NdBCO sample surpassed that of SmBCO, as expected from the higher J_c.

Figure 7 compares the temperature dependence of the trapped field of several NdBCO and SmBCO samples [54]. The temperature dependence of the trapped field of NdBCO was similar to that of SmBCO, and reached 7.0 T at 42 K. However, the maximum trapped field that the NdBCO samples could trap without a fatal destruction due to the magnetic pressure was lower than that of SmBCO. This suggests that the mechanical strength of NdBCO is lower than SmBCO. It is possible to evaluate the stress that caused the destruction of the sample using a theoretically derived equation [43], and Fig. 8 compares the stress thus estimated. It should be noted that this figure serves only as a relative comparison between the two materials, because all samples shown in the figure were embedded in 2 mm thick stainless-steel rings with epoxy resin for reinforcement. Nevertheless, it can be seen from the figure that the maximum stress that NdBCO sustained is indeed less than that of SmBCO.

Matsui *et al.* reported that a pre-sintering of the precursor pellet in pure oxygen before conducting the melt process is effective in minimizing the liquid loss and the shape change during the growth of the sample [55]. As a result, a pre-treated sample exhibited less inhomogeneity in the superconducting properties and misorientation was less significant. Accordingly, the trapped field was enhanced, and 1.4 T at the top surface of the sample was recorded at liquid nitrogen temperature with a sample that was 30 mm in diameter.

3.3 Gd-Ba-Cu-O

While the fabrication of a single domain bulk superconductor confronted many difficulties when Sm was replaced with a rare earth element that has a larger ionic radius, Nd, going to the opposite direction on the periodic table had been proved to be more successful. Nariki *et al.* first studied the effect of changing the calcination temperature of Gd_2BaCuO_5 (Gd211) and found that the particle size increases with the heat-treatment temperature [56]. A careful adjustment of the temperature yielded fine Gd211 single phase particles with a size smaller than 1 μm.

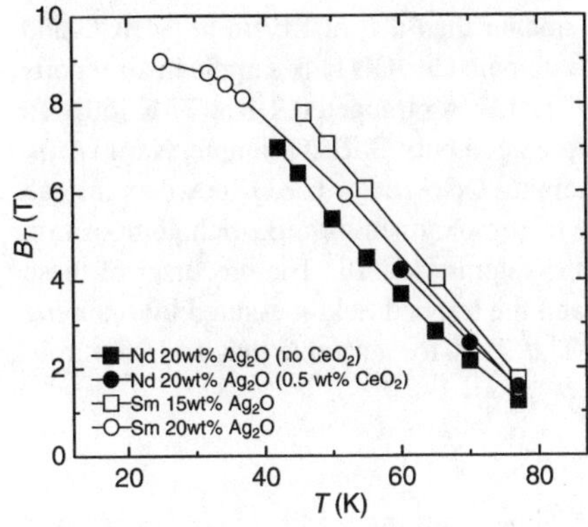

Figure 7: The temperature dependence of the trapped field of several NdBCO and SmBCO samples which are 30 mm in diameter [54]. The content of Ag_2O in the starting powders is indicated in the figure.

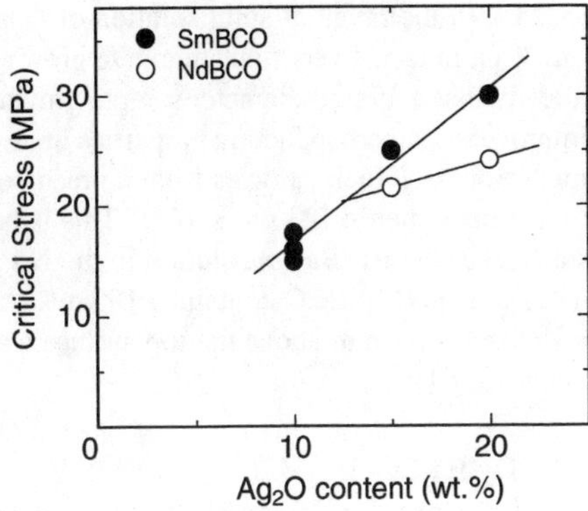

Figure 8: The stress that led to the destruction of SmBCO and NdBCO plotted as a function of the Ag_2O content [54].

Employing these fine Gd211 powders resulted in a reduction of the secondary phase inclusions in the melt-processed bulk GdBCO. As expected, J_c in the low-field region increased with this reduction in the size of Gd211 inclusions. The trapped field of a Ag-added GdBCO sample that is 32 mm in diameter was 1.5 T at 77 K, which was measured by scanning a Hall sensor 0.5 mm above the sample surface, about 30 min after the removal of the magnetization field [56].

The trapped field at 77 K of a similarly prepared GdBCO sample increased recently to 2.0 T when 20 wt.% Ag_2O was added to the precursor powders [57]. Here, the trapped field was measured with a total gap of 1.2 mm between the sample surface and the active area of the Hall sensor. As for NdBCO and SmBCO, adding Ag was important because the trapped field distribution of a similarly prepared Ag-free GdBCO sample revealed an irregular shape, indicating the presence of weak-links. The study on the temperature dependence of the trapped field reports 6.7 T at 55 K before the sample was fractured at a lower temperature, which was measured with a Hall sensor that was directly glued on the sample surface so that the gap to the active area of the sensor was 0.7 mm [57].

Nariki and co-workers refined further the average diameter of the Gd211 powders to 0.2 μm using a ball-milling technique, and prepared Ag-added bulk samples with employing these ultrafine powders [58]. The trapped fields at 77 K were 2.05 T and 2.54 T for 32 mm and 50 mm diameter samples, respectively, when measured 1.2 mm above the sample surface. The trapped field of the latter sample was 2.7 T when the Hall device was made to directly contact the sample surface, reducing the gap between the surface and the active area to 0.7 mm [58]. The trapped field measured between two 48 mm diameter GdBCO bulk samples prepared using Gd211 powders with an average diameter of 1 μm was 4.0 T when measured immediately after the removal of the external field, and 3.34 T after 30 min [59].

Because of the general trend that the RE/Ba substitution is less susceptible to the partial pressure of oxygen with increasing the atomic number, or decreasing the ionic radius, of RE

[5, 28, 29], the range of solid solution of Gd/Ba is smaller than that of RE/Ba in NdBCO and SmBCO. In fact, a very first attempt to grow a single-domain GdBCO bulk sample in air reports already that a 45 mm diameter sample grown with 10 wt.% Ag trapped 0.9 T at 77 K [60]. To improve the superconducting properties of an in-air prepared bulk GdBCO sample, Nariki *et al.* melt-processed their samples from a precursor containing $Gd_{0.95}Ba_{2.05}Cu_3O_y$ powders instead of a stoichiometric 123 phase [61]. This is because the employment of a Ba-rich composition suppresses the RE/Ba substitution in the Nd and Sm systems [49, 50]. The precursor of these in-air prepared GdBCO contained 10 wt.% Ag_2O, and the trapped field measured by scanning a Hall sensor 1 mm above the top surface was 1.2 T at 77 K for a sample that was 25 mm in diameter [61].

3.4 Dy-Ba-Cu-O

While it is necessary for a REBCO superconductor with a light rare earth element to conduct the melt processing in a reduced oxygen atmosphere to avoid the RE/Ba substitution that deteriorates the superconductivity, DyBCO can be processed in air without worrying about the substitution effect. On the other hand, the absence of RE/Ba substitution means that the field-induced pinning mechanism based on the presence of small RE/Ba substituted region would not be effective, and J_c would not high at large magnetic fields. Consequently, the trapped field might not be as large as in the light rare earth systems.

In fact however, the trapped field reported for DyBCO by Nariki *et al.* was beyond such expectations [62]. They first prepared fine Dy211 powders similarly as their work on GdBCO mentioned in the preceding subsection, and used these powders for the melt processing. As a result, they observed very fine Dy211 particles in the bulk sample, the average diameter of which was 0.6 μm. Larger grain samples were prepared with the addition of 10 wt.% Ag_2O in air, and the trapped fields measured at 77 K with a gap of 1.2 mm between the Hall sensor and the sample surface were 1.4 T and 1.7 T for samples that were 32 and 48 mm in diameters, respectively. This high trapped field was attributed to the large J_c values in zero magnetic field, which was brought about by the fine Dy211 particles embedded in the Dy123 matrix [62].

As mentioned above, no peak effect due to the RE/Ba substitution effect is expected for DyBCO. Nevertheless, Nariki *et al.* observed a peak in the J_c-B curves of melt-processed DyBCO. The peak was most pronounced for the sample prepared from a precursor whose Dy123:Dy211 ratio was 100:5 [63]. When the amount of Dy211 phase was increased relative to Dy123, J_c at zero magnetic field increased, but the peak effect gradually disappeared. This is the same trend observed for NdBCO prepared by changing the ratio of Nd123:Nd422 phases [64]. For NdBCO, the observation was attributed to the change in the volume fraction of the Nd123 phase in which the Nd/Ba partially substituted regions are distributed and acting as field-induced pinning centers [64]. Unlike NdBCO, however, the RE/Ba substitution would be very small in DyBCO, if any exists. On the other hand, a similar peak effect had been observed in Y123 for both single crystals and melt-processed samples [65–67] and is often related to the oxygen deficiency. Therefore, Nariki *et al.* explained the peak effect of DyBCO by oxygen deficient clusters [63]. In fact, the peak effect was very sensitive to the annealing temperature. It was argued that the presence of Dy211 phase in the Dy123 matrix assists the diffusion of oxygen and fewer oxygen deficient regions would exist in the sample when the amount of Dy211 is large, and this had caused the disappearance of the peak effect with increasing the relative amount of Dy211 phase. It was also shown that the peak effect of DyBCO is much pronounced than that of YBCO by comparing the J_c-B curves of similarly prepared DyBCO and YBCO samples [63].

Because of the peak effect in J_c-B curves of samples with a small content of Dy211, bulk samples with the size of 32 mm in diameter were prepared in air by adding 10 wt.% Ag_2O to precursor powders which had a composition of Dy123:Dy211=100:5 [68]. The trapped field of this sample was 0.9 T at liquid nitrogen temperature when measured 1.2 mm above the top surface. This is, however, lower than the trapped field of the above mentioned sample with the same size, 1.4 T, which was prepared from a starting composition of Dy123:Dy211=100:40, and for which no peak effect of J_c was observed. In spite of the lack of the peak effect of J_c, the sample with the higher content of Dy211 revealed a larger J_c than the other one at low magnetic fields, and this was more effective to increase the trapped field at liquid nitrogen temperature.

Interestingly, however, the peak effect of J_c affected the way how the magnetic flux density changed during decreasing the external field. Nariki *et al.* measured the field distribution *in-situ* of a superconducting magnet that generated the external field, and plotted the difference between the peak value of the field distribution and the external field [68]. While this value was 0.9 T when the external field was totally removed for the sample prepared with a small amount of Dy211 as mentioned above, a larger field difference of 1.2 T was observed when the external field was 1.5 T. This is in an interesting contrast with the sample that contains a larger amount of Dy211, the one that trapped 1.4 T at zero external field, because the latter sample showed a monotonous decrease for the difference between the field at the sample surface and the external one during decreasing the magnetization field. As a result, the maximum of the field difference is larger for the sample prepared with the smaller amount of Dy211. This might be an interesting observation, if the bulk superconductors will be used in an application that is operated in a non-zero external field.

3.5   Y-Ba-Cu-O

Until 1997, YBCO was the only material that possessed the capability of trapping large enough magnetic fields making applications worth considering. The trapped fields reported for YBCO were more or less the same if the difference in the sizes and the geometrical effect, as well as the general trend that it is more difficult to grow a homogeneous sample when the diameter increases, were taken into account. Typical values of the trapped field at the surface of a YBCO bulk superconductor at liquid nitrogen temperature were, for example, 0.72 T [18] and 0.56 T [69] for samples which were 45 mm and 24 mm in diameter, respectively, and 8.5 T at 51.5 K was reported for the trapped field between two YBCO samples, each 24 mm in diameter [69].

It was also already well known that irradiation techniques are effective for increasing the trapped field. A very large magnetic field, 3.1 T at 77 K and 10.1 T at 42 K, was reported for stacks of four 20 mm diameter YBCO samples [70]. Admixing uranium with the precursor powders of YBCO and irradiating the sample with thermal neutrons after melt processing was also reported to increase the trapped field [71]. The trapped field in a stack of four 20 mm diameter samples prepared in this way was 3.2 T at 77 K and 8.1 T at 62 K. While these studies show that the increase in trapped field by irradiation is significant, we will not include the recent development of these and related techniques in our review, because these techniques require special facilities in order to apply, which is probably a disadvantage for widely use irradiated materials in applications.

As had been proved to be effective for other rare earth element based superconductors, increasing the strength of the material with the addition of Ag pushed up the record of the trapped field of (non-irradiated) YBCO at low temperature by preventing the sample to be cracked due to the magnetic pressure [72, 73]. As a result, the trapped field measured above the surface of a YBCO sample increased to 11.4 T at 17 K [73]. The sample was 26 mm in diameter, and was reinforced by the addition of Ag and by embedding it in a Cr-Ni steel tube. The magnetic field in a 1 mm gap between two such YBCO samples was 14.35 T at 22.5 K. This value is to be compared with the maximum trapped field of 8.5 T at 51.5 K observed between two YBCO samples with no reinforcement [69], illustrating the effectiveness of Ag and the use of steel tubes.

The performance of YBCO was recently further improved with a new idea. This was made by doping the superconductor with a small amount of Zn [74–76]. It is more or less the general consensus that the peak effect observed in REBCO with a light rare earth element is caused by small regions where the RE element is substituted for Ba, creating spatial variation of T_c in the sample. The superconductivity of such region is weaker than the other regions, and therefore acts as a pinning center under a magnetic field. On the other hand, it is very well known that partially replacing Cu by Zn strongly deteriorate the superconductivity. The idea of the Dresden's group is, a substitution of Zn for Cu may form local regions of reduced T_c in Y123 that act as pinning centers under a magnetic field [74–76]. In fact, Krabbes $et\ al.$ have observed a peak effect near 3 T in the field dependence of J_c at 77 K when a slight amount of Zn was substituted for Cu [74].

Because of the improved J_c in the field range below 5 T, the trapped field of a 25 mm diameter bulk sample of Zn-doped YBCO increased to 1.12 T at 77 K, which is significantly larger than the corresponding value of 0.75 T of a reference sample to which Zn was not doped [74]. The difference between the trapped fields of Zn-doped and undoped samples increased with lowering the temperature, and the highest trapped field of the doped sample was 9 T that was recorded at 42.5 K. When measured in a 2.2 mm gap between two YBCO that had been doped with 0.12 wt.% Zn, the trapped field recorded at 47 K before cracking at a lower temperature was 11.2 T [75]. By further reinforcing the sample with adding Ag (10 wt.%), the maximum trapped field in the gap between two samples increased to 16 T recorded at 24 K [75, 76].

Detailed studies of the influence of varying the Zn content were reported by several authors [77–80]. These reports consistently show that there exists an optimum doping level for Zn where the largest J_c, and hence the largest trapped field, is observed. This is because the Zn-doped regions overlap if the doped amount is too large, and reduces T_c of the sample globally rather than acting as pinning sites. The optimum level of Zn doping seems to be considerably small, for instance, González $et\ al.$ reports that the largest enhancement in the field-dependent J_c was observed when 0.03% of the Cu site was replaced by Zn [80].

Another dopant that is proved to be effective in increasing the trapped field of YBCO is CeO_2. Walter $et\ al.$ reported the trapped field of a CeO_2 added YBCO that is square in shape with sizes of $40 \times 40 \times 14$ mm^3 [81]. The trapped field at liquid nitrogen temperature of this sample was 1.33 T when measured with a distance of 0.3 mm between the Hall probe and the sample surface. The enhancement of the trapped field with the addition of CeO_2 was attributed to the reduction in the size of Y211 particles and to the corresponding increase in J_c [24, 82–84]. While Pt seems to be more often used for the purpose of refining the 211 phase particles in the superconducting matrix, it is worthwhile mentioning that the trapped field reported by Walter $et\ al.$ is probably the largest one so far for YBCO prepared with no Ag or Zn addition.

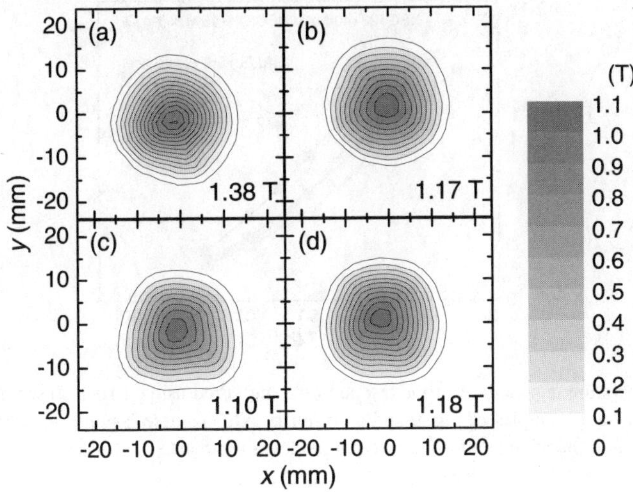

Figure 9: Contour plots of the trapped magnetic flux density at liquid nitrogen temperature of (NEG)BCO bulk samples that are 30 mm in diameter. The atomic molar ratios Nd:Eu:Gd of the RE211 powders in the precursors were (a) 1:1:0 (sample B), (b) 0:1:1 (sample C), (c) 0:1:0 (sample D), and (d) 0:0:1 (sample E), respectively [95].

3.6 Ternary system

Mixing two or more rare earth elements has also been widely studied. Among them, ternary systems studied extensively by Muralidhar and co-workers, particularly the (Nd,Eu,Gd)-Ba-Cu-O ((NEG)BCO) system, have been reported to exhibit superior magnetic-flux pinning properties [85–88]. According to them, J_c is extremely high, and exceeded 75,000 A/cm^2 in zero magnetic field and 68,000 A/cm^2 at 2.5 T for a $(Nd_{0.33}Eu_{0.33}Gd_{0.33})Ba_2Cu_3O_7$ (NEG123) sample prepared by the oxygen-controlled melt-growth process with adding 40 mol% excess $(Nd_{0.33}Eu_{0.33}Gd_{0.33})_2BaCuO_5$ powders [88].

However, the trapped fields of melt-processed (NEG)BCO superconductors are so far not yet extraordinarily large in spite of the large J_c values, and many efforts are still devoted to improve the performance of a bulk form sample. The first reported trapped field of a bulk (NEG)BCO superconductor was 0.7 T at 77 K for a sample that is 22 mm in diameter [89]. A somewhat higher trapped field was reported from the same group for another ternary system, (Nd,Sm,Gd)-Ba-Cu-O, which trapped 1.2 T at 77 K with a 30 mm diameter sample [90]. Recently, Yamada *et al.* successfully prepared single-grain samples of the (NEG)BCO system with diameters up to 38 mm [91]. The influence of changing the amount of Ag_2O in the starting composition on the field-trapping capability and on the mechanical strength was studied and a trapped field of 1.38 T was reported, increasing the record of the trapped field of (NEG)BCO by a factor of almost two [91].

An interesting aspect of the ternary system is that it provides a possibility to vary the pinning properties with changing the chemical ratio of the rare earth elements both of the 123 and 211 phases [92, 93]. For instance, by varying the Nd:Eu:Gd ratio of the 123 phase, Muralidhar and

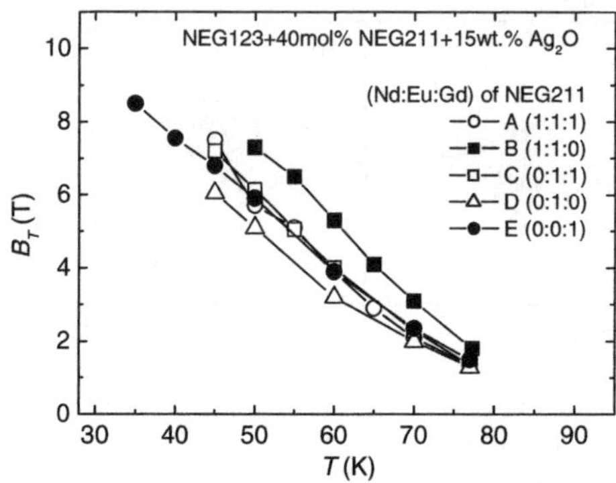

Figure 10: Temperature dependence of the trapped field measured using a Hall sensor mounted on the top surface of the same samples of Fig. 9 and an ordinary NEG sample (sample A) [95]. The trapped fields were recorded immediately after the external field was removed.

co-workers concluded that Nd mainly enhances flux pinning at low fields, Gd the intermediate and high-field region, and Eu control the position of the peak in the J_c-B curve. Furthermore, an enormously large irreversibility field of 14 T at 77 K for applied field parallel to the c-axis direction was reported quite recently by adjusting carefully the ratio of Nd:Eu:Gd of the 123 phase [94].

Attempts of growing c-axis oriented, large samples with varied chemical ratios of rare earth elements have also been started recently. Yamada *et al.* changed the molar ratio of the rare earth elements of the secondary phase, and investigated the magnetic field trapping capability [95]. Figure 9 shows the distribution of the magnetic flux density trapped by samples that are 30 mm in diameter for which the Nd:Eu:Gd ratio of the secondary phase was varied as indicated in the caption. The trapped field at the peak position measured by minimizing the distance between the Hall sensor and the sample surface is indicated in each profile. As shown in Fig. 9, the sample that was prepared by adding 40 mol% nominal $(Nd_{0.5}Eu_{0.5})_2BaCuO_5$ powders recorded a trapped field of 1.38 T at 77.3 K, which is as high as the trapped field of the above mentioned larger NEG sample prepared with nominal $(Nd_{0.33}Eu_{0.33}Gd_{0.33})_2BaCuO_5$ powders. On the other hand, the sample prepared from a precursor that only contained Gd211 as the RE211 phase showed a significant enhancement in the mechanical strength [95]. As a result, this sample recorded the largest trapped field among the samples studied by Yamada *et al.* before cracking at lower temperature, as shown in Fig. 10. The maximum trapped field of this 30 mm NEG sample was 8.5 T at 35 K [95].

3.7 Reinforcement

As many studies mentioned in the preceding subsections have shown, the trapped field of melt-processed bulk materials at low temperature is limited by the low mechanical strength of the material. The increase in the mechanical strength is therefore highly desirable. Adding Ag is effective and even crucial for some of the REBCO materials for increasing the mechanical strength as we have already seen. Some groups use a metal ring to compensate the tensile stress acting on the sample during magnetization by producing a compressive stress due to the difference in the thermal expansion between the metal and the superconductor [34, 73]. A theoretical analysis also showed the effectiveness of clamping the sample with a ring [96].

Another method to increase the mechanical strength of melt-processed bulk superconductors is impregnation of epoxy resin [97]. By impregnating phenol-formaldehyde to which polyamideimide was added as hardening material, a successful filling up of microcracks and pores of bulk SmBCO and GdBCO samples was reported. The epoxy resin was shown to be able to penetrate along the microcracks, and fill the structural defects in the region within 2 mm from the surface. The improvement in the mechanical properties was demonstrated by showing that no degradation in the trapped field was observed for the impregnated samples after cycling the temperature between 77 K and room temperature and repeating the magnetization experiments, while the samples without epoxy resin showed a fast decrease in the trapped field. A theoretical work reports that it is the upper and lower layers that affect the stress distribution of a resin impregnated sample rather than the region near the side surface [98].

This impregnation technique was soon improved by adding quartz as fillers in the resin and wrapping the sample with a fabric made of glass fiber prior to the impregnation of epoxy resin [99]. The filler decreases the expansion coefficient of the resin, which is significantly larger than that of the bulk sample. Because of the closer thermal expansion, it is expected that cracking due to thermal fatigue can be minimized. The measurement of the tensile strength at room temperature of samples reinforced with resin impregnation revealed a 50% strength increase [100]. The effect of wrapping with carbon fiber fabric prior to resin impregnation was more dramatic, the tensile strength was about ten times larger than that of a non-reinforced sample. Furthermore, it was reported that the stresses acting on the sample wrapped with carbon fiber fabric before impregnation was more homogeneous and smaller than that exerted on a non-impregnated sample from monitoring the stresses using distortion gauges while sweeping down an external magnetic field at 65 K. These results demonstrate the effectiveness of wrapping bulk superconductors in carbon fiber fabric for improving the mechanical properties.

Recently, a record high trapped field was reported for samples reinforced with the impregnation technique [101]. In this study, a YBCO sample was wrapped with carbon fiber and resin impregnated, leaving the top and bottom surfaces uncovered. A hole of 1 mm diameter was then drilled at the center of the sample. An 0.9 mm diameter Al wire was inserted into the hole, and this hole was filled with molten Bi-Pb-Sn-Cd alloy. The aim of embedding the Al wire with the metal alloy was to effectively remove the heat generated during magnetization, and to prevent flux jumping that caused cracking of the samples in an earlier experiment. The magnetic field between two such reinforced YBCO samples of 26.5 mm diameter stacked together was measured using Hall sensors. When the couple of superconductors was magnetized using a magnetic field of 17.9 T at 29 K, the trapped field was 17.24 T, a record-breaking value for the in-gap trapped field with an increase of about 1 T from the earlier record. Moreover, the authors claim that the trapped field was limited only by the available external field, and could be further increased if activation with a larger magnetic field would be possible.

4 MAGNETIZATION

As reviewed in the preceding section, the trapped field of melt-processed bulk superconductors well exceeds the magnetic field produced by conventional permanent magnets or by non-superconducting solenoid coils. Therefore, one of the expected applications of these superconductors is using them as strong magnets. Once these superconductors are magnetized and kept at low temperature, they may be considered as a new type of permanent magnet. Because the volume of a bulk superconductor is not large, cooling it with a refrigerator is not a big problem, thanks to the recent progress in refrigeration techniques. A more significant problem might be the magnetization. A superconducting "permanent" magnet has to be magnetized below the superconducting transition temperature, similarly to conventional permanent magnets, for which it is necessary to align the magnetic domains below the Curie temperature. However, the magnetic field that is required to fully magnetize the superconductor is now extremely large, because a field larger than at least the field to be trapped must be applied. Therefore, studying the magnetization techniques are of significant importance for practical applications of bulk superconductors.

One may classify the magnetization methods as quasi-static or pulsed-field magnetization. In the former method, the superconductor is subjected to a static magnetic field, and the external field is then slowly changed. In the latter technique, a magnetic pulse, or several pulses, is applied to the sample. As will be discussed in detail in the following, quasi-static magnetization is the most efficient way to magnetize the superconductor, and the result corresponds to the full ability of field trapping of the tested sample. The trapped fields reported in the previous section are those determined by this method. However, for practical applications, the pulse method has several advantages, and we will discuss both techniques, their merits and drawbacks, in detail in this section.

4.1 Quasi-static magnetization

The most effective way to magnetize a superconductor is to use a superconducting solenoid coil and change the external field quasi-statically. Magnetizing a superconductor with a static field is an already well studied and well understood issue since Bean's work of the critical state model [21]. Bean discussed the simplest model of the magnetization problem by assuming that the critical current density does not depend on the field. The field penetrated into the superconductor then depends linearly on the distance from the edge of the sample, and the slope is proportional to J_c, if the geometry effect can be ignored. Figure 11 depicts the magnetization process of a cylindrical sample by neglecting the demagnetization effect. Here (a) shows how the magnetic fluxes penetrate a superconductor when ascending the external field if the sample were cooled in the absence of a magnetic field, while (b) corresponds to the descending field stage. When we apply an external field large enough, and then remove it, a field distribution peaked at the center remains inside the sample as shown in Fig. 11(b) by hatching. This magnetization method is called zero field cooling (ZFC), because the sample was cooled to the target temperature before applying an external field. The trapped field is proportional to J_c and to the radius of the sample.

As it is obvious from Fig. 11, the field necessary to fully magnetize the sample with the ZFC method is twice as large as the trapped field at the center of the sample. Alternatively, the sample can be magnetized by applying the magnetic field above the superconducting critical temperature, cool the sample to the desired temperature in the field, and then decrease the external field to zero. This is called the field cooling (FC) method, and the change in the magnetic flux density inside the superconductor is schematically drawn in Fig. 11(c). Because the field is first applied above T_c, the magnetic flux density equals almost to the external field everywhere inside the sample before starting the external field to decrease. Therefore, the field

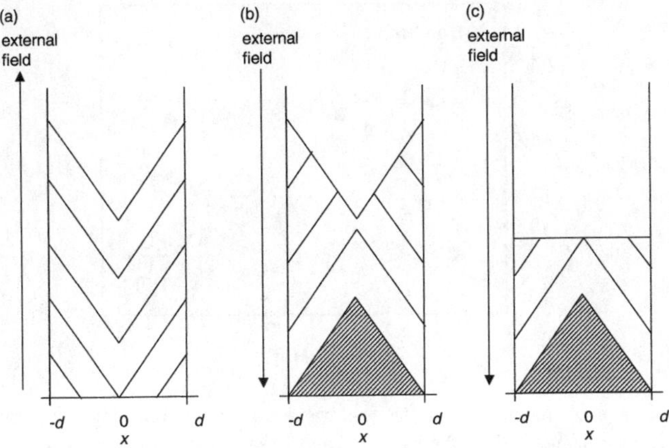

Figure 11: The change in the local magnetic flux density inside a superconductor during the magnetization process; (a) the field ascending and (b) field descending branches of ZFC magnetization, and (c) the FC magnetization. The shaded field distributions are trapped after magnetization.

necessary to fully magnetize the sample is ideally equal to the trapped field at the center of the sample. In other words, the field that is required is half of that of the ZFC method. Nevertheless, it is now necessary to use a 10 T class superconducting magnet to extract the full potential of melt-processed superconductors. Even at the liquid nitrogen temperature, the trapped field is larger than the magnetic field that an ordinary electromagnet can generate as we have seen, and one needs to use a large superconducting coil.

Unfortunately, a superconducting magnet is not a familiar equipment outside of research laboratories. It should also be pointed out that bulk superconductors might be often warmed up above T_c unlike a conventional permanent magnet, and magnetization may have to be conducted frequently. Furthermore, the bore space of a superconducting solenoid coil is usually small, which will make the designing of an application very difficult, because the whole apparatus has to be put into the narrow bore space at the event of magnetization, or a mechanism of an easy re-installation of magnetized superconductors is necessary.

For these reasons, a pulsed field magnetization (PFM) method is considered to be more feasible for many practical applications. The advantage of this technique is that the magnetizing coil can be made small since large currents are fed only instantly. Therefore, the magnetizing coil may be installed into the apparatus together with a bulk superconductor, allowing an *in-situ* magnetization. It is also possible to magnetize several superconductors in such a way that their magnetic poles direct to different directions. To demonstrate the feasibility of the PFM technique, Itoh *et al.* constructed a superconducting motor, in which twenty melt-processed YBCO superconductors were built-in and magnetized by the PFM method [102, 103].

4.2   PFM of YBCO

Figure 12 shows the amount of the trapped magnetic fluxes (Φ_T) of a YBCO bulk superconductor that was 34 mm in diameter and magnetized by the PFM method [104]. Φ_T was calculated

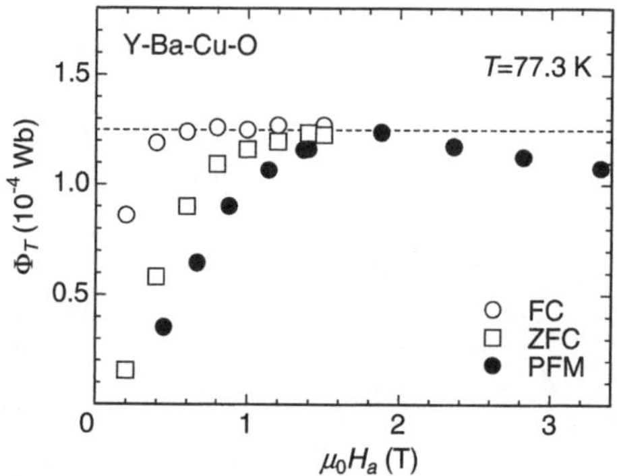

Figure 12: The amount of magnetic flux that was trapped by a 34 mm diameter YBCO bulk superconductor when it was magnetized by the FC, ZFC and PFM (single pulse) methods. The dashed line corresponds to the saturation value of Φ_T by the FC and ZFC methods. Data from Itoh and Mizutani [104].

by integrating the magnetic flux density that was measured by scanning a Hall sensor over the sample surface after magnetization was conducted. For comparison, Φ_T of the FC and ZFC methods are also plotted in the figure. The applied field μ_0H_a is the maximum field that was applied during magnetization. As expected, the amount of trapped magnetic flux of the FC method increased with the applied field, and saturated at about 0.6 T. The saturation value of Φ_T is shown by the dashed line in the figure, and corresponds to the full capability of trapping magnetic fluxes of the present sample. As usual, Φ_T of the ZFC method is lower than that of the FC method at low applied field, and the external field necessary to fully magnetize the sample is about twice as large as that of the FC method.

The maximum value of Φ_T of the PFM technique was similar to that of quasi-static magnetization techniques. This is what was expected, because PFM is nothing but a ZFC magnetization in which the external field is swept with a speed that is orders of magnitude larger than a usual ZFC magnetization. However, the applied field dependence of Φ_T was not the same for the ZFC and PFM methods as is evident from Fig. 12. The increase in Φ_T as a function of μ_0H_a is more gradual for PFM than ZFC in the low field region indicating that the penetration into the sample is more difficult for the flux lines with the pulse method. This less penetrability of the vortices at low applied fields was accounted for by the large viscous force that is exerted on flux lines when they fast move.

The viscous force F_v originates from the movement of the vortex core that is essentially in the normal state, and opposes to the direction to which the flux line moves [105–107]. F_v is to be balanced with the Lorentz and pinning forces, and is given by

$$F_v = \eta \frac{B}{\phi_0} v,$$

where B is the magnetic flux density, ϕ_0 the flux quantum, and v the velocity of the flux line. η is the viscosity coefficient which is related to the flux flow resistivity ρ_f by $\eta = \phi_0 B / \rho_f$, while ρ_f is essentially proportional to the normal state resistivity. As evident from the above equation, a large resistive force is exerted on moving flux lines when the superconductor is magnetized by the pulse method because the viscous force is proportional to the velocity of the vortices. This force blocks the free entrance of flux lines into the sample in addition to the pinning force, leading to the pbserved difference in the trapped field when the superconductor is magnetized by the ZFC and PFM methods.

It is also interesting to note that Fig. 12 shows a falloff of the total flux trapped by the PFM process after passing through a peak when the amplitude of the pulse is increased. This behavior was attributed to a temperature increase of the sample because the flux lines move against resistive forces, *i.e.*, pinning and viscous forces, which causes dissipation [104, 108, 109]. Consequently, the trapped field decreased because of the smaller J_c at higher temperature, and this brought about the observed peak of Φ_T as a function of the amplitude of the magnetic pulse.

The importance of accounting for the heat generation was demonstrated by studying the dynamic motion of flux lines using a pickup coil technique [110]. Five pickup coils were put concentrically in a 2 mm gap between two YBCO bulk superconductors to probe the flux motion through the superconductors, and two were used to measure the change in the magnetic flux density outside the samples. The experimental results were converted to the magnetic flux density, examples of which are shown in Fig. 13. Note that here the evolution of magnetic flux density during the ascending field branch is depicted in the left half of each figure, and the right half is for the descending branch. It is obvious from the figure that the flux gradient is steep near the edge of the sample and gradually reduces near the center during the early stage of the ascending branch. Usually, J_c is a decreasing function of the magnetic field, and the field gradient during ascending the external field is expected to decrease by going apart from the center of the sample, because the local magnetic flux density is larger at the edge of the sample. The observed field gradient, however, opposes to this expectation. On the other hand, the observation can be well explained by accounting for the viscous force, because this force is larger near the sample edge where a flux line enters the sample with a large velocity, and decreases with the deceleration of the flux line when it penetrates deeper into the sample. As a result, a concave field distribution is expected, consistent with the experimental results shown in Fig. 13.

The velocity of the vortices as a function of position and time was also estimated from a further analysis of the experimental data, which revealed an interesting non-monotonous time dependence with a peak structure, when the amplitude of the magnetic pulse was large [110, 111]. This behavior was accounted for by an increase in the sample temperature due to the heat generated by the motion of vortices against resistive forces. It was found that the viscous force has a significant importance for the heat generation. For field-pulses with an amplitude larger than a few tesla, the loss caused by the viscous force dominates over that of the pinning force. Under the assumption of an adiabatic condition, it was concluded that the local temperature may have been increased more than 5 K in some locations of the sample that was 35 mm in diameter when a magnetic pulse of 4.4 T with a rising time of 1 ms was applied [111].

The influence of the heat generation due to the fast motion of flux lines was also demonstrated by theoretical calculations. Tsuchimoto *et al.* have numerically simulated the PFM process [112–114], and the experimentally observed peak of Φ_T as a function of applied field (Fig. 12) was quite well reproduced by accounting for the heat generation due to the flux motion [113]. The same problem was recently solved analytically, giving the same conclusion [115].

102

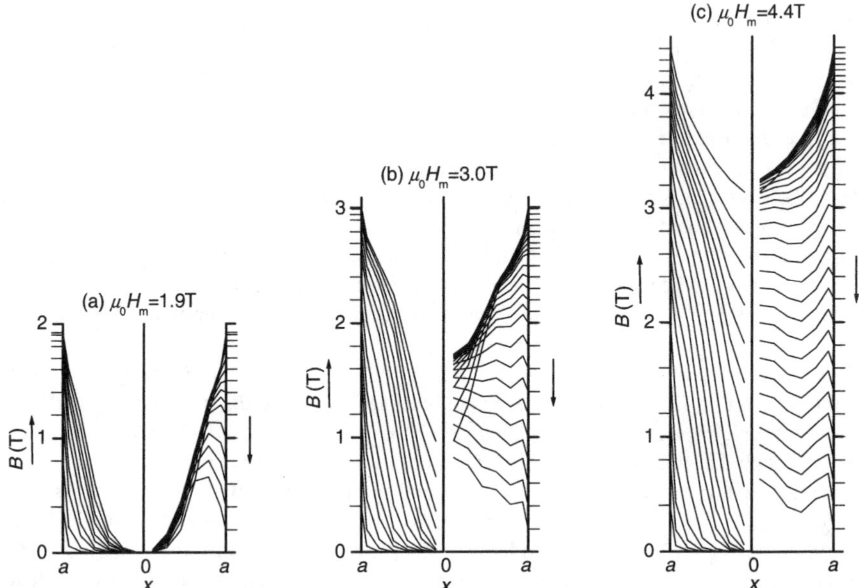

Figure 13: The time evolution of the magnetic flux density within the sample during PFM for magnetic pulses of different amplitudes [110]. The ascending field branch is depicted on the left half of each figure, while the right half is for the descending branch. Here 0 and a indicate the center and the edge of the sample, respectively.

The magnetic flux distribution during the PFM process (Fig. 13) was also quite well reproduced by a numerical calculation that takes into account the viscosity, or the flow resistivity [114]. For instance, the simulation resulted in a field gradient that is larger near the edge during the early stage of PFM similar to the experimental results, and the magnetic flux density continued to increase a little while at the center of the sample even after the external field had started to decrease, which is most obvious in the experimental results in the right half of Fig. 13(b). Furthermore, the increase in the local temperature during the PFM process calculated by using the material parameters appropriate for YBCO showed that the temperature near the edge of the sample may increase locally for more than 7 K when a 4 T magnetic pulse is applied at 77 K [114]. This is in a fairly good agreement with the experiments [111]. A quantitative calculation of the temperature increase was also conducted using numerical simulation for a ring sample, the outer and inner diameters being 46 mm and 16 mm, respectively, and it was concluded that the temperature may locally increase to 83 K when a magnetic pulse of 1.2 T is applied at 77 K [116].

Yanagi *et al.* monitored the temperature of a YBCO bulk sample during PFM experiments using a thermometer glued to the top surface of the superconductor that was cooled using a refrigerator [117]. A large increase in the temperature was observed, which was more significant at lower temperature, *i.e.*, when the pinning effect is larger. The temperature increased more than 15 K during PFM when a pulse with an amplitude larger than 5 T was applied at 42 K

[117].

While the influence of heat generation is much more serious for PFM, it is interesting to point out that some authors report that the temperature of the sample changed even in FC magnetization. Gruss *et al.* put a temperature sensor on top of a YBCO sample, and recorded the temperature change during decreasing the external field with a slow rate of 0.1 T/min [75]. Although, the sample was in a thermal equilibrium before the start of decreasing the external field, they observed that the temperature increased for 0.2 K when FC magnetization was conducted at 60 K and 1.8 K at 25 K, because of the dissipation brought about by the motion of flux lines.

Because of the heat generation, the trapped field by PFM is determined by the J_c value of an elevated temperature, not that of the temperature at which the magnetic pulse was applied. As a result, the sample is only incompletely magnetized. However, this in turn helps to suppress the creep effect after the application of the magnetic pulse, as was shown by Itoh and co-workers [104, 109]. When a YBCO sample immersed in liquid nitrogen was magnetized by FC or ZFC methods, the trapped field monitored using Hall sensors mounted on the top surface of the sample decreased as a logarithmic function of time as expected. The same time dependence was observed when a rather small magnetic pulse was applied. However, for large magnetic pulses, the creep rate was significantly suppressed [109]. This is because the trapped field was less than that expected from the full capability of the sample at liquid nitrogen temperature due to the heat generated during PFM. Consequently, the field gradient that the pinning force had to sustain was less than its limit, leading to a suppression of flux creep. This is the same mechanism that reduces the creep rate with post-cooling the sample after magnetization, as we have seen in Fig. 3.

The data shown in Fig. 12 indicate that the peak value of Φ_T by the pulse method was almost identical to the saturation value of the FC method. This means that the sample was fully magnetized when an optimized magnetic pulse was applied at 77 K. However, this is not true when the same experiment was conducted at lower temperatures, and Φ_T by PFM was smaller than that of FC magnetization [118, 119]. This finding was attributed to the larger pinning effect at lower temperatures. Furthermore, the falloff of Φ_T after passing through the peak was more serious when the temperature was decreased, indicating that the effect of heat generation is more significant at lower temperatures. A multi-pulse magnetization method that will be discussed in Section 4.4 was effective to increase the trapped field by the PFM method at low temperatures [118, 119].

4.3 PFM of SmBCO

Figure 14 shows the results of PFM experiments performed on SmBCO at liquid nitrogen temperature [111]. The data of SmBCO and YBCO are compared in this figure, both samples having the same diameter. Obviously, the amount of trapped magnetic flux of SmBCO also exhibited a peak in the applied field dependence. The peak Φ_T value was larger for SmBCO than YBCO reflecting the larger capability of trapping magnetic flux lines. However, Φ_T of SmBCO was smaller than that of YBCO for small magnetic pulses. This means that although the trapped field of SmBCO is larger than YBCO when both samples are optimally magnetized, penetrating into SmBCO is more difficult for the flux lines and requires a larger pulse. Furthermore, the amount of trapped magnetic flux by PFM started to decrease before reaching the FC value for SmBCO, indicating that the sample could not be fully magnetized by PFM even at liquid nitrogen temperature.

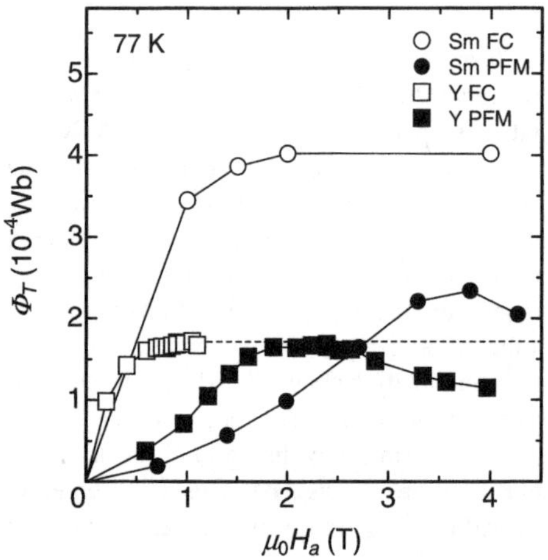

Figure 14: The amount of magnetic flux trapped by magnetizing YBCO and SmBCO using the FC and PFM methods. Both samples are 30 mm in diameter, and the magnetization was conducted at liquid nitrogen temperature [111].

The dynamic motion of flux lines inside SmBCO was extensively studied using the pickup coil method similar to the one applied to YBCO [111]. The flux distribution in SmBCO during PFM was qualitatively similar to that of YBCO, but the local magnetic flux density near the center of the sample was significantly less than that of YBCO when a magnetic pulse with a similar amplitude was applied. Consequently, the increase in the local temperature was smaller for SmBCO, which was estimated using the velocity of the flux lines determined from the output of the pickup coils. These observations were attributed to the larger pinning effect of SmBCO. Because of this less penetrability of flux lines into the sample, the trapped field of SmBCO is expected to be smaller than that of YBCO for relatively small magnetic pulses, consistent with the results shown in Fig. 14. Flux lines start to penetrate SmBCO at large applied fields, and Φ_T exceeds that of YBCO. However, the heat generated by the flux motion increases as well, which causes the peak in the applied field dependence of Φ_T. Accordingly, the amount of trapped magnetic flux by PFM was smaller than that of the full capability of the SmBCO sample [111].

The measurements of the field distributions after the application of magnetic pulses with various amplitudes revealed interesting results [120, 121]. Figure 15 shows the results for a SmBCO sample 36 mm in diameter, magnetized with a pulsed field at liquid nitrogen temperature [121]. It was found that no magnetic flux was trapped at the central region of the sample, because magnetic flux could not penetrate into this region with a small pulse. For larger pulses, the penetration front reached the sample center. However, the field distribution exhibited four distinct peaks locating diagonally instead of a single peak at the center, as it is most evident in Fig. 15(d). These four peaks are located on the growth sector boundaries (GSBs) of the sample, which is formed because the sample grows in a square shape. By measuring the field distribution of the same sample magnetized by the FC method after all PFM experiments had been finished, it was confirmed that the multi-peaked profile of Fig. 15(d) is not due to any

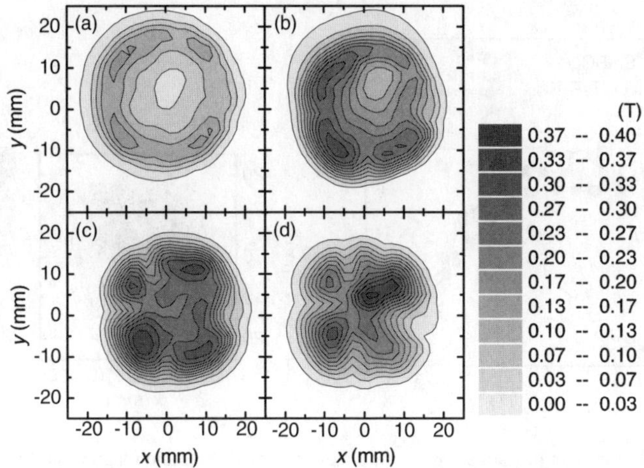

Figure 15: The distribution of the magnetic flux density trapped by a 36 mm diameter SmBCO sample when a single magnetic pulse was applied with an amplitude of (a) 1.3 T, (b) 2.4 T, (c) 3.0 T and (d) 3.9 T [121].

imperfection of the sample such as cracking or weak-links. Therefore, it was concluded that the GSB regions possess a comparatively stronger pinning effect than the regions between the boundaries, which brought about the multi-peaked field profile [121]. A similar splitting of the field profile was also observed in low-temperature experiments of PFM of SmBCO [122].

4.4 Multi-pulse magnetization

As discussed in the preceding sections, the magnetization of a bulk sample with a single magnetic pulse becomes increasingly difficult when the pinning effect increases. However, it is reported that the amount of magnetic flux can be increased by an iterative application of magnetic pulses with gradually reducing the amplitude, a technique that was named as IMRA method [118, 119]. Figure 16(a) shows the result of magnetizing a SmBCO sample at liquid nitrogen temperature with the use of this technique [121]. A large magnetic pulse was first applied, and thereafter, field pulses were repeatedly applied with decreasing the amplitude without demagnetizing the sample by heating it above the superconducting transition temperature. After the application of each magnetic pulse, the distribution of magnetic flux density was measured, from which Φ_T was calculated. The arrow in Fig. 16(a) indicates how Φ_T had changed with the iterative application of magnetic pulses, and the result is compared with that of FC magnetization and PFM with a single pulse. It can be seen that the amount of trapped magnetic flux increased with stepwise decreasing the amplitude of the magnetic pulse down to about 2 T. Compared to the peak value of single-pulse magnetization, Φ_T increased by a factor of about 1.4, and reached about 80% of that achieved with FC magnetization.

The field distribution after the first magnetic pulse of the experiment of Fig. 16(a), which was 4.1 T in amplitude, split in four peaks, because of the same reason already discussed (see Fig. 15(d)). During the iterative application of magnetic pulses, the amount of magnetic flux

(a)

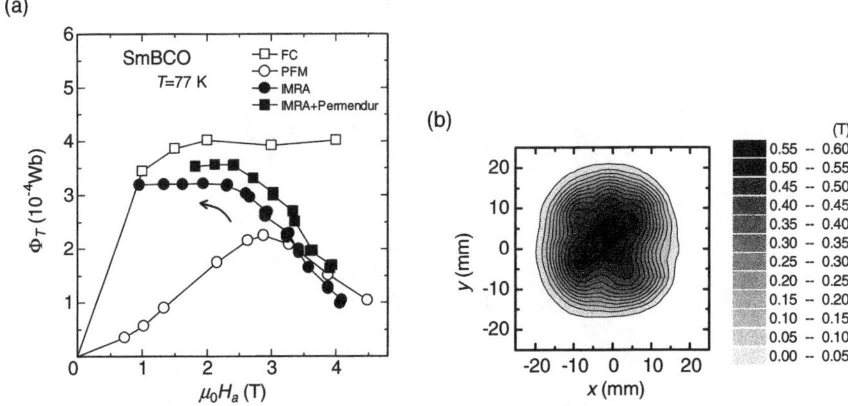

Figure 16: (a) The amount of trapped magnetic flux of a SmBCO sample magnetized by the FC method, the single-pulse magnetization (PFM), the IMRA method, and the IMRA method with sandwiching the sample by a pair of yoke pieces (Permendur alloy). (b) Field profile measured after an iterative application of magnetic pulses with reducing the amplitude from 4.1 T to 2.3 T [121].

increased progressively in all regions of the sample, but the increase at the regions between GSBs was larger than the boundary regions. As a result, the field distribution measured after decreasing the amplitude of the magnetic pulse to 2.3 T revealed a fairly concentric shape as shown in Fig. 16(b) [121]. The flux density at the GSB regions is still larger than that between them, but the difference had reduced, and the peaks are much less distinct compared to the field distribution after the first magnetic pulse.

Another method that was reported to be effective in enhancing the amount of trapped magnetic flux by PFM is sandwiching the superconductor with a pair of ferromagnetic yoke pieces [121, 123]. The increase in the trapped field was attributed to the field exerted from the ferromagnet that was magnetized by the trapped field of the superconductor [121]. Figure 16(a) includes the data of the combined magnetization of the IMRA and yoke methods. The yoke pieces used in this experiment were Permendur alloy, a well-known soft ferromagnetic material. As before, Φ_T increased with the iterative application of magnetic pulses, and it can be seen that the saturated value increased by combining the two techniques. The amount of trapped magnetic flux reached about 90% of that of FC magnetization for the particular sample of Fig. 16.

While the IMRA method is a multi-pulse technique in which successive magnetic pulses are applied at the same temperature with reducing the amplitude, an alternative method was proposed by Sander *et al.*, a multi-pulse technique with stepwise cooling [124, 125]. These authors reported that while their sample was not fully magnetized with a 1.9 T pulse at 70 K or below, the application of several magnetic pulses with the same amplitude but by decreasing the temperature from 75 K with a 5-10 K step increased the trapped field at the center of the sample. This is a very interesting result when compared to the IMRA method. A magnetic pulse with the same amplitude has a smaller effect on a superconductor at lower temperature because J_c increases. Hence, reducing the temperature while applying multi-pulses with the same amplitude is essentially the same as applying several pulses at the same temperature with decreasing the amplitude like in the IMRA method. Therefore, both techniques are probably closely related, and hence both were similarly effective in increasing the trapped field of a bulk sample that cannot be fully magnetized with a single pulse.

5 SUMMARY

Significant progress had been made in the trapped field achievable with a bulk REBCO super-conductor in the last 5-6 years. While YBCO was the only material worthwhile considering practical applications until 1997, there is now a variety of choice as a result of recent extensive studies. The addition of Ag into the starting composition was critically important to improve the mechanical strength of the material, and boosted up the amount of magnetic flux that the sample can trap without being cracked due to the enormous magnetic pressure. Without the addition of Ag, it was not possible to make full use of the superior J_c of REBCO with a light rare earth element that was melt-processed in a reduced oxygen atmosphere. The addition of Ag was, however, also important for YBCO to push up the limit of the magnetic field that the sample can safely trap. The addition of Zn to YBCO was another milestone, which showed that it is possible to incorporate the peak effect into an already well studied material. Another important achievement is the extraordinary large pinning effect of the ternary systems, which was shown to be able to be controlled by changing the chemical ratio of the rare earth elements, although the real ability of this material in a bulk form is yet to be decided. Despite the improvement in the mechanical strength of the material by the addition of Ag and/or the use of metal rings for reinforcement, it is still the mechanical properties that set a limit in the magnetic fields that a melt-processed bulk superconductor can trap. The recent study on resin impregnation is hence very fascinating as it might have finally removed away the cracking problem.

In exchange of the improved performance of the materials, the difficulty of fully magnetize bulk superconductors increase. We now need to apply a magnetic field larger than 10 T to fully magnetize the superconductor at low temperatures. Even at liquid nitrogen temperature, the field that is necessary is beyond the level that an ordinary non-superconducting electromagnet can generate. It would also not very easy to magnetize a bulk superconductor that is installed in an equipment. The magnetic pulse technique is important because it provides a flexible and a comparatively inexpensive means for the magnetization of bulk superconductors. For instance, an *in-situ* magnetization is possible with this method. It was shown that a large viscous force is exerted on flux lines that move very fast through the sample during pulsed-field magnetization. This force, in addition to the pinning force, is the source of dissipation, which leads to a heating of the sample. As a result, the trapped field is less than that of the ability of the sample. To increase the trapped field by the pulsed-field technique, it was effective to repeatedly apply magnetic pulses with either reducing the amplitude at the same temperature or decreasing the temperature while keeping the amplitude the same. These results are very encouraging, and deserve for further extensive studies.

ACKNOWLEDGEMENTS

The author is particularly indebted to Prof. Uichiro Mizutani of Nagoya University for many important discussions and collaborations. He is also grateful to his following collaborators and past and current students who works or worked with him on the melt processing and magnetization studies of bulk superconductors: T. Oka, Y. Itoh, Y. Yanagi, M. Yoshikawa, B. Latha, A. Takagi, S. Ikeda, A. Terasaki, S. Iwata, A. Mase, T. Hosokawa, H. Ishihara, K. Tazoe, T. Yamada and T. Iwasaki. Financial supports by Grant-in-Aids for Scientific Research from the Ministry of Education, Culture, Sports, Science and Technology of Japan are gratefully acknowledged.

REFERENCES

[1] S. Jin, T. H. Tiefel, R. C. Sherwood, R. B. van Dover, M. E. Davis, G. W. Kammlott and R. A. Fastnacht, Phys. Rev. B **37** (1988) 7850

[2] K. Salama, V. Selvamanickam, L. Gao and K. Sun, Appl. Phys. Lett. **54** (1989) 2352

[3] M. Murakami, M. Morita, K. Doi, K. Miyamoto, Jpn. J. Appl. Phys. **28** (1989) 1189

[4] M. Murakami, *Melt Processed High-Temperature Superconductors* (World Scientific, Singapore, 1992)

[5] M. Murakami, N. Sakai, T. Higuchi and S. I. Yoo, Supercond. Sci. Technol. **9** (1996) 1015

[6] D. A. Cardwell, Mat. Sci. Eng. B **53** (1998) 1

[7] C. Kim and G. Hong, Supercond. Sci. Technol. **12** (1999) R27

[8] J. R. Hull, Supercond. Sci. Technol. **13** (2000) R1

[9] M. Murakami, K. Yamaguchi, H. Fujimoto, N. Nakamura, T. Taguchi, N. Koshizuka and S. Tanaka, Cryogenics **32** (1992) 930

[10] G. Krabbes, P. Schätzle, W. Bieger, U. Wiesner, G. Stöver, M. Wu, T. Strasser, A. Köhler, D. Litzkendorf, K. Fischer and P. Görnert, Physica C **244** (1995) 145

[11] N. Ogawa, I. Hirabayashi and S. Tanaka, Physica C **177** (1991) 101

[12] M. Morita, M. Tanaka, S. Takebayashi, K. Kimura, K. Miyamoto and K. Sawano, Jpn. J. Appl. Phys. **30** (1991) L813

[13] T. Izumi, Y. Nakamura, T.-H. Sung and Y. Shiohara, J. Mater. Res. **7** (1992) 801

[14] Chan-Joong Kim, Ki-Baik Kim, Gye-Won Hong, Physica C **232** (1994) 163

[15] N. Ogawa and H. Yoshida, *Advances in Superconductivity IV* (Springer-Verlag, Tokyo, 1992) p. 455

[16] M. Kambara, Y. Watanabe, K. Miyake, A. Endo, K. Murata, Y. Shiohara and T. Umeda, J. Mater. Res. **12** (1997) 2873

[17] M. Morita, S. Takebayashi, M. Tanaka, K. Kimura, K. Miyamoto and K. Sawano, *Advances in Superconductivity III* (Springer-Verlag, Tokyo, 1991) p. 733

[18] K. Sawano, M. Morita, M. Tanaka, T. Sasaki, K. Kimura, S. Takebayashi, M. Kimura and K. Miyamoto, Jpn. J. Appl. Phys. **30** (1991) L1157

[19] T. B. Lindemer, F. A. Washburn and C. S. MacDougall, Physica C **196** (1992) 390

[20] P. J. Smith, D. A. Cardwell, N. Hari Babu, M. Kambara, and Y. Shi, Supercond. Sci. Technol. **14** (2001) 624

[21] C. P. Bean, Rev. Mod. Phys. **36** (1964) 31

[22] R. Weinstein, In-Gann Chen, J. Liu, D. Parks, V. Selvamanickam and K. Salama, Appl. Phys. Lett. **56** (1990) 1475

[23] P. W. Anderson, Phys. Rev. Lett. **9** (1962) 309

[24] Y. B. Kim, C. F. Hempstead and A. R. Strnad, Phys. Rev. **131** (1963) 2486

[25] N. Sakamoto, T. Akune and T. Matsushita, Jpn. J. Appl. Phys. **31** (1992) L1470

[26] H. Fukai, M. Tomita, M. Murakami and T. Nagatomo, Physica C **357-360** (2001) 774

[27] H. Fukai, M. Tomita, M. Murakami and T. Nagatomo, Supercond. Sci. Technol. **15** (2002) 1054

[28] K. Zhang, B. Dabrowski, C. U. Segre, D. G. Hinks, I. K. Schuller, J. D. Jorgensen and M. Slaski, J. Phys. C **20** (1987) L935

[29] W. Wong-Ng, B. Paretzkin and E. R. Fuller, J. Solid State Chem. **85** (1990) 117

[30] S I. Yoo, N. Sakai, H. Takaichi, T. Higuchi and M. Murakami, Appl. Phys. Lett. **65** (1994) 633

[31] S. Yoo, M. Murakami,N. Sakai, T. Higuchi and S. Tanaka, Jpn. J. Appl. Phys. **33** (1994) L1000

[32] S. I. Yoo and R. W. McCallum, Physica C **210** (1993) 147

[33] H. Ikuta, S. Ikeda, A. Mase, M. Yoshikawa, Y. Yanagi, Y. Itoh, T. Oka and U. Mizutani, Appl. Superconduct. **6** (1998) 109

[34] H. Ikuta, A. Mase, Y. Yanagi, M. Yoshikawa, Y. Itoh, T. Oka and U. Mizutani, Supercond. Sci. Technol. **11** (1998) 1345

[35] D. F. Lee, X. Chaud and K. Salama, Physica C **181** (1991) 81

[36] M. Mironova, D. F. Lee and K. Salama, Physica C **211** (1993) 188

[37] S. Haseyama, S. Kohayashi, M. Satoh, H. Miyairi, H. Nakane and S. Nagaya, *Advances in Superconductivity X*, eds. K. Osamura and I. Hirabayashi (Springer-Verlag, Tokyo, 1998) p. 653

[38] S. Kohayashi, H. Miyairi, S. Yoshizawa, S. Haseyama, S. Nagaya, M. Satoh and H. Nakane, *Advances in Superconductivity X*, eds. K. Osamura and I. Hirabayashi (Springer-Verlag, Tokyo, 1998) p. 693

[39] H. Ikuta, A. Mase, U. Mizutani, Y. Yanagi, M. Yoshikawa, Y. Itoh and T. Oka, IEEE Trans. Appl. Superconduct. **9** (1999) 2219

[40] H. Ikuta, A. Mase, T. Hosokawa, Y. Yanagi, M. Yoshikawa, Y. Itoh, T. Oka and U. Mizutani, *Advance in Superconductivity XI*, eds. N. Koshizuka and S. Tajima (Springer-Verlag, Tokyo, 1999) p. 657

[41] U. Mizutani, A. Mase, H. Ikuta, Y. Yanagi, M. Yoshikawa, Y. Itoh and T. Oka, Mat. Sci. Eng. B **65** (1999) 66

[42] H. Ikuta, N. Hirota, Y. Nakayama, K. Kishio and K. Kitazawa, Phys. Rev. Lett. **70** (1993) 2166

[43] Y. Ren, R. Weinstein, J. Liu, R. P. Sawh and C. Foster, Physica C **251** (1995) 15

[44] T. H. Johansen, Supercond. Sci. Technol. **13** (2000) 830

[45] T. H. Johansen, Supercond. Sci. Technol. **13** (2000) R121

[46] H. Takashima, M. Tsuchimoto and T. Onishi, Jpn. J. Appl. Phys. **40** (2001) 3171

[47] S. Kohayashi, S. Haseyama and S. Nagaya, *Advances in Superconductivity XII*, eds. T. Yamashita and K. Tanabe (Springer-Verlag, Tokyo, 2000) p. 500

[48] H. Ikuta *et al.* submitted to Supercond. Sci. Technol.

[49] Y. Watanabe, K. Miyake, A. Endo, K. Murata, Y. Shiohara and T. Umeda Physica C **280** (1997) 215

[50] H. Kojo, S. I. Yoo and M. Murakami, Physica C **289** (1997) 85

[51] M. Kambara, K. Miyake, K. Murata, T. Izumi, Y. Shiohara and T. Umeda, Physica C **330** (2000) 191

[52] N. Hari Babu, W. Lo, D. A. Cardwell and Y. H. Shi, Supercond. Sci. Technol. **13** (2000) 468

[53] D. A. Cardwell, N. Hari Babu, W. Lo and A. M. Campbell, Supercond. Sci. Technol. **13** (2000) 646

[54] H. Ikuta, T. Hosokawa, M. Yoshikawa and U. Mizutani, Supercond. Sci. Technol. **13** (2000) 1559

[55] M. Matsui, S. Nariki, N. Sakai and M. Murakami, Supercond. Sci. Technol. **15** (2002) 781

[56] S. Nariki, S. J. Seo, N. Sakai and M. Murakami, Supercond. Sci. Technol. **13** (2000) 778

[57] S. Nariki, N. Sakai, M. Matsui and M. Murakami, Physica C **378-381** (2002) 774

[58] S. Nariki, N. Sakai and M. Murakami, Physica C **378-381** (2002) 631

[59] S. Nariki, N. Sakai and M. Murakami, Supercond. Sci. Technol. **15** (2002) 648

[60] S. Haseyama, S. Kohayashi, J. Ishiai, S. Nagaya and S. Yoshizawa, *Advances in Superconductivity XI*, eds. N. Koshizuka and S. Tajima (Springer-Verlag, Tokyo, 1999) p. 701

[61] S. Nariki, H. Hinai, N. Sakai, M. Murakami and M. Otsuka, Physica C **378-381** (2002) 764

[62] S. Nariki, N. Sakai and M. Murakami, Physica C **357-360** (2001) 814

[63] S. Nariki and M. Murakami, Supercond. Sci. Technol. **15** (2002) 786

[64] S. Ikeda, T. Oka, Y. Yamada, M. Yoshikawa, Y. Yanagi, Y. Itoh and U. Mizutani, Jpn. J. Appl. Phys. **36** (1997) L345

[65] M. Daeumling, J. M. Seuntjens and D. C. Larbalestier, Nature **346** (1990) 332

[66] M. Ullrich, D. Müller, K. Heinemann, L. Niel and H. C. Freyhardt, Appl. Phys. Lett. **63** (1993) 406

[67] T. Matsushita, D. Yoshimi, M. Migita and E. S. Otabe, Supercond. Sci. Technol. **14** (2001) 732

[68] S. Nariki and M. Murakami, Physica C **378-381** (2002) 759

[69] G. Fuchs, G. Krabbes, P. Schätzle, S. Gruß, P. Stoye, T. Staiger, K.-H. Müller, J. Fink and L. Schultz, Appl. Phys. Lett. **70** (1997) 117

[70] R. Weinstein, J. Liu, Y. Ren, R-P. Sawh, D. Parks, C. Foster and V. Obot, *Proceedings of the 10th Anniversary HTS Workshop on Physics, Materials and Applications* eds. B. Batlogg, C. W. Chu, W. K. Chu, D. U. Gubser, and K. A. Müller (World Scientific, Singapore, 1996) p. 625

[71] Y. Ren, R. Weinstein, R. Sawh and J. Liu, Physica C **282-287** (1997) 2301

[72] P. Schätzle, G. Krabbes, S. Grußand G. Fuchs, IEEE Trans. Appl. Supercond. **9** (1999) 2022

[73] G. Fuchs, P. Schätzle, G. Krabbes, S. Gruß, P. Verges, K.-H. Müller, J. Fink and L. Schultz, Appl. Phys. Lett. **76** (2000) 2107

[74] G. Krabbes, G. Fuchs, P. Schätzle, S. Gruß, J. W. Park, F. Hardinghaus, G. Stöver, R. Hayn, S.-L. Drechsler and T. Fahr, Physica C **330** (2000) 181

[75] S. Gruss, G. Fuchs, G. Krabbes, P. Verges, G. Stöver, K.-H. Müller, J. Fink and L. Schultz, Appl. Phys. Lett. **79** (2001) 3131

[76] G. Krabbes, G. Fuchs, P. Verges, P. Diko, G. Stöver and S. Gruss, Physica C **378-381** (2002) 636

[77] Y. X. Zhou, W. Lo, T. B. Tang and K. Salama, IEEE Trans. Appl. Supercond. **11** (2001) 3700

[78] W. Lo, Y. X. Zhou, T. B. Tang and K. Salama, Physica C **354** (2001) 152

[79] Y. X. Zhou, W. Lo, T. B. Tang and K. Salama, Supercond. Sci. Technol. **15** (2002) 722

[80] M. T. González, N. Hari-Babu and D. A. Cardwell, Supercond. Sci. Technol. **15** (2002) 1372

[81] H. Walter, M. P. Delamare, B. Bringmann, A. Leenders and H. C. Freyhardt, J. Mater. Res. **15** (2000) 1231

[82] S. Piñol, F. Sandiumenge, B. Martínez, V. Gomis, J. Fontcuberta, X. Obradors, E. Snoeck and Ch. Roucau, Appl. Phys. Lett. **65** (1994) 1448

[83] M. P. Delamare, I. Monot, J. Wang, J. Provost and G. Desgardin, Supercond. Sci. Technol. **9** (1996) 534

[84] S. Yeung, A. Banerjee, J. Fultz and P. J. McGinn, Mater. Sci. Eng. B **53** (1998) 91

[85] M. Muralidhar, H. S. Chauhan, T. Saitoh, K. Kamada, K. Segawa and M. Murakami, Supercond. Sci. Technol. **10** (1997) 663

[86] M. Muralidhar, M. R. Koblischka, T. Saitoh and M. Murakami, Supercond. Sci. Technol. **11** (1998) 1349

[87] M. Muralidhar and M. Murakami, Physica C **309** (1998) 43

[88] M. Muralidhar, K. Segawa and M. Murakami, Mater. Sci. Eng. B **65** (1999) 42

[89] M. Muralidhar, S. Nariki, M. Jirsa and M. Murakami, Physica C **372-376** (2002) 1134

[90] M. Muralidhar, S. Nariki, M. Jirsa, Y. Wu and M. Murakami, Appl. Phys. Lett. **80** (2002) 1016

[91] T. Yamada, H. Ikuta, Y. Yanagi, M. Yoshikawa, Y. Itoh and U. Mizutani, Supercond. Sci. Technol. (*submitted*)

[92] M. Muralidhar, M. R. Koblischka and M. Murakami, Supercond. Sci. Technol. **12** (1998) 555

[93] M. Muralidhar, M. Jirsa, N. Sakai and M. Murakami, Appl. Phys. Lett. **79** (2001) 3107

[94] M. Muralidhar, N. Sakai, N. Chikumoto, M. Jirsa, T. Machi, M. Nishiyama, Y. Wu and M. Murakami, Phys. Rev. Lett. **89** (2002) 237001

[95] T. Yamada, H. Ikuta, M. Yoshikawa, Y. Yanagi, Y. Itoh and U. Mizutani, Physica C (*in press*)

112

[96] T. H. Johansen, Q. Y. Chen and W.-K. Chu, Physica C **349** (2001) 201

[97] M. Tomita and M. Murakami, Supercond. Sci. Technol. **13** (2000) 722

[98] M. Tsuchimoto, H. Takashima, M. Tomita and M. Murakami, Physica C **378-381** (2002) 718

[99] M. Tomita and M. Murakami, Physica C **354** (2001) 358

[100] M. Tomita, M. Murakami and K. Yoneda, Supercond. Sci. Technol. **15** (2002) 803

[101] M. Tomita and M. Murakami, Nature **421** (2003) 517

[102] Y. Itoh, Y. Yanagi, M. Yoshikawa, T. Oka, S. Harada, T. Sakakibara, Y. Yamada and U. Mizutani, Physica C **235-240** (1994) 3445

[103] Y. Itoh, Y. Yanagi, M. Yoshikawa, T. Oka, S. Harada, T. Sakakibara, Y. Yamada and U. Mizutani, Jpn. J. Appl. Phys. **34** (1995) 5574

[104] Y. Itoh and U. Mizutani, Jpn. J. Appl. Phys. **35** (1996) 2114

[105] Y. B. Kim, C. F. Hempstead and A. R. Strnad, Phys. Rev. **139** (1965) A1163

[106] Y. B. Kim and M. J. Stephen, *Superconductivity* ed. R. D. Parks, (Marcel Dekker, Inc., New York, 1969) Chap. 19

[107] M. Thinkham, *Introduction to Superconductivity* (McGraw-Hill, Inc., New York, 1975)

[108] Y. Itoh, Y. Yanagi, M. Yoshikawa, T. Oka, Y. Yamada and U. Mizutani, Jpn. J. Appl. Phys. **35** (1996) L1173

[109] Y. Itoh, Y. Yanagi and U. Mizutani, J. Appl. Phys. **82** (1997) 5600

[110] A. Terasaki, Y. Yanagi, Y. Itoh, M. Yoshikawa, T. Oka, H. Ikuta and U. Mizutani, *Advances in Superconductivity X* eds. K. Osamura and I. Hirabayashi (Springer-Verlag, Tokyo, 1998) p. 945

[111] H. Ikuta, H. Ishihara, T. Hosokawa, Y. Yanagi, Y. Itoh, M. Yoshikawa, T. Oka and U. Mizutani, Supercond. Sci. Technol. **13** (2000) 846

[112] M. Tsuchimoto, H. Waki, T. Honma, Y. Itoh, Y. Yanagi, M. Yoshikawa, T. Oka, Y. Yamada and U. Mizutani, IEEE Trans. Magn. **32** (1996) 1168

[113] M. Tsuchimoto, H. Waki, Y. Itoh, Y. Yanagi, M. Yoshikawa, T. Oka, Y. Yamada and U. Mizutani, Cryogenics **37** (1997) 43

[114] M. Tsuchimoto and K. Morikawa, IEEE Trans. Appl. Superconductivity **9** (1999) 66

[115] S. Bræck, D. V. Shantsev, T. H. Johansen and Y. M. Galperin, J. Appl. Phys. **92** (2002) 6235

[116] H. Ohsaki, T. Shimosaki and N. Nozawa, Supercond. Sci. Technol. **15** (2002) 754

[117] Y. Yanagi, Y. Itoh, M. Yoshikawa, T. Oka, Y. Yamada and U. Mizutani, *Advances in Superconductivity IX* (Springer-Verlag, Tokyo, 1997) p. 733

[118] Y. Yanagi, Y. Itoh, M. Yoshikawa, T. Oka, A. Terasaki, H. Ikuta and U. Mizutani, *Advances in Superconductivity X* eds. K. Osamura and I. Hirabayashi (Springer-Verlag, Tokyo, 1998) p. 941

[119] U. Mizutani, T. Oka, Y. Itoh, Y. Yanagi, M. Yoshikawa and H. Ikuta, Appl. Superconductivity **6** (1998) 235

[120] A. B. Surzhenko, S. Schauroth, D. Litzkendorf, M. Zeisberger, T. Habisreuther and W. Gawalek, Supercond. Sci. Technol. **14** (2001) 770

[121] H. Ikuta, H. Ishihara, Y. Yanagi, Y. Itoh and U. Mizutani, Supercond. Sci. Technol. **15** (2002) 606

[122] Y. Yanagi, Y. Itoh, M. Yoshikawa, T. Oka, T. Hosokawa, H. Ishihara, H. Ikuta and U. Mizutani *Advances in Superconductivity XII* eds. T. Yamashita and K. Tanabe (Springer-Verlag, Tokyo, 2000) p. 470

[123] H. Ikuta, Y. Yanagi, M. Yoshikawa, Y. Itoh, T. Oka and U. Mizutani, Physica C **357-360** (2001) 837

[124] M. Sander, U. Sutter, R. Koch and M. Kläser, Supercond. Sci. Technol. **13** (2000) 841

[125] M. Sander, U. Sutter, M. Adam and M. Kläser, Supercond. Sci. Technol. **15** (2002) 748

**COATED CONDUCTORS AND HTS MATERIALS
BY CHEMICAL DEPOSITION PROCESSES**

Sandrine BEAUQUIS, Carmen JIMENEZ and François WEISS,
Laboratoire des Matériaux et du Génie Physique, LMGP-ENSPG-INPG, CNRS-UMR5628,
BP 46, F-38402 Saint Martin d'Hères, FRANCE

1   INTRODUCTION.

In recent years, research efforts have been made to develop new technologies to elaborate tape-shaped 'high temperature superconductors (HTS) coated conductors' essentially based on the $YBa_2Cu_3O_7$ (YBCO) compound. The main interest of YBCO is its strong pinning properties in high magnetic fields at 77 K. It is not easy to fabricate YBCO based conductor cables with high critical-current density (J_c), because the grain-boundaries misorientation angle needs to be lower than few degrees due to the intergranular weak links of this material (Figure 1) [1, 2]. Whereas, epitaxially grown films show high J_c typically $> 10^6$ A/cm^2 when the grain-boundaries misorientation is inferior to 5 degrees.

Thus YBCO conductors have been developped as tape-shaped coated conductors on different metallic substrates by using deposition techniques of the superconducting material. It is generally necessary to deposit one or several buffer layers, either to improve the matching between the substrate and YBCO, or as chemical barrier (Figure 2).

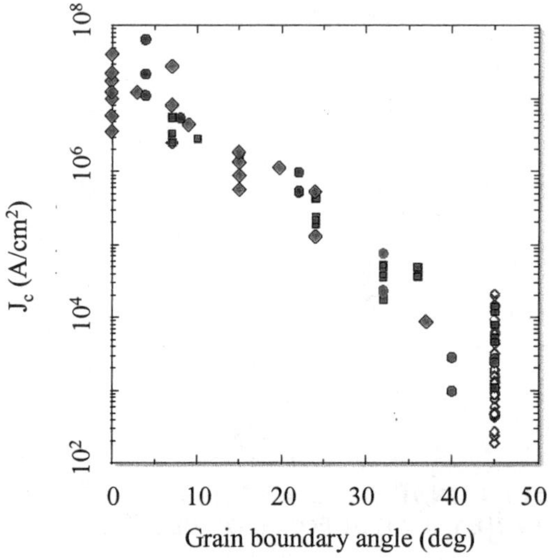

Figure 1: Orientation dependence of grain-boundary critical currents
in YBCO bicrystals [1, 2]

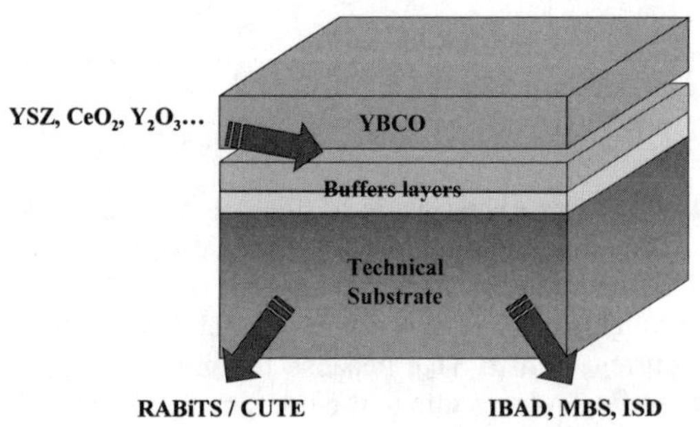

Figure 2: Typical architecture of a tape-shaped coated conductor on a metallic substrate

There are several approaches to elaborate YBCO coated conductors:
    (1)   Deposition of YBCO layers on polycrystalline metal substrates with artificially textured buffer layer deposited by Ion Beam Assisted Deposition (IBAD) [3, 4], Modified Bias Sputtering (MBS) [5] or Inclined Substrate Deposition (ISD) [6].

(2) Deposition of YBCO films on textured metal substrate like Cube Texture (CUTE) or Rolling Assisted Biaxially Textured Substrates (RABiTS) [7] with or without buffer layer.

In the first part of the present paper, we discuss the preparation of flexible metallic substrates. In the second part, we detail the chemical engineering of several Chemical Deposition Processes used for the synthesis of thin or thick superconducting oxides. The deposition techniques considered are:

i) Metal Organic Chemical Vapor Deposition (MOCVD),

ii) Spray pyrolysis

iii) Chemical Solution Deposition, essentially, Metal-Organic Decomposition (MOD), and Sol-gel deposition.

These chemical routes are very promising for coated conductor production, since they are considered as low cost techniques that can be extrapolated to a large-scale production. We present for each of these techniques the basic principle involved, the main results obtained nowadays for the fabrication of HTS thin and thick films, and their development for the realisation of coated conductors in reel-to-reel systems.

To form the complex architecture of coated conductors, different compounds have been synthesised by chemical deposition processes: from buffer layers (CeO_2, YSZ, Y_2O_3...) to the superconducting phases (Y-123, Hg-1223, Tl-1223). In the present review, the basic discussion concerning film preparation is concentrated on YBCO based conductors. Other families of superconducting oxides (Tl- and Hg-based superconductors) are also investigated; the most relevant results concerning these materials will be also presented.

2 BIAXIALLY TEXTURED SUBSTRATE PREPARATION

For HTS coated conductors, the choice of the metallic flexible substrate is critical. The basic requirements for substrates are:

 - No chemical interaction with the superconducting material
 - Heat durability in oxygen pressure during the deposition
 - Sufficient flexibility, proper mechanical tolerance
 - Suitable hardness to get a mirror-like flat surface
 - Low magnetism
 - Thermal expansion nearly equal to the superconductor from deposition temperature to room temperature

Last requirement is predominant: if the thermal expansion of the substrate is very far of the superconductor thermal expansion, the constraints induced in the superconducting material provoke cracks that affect its superconducting properties [8]. Figure 3 shows the thermal expansion coefficient for different metals, compared to this of YBCO. From this point of view, nickel or copper-based substrates match well with YBCO and are suitable for coated conductors elaboration.

118

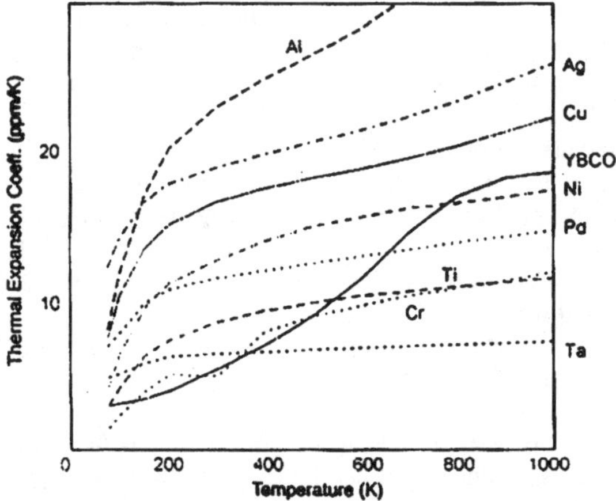

Figure 3: Average thermal expansion coefficient for various metals and for YBCO versus temperature (graph adapted from unpublished graph of A. Usoskin and H Freyhardt, reproduced from [8])

At present, two independent techniques are used for preparing biaxially textured substrates. The first one consists of the deposition of textured buffer layer on a polycrystalline metallic substrate; in the second one, the metallic substrate is mechanically worked and annealed to promote a textured surface.

2.1 Polycrystalline metal substrates with artificially textured buffer layer

2.1.1   Ion beam assisted deposition method (IBAD)

Ion beam assisted deposition (IBAD) technique achieves biaxial texture by means of an ion gun (Ar + O), which orients an oxide film buffer layer during its growth onto the polycrystalline metallic substrates (Figure 4). When using another ion beam (Ar) for the ablation of a target, the method is called 'dual ion beam sputtering system'. The use of this technique for the preparation of buffer layer as template for YBCO deposition, was proposed by Fujikura Ltd in 1991[9]. The buffer layers obtained by this method are mainly MgO and YSZ.

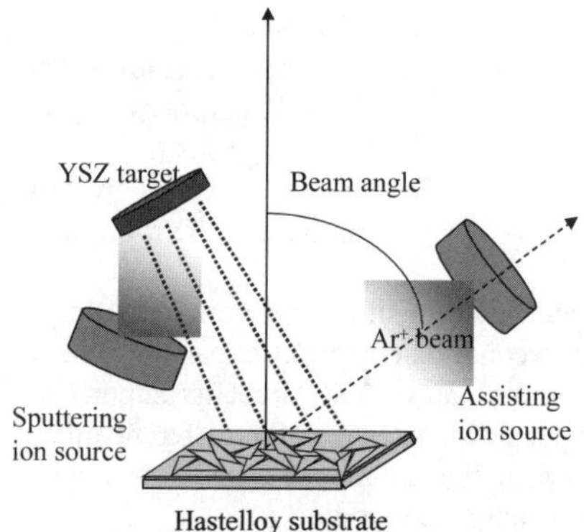

Figure 4: Schematic diagram of the ion-beam-assisted deposition system

Figure 5 shows that the sharpest in-plane texturing is obtained when the ion incident angle is 55° from the substrate normal, which correspond to the (111) axis of the YSZ unit lattice as referred to the (h00) direction [10]. The thickness dependence of the FWHM for both rocking curves of YSZ (200) peak and YSZ (111) ϕ–scan is shown in Figure 6 [11]; a FWHM of about 20° is obtained for a thickness of 600 nm. The deposition rate of this method is about 1nm/min. So the long time required to obtain a good YSZ layer is not favourable for a mass production of coated conductors.

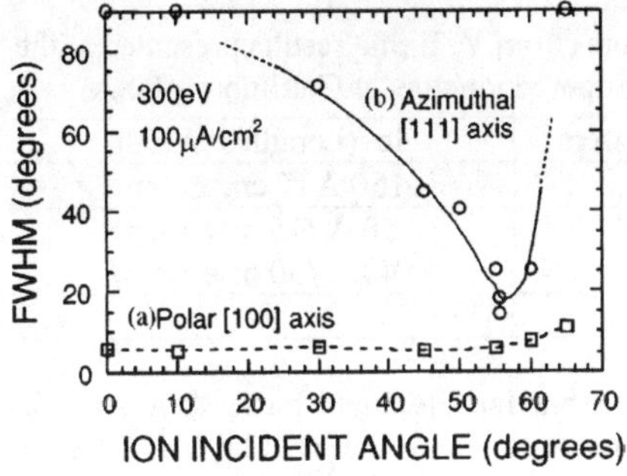

Figure 5: FWHM values for (a) rocking curves of YSZ (200) reflection and for (b) ϕ-scan of YSZ(111), as a function of the Ar^+ ion incident angle (reproduced from [10]).

Figure 6: FWHM values for (a) rocking curves of YSZ (200) reflection and for (b) ϕ-scan of YSZ(111) as a function of film thickness (reproduced from [11]).

An alternative to YSZ is MgO [12, 13]; the same degree of texture can be reached by IBAD on very thin films (10nm) (Figure 7, FWHM ϕ–scan MgO (111) <7°).

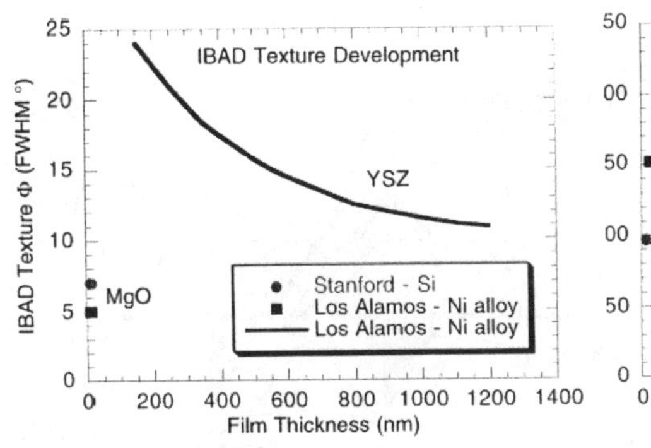

Figure 7: FWHM values of YSZ (111) ϕ–scan and MgO(111) ϕ–scan as a function of film thickness (reproduced from [13])

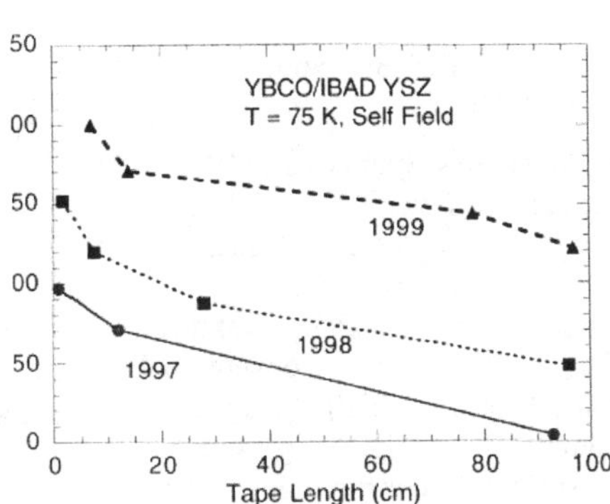

Figure 8: Improvement of the evolution of critical current of IBAD-based YBCO films with length (reproduced from [14]).

The improvement in substrate quality is reflected by a J_c enhancement; Figure 8 shows the evolution of critical current for IBAD-based YBCO coated conductors [14].

IBAD technique has been scaled-up to long length by Fujikura Ltd, by H.C. Freyhardt (Univ. of Goettingen) and by Los Alamos National Laboratory (LANL).

Table 1 shows recent J_c results obtained for IBAD coated conductors (presented by Y Iijima at the International Workshop on Processing and Applications of Superconductors at Gatlinburg-USA in July 2002). High values of J_c ($\sim 10^6$ A/cm^2) have been obtained in long length Hastelloy/ YSZ/GZO/YBCO coated conductors (GZO = Gd$_2$Zr$_2$O$_7$).

Table 1: Critical current of IBAD coated conductors (from Y. Iijima results presented at the International Processing and Applications of Superconductors at Gattlinburg-USA)

Architecture	Critical currents A/cm^2	Ic (Length x Width)
Hastelloy/YSZ/GZO//YBCO	1.7×10^6	150 A (7 cm x 1 cm)
	0.8×10^6 - 10^6	50 A (15 m x 1 cm)
		40 A (30 m x 1 cm)
Hastelloy/GZO//CeO$_2$/YBCO	3×10^6	-

H.C. Freyhardt (ZFW) has proved the deposition of biaxially textured buffer layers on flat and on cylindrically curved substrates [15]. The deposition rate of the YSZ layer is 70 nm/h. The reproducible performances of IBAD substrates covered with YBCO by PLD are 145 A measured end-to-end for 2-meter length and 10-mm width tape, and 70 A for 10-meter length and 4-mm width tape. Current is proportional to the cross section of the HTS coating. Another way to present the performance is using the figure of merit of critical current divided by effective tape width (Ic/w), because the width is the primary determinant of the cross section of the HTS coating, i.e. 145A/cm-width in 2 meters and 175 A/cm-width in 10 meters. Taking into account this parameter, the best values obtained by this group are 391 A/cm-width, i.e. Ic= 137A, for 20 cm, and 223 A/cm-width for 10m, i.e. a Jc of 2MA/cm^2. These results were presented during ASC 2002 in Houston

2.1.2   Modified Bias Sputtering (MBS)

Modified Bias Sputtering (MBS) is a technique invented by Fukutomi *et al* (National Research Institute of Metals), which permits the formation of biaxially textured films using a conventional radiofrequency (RF) magnetron sputtering [16]. The Ar$^+$ plasma has a particular configuration. Two electrodes are used to control the energy and the direction of impinging Ar$^+$ ions on the growing surface of the YSZ layer (Figure 9).

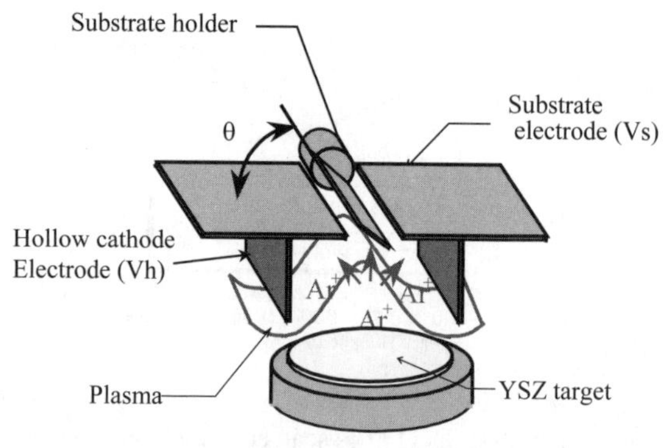

Figure 9: Schematic representation of Modified Beam Sputtering (MBS) process.

At present, the sharpness of the in-plane alignment is inferior to that obtained by IBAD, but this method is much simpler with lower costs performance for industrial applications.

2.1.3 Inclined Substrate Deposition (ISD)

Inclined Substrate Deposition (ISD) method permits to obtain biaxially textured YSZ films using pulsed laser deposition (PLD). As shown in Figure 10, YSZ in-plane ordering is obtained by inclining the substrate during the deposition [17]. In ISD, the YSZ <110> axis is favoured to align towards the direction of ablation plume, contrary to the IBAD for which the <111> axis align towards the direction of the Ar^+ beam. The origin of in-plane texturing seems to be a 'shadowing effect' that had been reported for several thin films grown by obliquely arranged vacuum evaporations. A similar biaxial alignment was observed in CeO_2 and MgO by the ISD process.

In small samples, Quinton *et al* [18] have obtained critical current value of $1.2x10^5$ A/cm^2 with YBCO films deposited on 1-μm YSZ ISD layer, Hasegawa *et al* [19] have obtained 4.3 10^5 A/cm^2 in equivalent conditions and Bauer *et al* [20] have obtained $7.9x10^5$ A/cm^2 for the Hastelloy®/MgO/YBCO architecture.

Recently, at the International Workshop on Processing and Applications of Superconductors at Gatlinburg 2002, B. Balachandran from Argonne National Laboratory (ANL) presented results of $3.5x10^5$ A/cm^2 for 1-meter long MgO ISD-buffered YBCO coated conductors. In the same workshop, K. Ohmatsu from Sumitomo Elec. Ind. Ltd. presented production of 50-m length of YSZ and CeO_2 layers; they used Reverse ISD, which is a variant of ISD, based on two steps deposition, by changing the inclination of the substrate in 180°.

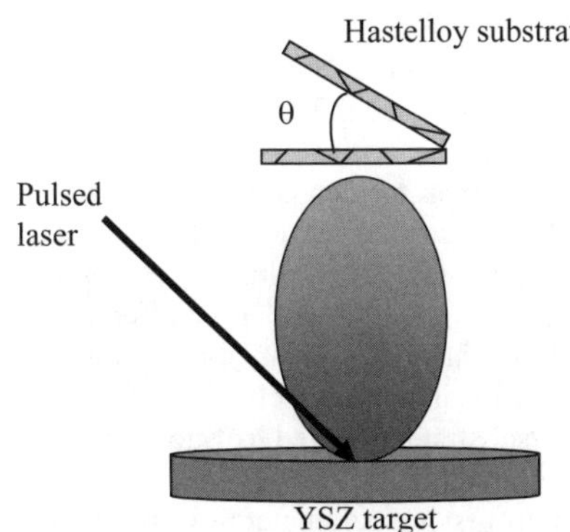

Figure 10: Schematic representation of the Inclined Substrate Deposition (ISD) method

2.1.4 Ion-beam nanotexturing (ITEX)

This technique has been recently developed (2001) by LBNL. ITEX method produces a biaxially textured layer by oblique ion irradiation on an amorphous film surface [21]. The ion beam parameters are similar to those of the IBAD method but the texturing process is separated from the layer deposition: an amorphous layer of buffer material is deposited by a rapid technique and then exposed to oblique irradiation. Only a thin layer near the surface is directly modified (1-2 nm of an YSZ amorphous layer for a 300 eV Ar^+ ion bombardment). The results obtained by this method are quite similar to those obtained with early IBAD: 2.5×10^5 A/cm^2 for HA230/YSZ/YBCO architecture, with YSZ and YBCO deposited by PLD [21] (HA230 is nickel-based alloy called Haynes Alloy N° 230).

2.2 Biaxially Textured Substrates

To elaborate coated conductors, it is interesting to achieve texture in the metal itself. It is possible by mechanical rolling of a face-centred cubic (FCC) metal and subsequent heat treatment. FCC metals such as silver and nickel are suitable for YBCO based coated conductors because their lattice parameters are nearly to the lattice a parameter of YBCO.

2.2.1 Cube-textured silver tapes (CUTE tapes)

Ag is the only metal tape on which superconducting oxides can grow without any chemical reaction, so it is not necessary to cover the Ag tape with a buffer layer. Texture of Ag was studied to determinate the stable structure after recrystallization. The cubic {100}<001> texture takes place for warm rolling and subsequent annealing at high temperature (between 700°C and 850°C) [22], and the bronze type structure {110}<211> is stable for lower temperature rolling (< 100°C). On another hand, studies on YBCO deposition on Ag single crystal [23] have shown that one excellent in-plane orientation of YBCO is obtained on the Ag (110) plane, whereas 2 types of in-plane orientations are obtained on Ag (100) plane. According to these results, it seems that the most adapted Ag textures for YBCO deposition are {110}<011> and {110}<112>. The processing of long length tapes with these two types of textures has been already carried out [24].

YBCO films deposited on Ag (001) present a c-axis texture, but with two four-fold symmetries in the ab plane shifted at 45°. High grain-boundaries misorientation induced by these two families of crystallites decreases dramatically the critical current. Attempts to improve the in plane texture of YBCO have been performed by depositing thin CeO_2 [25] and Y_2O_3 [26] layers on Ag (001) substrates by Pulsed Laser Deposition (PLD) and Metal Organic Chemical Vapor Deposition (MOCVD), respectively. Nevertheless, the epitaxial quality of YBCO is better when deposited directly on Ag (110), and no buffer layer is required. A 1-meter long YBCO sample on Ag (110) tape has been obtained by Yamazaki et al [24] with a critical current of $6\ 10^4$ A/cm^2. The best critical current, 2×10^5 A/cm^2, has been obtained by MOCVD [27], on the CUTE {110}<011> substrate prepared by Suo $et\ al$ [28].

The texture quality of Ag tapes is sensitive to a lot of processing parameters like purity, grain size, rolling conditions (preheating temperature, recrystallisation temperature…) [28]. Therefore, it is difficult to obtain textured Ag tapes reproducibly. In addition, the texture strongly depends on the ribbon thickness, and no sharp {110}<011> texture is obtained with tapes thinner than 300 µm. Furthermore, the mechanical strength of pure Ag is insufficient for handling, and so, it is difficult to deposit YBCO films on long tapes or to make coils. Silver alloys have been studied in order to improve mechanical strength and crystal orientation by alloying or by cladding [29]. Yoshino $et\ al$ have reported that Ag /Ni or Ni alloy clad tapes are alternatives as a means of strengthening Ag tape, and that the addition of 0.1% of Cu to Ag is effective for obtaining high J_c value of about 10^5 A/cm^2 by increasing surface smoothness and in plane alignment [30, 31]. Suo et al have also proposed new Ag/Ni composites, which allow to reinforce the substrate and to reduce the amount of silver used for their fabrication by 40% [29].

2.2.2 Rolling-Assisted Biaxially Textured Substrate (RABiTS)

The term Rolling-Assisted Biaxially Textured Substrates (RABiTS) has been invented by Goyal *et al* [32, 33, 7], and it refers to *"substrates with biaxially textured, chemically compatible surfaces for epitaxial growth of superconducting or other electronic devices"* [32]. The method to texture the substrate employs thermomechanical processing of base materials such as Cu or Ni. This is followed by deposition of appropriate chemical and structural buffer layers on the textured base metal.

2.2.2.1 Textured nickel-based substrates

A biaxially textured tape can easily be obtained on nickel or nickel alloy with {100}<001> orientation at high processing speed. The fabrication of biaxially textured nickel tape takes place by consecutive rollings of polycrystalline randomly oriented high purity (>99.99%) bars to total deformation greater than 90%. Subsequent annealing of the substrate is then performed in a controlled atmosphere. Tapes prepared in this way present single cube texture, with FWHM values of 6° from Ni (111) ϕ–scan.

High J_c values of over 10^6 A/cm^2 were rapidly reported on these substrates [34, 35]. However it is necessary to deposit buffer layers on the metal tape to improve the crystallinity of YBCO and to avoid the chemical reaction between the nickel and the superconducting layer. Typical buffer layers are oxides, which should be deposited without growth of a non-epitaxial NiO layer; this is possible using low temperature techniques or depositing in non-oxidizing atmosphere or high vacuum.

To date, the combination of YSZ (used as chemical barrier), and CeO_2 (for lattice matching) has been mainly studied. Instead of this multilayered architecture, Ichinose *et al* showed that a Y_2O_3 single layer could act as chemical barrier and for lattice matching [36]. Y_2O_3 single layer deposited on the {100}<001> textured nickel tape using electron beam (EB) evaporation presented a columnar growth, very sensitive to deposition conditions [37]. To improve the surface morphology, a dense Yb_2O_3 cap layer was deposited on the Y_2O_3 surface by sputtering. YBCO films produced on this structure by PLD had J_c values exceeding 10^6 A/cm^2 [38]. At almost the same time, M. Paranthaman et al [39] investigated the used of biaxially textured RE_2O_3 (RE=Rare Earth) buffer layers obtained by reactive evaporation. They deposited Y_2O_3, Gd_2O_3 and Yb_2O_3 on Ni tapes at temperatures around 650°C. These films were deposited under partial pressure of H_2O as source for oxygen, and at the same time, to prevent the formation of NiO.

Nevertheless, the "standard" buffer architecture for PLD, which lead to the best values of J_c until now, is CeO_2/YSZ/CeO_2. Recently, other architectures have been proposed, as for example LaNiO$_3$/YSZ/CeO$_2$, with Tc=90.5 K and Jc= 1x10^{6}A/cm^2 [40]. Other proposed

124

buffer layers are Lanthanum Zirconium oxide ($La_2Zr_2O_7$), Lanthanum Manganese Oxide ($LaMnO_3$) or Lanthanum Strontium Manganese oxide ($LaSrMnO_3$) [41].

Concerning the evaporation methods, J_c values exceeding 10^6 A/cm^2 have been obtained by thermal evaporation with the CeO_2/YBCO architecture [42, 43] in 20-cm length samples. In contrast to PLD, a thin (100 nm) CeO_2 layer is enough to prevent diffusion. Multilayers architectures (CeO_2/MgO, Y_2O_3/CeO_2 and thicker $Ce_{1-x}Gd_xO_2$) have been also tested, but they did not lead to a significant improvement. A reel-to-reel thermal evaporation system developed by THEVA was presented during ASC 2002 [44]. The long length deposition of CeO_2 is already a standard process, yielding up to 25 m of buffered tape. First attempts to deposit YBCO in a dynamic mode have been already performed.

As mentioned, one difficulty with RABiTS nickel-based coated conductors is the formation of polycrystalline or non-epitaxial NiO on the surface of the nickel tape during the deposition of oxide buffer layers. The oxidation of nickel has been controlled and used to grow biaxially textured NiO on a nickel tape surface. This technique named Surface Oxidation Epitaxy (SOE) was invented by Matsumo *et al* [45, 46]. This method seems to be useful for speeding up tape manufacturing because the SOE batch process in an electric furnace can easily form the biaxially textured NiO layer. This method is advantageous from the viewpoint of the simplification of the buffer layer formation.

In SOE processing, a cube-textured nickel tape is inserted into an electric furnace heated to 1000-1300°C for metal oxidation (Figure 11). Nickel oxide can epitaxially grow on the nickel surface if the oxidation temperature and the pressure are optimized. The sharpness of the cube texture of these NiO layers is strongly dependent on the substrate texture and on the grain size of the substrate material, as well as the processing conditions during the oxidation [47].

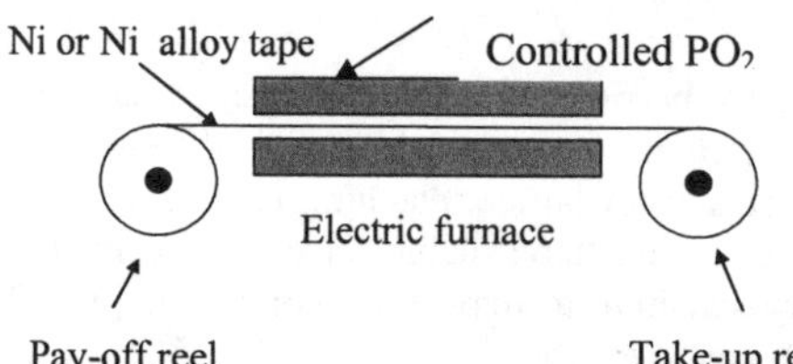

Figure 11: Schematic illustration of the SOE process for long length tapes

Figure 12 shows the cross section of NiO respectively on a) Ni 0.1 at % W and b) Ni 0.1 at % Mo [47]. Long length biaxially textured SOE tapes up to 50 m have been already fabricated [48]; however, YBCO directly deposited on SOE tapes by PLD had very low J_c values (0.05 10^6 A/cm^2). This is due to the existence of grooves on the grain boundaries of NiO. The NiO grooves are transferred from thermal grooves of underlying Ni and produce superconducting weak couplings which suppress the transport critical current. The deposition of a MgO cap layer flatten the NiO surface and J_c value of 0.3 10^6 A/cm^2 was obtained on a short sample of YBCO/MgO/NiO/Ni tape [48]. Other cap layers as $SrTiO_3$ and CeO_2 have been investigated because of their smaller lattice misfit with YBCO than MgO, but no higher J_c have been obtained with these oxides [49].

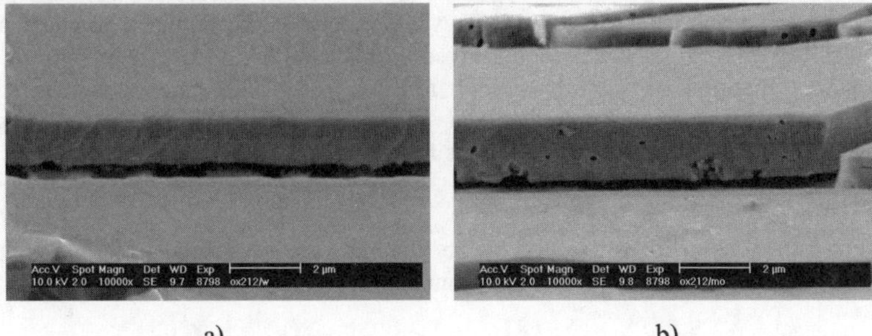

a)   b)

Figure 12: Cross section of NiO on a) Ni 0.1at%W, and b) Ni0.1at%Mo.

Further studies about the influence of the surface structure and chemistry of the biaxially textured Ni (001) substrates on the seed-layer growth are currently been performed at Oak Ridge National Laboratory (ORNL) [50]. They claimed that the formation of a $c(2x2)$ two-dimensional superstructure is due to the surface segregation of sulfur, initially contained in the metal. The role of this surface structure seems crucial during the first step of growth of buffer layers, mainly for oxides with perovskite or fluorite structures.

The technological issues using nickel tape are related to the mechanical strength and magnetism of nickel. The substrate must be thin because a very important consideration in most applications is a high the engineering critical current density value J_{eng}, defined as a critical current carried by the superconducting layer, divided by the total cross section of the conductor, which includes the substrate and all the functional layers. The substrate must be thin; nevertheless, the mechanical strength must be sufficient. Various approaches to the fabrication of stronger substrates with reduced magnetism have been suggested including alloys and composite structures.

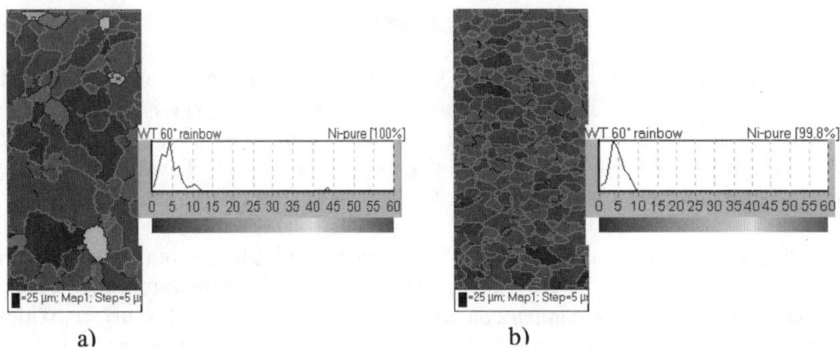

a)   b)

Figure 13: EBSD mappings of (a) Ni+ 0.1 at% W annealed over 30 min at 850 °C and (b) Ni+ 5.0 at.% W annealed over 45 min at 1000 °C.

The yield strength (YS) of the tape at room temperature can be improved by adding solutes like Cr, V, Cu, Fe or W to Ni, without deteriorating the cube texture. Eickemeyer *et al* have shown that nickel-refractory metal alloys (Mo, W, Nb, or Ta) are promising material group for highly textured substrate tapes [51]. They reported that the texture sharpness and the thermal stability in Ni tapes could be improved by micro-allowing with Mo, W, Nb, or Ta. The orientation mapping obtained by EBSD from Ni-W RABiTS tapes is shown in Figure 13, together with the associated grain boundary angle distribution. The very strong alignment of the grains in the cube orientation is obvious from Figure 13 b), especially for the Ni 5 at.% W.

Micro-allowing nickel with refractory metals increase the resistance of grain boundaries against thermal etching processes and thus grain boundary grooving [51] (Figure 14).

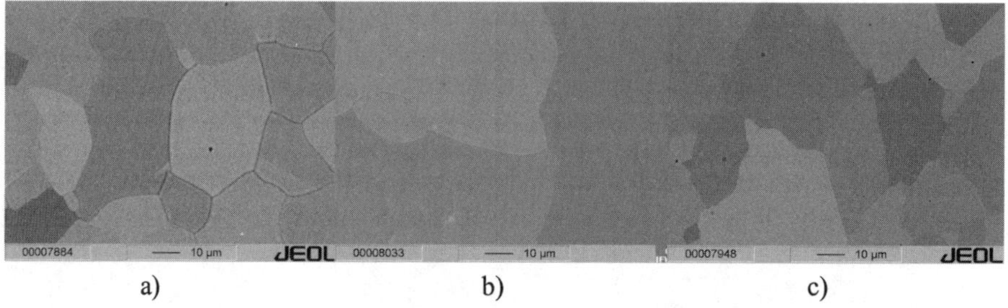

Figure 14: Scanning electron micrographs of RABiTS surfaces to manifest different degrees of grain boundary grooving: a) Ni (99.99 at-%) b) Ni(99.94 at-% purity)+ 0.1 at-% W annealed at 850°C in hydrogen gas over 30 min, and c) Ni (99.94 at-% purity) + 5 at.% W after annealing for 45 min at 1000°C in hydrogen gas [51].

A comparative study on these nickel and nickel alloys substrates recovered with CeO_2/YBCO layers by thermal evaporation demonstrated that superconductors on Ni5%W alloys routinely reproduced J_c values of 1 MA/cm^2. The best result was obtained in a 12 cm of 1cm-wide sample, conducting 135 A through a 2-µm YBCO layer [42].

Goyal *et al* reported that Ni substrate with minor alloying additions of W and Fe is significantly stronger than the pure Ni substrate (YS of Ni -3at%W-1.7%Fe is around 3 *YS of Ni). They measured a critical current value of 1.9×10^6 A/cm^2 in a YBCO/CeO_2/YSZ/Y_2O_3/Ni-3at%W-1.7at%Fe sample. J_c at 0.5 T reduced by only 21% indicating strongly-linked grain boundaries in the YBCO film on this substrate [52].

Highly biaxially textured substrates have been strengthened by introducing Al_2O_3 particles with internal oxidation [53]. Ni-Al tapes possess a good cube texture and the strength is increased by a factor of 5 in comparison with pure Ni. Nevertheless the surface oxides of the tape are mixed oxides of Ni and Al. A solution proposed by De Boer *et al* is a designed composite consisting of a pure Ni (or alloyed) surface region and an Al_2O_3 strengthened core [53].

There is a range of Ni-based binary or ternary alloys that can be used as a substrate. Non-magnetic textured metal substrates or with substantially reduced magnetism at device-operating temperature are desirable. NiCu, NiV, NiCr, NiW are non-magnetic depending on the alloy composition and the temperature of the application envisaged. Thompson *et al* studied a series of biaxially textured $Ni_{1-x}Cr_x$ substrates [54]. Ni-Cr alloys with Cr content as high as 13at-%Cr can be thermo-mechanically processed to form substrate with very sharp and fully formed cube texture. The Curie temperature of this alloy is inferior to 10 K (625 K for Ni), the alloy is paramagnetic with a saturation magnetization of only 0.4 G cm^3/g (57.5 G cm^3/g for Ni).

The NiO approach has been also tested in some non-magnetic biaxially textured binary alloys as Ni-11%V; the $YBCO/CeO_2/NiO/Ni$-V architecture was fabricated by PLD, giving J_c values of 0.6 10^6 A/cm^2 in small samples [55].

Ni/NiC (Ni-20wt%Cr) and Ni/SS (stainless steel) composites have been also recently studied [56, 57]. The main technical issues are the reduction of ferromagnetism, as a consequence of the reduction of volume fraction of nickel, and the increase of mechanical strength. The same cold rolling and recrystallisation process than used for Ni tapes lead to a cube textured at the surface of these composites. The SOE process gives a biaxially textured NiO (100) layer on the top of these composite tapes. MgO and YBCO layers were deposited by PLD on these samples, which revealed Jc values of $0.1x10^6$ A/cm^2.

2.2.2.2   Textured copper substrates

A possible low cost alternative for fabrication of coated conductor tapes is to use Cu as the substrate. The advantages of Cu and Cu alloy tapes are the easy formation of a sharp texture, they are non-magnetic cold workable, and there is no chemical interaction with YBCO. Piñol *et al* have obtained sharp biaxially cube Cu {100}<001> textured tape [58]. The main disadvantage is the poor resistance to the oxidation in an oxygen atmosphere at high temperature during the synthesis of superconducting oxide. Nevertheless, it is possible to grow one protective metallic biaxially cube textured layer. Diaz *et al* have deposited protective film of Ag, Au and Ag-Au by electro-epitaxial deposition technique on Cu substrates in order to avoid the oxidation of the Cu surface [59].

Investigations on coated conductors with copper-based substrate are rare up to now. During ASC 2002, C. Cantoni presented high J_c value of about 10^6 A/cm^2 for a Cu/TiN/MgO/LMO/YBCO architecture prepared at Oak Ridge National Laboratory, using 2mm-thick Cu deposited on single crystal substrate. In this architecture, TiN is used as diffusion and oxidation barrier and MgO as oxygen diffusion barrier.

3 CHEMICAL DEPOSITION PROCESSES FOR COATED CONDUCTORS: BASIC PRINCIPLES AND MAIN RESULTS

Chemical Deposition processes, used for the formation of HTS films, are based on chemical reactions between inorganic or organic metal-containing species, leading to the formation of a layer on an appropriate substrate.

These processes can be classified in two main groups:

1. **One step processes**, chemical vapor deposition (CVD) or metalorganic chemical vapor deposition *(MOCVD)* using volatile chemical precursors, where the superconducting oxide phase is formed directly after chemical reactions from the gas phase on the surface of a determined substrate.

2. **Two step processes**, (spray pyrolysis, chemical solution deposition) which are based on the deposition of an unreacted precursor film, followed by a recrystallisation step, where solid state reactions induce the formation of the desired superconducting phase.

All these processes are in general simple and less expensive than physical deposition processes. They are commonly used in the production of various coatings and thin films with the main advantage that they allow an easy scaling up to large size and a conformal coverage of complex shapes.

3.1 Metalorganic Chemical Vapor Deposition (MOCVD) of YBCO

Chemical Vapor Deposition (CVD) or Metal Organic Chemical Vapor Deposition (MOCVD), depending on the nature of the chemical precursor used, offers several advantages as compared to physical methods: conformal step coverage and selective growth. In addition, CVD is a cheap and versatile method. In CVD or MOCVD processes, the gaseous precursor molecules are transported to the deposition surface and react to form a thin film by chemical reaction [60]. The pressure used in this process is generally in the range of a few Torr. This pressure is easily attained with classical one-stage vacuum pumps and does not need ultra high vacuum equipment, which considerably increase the equipment and maintenance costs. On the other hand, as a large variety of metalorganic precursors are available for MOCVD, buffer layers as well as superconducting layers can be deposited.

These precursors have to fulfil several basic properties like purity (over 95%), high yield synthetic routes, easy handling, low hazard, and low cost. However, essential requirements of suitable precursors for MOCVD processes are:

1. Adequate volatility to achieve acceptable growth rates at moderate evaporation temperature

2. Enough temperature gap between evaporation and decomposition

3. Clean decomposition with no residual impurities

4. Good compatibility with co-precursors during the growth of complex oxides (Y, Ba and Cu, for example)

5. Good solubility in a vast range of organic solvents (for instance, 1-2 dimethoxyetane, xylene, octane), and stability in solution

The precursors generally used in MOCVD for the synthesis of buffer layers and superconductors, are metal alkoxides $M(OR)_n$ and β-diketonates $M(RCOCHCOR')_n$, β-diketonatoalkoxides $M(OR)_{n-x}(RCOCHCOR')_n$ (with R and R' = Me, Et, ^tBu, for instance), and to a lesser extend, alkyl or cyclopentadienyl derivates. [61]. These compounds are fluorine-free and most of them are solids. To obtain the precursor vapours, classical methods like direct sublimation or gas-bubbler methods can be used, but these evaporation methods require precursors with a very good stability under repeated cycles. Well-controlled, high-temperature delivery lines are also required for the transportation in gas phase. When depositing multilayers or multicomponents materials, the complexity of the delivery system increases. These aspects are crucial to obtain good quality HTS films.

Precursors mainly used for HTS synthesis by thermal MOCVD are β-diketonates. The vapour pressures are satisfactory for yttrium and copper tetramethylheptanedionates (tmhd or thd) ($Y(thd)_3$, $Cu(thd)_2$). Nevertheless, the barium complexes of 2,2, 6,6 tetramethy 3,5-heptanedionate show limited volatility at elevated temperatures, and unfortunately, the compounds are reported to decompose near the temperature required for sublimation. The large ionic radius of Ba (~ 1.96 Å), and the large coordination number c (up to 12) tend to form low volatility oxygen-bridged polynuclear cluster in order to fulfil the saturation requirement of the central metal atom. For example, $Ba(thd)_2$ is tetrameric [61, 62, 63]. In addition, the coordination depends on the synthesis method, mainly on the nature of the solvent, aqueous or non-aqueous, on the barium source, and on the purification method. The synthesised "$Ba(thd)_2$" can display quite different structures and properties in term of volatility and stability. A decrease in the volatility with aging is also common for these molecules. Thus, there has been a large research effort aimed to increase the vapour pressure of these complexes for MOCVD application. Addition of a Lewis base, such as polyethers [64] (tetraglyme, triglyme) or polyamines [65] (pentamethyldiethylenetriamine), leads to a range of monomeric complexes. The adduction of a ligand is undoubtedly useful to protect the [Ba(thd)2] complex from hydrolysis and in stabilising the complex in solution during liquid injection MOCVD.

3.1.1 Single-liquid sources for MOCVD systems

A lot of efforts to overcome this problem with barium precursors have been performed. Solutions based on the use of single-source have been proposed. Several systems have been developed using a single solid source or a single liquid source. We will present here only the systems that use a mixture of precursors in a single liquid solution. These single liquid sources are based on the chemical compatibility of the precursors in the solution, i.e. no formation of complexes or precipitates. Some examples of these systems are:

- Generation of droplets from solution containing the precursor by ultrasonic nebulization and transport through a heated zone where the evaporation of the whole mist takes place [66, 67]. This system is known as Aerosol Assisted CVD (AACVD).
- A band is continuously wet with the precursor solution. This impregnated tape crosses a heated zone, where the solvent is first evaporated and evacuated. Then the band with the dry precursors moves into a zone heated at a higher temperature,

130

where the precursors completely evaporate [68, 69]. This method is known as Band Evaporator MOCVD (BEMOCVD)

- The solution is injected via an injector valve into a heated zone, where it is evaporated without contact [70, 71, 72]. The solution enters in the evaporator as small droplets that are easily dispersed and flashed evaporated. This system is known as Pulsed injection MOCVD (PIMOCVD) or Injection MOCVD (IMOCVD).

The main advantages of these liquid delivery systems are:

- The evaporation rate does not depend on precursor's volatility. Very high growth rate can be obtained depending only on the feeding rate.
- Only a small amount of solution is evaporated at once, the rest of the solution is kept at room temperature under inert atmosphere. This leads to a better stability and reproducibility of the precursor vapours.

There are several parameters which depend on the evaporator system, for example, as commented, the evaporation rate.

In AAMOCVD, the evaporation rate depends on the carrier gas flow, and on the power used for the generator connected to the ultrasonic element. The frequency used for the aerosol generation defines the distribution of the droplets size in the aerosol. Smaller droplets are easily evaporated.

In BEMOCVD, the solution feeding rate is controlled by an external liquid massflow controller. The speed of the band carrying the solution has to be adapted to the feeding rate in order to have a linear evaporation rate [73].

In PIMOCVD, the solution is injected directly by a micro-valve (Figure 15). The frequency, the opening time and the solution concentration are the main parameters controlling the growth rate. Other parameters are the viscosity of the solution, and the gas pressure used to pressurize the solution container.

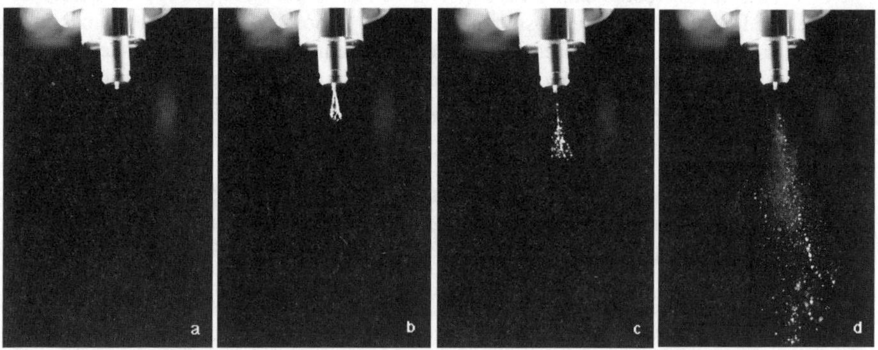

Figure 15: Injection of a liquid solution as used in PIMOCVD

Solvents have to be adapted to each system. The principal requirements are viscosity and vapour pressure. For example, in AACVD, the solvent has to present a low viscosity, and a low vapour pressure. Alcohols and acetylacetonate are common solvents for this technique. In BEMOCVD, the solvent has to present higher vapour pressure to ensure the solution reaches the band in a liquid state. An early evaporation of the solvent before reaching the tape would clog the feeding line. Typical solvent for this system are diglyme (bis (2-methoxyethyl) ether) or xylene. In PIMOCVD, a larger range of vapour pressure is possible. Nevertheless, if the vapour pressure is too low, the injector can be also clogged up. Usual solvents are monoglyme (1-2 dimethoxyethane) or saturated hydrocarbons (for instance, octane).

Finally, these vapour sources have to be connected to the reaction chamber by heated lines to ensure the transport of the vapour precursors. The temperature of these lines has to be at least equal to the temperature for evaporation to avoid recondensation of vapours. These temperatures are typically in the range of 200-300°C. A carrier gas is normally provided to optimise the vapour transport; usually inert gas (Ar, He) is used to avoid early decomposition of the precursors, but in some cases it is also possible to mix the precursor vapours with reactive gases (O_2, N_2O). The transfer time, between the evaporator and the deposition zone, is very short (~1 s), which means gas speed of a few m/s.

3.1.2   Deposition of buffer layers and YBCO by MOCVD in static mode

High quality YBCO films have been deposited, in static mode, using the single source systems described in 3.1.1 on single crystals [74, 75, 76, 77, 78] and on metallic substrates [79, 80]. These systems have also been used to deposit high quality buffer layers [81, 82, 83, 84].

Several oxide buffer layers have been synthesised by PIMOCVD, essentially Y_2O_3, YSZ, Gd_2O_3, GdSZ, MgO, $SrTiO_3$ and CeO_2. Precursors for these materials are commercially available. The deposition conditions have been optimised for these materials from the HTS application point of view (biaxial texture, dense film, optimal thickness), and as a function of substrates [82]. Typical deposition parameters for the usual buffer layers and for superconducting layers are summarized in Table 2. The deposition rates depend on the deposition parameters and typically vary from 1 μm/h to 6μm/h.

The use of HTS films grown by MOCVD in others applications than coated conductor, has been also tested. A PIMOCVD reactor scaled-up to 3" substrates has been already developed [85] at the Vilnius University, and a 4" PIMOCVD system is currently under process optimization [86]. Depositions of 200nm YBCO films on 3" $LaAlO_3$ substrates under optimal conditions leads to Jc values of 4-6 MA/cm^2 at 77 K over almost all the surface (Figure 16). The films showed also a low-field surface resistance, Rs, of about 0.35mΩ (at 8.5GHz and 77k) which remained constant up to 4 mT.

Table 2: Typical deposition parameters for some buffer layers and YBCO by PIMOCVD.

Parameters	Y_2O_3 [84]	YSZ [82]	CeO_2 [82]	MgO[79]	YBCO [75]
Precursor	$Y(thd)_3$	$Y(thd)_3$, $Zr(thd)_4$	$Ce(thd)_4$	$Mg(thd)_2$	$Y(thd)_3$ $Cu(thd)_2$ $Ba(thd)_2$
Solvent	monoglyme	monoglyme	heptane	monoglyme	monoglyme
Injection frequency (Hz)	2-3	3	4	1-2	1
Injection time (ms)	1-2	2	1	1	4
Evaporation temperature °C)	250	300	230	280	300
Deposition temperature (°C)	600-800	600-700	800-850	700-800	775-825
Deposition pressure (Torr)	5	5	5	5	5
% O_2 in $Ar+O_2$	40	42	42	30	41

PIMOCVD was also used for the elaboration of $YBa_2Cu_3O_7/SrTiO_3/La_{0.67}Sr_{0.33}MnO_3$ and $YBa_2Cu_3O_7/La_{0.67}Sr_{0.33}MnO_3$ heterostructures. These multilayers were prepared and tested as SPQIDs devices, based on spin-polarized carrier injection effect [87, 88].

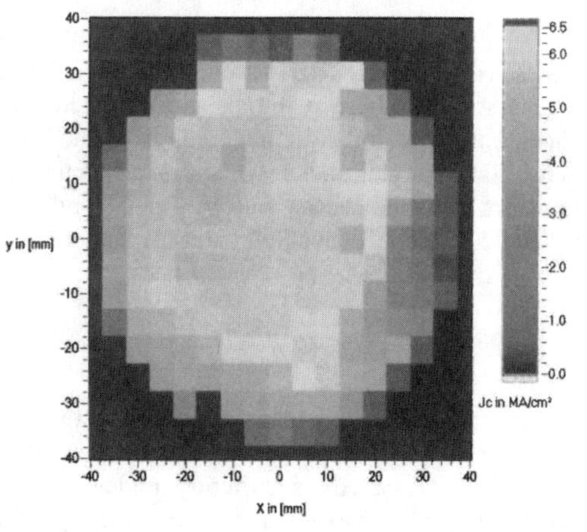

Figure 16: Distribution of Jc values (at 77k) in a 200nm YBCO film, as determined by AC susceptibility third harmonic measurements [85]

Several architectures on metallic substrate have been tested using static PIMOCVD systems. It is important to notice that one limiting factor for the substrate choice in PIMOCVD process is the used of oxidizing atmosphere at high temperature.

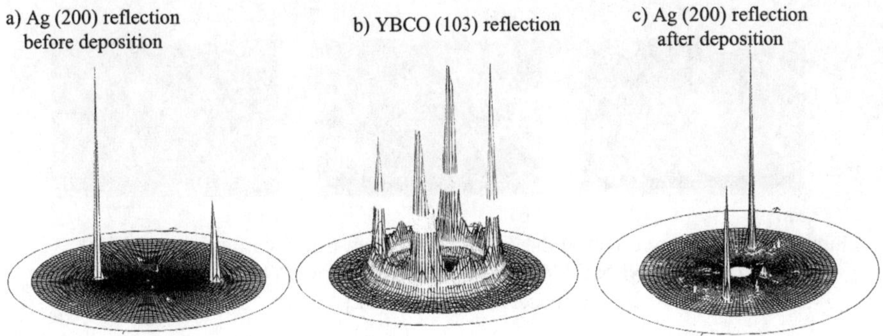

Figure 17: Pole figures of reflections corresponding to: a) silver substrate before deposition, b) YBCO layer on this substrate, and c) silver substrate after deposition [27]

Direct deposition on silver {110}<011> textured substrates was possible, thanks to the chemical stability of silver with oxygen [27]. YBCO films were biaxially textured (Figure 17), but the in-plane texture was not complete. The Jc value obtained for this sample was $2\times10^5 A/cm^2$.

The direct deposition of YBCO on nickel-based RABiTs substrates is not possible by MOCVD; instead, SOE substrates are more suitable. YBCO deposited on biaxially textured NiO lead to superconducting films, but the quality is limited by the nickel diffusion in YBCO. Several buffer layers architectures have been studied to optimise the network matching and to solve the interdiffusion phenomena.

Ni/NiO substrates prepared by SOE at IFW as described by D. Selbmann [89, 90] were used. The nickel-based tapes were microalloys of 0.1%at Mo or W. MgO(400nm)/YBCO(700nm) layers [79] were grown by PIMOCVD on the top of these substrates. The buffer layers were deposited at a growth rate of 1.5 µm/h, and the YBCO films at 5 µm/h. A T_c(onset) of 87K and a critical current density $J_c=10^4 A/cm^2$ at 77K were measured by susceptibility.

YSZ(200nm)/YBCO(750nm) structures [91] were also deposited on these Ni/NiO substrates. The transition was rather sharp (Tc~88-89K) but the Jc values were low ($J_c\sim10^4 A/cm^2$), due to the lattice mismatch between YSZ and YBCO (~6%), and because of the Zr diffusion in the YBCO film. Adding a second buffer layer of Y_2O_3 leads to Jc values of $5\times10^5 A/cm^2$. The beneficial effect of the Y_2O_3 layer, in addition to the diffusion barrier role, is the surface flattening effect [92]. In fact, NiO thermally grown on Ni substrates presents some grooves, which are transferred to the subsequent layers. The YSZ layers reproduce the morphology of the NiO surface (Figure 18 a). The second buffer layer fills the grooves and improves the surface roughness (Figure 18 b).

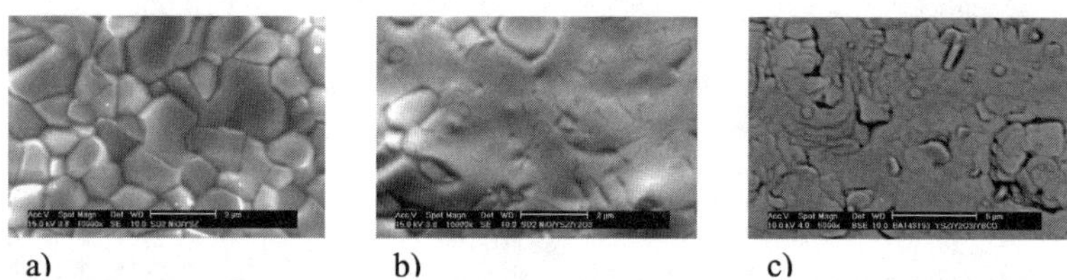

a)   b)   c)

Figure 18: Flattening effect of the Y_2O_3 buffer layer. Surface morphology of (a) Ni/NiO//YSZ, (b) Ni/NiO//YSZ/Y_2O_3, and (c) Ni/NiO//YSZ/Y_2O_3/YBCO as observed by SEM [92]

YBCO layers were also deposited on the top of Ni substrates ex-situ buffered with CeO_2/YSZ/CeO_2 by PLD, and provided by Cryoelectra GmbH [91]. Jc values close to 1 MA/cm^2 were obtained on these samples. When depositing the CeO_2(150nm)/YSZ(200nm)/CeO_2(150nm)/YBCO structure by PIMOCVD on Ni/NiO substrate, Jc values of 6×10^5 A/cm^2 were obtained.

Some depositions on YSZ-IBAD substrates (YSZ/Stainless Steel, YSZ/SS) [93] were also performed in static mode. CeO_2(300nm)/YBCO(750nm) structures grown on the top of YSZ/SS substrates leads to J_c values of 1.3 MA/cm^2.

3.1.3 Deposition of buffer layers and YBCO by MOCVD in reel-to-reel mode

In order to deposit HTS thin or thick films on moving tapes special reel-to-reel deposition systems, based on single solution source, have been developed at IOPW [69, 94], at LMGP [92], and at JIPELEC, in the context of the European READY project [95]. In a general way, these reactors (Figure 19) consist of a process chamber (1) containing all the necessary elements for the MOCVD process: an evaporator, a resistive furnace (2), a showerhead for precursors vapours distribution (3), a pumping system (4), a pressure control system, a gas panel, and a reel-to-reel system (5). All the system is under vacuum. The geometry and deposition parameters (essentially gas distribution) have to be optimised to improve the deposition yield in the reaction zone. The temperature homogeneity in the deposition zone has to be better than ± 3 °C. Using lateral furnace allows to improve the thermal profile in the deposition zone, and to perform a preheating of the substrates.

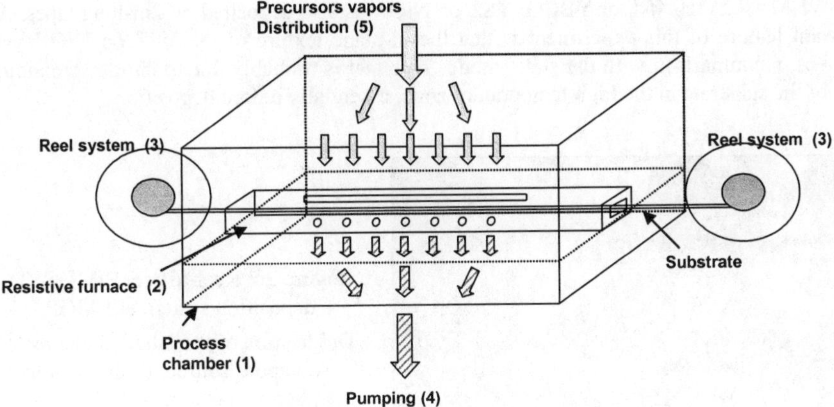

Figure 19: Schematic representation of a MOCVD reactor

These prototypes consist of 1 deposition chamber, and buffer layers and YBCO films depositions are performed sequentially. A reel and de-reel system is provided to transfer tapes in two ways. After YBCO deposition, the lateral furnace can be also used as post deposition annealing furnace to perform the oxygen loading of the films. The travelling velocity of the tape can vary from 0.5m/h to 20m/h, depending on the deposition rate and on the film thickness. When depositing on small samples, a stainless steel or a nickel alloy at 5%W tape is used as substrate holder; the samples are glued or mechanically fixed to this holder, and the deposition takes place in a reel-to-reel mode.

Thermodynamics calculations allow simulating the CVD process, which is a very practical tool for reactor design. For instance, modelling of temperature distribution, gas flow distributions, and deposition rate of YBCO was calculated by the computer program FLUENT; these calculation has allowed to optimised the reactor geometry at IOPW [96, 97, 98]. In this reactor, 300nm-thick YBCO films were deposited on LAO single crystal travelling at a speed of 4m/h. The crystalline quality of these films was excellent: FWHM of the YBCO (102) reflection in ϕ-scan was 0.65°, while the FWHM of the YBCO (005) in the rocking curve was 0.098°. The J_c value was of 4.8 MA/cm^2 at 77 K. On IBAD YSZ samples, YBCO deposited at the same conditions leads to a Jc value of 1 MA/cm^2, and for 500nm-thick YBCO layers deposited on CeO$_2$/YSZ/CeO$_2$ buffered nickel, a J_c value of 0.25 MA/cm^2 was achieved [94]. This reactor has recently been improved to sequentially deposit buffer, superconducting, and protective layers in three in-line chambers [98]. More experiments are currently performed.

The reactor developed at LMGP is presented in Figure 20. The experience obtained in static substrates deposition has been transferred to this continuous system. Buffer layers and superconducting films have been successfully deposited in this reel-to-reel system.

CeO$_2$ and YBCO layers were deposited on 2m-long tape in a continuous mode. The average value of the in-plane texture from ϕ-scan measurements was FWHM= 9.23°±0.48° for CeO$_2$

136

and FWHM= 8.26°±0.45° for YBCO. YSZ on NiO was also deposited on 2m-long tapes. An important feature of this experiment is that the in-plane texture of the YSZ on NiO layers improves in comparison with the static mode. This fact is probably due to shorter remaining times of the substrate in the high temperature zone, essentially before deposition.

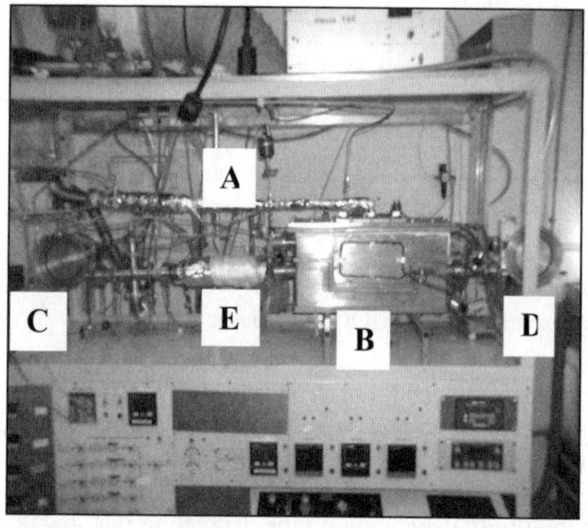

Figure 20: Reel-to-reel PIMOCVD deposition system at LMGP.

Description of the labelled elements: A: vapour source, B: deposition chamber, C: take-up reel, D: play-out reel, E: annealing chamber

CeO_2 and YSZ films were deposited by PIMOCVD, in continuous mode, on the top of YSZ/SS IBAD substrates. The in-plane texture versus thickness of the CeO_2 layer follows an exponential decay [99]. This result was previously reported by J. Wiesmann [100], when increasing thickness of YSZ layer by IBAD: a saturation of the in-plane texture was observed at a FWHM = 7° for 5.5µm-thick films. In our study, we started working with a YSZ(800nm)/SS structure prepared by IBAD. The in-plane texture of the initial YSZ layer showed a FWHM = 13.5° value from the YSZ(111) reflection in φ-scan mode. The saturation in texture was found at a FWHM=8.7° for a 600nm-thick CeO_2 film, and at a FWHM=10° for a 300nm-thick YSZ film. In addition to these lower saturation thickness values, the growth rate of the films by MOCVD was around 10 µm/h, as compared to 0.3-0.7 µm/h by IBAD.

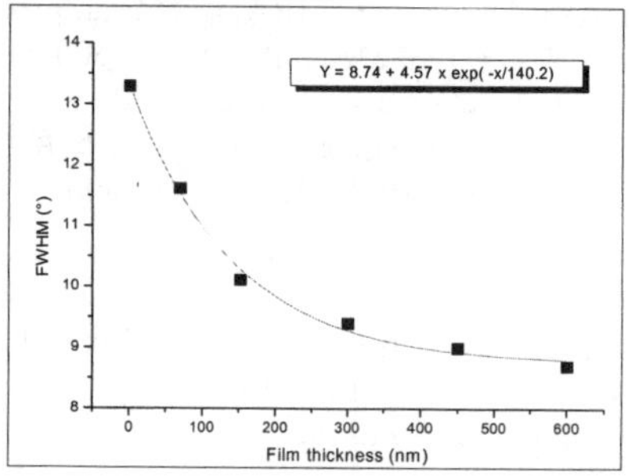

Figure 21: Influence of thickness on the in-plane texture of the CeO_2 films deposited on YSZ/SS IBAD substrates. FWHM values were obtained from the CeO_2(111) reflection in φ-scan mode [99]

On the other hand, AFM measurements of the surface roughness showed lower values for CeO_2 films than for YSZ; denser YBCO films were consequently obtained on these samples.

Firstly, the optimal architecture found up to now on YSZ IBAD substrates is YSZ(200nm) /CeO_2 (150nm)/YBCO. The tape velocity used for the deposition of the buffer layer ranged between 3-4 m/h, and for the superconducting layer between 1-2 m/h. Structures deposited in these conditions lead to critical current densities of $0.8\ 10^6$ A/cm^2 at 77K with a Tc= 90K.

Secondly, YSZ(200nm)/CeO_2(150nm)/YBCO(750nm) structures were deposited on Ni/CeO_2 samples supplied by Theva [43]. Tc(onset)= 89K with transition of 1K, and Jc values of 0.5 MA/cm^2 have been obtained.

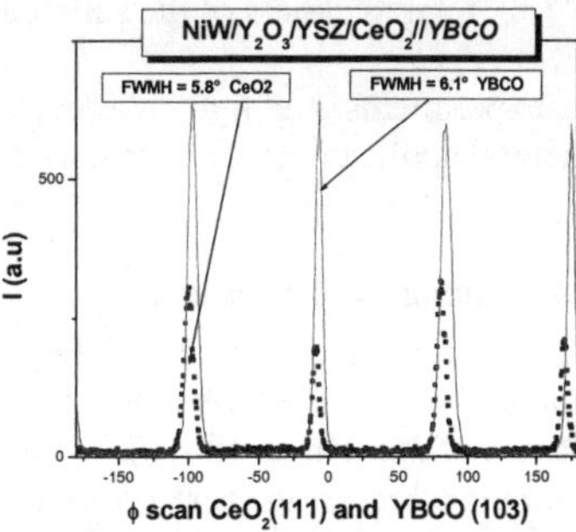

Figure 22: ϕ-scan of the CeO_2 (111) and YBCO (103) reflections for the NiW/Y_2O_3/YSZ/CeO_2// YBCO [101] heterostructure.

Recently, YBCO films were deposited on buffered NiW tapes supplied by American Superconductor. The whole structure consisted of NiW/Y_2O_3/YSZ/CeO_2//YBCO [102]. Early experiments gave T_c (onset) = 89.5K with a sharp transition (0.8 K) , and Jc values of 0.6 MA/cm^2. The in-plane texture of the YBCO layer has a FWHM=6.1° (Figure 22).

Table 3: Comparison of the Jc values obtained on static and on reel-to-reel mode

Architecture	Static mode:		In reel-to-reel mode:	
	Jc (A/cm^2)	Tc onset (K)	Jc (A/cm^2)	Tc onset (K)
Ni/NiO//YSZ/Y_2O_3/YBCO	5×10^5	90	5×10^5	91
Ni/CeO_2/Y_2O_3/CeO_2//YBCO	1×10^6	89.5	-	-
Ni/NiO//CeO_2/Y_2O_3/CeO_2/YBCO	6×10^5		6×10^5	89.8
NiW/CeO_2//YSZ/CeO_2/YBCO	-	-	5×10^5	89
SS/YSZ//YSZ/CeO_2/YBCO	1.3×10^6	90	8×10^5	90

Up to now, the longest length reported for YBCO films deposited by MOCVD corresponds to the results obtained at Intermagnetic General Corporation (IGC)-SuperPower, USA. They reported during the MRS superconductivity Workshop 2003 critical currents of 90 A over 1-

meter length per cm-wide IBAD tape. More results can be found in [103]. On the other hand, N. Kashima from Chubu Electric Power Co., Japan, reported the successful deposition of 100 meters length of YBCO on rolled non-textured Ag tape using a six-stage MOCVD system.

3.1.4 Alternatives to the proposed coated conductor architectures

Others experiences performed with these deposition methods rely on:

> Alternative buffer layer to decrease the surface roughness and to reduce the weak link behaviour of films, high angle grain boundaries (NiO, GdSZ, Gd_2O_3) or conductive buffer layers.
> Multilayers of superconducting REBCO materials, ie: SmBCO, to increase the total thickness of the superconducting layer keeping the same J_c values.

These experiments highlight alternatives to the fabrication of HTS structures for coated conductors' applications. The results have to be considered as preliminary ones.

3.1.4.1 Alternative buffer layers:

In order to reduce the effect of the NiO surface roughness and grooves on the stacking structures, the NiO layer obtained by SOE can be used as a seed layer. We have deposited NiO layers of 350 nm by PIMOCVD at 650°C. The in-plane texture was improved in 2°, and a smoothing effect on the surface is shown in Figure 23.

a) 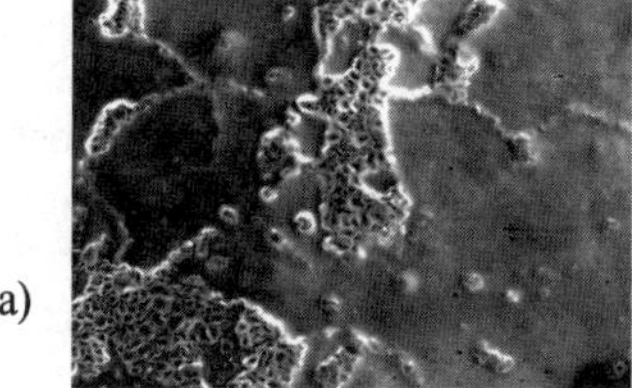   b) 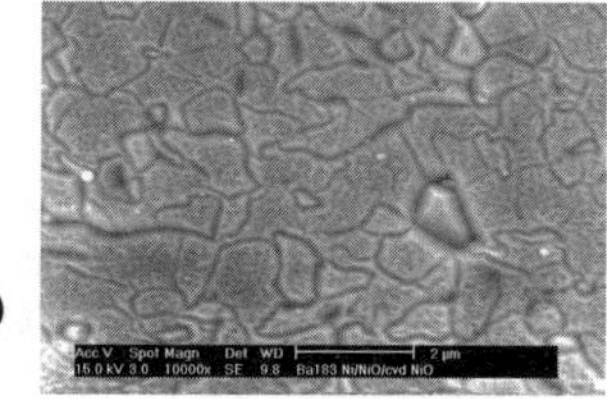

Figure 23: Surface morphology of the Ni/NiO substrates a) as obtained by SOE, b) after deposition of a NiO layer by MOCVD.

A very attractive approach is the use of conductive buffer layers, in order to be electrically coupled to the superconducting film to minimise the effect of the transition to the dissipative regime. In addition, coupling to a good metal increases the thermal conductivity of the entire tape, and in particular improves the cooling rate of the conductor.

Mainly $LaNiO_3$ (LNO) and $SrRuO_3$ (SRO) have been proposed until now [104]. Both are perovskite-type conductive metallic oxides with pseudo-cubic lattice parameters of 3.96 Å and 3.86 Å, and room temperature resistivities of about 300 $\mu\Omega$.cm and 600 $\mu\Omega$.cm. Films prepared by PLD on LAO and STO show that the better structure is YBCO/SRO/LNO/LAO, which gives Jc of 4.4. MA/cm^2.

LaNiO$_3$ films have been grown by PIMOCVD [105] on LAO and STO substrates as electrodes for ferroelectrics. Films are epitaxially (00l) oriented on both substrates. The deposition temperature ranges between 750°-800°C.

Recent results obtained by PIMOCVD show that YBCO and LNO films grown on NiO(SOE)/ RABiTS Ni0.1%Mo substrates are well-textured (Figure 24 a). Nevertheless, an intermediate buffer layer has to be used to avoid nickel diffusion in YBCO, as for example SrRuO$_3$.

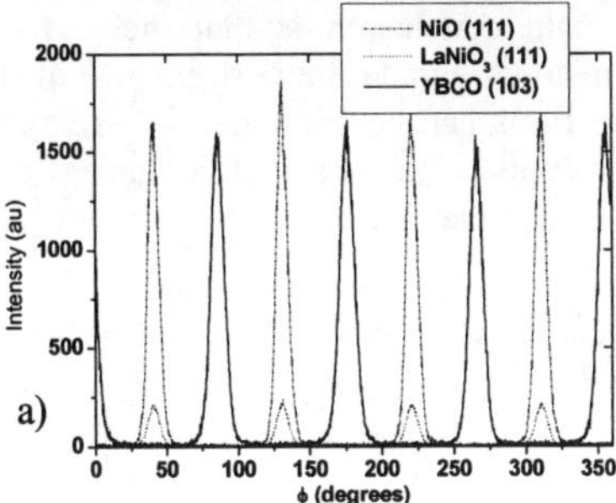

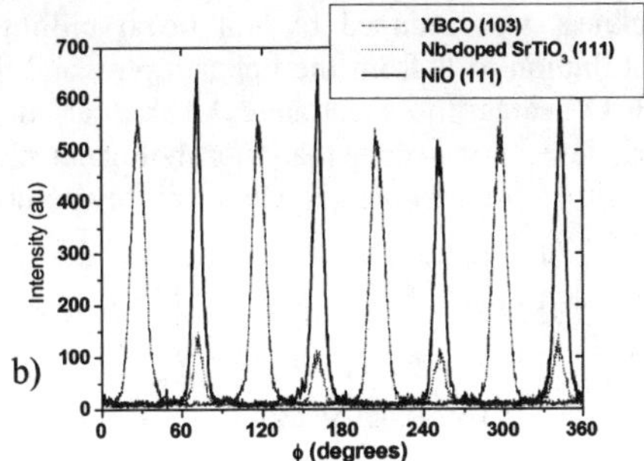

Figure 24: Φ-scan of two heterostructures fabricated by PIMOCVD using conductive buffer layers: a) YBCO/LaNiO$_3$/NiO/Ni and b) YBCO/Nb-doped SrTiO$_3$/NiO/Ni

SrRuO$_3$ films have also been prepared by PIMOCVD [106]. Films were in-plane textured on perovskite substrates (SrTiO$_3$, NdGaO$_3$, LaAlO$_3$) and on MgO. On YSZ the (110) plane is twinned along two directions of substrate structure. The heterostructure YBCO/SRO in STO and in LAO showed T$_c$ close to 91 K, with a transition smaller than 1 K, and Jc~1.5 MA/cm^2. In this work, the SRO films were grown at 825°C, to deposit the heterostructure in the same deposition run than the other layers, but other work [107] has reported the deposition of SRO at lower temperatures (500-750°C). Nevertheless, any work has been reported until now about the synthesis of these structures on metallic substrates by MOCVD.

La$_{0.7}$Sr$_{0.3}$MnO$_3$ (LSMO) has been also proposed as conductive buffer layer. In fact, this material has been intensely studied due to its colossal magnetoresistance properties [108]; in addition, LSMO is also an electrically conductive oxide with good thermal stability. The structure YBCO/LSMO/Ni has been successfully produced by the combination of PLD for YBCO deposition and rf-magnetron sputtering for LSMO [41].

This material has been also grown using PIMOCVD. The reproducibility of this process and its capability in growing ultra-thin interlayers (few monolayers) has been proved by the fabrication of (La$_{0.7}$Sr$_{0.3}$MnO$_3$/SrTiO$_3$)$_{15}$ superlattices on SrTiO$_3$ substrates, with thickness of (3.1 nm/1.2 nm)$_{15}$ [109]. This material has been also used in the fabrication by PIMOCVD of YBCO/STO/LSMO heterostructure for SPQID devices [87].

140

Another conductive layer recently tested by PIMOCVD is Nb-doped SrTiO3 (Nb-STO). Experiments performed until now show that the Nb-STO/YBCO structures grows epitaxially on NiO(SOE)/ RABiTS Ni0.1%Mo substrates (Figure 24 b).

3.1.4.2 Multilayers of superconducting REBCO

The decrease of the J_c when increasing YBCO thickness has been reported for single-crystal oxides substrates [110], and has been corroborated for flexible metal substrates [111]. Superconducting behaviour of thick samples were analysed as a function of depth, i.e. the thickness was reduced by ion beam milling. The main conclusion is that there is not contribution to I_c from the upper layers, and they somehow degrade the performance of the YBCO nearest the substrates. All the current in these films can be contained in ~1.25μm-thick layer.   It seems that morphological changes, essentially increase of roughness and porosity; with thickness is one of the key points in the J_c degradation.

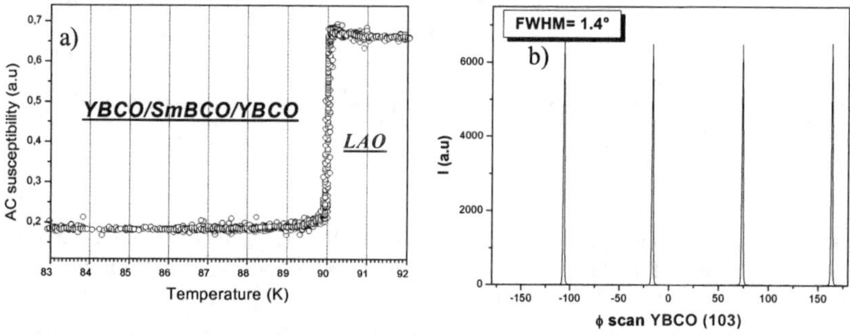

Figure 25: Characterization of the stacking structure YBCO/SmBCO/YBCO on LAO obtained by PIMOCVD with a total thickness of 2 μm: a) superconducting transition, b) φ-scan of the YBCO (103) reflection.

A solution has been proposed which consists in the growth of multilayer stacking of two superconducting materials: $YBa_2Cu_3O_7$ and $SmBa_2Cu_3O_7$. This multilayer has been obtained by PLD on IBAD-YSZ substrates at LANL [112], carrying more than 200 A/cm on 1-meter tape.

Recent results have shown that it is possible to reproduce by PIMOCVD the stacking structure Y123/Sm123/Y123, leading to 2μm-thick HTS layers with a J_c value of 1.2 10^6 A/cm^2. The texture of the YBCO layer seems to be as good as for thinner samples, and the Tc transition is very abrupt with a Tc of 90 K (Figure 25).

3.2 *Spray pyrolysis deposition*

Aerosol routes offer a variety of approaches for film generation. These processes have been used to deposit many materials including ceramic superconductors, simple metal oxides, non-oxide ceramic, etc. at high rates [113]. Among the different experimental procedures, the spray pyrolysis relies on the decomposition on a heated substrate of the droplets produced by an aerosol of non-volatile liquid source precursors.

As mentioned above, the spray pyrolysis process the deposition of an unreacted precursor film, followed by a recrystallisation step, where solid-state reactions induce the formation of the superconducting phase. In the spray pyrolysis, the aerosol produced by pulverisation is transported at nearly room temperature, and at atmospheric pressure, from the source to the reaction zone and the deposition surface by convection. This deposition stage leads to a non-completely reacted film (precursor film), which should be followed by one or several annealing treatments: first, to control the film recrystallisation, and then, a final oxidation step to improve the superconducting properties.

The experimental set-up (Figure 26) consists of a high frequency ultrasonic generator coupled to a piezoelectric element to generate the aerosol. The droplets are carried by a flow gas of inert gas or oxygen to the deposition region. The substrate is heated at deposition temperature. The residual products are evacuated by the exhaust.

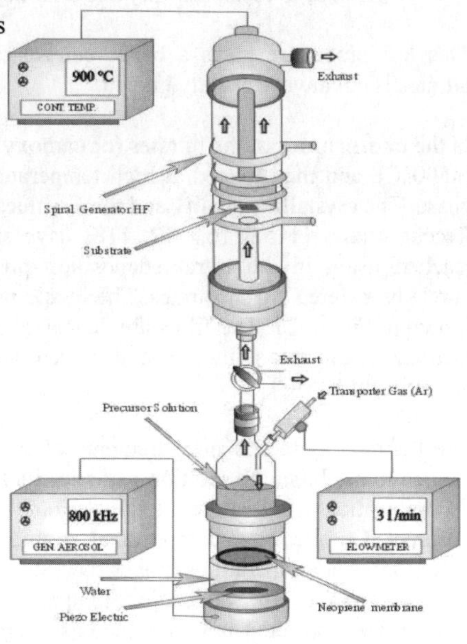

Figure 26 : Schematic representation of the spray pyrolysis deposition system.

The main features of this process in the deposition of HTS films are that the growth rate can achieve several µm/min, and that it is possible to work with aqueous solutions presenting a low toxicity [114]. The formation of YBCO films can take place in-situ or ex-situ, depending on the substrate temperature. For the spray pyrolysis of HTS materials (YBCO, Tl and Hg based superconductors) aqueous nitrate solutions have been used in most of the cases but sometimes also carboxylates. Nitrate solutions are generally preferred, especially in the case of Ba based superconductors, because they allow to circumvent problems of carbon contamination in the layer.

3.2.1 Deposition of YBCO films by spray pyrolysis

All-nitrates approach is chosen for easy preparation, low cost of raw materials, non-toxic content and good conditions to avoid Ba carbonate formation [113]. Aqueous nitrate solutions can be prepared:

i) by dissolving $Y(NO_3)_3 \cdot 3H_2O$, $Ba(NO_3)$, and $Cu(NO_3) \cdot 6H_2O$, or $Ca(NO_3)_2 \cdot 4H_2O$ in highly distilled and deionized water to give the suitable cationic ratio for deposition.

142

ii) from raw materials (Y_2O_3, $BaCO_3$, and CuO) dissolved in the appropriate molar ratio of diluted HNO_3. This nitrate solution is prepared on a hot magnetic stirrer, at 100°C, to reach stability and a homogenous mixture.

This last procedure gives a better control of stoichiometry because the water content of nitrates is not always exactly known.

In the *ex-situ* process, the nitrates (or carboxylates) are deposited at relative low temperatures (<500°C), and then treated at high temperature to react the precursor and to form the right phase. The crystalline quality and the J_c values obtained by this method are not very good.
Recent studies [115, 116, 117, 118] have shown that biaxial texturing of YBCO can be reached using in-situ nitrate deposition process on single crystalline substrates and on biaxially textured Ag substrates. The in-situ process can be summarized in the thermal profile shown in Figure 27. The films are deposited at 850°C. This deposition is followed by in-situ thermal treatments (sintering in oxygen at the same temperature than deposition, and annealing at 525°C)

The thermogravimetric measurements of the nitrates decomposition have been intensively studied to establish the reaction paths for individual and mixed components (Figure 28). The decomposition temperature of the various nitrates used for HTS films formation is very different. Cu nitrates decomposes at around 200°C, $Y(NO)_3$ around 400°C, and Ba nitrate decomposes at temperatures higher than 800°C. The solution concentration has to be adapted as a function of the deposition temperature: the Ba/Y ratio remains constant for all temperatures, but the Cu/Y ratio increases rapidly at deposition temperatures of above 300°C [116], due to the low decomposition temperature of Cu nitrate.

An all-nitrate precursors approach has allowed to grow epitaxial HTS oxide film on $SrTiO_3(001)$ substrate at a high deposition rate and at high temperature[119, 120]. 1µm-thick YBCO films were prepared in-situ at 850°C, form a precursor solution with a final molarity of 0.3 mol/l, reaching very high local deposition rate, of around 12 µm/hr [121]. X-ray diffraction results show that the films do not have only a strong c-axis texture, but also a sharp in-plane biaxial orientation. FWHM values for YBCO (102) and YBCO (005) diffraction lines can be as low as $\Delta\varphi\sim1.4°$, and $\Delta\omega\sim0.4°$, respectively. Furthermore, the critical temperature (T_c) and the critical current density (J_c) of this film are 91K and 1.4 x $10^6 A/cm^2$ (at 77K, 0T), as presented in Figure 29. The critical current map included in this figure, shows well-connected grains, but also inhomogeneities on a longer grain distance (~2000 µm). Average J_c value for this film (1-µm thick) is 1.1 MA/cm², while local J_c can be as high as 3.6MA/cm².

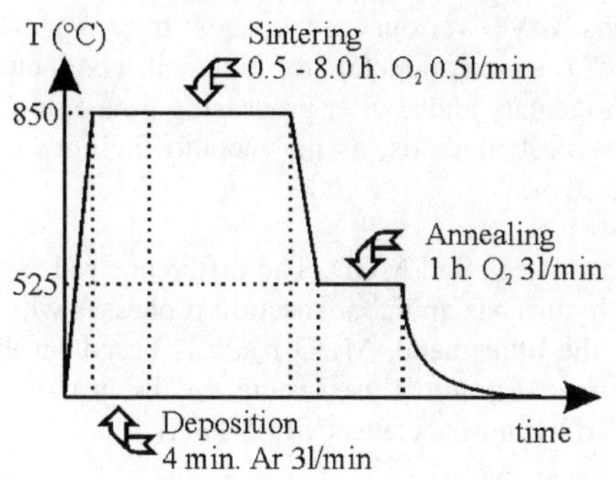

Figure 27: Conditions for the fabrication of YBCO films by in situ spray pyrolysis. Thermal profile for the deposition step and post-deposition annealing.

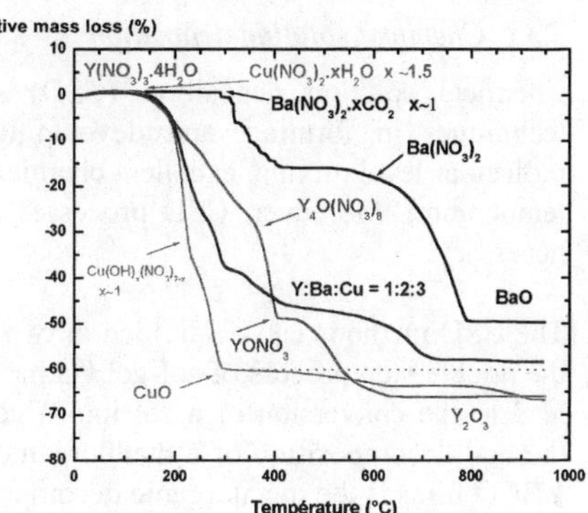

Figure 28: Thermogravimetric analysis of the nitrates used for YBCO deposition in the all-nitrates spray pyrolysis process.

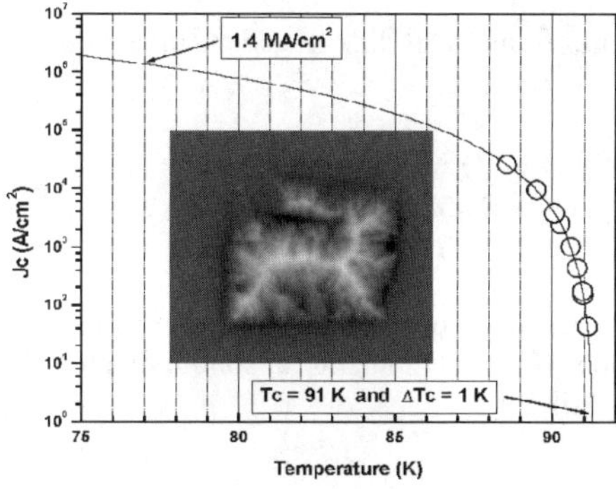

Figure 29: Tc, Jc and critical current profile of YBCO film obtained by all-nitrate spray pyrolisys on single-crystal SrTiO₃.

YBCO thick films have been grown on different single crystal substrates with high current performance (J_c >1 MA/cm²) and on Ag single crystals or biaxially textured tapes. The deposition conditions are being currently adapted to buffered coated conductor substrates. First attempt to deposit Y_2O_3 layers has shown that the growth rate is an important factor on the crystalline quality of the films [119].

The scaling up of this process for producing long length-coated conductors in a reel-to-reel equipment is currently under development.

144

3.3   Chemical solution deposition

Chemical solution deposition (CSD) methods have various advantages over vacuum techniques in forming, and developing YBCO superconductor to practical conductor: molecular level mixing, excellent chemical homogeneity and shorter processing time at lower temperature. Moreover, CSD processes are low cost methods, as no vacuum facilities are necessary.

The CSD methods can be divided in two routes: sol-gel, and MOD. The difference relies on the double-step process of sol-gel via based on hydrolysis and condensation processes which lead to the conversion of a sol into a gel. On the other hand, MOD route is based on the thermal decomposition of a metalorganic precursor. The most used route for the growth of YBCO films is the metalorganic decomposition of trifluoroacetates (MOD-TFA).

3.3.1   Sol-gel method

The term sol-gel processing designs the preparation of oxide material by a wet-chemical route. In sol-gel processes, the precursors consist of a metal surrounded by ligands. The ligands are organic molecules; the most widely used in sol-gel synthesis are alkoxides. The solution converts to a colloidal suspension of particles (with size between 1nm and 1μm) in a liquid named sol. The following reactions take place during the solution to sol conversion:

$$M(OR)_4 + H_2O \rightarrow HO\text{-}M(OR)_3 + ROH \qquad \text{(hydrolysis)}$$

The hydrolysed or partially hydrolysed molecules can link together in a condensation reaction:

$$(OR)_3M\text{-}OH + HO\text{-}M(OR)_3 \rightarrow (OR)_3M\text{-}O\text{-}M(OR)_3 + H_2O$$
$$(OR)_3M\text{-}OR + HO\text{-}M(OR)_3 \rightarrow (OR)_3M\text{-}O\text{-}M(OR)_3 + ROH$$

Successive hydrolysis and condensation reactions form oligomer and finally inorganic polymers during the sol to gel conversion. The gel contains a solid polymer skeleton enclosing a continuous liquid phase (Figure 30).

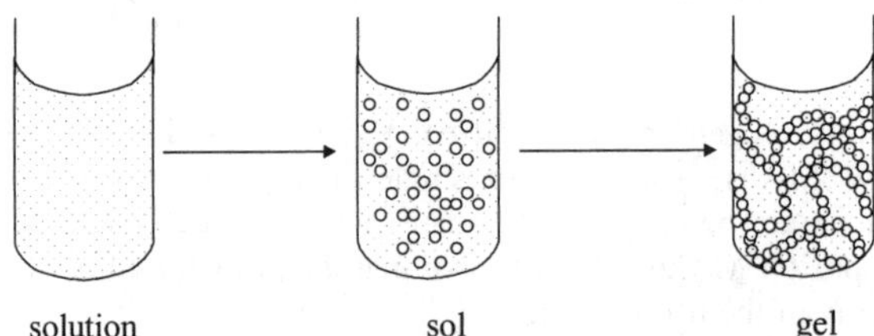

Figure 30: schematic representation of sol and gel formation.

In sol-gel route, the precursor is applied to the substrate surface by dip coating or spin coating. Then, the evaporation of solvent takes place or even is promoted to condensate the film, while this densification approaches the reactive species. For further densification and crystallisation, samples are usually treated at high temperature in a controlled atmosphere. To increase the final thickness, sequential wetting and drying cycles can be performed.

When using dip coating method in continuous reel-to-reel system, the process can be dividing into three steps: dipping, withdrawal, and heating. The film thickness depends on concentration, viscosity of solution, and withdrawal rate [122].

Textured dense and crack free RE_2O_3 (RE= La, Pr, Sm, Er, Ce, Er, Gd, Tb, Ho, Yb, and Lu) buffer layers, and perovskites such as $SrTiO_3$, $LaAlO_3$, $La_2Zr_2O_7$, $PbTiO_3$ and $BaZrO_3$ have been successfully grown on nickel substrates by sol-gel processing [123, 124]. Paranthaman et al have produced 1-2meter lengths of epitaxial Eu_2O_3 and $La_2Zr_2O_7$ buffer layers on nickel tapes [125, 126]. Critical current values of $J_c = 4.8 \times 10^5$ A/cm^2 and $J_c > 10^6$ A/cm^2 have been obtained in respectively LZO(sol-gel) /YSZ(sputtered) /CeO$_2$(sputtered) /YBCO(e-beam coevaporation) and Eu_2O_3(sol-gel) /YSZ(sputtered) /CeO$_2$(sputtered) /YBCO(e-beam coevaporation). This process has been scaled-up at American Superconductors for the production of 30 nm Gd_2O_3 seed layer on nickel substrates.

For YBCO compounds, some difficulties arise due to the formation of barium carbonates ($BaCO_3$) instead of oxides during the firing process. Moreover in the case of YBCO, copper alkoxides are not very soluble in organic solvents and Yttrium alkoxides (and the other rare earth alkoxides) are readily hydrolysed by very small amount of water. Thus the preparation of homogenous solution with the required formula is difficult.

Yttrium trimethyl acetate, barium hydroxide, and copper trimethyl acetate dissolved in a mixture of propionic acid and amine. Xylenes of alcohols were used for dilution and for controlling solution viscosity in [127]. Spin-coated films were dried at 200-250°C for several minutes. This process was repeated. The final film was heated at 800-920°C under controlled atmosphere, and O$_2$-annealed at 400°C for 24 h. Films deposited on YSZ and LAO exhibited sharp Tc values near 90 K, with J^c values above 10^5 A/cm^2. D. Marguillier *et al* used Y, Ba, and Cu acetates in acetic solution, adjusting the pH with ammonia [128].

Mutlu et al [122] proposed another sol-gel via based on Yttrium isoproxide, Ba2-ethylhexanoate, and Cu (II) metoxide precursors separately dissolved in glacial acetic acid and triethanolamine. Methanol and ethanol were added to dilute the solution. A continuous reel-to-reel system was used to coat YBCO films on sol-gel $SrTiO_3$ buffered nickel tapes. After dipping the tapes, a drying step at 200°C and a heat treatment at 500°C take place. These steps are repeated up to 15 times. Then samples are annealed at high temperature and oxygenated. No J$_c$ values are given in this work.

3.3.2 Metalorganic decomposition (MOD)

The metalorganic decomposition (MOD) method consists on combining precursors of the desired metals in the corresponding ratio, which are deposited on substrates, and this precursor film is transformed in the final material by high temperature thermal decompositions. The decomposition then yields an atomic level mixture of the oxides/carbonates. MOD uses as precursors a metal linked to organic groups, most commonly metal carboxylate and modified metal alkoxides. The MOD ligands are separated from the precursor by thermolysis; therefore, compounds that sublime, and evaporate are not suitable as MOD precursors.

It has been shown that the MOD technique has a very high flexibility in preparing simple oxides buffer layers (CeO_2, YSZ) and binary compounds such as $LaAlO_3$, $SrTiO_3$ or $BaZrO_3$ [129, 130, 131].

3.3.3 Metal organic decomposition of trifluoroacetates (MOD-TFA)

Among the chemical solution processes, metal organic decomposition process using trifluoroacetates (TFA-MOD), first developed by Gupta *et al* [132] in 1988, appeared recently as very appealing alternatives to grow biaxially textured YBCO films in combination with textured metallic substrates [133].

Usually, the thermal decomposition of the MO precursors leads to the formation of metal carbonates and the corresponding ketones, both of which decompose further following their own thermodynamic characteristics. In Y-Ba-Cu-O-C system, relatively stable $BaCO_3$ forms as intermediate compound during the Y-123 synthesis. In order to avoid this problem, trifluoacetate are used as a metal precursor. The barium, yttrium, and copper trifluoracetates decompose into oxygen and oxy-fluorides, which are later converted to their oxides by reaction with a moist atmosphere. The chemistry of this system is similar to that present in the electron-beam-coevaporated of BaF_2, Y, and Cu used in the barium fluoride process [134]. McIntyre et al have extensively studied the kinetics of transformation on (001) $SrTiO_3$ and (001) $LaAlO_3$ substrates [135, 136, 137].

Recently a lot of work has been performed in the application of the MOD-TFA method to fabricate YBCO-based coated conductors. This process involves several steps: solution preparation and deposition, the pyrolysis of the metalorganic precursors leading to nanometric phases, and the growth process of the epitaxial crystalline phase. The thermal process requires water vapour treatments during pyrolysis (or calcining), and growth (or annealing at high temperature) processes. Water vapour suppresses the volatilisation of Cu trifluoracetate in the calcining process, and it decomposes the oxyfluoride precursors in the firing process [135]. A typical schedule of YBCO films preparation by the TFA-MOD method is shown in Figure 31 [138]. A lot of efforts [139, 140] have been made until now to understand the influence of the large number processing parameters on the final microstructure and superconducting properties. Nevertheless, a complete understanding of these factors is not established yet.

Araki *et al* have worked on a purification method of the TFA solution called the solvent-into-gel method. Figure 32 shows the process for the preparation of a pure coating solution using this method. By replacing impurities in blue viscous solution with solvent, highly purified coating solution is obtained. Impurity rate decreases from about 5% to 0.25% [141]. With this coating solution, critical current density values of $Jc=7X10^6$ A/cm^2 have been obtained for very thin films (200 nm) on single crystalline $LaAlO_3$ substrates.

Araki *et al* have also studied the effect of the time of the calcining process (Figure 33). They showed that long time heating between 200 and 250°C induces CuO formation in the YBCO film which deteriorates J_c. They propose a growth model based on the existence of quasi-liquid (Figure 34). They define the quasi as an intermediate material having a large mobility derived from CuO, amorphous matrix of Y-Ba-Cu-O-F and water vapour [142].

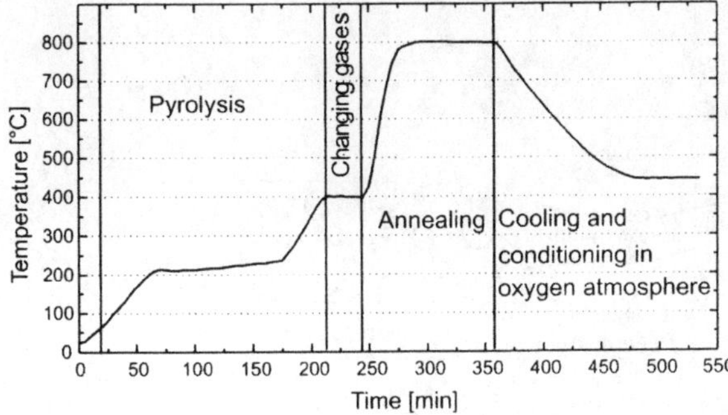

Figure 31: typical heating schedule of the YBCO annealing. The heating starts with dry oxygen. At 60°C water vapour is added and the atmosphere is kept humid until the end of the dwell time at 800°C from [138].

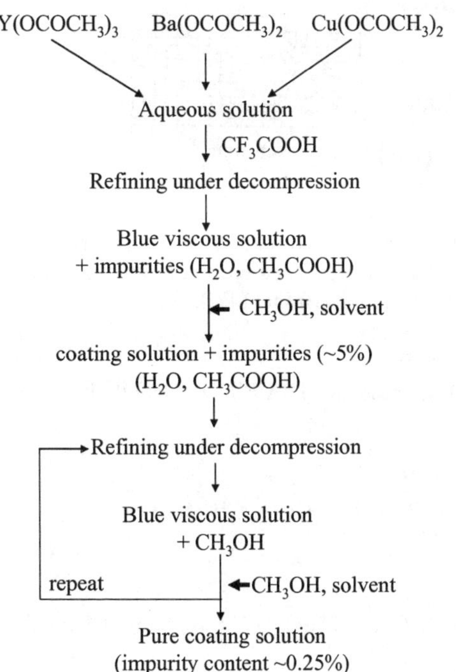

Figure 32 : process for preparation of a coating solution using the Solvent-into-gel (SIG) method (from [140])

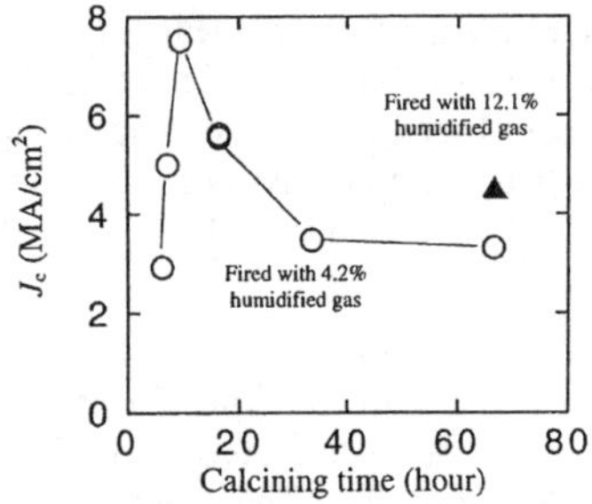

Figure 33: Dependence of calcining time at 200-250°C on J_c of the fired films (open circles) (from [142])

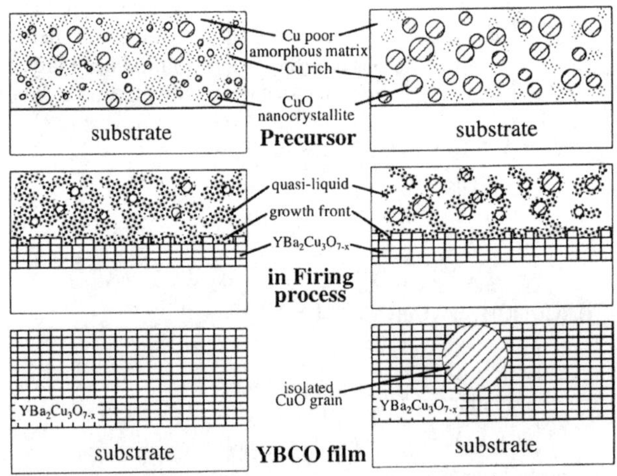

Figure 34: Growth mechanism of YBCO grains prepared by TFA-MOD in the firing process. (a) Proper time calcined precursor and (b) long time calcined precursor. Precursor (b) has large and much CuO crystallites (from [142]).

During the formation of the $YBa_2Cu_3O_{7-x}$ phase on TFA-MOD method, two chemical reactions are considered to take great roles:

$$BaF_2 + H_2O \rightarrow BAO + HF$$
$$2\,BAO + \tfrac{1}{2}\,Y_2Cu_2O_5 + 2\,CuO + x\,O_2 \rightarrow Y\,Ba_2Cu_3O_{7-\delta}$$

which can be expressed in only 1 reaction

$$2BaF_2 + \tfrac{1}{2}\,Y_2Cu_2O_5 + 2\,CuO + 2H_2O \rightarrow Y\,Ba_2Cu_3O_{6.5} + HF\ [139]$$

The advancement of this reaction is mainly controlled by temperature and water pressure P_{H_2O}, while oxygen pressure P_{O_2} shifts the stability limits of the $YBa_2Cu_3O_{7-x}$ phase. It is necessary that BaF_2 reacts with water vapour entirely to have no residual BaF_2 in films and good J_c [143]. Castaño *et al* [144] found, for this step, a strong influence of the temperature on the porosity of the films, related to c-axis grains concentration.

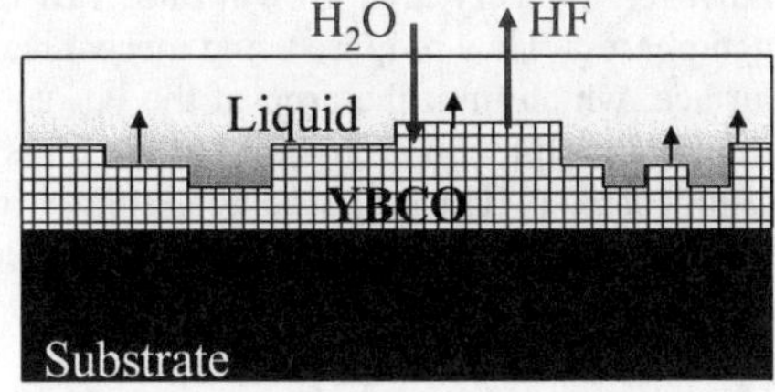

Figure 35: Schematic representation of YBCO formation during heating in the TFA-MOD process

Difficulties for H_2O and HF diffusion increase with the thickness of the layer (Figure 36), leading to a decrease of critical current density [145]. Solovyov *et al* have studied the kinetics and the growth rate limiting mechanisms on the ex-situ processing [146, 147]. As previously commented, the chemistry of these two systems is similar, because the essential factor is the presence of BaF_2, and consequently the release of HF. They found that the growth rate of the superconducting phase remains constant during heat treatment, and it is proportional to the square root of P_{H_2O} in the processing atmosphere. The decomposition of an oxifluride compound in the precursor film occurs at the liquid layer-precursor film interface. The rate of this decomposition reaction defines the growth rate of $YBa_2Cu_3O_7$. The rate of this reaction, for the usual water vapour pressure (25-200 Torr), is limited by the out-diffusion of the decomposition product HF, and in more extent, by the rate of removal of HF from the surface of the precursor film into the reaction chamber.

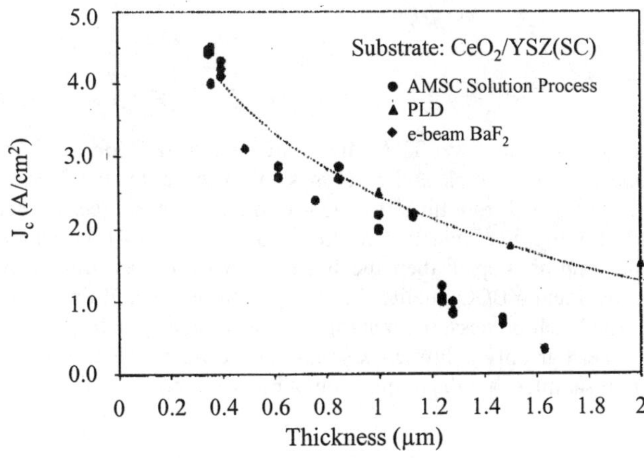

Figure 36: Jc (77K, 0T) of YBCO grown from MOD-TFA process on a CeO$_2$/YSZ/single creistal substrate as a function of film thickness. Data for YBCO films deposited by PLD and e-beam BaF$_2$ are shown for comparison (from [145]).

150

Solovyov *et al* have also studied the influence of the sample size, essentially for long tapes. [148]. The model used predicts that processing in a tubular reactor at atmospheric pressure will require inordinately large gas flows. They have also calculated the limitation in a partial vacuum reactor, compared to experimental results. At lower pressure, precursor conversion reaction can be more efficient and homogeneous. In fact, in this case it is possible to achieve very high growth rate at very low P_{H2O} values; however, for very high growth rate, YBCO nucleates as a mixture of random and a-axis oriented phases. They propose using mechanical restrictions (as ceramic barriers) under the tape surface, which limit the drop of the P_{HF} and control the growth rate. The compromise between time of production and quality of the films lead to growth rates of around 0.2 nm/s. Yoshizumi *et al* (MIT) proposed at Gattlinburg 2002, seeding first the substrate at low P_{H2O}, and then increase the growth rate using higher P_{H2O}, which would increase the production rate in a factor of 12.

Araki *et al* have also worked on the optimisation of the deposition conditions on metallic substrate. On IBAD substrates with a CeO_2 cap layer, they obtained a J_c=1.7 x 10^6A/cm^2 (77k, 0T) [149, 150] for 0.15 µm YBCO layer. More recently Araki *et al* showed the presence of a $BaCeO_3$ layer between YBCO and the CeO_2 buffer layer. Controlling the firing process it is possible to avoid the formation of this layer and to obtain a J_c of 2.5 x10^6 A/cm^2 in 0.24 µm thick YBCO films on CeO_2 and YSZ (IBAD) buffered metal (hastelloy-C) tapes [151].

Siegal *et al* reported J_c values of 1.3 x10^6 A/cm^2 in coated conductors fabricated with solution chemistry for both buffer and superconducting layers deposited on nickel RABiTS [152]. J_c values of 2 x10^6 A/cm^2 have been achieved in 0.4 µm thick YBCO films on $Ni/CeO_2/YSZ/CeO_2$ substrates [145].

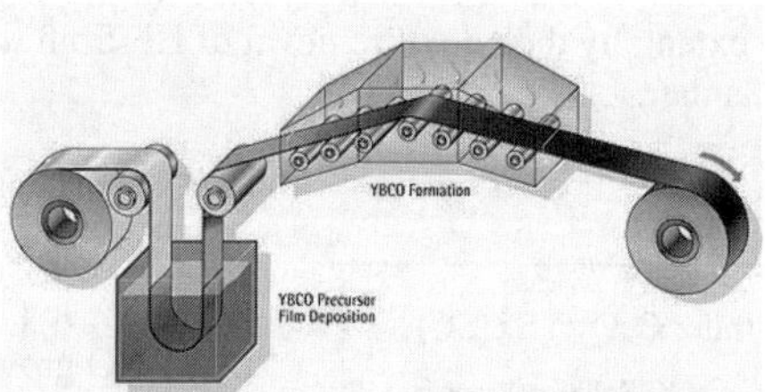

Figure 37: Reel-to-reel processing scheme for fabrication of continuous lengths of YBCO tapes by MOD from [153].

Currently, continuous production systems by TFA-MOD are being developed. Two configurations have been proposed: reel-to-reel, and batch system. The volumetric YBCO growth rate is defined as the linear growth rate times the deposition area. In the reel-to-reel process the tape moves linearly through the annealing reactors, so only a part of the whole sample is treated at once. The limiting step is then the linear growth rate, as previously explained, related with HF removal and YBCO quality. In this system there is not limitation of the length to be treated. For the batch process, the material is wound on a cylinder, and all the material is elaborated at once, but only a limited amount can be treated. The time of process is the same than for small samples, but the output cannot be increased.

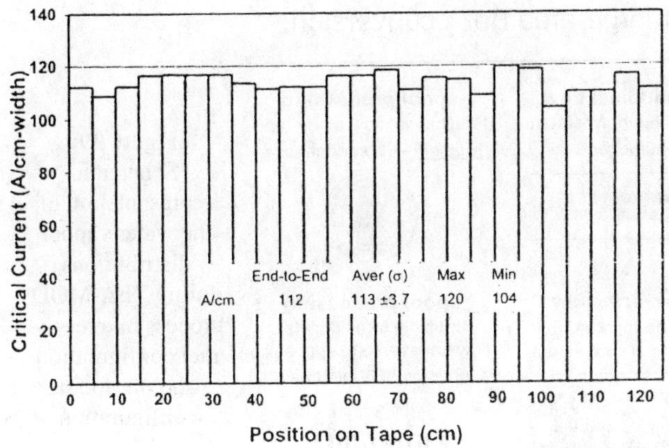

Figure 38: Critical current vs. length for 1.25 meter YBCO tapes with Ni-5at%W substrate. Statistical data is included for Ic measurements at 77K, self field taken at 5 cm intervals from [153].

Reactors in reel-to-reel configuration have been already constructed at American Superconductors (AMSC); Figure 37 shows a scheme of this kind of reactor. Rupich *et al* have obtained uniform current densities on 1.25-meter long (TFA-MOD) YBCO coated conductor with the architecture: RABiTS Ni alloy/Y_2O_3/YSZ/CeO_2/YBO/Ag [153] (Figure 38). Tests on 10-meter-long wires produced at AMSC gave performance levels over 100 A/cm-width (http://www.amsuper.com/wirefact.htm).

The batch or spool approach for continuous production has been adapted at Oak Ridge National Laboratory (ORNL) and at ISTEC. This type of process allows homogenising the atmosphere in order to optimise the HF removal. As it can be seen in Figure 39, the water vapour distribution on the reel-to-reel system depends on the flow stream configuration. When using a longitudinal configuration, the concentration of water vapour decreases and HF concentration increases with length; the growth conditions changes during processing, leading to low quality films. When using a transversal configuration, multiple inlets for water vapour and outlets for HF vapours are needed in all the processing area. On the other hand, tape width should be smaller enough to ensure a homogeneous lateral distribution.

When using spool processing, multiple inlets are used, longitudinal to the tape length, but the rotation movement allows homogenizing the processing atmosphere. At ORNL, 6-m tapes have been processed in this system. Another advantage of the batch configuration is that vacuum or low pressure processes are easily adapted.

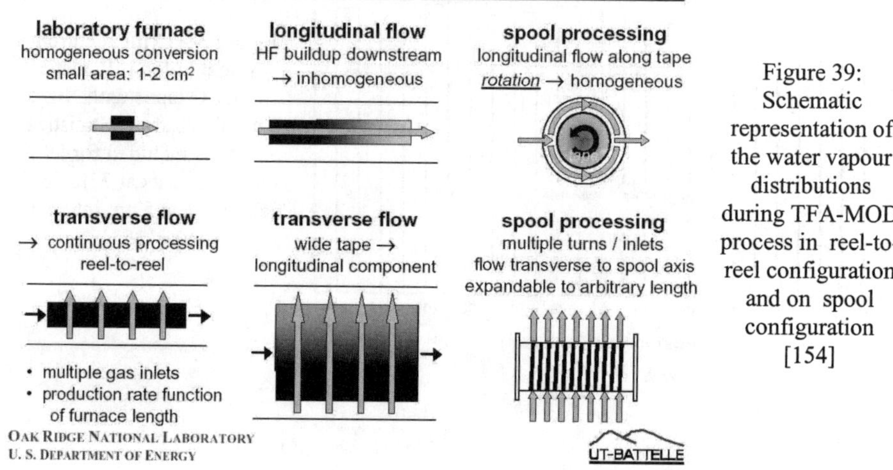

Figure 39: Schematic representation of the water vapour distributions during TFA-MOD process in reel-to-reel configuration and on spool configuration [154]

Many improvements have been made last four years, and the MOD-TFA process is considered to be one of the promising methods for elaboration of coated conductors. Figure 40 shows historical improvement in critical current density of YBCO films grown from a TFA-MOD process as a function of film thickness.

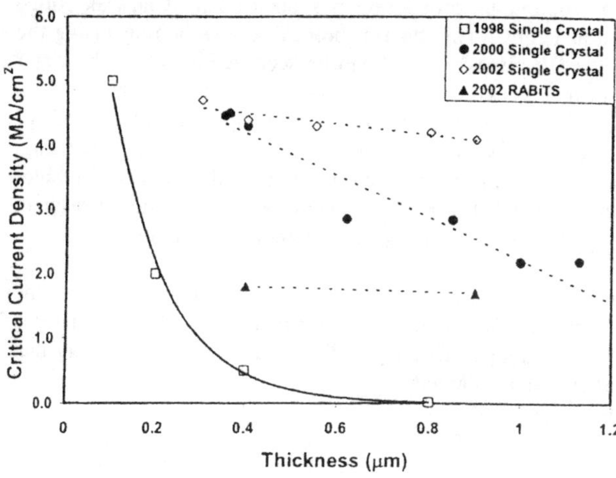

Figure 40: Historical improvement in critical current density of YBCO films grown by TFA-MOD on single crystal and RABiTs as a function of film thickness (from [153]).

3.4   Chemical Deposition of other HTS

Tl and Hg based oxide superconductors like $TlBa_2Ca_2Cu_3O_z$ (named Tl-1223) with T_c =115 K, and $(Hg, Re)Ba_2Ca_2Cu_3O_{8+x}$ (named (Hg, Re)-1223), with T_c= 134 K, are potential candidates for high current applications. Interested readers can find more information about Tl-based HTS phase in [155, 156, 157] and Hg-based HTS phase in [158, 159]

3.4.1   Tl-BCCO films:

The fabrication of thin films of the Tl-based HTS materials involves a compromise between the high temperatures needed to form the superconducting phases [160], and the high volatility of thallium at these temperatures. To circumvent this problem, the TBCCO film fabrication is performed by a two-step process:

i)   Deposition of a precursor layer of Ba, Ca and Cu by usual deposition methods. Sometime this precursor layer contained already thallium [161].

ii)   Ex-situ thallination anneal in the presence of thallous oxide vapour.

The Tl-1223 phase is of interest for power applications due to its current-carrying capability in high magnetic fields. Concerning the double TlO layer phase, high performance Tl-2212 films [162] (T_c=110 K) have been prepared for applications in passive microwave devices. However, the higher Tc of the Tl-2223 phase (Tc=125 K) suggests that devices with Tl-2223 films could operate at higher temperature.

Tl-1223 and Tl-2223 films [163, 164] have also been synthesized using spray pyrolysis (see section 3.2) from an all-nitrates solution, followed by an ex-situ thallination based on the conditions established for the preparation of bulk samples [160]. This method has deposition rate faster than vacuum techniques. For the precursor layers with a composition Ba:Ca:Cu=2:2:3 were obtained from a solution with a composition 2:1.8:0.4 [164]. As already mentioned, this difference in composition is due to the differences in decomposition temperatures of the nitrates: Cu between 150 and 300°C, Ca between 600 and 650°C, and Ba just below 800°C. These decomposition temperatures lead to deposition at temperatures above 800°C. The ex-situ thallination was carried out in a sealed quartz tube under an oxygen pressure of 0.5 bar at 900°C for Tl-1223, and 850°C for Tl-2223. The Tl source is a pellet with a nominal composition Tl-1223 or Tl-2223, placed in contact with the layer. In these conditions, Tl-1223, could be obtained relatively as a pure phase (fig 41a). The formation pathway in thin films follows the sequence Tl-2212 =>Tl-2223 => Tl-1223, as for bulk samples [165]. The best superconducting properties obtained so far are T_c = 113 K, J_c = 7.5 x 10^5 A/cm^2 (77 K, 0 T), (fig 41b).

A recent study relies on the effects of fluorine addition: it improves the crystalline quality, but further experiments have to be carried out to improve the homogeneity of fluorine content. Anyway, transport critical current density reached 7 x 10^5 A/cm^2 in these early results [164].

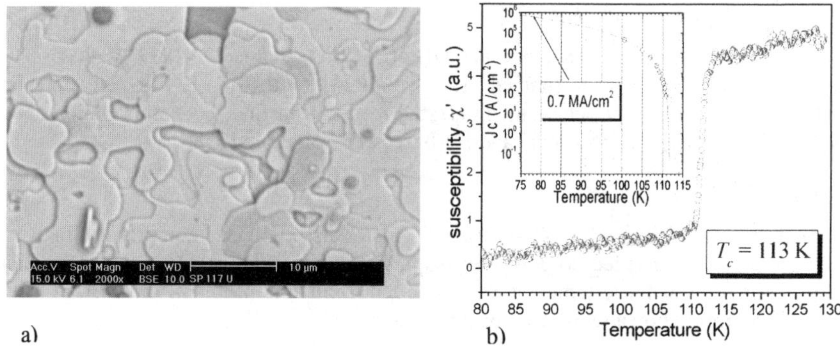

Figure 41: Results from the Tl-1223 film obtained by spray pyrolysis a) SEM picture from backscattered electrons, b) T_c and J_c values obtained by susceptibility measurements

3.4.2   Hg-BCCO films

Mercury-containing cuprates presents the highest Tc at ambient pressure among the superconductors known, in particular $HgBa_2Ca_2Cu_3O_{8+x}$ (Hg-1223) phase has a Tc ~135 K. This makes it a very promising material. High performance can be reached by a proper choice of constituting cations, which should take into account various characteristics of superconductivity as well as sufficient chemical stability to CO_2 and moisture in air. In this aspects, Ba site are normally substituted by Sr or La, and Ca site by Y, Sm, Eu, Gd, Dy, Ho, Er, Tm Yb. For example, the substitution of Sr for Ba ($HgSr_{2-y}BaYCa_2Cu_3O_Z$) increase the stability of the precursor film, but Tc decreases with Sr content. The substitution of the Hg-1223 phase have been studied for the Hg-site mainly with Tl or Pb [166], and in a less extend with Bi and Re. These cations facilitate the stabilisation of the Hg-1223 phase, allowing a diminution of the synthesis pressure. In particular, the substitution with Re is very efficient in preventing Ba carbonatation of the precursor film [167]. On the other hand, it optimises the hole doping by supplying the right amount of oxygen. In thin films, Re substitutions enhances the superconducting properties, which probably relies on a improved metallization of the charge transfer block (Hg, Re)$O_δ$ [168].

S.O. Klimonsky et al [166], have synthesized (Hg, Pb)(Sr, Ba)$_2$Ca$_2$Cu$_3$O$_z$ films in a two step procedure using alternatively PLD or MOCVD techniques. The precursor film has to be deposited at relative high temperature (600°C), and MOCVD technique appears to be more appropriated for such high temperature deposition. 1μm-thick film of Sr-rich (Hg,Pb,)-1223 phase with J_c of 1 MA/cm^2 (77 K, 0T), and 4 x10^{4}A/cm^2 (77 K, 4 T) were reproducibly prepared.

Recently, thick films of (Hg,Re)Ba$_2$Ca$_2$Cu$_3$O$_{8+d}$ have been produced using an aerosol technique for the spray pyrolysis of the precursor layer [169, 170].

The films are processed in three main steps:

i)   Preparation of the precursor film by the aerosol technique (or any other thin film deposition method)

ii)   Preparation of the Hg-source

iii)   Formation of the Hg-1223 film through reaction of the precursor with Hg (vapour) by gas phase diffusion in a closed quartz tube at a total pressure close to 30 bars.

An aqueous solution containing Re, Ba, Ca, and Cu nitrates with a concentration fixed to 0.3 M is used for the precursor layer deposition. After deposition, no special precautions are required to store the precursor film due to the chemical stabilization of the Ba against CO_2 contamination owing to rhenium activity. The source of mercury is a un-reacted pellet of (Hg,Re)-1223. The precursor film is placed in face-contact with the un-reacted pellet and wrapped all together in a gold foil. This is inserted in a quartz tube connected to a thermobaric analyser technique (TBA) and closed under vacuum. The principle of the TBA sensor is based on the elastic deformation of the quartz as a function of pressure.

Films of 1-2 μm thickness were epitaxially grown by this technique on MgO, and presented a $T_c{\sim}130$ K. X ray diffraction data show a good c-axis orientation. In these films, the critical current density can be as high as $J_c{\sim}4.4\,x10^5$ A/cm^2 (77 K, 0 T) [171].

Thin films (~350 nm) on MgO presented also a c-axis orientation ($\Delta\omega = 0.7°$) and in plane texture. The intragrain critical current density determined from d.c. magnetisation reaches 1.6 MA/cm^2 at 77 is self-field [169]. The surface morphology of one of these thin films is presented in Figure 42.

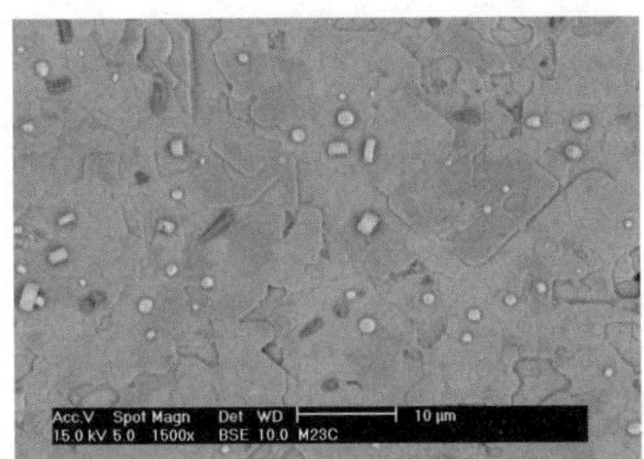

Figure 42: SEM micrograph of a Hg-1223 thin films prepared by spray pyrolysis

(Courtesy of D. De Barros)

4   PROCESS-RELATED TECHNICAL ISSUES

All the Chemical Deposition Methods presented in this study have shown that they are able to produce high quality HTS films either on single-crystalline (LAO,STO…) substrates or on technical metallic substrates with J_c over 1MA/cm^2 in most of the cases.

The major issues for the economical manufacturing of long length YBCO coated conductors pertain to the capability of uniform, epitaxial deposition at high rates over a large area with a continuous process. Such capability has to meet the other important requirements of proper stoichiometric composition and epitaxial structure. The volumic growth rate Vr (growth speed x covered surface aera A) is a major parameter splitting the different deposition techniques. Vr will fix the whole production time of a given coated conductor as well as its average effective cost sums-up the performance of various deposition techniques, extrapolated to the largest deposition zones actually covered, and gives a comparison between Physical deposition techniques and Chemical deposition techniques.

Table 4: Deposition technique and deposition rate

	r (µm/hr)	Vr (µm.cm^2/hr)	A (covered surface area cm^2)
Pulse Laser Deposition	3 - 7	25 - 60	10-20
Magnetron Sputtering	0.3	25 - 100	80 (4")
Thermal Evaporation	1.2	400	320 (8")
PLD - IBAD for buffer		36	
MOCVD	5 -10	250 - 500	50
MOCVD (extrapolated)	40	15000	400
MOD, spray pyrolysis	>40	> 15 000 !!!	> 400 !!

In Chemical Processes, the covered surface area can normally be extended to large sizes by simple chemical engineering techniques and a proper gas phase distribution. The local growth rate can easily be a factor 10 higher than in physical deposition processes. Chemical Processes are therefore much faster than Physical processes and are certainly better adapted for mass production of long length coated conductors. As an example, a 400cm^2 MOCVD reactor (as shown in Table 4 will be able to produce a 1cm wide coated conductor tape transporting 100A (1µm YBCO with $J_c=10^6$A/cm^2) at a speed of 150m/hr.

For each of the processing option presented, it is finally very important to assess clearly the cost of chemicals, the mass transfer (reactor yield) and reaction kinetic data, the need and cost of diagnostic and control elements and the environmental issues, in order to select the best process for coated conductor production.

Concerning the cost of chemicals, it appears in Table 5 that the nitrate route is certainly the cheapest one. Equipment (no vacuum), production and maintenance costs of this "green" process strengthen the assess that, among the Chemical deposition techniques, it is actually the only one able to reach the 10€/KA.m criterion fixed for coated conductors.

Table 5: Cost comparison of chemicals (raw estimations) for three 1-µm-thick layers (YSZ, CeO$_2$ and YBaCuO) on a 1cm wide and 1-meter-long tape, with a reactor yield of 50%.

COSTS in €	Nitrate spray pyrolysis	MOCVD (tmhd)	MOD (TFA)
Research cost	0.052	31.3	75
Production cost	0.018	3.1	7.5
Cost / KA.m	0.18	31	75

5 SUMMARY

The main aim of this chapter is to present the state of art of the chemical deposition process for the elaboration of HTS coated conductors. Nowadays, two kind of technical substrates are available: polycrystalline metal substrates with artificially textured buffer layer (obtained by IBAD, ISD, MBS and ITEX), and biaxially textured substrates (CUTE and RABiTS), essentially Ni and Ag-based tapes. An overview of the current status of substrates fabrication in terms of long length and quality is given. At present, the best results in terms of J_c are obtained on substrates prepared with the IBAD technique. Description of some limiting factors in the use of these technical substrates for YBCO deposition, as well as the current proposed solution, are also given. Some of these factors are chemical interaction or interdiffusion with YBCO, compatibility of thermal expansion coefficient, or lattice matching. SOE route, i.e. epitaxial growth of a NiO layer on biaxially textured Ni, is an alternative. From the application point of view, mechanical strength and magnetism are also crucial parameters to take into account; the use of nickel-refractory metals microalloys or alloys, and other composites has been also envisaged.

The main chemical deposition processes involved in the deposition of buffer layers and YBCO have been described. These processes can be classified in two main groups:
> - one-step processes, as Metal Organic Chemical Vapor Deposition (MOCVD) for which the superconducting phase is formed directly after chemical reactions from the gas phase on the surface of the substrate,
> - two-step processes, as spray pyrolysis and chemical solution deposition (sol-gel and metal organic decomposition), which are based on the deposition of an unreacted precursor film, followed by a recrystallisation step where solid state reactions induce the formation of the superconducting phase.

The main technical issue in MOCVD is stable precursors sources. Several single-liquid sources are presented with their specifications. More important results on buffer layers and YBCO depositions are given to show the performance of these systems. Advances of MOCVD route on the continuous deposition mode are summarized.

A brief description of the basic principles involved in spray pyrolysis is presented, together with the results obtained up to now. For the chemical deposition methods, we give a fast review of the main results obtained by sol-gel methods. The metal organic decomposition of trifluoroacetates is at the moment one of the most studied route for coated conductor fabrication. Some parameters are currently well known, but several aspects need further investigation to be controlled. Some problems, as for example the decrease of J_c with films thickness related to fluorine diffusion, are discussed. These physical aspects are also related to technological aspects; as for example the two approaches used in continuous mode production.

Some results concerning the chemical deposition processes of other families of superconducting oxides (Tl- and Hg-based superconductors) are also presented.

All the Chemical Deposition Methods presented in this study have shown that they are able to produce high quality HTS films, either on single-crystalline substrates or on technical substrates, with J_c over 1MA/cm^2 in most of the cases. At present, only the MOCVD and MOD method are able to produced long length tapes; nevertheless spray pyrolysis nitrate route seems also very promising as the cheapest one.

GLOSSARY

AAMOCVD Aerosol assisted Chemical Vapour Deposition
BEMOCVD Band Evaporator Chemical Vapour Deposition
CUTE: Cube texture
CVD: Chemical vapour deposition
Hg-BCCO: mercury barium calcium copper based HTS
HTS: High temperature superconductor
IBAD: Ion beam assisted deposition
IMOCVD: Injection Chemical Vapour Deposition
ISD: Inclined substrate deposition
ITEX: Ion beam nanotexturing
MBS: modified bias sputtering
MOCVD: Metalorganic Chemical Vapour Deposition
MOD: Metalorganic Decomposition
PIMOCVD: Pulsed injection Chemical Vapour Deposition
PLD: Pulsed laser Deposition
RABiTS: Rolling assisted Biaxially textured substrates
RISD Reverse inclined substrate deposition
SOE: Surface oxidation epitaxy
TFA: Trifluoracetates
Tl-BCCO: thallium barium calcium copper based HTS
Tmhd: tetramethylhepanedionates
Thd: see Tmhd
YBCO: $YBa_2Cu_3O_{7-x}$
YSZ: Yttrium Stabilised Zirconia

REFERENCES

[1] D. Dimos, P. Chaudhari, J. Mannhart and F. K. LeGoues, Phys.Rev.Lett. 61 (2), [1988] 219

[2] H.Hilgenkamp and J. Mannhart, Rev. Mod. Phys 74, [2002] 485.

[3] Y. Iijima, N. Tanabe, Y. Ikeno and O. Kohno, Physica C 185 [1991]1959.

[4] Y. Iijima, K.Onabe, N. Futaki et al. J.Appl.Phys. 74 (3) [1993] 1905.

[5] M. Fukutomi, S. Aoki, K. Komari et al., Physica C 219 [1994] 333.

[6] M. Bauer, R. Semerad and H. Kinder, IEEE Trans.Appl.Supercond. 9 (2) [1999] 1502.

[7] A.Goyal, D.P. Norton, D.M. Kroeger, D.K. Christen, M. Paranthaman, E.D. Specht, , J.D. Budai , Q. He, S. Saffian, F.A. List, D.F. Lee, E. Hatfield, P.M. Martin, C.E. Klabunde, J. Mathis, C. Park J. Mater. Res. 12 [1997] 2924.

[8] J.L. MacManus-Driscoll, Annu. Rev .Mater. Sci. 28, [1998] 421.

[9] Y. Ijjima, N. Tanabe, Y. Ikeno O. Kohno. Physica C vol 185-189 (1991) 1959-1960.

[10] Y. Iijima, K. Onabe, N. Futaki, N. Tanabe, N Sadakata, O. Kohno and Y. Ikeno, IEEE Trans. Appl. Supercond. 3, [1993] 1510.

[11] Y. Iijima, M. Hosaka, N. Tanabe, N Sadakata, T. Saitoh, O. Kohno and K. Takeda, J. Mtaer. Res. 12, [1997] 2913.

[12] C. P Wang, K. B Do, M. R Beasley, T. H. Geballe, and R. H Hammond, Appl. Phys. Lett. 71, [1997] 2955.

[13] J. R. Groves, P. N. Arendt, S. R. Foltyn, Q. X. Jia, T.G. Holesinger, H. Kung, E.J. Peterson, R. F. DePaula, P.C. Dowden, L. Stan, L.A. Emmert, J. Mater. Res. 16, [2001] 2175.

[14] J.O. Willis, P.N. Arendt, S.R. Foltyn, Q.X. Jia, J.R. Groves, R.F. DePaula,P.C. Dowden, E.J. Peterson, T.G. Holesinger, J.Y. Coulter, M. Ma, M.P. Maley, D.E. Peterson, Physica C 335, [2000] 73.

[15] J. Dzick; J. Wiesmann; J. Hoffmann; K. Heinemann; A. Usoskin; F. Garcia-Moreno; A. Isaev; HC. Freyhardt. Applied Superconductivity 1997. Proceedings of EUCAS 1997 3rd European Conference on Applied Superconductivity. (1997); vol 2.1001-1004.

[16] M. Fukutomi, S. Aoki, K. Komari R. Chatterdjee and H. Maeda, Physica C 219, [1994] 333

[17] K. Hasegawa, N Yoshida, K Fujino, H. Mukai, K. Hayashi, K. Sato, T. Okhuma, S. Honjo, H. Ishii and T. Hara advances in superconductivity 9, [1997] 745.

[18] W. A. J. Quinton and F. Baudenbacher, Physica C 292, [1997] 243.

[19] K. Hasegawa, N. Yoshida, K. Fujino et al. Proc. of 16th ISEC/ICMC, Amsterdam : Elsevier Science, [1997] 1413 .

[20] M. Bauer, R. Semerad and H. Kinder, IEEE Trans.Appl.Supercond. 9 (2), [1999] 1502.

[21] R.P. Reade, P. Berdhal, and R.E. Russo, Applied Physics Letter 80 (8), [2002] 1352.

[22] T. Doi, N. Sugiyama, T. Yuasa, T. Ozawa, K. Higashiyama, S. Kikuchi and K. Osamura, Advances in Superconductivity 8, ed H. Hayakawa and Y. Enomoto (Tokyo springer) [1996] 903

[23] J. D. Budai, R. T. Young and B. S. Chao Appl. Phys. Lett 62, [1993] 1836.

[24] M. Yamazaki, T. D. Thanh, H. Kubota, Y. Kudo, H. Yoshino and H. Nagamura Advances in Superconductivity 11, ed N. Koshizuka and S. Tajima (Tokyo: Springer) [1999] 789.

[25] Y. Takahashi, K. Matsumoto, S. B. Kim and I. Hirabayashi, IEEE Trans. Appl. Supercond. 9, [1999] 2272.

[26] M. Hasegawa, Y. Yoshida, K. Matsumoto, I. Hirabayashi, M. Iwata, Y. Takai, H. Akata and K. Higashiyama, IEEE Trans. Appl. Supercond. 9, [1999] 2240.

[27] A. Teiserskis, A. Abrutis, Z. Saltyte, C. Jiménez, F. Weiss, J.-P. Sénateur, Hong Li Suo, J.Y. Genoud, R. Flükinger. Proced. Intern. Conf. 'Thin films deposition of oxide multilayers. Industrial-scale processing", TFDOM [2000] 57.

[28] H. Suo, J.Y. Genoud, G. Triscone, E. Walker, M. Schindl, R. Passerini, F. Cléton, M. Zhou and R. Flükiger, Supercond. Sci. Technol. 12 [1999] 624.

[29] H. Suo, J.Y. Genoud, M. Schindl, E. Walker and R. Flükiger, Supercond. Sci. Technol 14 [2001] 854.

[30] H. Yoshino, H. Kubota, M. Yamazaki, Y. Kudo and T. D. Thanh, Physica IEEE Trans. App Supercond 11 [2001] 3142.

[31] H. Yoshino, M. Yamazaki, T. D. Thanh, H. Kubota, , and Y. Kudo, Physica C 357-360 [2001] 923.

[32] A. Goyal, D. P. Norton, J. D. Budai, M. Paranthaman, E. D. Specht, D. M. Kroeger, D. K. Christen, Q. He, B. Saffian, F. A. List, D. F. Lee, P. M. Martin, C. E. Klabunde, E. Hartfield. V. K. Sikka. Appl. Phys. Lett. 69,(12) 1996, 1795-1797.

[33] D. P. Norton, A. Goyal ; J. D. Budai, D. K. Christen , D. M. Kroeger, E. D. Specht, Q. He, B. Saffian, M. Paranthaman, C. E. Klabunde, D. F. Lee, B. C. Sales, F. A. List. Science, Vol 274, (1996), 755.

[34] J.E. Mathis, A.Goyal, D.F. Lee, F.A. List, M. Paranthaman, E.D. Specht, D.M. Kroeger, P.M. Martin.Japan. J. Appl. Phys 37 [1997] L1379.

[35] C. Park, D.P. Norton, D.K. Christen, D. T. Verebelyi, R. Feenstra, J.D. Budai, A.Goyal, D.F. Lee, E.D. Specht, D.M. Kroeger, M. Paranthaman, IEEE Trans. App Supercond 9 [1999] 2276.

[36] A. Ichinose, A. Kikuchi, K. Tachikawa and S. Akita, Physica C 302 [1998] 51.

[37] A. Ichinose, G. Daniels, C. Y. Yang *et al.*, IEEE Trans. Appl. Supercond. 9 [1999] 2280.

[38] D.F. Lee, M. Paranthaman, J.E. Mathis, A. Goyal, D.M. Kroeger, E.D. Specht, R.K. Williams, F.A. List, P.M. Martin, C. Park, D.P. Norton, D.K. Christen Japan. J. Appl. Phys. 38 [1999] L178

[39] M. Paranthaman, D.F. Lee, A.. Goyal, E.D. Specht, P.M. Martin, X. Cui, J.E. Mathis, R. Feenstra, D.K. Christen, D.M. Kroeger. Supercond. Sci. Technol. 12 (1999) 319-325.

[40] T. Aytug, J.Z. Wu, B.W. Kang, D.T. Verebelyi, C. Cantoni, E.D. Specht, A. Goyal, M. Paranthaman, D.K. Christen, Physica C 340 (2000) 33-40

[41] T. Aytug, M .Parnathaman, B.W. Kang, S. Sathyamurthy, A. Goyal, D.K. Christen. Appl. Phys. Lett. 79, No.4 [2001] 2205.

[42] J. Knauf, R. Semerad, W. Prusseit, B. DeBoer, J. Eickemeyer, IEEE Trans. Appl. Supercond 11 [2001] 2885.

[43] R. Nemetschek, W. Prusseit, B. Holzapfel, J. Eickmeyer, B. DeBoer, U. Miller and E. Maher, Physica C 372-376 [2002] 880.

[44] R. Nemestchek, W. Prusseit, B. Holzapfel, J. Eickemeyer, U. Miller, E. Maher. Submitted to ASC2002.

[45] K. Matsumoto, Y. Niiori, I. Hirabayashi, N. Koshizuka, T. Watanabe, Y. Tanaka, and M. Ikeda, Advances in Superconductivity. Proceedings of the 10th International Symposium (Tokio). Vol 2 [1998] 611.

[46] K. Matsumoto, S. B. Kim, J. G. Wen, I. Hirabayashi, S. Tanaka, N. Uno, and M. Ikeda, IEEE Trans. Appl. Supercond 9 [1999] 1539.

[47] D. Selbmann, J. Eickmeyer, H. Wendrock, C. Jimenez, S. Donet, F. Weiss, U. Miller and O. Stadel, Journal de Physique IV. 11 [2001] Pr11-239.

[48] K. Matsumoto, T. Watanabe, T. Tanigawa, T. Maeda, S. B. Kim and I. Hirabayashi, IEEE Trans. App Supercond 11 [2001] 3138.

[49] T. Watanabe, K. Wada, Y. Ohashi, M. Ozaki, K. Yamamoto, T. Maeda and I. Hirabayashi, Physica C 378-381 [2002] 911.

[50] C.Cantoni, D.K. Christen, R. Feenstra, A. Goyal, G.W. Ownby, D.M. Zehner, D.P. Norton. Appl. Phys. Lett. Vol 79, No. 19 (2001) 3077- 3079

[51] J. Eickemeyer, D. Selbman, R. Opitz, B. de Boer, B. Holzapfel, L. Schultz and U. Miller, Supercond. Sci. Technol 14 [2001] 152.

[52] A.Goyal, R. Feenstra, M. Paranthanam, J. R. Thompson, B. Y. Kang, C. Cantoni, D. F. Lee, F. A. List, P. M. Martin, E. Lara-Curzio, C. Stevens, D. M. Kroeger, M. Kowalewski, E. D. Specht, T. Aytug, S. Sathyzmurthy, R. K. Williams and R. E. Ericson, Physica C 382 [2002] 251.

[53] B. de Boer, V. Sarma, N. Reger, J. Eickemer and B. holzapfel, Physica C 372-376 [2002] 798.

[54] J. R. Thompson, A. Goyal, D.K. Christen and D.M. Kroeger, Physica C 370 [2002] 169.

[55] V. Boffa, T. Petrisor, G. Gelentano, F. Fabbri, C. Annino, S. Ceresara, V. Galluzzi, U. Gambardella; G. Grimaldi and A. Mancini, Supercond. Sci. Technol 13 [2000] 1467.

[56] T. Watanabe, K. Matsumoto, T. Tanigawa, T. Maeda, I. Hirabayashi. IEEE. Transc. On applied supercond. Vol 11, No 1 [2001] 3134.

[57] T. Watanabe, K. Matsumoto, T. Maeda, T. Tanigawa, I. Hirabayashi. Physica C 357-360 [2001] 914.

[58] S. Pinol, J. Diaz, M. Segarra and F. Espiell, Supercond. Sci. Technol 14 [2001] 11.

[59] J. Diaz, M. Segarra, F. Espiell and S. Pinol, Supercond. Sci. Technol 14 [2001] 576.

[60] C.E. Morosanu, "Thin Films by Chemical Vapour Deposition" Thin Films science and technology collection, Ed. Elsevier [1990].

[61] L.F. Hubert-Pfalzgraf, H. Guillon. Appl. Organometallic Chemistry 12 [1998] 221.

[62] J.F. Roeder, T.H. Baum, S.M. BiLodeua, G.T. Stanf, C. Rogaglia, M.W. Russell, P.C. van Buskivk. Advanc.Mater. Optoelecton. 10 (2000) 145.

[63] G. Wahl, F. Weiss, O. Stadel. Deposition technique: chemical deposition. to be published

[64] K. Timmer, C.I.M.A. Spee, A. Mackor, H.A. Meinema, A.L. Spek, P. van Sluis. Inorg. Chim. Acta 190 [1991] 109.

[65] S.R. Drake, M.B. Hursthouse, K.M. Abdul Malik, D.J. Otway. J. Chem. Soc. Dalton Trans. [1993] 2883.

[66] F. Weiss, K. Fröhlich, R. Haase, M. Labeau, D. Selbmann, J.P. Senateur, O. Thomas. J. Physique IV C3 3 [1993] 321.

[67] W. Meffre, J.L. Deschanvres, M.F. Joubert, L. Albello, J.C. Joubert, B.Jacquier. J. de Physique IV Vol 9 8 [1999] 583.

[68] G. Wahl, M. Pulver, W.Decker, L .Klippe. Surf. and coatings Tehcnol. 100 [1998] 132.

[69] O. Stadel, J. Schmidt, G. Wahl, F. Weiss, D. Selbmann, J. Eickemeyer, O. Yu. Gorbenko, A. R. Kaul, C. Jimenez, Journal de Physique IV. 11, [2001] Pr11-233.

[70] J.P. Sénateur, R. Madar, F. Weiss, O. Thomas, A. Abrutis, 93/08838, France, [1993], extended : FR94/0000858, Europe/USA, [1994].

[71] F. Felten, J.P. Senateur, F. Weiss, R. Madar, A. Abrutis, J. de Physique IV C5 [1995] 1079 .

[72] J.P. Senateur, F. Felten, S. Pignard, F. Weiss, A. Abrutis, V. Bigelyte, A. Teiserskis, Z. Saltyte, B. Vengalis, Journal of Alloys and Compounds, 251 [1997] 288.

[73] C. Jimenez, H. Guillon, B. Pierret, O. Stadel, J. Schmidt, U. Krause,G. Wahl. J. Physique IV Vol 11 [2001] Pr3-669.

[74] K . Fröhlich, J .Souc, D. Machajdik, F. Weiss. Appl. Supercond. 158 [1997] 233.

[75] A. Abrutis, J.P. Senateur, F. Weiss, V. Bigelyte, A. Teiserskis, V. Kubilius, V. Galindo, S. Balevicius. Journal of Crystal Growth 191 [1998] 79.

[76] L. Klippe, G. Wahl. Journ. of Alloys and Compounds 251 [1997] 249.

[77] F. Weiss, U. Schmatz, A. Pisch, F. Felten, S. Pignard, J.P. Senateur, A. Abrutis, K. Fröhlich, D. Selbmann, L. Klippe, Journal of Alloys and Compounds 251, [1997] 264.

[78] J. Lindner, F. Weiss, J.P. Senateur, A. Abrutis, Integrated Ferroelectrics, 30 [1-4] [2000] 301.

[79] C. Jimenez, F. Weiss, J.P. Senateur, A. Abrutis, M. Krellmann, D. Selbmann, J Eickemeyer, O. Stadel, G. Wahl, IEEE Trans. App Supercond., 11 [1] Part 3 [2001] 2905.

[80] U. Schmatz, F. Weiss, L. Klippe, O. Stadel, G. Wahl, M. Krellmann, .D Selbmann, L. Hubert-Pfalzgraf, H. Guillon, J. Peña, M. Vallet-Regi. EUROCVD Proced. Vol 97-25 Electrochem. Soc. [1997] 1005.

[81] M . Krellmann, D. Selbmann, U. Schmatz and F. Weiss. J. of Alloys and Compounds 251 [1997] 307.

[82] A. Abrutis, V.Plausinaitienë, A.Teiserskis, V.Kubilius, Z.Saltytë, J.P.Sénateur, L.Dapkus. J. de Physique.IV 9 [1999] 689.

[83] A.Abrutis, V. Plausinaitiene, A. Tesarskis, V. Kubilius, J.P. Senateur, F. Weiss. Chemical Vapor Deposition 5, No 4 [1999] 171.

[84] W. Meffre, PhD Thesis INPG [1999] 85.

[85] A. Teišerskis, A. Abrutis, V. Kubilius, Z. Šaltytė, J.P. Sénateur, F. Weiss. Proced. Intern. Conf. 'Thin films deposition of oxide multilayers. Industrial-scale processing", TFDOM, Sep 2000 19

[86] Internal communication: agreement between LMGP, University of Vilnius and JIPELEC.

[87] V. Plausinaitiene, A. Abrutis, B. Vengalis, R. Butkute, J.P. Sénateur, Z. Saltyte, V. Kubilius. Physica C 351 [2001] 13.

[88] B. Vengalis, V. Plausinaitiene, A. Abrutis, Z. Saltyte, R. Butkute, V. Petrauskas, J. Maria, G. Bonfait. J. Phys IV 11 [2001] Pr11-53.

[89] J. Eickemeyer, D. Selbmann, R. Opitz H. Wendrock, E. Maher, U. Miller, W. Prusseit. Physica C 372-376 [2002] 814.

[90] D. Selbmann, J. Eickemeyer, H. Wendrock, C. Jimenez, S. Donet, F. Weiss, U. Miller, O. Stadel, J. de Physique IV, 11 [2001] Pr11-239.

[91] S. Donet, F. Weiss, J.P. Sénateur, P. Chaudouet, A.Abrutis, A. Teiserskis, Z. Saltyte, D. Selbmann, J. Eickemeyer, O. Stadel, G. Wahl, C. Jimenez and U. Miller. Physica C 372-376 [2002] 652.

[92] S. Donet, F. Weiss, J.P. Senateur, P. Chaudouet, A. Abrutis, A. Teiserskis, Z. Saltyte, D. Selbmann, J. Eickemeyer, O. Stadel, G. Wahl, C. Jimenez and U. Miller, Journal de Physique IV, 11 [2001] Pr11-319.

[93] J. Dzick, S. Sievers, J. Hoffmann, K. Thiele, F. Garcia-Moreno, A. Usoskin, C. Jooss, H.C. Freyhardt. MRS Proceedings Vol 585, 55-66.

[94] O. Stadel, J. Schmidt, G. Wahl, F. Weiss, D. Selbmann, J. Eickemeyer, O. Yu. Gorbenko, A.R. Kaul, C. Jimenez, Physica C 372-376 [2002] 751.

[95] Brite Euram READY Project, BRPR-CT98-076

[96] O. Stadel, L. Klippe, J. Schmidt, G.Wahl, S.V. Samoylenkov, O-Yu Gorbeko, A.R. Kaul
J. Phys. IV 9, [1999] Pr8-561.

[97] O. Stadel, J. Schmidt, G. Wahl, F. Weiss, D. Selbmann, J. Eickemeyer, O.Yu Gorbenko, A.R. Kaul, C. Jimenez. J. Phys. IV 11 [2001], Pr11-233.

[98] O .Stadel, J. Schmidt, M. Liekeffet, G. Wahl, O-Yu Gorbenko, A.R. Kaul. Submitted to ASC2002.

[99] S. Donet, F. Weiss, P. Chaudouet, S. Beauquis, A. Abrutis, H.C. Freyhardt, A. Usokin, D. Selbmann, J. Eickemeyer, C. Jimenez, C.E. Bruzek, J.M. Saugrain. Submitted to ASC2002.

[100] J. Wiesmann, J. Dzick, J. Hoffman, K. Heinemann, H.C. Freyhardt. J. Mat. Res. Vol 13 No 11 (1998) 3149-3152.

[101] S. Donet, F. Weiss, P. Chaudouet, D. Selbmann, W. Prusseit, C.E. Bruzek, J.M. Saugrain. Submitted to EUROCVD 14, hold during ECS2003.

[102] S. Donet, F. Weiss, P. Chaudouet, D. Selbmann, W. Prusseit, C.E. Bruzek, J.M. Saugrain. Submitted to EUROCVD 14, hold during ECS2003.

[103] L.R. Motowidlo, V. Selvamanickam, G. Galinski, N. Vo, P. Haldar, R.S. Sokolowski. Physica C 333 [2000] 44.

[104] T.Aytug, J.Z. Wu, C. Cantoni, D.T. Verbelyi, E.D. Specht, M .Paranthaman, D.P. Norton, D.K. Christen, R.E. Ericson, C.L. Thomas. Appl. Phys. Lett. Vol 76, No 6 [2000] 760.

[105] M. A. Novozhilov, A.R. Kaul, O.Yu. Gorbenko, I.E. Graboy, G. Wahl, U. Krausse. J. Phys IV 9 [1999] Pr8-629.

[106] A. Abrutis, V. Plausinaitiene, S. Pasko, A. Tesierskis, V. Kubilius, Z. Saltyte, J.P. Senateur. J. Phys IV 11 [2001] Pr3-1169.

[107] N. Okeda, K. Saito, H. Funakubo. Jap. J. Appl. Phys. 39 [2000] 572.

[108] Y. Yu, W.X. Li, Q.G. Gon, G. Xiao, A. Gupta, P. Lecoeur, J.Z. Sun, Y.Y. Wang, V.P. Dravid. Phys. Rev. B 54 [1996] R8357.

[109] C. Dubourdieu, M. Rosina, H. Roussel, F. Weiss, J.P.Sénateur. Appl. Phys. Lett. Vol 79, No. 9 [2001] 1246.

[110] S.R. Foltyn, P. Tiwari, R.C. Dye, M.Q. Le, X.D. Wu. Appl. Phys. Lett. 63 [1993] 1848.

[111] S.R. Foltyn, Q.X. Jia; P.N. Arendt, L. Kinder, Y. Fan, J.F. Smith. Appl. Phys. Lett. 75, No. 23 [1999] 3692.

[112] Results presented by D. Peterson (LANL) at the MRS Workshop on "Processing and Applications of superconductors" hold in Gatlinburg in August 2002.

[113] M. Jergel. Supercond. Sci. Technol. 8 [1995] 67.

[114] G. B. Blanchet. Supercond. Sci. Technol. 8 [1991] 69.

[115] A. Ferreri, J.J. Wells and J.L. MacManus-Driscoll, IEEE Trans. App Supercond.11 [1] Part 3, [2001] 2742.

[116] G.B. Blanchet, C.R. Fincher Jr., Supercond. Sci. Technol. 4 2 [1991] 69.

[117] J.L. MacManus-Driscoll, A. Ferreri, J.J. Wells, J.G.A. Nelstrop, Supercond. Sci. Technol 14 2 [2001] 96.

[118] T.C. Shields, K. Kawano, T.W. Button, J.S. Abell. Supercond. Sci. Technol. 15 [2002] 99.

[119] A. Sin, Z. Supardi, A. Sulpice, P. Odier, F. Weiss, L. Ortega, M. Nunez-Regueiro, IEEE Trans. Appl. Supercond. 11 1 Part 3, [2001] 2877.

[120] P. Odier, F. Weiss, Z. Supardi, patent pending Fr0202217.

[121] Z. Supardi , G. Delabouglise, C. Peroz, A. Sin, C. Villard, P. Odier, F. Weiss. Physica C, in press.

[122] I.H. Mutlu, E. Celik, M.K. Ramazanoglu, Y. Akin, Y.S. Hascicek. IEEE Transc. On Appl. Supercond. Vol 10, No 1 [2000] 1154.

[123] H. Okuyucu, E. Celic, M. K. Ramzanoglu, Y. Akin, I. H. Mutlu, W. Sigmund, J. E. Crow and Y. S. Hascicek, IEEE Trans. Appl. Supercond. 11 [2001] 2927.

[124] E. Celic, H. Okuyucu, I. H. Mutlu, M. Tomsi, J. Schwartz and Y. S. Hascicek, IEEE Trans. Appl. Supercond. 11 [2001] 3162.

[125] T. G. Chirayil, M. P. Paranthanam, D. B. Beach, D. F. Lee, A. Goyal, R. K. Williams, X. Cui, D. M. Kroeger, R. Feenstra, D. T. Verebelyi and D.K. Christen, Physica C 336 [2000] 63.

[126] M. P. Paranthanam, T. G. Chirayil, F. A. List, X. Cui, A. Goyal, D. F. Lee, E. D. Specht, P. M. Martin, R. K. William, D. M. Kroeger, J. S. Morrell, D. B. Beach, R. Feenstra and D. K. Christen J. Am. Ceram. Soc 84 (2) [2001] 273.

[127] D. Shi, Y. Xu, S.X. Wang, J. Lian, L.M. Wang, S.M. McClellan, R. Buchanan, K.C. Goretta. Physica C 371 (2002) 97.

[128] D. Marguillier, R .Cloots, A. Rulmont, J.F. Fagnard, Ph. Vanderbemden, M. Ausloos. Physica C 372-376 (2002) 715.

[129] S. Sathyamurthy, K. Salam, Physica C 377 [2002] 208.

[130] M. W. Rupich, W. Palm, E. Siegel, S. Annavarapu, L. Fritzemeier, M. D. Teplitsky, C. Thieme and M. Paranthanam, IEEE Trans. Appl. Supercond. 9 [1999] 1527.

[131] O. Castaño, A. Palau, J. C. Gonzalez, S. Pinol, T. Puig, N. Mestres, F. Sandiumenge, X. Obradors, Physica C 372-376 [2002] 806.

166

[132] A. Gupta, R. Jagannathan, E. I. Cooper, E. A. Giess, J. I. Landman and B. W. Hussey, Appl. Phys. Lett. 52 [1988] 2077.

[133] M. W Rupich, Q. Li, S. Annavaparu, C. Thieme, W. Zhang, V. Prunier, M. Paranthaman, A. Goyal, D. F. Lee, E. D. Specht and F. A. List, IEEE Trans. Appl. Supercond. 11 [2001] 2927.

[134] R. Feenstra, T.B. Lindemer, J.D. Budai, M.D. Galloway .J. Appl. Phys. 69 [1990] 6569.

[135] P.C. McIntyre, M.J. Cima, M.F. Ng. J. Appl. Phys. 68 (8) [1990] 4183.

[136] P.C. McIntyre, M.J. Cima, J.A. Smith, R.B.Hallock, M.P. Siega, J.M. Phillips. J. Appl. Phys. 71 (4) [1992] 1868.

[137] P.C. McIntyre, M.J. Cima. J. Appl. Phys. 77 (10) [1995] 5263.

[138] M. Falter, W. Hässler, B. Schobach, B. Holzapfel, Physica C 372-376 [2002] 46.

[139] J.A. Smith, M.J. Cima, N. Sonnenberg. IEEE Transactions on applied Superconductivity Vol9, No 2 [1999] 1531.

[140] T. Araki, Y. Takahashi, K. Yamagiwa, Y. Iijima, K. Takeda, Y. Yamada, J. Shibata, T. Hirayama and I. Hirabayashi, Physica C 357-360 [2001] 991.

[141] T. Araki, K. Yamagiwa, I. Hirabayashi, K. suzuki and S. tanaka, Supercond. Sci. Technol. 14 [2001] L21.

[142]T. Araki, I. Hirabayashi, J. Shibata, Y. Ikuhara, Supercond. Sci. Technol. 15 [2002] 913.

[143] T. Araki, Y. Takahashi, K. Yamagiwa, T. Yuasa, Y. Iijima, K. Takeda, S. B. Kim, Y. Yamada and I. Hirabayashi, IEEE Trans. App Supercond. 11 (1) [2001] 2869.

[144] O. Castaño, A. Cavallaro, A. Palau, J.C. Gonzalez, M. Rossell, T. Puig, F. Sandiumenge, N. Mestres, S. Piñol, A. Pomar, X. Obradors. Supercond. Sci. Technol. 16 (2003) 45-53.

[145] M. W. Rupich, Q. Li, S. Annavarapu, C. Thieme, W. Zhang, V. Prunier, M. Paranthaman, A. Goyal, D. F. Lee, E. D. Specht and F. A. list, IEEE Trans. App Supercond. 11 (1) [2001] 2927.

[146] V.F. Solvyov, H.J. Wiesmann, M. Suenaga. Physica C 353 [2001] 14.

[147] V.F. Solvyov, H.J. Wiesmann, L-. Wu, Y. Zu, M. Suenaga. Appl. Phys. Lett. Vol 76 No144 [2000] 1911.

[148] V.F. Solovyov, H.J. Wiesmann, L.-J. Wu, Y. Zhu, M. Suenaga. IEEE Transac. On appl. Supercond. Vol 11. No. 1 [2001] 2939.

[149] Y. Takahashi, T. Araki, K. Yamagiwa, Y. Yamada, S.B. Kim, Y. Iijima, K.Takeda, T. Hirayama, I. Hirabayashi, Physica C 357-360 [2001] 1003.

[150] Y. Yamada, S.B. Kim, T. Araki, Y. Takahashi, T. Yuasa, H. Kurosaki, I. Hirabayashi, Y. Iijima, K. Takeda, Physica C 357-360 [2001] 1007.

[151] T. Araki, T. Yuasa, H. Kurosaki, Y. Yamada, I. Hirabayashi, T. Kato, T. Hirayama, Y. Iijima and T. Saito, Supercond. Sci. Technol. 15 [2002] L1.

[152] M. P. Siegal, P. G. Clem, J. T. Dawley, R. J. ONG, M. A. Rodriguez and D. Overmyer, Appl. Phys. Lett. 80 (15) [2002] 2710.

[153] M. W. Rupich, U. schoop, C. Thieme, W. Zhang, X. Li, T. Kodenkantath, N. Nguyen, E. Siegal, D. Buczek, J. Lynch, M. Jowett, E. Thompson, J. S. wang, J. scudiere, A. P. Malozemoff, Q. Li, S. Annavarapu, S. Cui, L. Fritzmeier, B. Aldrich, C. Craven, F. Niu, A.

Goyal and M. Paranthaman, submitted to ASC 2002, to be published in IEEE Trans. App Supercond.

[154] "YBCO processing and processing- strategic research" ORNL presentation available at the web site (http://www.ornl.gov/HTSC/fy02peer.htm)

[155] A. P. Bramley, J.D. O'Connor, C.R.M. Grovenor. Supercond.Sci. Technol. 12 [1999] R57-R74.

[156] S.S.P. Parkin, V.Y. Lee, A.I. Nazzal, R. Savoy, R.Huang, T.C. Gorman, R. Beyers. Phys. Rev. B 38 [1988] 6531.

[157] Z.Z. Sheng, A.M. Hermann, Nature 332 [1988] 55.

[158] A. Schilling, M. Cantoni, J.D. Guo, H.R. Ott. Nature 363 [1993] 56.

[159] S. N. Putilin, I. Bryntse, E.V. Antipov .Mat. Res. Bull. 26 [1991] 1299.

[160]Th. Hopfinger, M. Lomello-Tafin, J.L. Jorda, Ph. Galez, M. Couach, R.E. Gladyshevskii, J.L. Soubeyroux. Physica C 351 [2001] 53.

[161] W. Li, D.Z. Wang, J.Y. Lao, Z.F. Ren, J.H. Wang, M. Paranthaman, D.T. Verebelyi, D.K. Christen. Supercond. Sci. Technol. 12 [1999] L1.

[162] H. Schneidewind, M. Manzel, G. Bruchlos, K. Kirsch. Supercond. Sci. Technol. 14 [2001] 200.

[163] S. Phok, Z. Supardi, A. Sin, Ph. Galez, J. L. Jorda and F. Weiss, Journal de physique IV 11 [2001] Pr11-157.

[164] S. Phok, P. Galez, J. L. Jorda, C. Peroz, C. Villard, D.De Barros and F. Weiss, IEEE transactions on Applied Superconductivity, [2003] to be published.

[165] Ph Galez, J.L. Soybeyroux, Th. Hopfinger, Ch. Opagiste, M. Lomello-Tafin, Ch. Bernard, J.L. Jorda. Supercond. Sci. Technol. 14 [2001] 583.

[166] S.O. Klimonsky, S.V. Samoilenkov, O.Yu. Gorbenko, D.A. Emelianov, A.V. Lyashenko, S.R. Lee, A.R. Kaul, Yu.D. Tretyakov, D.G. Andrianov, A.V. Kalinov, I. V. Voloshin. Physica C 383 [2002] 37.

[167] A. Sin, P. Odier, M. Nuñez-Rigueiro. Physica C 330 [2000] 9.

[168] K.Kishio, J. Shimoyama, A. Yoshikawa, K. Kitazawa, O. Chmaissem, J.D. Jorgensen. Low Temp. Phys. 105 [1996] 1359.

[169] A. Sin, Z. Supardi, P. Odier, F. Weiss, L. Ortega, A. Sulpice, M. Nunez-Regueiro, Thin solid film 388 1-2 [2001] 251.

[170] A. Sin, F. Weiss, P. Odier, Z. Supardi, M. Nuñez-Regueiro, Physica C 341 Part 1, [2000] 399.

[171] A. Sin, Z.I. Supardi, P. Odier, F. Weiss, M. Nuñez-Regueiro. Supercond. Sci. Technol. 13 [2000] 617.

TlBaCaCuO SUPERCONDUCTING THIN FILMS BY AN EX-SITU MOCVD BASED APPROACH: FROM BaCaCuO(F) MATRICES TO DEVICES

Graziella MALANDRINO and Ignazio L. FRAGALA'.

Dipartimento di Scienze Chimiche, Universita' di Catania, and I.N.S.T.M. UdR di Catania, Viale Andrea Doria 6, I-95125 Catania, ITALY

1. INTRODUCTION

Since the discovery of superconductivity in the TlBaCaCuO system [1,2], enormous efforts have been made to synthesise reproducibly and selectively single phases both as bulk [3-11] and thin film [12-18] materials. In particular, nine well defined phases have been reported for the homologous series $Tl_mBa_2Ca_{n-1}Cu_nO_{2n+m+2}$ (where for m = 1, n = 1-5 and for m = 2, n = 1-4). Thus, five phases with m=1 have a single TlO layer [3-7], while four with m =2 adopt a double TlO layer structure [8-11]. The single and double TlO layer phases may be formulated as alternating rocksalt-like single TlO layers and perovskite $Ba_2Ca_{n-1}Cu_nO_{2n+2}$ layers for the $TlBa_2Ca_{n-1}Cu_nO_{2n+3}$ (hereafter Tl-1,2,n-1,n) (Fig. 1), and alternating rocksalt-like double TlO layers and perovskite layers for the $Tl_2Ba_2Ca_{n-1}Cu_nO_{2n+4}$ (hereafter Tl-2,2,n-1,n) (Fig. 2) phases. For a general view on thallium based high-temperature superconductors see ref. [19,20], while for a review focusing on TlBaCaCuO films see ref. [21].

Among the TlBaCaCuO phases, the $Tl_2Ba_2Ca_2Cu_3O_{10}$ (Tl-2223) and $Tl_2Ba_2CaCu_2O_8$ (Tl-2212) phases represent the best candidates due to the remarkably high critical transition temperatures (T_c) of 125 K and 110 K, respectively. The same phases also show appealing microwave properties [22, 23]. Challenging surface resistance R_s and critical current J_c values have been reported for Tl-2212 thin films on $LaAlO_3$ (100) at 77 K [24].

All the Tl-based superconductors are inherently difficult to produce in form of thin films because of the high volatility of Tl oxide. In addition, several problems are associated with the reproducible fabrication of Tl-based films due to the intriguing interplay among the various TlBaCaCuO superconducting phases, whose properties considerably vary according to their stoichiometry and structure. The synthetic control over the desired phase is, therefore, of paramount importance for both fundamental studies and technological applications. In addition, the close similarity of the a-axis lattice constants (≈ 3.85 Å) for both single and double TlO layer phases [3,4,8-10] gives rise to the formation of intergrowths in the TlBaCaCuO system. Two different types of intergrowths have been reported: i) the intergrowths containing phases with differing numbers of Cu and, consequently, Ca layers (i.e. intergrowths due to the presence of two sorts of perovskite units) [16, 25] and ii) the intergrowths containing phases with different number of TlO layers (i.e. due to the presence of two sorts of rocksalt-like units) [26, 27]. Such phenomena have been observed both on a microscopic and macroscopic scale depending upon the relative amount of intergrowths. The presence of randomly dispersed minor intergrowths of other phases in the domains of the major phase [16, 25] has no consequence in the X-ray diffraction pattern, even though it influences the linewidth of the 00ℓ reflections of the major phase [25]. These intergrowths can only be detected by high resolution electron microscopy and transmission electron microscopy studies [16, 25, 28]. On a macroscopic scale, we distinguish two cases. When distinct crystalline domains of different phases are present, the X-ray diffraction patterns are indicative of each distinct phase. By contrast, different phases present in the same domain may result in intergrowths with characteristic $\theta/2\theta$ diffraction patterns [27,29].

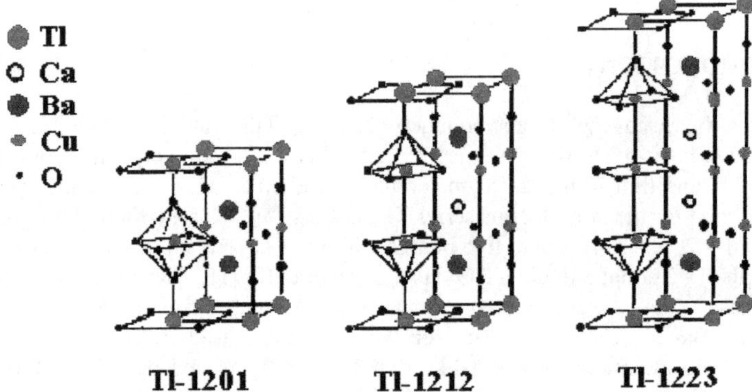

Fig. 1 Ideal structures of $TlBa_2Ca_{n-1}Cu_nO_{2n+3}$ (n = 1, 2, and 3).

To date, efforts to produce single-sided (SS) [17,18,30-32] and double-sided (DS) [33-37] thin films of Tl-based superconductors have been largely focused on physical vapour deposition (PVD) techniques. Despite the top-level intrinsic properties, fabrication of Tl-based superconducting materials is often confined to laboratory scale, which is expensive and hardly scalable to manufacturing of large area films, while any perspective applications require scalable and reproducible processes suited for high-quality films.

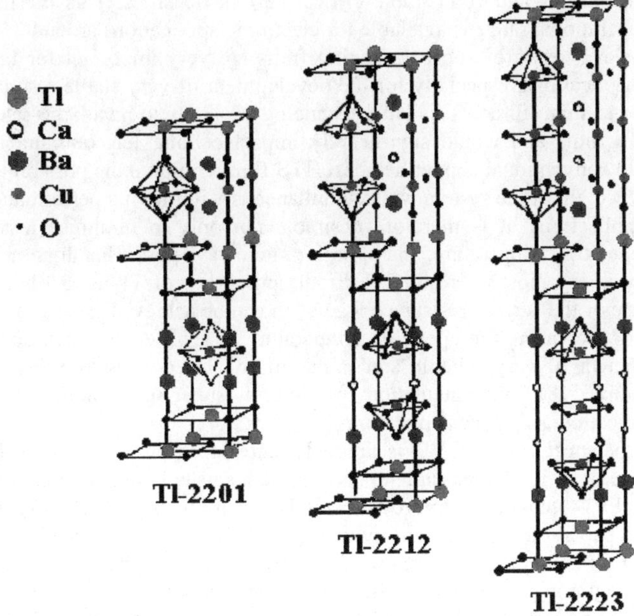

Fig. 2 Ideal structures of $Tl_2Ba_2Ca_{n-1}Cu_nO_{2n+3}$ (n = 1, 2, and 3).

Therefore, the development of low-cost processes represents a crucial step for large-scale applications.

Metal-organic chemical vapour deposition (MOCVD) has proved to be a challenging alternative due to the use of simplified apparatus, lower deposition temperatures, capability to coat complex shapes, and adaptability to large scale processing [38]. MOCVD has been successfully applied to the preparation of TlBaCaCuO phases containing double TlO layer phases, such as Tl-2212 and Tl-2223, single TlO layer phases, such as Tl-1223 and Tl-1234, and intergrowth phases [39-48]. In particular, pioneeristic has been, in this field, the work of T. J. Marks et al. beginning from the first report of TlBaCaCuO films obtained using an MOCVD based approach [41], up to the most recent publications on the precursor performance issues [43] and the fluoride effect on the phase-selective growth of Tl-1223 films [44].

In regard to applications, high critical temperature Tc superconducting (HTS) materials have stimulated considerable interest in terms of potential applications/benefits, and microelectronic devices are expected to have a major commercial impact [49,50]. In particular, there has been great interest in applications ranging from active Josephson-effect devices [51,52] and interconnects in multichip modules to passive microwave devices such as delay lines and filters [53]. In spite of Tl toxicity, thallium cuprate thin films offer several advantages compared to $Y_1Ba_2Cu_3O_{7-x}$ thin films, including higher critical temperatures and comparable critical current densities. This means that a wide margin of operational range is available for electronic applications around 77 K.

Low microwave losses of superconducting films are very important for high frequency electronic applications, especially for the development of very stable resonators, lossless delay lines and sharp filters. This latter technology is likely to be a large-scale application of HTS thin films and would significantly impact cellular telecommunication [50,54]. Compared to conventional copper devices, HTS films combine the possibility of reducing the size and weight of the system while simultaneously improving performances. For these potential applications, it is therefore desirable not only to minimise loss, but also to maximise operating temperature, and optimise microwave power handling in HTS films.

In this respect, the most interesting electronic properties of Tl-based films is, probably, the low value of the surface resistance R_s, which can be achieved up to nearly 100 K. The possibility of increasing the operation temperature of HTS devices can open the road to the development of very reliable and miniaturised closed-cycle coolers, making much more attractive the implementation of hybrid super-semiconducting systems for telecommunication and space applications.

In this perspective, MOCVD is uniquely suited to grow DS films to be applied as passive filters. It is, in fact, possible to coat simultaneously both sides of substrates, whilst line of sight PVD techniques require sequential deposition steps of each of the two sides.

In this paper, we report an overview on the *ex-situ* fabrication processes of TlBaCaCuO films, both SS and DS, produced by an MOCVD based approach on various substrates, beginning from matrix deposition and thallium vapour diffusion step, to electrical characterisation and potential devices. In particular, the paper first describes the matrix deposition processes for SS films. In the second section, the importance of processing parameters during the thallium vapour diffusion step is assessed in the perspective of reproducible and selective fabrication of single phases and of the formation of new metastable $Tl_{1+x}Ba_2Ca_2Cu_3O_{9+x}$ and $Tl_{1+x}Ba_2CaCu_2O_{7+x}$ intergrowth phases. Thirdly, the formation of *a-axis* oriented TlBaCaCuO films is reported and commented in function of the substrate nature. The processes for the preparation of DS films and their morphological and structural characterisation are described in detail in the following. Finally, the transport and microwave properties are reported and the realisation of dual mode microwave filters based on Tl-2212 DS films is discussed.

2. EXPERIMENTAL PROCEDURES

MOCVD reactors. A low pressure, horizontal, cold wall multicomponent reactor [55, 56] with a resistive substrate heater was used to deposit SS-BaCaCuO(F) precursor films, while an horizontal monocomponent hot-wall reactor was used to deposit DS-BaCaCuO(F) films [57]. Mass flows of carrier gas (Ar) and reactant gas (O_2) were controlled with 1160 MKS flowmeters using an MKS 147 electronic control unit (±1 sccm accuracy). Depositions were carried out at 4-5 torr total pressure.

Thallium vapour diffusion process. The crucible geometry in an "open" reactor has been used for the thallium vapour diffusion step (from now on annealing). In the crucible geometry the matrix is placed inside a crucible together with a source of thallous oxide vapour typically a powder of the Tl_2O_3, BaO, CaO, CuO oxides with an appropriate stoichiometry. Due to the thallium toxicity [58], particular attention has to be devoted to this process. The annealing reactor has to be gasproof, so that the unique outlet gas flow may be bubbled in a series of three traps containing NaI solution (10 % in H_2O) to convert any Tl exhaust to unsoluble yellow TlI. The third trap solution has to be colorless. The entire system, reactor and traps, was allocated inside a totally filtered hood.

Compositional and structural characterisation. The chemical composition of the films was analysed by Energy Dispersive X-Ray Analysis (EDX) using a windowless IXRF solid state detector and the surface morphology was examined by Scanning Electron Microscopy (SEM) using a LEO Iridium 1450. θ-2θ X-ray diffraction (XRD) patterns were recorded on a θ-θ D5005 Bruker diffractometer using a Cu Kα radiation operating at 40 kV/30 mA over a $4°<2\theta<80°$ angular range. The rocking curves and pole figures were recorded on a θ-2θ D5005 Bruker diffractometer equipped with an Eulerian cradle.

Transport and microwave properties. A standard four probe technique was employed to perform resistive transition measurements using sputter deposited Au contacts. Samples were cooled in N_2 vapours. The supply currents ranged between 1 and 10 μA in order to rule out the possibility of critical current effects. Transport and superconducting properties of Tl SS and DS films grown on 10×10 mm^2 $LaAlO_3$ substrates were investigated using an inductive, non destructive, method, which measures (depending on the temperature and on the driving current) the third harmonic component voltage across a small sensor coil mounted very close to the film surface. The surface impedance R_s was measured at a frequency of 87 GHz using a cylindrical OFHC copper cavity. The transmitted power P is measured as a function of the stepwise swept frequency f of the input signal and detected with a diode voltage. The loaded quality factor Q_L of the cavity is then converted into the surface resistance value. A detailed description of the experimental set up, including a discussion of the accuracy and limits of the technique, is contained in ref. [59].

3. RESULTS AND DISCUSSION

TlBaCaCuO films have been prepared through an *ex-situ* process using MOCVD to deposit the precursor films combined with a thallium vapour diffusion step. *Ex-situ* growth processes produce highly textured films, whose grains nucleate and grow from sites randomly distributed over the substrate surface. When all the nuclei have very similar orientations, as a result of a strong interaction with a well-lattice-matched substrate, these films may be considered "heteroepitaxial", even though they may contain a high density of low-angle grain boundaries. This type of epitaxy is called graphoepitaxy or artificial epitaxy [60]. In this case, the crystallographic relations between film and substrate are independent of any bonding that exists across the film-substrate interface, but are entirely dependent on features, such as cleavage steps, that exist on the surface of the substrate. Of course heteroepitaxial growth may only occur when there is a small misfit between the lattice parameters of the unit cells of film and substrate. When there is a large misfit, the film grows with a fibre texture, i.e. the grains are out of plane aligned, but no alignment occurs in the *a-b* plane. In fact, *c-axis* oriented films have been obtained not only on well matched substrates but also on substrates with a very high mismatch, such as YSZ (100), and even on randomly oriented YSZ substrates (i.e. substrates cut from YSZ single crystals without any particular alignment).

c-axis films have been obtained on a variety of commercially available substrates such as $SrTiO_3$, $LaAlO_3$, $NdGaO_3$, YSZ (yttria stabilised zirconia), MgO, or on buffer layers such as CeO_2, Sr_2AlTaO_6, and $LaAlO_3$ [21]. The best combination of HTS material and substrate is usually very dependent on the end-use application, therefore the substrate choice is obviously dictated by the final application.

3.1. Single sided films

3.1.1. *c-axis* oriented films

The most typical orientation reported for TlBaCaCuO films is along the *c-axis* (i.e. the film grains have their c axes perpendicular to the substrate) independently of the operating conditions used during the annealing step. This means that at variance with the $YBa_2Cu_3O_{7-\delta}$ system, in which case the *a-axis* growth is observed for particular deposition temperatures, the driving force of *c-axis* orientation for TlBaCaCuO films is generally due to a kinetic factor since the growth along the <001> is much faster than along the <100> direction.

3.1.1.1. Nature of MOCVD precursors. The quality of any MOCVD process crucially depends on the apparatus as well as on prerequisites of adopted precursors [38]. Thus, reproducible results are always associated with a careful control of operational deposition parameters and, even more importantly, with the use of highly pure precursors with clean decomposition pathways and constant mass-transport properties.

The precursors used to fabricate the Ba-Ca-Cu matrices were "second generation" metal-organic complexes of the alkaline-earth metals $Ba(hfa)_2 \cdot$tetraglyme, $Ca(hfa)_2 \cdot$tetraglyme (Hhfa = 1,1,1,5,5,5-hexafluoro-2,4-pentanedione, tetraglyme = 2,5,8,11,14-pentaoxatetradecane) prepared as reported elsewhere [61], and commercially available precursors for copper, $Cu(acac)_2$ (Hacac = acetylacetone) and $Cu(tmhd)_2$ (Htmhd = 2,2,6,6-tetramethyl-3,5-heptandione).

It is well known that these "second-generation" precursors represent, to date, the best available source of alkaline-earth metals, even considering the drawback of decomposition side-processes of the Hhfa ligand, which produces fluoride phases. Nevertheless, note that the fluoride phases are easily removed during the annealing step (see section 3.1.1.3). On the other hand, it has recently been reported that BaCaCuO(F) precursor matrices show increased chemical resistance against the influence of atmospheric surroundings in comparison with fluorine free precursor films [62].

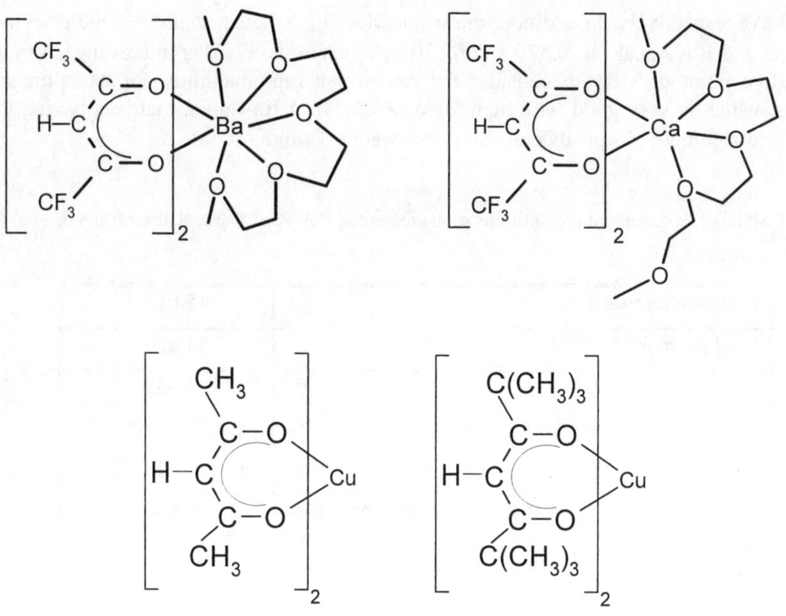

Cu(acac)₂

Cu(tmhd)₂

Fig. 3 Precursor molecules utilised in the MOCVD deposition of BaCaCuO(F) thin films.

3.1.1.2. MOCVD fabrication of SS BaCaCuO(F) matrices. BaCaCuO(F) precursor matrices have been deposited on randomly oriented and single crystal YSZ (100) (YSZ = yttria stabilised zirconia), $LaAlO_3$ (100), and MgO (100) substrates from Ba(hfa)$_2$•tetraglyme, Ca(hfa)$_2$•tetraglyme and Cu(acac)$_2$ precursors. Details of operating conditions are reported in Table 1.

The as-deposited films are homogeneous in composition, thickness and morphology. They have mirror-like surface and a yellowish-brown colour due to the presence of Cu(I) and/or Cu(II) oxide phases (*vide infra*), and their thicknesses range from 0.9 μm to 1.1 μm.

Particular attention has been devoted to the optimisation of precursor sublimation temperatures to obtain the correct stoichiometry of the matrix. An appropriate tuning of precursor sublimation temperatures within the ranges reported in Table 1 has allowed to obtain Ba:Ca:Cu stoichiometries in a 2:1:2 or 2:2:3 ratio. EDX analyses of several selected areas (100 μm × 100 μm) indicate that the 2±0.2:1±0.1:2±0.2 or the 2±0.2:2±0.2:3±0.25 stoichiometries are found over the whole 10 × 10 mm^2 surface (Fig. 4). The Ca K_α and K_β peaks are observed at 3.690 and 4.010 keV, the Ba L lines are spread in the 4.400-5.800 keV range, and the Cu K_α and K_β peaks are found at 8.040 and

8.900 keV, respectively. In addition, the windowless EDX detector allowed the detection of the O and F K_α peaks at 0.520 and 0.670 keV, respectively. It is interesting, from an applicative point of view to consider the run to run reproducibility for an optimised process, which is very good with high yield (> 85 %) of Ba-Ca-Cu matrices having the correct stoichiometry. Typical deposition rates were ~ 1 μm/h.

Table 1 MOCVD experimental conditions of as-deposited BaCaCuO(F) precursor films.

Substrate temperature			450°C
Total pressure			5 torr
	↗	Ca	90-95°C
Precursor sublimation temperatures	→	Ba	120-130°C
	↘	Cu	125-135°C
Flow rate of the carrier gas (Ar) for each precursor			80 sccm
Flow rate of the reactant gas (O_2)			500 sccm
Growth time			60 min

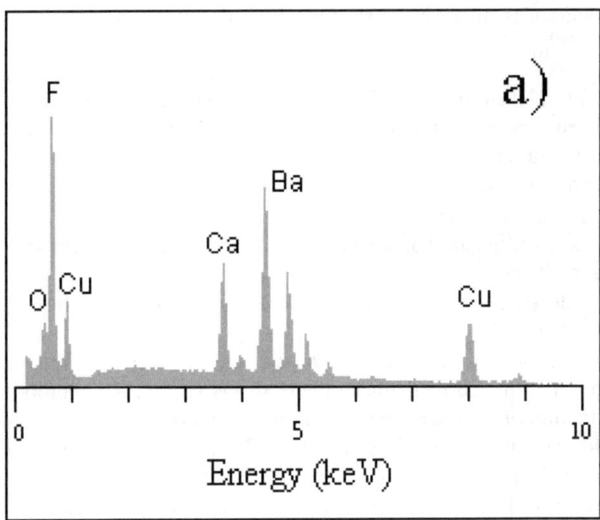

Fig. 4 EDX analysis of an as-deposited BaCaCuO(F) matrix.

Moreover, they have poor crystalline structures. Thus, low intensity peaks can be detected in glancing incidence diffraction (GID) experiments. The associated XRD pattern (Fig. 5), only shows small peaks due to the barium fluoride (BaF_2) and copper(II) oxide (CuO) phases. No crystalline phases containing Ca are evident.

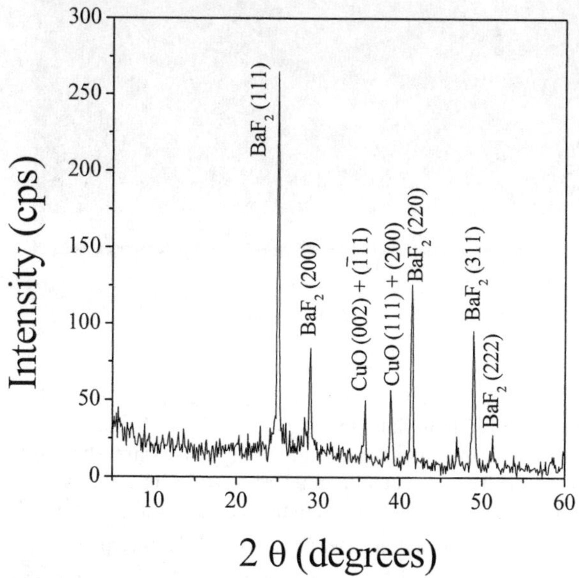

Fig. 5 θ-2θ XRD pattern of an as-deposited BaCaCuO(F) matrix.

It is well known [12] that the formation of the fluoride phases is due to side-processes accompanying the decomposition of the fluorinated precursors. The direct annealing of the as-deposited matrices under thallium vapour diffusion, allows an easy fluorine removal through the reaction:

$$BaF_2 + Tl_2O \Leftrightarrow 2\ TlF\uparrow + BaO \qquad (1)$$

(TlF is a gas at the operating temperature).

The SEM images of as-deposited BaCaCuO(F) films always show smooth surfaces with very fine grains (100 nm) (Fig. 6).

To conclude, the used MOCVD process yields, reproducibly, precursor films, of almost identical characteristics, which are homogeneous and possess comparable thickness and morphology. This is an important point since it is known that the nature of the precursor matrices affects the nature of the final thallium superconducting films [63].

178

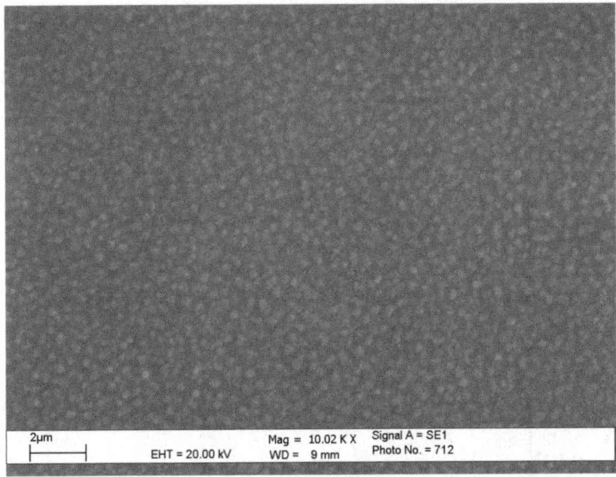

Fig. 6 SEM image of an as-deposited BaCaCuO(F) matrix.

3.1.1.3. Phase formation of c-axis oriented TlBaCaCuO films. The fabrication of thin films of the Tl-based HTS materials involves a delicate balance between the high temperatures required to form the superconducting phase and the high volatility of thallium at these temperatures. Various parameters have to be optimised to reproducibly and selectively yield a monophasic film, namely Tl_2O partial pressure (P_{Tl_2O}), oxygen partial pressure (P_{O_2}) and substrate temperature. A subtle interplay between all these parameters occurs since Tl_2O, which is actually the thallium source in the diffusion process, is produced through the reaction:

$$Tl_2O_3 \Leftrightarrow Tl_2O \uparrow + O_2 \qquad (2)$$

Therefore, P_{O_2} together with temperature determines the P_{Tl_2O}. In addition, P_{Tl_2O} and P_{O_2} play a crucial role in the formation of a TlBaCaCuO phase, for e.g in the case of Tl-2212 their role can be understood from the simplified expression:

$$2\ BaF_2 + CaF_2 + 2\ CuO + 4\ Tl_2O + O_2 \Leftrightarrow Tl_2Ba_2Ca_1Cu_2O_8 + 6\ TlF \uparrow \qquad (3)$$

An obvious consequence is that the direct procedure requires suitable oxide mixtures to warrant enough P_{Tl_2O} for simultaneous elimination of fluoride phases as well as for the formation of the superconducting film.

Finally, it is well known that a fine balance occurs between temperature and P_{O_2}, since it has been shown that lower P_{O_2} allow formation of a TlBaCaCuO phase at lower temperature [64].

Two different strategies have been used to produce superconducting TlBaCaCuO films from the first MOCVD based reports to date. Previous treatments were carried out making an intermediate hydrolytic step (three step process) to eliminate fluoride phases arising from the alkaline-earth precursor decomposition [12,29]. Afterwards, the direct annealing of the as-deposited matrices has been shown to eliminate fluoride phases through the methathetic reaction (1) giving rise to thallium based superconducting phases. The reaction atmosphere, during the annealing step, has been controlled by mixing different flows (F) of oxygen and argon gasses. The oxygen partial pressure (P_{O_2}) has been evaluated through the following relation:

$$P_{O_2} = \frac{F_{O_2}}{F_{O_2} + F_{Ar}} \times P_{Tot}$$

where P_{Tot} is the total pressure (1 atm).

Adopting the three step process, a careful selection of the discussed parameters allowed the selective and reproducible formation of Tl-2223 and Tl-2212 films [12]. In that case, the annealing step was carried out, in both cases, under oxygen at 890°C and selectivity was obtained by adjusting the P_{Tl_2O}, varying the stoichiometry of the oxide mixture.

Analogously, considering the correlation between P_{Tl_2O}, P_{O_2} and temperature [65], the selective formation of monophasic films may be achieved by varying temperature or P_{O_2}, keeping constant the other parameters. A fourth important variable not discussed so far is the reaction time of the thallium diffusion process. In a previous work [29], we investigated, through the three step process, the phase formation, the morphologic and the transport properties of MOCVD-derived TlBaCaCuO films on random YSZ substrates with respect to all of the variables of the annealing step. Most relevant results are summarised in Table 2.

Table 2 Experimental conditions and results of films grown on random YSZ substrates.

Sample	T_{ann} (°C)	P_{O_2} (atm)	Annealing Time	Oxide mixture Tl_2O_3:BaO:CaO:CuO	TlBaCaCuO phase film	Tc(K)
A	900	1	5	1:2:2:3	Tl-2212	99
B	860	0.13	15	1:2:3:4	Tl-2223	109
C	850	0.08	15	1:2:3:4	Tl-1223/2223	102
D	820	0.13	60	1:2:2:3	Tl-1234, Tl-1223	

180

The films A and B exhibit XRD patterns indicative of the Tl-2212 (Fig. 7 a) and Tl-2223 (Fig. 7 b) phases, respectively. The XRD scans show only the (00ℓ), thus indicating that crystallite c-axes are preferentially oriented perpendicular to the substrate surface. The samples C and D consist of two rather uncommon phases: the Tl-1223/2223 (Fig. 7 c) intergrowth structure and the Tl-1234 (Fig. 7 d), respectively. The XRD pattern of the Tl-1234 phase indicate that also this film is *c-axis* oriented. Note that all the reflections are observed for the single TlO phases, which possess a primitive tetragonal structure, while only the even reflections are observed for the double TlO phases, since they possess a body centred tetragonal cell (i.e. the $h+k+\ell = 2n$ symmetry rule applies).

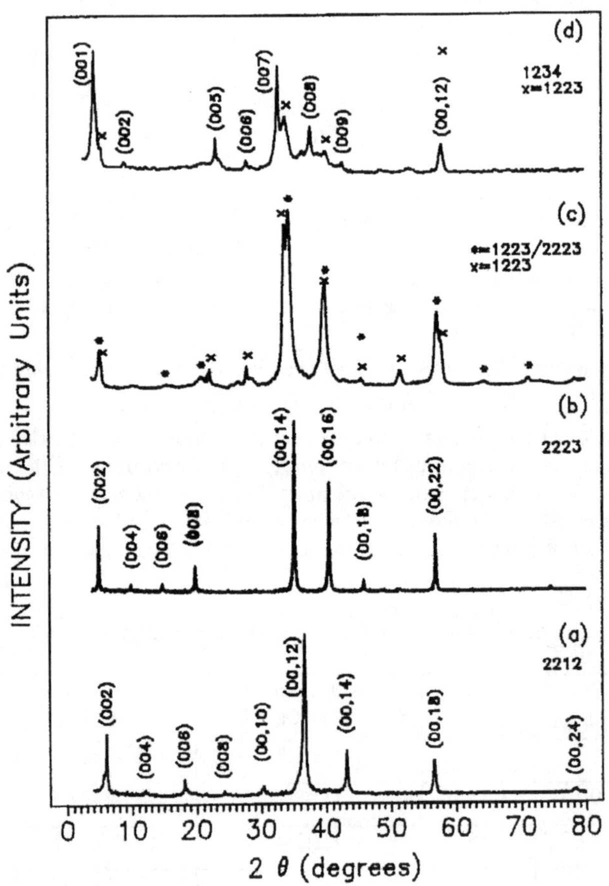

Fig. 7 θ-2θ XRD patterns of (a) Tl-2212, (b) Tl-2223, (c) Tl-1223/2223, and (d) Tl-1234 grown on random YSZ substrates.

The dependence of the electrical resistance of A, B and C films as a function of the temperature has been studied by using the four-point-probe technique (Fig. 8). Resistance measurements show a good normal state metallic behaviour with zero resistances occurring at 99 K, 102 K, and 109 K for Tl-2212, Tl-1223/2223, and Tl-2223 films, respectively.

These samples show critical current densities (J_c) of about 10^3 A/cm^2 at 77 K. Note that, despite the random orientation of adopted substrates, the films are *c-axis* oriented and present good T_c values. These data indicate that good lattice matched substrates are not essential to grow *c-axis* oriented TlBaCaCuO films. Nevertheless, low J_c values are observed as a consequence of the absence of any in-plane alignment.

These results point to the possible use of TlBaCaCuO phase films (not epitaxial, but only *c-axis* oriented) for those applications such as coated conductors, which are of great interest, to date, and do not require excellent performance in terms of J_c.

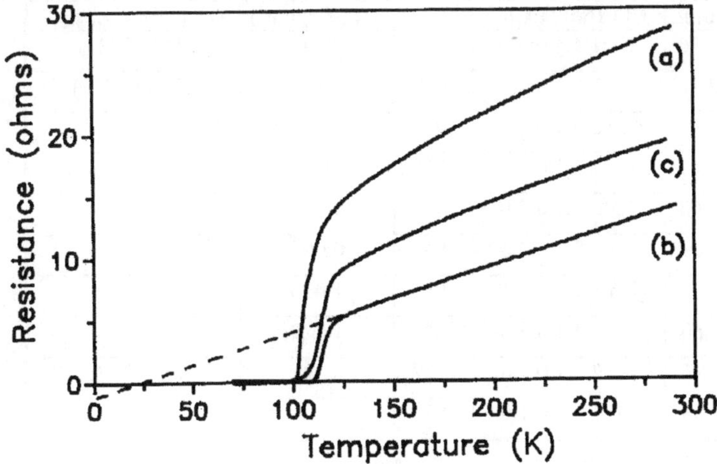

Fig. 8 Variable-temperature resistance data of (a) Tl-2212, (b) Tl-2223, and (c) Tl-1223/2223 grown on random YSZ substrates.

The best performance in terms of T_c, J_c and surface resistance are only encountered for films grown on well matched substrates such as SrTiO$_3$ (100) and LaAlO$_3$ (100) films.

SrTiO$_3$ is a cubic perovskite substrate which has an excellent lattice match with TlBaCaCuO films (a_{SrTiO_3} = 3.91 Å). Nevertheless, there are various opinion on its application because of discordant data on its reactivity with TlBaCaCuO films depending on the processing temperature [21]. LaAlO$_3$ (100), which may be considered a pseudo-cubic perovskite structure (a_{LaAlO_3} = 3.79 Å), has been widely used due to its excellent match with the *a-b* plane of the TlBaCaCuO unit cells. Actually, LaAlO$_3$ has a 0.2% rhombohedral (r) distortion [66], therefore the (100) and (200) lines of the pseudo-cubic structure should be referred to as (012) and (024), respectively, in the correct r indexing (according to ICDD card #31-22 [67]). Nevertheless, the simpler pseudo-cubic indexing will be used in the following. This substrate promotes heteroepitaxial growth even in *ex-situ* processes because of a good in plane alignment, which results in films with very low angle grain boundaries and in turn high values of J_c and low values of R_s.

An optimised process using the direct annealing of as-deposited matrices deposited on $LaAlO_3$ (100) (see equation (3)) yielded high quality epitaxial Tl-based superconducting films [48,68]. In particular, films annealed at a temperature of 800-820 °C using an oxygen partial pressure of 0.15-0.2 atm formed Tl-2212 epitaxial samples. The thickness of the films after the thallium vapour diffusion step was about 1-1.3 μm.
The stoichiometric ratios of the elements for a batch of 10 samples are reported in Table 3.

Table 3. The structural and transport properties of Tl-2212 films grown on $LaAlO_3$ (100) substrates.

Sample	Tl:Ba:Ca:Cu	Tc(K)	$Jc(10^6 A\backslash cm^2)$	$I_{Tl-2212}/I_{Impurity\ phase}$ [a]
a, 50	1.8:1.4:1.0:2.0	101	0.4	2 (Tl-2223)
b, 51	2.0:1.6:1.0:1.7	101	0.1	12 (Tl-1212)
c, 55	2.0:1.9:1.0:1.8	104	0.25	∞
d, 67	1.3:1.2:1.0:1.2	91	0.462	∞
e, 68	2.5:2.0:1.0:2.0	97	0.547	∞
f, 69	2.5:2.0:1.0:2.3	98	0.009	∞
g, 75	2.6:1.7:1.0:2.1	96	1.036	∞
h, 80	1.8:1.3:1.0:2.6	83	1.122	∞
i, 82	2.1:2.2:1.0:1.8	< 77	1.2	6 (Tl-2201)
l, 83	2.0:1.6:1.0:1.9	93	0.46	∞

[a] The impurity phase is reported between parenthesis

The observed stoichiometries point to a good reproducibility in terms of composition both in regard to the MOCVD deposition as well as to the annealing step.

The XRD patterns of these samples indicate that films are highly *c-axis* oriented since only the 00ℓ reflections are observed. The XRD patterns of the sample batch reported in table 3 indicate that most of the films are monophasic and only in few cases other TlBaCaCuO phases are present as minor impurities. The amount of impurity phase has been approximately estimated by determining the $I_{Tl-2212}/I_{Impurity\ phase}$ ratios of the heights of the lower angle XRD peaks [32], $2\theta \cong 6°$, $2\theta \cong 7°$ and $2\theta \cong 7.7°$ for the Tl-2212, Tl-1212 and Tl-2201 phases respectively. Great care has been taken in positioning the samples for the XRD measurements to have a precise film alignment.

In Fig. 9 are reported two typical XRD patterns, one for the monophasic Tl2212 film *l*, the other for film *a* which shows the presence of the Tl-2212 as the major phase and of the Tl-2223 as the impurity phase. The inset in Fig. 9 shows the θ rocking curve of the (00,12) reflection for film *l*. The full width half maximum value (FWHM) of 1.28° indicates a reasonable alignment of the grain c-axes perpendicular to the substrate.

The XRD patterns clearly indicate a c-axis preferred orientation of the Tl-2212 films, even though no further information on the in-plane growth can be obtained from this measurement. To gain further insight into the in-plane orientation, a pole figure of the Tl-2212 (107) reflection has been collected using an Eulerian cradle (4-circle goniometer). The (107) pole figure of a Tl-2212 film on LaAlO$_3$ is reported in Fig. 10. In the polar plot, the radial dimension, indicated as χ or Ψ, refers to the angle between the normal to the film surface and the plane of the X-ray beam. The azimuth, φ, corresponds to the rotation of the substrate about the surface normal. The peaks at $\Psi=47°$ are the (107) reflections from <001> oriented Tl-2212 films. There are four peaks at 90° intervals as expected for this tetragonal material.

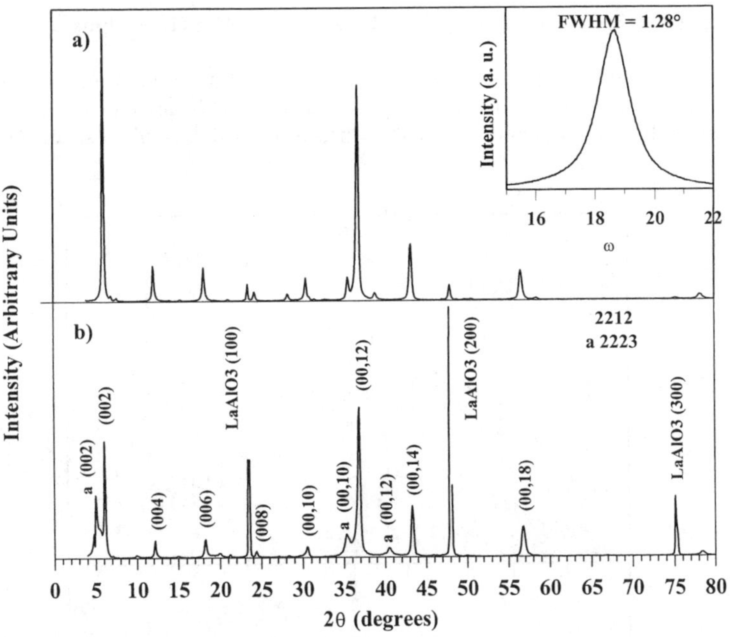

Fig. 9 θ-2θ XRD patterns of a) the Tl-2212 film *a* and b) the Tl-2212 film *l*, both grown on LaAlO$_3$ (100).

184

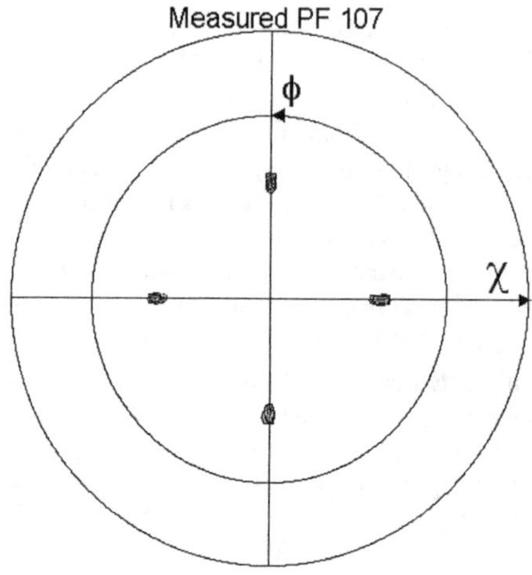

Fig. 10 (107) pole figure of a Tl-2212 film grown on LaAlO₃ (100) substrate.

The EDX spectrum (Fig. 11) shows, in addition to the Ba, Ca and Cu peaks, the Tl M lines centered at 2.270 keV. Moreover, Auger electron spectroscopy (AES) depth profiles have shown no fluorine contamination both in the surface and the bulk of superconducting films.

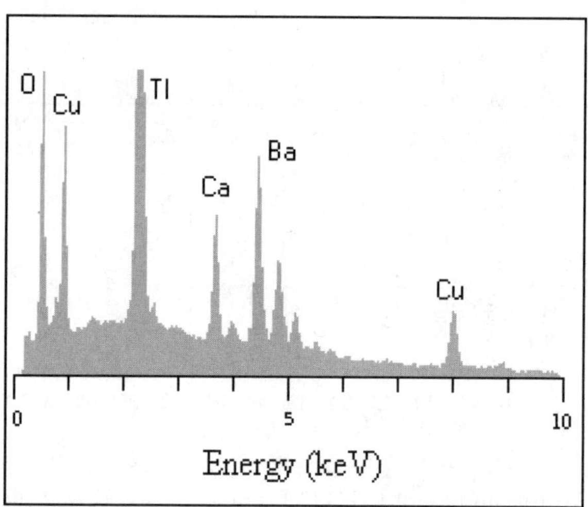

Fig. 11 EDX analysis of a Tl-2212 film grown on LaAlO₃ (100) substrate.

The SEM images show that pinholes are usually present on the film surface while outgrowths are observed only when a slight different stoichiometry is observed for Ba or Cu elements. In Fig.12 two SEM images related to sample *a* and *l* respectively are reported. Note that in the case of sample *a* the surface is very smooth with 10-15 µm grains and presents several pinholes. A slightly rougher morphology has been observed for sample *l* even though no outgrowths are visible on the surface.

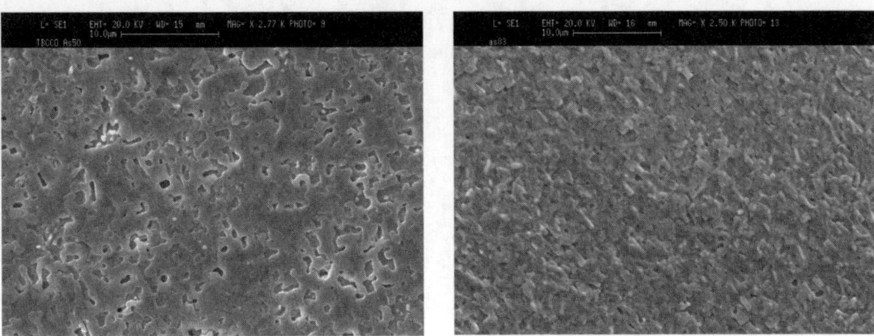

Fig. 12 SEM images of a) the Tl-2212 film *a* and b) the Tl-2212 film *l*.

The microwave properties were studied using an inverted microstrip meander-line resonator. The quality factor Q of the fundamental resonant mode was measured as a function of temperature and input power by means of a network analyser. The technique has been carefully optimised to minimise dielectric, radiation and other external contributions to the losses of the resonator [69]. In this configuration, the surface resistance R_s of the superconducting material is directly related to the Q value through the relation $R_s = \Gamma/Q$, where Γ is a geometrical factor, depending only on the electromagnetic field distribution inside the device. Γ can be analytically evaluated using standard electromagnetic techniques [70].

Each resonator is formed by a couple of films placed on the two sides of a thin sapphire, of thickness h = 100 µm. One of the films is patterned by standard wet photolithography in a meander-line geometry of variable line width w and fixed length L = 30 mm, yielding a fundamental mode f_1 at 1.2 GHz. Two samples were measured: film *a* and *l* were patterned to 150 µm wide and 50 µm wide transmission lines, respectively, while sample *b* was employed as ground plane in both cases. Due to the low aspect ratio w/h, the conductive losses are mainly ($\geq$ 90 %) concentrated in the microstrip, allowing a direct comparison of the microwave properties of the two meandered lines. In spite of the presence of a quite large 2223 impurity phase, the former film showed better microwave properties, both in terms of temperature and field dependence. The best surface resistance values, measured at 4.2 K, were 160 µΩ and 350 µΩ for sample *a* and *l* respectively. In Fig.13 the surface resistance of film *a* as a function of the total current I carried in the

186

transmission line resonator is shown at different temperatures. I is related to the microwave input power P_{in} through the relation [70]:

$$I = \sqrt{\frac{4QP_{in}S_{12}}{\pi Z_0}}$$

S_{12} is the power transmission coefficient at the resonance, and Z_0 is the intrinsic impedance of the transmission line. The maximum current corresponds to a microwave input power of 10 mW.

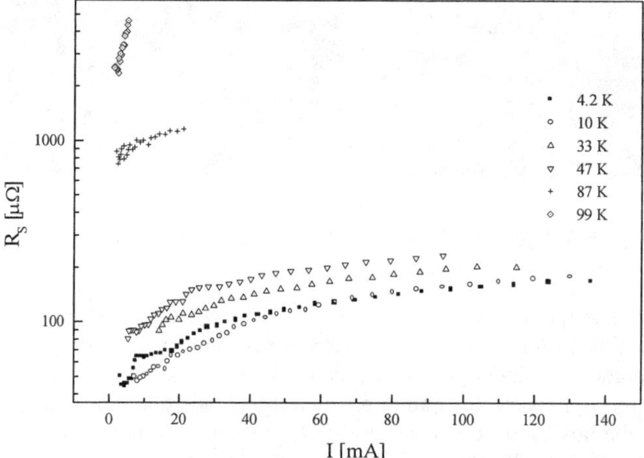

Fig. 13 Surface resistance R_s of sample *a* vs the total current I circulating in the microstrip at various temperatures.

For temperatures below 50 K the dependence was quite flat while R_s values were more than one order of magnitude better than copper up to the maximum power available. It is worth noting that at 87 K the Tl-2212 sample still showed R_s values below 1 mΩ.
Film *I* had a similar dependence, even if, due to the lower quality factor, the maximum current circulating in the microstrip was smaller. The larger values of the surface resistance is likely to be ascribed to the smaller microstrip width, which enhances the effects of the film's edges, degraded by patterning, on the overall microwave response.

In regard to electrical characterisation, the measurements were all performed by the inductive method on the unpatterned 10×10 mm^2 samples. The plot of the third harmonic component voltage U_{31} as a function of the temperature is reported in Fig.14. T_c and J_c values measured for a typical sample are respectively, T_c= 100.5 K with ΔT_c = 1.5 and J_c

$(77 \text{ K}) = 2 \times 10^5$ A/cm^2. The critical current density is evaluated by considering the calibration factor of the coil and the thickness of the film. Four probe resistive measurements on patterned films showed that the etching process did not affect the values of T_c and J_c.

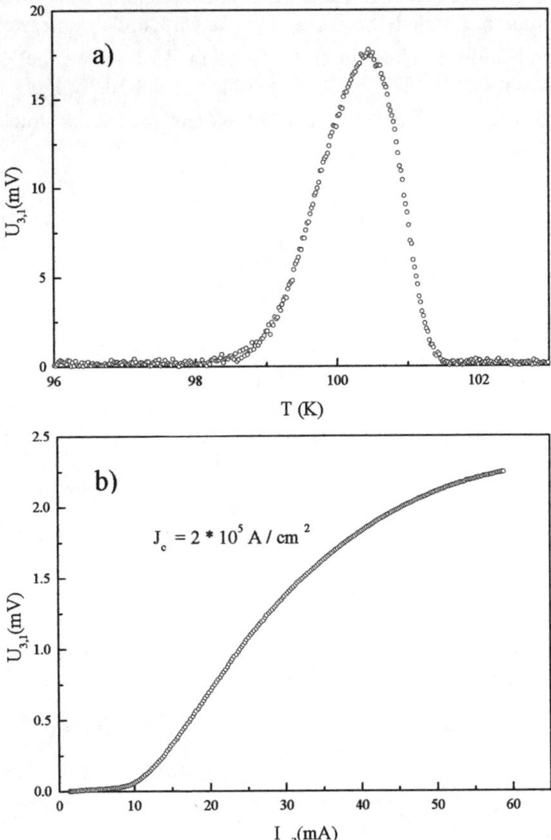

Fig. 14 Third harmonic component of the coil voltage $V_{3,1}$ as a function of the temperature a) and as a function of the coil current b) of a typical Tl-2212 film grown on LaAlO$_3$ (100) substrate.

3.1.1.4 Nature and characterisation of intergrowth phases. An interesting consequence of systematic studies carried out changing all of the variables may be the singling out of particular ranges of P_{O_2} and temperature which allow the formation of new metastable phases such as intergrowth phases.

We have in Catania identified, through XRD i.e. on a macroscopic scale, two different phases namely, the $Tl_{1+x}Ba_2Ca_2Cu_3O_{9+x}$ and $Tl_{1+x}Ba_2CaCu_2O_{7+x}$ intergrowth structures. Two observations were of importance for the identification of the intergrowth phases [46]. The intergrowth phases i) follow the formation of the double TlO layer phases and precede those of the single-thallium layer 1212 and 1223 phases, and ii) the diffraction pattern of one of this structure resembled that of a $Tl_{1+x}Ba_2Ca_2Cu_3O_{9+x}$ integrowth phase film obtained via laser ablation by Holstein et al. [27]. Fig. 15 a shows the X-ray diffraction pattern of a representative (Tl-1223/2223) phase obtained at 850 °C, P_{O_2} = 0.13 atm, 5 min annealing time and an oxide mixture Tl_2O_3:BaO:CaO:CuO of 1:2:3:4 stoichiometry.

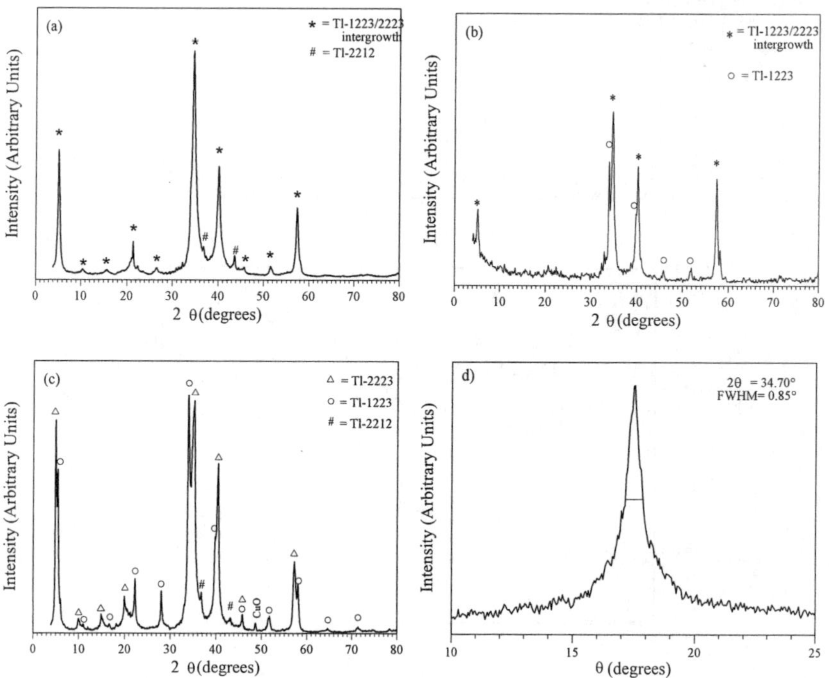

Fig. 15 θ-2θ XRD patterns of (a) a Tl-1223/2223 intergrowth film (850°C, P_{O_2} = 0.13 atm, 5 min), (b) a Tl-1223/2223 intergrowth film containing distinct domain of Tl-1223 (860°C, P_{O_2} = 0.13 atm, 15 min), and (c) a film containing distinct Tl-1223 and Tl-2223 domains (900°C, P_{O_2} = 1 atm, 5 min); (d) rocking curve of the peak at 2θ = 34.7° of a Tl-1223/2223 intergrowth film.

Reflection values of this intergrowth are reported in Table 4 and compared with those of Tl-1223 (ICDD card #44-0476) and Tl-2223 (ICDD #41-1334). Analogously, reflections of the $Tl_{1+x}Ba_2CaCu_2O_{7+x}$ are reported in the same table and compared with the pure Tl-1212 (ICDD #41-0123) and Tl-2212 (ICDD #39-1482) phase reflections. No unidirectional or "constant" shift are observed in both cases, thus indicating that they represent completely different structures from the pure thallium phases. Note, in this context, that when distinct crystalline domains of Tl-1223 are present in the intergrowth $Tl_{1+x}Ba_2Ca_2Cu_3O_{9+x}$ film, the Tl-1223 reflection peaks are detectable in the XRD pattern (Fig. 15 b). When distinct crystalline domains of the Tl-1223 and Tl-2223 phases are present in the film, the X-ray diffraction pattern is indicative of each distinct phase (Fig. 15 c).

The rocking curve full width half-maximum (FWHM) value of 0.75° (obtained for better oriented samples) for the peak at $2\theta = 34.70°$ (Fig. 15 d) shows that the intergrowth phase is *c-axis* oriented.

Table 4 X-ray reflections of the Tl-1223/2223 and Tl-1212/2212 intergrowth phases compared to ICDD data of the parent phases.

Tl-1223 2θ (hkl)	Tl-2223 2θ (hkl)	Tl-1222/2223 2θ	Tl-1212 2θ (hkl)	Tl-2212 2θ (hkl)	Tl-1212/2212 2θ
5.56 (001)	4.96 (002)	5.25	6.99 (001)	6.01 (002)	6.32
11.20 (002)	9.93 (004)	10.45	13.98 (002)	12.03 (004)	12.77
16.74 (003)	14.92 (006)	15.75	21.03 (003)	18.09 (006)	19.30
22.39 (004)	19.93 (008)	21.05	28.18 (004)	30.27 (00,10)	
28.09 (005)		26.05	35.36 (005)	36.70 (00,12)	36.28
33.86 (006)	35.30 (00,14)	34.70	42.72 (006)	43.20 (00,14)	43.00
39.72 (007)	40.60 (00,16)	40.20	50.41 (007)	49.59 (00,16)	50.14
45.69 (008)	45.78 (00,18)	45.75	58.03 (008)	56.30 (00,18)	57.21
58.07 (00,10)	56.77 (00,22)	57.40			

Close-ups of experimental X-ray θ-2θ diffraction patterns for Tl-1223/2223 intergrowth, Tl-1223 and Tl-2223 phases in the 2θ range 4°-8° and 32°-37° are reported in Fig. 16. The intergrowth phase peak at $2\theta = 5.25°$ lies intermediate (Fig. 16 a) between those of (002) Tl-2223 ($2\theta = 5.05°$) and (001) Tl-1223 ($2\theta = 5.50°$). In the 2θ range 32°-

37° (Fig. 16 b), where the pure phase peaks are closely spaced, a well defined, although slightly broadened, intergrowth peak is observed at $2\theta = 34.70°$.

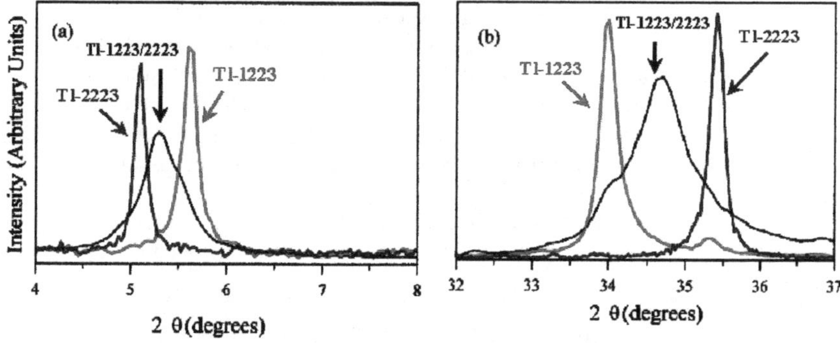

Fig. 16 Close-ups of experimental θ-2θ XRD patterns over the $4°< 2\theta <8°$ (a) and $32°< 2\theta < 37°$ (b) ranges for the Tl-1223/2223, Tl-1223 and Tl-2223 phase films.

Note that this experimental profile compares very well to the computational pattern based on a statistical model of X-ray diffraction by one-dimensional randomly disordered Tl-1223 and Tl-2223 layer thin films, reported by Holstein et al. [71]. This type of intergrowths, in fact, are not ordered, even though they give rise to a particular X-ray diffraction pattern that could no longer be assigned to crystalline $Tl_mBa_2Ca_{n-1}Cu_nO_{2n+m+2}$ phases. Therefore, the $Tl_{1+x}Ba_2Ca_2Cu_3O_{9+x}$ intergrowth phase can be confidently identified as a random disordered sequence of 50% Tl-1223 and 50% Tl-2223 layers (i.e. $x \approx 0.5$) [46]. The superconducting transition onset of a typical Tl-1223/2223 intergrowth film on YSZ substrate is at $T_{c_{onset}} = 118$ K, and zero resistance is achieved at $T_{c_0} = 101$ K. The observed T_{c_0} is slightly lower than that reported (107 K) for the Tl-1223/2223 intergrowth phase film on $LaAlO_3$ [27]. This difference might be related to the different substrate nature, YSZ vs. $LaAlO_3$, as well as to the nature of present discrete grain films since the poor connectivity between adjacent grains is responsible for the broad transition [46].

A model diagram of a plausible architecture of this intergrowth structure is pictured in Fig. 17. The model adopts an hypothetical ten layer sequence of randomly disordered Tl-1223 (c= 15.913 Å) and Tl-2223 (c/2= 17.94 Å) layers.

A similar assignment can be done for the Tl-1212/2212 intergrowth phase. This structure has been obtained at 850 °C, $P_{O_2} = 0.67$ atm, 5 min annealing time and an oxide mixture Tl_2O_3:BaO:CaO:CuO of 1:1:2:3 stoichiometry.

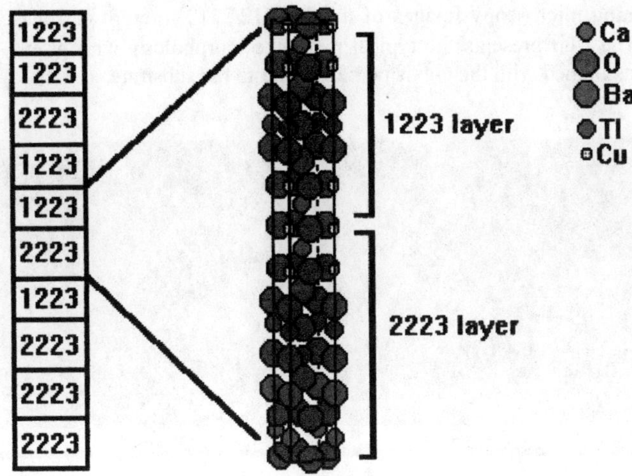

Fig. 17 Model diagram of the Tl-1223/2223 intergrowth phase showing a hypothetical ten layer sequence of randomly disordered 1223 (50%9 and 2223 (50%) layers.

In Fig. 18 are reported the close ups of significant 2θ regions for the new Tl-1212/2212 intergrowth structure. The reflection at $2\theta = 6.30°$ lies intermediate (Fig. 16 a) between those of (002) Tl-2212 and (001) Tl-1212, while that observed at 36.2° lies exactly in between those of the (005) Tl-1212 and (00,12) Tl-2212 pure phases. Therefore, also in this case, a random disordered sequence of 50% Tl-1212 and 50% Tl-2212 layers (i.e. $x \approx$ 0.5) can be hypothesised.

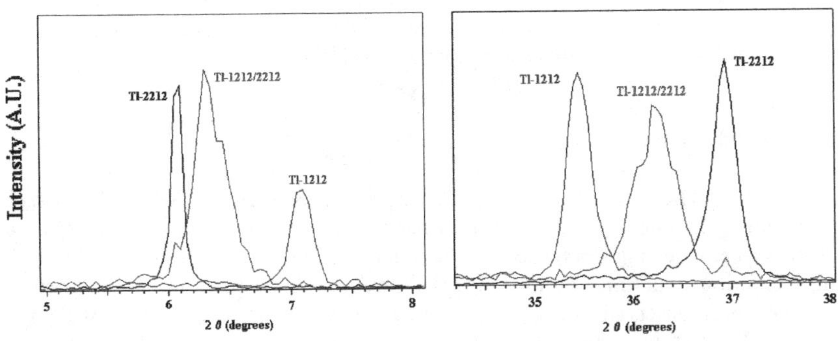

Fig. 18 Close-ups of experimental θ-2θ XRD patterns over the 5°< 2θ <8° (a) and 34°< 2θ < 38° (b) ranges for the Tl-1212/2212, Tl-1212 and Tl-2212 phase films.

192

Scanning microscopy images of the Tl-1212/2212 intergrowth phase is reported in Fig. 19. The film presents the typical plate-like morphology with grains, tens of μm in dimension, aligned with the c-axis perpendicular to the substrate.

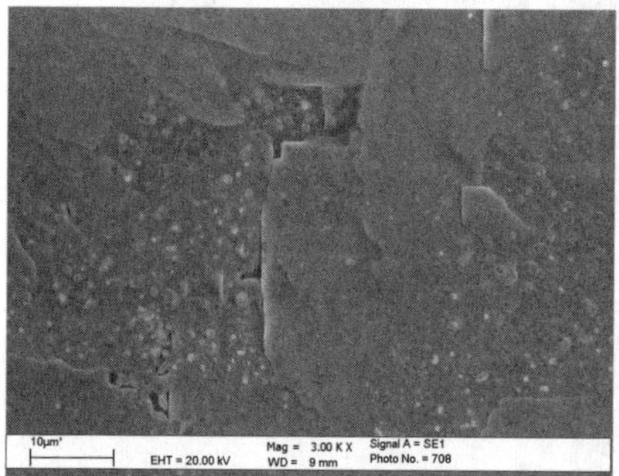
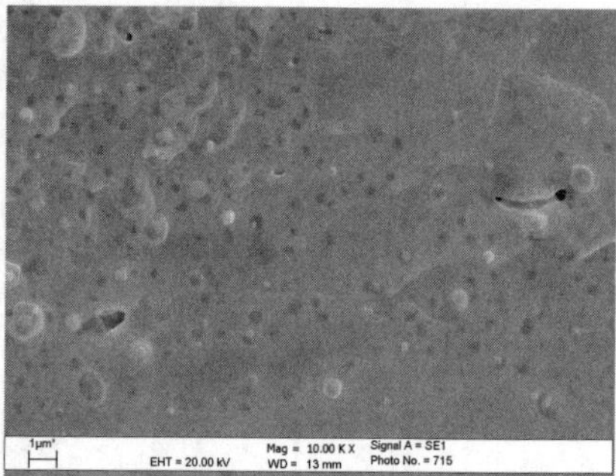

Fig. 19 SEM images of the Tl-1212/2212 film at low (left) and high (right) magnification.

To our knowledge, only one report has been reported, to date, on an ordered intergrowth between phases with a single TlO layer and different perovskite-like $Ba_2Ca_{n-1}Cu_nO_{2n+2}$ units [72]. Very few examples of intergrowths containing the same perovskite layers and both the rocksalt-like single and double TlO layers [26,27,73] have been reported. In a recent paper, Zhang et al. [74] have identified by selected-area electron diffraction and high-resolution electron microscopy several superstructures related to regular *c-axis* intergrowths between Tl-1223 and Tl-2223 layers. In function of the ratios between the Tl-1223 and Tl-2223 layers the c-axis parameters associated with the various superstructures have been determined. Nevertheless, the XRD pattern of the Tl-1223/2223 intergrowth structure does not fit univocally any of the calculated patterns based on the unit cell parameters of the reported superstructures [74,75].

3.1.2. Single sided *a-axis* oriented films

All of the studies reported, up to date, on the preparation of TlBaCaCuO *c-axis* oriented films. The *c-axis* growth comes in handy for many applications, and, since it shows the superconducting CuO_2 sheets parallel to the substrate, it is associated with films having high critical currents and low surface resistances. In some cases, *a-axis* oriented crystals have been observed as outgrowths with a minor or major extension versus *c-axis* orientation. Nevertheless, to our knowledge, our previously reported example of TlBaCaCuO *a-axis* oriented films [76,77], whose nature was assessed by X-ray diffraction, is still unique.

a-axis oriented films may be applied in the preparation of a particular type of Josephson's junctions, the bi-epitaxial junctions [78]. In this case the tunneling effect is observed between two superconducting films having different epitaxial relation with the same monocrystalline substrate. The different epitaxial growth is observed using a buffer layer that alters the usual epitaxial relation between the superconducting film and the substrate.

3.1.2.1. Nature of the buffer LaAlO₃ layer. High quality $LaAlO_3$ (100) buffer layers have been grown by MOCVD on $SrTiO_3$ (100) substrates from the La(hfa)$_3$•diglyme (diglyme = bis(2-methoxyethyl)ether) complex and Al(acac)$_3$. The novelty of the used approach relies upon the use of the La(hfa)$_3$•diglyme which acts as a solvent for the aluminum precursor, thus affording a two-component liquid source. Further details on the single precursor source and deposition conditions may be found in ref. [79].

A short comment deserves the nature of the $LaAlO_3$ buffer layer. Fig. 20 shows the θ-2θ XRD pattern of a film deposited at 1000 °C.

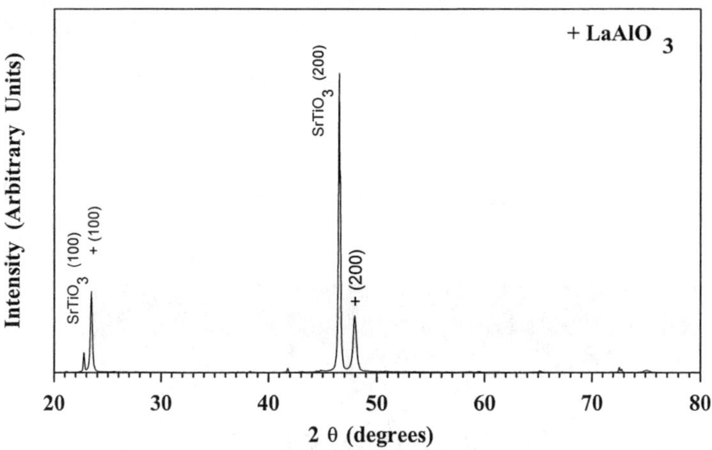

Fig. 20 θ-2θ XRD pattern of an $LaAlO_3$ (100) film grown on $SrTiO_3$ (100).

The two lines observed in addition to the expected (100) and (200) $SrTiO_3$ reflections can be identified as (100) and (200) of the $LaAlO_3$ perovskite structure. The θ-2θ pattern clearly indicates a c-axis preferred orientation of $LaAlO_3$, even though no further information on the in-plane growth can be obtained from this measurement. To gain further insight into the in-plane crystallography, the (111) $LaAlO_3$ pole figure has been collected using an Eulerian cradle. Four distinct poles at ψ=54° and φ=0, 90, 180 and 270°, respectively, have been found (Fig. 21 a) in accordance with the typical cube-on-cube growth:

$$[001]\ SrTiO_3\ ||\ [001]\ LaAlO_3, \qquad [100]\ SrTiO_3\ ||\ [100]\ LaAlO_3$$

194

Further information has been obtained from the rocking curve of the (100) LaAlO$_3$ reflection. The rocking curve shown in Fig. 21 b (FWHM$\cong$0.21°) confirms the low mosaicity of the LaAlO$_3$ thin film. SEM images (Fig. 22) show that films are highly smooth and consists of ~80 nm grains uniformly distributed.

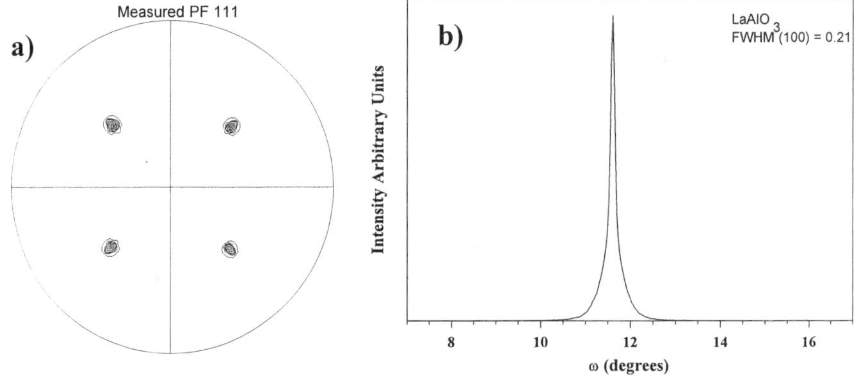

Fig. 21 (111) pole figure a) and rocking curve of the (100) reflection b) of an LaAlO$_3$ (100) film grown on SrTiO$_3$ (100) substrate.

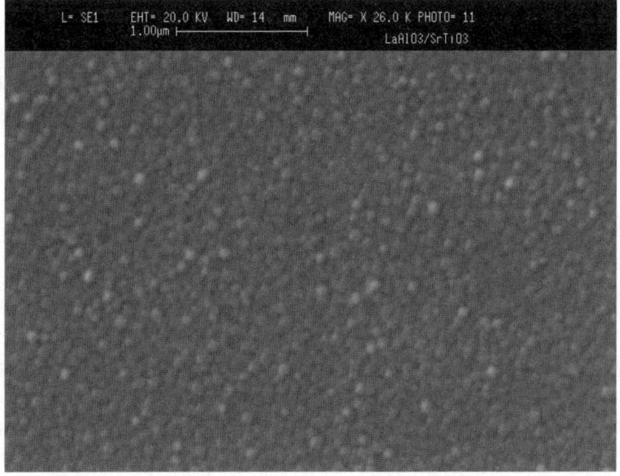

Fig. 22 SEM image of an LaAlO$_3$ (100) film grown on SrTiO$_3$ (100) substrate.

3.1.2.2. Phase formation of a-axis oriented TlBaCaCuO films. The unusual *a-axis* orientation has been observed for TlBaCaCuO films grown on the $LaAlO_3$ buffer layer deposited in our labs. The films were prepared following the same procedure as described therein: i) deposition of a BaCaCuO(F) matrices, and ii) *ex-situ* annealing in a thallium environment. In all experiments the BaCaCuO(F) precursor matrices have been, simultaneously, deposited on substrates of commercial $LaAlO_3$ <100> oriented, (hereafter C-$LaAlO_3$) and on buffer layer of $LaAlO_3$ (100)/ $SrTiO_3$ (100) (hereafter B-$LaAlO_3$).
The superconducting thallium based films have been obtained under the following conditions: i) 760-900°C annealing temperature, ii) 5-80 minutes annealing time, iii) 0.67 1.0 atm P_{O_2}, iv) oxide mixture Tl_2O_3, BaO, CaO and CuO in the ratio 1:1:2:3.

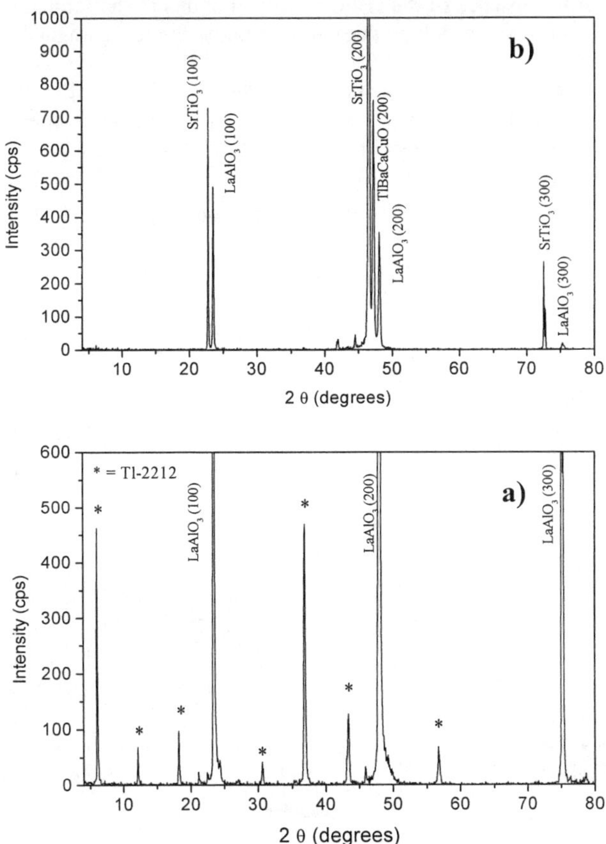

Fig. 23 θ-2θ XRD patterns of a) a Tl-2212 film grown on C-$LaAlO_3$ and b) a Tl-2212 film grown on B-$LaAlO_3$.

196

The Fig. 23 a shows a typical x-ray diffraction pattern of the sample deposited on C-LaAlO$_3$ substrates. The XRD pattern shows all the (00ℓ) reflections of the Tl-2212 phase besides to the peaks at $2\theta = 23.45°$ and $2\theta = 48.15°$ associated with the C-LaAlO$_3$ (100) and (200) reflections, respectively. This indicates that the film is oriented with the c-axis perpendicular to the substrate surface.

In Fig. 23 b is reported the XRD pattern of a TlBaCaCuO film deposited on B-LaAlO$_3$. The pattern shows a unique peak at 47.20° associated with the (200) reflection of the thallium based phase. In fact, all the other peaks are attributable to the SrTiO$_3$ substrate (100) and (200) reflections, peaks observed at 22.75° and 46.50°, and to the LaAlO$_3$ buffer layer (100) and (200) reflections, peaks at 23.50, and 48.10°. This indicates that the superconducting film grains are oriented with their a-axes perpendicular to the substrate surface. Note that the absence of the (100) reflection points to the formation of a double TlO layer phase, since these phases due to their body centred tetragonal cells, do not present this reflection.

Due to the similar values of a-axis lattice constants for all the TlBaCaCuO phases, it is not possible to determine the superconducting phase from the θ-2θ pattern. To this aim, it is mandatory to find reflections related also to the c-axis parameters.

This information can be obtained by recording a series of patterns from $2\theta = 25°$ to $2\theta = 34°$, where the most intense (hkℓ) reflections fall, on varying the Ψ angle from 35° to 55°.

The θ-2θ scan recorded at $\Psi = 43°$ is reported in Fig. 24.

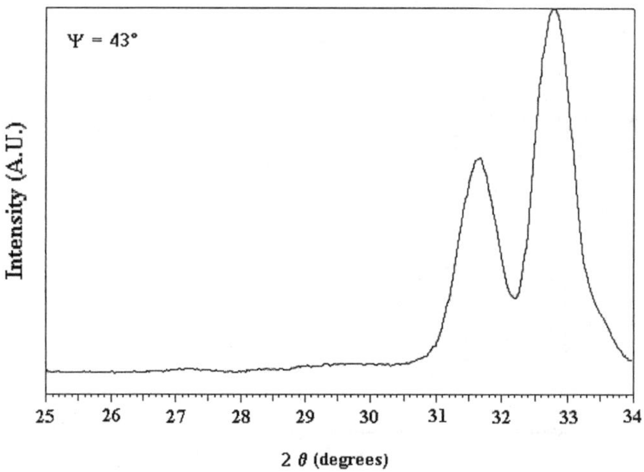

Fig. 24 θ-2θ XRD pattern at $\Psi = 43°$ of an *a-axis* oriented Tl-2212 film grown on B-LaAlO$_3$.

While the peak at 32.90° is not indicative of any particular phase, being associated with the (110) reflection, the peak at 31.45° is attributable to th-e (107) reflection of the Tl-2212 phase. This unequivocally indicates that the *a-axis* oriented films are formed by the Tl-2212 phase, analogously to the *c-axis* films simultaneously annealed.

To confirm the out of plane alignment, rocking curves have been recorded in both cases. The rocking curve of *c-axis* oriented Tl-2212 film (00,12) reflection (Fig. 25 a) presents a FWHM of 0.72°, which indicates that the grains have a low dispersion. The rocking curve of *a-axis* oriented Tl-2212 film (200) reflection (Fig. 25 b) presents a FWHM of 0.98°. This is indicative of a similar grain dispersion of *a-axis* vs. *c-axis* oriented films.

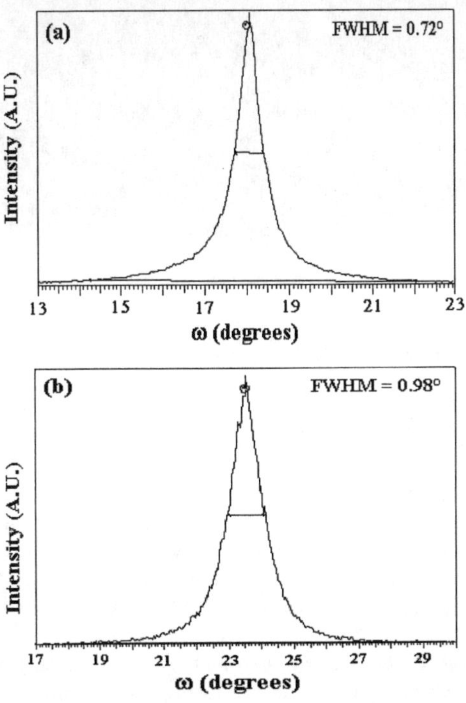

Fig. 25 Rocking curves of the (a) (00,12) Tl-2212 and of the (b) (200) Tl-2212 films.

The morphology has been investigated by scanning electron microscopy. The SEM image of an *a-axis* oriented film (Fig. 26) shows a particular texture with long and narrow grains of the superconducting phase that intersect to each other at right angles. This morphology is typical of a-axis grown grains.

These results, considered the wide temperature and P_{O_2} ranges, confirm that the formation of *a-axis* oriented films is independent on the annealing temperature, but is affected by the morphological nature of the B-LaAlO$_3$ substrate. In addition, also the chemical nature can be ruled out, since different orientation have been obtained simultaneously on substrates with the same composition.

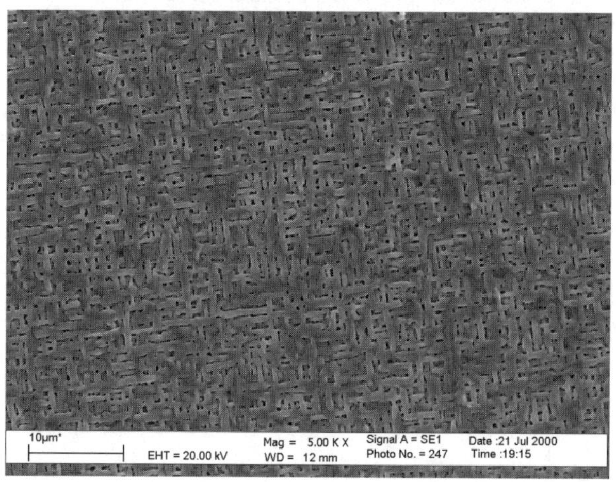

Fig. 26 SEM image of an *a-axis* oriented Tl-2212 film grown on B-LaAlO$_3$ (100) substrate.

3.2. Double sided films

Microwave applications, such as bandpass filters for cellular communications, require films with low microwave surface resistance up to high power levels to minimise the filter loss and intermodulation distortion products. Substrates with low microwave losses such as sapphire, LaAlO$_3$ and MgO are essential for microwave applications. In the following paragraphs, the fabrication and characterisation of DS films on LaAlO$_3$ (100) [80] and MgO (100) are reported.

3.2.1. MOCVD fabrication of DS BaCaCuO(F) matrices. The present MOCVD process relies upon a novel approach based on the use of a single source multimetal mixture in a monocomponent hot-wall reactor whose geometry allows uniform coating of both sides of 10×10 mm^2 LaAlO$_3$ (100) or 10×10 mm^2 MgO (100) substrates. The multicomponent precursor consists of a homogeneous mixture of Ba(hfa)$_2$•tetraglyme, Ca(hfa)$_2$•tetraglyme and Cu(tmhd)$_2$ in an appropriate stoichiometric ratio. The Ca(hfa)$_2$•tetraglyme precursor likely acts, upon melting, as a solvent for the other species and gives a homogenous mixture of all the three precursors. The multimetal mixture precursor has improved characteristics/advantages over the separate management of singular sources, with different operational parameters.

The single source behaviour of the three precursor mixture has been assessed by a differential scanning calorimetric (DSC) study (Fig. 27). The DSC scans of the single Ba,

Ca and Cu precursors, show evidence of the endothermic peaks due to melting (94.9 °C for Ca(hfa)$_2$•tetraglyme, 153.6 °C for Ba(hfa)$_2$•tetraglyme, and 196.3 °C for Cu(tmhd)$_2$) and to evaporation from melts in the 250-280 °C, 260-290 °C, and 210-260 °C temperature ranges, respectively. In the DSC scan of the multicomponent precursor, the lower temperature peak (89.9 °C) represents melting of the Ca(hfa)$_2$tetraglyme component, while the endothermic broad peak in the 110-160 °C temperature range is likely associated with Ba(hfa)$_2$•tetraglyme and Cu(tmhd)$_2$ dissolution. The highest temperature (190-270°C) endothermic process is associated with evaporation of the melt mixture. Note, in this context, that the endothermic peaks expected for the Ba(hfa)$_2$•tetraglyme and Cu(tmhd)$_2$ melting, at 153.6 °C and 196.3 °C respectively, are not observed, thus confirming that these precursors loose their chemical identity forming a single source.

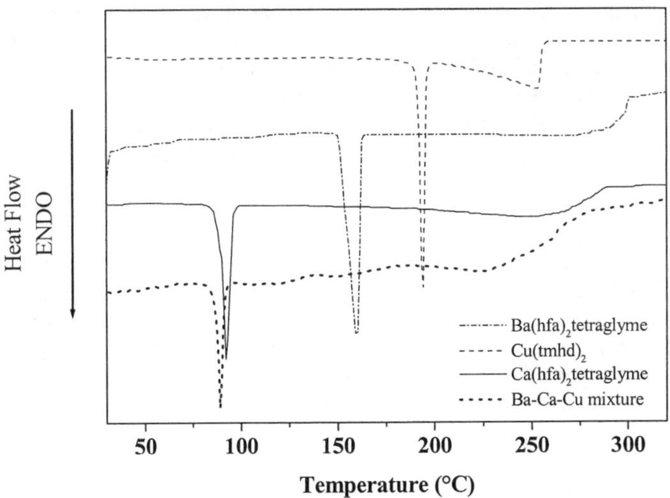

Fig. 27 DSC curve of the 3:1:0.5 source mixture compared with data of individual precursors.

Various stoichiometric mixtures and parameter conditions were tested to fix the best suited stoichiometry to grow the 2:1:2 (Ba-Ca-Cu) matrices and the best ratio was found to be 3:1:0.5 for the three precursors. In order to obtain DS matrices, the substrate was positioned in the center of the 25 mm diameter reactor using a special stainless steel substrate holder. Also in this case the as-deposited matrices have poor crystalline structures, since low intensity peaks associated with reflections of BaF$_2$ and CuO phases have been observed. EDX analyses of matrix films show some minor compositional inhomogeneity over the whole area (10×10 mm^2) with 2±0.2:1±0.1:2±0.2 Ba:Ca:Cu stoichiometries. Identical compositions are found in corresponding points of the two LaAlO$_3$ sides. The SEM plane view image points to homogeneous films with 1μm grains.

The advances/advantages of the present synthetic strategy to simultaneously coating both sides of the $LaAlO_3$ substrates can be compared with limitation of PVD techniques. The intrinsic line of sight nature of the latter methodologies requires sequential depositions, thus often resulting in contamination/alteration of the first deposited side.

Finally, it is of interest to compare this horizontal hot-wall MOCVD approach with the vertical hot-wall MOCVD apparatus proposed by Ito et al. [81] for the preparation of DS-YBaCuO thin films. This earlier procedure has an 80 °C temperature differential along a 10 mm substrate and requires a pulling up procedure to balance the thermal inhomogeneities. By contrast, the present approach allows uniform temperatures in a static substrate, and, therefore, an easier scaling to larger area films.

3.2.2. Thallium vapour diffusion of DS matrices. BaCaCuO(F) matrices have been cleanly converted into pure TlBaCaCuO superconducting films adopting the thallium vapour diffusion process described in the previous section, using the crucible geometry.

Note that, this procedure requires a fine tuning of the Tl_2O vapor pressure over both sides of the processed matrices to cleanly convert fluoride phases and to drive the transformation into the desired Tl-2212 superconducting pure phase.

Several experiments have been made to produce the most suitable Tl_2O uniform partial pressure over the entire sample. This is crucial to have synthetic control over a desired phase, and becomes even more difficult in the case of DS films. The BaCaCuO(F) matrices have been exposed to Tl_2O vapours in alumina crucibles with optimised geometry for 10 x 10 mm^2 substrates and tightly covered with a thin gold foil. The geometry of the crucible assembly has proven crucial to maintain uniform Tl_2O pressures over both sides of the matrices and, hence, to form homogeneous DS Tl-2212 films. DS matrices placed on the bottom of the boat produced DS films with different phases on the two sides, namely the Tl-2212 phase formed on the top and the Tl-1212 phase on the bottom side. This indicates a Tl_2O vapour pressure gradient along the vertical of the boat. Identical Tl_2O vapour pressures, hence identical Tl-2212 phases, have been obtained on both sides upon suspending, over a finally grounded oxide mixture, the 10 x 10 mm^2 matrices horizontally and equally spaced between the bottom of the boat and the gold covering foil. This geometry yielded best results, in terms of identical Tl-2212 phase formation on both sides as well as minimal failures.

In terms of best operational annealing parameters, monophasic DS Tl-2212 films have been obtained adopting: i) 800-820°C annealing temperature, ii) 30-40 min annealing time, iii) 0.7-1 atm oxygen partial pressure, and finally iv) an oxide mixture Tl_2O_3: BaO: CaO: CuO in the 1: 1: 2: 3 ratio. The resultant DS Tl-2212 films grown on $LaAlO_3$ (100) exhibit XRD patterns (Fig. 28) indicative of well-crystallised *c-axis* oriented Tl-2212 phase. The (00,12) rocking curve with a full width half maximum (FWHM) of 0.9° points to an high out of plane alignment. The in-plane alignment has been confirmed by recording the (107) pole figures of both sides of the films. In Fig. 29 are reported the pole figures, in both bidimensional and tridimensional forms. The presence of four poles at $\Psi =$ 47.8° every 90° of φ confims the epitaxial nature of double sided films grown on $LaAlO_3$ (100). The Tl-2212 phase films are fluorine free, as confirmed from the EDX and WDX data.

DS Tl-2212 films grown on MgO (100) show analogous XRD patterns and good out-of- plane alignment, but no alignment is observed in the *a-b* plane.

SEM micrographs of both sides of DS Tl-2212 films on LaAlO$_3$ (100) (Fig. 30) show dense and homogeneous structures consisting of smooth overlapping platelets up to roughly 10 µm diameter.

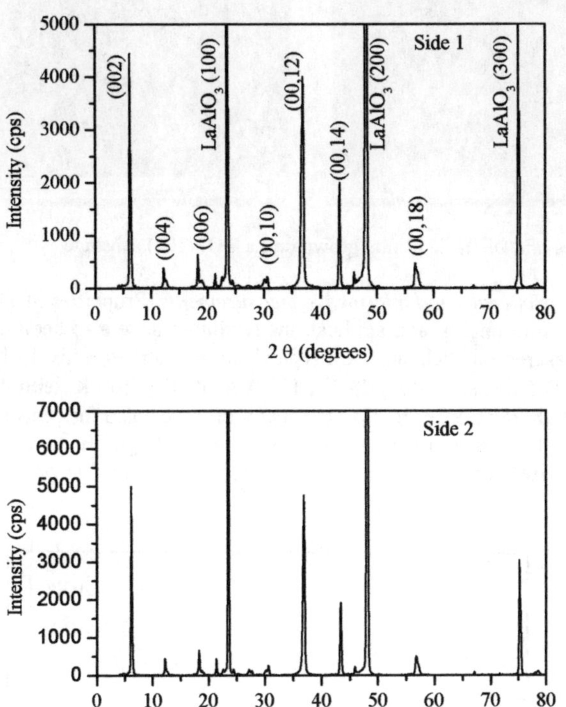

Fig. 28 XRD patterns of a DS Tl-2212 film grown on LaAlO$_3$ (100) substrate.

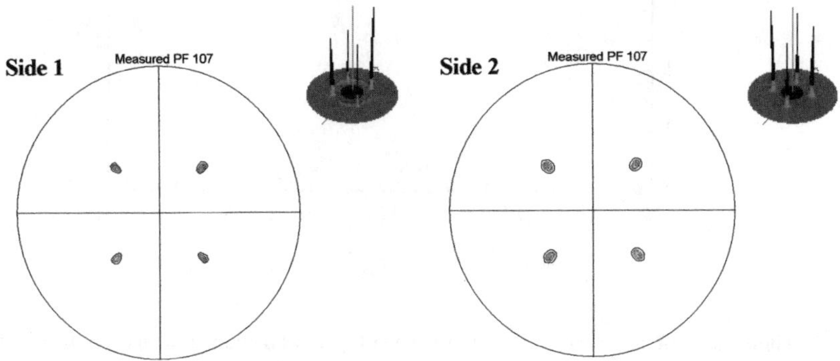

Fig. 29 (107) pole figures of a DS Tl-2212 film grown on LaAlO$_3$ (100) substrate.

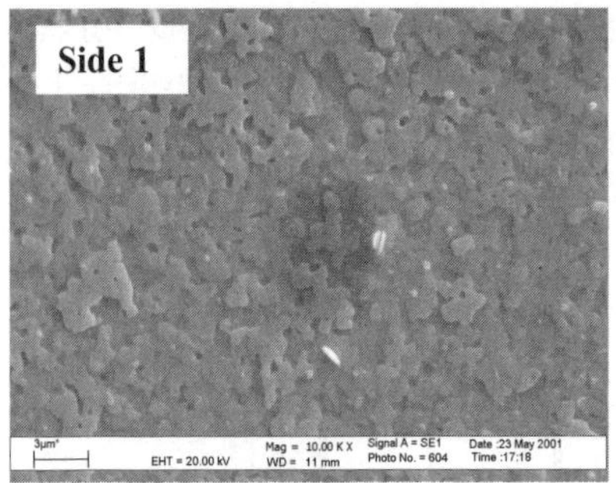

Fig. 30 SEM images of a DS Tl-2212 film grown on LaAlO$_3$ (100) substrate.

3.2.2.3. Transport properties and microwave measurements. Properties of Tl films that are key diagnostics to film quality and applications readiness have also been investigated. J_c and T_c values measured on each side of a typical sample are respectively $J_{c,1}$ (77 K)=3 x 10^4 A /cm^2, $T_{c,1}$=98.5 K and $J_{c,2}$ (77 K)=7 x 10^4 A /cm^2, $T_{c,2}$=100 K, with $\Delta T_c = 1$ in both cases. Note that slight differences in T_cs are usually observed also for films grown by PVD techniques. The plot of the third harmonic component voltage U_{31} as a function of the temperature is reported in Fig. 31. The peak represents the T_c value of a typical double-sided 2212 film.

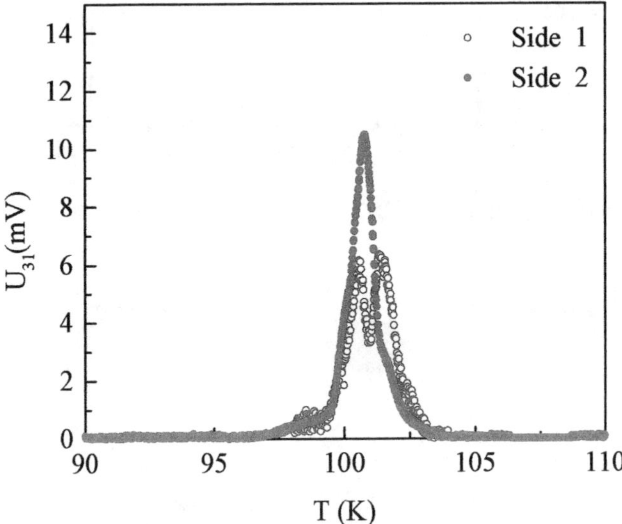

Fig. 31 Third harmonic component of the coil voltage $V_{3,1}$ as a function of the temperature of a DS Tl-2212 film grown on LaAlO$_3$ (100) substrate.

Surface impedance measurements have been performed at 87 GHz employing an end wall cavity method. The surface resistance as a function of temperature shows a residual term R_{res} ≈ 10 mΩ for the side 2, while for the side 1 the R_{res} term is about one order of magnitude higher. The former R_{res} value is comparable to the surface resistance values measured on SS Tl-2212 film [68], and when scaled for the usual quadratic frequency law, it is even better than results reported by Willemsen *et al.* for SS films ($R_{s,min} \sim 50$ $\mu\Omega$ at 3.7 GHz) [24].

3.3. Devices

Dual mode devices based on HTS films represent an interesting class for telecommunication applications since they combine a miniaturised size with a good power handling. On the basis of the transport and microwave properties, DS Tl-2212 films grown by MOCVD are very promising for the development of passive microwave planar devices for telecommunication applications.

Some of the DS Tl-2212 phase films grown on LaAlO$_3$ (100) substrates have been tested for the realisation of dual mode microwave filters [82]. Both sides of the used films are smooth, homogeneous in thickness and composition and show epitaxially grown Tl-2212 phases. Typical values of T_c =100 K and J_c = 0.5 MA/cm^2 at 77 K have been measured for both sides. Microwave measurements have been performed at 20 GHz by using dielectric resonators.

Two planar superconducting single stage dual mode filters operating in C-band with 1% and 10% fractional bandwidth have been realised. The basic element of the device is a square patch resonator diagonally crossed by unequal slots providing both a size reduction and a simple and controllable way to couple the two degenerate modes. Device response has been studied at different temperatures performing power and intermodulation measurements. Fig. 32 shows the layouts of the dual mode filters realised. The basic resonator is a 6x6 mm^2 squared patch crossed by a pair of diagonal cuts L_{slot} = 6 mm long and w = 0.17 mm wide.

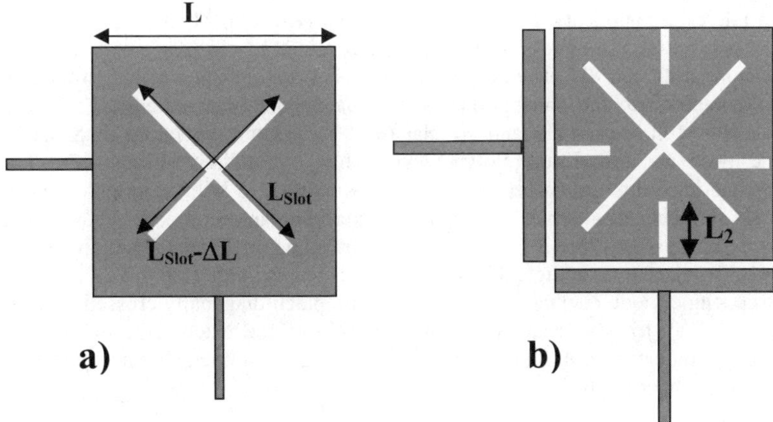

Fig. 32 Geometry of a) cross type and b) star type dual mode filter with a direct and capacitive external connection, respectively.

This first typology is named *cross type* (Fig. 32 a); in the second one, named *star type* (Fig. 32 b), four transverse cuts, of length $L_2 = 2$ mm, are further introduced. For a detailed description of the principle of operation of these typologies see ref. [83]. Briefly, the cuts increase the path of r.f. current and hence reduce the frequency of each resonance. It was found that the resonant frequencies are decreased by more than 30% if $L_{slot} = L$, and that a further 15% reduction is achieved by introducing the four symmetric transverse cuts with $L_2 = L/3$. Direct and capacitive external connections have been considered for the *cross type* and *star type* configuration, respectively. In the former case, feed lines 0.17 mm wide, having 50 Ω internal impedance, are directly attached to the resonator, while in the latter one a gap of 0.1mm is present between the stubs (0.6 mm width and 6 mm length) and the patch resonator. They present a frequency fractional bandwidth of 10% and 1% for the external direct and capacitive couplings, respectively. We focused on these two configurations since they allowed the investigation of either the largest power handling (direct coupled *cross type*) or the best miniaturisation (capacitive coupled *star type*) previously obtained with this class of filter by using YBCO films [83].

The scattering parameter S_{12} measured as a function of the frequency for the capacitive *star type* filter shows a good agreement with the simulation, even if the response showed at low temperatures, T<50K, an enhanced in-band ripple due to the temperature dependence of the coupling between the two modes. The simulations have been performed with ENSEMBLE, a 2.5D EM simulator, based on the moment method assuming the resonators as perfect conductors on a substrate with a dielectric constant $\varepsilon_r = 23.6$.

The insertion losses *IL* as a function of the temperature and of the input power have been studied and compared with those obtained on the YBCO filter [82]. It was found that the losses of Tl-2212 device are lower than the YBCO device in case of $T \geq 85$ K. It is worth mentioning that at 90 K the *IL* of the Tl filter are still lower than those obtained for the filter realised with 1μm gold (*IL*=7dB) at the same temperature, while the YBCO filter is already in the normal state. The Tl-2212 filter presents a power handling lower than the YBCO filter, which can be ascribed to the presence of weak links in these films. Further details on the filter characteristics may be found in ref. [82].

An interesting application of DS Tl-2212 films grown on MgO (100) substrates has been the realisation of a novel compact antenna obtained by crossing a square patch with two or more slots [84, 85]. The proposed design has an antenna size reduction of about 40% as compared to the conventional square patch microstrip antennas. Single patch antenna both with linear (LP) and circular (CP) polarisation operating in X-band have been designed and tested at prototype level. They are realised by using DS Tl-2212 superconducting films grown on MgO substrates and tested with a portable cryocooler. The films have been patterned by standard photolithographic process. They showed at T=77 K return loss less than 25 db and a power handling of 23 dbm. The top view of the basic resonator is shown in Fig. 33 and described in details in ref. [83].

The basic single mode resonator is a L x L square patch diagonally crossed by slots (L_1, L_2) long and by transverse cut (L_C) on each side of the patch. Due to the cuts, the fundamental frequency is shifted downward increasing the slot length since the r.f. current path increases accordingly.

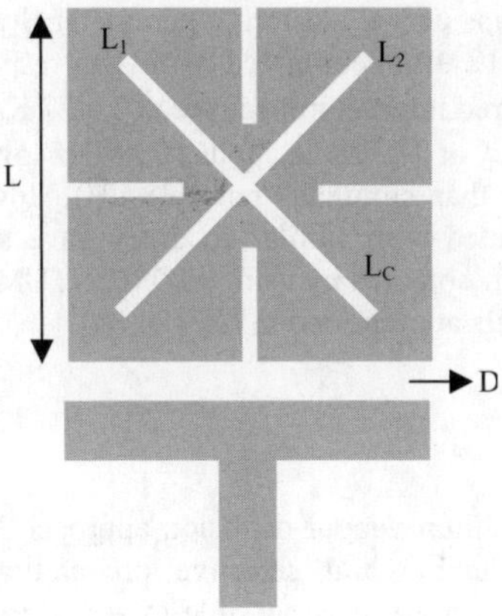

Fig. 33 Layout of gap-coupled-fed miniaturized antenna.

Since the cross slots introduce also a phase shift in the modes, circularly polarised (CP) antennas can be easily obtained with a single fed by introducing an opportune difference $\Delta L = L_2 - L_1$ between the slots.

The antenna performances have been tested using a compact Stirling cryocooler equipped with a quartz glass window near to the antenna at Yamagata University [86]. The distance between source and patch antenna is 1.5 m.

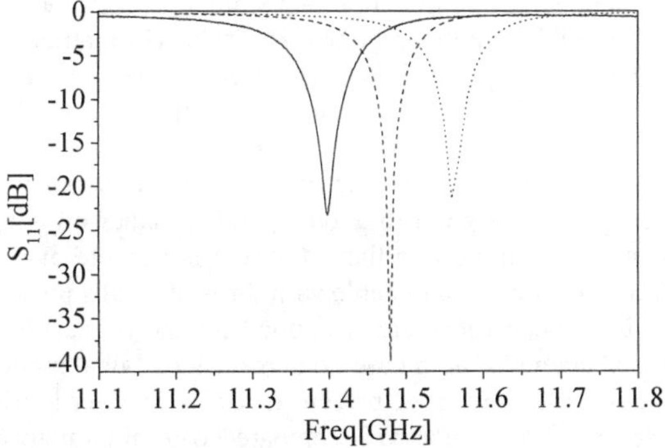

Fig. 34 Experimental scattering parameters S_{11} (continuous curve) of the LP Tl-2212 antenna measured at T = 77 K compared with ENSEMBLE (dashed curve) and MOMENTUM (dotted curve) simulation.

Fig. 34 shows the measured (continuos curve) scattering parameter S_{11} at T= 77 K of the LP antenna realised using a Tl-2212 film compared with the response of Ensemble simulator (dashed curve). The measured internal impedance at T= 77 K is Z= 47 Ω and Z= 52 Ω at T<60 K. The centre frequency at T=77K is about 100 MHz lower than simulation and S_{11} value (-23 dB) is also lower than simulated one (-38 dB). However, it should be noted that different simulators operated with similar accuracy gave somewhat different results. For comparison the simulation obtained by using MOMENTUM is also shown.
In regard to CP antennas further details are reported in ref. [84,85].

4. CONCLUSIONS

The combined MOCVD and thallium vapour diffusion approach has proved to be a very valuable route to the reproducible and selective preparation of TlBaCaCuO monophasic films. From this brief overview it is possible to assess the versatility of the used approach, which easily allows not only the formation of single- but also of double-sided films.

In particular, the reported results indicate that this synthetic strategy can be successfully used to prepare metastable structures such as $Tl_{1+x}Ba_2Ca_2Cu_3O_{9+x}$ and $Tl_{1+x}Ba_2CaCu_2O_{7+x}$ intergrowth phases. It is also important to note that through a full MOCVD route it is possible to prepare *a-axis* oriented TlBaCaCuO films, that could be applied to fabricate bi-epitaxial Josephson junctions.

Nevertheless, a major advantage of the presently adopted *ex-situ* MOCVD based approach remains the possibility to easily produce films on both sides of a substrate. The proposed method for the fabrication of DS Tl-2212 films represents a challenging application of a multimetal single source in a monocomponent, simple hot-wall MOCVD reactor for the deposition of DS multicomponent BaCaCuO(F) matrices. It offers the best promise for a scalable and reproducible deposition technology due to the overall simplification of the MOCVD process. Besides this, it offers prerequisites of good selectivity and reproducibility in the formation of DS Tl-2212 phase films, mostly due to the accurate control of the matrix stoichiometry and of the Tl vapour diffusion process. The electrical transport properties point to good T_c and J_c values on both sides of the Tl-2212 films. The microwave data outline that MOCVD is capable of producing samples with a very low surface resistance, comparable with the best results present in literature.

The DS Tl-2212 have been successfully applied for the fabrication of passive filter devices and miniaturised antennas, both based on cross-slotted dual mode patch resonator. The prototype dual mode filters realised present in terms of insertion losses and power handling, a slightly degraded performance if compared with high-quality YBCO devices. However, they allow the development of devices operating at lower reduced temperatures where the superconducting properties are better established or, alternatively, the use of simpler and lighter cooling systems working at higher temperatures without compromising device performance. X-band LP and CP antennas have been designed and tested using HTS films in a small size cryocooler.

In summary, the proposed approach certainly represents a promising low-cost route to large, high-quality, single- and double-sided superconducting Tl-Ba-Ca-Cu-O films, since both the MOCVD deposition and the annealing appear well suited for fabrication of larger-area films.

ACKNOWLEDGEMENTS

The authors thank Prof. A. Andreone, Dr. A. Cassinese, Dr. G. Pica and Prof. R. Vaglio for their invaluable contribution relative to transport properties, microwave characterisation and device fabrication, and Dr. L. M. S. Perdicaro for her synthetic contribution. Finally, the authors thank the Ministero dell'Università della Ricerca Scientifica e Tecnologica (MURST, Rome) for financial support.

REFERENCES

[1]   Z. Z. Sheng and A. M. Hermann, Nature 332 [1988] 55.
[2]   Z. Z. Sheng and A. M. Hermann, Nature 332 [1988] 138.
[3]   R. Sugise and H. Ihara, Jpn. J. Appl. Phys. 28 [1989] 334.
[4]   M. A. Subramanian, J. B. Parise, J. C. Calabrese, C. C. Torardi, J. Gopalakrishnan and A. W. Sleight, J. Solid State Chem. 77 [1988] 192.
[5]   H. Ihara, R. Sugise, M. Hirabayashi, N. Terada, M. Jo, K. Hayashi, A. Negishi, M. Tokumoto, Y. Kimura and T. Shimomura, Nature 334 [1988] 510
[6]   R. Sugise, M. Hirabayashi, N. Terada, M. Jo, T. Shimomura and H. Ihara, Physica C 157 [1989] 131.
[7]   A. Iyo, Y. Tanaka, Y. Ishiura, M. Tokumoto, K. Tokiwa, T. Watanabe and H. Ihara, Supercon. Sci. Technol. 14 [2001] 504.
[8]   Y. Shimakawa, Y. Kubo, T. Manoko, Y. Nakabayashi and H. Igarashi, Physica C 156 [1988] 97.
[9]   R. S. Liu and P. P. Edwards, Physica C 179 [1991] 353.
[10] C. C. Torardi, M. A. Subramanian, J. C. Calabrese, J. Gopalakrishnan, K. J. Morissey, T. P. Askew, R. B. Flippen, U. Chodry and A. W. Sleight, Science 240 [1988] 631.
[11] Z. Y. Chen, Z. Z. Sheng, Y. Q. Tang, Y. F. Li, L. M. Wang and D. O. Pederson, Supercond. Sci. Technol. 6 [1993] 261.
[12] G. Malandrino, D. S. Richeson, T. J. Marks, D. C. DeGroot, J. L. Schindler and C. R. Kannewurf, Appl. Phys. Lett. 58 [1991] 182.
[13] A. Piehler, N. Reschauer, U. Spreitzer, J. P. Strobel, R. Schonberger, K. F. Renk and G. Saemannischenko, Appl. Phys Lett. 65 [1994] 1451.
[14] S. L. Yan, L. Fang, M. S. Si, H. L. Cao, Q. X. Song, J. Yan, X. D. Zhou and J. M. Hao, Supercond. Sci. Technol. 7 [1994] 681.
[15] L. Y. Su, C. R. M. Grovenor, M. J. Goringe, A. P. Jenkins and D. Dew-Hughes, Appl. Phys. Lett. 66 [1995] 1542.
[16] B. J. Hinds, D. L. Schulz, D. A. Neumayer, B. Han, T. J. Marks, Y. Y. Wang, V. P. Dravid, J. L. Schindler, T. P. Hogan and C. R. Kannewurf, Appl. Phys. Lett. 65 [1994] 231.

[17] J. D. O'Connor, D. Dew-Hughes, N. Reschauer, W. Brozio, H. H. Wagner, K. F. Renk, M. J. Goringe, C. R. M. Grovenor and T. Kaiser, Physica C, 302 [1998] 277.

[18] Š. Chromik, M. Jergel, Š. Gaži, V. Štrbìk, F. Hanic, C. Falcony, M. Vaško and Š. Beňačka, Physica C 354, [2001] 429.

[19] A. M. Hermann, J. V. Yakhmi, Thallium Based High Temperature Superconductors, Marcel Dekker, Inc., New York, NY, USA, [1994] and reference therein.

[20] M. P. Siegal, E. L. Venturini, B. Morosin and T. L. Aselage, J. Mater. Res. 12 [1997] 2825.

[21] A. P. Bramley, J. D. O'Connor and C. R. M. Grovenor, Supercond. Sci. Technol. 12, [1999] R57.

[22] S. Huber, M. Manzel, H. Bruchlos, S. Hensen and G. Müller, Physica C 244 [1995] 337.

[23] A. P. Jenkins, A. P. Bramley, D. J. Edwards, D. Dew-Hughes and C. R. M. Grovenor, J. Supercond. 11 [1998] 5.

[24] B. A. Willemsen, K. E. Kihlstrom, T. Dham, D. J. Scalapino, B. Gowe, D.A. Bonn and W. N. Hardy, Phys. Rev. B Condens. Matter 58 [1998] 6650.

[25] B. Morosin, M. Grant Norton, C. Barry Carter, E. L. Venturini and D. S. Ginley, J. Mater. Res. 8 [1993] 720.

[26] M. Verwerf, G. Van Tendeloo and S. Amelinckx, Physica C 156 [1988] 607.

[27] W. L. Holstein, L. A. Parisi, C. R. Fincher and P. L. Gai, Physica C 212 [1993] 110.

[28] D. J. Werder and S. H. Liou, Physica C 179 [1991] 430.

[29] G. G. Condorelli, I. L. Fragalà, G. Malandrino, F. Miletto Granozio and M. Valentino, Superconductivity and Superconducting Materials Technologies Ed. P. Vincenzini, TECHNA, Faenza, Italy [1995] 579.

[30] J. Y. Lao, J. H. Wang, D. Z. Wang, S. X. Yang, Y. Tu, J. G. Wen , H. L. Wu, Z. F. Ren, D. T. Verebelyi, M. Paranthaman, T. Aytug, D. K. Christen, R. N. Bhattacharya and R. D. Blaugher, Supercond. Sci. Technol. 13 [2000] 173.

[31] H. Q. Chen,. L. G. Johansson Q. H. Hu, Z. G. Ivanov, D. Erts and E. Stepantsov, J. Supercond. 11 [1998] 73.

[32] J. D. O'Connor, A. P. Jenkins, C. R. M. Grovenor, M. J. Goringe and D. Dew-Hughes, Supercond. Sci. Technol. 11 [1998] 207.

[33] S. Huber, M. Manzel, H. Bruchlos, F. Thrum, M. Klinger and A. Abramowicz, Applied Superconductivity 1997. Proceedings of EUCAS 1997, Third European Conference on Applied Superconductivity, Institute of Physics Publishing, Bristol, UK. 1, [1997] 351.

[34] H. Schneidewind, M. Manzel, G. Bruchlos and K. Kirsch, Supercond. Sci. Technol. 14 [2001] 200.

[35] K. S. Kale, A. P. Jenkins, K. L. Jim, J. D. O'Connor, D. Dew-Hughes, D. J. Edwards, A. P. Bramley, B. J. Glassey, C. R. M. Grovenor, M. J. Goringe and B. Pecz, High Temperature Superconductors: Synthesis, Processing, and Large-Scale Applications. Proceedings TMS, Miner. Metals & Mater. Soc., Warrendale, PA, USA [1996] 345.

[36] A. P. Jenkins, K. S. Kale, D. Dew-Hughes, D. J. Edwards, A. P. Bramley and C. R. M. Grovenor, IEE Colloquium on Superconducting Microwave Circuits, IEE, London, UK [1996] 3.

[37] F. A. Miranda, S. S. Toncich and K. B. Bhasin, Microwave Opt. Technol. Letters 6 [1993] 752.

[38] M. L. Hitchman and K. F. Jensen, Chemical Vapour Deposition: Principles and Applications, Academic Press, London, [1993].

[39] N. Hamaguchi, R. Gardiner, P. S. Kirlin, R. Dye, K. M. Hubbard and R. E. Muenchausen, Appl. Phys. Lett. 57 [1990] 2136.

[40] J. A. Ladd, B. T. Collins, J. R. Matey, J. Zhao and P. Norris, Appl. Phys. Lett. 59 [1991] 1368.

[41] D. S. Richeson, L. M. Tonge, J. Zhao, J. Zhang, H. O Marcy, T. J. Marks, B. W. Wessels and C. R. Kannewurf, Appl. Phys. Lett. 54 [1989] 2154.

[42] X. F. Zhang, Y. S. Sung, D. J. Miller, B. J. Hinds, R. J. McNeely, D. B. Studebaker, and T. J. Marks, Physica C 274 [1997] 146.

[43] B. J. Hinds, R. J. McNeely, D. B. Studebaker, T. J. Marks, T. P. Hogan, J. L. Schindler, C. R. Kannewurf., X. F. Zhang and D. J. Miller, J. Mat. Res 12 [1997] 1214.

[44] R. J. McNeely, J. A. Belot, T. J. Marks, Y. Wang, V. P. Dravid, M. P. Chudzik and C. R. Kannewurf, J. Mater. Res. 15 [2000] 1083.

[45] G. Malandrino, G. G. Condorelli, I. L. Fragalà, F. Miletto Granozio, U. Scotti di Uccio and M. Valentino, Nuovo Cimento 16D [1994] 1953.

[46] G. Malandrino, G. G. Condorelli, I. L. Fragalà, F. Miletto Granozio, U. Scotti di Uccio and M. Valentino, Supercond. Sci. Technol., 9 [1996] 570.

[47] G. Malandrino, G. G. Condorelli, G. Lanza, I. L. Fragalà, U. Scotti di Uccio and M. Valentino, J. Alloys Comp. 251 [1997] 332.

[48] A. Andreone, A. Cassinese, F. Palomba, G. Pica, M. Salluzzo, G. Malandrino, V. Ancarani and I. L. Fragalà, Intern. J. Mod. Physics 13 [1999] 1321.

[49] A. Lauder, K. E. Myers and D. W. Face, Adv. Mater. 10 [1998] 1249.

[50] A. P. Jenkins, D. Dew-Hughes, D. J. Edwards, D. Hyland and C. R. M. Grovenor, IEEE Trans. Applied Supercond. 9 [1999] 2849.

[51] A. R. Kuzhakhmetov, O. S. Chana, D. M. C. Hyland, C. R. M. Grovenor, G. Burnell, M. G. Blamire and H. Schneidewind, Physica C 372-376 (Pt.1) [2002] 322.

[52] O. S. Chana, A. R. Kuzhakhmetov, D. M. C. Hyland, D. Dew-Hughes, C. R. M. Grovenor, Y. Koval, R. Kleiner, P. Muller and P. A. Warburton, Physica C 362 [2001] 265.

[53] D. W. Face, R. J. Small, M. S. Warrington, F. M. Pellicone and P. J. Martin, Physica C 357-360 [2001] 1488.

[54] H. Schneidewind, M. Manzel and T. Stelmer, Physica C 372 (Pt.1) [2002] 493.

[55] G. Malandrino, PhD Thesis [1994] Università di Catania, Italy.

[56] G. G. Condorelli, G. Malandrino and I. L. Fragalà, Chem. Mater. 6 [1994] 1861.

[57] G. Malandrino, A. M. Borzì, I. L. Fragalà, A. Andreone, A. Cassinese and G. Pica, J. Mater. Chem. 12 [2002] 3728.

[58] G. E. Myers, Thallium Safety p. 599 in A. M. Hermann, J. V. Yakhmi, Thallium Based High Temperature Superconductors, Marcel Dekker, Inc., New York, NY, USA, [1994].

[59] S. Hensen, G. Mueller, C. T. Rieck and K. Scharnberg, Phys. Rev. B 56 [1997] 6237.

[60] E. S. Machlin, Materials Science in Microelectronics (Gyro Press, Croton–on-Hudson N.Y. [1995].

[61] G. Malandrino, F. Castelli and I. L. Fragalà, Inorg. Chim. Acta 224 [1994] 203.

[62] J. C. Cheang-Wong, Mi Jergel, Š. Chromik, V. Štrbìk, J. Rickards, Ma Jergel, C. Falcony and E. Andrade, Supercond. Sci. Technol. 14 [2002] 90.

[63]   L. Y. Su, C. R. M. Grovenor and M. J. Goringe, Supercond. Sci. Technol. 7 [1994] 133.

[64] B. T. Ahn, W. Y. Lee and R. Beyers, Appl. Phys. Lett. 60 [1992] 2150.

[65] W. L. Holstein, J. Phys. Chem. 97 [1993] 4224.

[66] S. Geller and V. B. Bala, Acta Crystallogr. 9 [1956] 1019.

[67] ICDD-International Centre for Diffraction Data, Newton Square, PA 19073-3273.

[68] G. Pica, A. Andreone, F. Palomba, M. Salluzzo, R. Vaglio, G. Malandrino, V. Ancarani, I. L. Fragalà, A. Cassinese and G. Müller, Eur. Phys. J. B 18 [2000] 405.

[69] A. Andreone, A. Cassinese, A. Di Chiara, M. Iavarone, F. Palomba, A. Ruosi and R. Vaglio, J. Appl. Phys. 82 [1997] 1736.

[70] M. J. Lancaster, Passive Microwave Devices Applications of High Temperature Superconductors, (Cambridge University Press, Cambridge) [1997].

[71] W. L. Holstein, J. Appl. Phys. 74 [1993] 4963.

[72] D. M. Ogborne, M. T. Weller, P. C. Lanchester, C. R. M. Grovenor and L. Romano, Supercond. Sci. Technol. 5 [1992] 78.

[73] Y. S. Sung, X. F. Zhang, P. J. Kostic and D. J. Miller, Appl. Phys. Lett. 69 [1996] 3420.

[74] X. F. Zhang, Y. S. Sung and D. J. Miller, Physica C 301 [1998] 221.

[75] Th. Hopfinger, O. O. Shcherban, Ph. Galez, R. E. Gladyshevkii, M. Lomello-Tafin, J. L. Jorda, and M. Couach, J. Alloys Comp. 333 [2002] 237.

[76] G. Malandrino, A. R. Frassica, G. G. Condorelli, G. Lanza and I. L. Fragalà, J. Alloys Comp. 251 [1997] 314.

[77] G. Malandrino, I. L. Fragalà, U. Scotti di Uccio and M. Valentino, Il Nuovo Cimento 19 D [1997] 1053.

[78] K. Char, M. AS. Colclough, S. M. Garrison, N. Newman and G. Zaharchuk, Appl. Phys. Lett. 59 [1991] 733.

[79] G. Malandrino, I. L. Fragalà and P. Scardi, Chem. Mater. 10 [1998] 3765.

[80] G. Malandrino, A. M. Borzì, I. L. Fragalà, A. Andreone, A. Cassinese and G. Pica, J. Mater. Chem. 12 [2002] 3728.

[81] Y. Ito, Y. Yoshida, M. Iwata, Y. Takai and I. Hirabayashi, Physica C 288 [1997] 178.

[82] A. Cassinese, A. Andreone, E. Di Gennaro, G. Pica, R. Vaglio, G. Malandrino, L. M. S. Perdicaro, I. L. Fragalà, and C. Granata, Supercond. Sci. Technol., 14 [2001] 406.

[83] A. Cassinese, F. Palomba, G. Pica, A. Andreone and G. Panariello, Appl. Phys. Lett. 77 [2000] 4407.

[84] A. Cassinese, M. Barra, I. Fragalà, M. Kusunoki, G. Malandrino, T. Nakagawa, L. M. S. Perdicaro, K. Sato, S. Ohshima and R. Vaglio, Physica C 372-376 [2002] 500.

[85] M. Barra, A. Cassinese, I. Fragalà, M. Kusunoki, G. Malandrino, T. Nakagawa, L. M. S. Perdicaro, K. Sato, S. Ohshima and R. Vaglio, Supercond. Sci. Technol. 15 [2002] 581.
[86] K. Ehata, K. Sato, M. Kusunoki, M. Mukaida, S. Ohshima, Y. Suzuki and K. Kanao, IEEE Trans. Applied Supercond. 11 [2001] 111.

Hg-BASED SUPERCONDUCTING THIN FILMS FOR DC SQUIDS BY IMPROVED FABRICATION PROCESS

N. INOUE[a*], A. TSUKAMOTO[b], Y. MORIWAKI[a**], T. SUGANO[a], X. –J. WU[a***], A. OGAWA[a], S. ADACHI[a], K. TAKAGI[b] and K. TANABE[a]

[a] Superconductivity Research Laboratory, International Superconductivity Technology Center, 1-10-13 Shinonome, Koto-ku, Tokyo 135-0062, JAPAN
[b] Advanced Research Laboratory, Hitachi, Ltd., Kokubunji, Tokyo 185-8601, JAPAN

1. INTRODUCTION

Hg-based superconductors have the highest critical temperature T_c values among the cuprate superconductors, for example, 127 K for the Hg-1212 phase ($HgBa_2CaCu_2O_y$) and 135 K for the Hg-1223 phase ($HgBa_2Ca_2Cu_3O_y$) [1-3]. Application of these materials to electronic devices such as a superconducting quantum interference device (SQUID) would lead to various advantages including improved performance at liquid nitrogen temperature (77 K) and a higher operating temperature. Especially, the higher operating temperature induces the reduction of cooling load and vacuum space for device systems. Consequently, for the SQUIDs with these materials, advancement of downsizing, portability, position-sensing resolution and so on can be expected. Therefore the fabrication of Hg-based superconducting thin films has been extensively studied so far [4-10]. However, few manufacturing trials for electronic devices have been reported [11,12].

We have been studying the synthesis and properties of the Hg-based superconductors since the discovery of these materials [13]. Recently, we have tried to fabricate high-T_c Hg-based superconducting thin films with the aim of applying them to electronic devices and confirmed the good properties of Josephson bicrystal junctions and dc-SQUIDs with these films [14-30]. In particular, some of their properties such as the I_cR_n products were found superior to those for the junctions with $YBa_2Cu_3O_y$ (Y-123) thin films. In this chapter, we review our thin film fabrication process and the properties of the obtained films [14,15,20-24,26,27]. Furthermore, we describe the transport properties of bicrystal grain boundary junctions [16-18,25,29] and dc SQUIDs [16-18,28,30].

2. THIN FILM FABRICATION

2.1 Two step process

To fabricate Hg-based superconducting thin films, the two-step process is usually employed, which consists of (1) the preparation of precursor films and (2) the annealing in Hg-vapor atmosphere, because of the high volatility of mercury [4-10]. By means of the two-step process several groups succeeded in fabricating Hg-1212 and Hg-1223 thin films. Some of them exhibited good superconducting properties, i.e., J_c above 1×10^6 A/cm^2 at 77 K in a self-field [6-9]. We also fabricated 200 - 500-nm-thick Hg-1212 and Hg-1223 thin films through the similar process. Shimoyama *et al.* [31] reported that the transport properties and chemical stability of Hg-1223 ceramics are significantly improved by doping Re to the Hg site. That seems quite beneficial to practical applications of this superconducting material. Maignan *et al.* [32] investigated the transition metal-doping effects on Hg-1212 ceramic samples. Their results suggested that Mo- and V-doping to Hg-1212 reduces the thickness of the so-called charge reservoir without appreciable T_c-depression. It can be expected that the reduction in the thickness probably improve its J_c properties. Here we describe the transition metal-doping effects on Hg-1223 and Hg-1212 thin films [14,15, 19,20].

2.1.1 Fabrication of 1223 phase

Both undoped and Re-doped Hg-1223 thin films were fabricated on SrTiO$_3$ (STO) substrates by the conventional two-step process. Precursor films consisting of alternate stack of Re$_x$Ba$_2$Ca$_2$Cu$_3$O$_y$ ($x = 0$, 0.1) and HgO (thickness ratio = 7 : 1, typical Re$_x$Ba$_2$Ca$_2$Cu$_3$O$_y$ thickness = 35 nm, total thickness = 500 - 800 nm) were prepared by pulsed laser deposition (PLD) using a KrF excimer laser (λ = 248 nm). HgO cap layers were also deposited on top of the multilayer films to protect it from moisture and CO$_2$. The absence of the cap-layer leads to the surface degradation and appearance of a substantial amount of impurities phases [15]. The Re$_x$Ba$_2$Ca$_2$Cu$_3$O$_y$ targets were prepared by heating a pressed powder mixture of ReO$_2$, BaO$_2$, CaO, and CuO at 750 - 920 °C for 30 h (total time) in flowing oxygen gas. The deposition was performed in vacuum at room temperature. The precursor film was sealed together with two unreacted pellets in an evacuated quartz tube and heated at 725 – 800 °C for 5 - 24 h. The pellets were pressed powder mixture of monoxide with Hg : Ba : Ca : Cu cation ratios of 1 : 2 : 3 : 4 and 0 : 2 : 2 : 3, respectively. Such a method using two pellets was previously employed to obtain a stable mercury vapor flow [33]. The precursor film was placed between the two pellets with its surface facing to the former pellet. The fabricated films were further posted-annealed in oxygen at 320 ℃ for 24 h to adjust their oxygen content.

Figures 1(a) and 1(b) show typical X-ray diffraction (XRD) θ - 2θ patterns for the undoped and Re-doped films, respectively. Major peaks can be assigned to the (00l) reflections of the Hg-1223 phase, indicating that Hg-1223 phase is dominant and c-axis orientation perpendicular to the substrate surface is achieved. Although weak peaks from impurity phases and the Hg-1212 phase are also observed, it is clear that more than 75 % phase-pure and highly c-axis oriented Hg-1223 films can be obtained both for undoped and Re-doped cases. However, the typical full width at half maximum (FWHM) value of the (006) rocking curve was approximately 1.0°. The typical c-axis lattice constants determined from the (006) peak of the Hg-1223 phase for as-grown undoped and Re-doped thin films

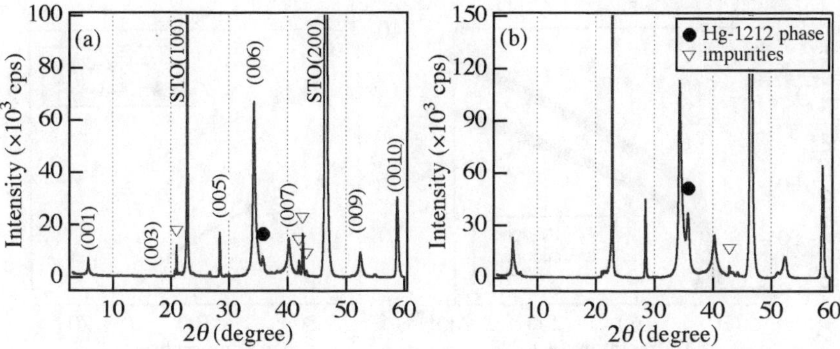

Fig. 1 Typical XRD θ - 2θ patterns for (a) undoped and (b) Re-doped Hg-1223 thin films fabricated by the conventional two-step process.

were 1.568 and 1.565 nm, respectively. These values are substantially smaller than those for undoped ceramics (1.585 nm)[3] and rather close to those reported for 10% Re-doped ceramics [31]. It is thus likely that the smaller c-axis lengths of the films are caused by internal stress.

Figures 2(a) and 2(b) show typical SEM pictures for the surface of undoped and Re-doped Hg-1223 thin films, respectively. Round-shape plate-like grains with a typical size of 5 μm are observed in the undoped film. On the other hand, rectangular-shape grains with a typical size of 3 - 5 μm are grown and appear to coalesce better for the Re-doped film, indicating that the Re-doping significantly improves the surface morphology.

Figures 3(a) and 3(b) show the temperature dependence of the electrical resistivity and the critical current density J_c determined with the 1 μV/mm criterion for post-annealed undoped and Re-doped films. The resistivity near room temperature less than 300 $\mu\Omega$cm for the Re-doped film compares well with those for high-quality Y-123 thin films. Although the undoped film seems to show even lower resistivity, the surface roughness as seen in the SEM

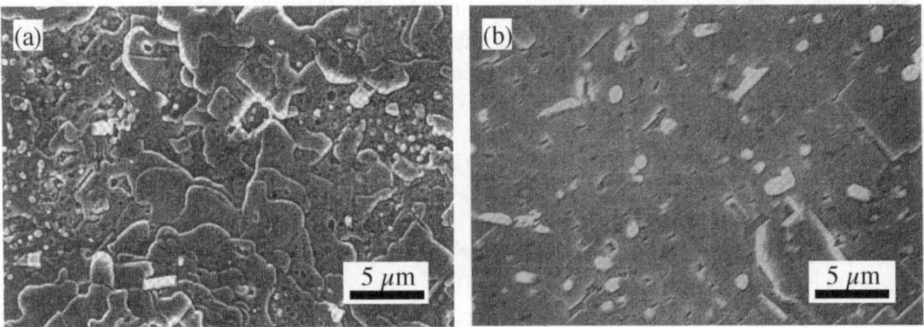

Fig.2 SEM pictures for the surfaces of (a) undoped and (b) Re-doped Hg-1223 thin films fabricated by the conventional two-step process.

216

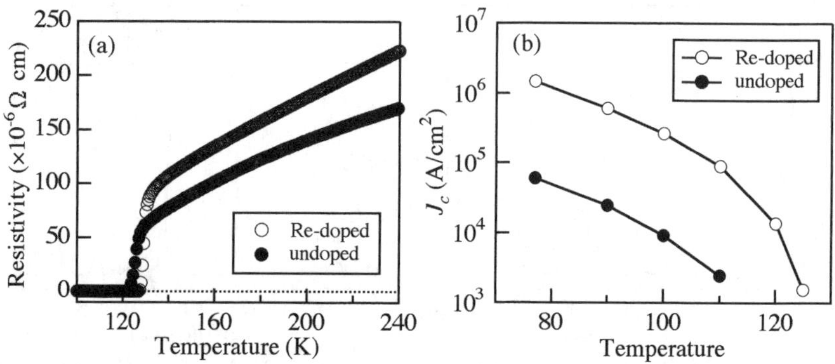

Fig. 3 Temperature dependence of (a) resistivity and (b) J_c for undoped and Re-doped Hg-1223 thin films fabricated by the conventional two-step process.

picture (Fig. 2(a)) makes accurate estimation of the value quite difficult. The undoped film exhibited a typical T_c of 120 - 123 K after post-annealing in oxygen, while slightly higher T_c values up to 127.5 K were observed for both as-prepared and annealed Re-doped films. This result indicates that the Re doping has effect to introduce more holes to the Hg-1223 phase as well as to improve the surface morphology. The Re-doped film exhibits a J_c at 77 K of 1.5 $\times 10^6$ A/cm^2 in a self-field, which is comparable to the highest J_c value ever reported for Hg-1223 thin films [7]. On the other hand, the undoped film shows a self-field J_c value less than 10^5 A/cm^2 at 77 K. This lower J_c value for the Re-free film is primarily attributed to the weak links which may exist in the as-prepared film or result from deterioration after the patterning process.

2.1.2 Fabrication of 1212 thin films

Re-doped Hg-1223 films exhibit superior film morphology and a T_c up to 127.5 K which is slightly higher than that of Re-free Hg-1223 films. They also show a transport J_c at 77 K of 1.5 × 10^6 A/cm^2 in a self-field. However, the crystallinity of the present Re-doped Hg-1223 thin films is still not so good. For instance, the FWHM value of XRD rocking curves for the Hg-1223 films is typically 1.0° and they contain a small amount of the 1212 phase. Crystallographic quality is important in fabricating reliable electronic devices. We prepared and characterized Re-, Mo- and V-doped Hg-1212 thin films with the aim of obtaining Hg-based films with higher crystallinity and better superconducting properties.

200 - 400-nm-thick $(Hg_{0.9},M_{0.1})Ba_2CaCu_2O_y$ ((Hg,M)-1212) thin films (M = Re, Mo, V) were fabricated on STO substrates, employing the two-step process similar to that used for the fabrication of (Hg,Re)-1223 thin films. A precursor film consisting of an alternate stack of $M_{0.1}Ba_2CaCu_2O_y$ and HgO was prepared by PLD at room temperature. An HgO cap layer was also deposited on top of the multilayer film. The precursor film was annealed at 700 - 800 °C for 5 h with unreacted two pellets prepared by mixing monoxides with compositions of Hg:Ba:Ca:Cu = 1:2:1:2 and 0:2:1:2. The fabricated films were further post-annealed in flowing oxygen at 300 °C for 6 h.

Figure 4 shows the XRD θ - 2θ patterns for the obtained films. Major peaks except intense peaks from the STO substrate can be assigned to the (00l) reflections of the 1212

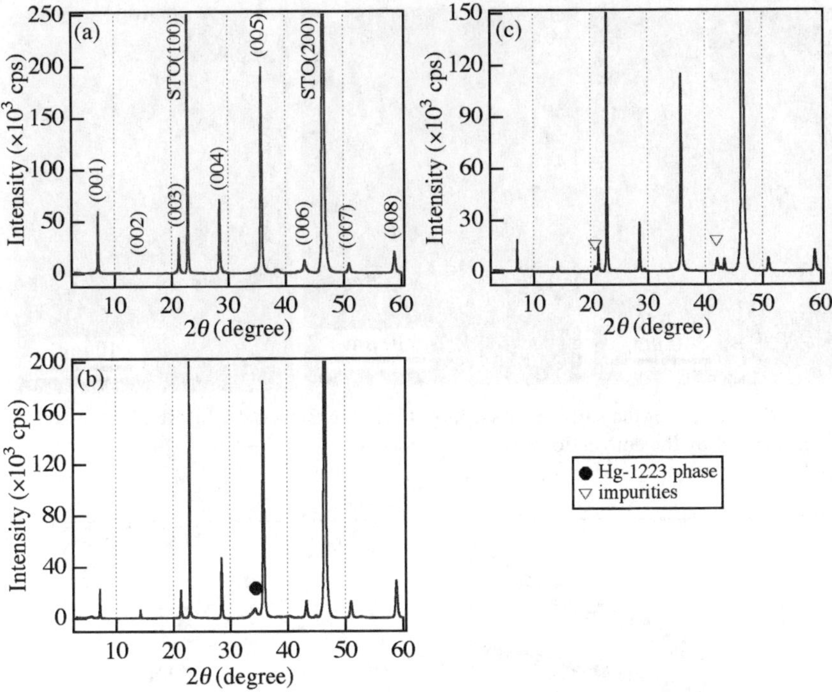

Fig. 4 XRD θ - 2θ patterns for (Hg,M)-1212 thin films (M = (a) Re, (b) Mo and (c) V) fabricated by the conventional two-step process.

phase, indicating that (Hg,M)-1212 phase is dominant and the c-axis orientation perpendicular to the substrate is achieved. Although peaks from impurities are also observed in the patterns, their intensities are quite weak. The typical value of FWHM for the (005) rocking curve for the 1212 film was less than 0.2°, while that for the (006) of the 1223 film was around 1.0° [14,15]. These results indicate that crystallinity and phase purity of the 1212 film is better than those for the 1223 film. In-plane alignment in the 1212 film was confirmed from the XRD pole figures of the (103) reflection. Moreover, the 1212 phase was easily obtained and reproducibility in production was excellent, as compared with the 1223 film.

Figure 5 shows SEM pictures for the film surfaces. In the cases of M = Re and Mo, it is seen that crystal grains with a few microns are well-connected to each other. However, many submicron-size defects are seen. A rather smooth surface is attained in the V-doped film. From the cross-sectional views for a fractured surface of the Re-doped film, it was found that plate-like grains crystallized on the substrates. Some defects such as pinholes, outgrowths and large steps $etc.$ were observed. Elimination of these defects seems important to improve the film quality.

Figure 6(a) shows the temperature dependence of the resistivity. The films exhibited superconducting transitions at temperatures around 120 K. The Re-doped film exhibits the highest T_c of 122 K. Figure 6(b) shows the temperature dependence of the critical current

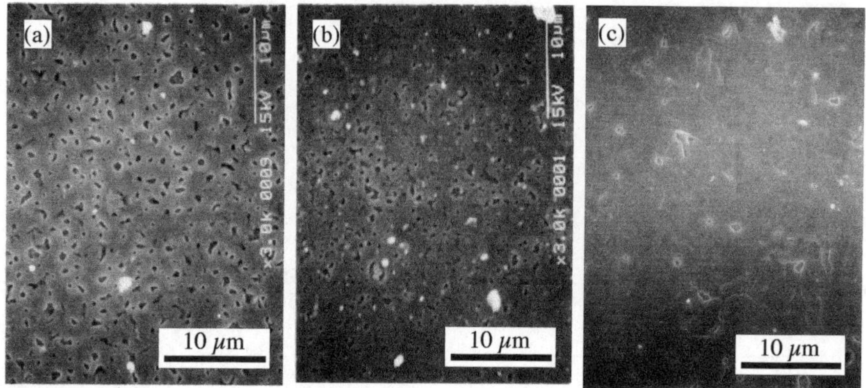

Fig. 5 SEM pictures for the surfaces of (Hg,M)-1212 thin films (M = (a) Re, (b) Mo and (c) V) fabricated by the conventional two-step process.

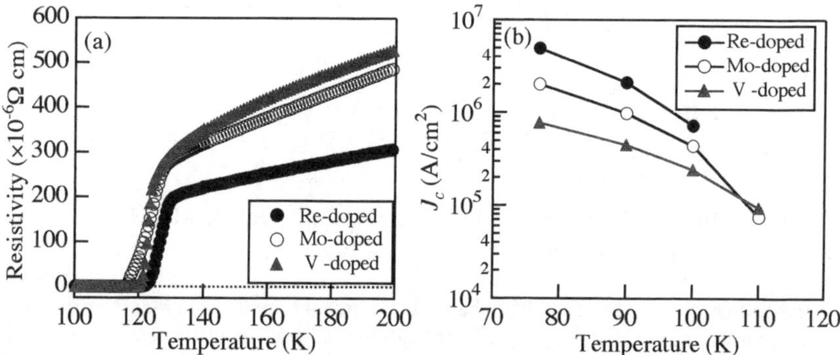

Fig.6 Temperature dependence of (a) resistivity and (b) J_c for (Hg,M)-1212 thin films (M = Re, Mo and V) fabricated by the conventional two-step process.

density J_c. Typical J_c values at 77 K were 4.9×10^6, 2.0×10^6, 7.6×10^5 A/cm^2 for M = Re, Mo, V, respectively. The J_c values for the (Hg,Re)-1212 films were found to be substantially higher than those for the (Hg,Re)-1223 films fabricated by a similar method [14,15].

2.2 Improved Fabrication Process

We succeeded in fabricating the (Hg,Re)-1212 thin films on STO substrates exhibiting a high-J_c value (4.9×10^6 A/cm^2) through the conventional two-step process, as mentioned above. However, careful observations of surfaces and cross sections for the films using SEM revealed the inclusion of submicron-size defects such as pinholes, outgrowths and large steps *etc.*. Since these defects may cause the degradation of device properties, the fabrication method for high-quality Hg-based superconducting thin films must be established for practical applications.

High-quality Y-123 thin films are generally fabricated through an *in situ* process by PLD or sputtering technique, which includes the deposition on pre-heated substrates. In this process Y-123 crystals nucleate at the substrate surface and grow epitaxially from the substrate. In the case of Hg-based thin films, however, it is extremely difficult for the materials to crystallize on pre-heated substrates because of their low evaporation temperature. In the two-step process which is commonly employed to fabricate Hg-based thin films, the crystal growth mechanism is considered as shown in Fig. 7. First, the precursor films are deposited on a substrate at room temperature. (Fig. 7(a)) During the heat treatment, most of the nucleation begins on the substrate, while even at arbitrary places in the precursor film nucleation can occur. (Fig. 7(b)) These crystal nuclei located at the regions far from the substrate surface grow freely in random directions. (Fig. 7(c)) As a result, they give rise to pinholes, outgrowths and large steps *etc.* in the films. (Fig. 7(d))

To decrease the number of randomly oriented crystals, we proposed the following improved fabrication processes.

(1) For very thin films (film thickness is less than 100 nm), rather thin precursor films are employed and annealed.

(2) For thicker films (film thickness is 100 nm or more), the very thin films are prepared by process (1) first and then the same deposition and annealing procedures are repeated.

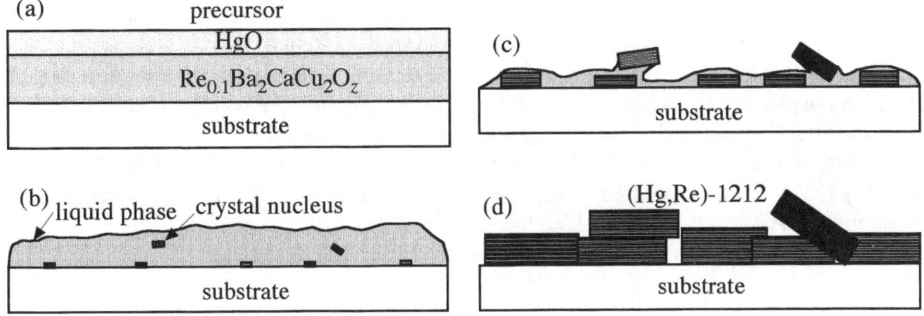

Fig. 7 Crystal growth mechanism in the conventional two-step process.

2.2.1 Very thin films (Film Thickness < 100 nm)

We considered that crystals in an Hg-based superconducting thin film may mostly grow epitaxially from the substrate when the precursor-film thickness is thinner than a certain critical thickness [21]. To confirm the availability of the idea, less-than-100-nm-thick (Hg,Re)-1212 thin films were fabricated on STO substrates, and their properties were characterized. As a precursor, an approximately 100-nm-thick $Re_{0.1}Ba_2CaCu_2O_z$ film was prepared by a PLD technique. A 25-nm-thick HgO cap layer was also deposited on top of the $Re_{0.1}Ba_2CaCu_2O_z$ film. The precursor film was sealed in a quartz tube together with unreacted two pellets prepared by mixing monoxides with compositions of Hg:Ba:Ca:Cu = 1:2:1:2 and 0:2:1:2 and annealed at 650 - 800 °C for about 5 h.

Figure 8(a) shows an SEM picture for the surface of a 75-nm-thick (Hg,Re)-1212 thin film obtained from a 100-nm-thick precursor film. A quite smooth surface is observed. Neither pinholes nor large outgrowth grains are seen. For comparison, surface pictures of

220

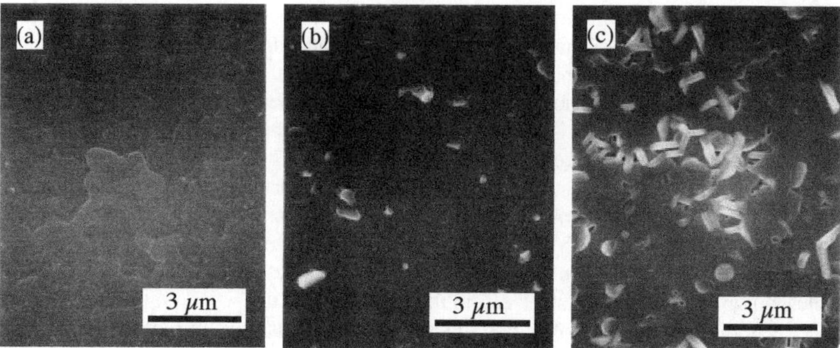

Fig. 8 SEM pictures for the surfaces of (Hg,Re)-1212 thin films. Film thickness t = (a) 75 nm, (b) 110 nm and (c) 150 nm.

110- and 150-nm-thick (Hg,Re)-1212 thin films which were fabricated by employing the 150- and 200-nm-thick precursor films are also shown in Figs. 8(b) and 8(c). For the case of film thickness larger than 100 nm, submicron-size defects such as outgrowths and pinholes appear remarkably in the films and their density seems to increase with the thickness. Thus the critical thickness is considered to be approximately 100 nm. Figure 9 shows an SEM picture for the fractured cross section of the 75-nm-thick (Hg,Re)-1212 thin film. In the cross-sectional view, the substrate is uniformly covered by the film without appreciable steps. No gap or void is observed at the interface between the film and the substrate.

Figure 10 shows the XRD θ-2θ pattern for the 75-nm-thick film. Peaks from the (Hg,Re)-1212 phase with the c-axis orientation perpendicular to the substrate are predominant. A typical value of FWHM for the (005) rocking curve was 0.3°, which is comparable or slightly larger than the minimum value of 0.2° for thicker (Hg,Re)-1212 films

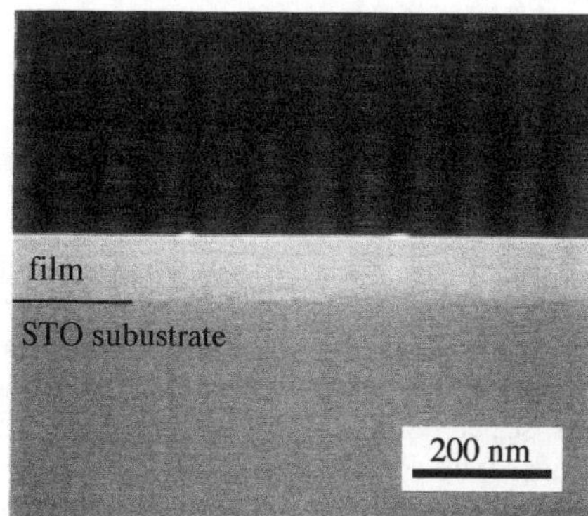

Fig. 9 SEM picture for the fractured cross-section of a 75-nm-thick (Hg,Re)-1212 thin film.

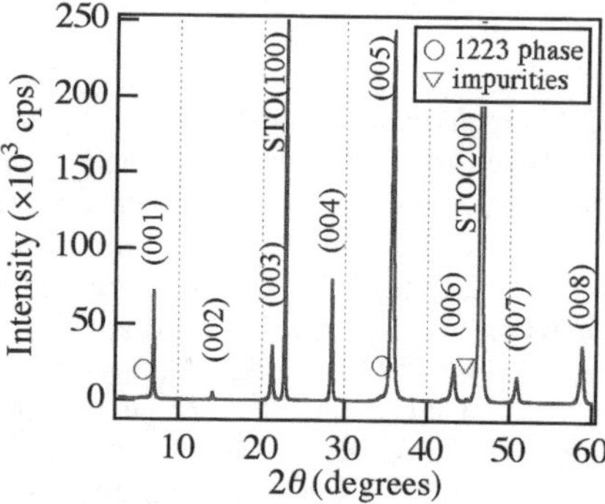

Fig. 10 XRD θ-2θ pattern for a 75-nm-thick (Hg,Re)-1212 thin film.

fabricated by the conventional two-step process [19,20]. By the XRD ϕ-scan measurement, four (103) reflection peaks from the (Hg,Re)-1212 film were observed at an interval of 90° and their positions coincide with the (101) reflection peaks from the STO substrate. This confirms that (Hg,Re)-1212 grains are in-plane aligned with their a-axis parallel to STO[100] and thus they are epitaxially grown on the substrate.

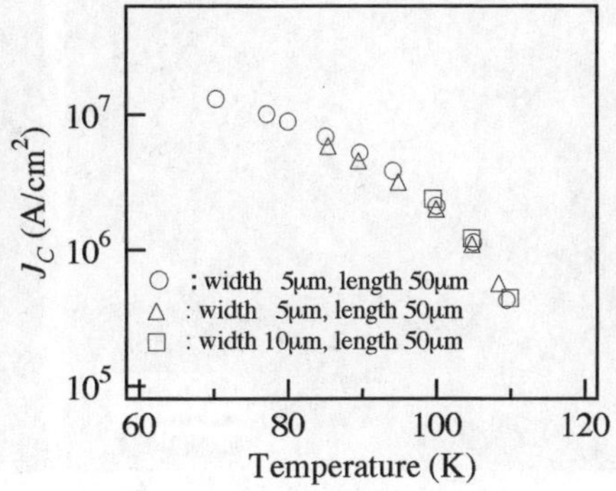

Fig. 11 Temperature dependence of J_c for one of the best 75-nm-thick (Hg,Re)-1212 thin film.

As-fabricated films exhibited a typical T_c of 116 K. Figure 11 shows the temperature dependence of the transport J_c in a self-field for one of our best films. We measured J_c for three bridges patterned on this film. The obtained results for them are in good agreement, as seen in the figure. In spite of the lower T_c than those for thicker films, the J_c is as high as 1.0×10^7 A/cm^2 at 77 K. To the best of our knowledge, this is the highest J_c value at 77 K ever reported for Hg-based superconducting thin films. The J_c values at 77 K of $3 - 10 \times 10^6$ A/cm^2 were reproducibly obtained.

2.2.2 Thicker Films (Film Thickness $\geq$ 100 nm)

For actual device applications, the thickness of films must be controlled at the request of specific designs. We proposed the following process for thicker films [24]. First, a homogeneous very thin film is prepared by above-mentioned process. Next, the same procedure of deposition and annealing is repeated on the firstly obtained film. The thickness of the second precursor film is adjusted so as to obtain a desired total thickness. During the second annealing process, crystal growth begins on a homogeneous seed layer, so that we can expect reduction of defects as compared with the conventional two-step process.

To confirm the availability of this process, thicker (Hg,Re)-1212 thin films were fabricated on STO substrates and their properties were characterized. First, about 75-nm-thick homogeneous (Hg,Re)-1212 thin films were prepared on STO substrates. Next, the 75- and 150-nm-thick-film processes were repeated on the firstly-obtained (Hg,Re)-1212 thin films which were used as seed layers and 150- and 225-nm-thick (Hg,Re)-1212 films were obtained.

Figure 12 shows SEM pictures for the surface of the obtained 150- and 225-nm-thick films. Both films have quite smooth surfaces. Neither pinholes nor large outgrowth grains are seen. SEM cross-sectional observation confirmed that the substrates were uniformly covered by the films without appreciable steps. When the seed layer included some defects, a smooth surface was not obtained, indicating that the surface condition of the seed layer is very important to make smooth surface. Figure 13 shows the XRD θ-2θ pattern for the 150-nm-thick film. It is clear that the (Hg,Re)-1212 phase with the c-axis orientation is predominant. Figure 14 shows a TEM picture for the cross section of the 150-nm-thick film. The interface caused by the repeated process in the (Hg,Re)-1212 thin film is hardly seen.

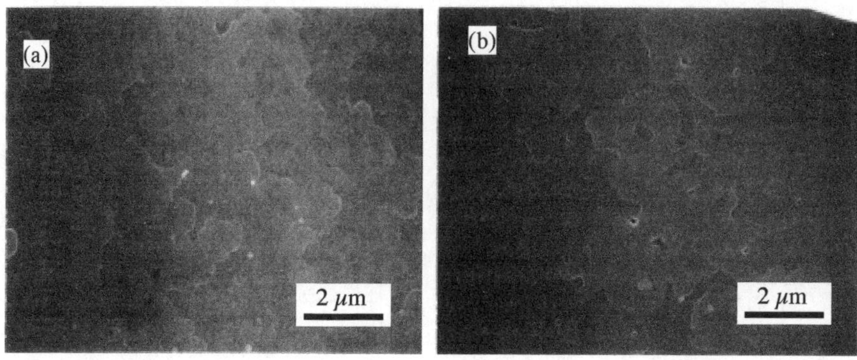

Fig. 12 SEM pictures for the surfaces of (Hg,Re)-1212 thin films fabricated by the improved process. Film thickness t = (a) 75 + 75 = 150 nm, (b) 75 + 150 = 225 nm.

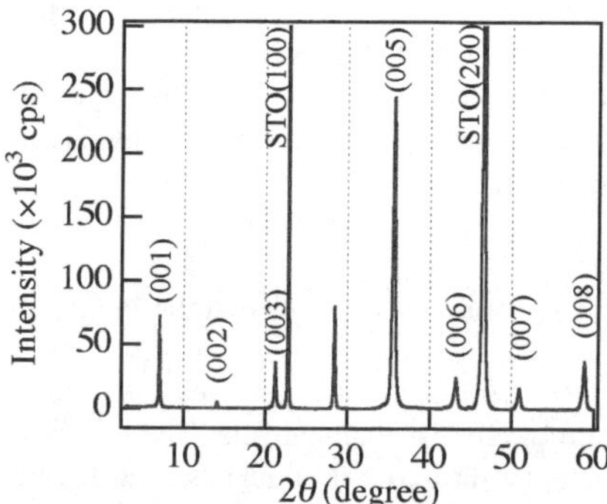

Fig. 13 XRD θ - 2θ pattern for an (Hg,Re)-1212 thin film. Film thickness t = 75 +75 = 150 nm.

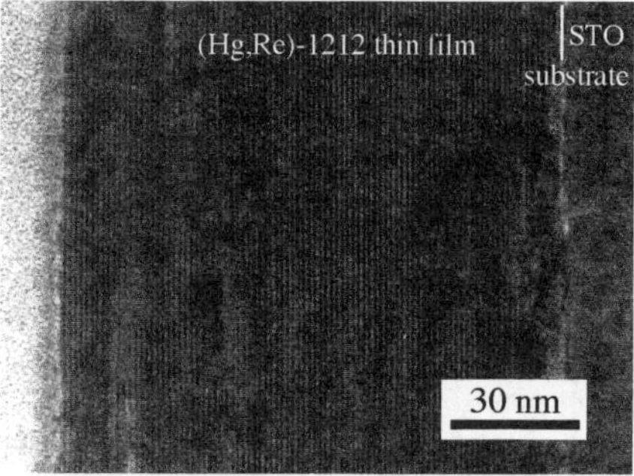

Fig. 14 TEM picture for the cross section of an (Hg,Re)-1212 thin film fabricated by the improved process. Film thickness t = 75 +75 = 150 nm.

This indicates that homoepitaxial growth is realized in this process.

2.3 (Hg,Re)-1212 film fabrication on various substrates

It is well known that superconducting properties of high-T_c films strongly depend on their microstructures. However, there have been few detailed investigations on the relationship between the microstructures of the Hg-based films on different substrates and their superconducting properties. Here we describe our results of microstructural examination for (Hg,Re)-1212 thin films grown on various substrates [22,23,26,27].

75-nm-thick (Hg,Re)-1212 thin films were fabricated on $LaAlO_3$ (LAO) and $(LAO)_{0.3}$-$(SrAl_{0.5}Ta_{0.5}O_3)_{0.7}$ (LSAT) substrates, by   the improved process similar to that used for the fabrication of 75-nm-thick film on STO substrate. Figure 15 shows the XRD θ-2θ patterns

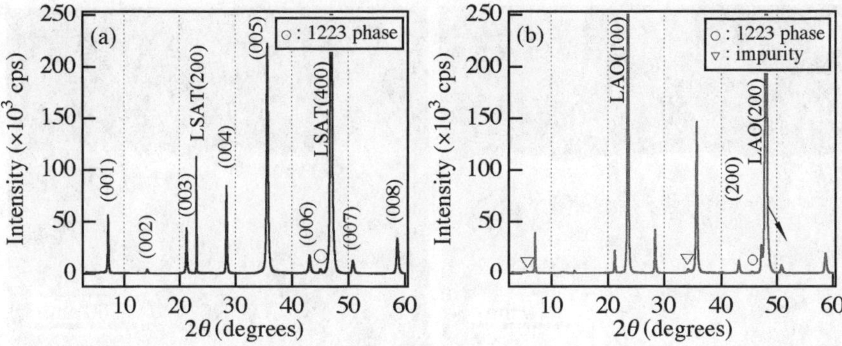

Fig. 15 XRD θ - 2θ patterns for (Hg,Re)-1212 thin films on (a) LSAT and (b) LAO substrates. Film thickness t = 75 nm.

for the obtained films. The patterns indicate that the (Hg,Re)-1212 phase with the c-axis orientation is dominant. Although we also tried fabricating films on MgO, NdGaO$_3$ and LaSrGaO$_3$ substrates, they did not exhibit strong peaks from the 1212 phase. The film on LAO substrate shows small peaks from a-axis-oriented grains, as seen in Fig. 15(b). The XRD ϕ-scan confirmed that the greater part of (Hg,Re)-1212 grains are in-plane aligned with their a-axis parallel to the substrate [100] direction and thus they are epitaxially grown on each substrate.

Figures 16 and 17 show SEM pictures for the surfaces and the cross-sections of the films. The film on STO has a quite smooth surface as already shown in Fig. 8(a) and 9, while submicron-size defects are observed in the films on LSAT and LAO. The density of defect regions in the film on LSAT substrate looks smaller than that of the film on LAO substrate.

Typical T_c values of 112 and 114 K for as-fabricated films on LSAT and LAO substrates, respectively, are comparable to that for the thin film on STO substrate. Figure 18 shows the temperature dependence of the transport J_c for one of our best films on each substrate. The

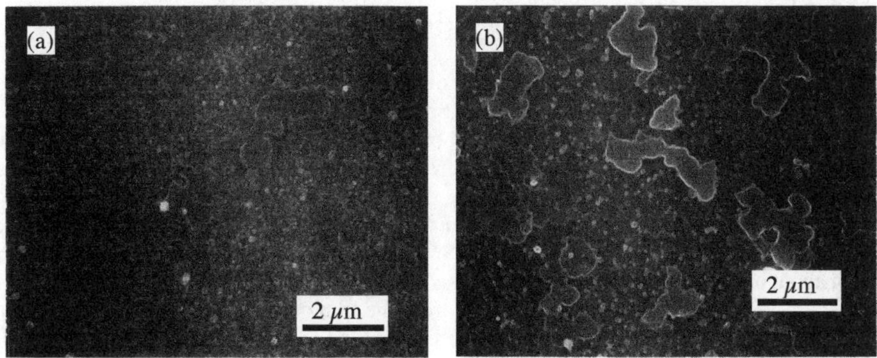

Fig.16 SEM pictures for the surfaces of (Hg,Re)-1212 thin films on (a) LSAT and (b) LAO substrates. Film thickness t = 75 nm.

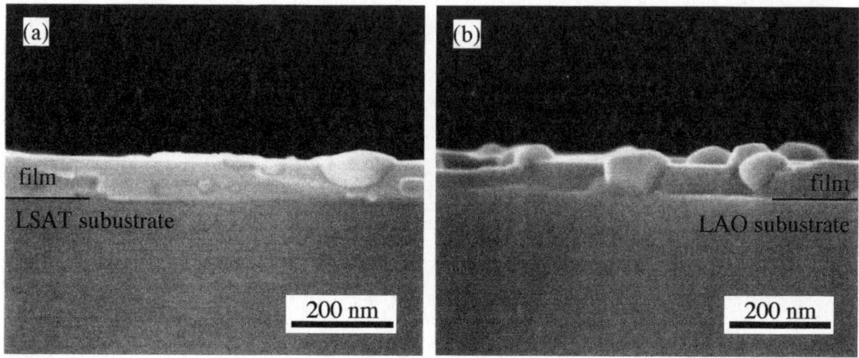

Fig. 17 SEM pictures for fractured cross-section of (Hg,Re)-1212 thin films on (a) LSAT and (b) LAO substrates. Film thickness t = 75 nm.

films on LSAT and LAO substrates exhibit typical J_c values of 4.5×10^6 and 2.7×10^6 A/cm^2 at 77 K, respectively, which are substantially lower than that on STO substrate, i.e. 1.0×10^7 A/cm^2 [21].

Figure 19(a) is a TEM micrograph taken from the film on STO substrate. The film is well c-axis-oriented, but many defect regions, can be found at the interface between the STO substrate and the film. It was found that a step-like region is formed at the top layer of STO substrate with a length of about 60 - 70 nm and a thickness of about 6 nm. A further study indicated that the lattice parameters in such step-like regions are different from either that of STO or that of (Hg,Re)-1212, suggesting that a chemical reaction occurred, and a new phase of $(Ba_x,Sr_{1-x})(Cu_y,Ti_{1-y})O_z$ was formed in these regions [22]. It is clear that this phase was produced by substituting Ba for Sr, and Cu for Ti in the STO substrate. The released Sr and Ti from the substrate diffused into the film. Such diffusion resulted in a variation in the lattice parameters of (Hg,Re)-1212, and then caused some long periodicity fringes in the film. From Fig. 19(a), it can be estimated that the dimension of the region containing diffused Sr and Ti in the film is about 10 nm in average, being about one-seventh of the film thickness.

Figure 19(b) is a TEM micrograph taken from the film on LSAT substrate. It is seen that the defect region does not contain any lattice fringes, and looks like an amorphous phase. It is likely that such defect regions come from precipitation of the precursor film, i.e., the chemical reaction may not be completed during the annealing treatment. However, it should be noted that even longer-time annealing resulted in no significant improvement in the

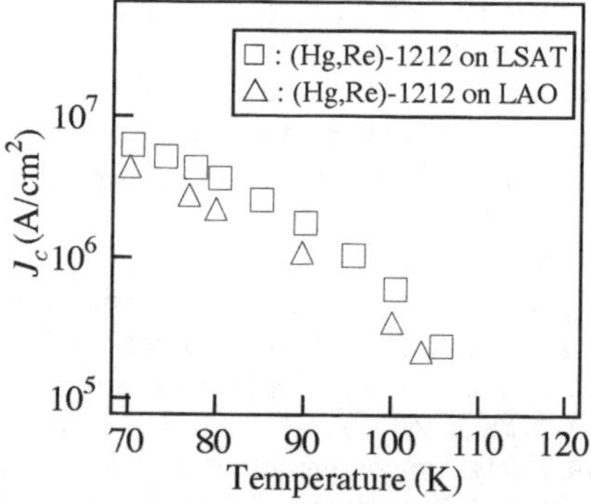

Fig. 18 Temperature dependence of J_c in a self-field for (Hg,Re)-1212 thin films on LSAT and LAO substrates. Film thickness t = 75 nm.

microstructure, suggesting that the existence of the amorphous defects is related to some difference in film growth mode.

Figure 19(c) is a TEM image taken from the film on LAO substrate. In this film, both a-axis- and c-axis-oriented grains are observed. Moreover, a grain with a random orientation is found. We can recognize that the film surface on the randomly oriented grain is not flat, and the dimension of the grain is about 0.05 μm, which is in the same order as the white contrast particles shown in Fig. 16(b). Thus, it is considered that the high density of particles in Fig. 16(b) indicates the presence of an amount of such defect regions.

Comparing the films on STO, LSAT and LAO substrates, we found that only in the film on STO substrate, a chemical reaction occurred between the precursor film and STO substrate. The influence of Sr and Ti substitution on the superconducting properties of (Hg,Re)-1212 is not clear. However, such defect regions are located in a limited area near the interface with a rather small thickness. Thus, the film on STO substrate has a good superconducting path, and the influence of the defect on J_c seems not so significant. In the films on LSAT and LAO substrates, there are no traces showing a reaction between the films and the substrates. In film on LSAT substrate, many precipitated particles are found. Because they are unreacted materials, such particles are non-superconducting. The presence of such particles directly leads to a decrease of the effective superconducting volume and thus a decrease of the J_c value of the whole film. In the film on LAO substrate,

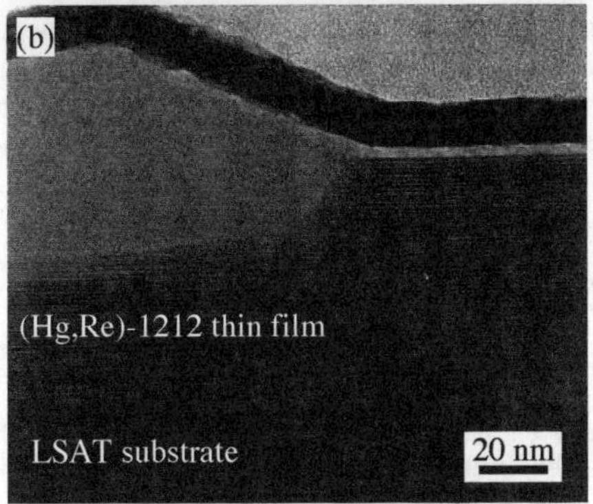

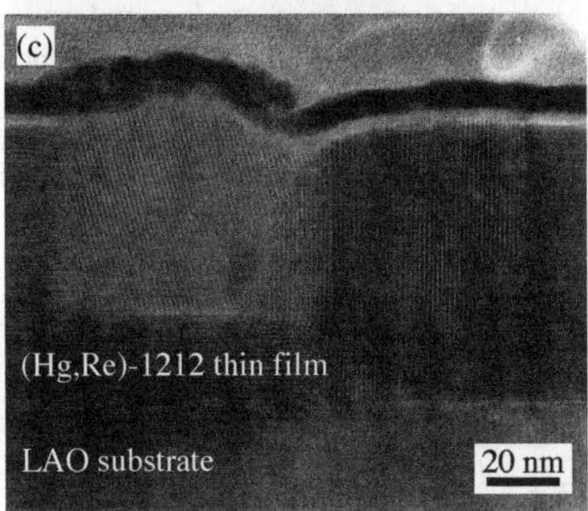

Fig. 19 TEM images taken from (Hg,Re)-1212 thin films on (a) STO, (b) LSAT and (c) LAO substrates. Film thickness t = 75 nm.

two types of defects such as the a-axis-oriented grains and the outgrowth particles are seen. The J_c along the c-axis is considered to be more than one order of magnitude lower than that in the a-b plane. This implies that J_c across the boundaries of such defects is similarly low. Because the a-axis-oriented grains can be hardly observed by SEM, the actual defect density in the film on LAO substrate should be much higher than that in the film on LSAT substrate. Thus, the total volume of defect regions in the film on LAO substrate is much higher than that in the film on LSAT, giving a reasonable explanation for the lower J_c of the former film.

3. BICRYSTAL JUNCTION

Bicrystal junctions obtained by a relatively simple process are useful for applications of high-T_c superconductors to electronic devices such as SQUIDs. Moreover, to understand the electrical transport across the grain boundaries in high-T_c superconductors, systematic studies have been carried out using Y-123 thin films fabricated on bicrystal substrates [34-40]. It was revealed that their transport properties significantly depend on the misorientation angle θ, and the J_c decreases nearly exponentially with increasing θ. From these results, the mechanisms controlling the transport properties were deduced. We have systematically investigated the transport properties of (Hg,Re)-1212 bicrystal junctions fabricated on both STO and LSAT bicrystal substrates with different misorientation angles to clarify the relationship between the junction properties and the misorientation angles [25,29].

3.1 Bicrystal junctions on STO substrates

About 160-nm-thick (Hg,Re)-1212 thin films were prepared by the improved process, i.e., repeating the 80-nm-thick-film process. STO [001]-tilt symmetric bicrystal substrates with different misorientation angles $\theta = 20°$, $24°$, $36.8°$ and $45°$ were used. Figure 20(a) shows an SEM plane-view picture of the bicrystal boundary for a (Hg,Re)-1212 thin film on a $\theta = 24°$ STO bicrystal substrate. Many particles are observed along the bicrystal boundary. SEM observations of the bicrystal boundaries were carried out for all the films with different misorientation angles. Almost all of them had similar particles. Figure 20(b) shows an SEM cross-sectional picture of the bicrystal boundary for a film on a $\theta = 24°$ STO bicrystal substrate. It is found that a large reacted region is formed around the bicrystal boundary. That region grows into the substrate with a thickness of about 150 nm and a width of about 300 nm. As we discussed in the previous section, the reaction or mutual diffusion between the film and the STO substrate without a bicrystal boundary occurred in a limited region with a

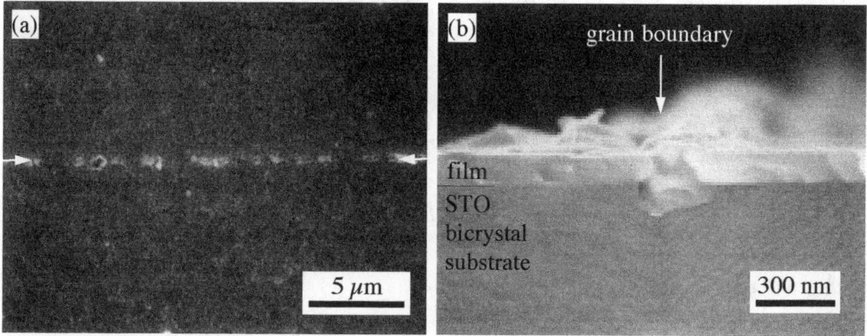

Fig. 20 SEM (a) plane-view and (b) cross-sectional pictures of the bicrystal boundaries in (Hg,Re)-1212 thin films on $\theta = 24°$ STO bicrystal substrates fabricated by the improved process. The bicrystal boundaries are indicated by the arrows.

thickness of about 10 nm [26,27]. The size of reacted regions at the bicrystal boundary is much larger than that without the bicrystal boundary, suggesting that the reaction is considerably accelerated at the bicrystal boundary. However, it is noted that there are also many narrow channels without precipitates across the boundary in Fig. 20(a).

The current-voltage (I - V) characteristics of the junctions were examined using a conventional four-probe method and the critical current I_c and the asymptotic junction resistance R_n were estimated. Figure 21(a) shows the I - V curve at 4.2 K for a junctions on a $\theta = 24°$ STO bicrystal substrate. The junctions with $\theta = 20 - 45°$ exhibited resistively-shunted-junction (RSJ)-like I - V characteristics, similar to that shown in Fig. 21(a). Figure 21(b) shows the magnetic-field H dependence of I_c for a typical junction. The field was applied perpendicular to the plane of the film. Though a Fraunhofer-like pattern was observed in some junctions as shown in Fig. 21(b), the curves for most of the junctions significantly deviate from the ideal diffraction pattern, indicating inhomogeneous current distribution. In particular, some junctions with higher tilt angles showed I_c - H curves with short periodic oscillations which are similar to those for SQUIDs. This seems consistent with the parallel array of narrow junctions expected from the SEM picture in Fig. 20(a). However, all the junctions showed a relatively large modulation of 80 - 90%, implying rather small excess current.

In Fig. 22, the J_c, R_nS and I_cR_n product at 4.2 and 77 K of the bicrystal junctions on STO substrates are plotted as a function of the misorientation angle. For comparison among junctions having different width, the I_c and R_n are normalized by the junction cross section S. The $\theta = 20°$ junction shows the I_cR_n products of 1.9 and 0.71 mV at 4.2 and 77 K, respectively. To the best of our knowledge, this value at 77 K is the highest I_cR_n product ever reported for Josephson junctions with Hg-based superconducting thin films. The J_c and R_nS do not show a systematic variation, while the I_cR_n product decreases monotonically with an increase in θ. Figure 23 shows the relationship between the J_c and R_nS at 77 K for $\theta = 24°$ (Hg,Re)-1212 junctions which were fabricated on STO bicrystal substrates by the conventional two-step process [17]. Many juctions have I_cR_n products larger than 0.2mV. However, both the J_c and R_nS of the junctions show a rather large spread. The largest J_c was almost 10 times larger than the smallest one. In contrast, a much smaller J_c spread was

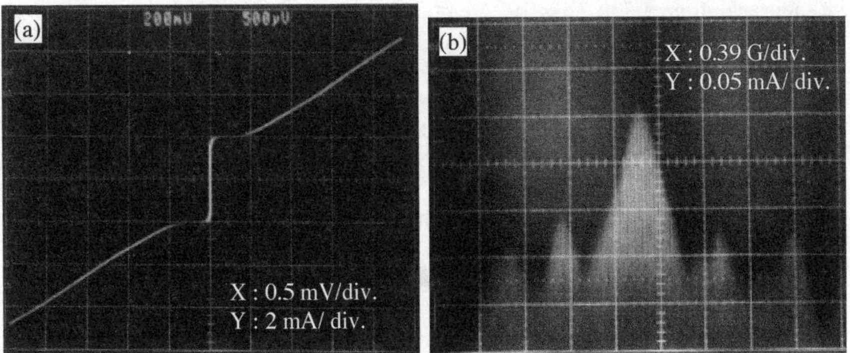

Fig. 21 (a) I - V characteristics and (b) magnetic-field dependence of I_c at 4.2 K for a 5-μm-wide (Hg,Re)-1212 junction on $\theta = 24°$ STO bicrystal substrate.

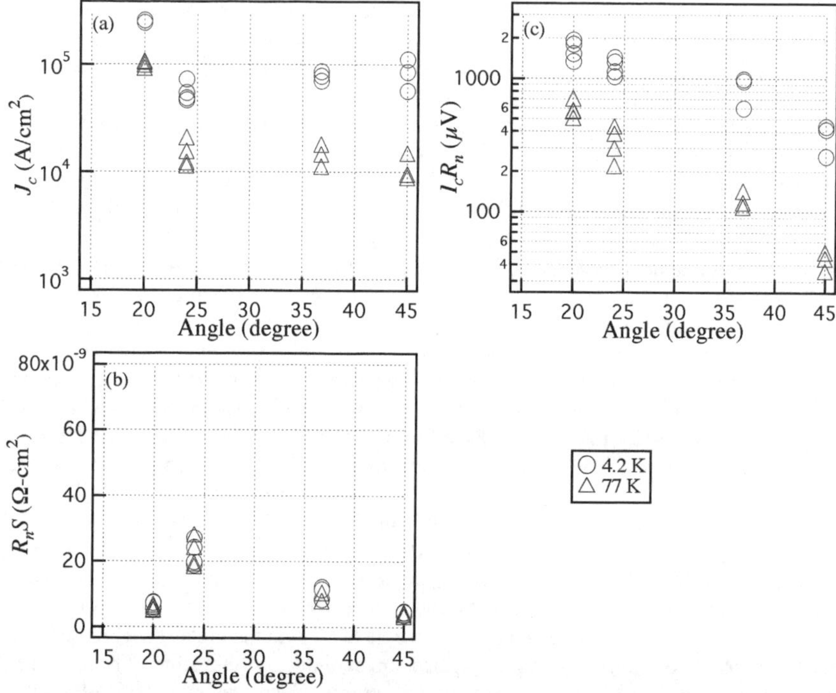

Fig. 22 Misorientation angle dependence of (a) J_c, (b) R_nS and (c) I_cR_n for the (Hg,Re)-1212 junctions on STO substrates measured at 4.2 and 77 K.

observed for the junctions fabricated by the improved process, i.e., the obtained J_c values for the $\theta = 24°$ junctions on the same substrate were in a range from 4.5 to 7.3×10^4 A/cm^2. This implies that the film quality is important to achieve small spreads of junction properties.

3.2 Bicrystal junctions on LSAT substrates

It was found that LSAT substrates did not react with the (Hg,Re)-1212 films [26]. Thus we expected preparation of bicrystal junctions with no significant precipitates by using LSAT bicrystal substrates. About 160-nm-thick (Hg,Re)-1212 thin films were prepared by the improved process, i.e., repeating the 80-nm-thick-film process. LSAT [001]-tilt

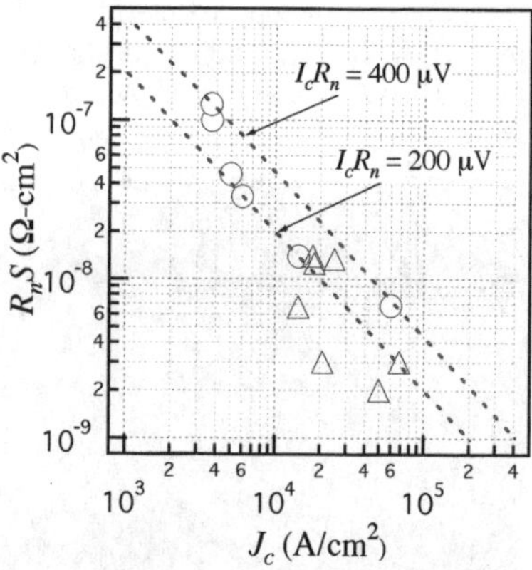

Fig. 23 Relationship between the J_c and R_nS at 77 K for $\theta = 24°$ (Hg,Re)-1212 junctions which were fabricated by the conventional two-step process.

symmetric bicrystal substrates with different misorientation angles $\theta = 24°$ and $36.8°$ were used. Figure 24(a) shows an SEM plane-view picture of the bicrystal boundary for a film on a $\theta = 36.8°$ LSAT bicrystal substrate. It is difficult to find the bicrytal boundary line in the SEM picture, while it could be seen by optical microscope observation. A pair of arrows indicates the location of the bicrytal boundary. It is found that precipitates aligned along the bicrytal boundary do not exist. Figure 24(b) shows an SEM cross-sectional picture of the bicrystal boundary for a film on a $\theta = 36.8°$ LSAT bicrystal substrate. Reaction between the film and the LSAT substrate is not observed even near the bicrytal boundary and the bicrystal junction looks homogeneous as expected.

The I - V characteristics of $\theta = 24°$ and $36.8°$ junctions exhibited a RSJ-like behavior with little excess current, as seen in Fig. 25(a). Figure 25(b) shows the magnetic-field dependence of I_c at 4.2 K for a 5-μm-wide junction on a $\theta = 24°$ LSAT bicrystal substrate. It

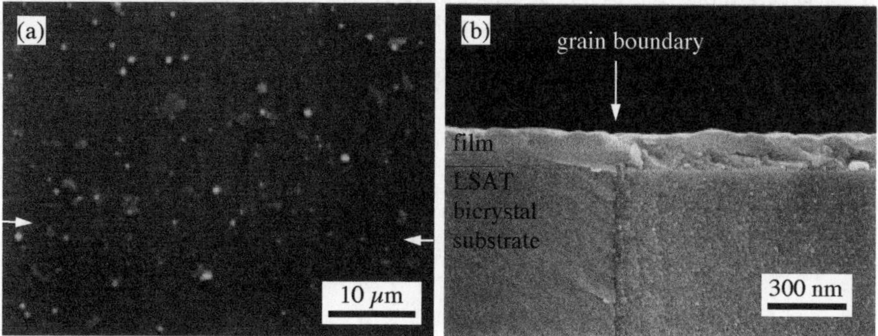

Fig. 24 SEM (a) plane-view and (b) cross-sectional pictures of the bicrystal boundaries in (Hg,Re)-1212 thin films on $\theta = 36.8°$ LSAT bicrystal substrates fabricated by the improved process. The bicrystal boundaries are indicated by the arrows.

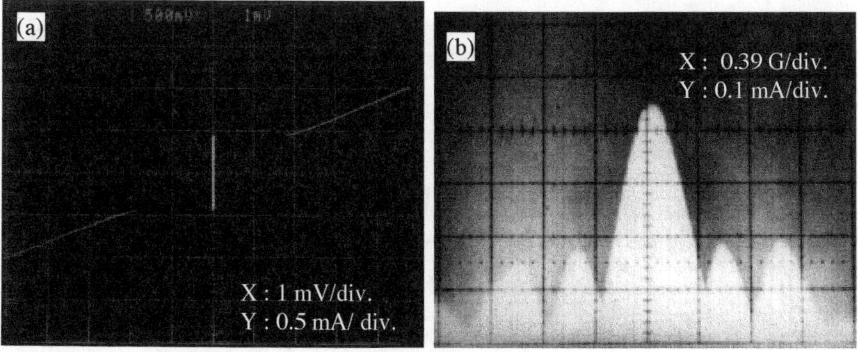

Fig. 25 (a) I - V characteristics and (b) magnetic-field dependence of I_c at 4.2 K for a 5-μm-wide (Hg,Re)-1212 junction on $\theta = 24°$ LSAT bicrystal substrate.

shows more than 95% modulation and indicates that the junctions on LSAT substrates have rather homogeneous current distribution.

Figure 26 shows the J_c, R_nS and I_cR_n product at 4.2 and 77 K of the bicrystal junctions on LSAT substrates with $\theta = 24°$ and 36.8°. The $\theta = 24°$ junction shows the I_cR_n products of 2.2 and 0.38 mV at 4.2 K and 77 K, respectively. The J_c values and the I_cR_n products for the $\theta = 24°$ junctions are larger than those for the $\theta = 36.8°$ junctions, while the R_nS slightly increases with an increase in θ. The spreads of J_c are even smaller than those for the junctions on STO, i.e., the obtained J_c values for the $\theta = 24°$ junctions are in a range from 4.4 to 4.8×10^4 A/cm^2.

3.3 Comparison between properties of junctions on different substrates

The morphologies of bicrystal boundaries on STO and LSAT substrates are quite different. Consequently, the junction properties especially for high-angle bicrytal boundaries may be largely influenced by the structural defects. Here, we focus our attention on the difference in the transport properties between the junctions on STO and LSAT substrates. On LSAT substrates, the J_c decreases and R_nS increases with an increase in θ. However, quite different tendencies of the J_c and R_nS are observed for the junctions on STO substrates.

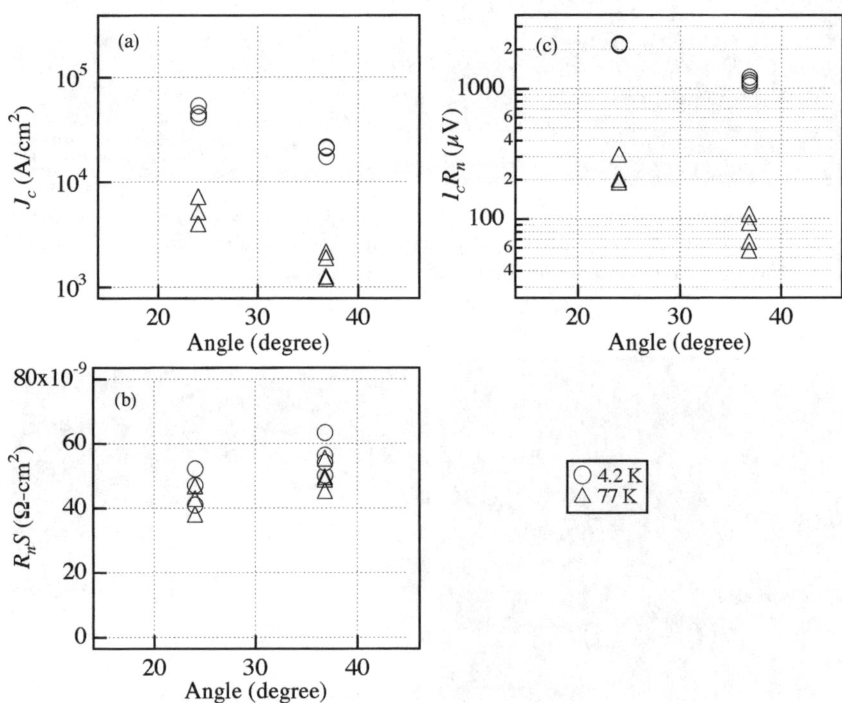

Fig. 26 Misorientation angle dependence of (a) J_c, (b) R_nS and (c) I_cR_n for the (Hg,Re)-1212 junctions on LSAT substrates measured at 4.2 and 77 K.

In previous studies on Y-123 bicrystal junctions, it was reported that the J_c decreases exponentially in many cases, with increasing θ [34-38,40]. This J_c variation has been mainly attributed to a change in the thickness of carrier depletion layer or a barrier near grain boundaries [41,42], though other origins such as the predominant d_{x2-y2}-symmetry of the order parameter [37] and faceting of bicrystal boundaries [43] may have some contribution. The observed J_c tendency for the junctions on LSAT is consistent with the results in Y-123 bicrystal junctions. For the junctions on STO substrates, precipitates were observed along the bicrytal boundaries, as seen in Figs. 20(a) and 20(b). The existence of precipitates and resultant inhomogeneous current distribution is considered to be a main reason for the observed unusual J_c behavior for the junctions. In contrast, the magnetic-field dependence of I_c for the junctions on LSAT indicates more homogeneous current distribution, as shown in Fig. 25(b), which is consistent with the monotonic J_c decrease with θ.

One of the important "figure of merits" of Josephson junctions is the I_cR_n product that determines the performance of the junctions at high frequencies. For the same misorientation angles θ, the I_cR_n products of the junctions on LSAT are clearly higher than those of the junctions on STO. This is also attributable to the homogeneous microstructure and current distribution in the former junctions. For both the junctions on STO and LSAT substrates, the I_cR_n products monotonically decrease with an increase in θ and this tendency agrees with that reported for Y-123 junctions [36]. For $\theta \geq 20°$, the junctions on STO exhibited RSJ-like I - V characteristics. There is a possibility of obtaining even higher I_cR_n products by using $\theta = 20°$ LSAT bicrystal substrates, if they are commercially available.

4. DC SQUID

One of the promising applications of high-T_c superconductors is the SQUIDs containing only one or two Josephson junctions. High-T_c SQUIDs are usually fabricated using Y-123 as a superconductor because high-quality epitaxial thin films can be obtained routinely. However, the I_cR_n products of Y-123 at 77 K are 10% or lower of those obtained at 4.2 K because the T_c of Y-123 is close to 77 K. The use of superconductors with a higher T_c is an apparently promising way of overcoming these problems. We have fabricated (Hg,Re)-1212 dc SQUIDs by both the conventional and the improved processes and evaluated their properties. These results are described in this section.

4.1 Properties of SQUIDs fabricated employing the conventional process

About 200-nm-thick (Hg,Re)-1212 thin films were fabricated on $\theta = 24°$ STO bicrystal substrates by the conventional two-step process [16-18]. Two dc-SQUIDs with stripline-type pattern for one and washer-type pattern for the other were prepared. Figures 27(a) and 27(b) show an SEM photograph of the stripline-type SQUID and an optical microscope image of the washer-type SQUID with a feedback coil on the same substrate.

Table 1 shows the parameters for the SQUIDs with two 5-μm-wide junctions and their properties measured at 77 K. For both of the patterns, the voltage modulation induced by applying the magnetic field was observed up to about 110 K. Figure 28 shows the voltage – flux (V–Φ) characteristics of the stripline-type SQUID measured at 77 K. The volgate modulation depth ΔV is 90 μV at 77 K. This value is comparable to that of SQUIDs made of high-quality Y-123 thin films. On the other hand, the washer-type SQUID showed smaller ΔV, because of its larger inductance and smaller R_n. The washer-type SQUID equipped with the feedback coil could be operated in a flux-locked loop (FLL) scheme with dc-bias current.

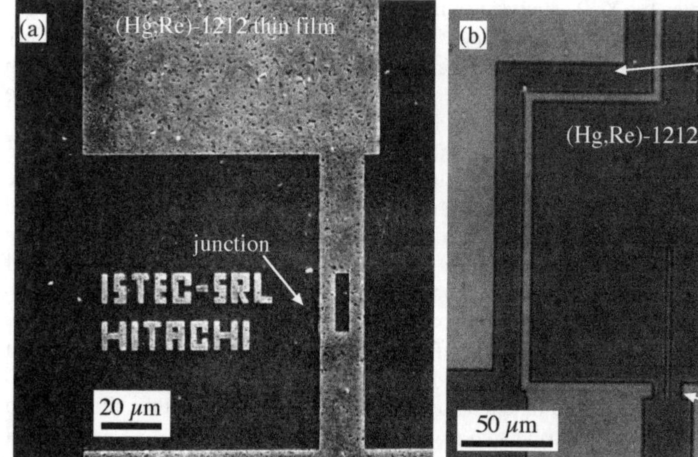

Fig.27 (a) SEM photograph of a stripline-type SQUID and (b) optical microscope image of a washer-type SQUID with a feedback coil. (Hg,Re)-1212 thin films were fabricated by the conventional two-step process.

Figure 29 shows the flux noise measured at 77 K for this SQUID. The flux noise with $1/f$ frequency dependence is dominant at low frequencies. The $S_\phi^{1/2}$ at 1 Hz is about 10^{-3} $\Phi_0/\mathrm{Hz}^{1/2}$, which is about one and two orders of magnitude larger than that for the best Y-123 SQUID measured in dc-bias and in ac-bias schemes, respectively [44]. In the case of Y-123, SQUIDs made of

Table 1 Parameters of (Hg,Re)-1212 SQUIDs

	stripline type	washer type
Outer dimensions	15 μm × 20 μm	150 μm × 150 μm
Slit size	5 μm × 20 μm	2 μm × 75 μm
Junction width	5 μm	5 μm
Film thickness	200 nm	200 nm
J_c (film)	5×10^6 A/cm^2	6×10^6 A/cm^2
Inductance	18 pH	33 pH
$2I_c$	100 μA	500 μA
$R_n/2$	2.3 Ω	0.65 Ω
$I_c R_n$	230 μV	325 μV
ΔV	90 μV	19 μV

low-quality films exhibit a large flux noise. (Hg,Re)-1212 thin film fabricated by conventional two-step process included submicron-size defects as described above. By using the improved process, it is expected that the noise properties of (Hg,Re)-1212 SQUIDs will be substantially improved.

4.2 Flux Noise for SQUIDs fabricated employing the improved process

About 160-nm-thick (Hg,Re)-1212 thin films were fabricated on $\theta = 36.8°$ STO and 24° LSAT bicrystal substrates by the improved process, i.e., repeating the 80-nm-thick-film process [28,30]. We measured the flux noise of washer-type SQUIDs with dc bias and ac bias current schemes at 77 K. Figure 30(a) shows the flux noise spectra at 77 K with a dc and ac bias for the SQUID fabricated on STO bicrystal substrate. The flux noise with the dc

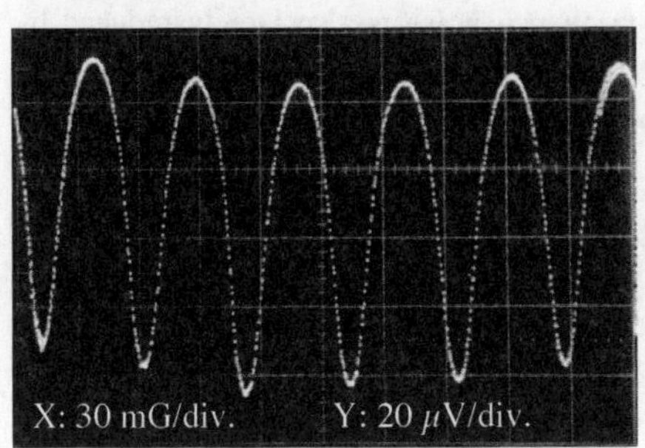

Fig. 28 Voltage-flux characteristics of the stripline-type SQUID measured at 77 K. (Hg,Re)-1212 thin film was fabricated by the conventional two-step process.

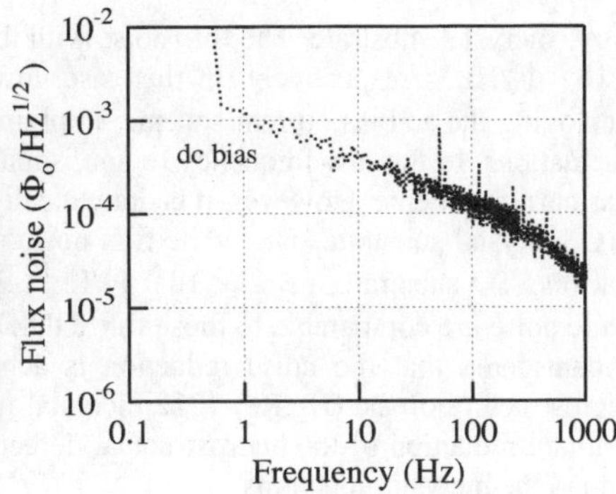

Fig. 29 Flux noise spectrum of the washer-type SQUID measured at 77 K with dc bias current. The (Hg,Re)-1212 thin film was fabricated by the conventional two-step process.

and ac bias at 1Hz and 77 K are 3×10^{-4} and 6×10^{-5} Φ_0/Hz$^{1/2}$, respectively. The low-frequency noise is dramatically reduced by employing the ac bias current scheme. There are generally two causes of $1/f$ noise. One is fluctuations in the junction I_c and the other is the thermally activated motion of vortices in the superconducting thin film that forms the SQUID loop. The $1/f$ noise caused by I_c fluctuations can be eliminated by the ac bias current scheme. Indeed the low-frequency noise is largely reduced by employing the ac bias current scheme, as shown in Fig. 30(a), implying that the $1/f$ noise is mainly generated by I_c fluctuations. The white noise of 2×10^{-5} Φ_0/Hz$^{1/2}$ is slightly higher than that of the SQUIDs made from high-quality epitaxial Y-123 thin films.

Figure 30(b) shows the flux noise spectra for the (Hg,Re)-1212 SQUID fabricated on

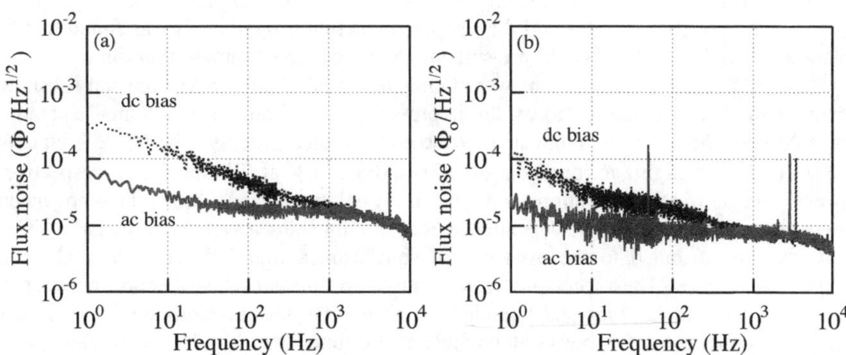

Fig. 30 Flux noise spectra of the (Hg,Re)-1212 SQUIDs on (a) $\theta = 36.8°$ STO bicrystal substrate and (b) $\theta = 24°$ LSAT bicrystal substrate measured at 77 K with dc and ac bias current. The films were fabricated by the improved process.

LSAT bicrystal substrate. The flux noise with the dc and ac bias at 1 Hz and 77 K are 8 and 2 $\times$ 10^{-5} $\Phi_0/\text{Hz}^{1/2}$, respectively. In this case, low-frequency noise is also largely reduced by employing the ac bias current scheme, implying that the $1/f$ noise is mainly generated by I_c fluctuations. In the low frequency region, some $1/f$ noise remains even by employing the ac bias current scheme. However, it is dramatically decreased as compared with the cases using STO bicrystal substrate. The white flux noise is 8×10^{-6} $\Phi_0/\text{Hz}^{1/2}$, which is also lower than that on STO substrate, i.e. 2×10^{-5} $\Phi_0/\text{Hz}^{1/2}$. Both the observed $1/f$ noise at 1 Hz and the white noise are comparable to those for SQUIDs made from high-quality Y-123 thin films. It is considered that the noise reduction is achieved by the improved homogeneity in the microstructure of the (Hg,Re)-1212 bicrystal junctions on LSAT bicrystal substrate and the resultant reduction of the microstructual defects which cause the motion of the trapped flux around the bicrystal junctions.

5. CONCLUDING REMARKS

In this chapter, we presented the fabrication process of high-quality Hg-based superconducting thin films, and its application to bicrystal junctions and dc-SQUIDs. From the investigation on the transition metal doping, it was found that the Re doping has a significant effect on the fabrication of Hg-1223 and Hg-1212 thin films with superior surface morphology and superconducting properties. The crystallinity of the (Hg,Re)-1212 thin film was better than that of the 1223 thin film. To decrease the density of defects in the films fabricated by the conventional two-step process, we developed the following improved processes.

(1) For very thin films (film thickness is less than 100 nm), rather thin precursor films are employed and annealed.

(2) For thicker films (film thickness is 100 nm or more), the very thin films are prepared by process (1) first and then the same deposition and annealing procedures are repeated.

75-nm-thick (Hg,Re)-1212 thin films with no appreciable defects were successfully obtained by process (1). 150- and 225-nm-thick (Hg,Re)-1212 thin films were also obtained without degrading the surface smoothness by process (2). These films exhibited high J_c value at 77 K of $(3 - 10) \times 10^6$ A/cm^2 with a small spread for several bridges on the same substrate. We have also investigated the relationships between the microstructures of the (Hg,Re)-1212 thin films on various substrates and their superconducting properties. It was found that two substrate materials, STO and LSAT are suitable to fabrication of smooth thin films.

(Hg,Re)-1212 bicrystal junctions were fabricated on STO and LSAT bicrystal substrates. Properties of junctions fabricated by the improved process show a rather small spread, as compared with the junctions fabricated by the conventional process. The $\theta = 20°$ junctions on STO exhibited high I_cR_n products of 0.71 and 1.9 mV at 77 and 4.2 K, respectively. However, the current distribution along the bicrystal boundary is quite inhomogeneous. Moreover, misorientation angle dependence of J_c did not show a systematic variation. These results can be attributed to the existence of significant amount of precipitates along the bicrystal boundaries. These precipitates are caused by the enhanced reaction between the films and STO near the bicrystal boundaries. In contrast, the junctions on LSAT with no trace of precipitates at the bicrystal boundaries exhibited the I_c - H curves close to the Fraunhofer pattern and a monotonic decrease of J_c with increasing θ. These results are consistent with the expectation that the microstructure of the junctions on LSAT substrates is much more homogeneous.

(Hg,Re)-1212 SQUIDs were fabricated on STO and LSAT bicrystal substrates. A strip-line SQUID made of the film by the conventional process exhibited ΔV of 90 μV at 77 K. This value is comparable to that of high-quality Y-123 SQUIDs. Low-frequency noise of washer-type SQUIDs made by the improved process on STO and LSAT substrates is largely reduced by employing the ac bias current scheme, implying that the $1/f$ noise is mainly generated by I_c fluctuations. The $1/f$ noise with ac bias at 1 Hz and the white flux noise at 77 K for the SQUID on LSAT substrate are 2×10^{-5} and 8×10^{-6} $\Phi_0/Hz^{1/2}$, respectively. These are substantially lower than those on STO substrate and comparable to those for SQUIDs made from high-quality Y-123 thin films. It is considered that the noise reduction is achieved by the improved homogeneity in the microstructure of the bicrystal boundary on LSAT bicrystal substrate and the resultant reduction of the microstructual defects which cause the motion of the trapped flux around the bicrystal junctions.

ACKNOWLEDGMENTS

The authors would like to thank T. Utagawa and Dr. Y. Tarutani for patterning of the thin films. This work was supported by the New Energy and Industrial Technology Development Organization (NEDO) as Collaborative Research and Development of Fundamental Technologies for Superconductivity Applications.

FOOTNOTE

* Present address: Technical Research Center, The Chugoku Electric Power Co., Inc., 3-9-1, Kagamiyama, Higashihiroshima 739-0046, Japan
** Present address: Okayama Service Office, The Chugoku Electric Power Co., Inc., 2-6-51, Aoe, Okayama 700-8507, Japan
*** Present address: Department of Materials Science, Fudan University, Handanlu 220, P. O. Box 200433, Shanghai, China

REFERENCES

[1]   S. M. Loureiro, E. V. Antipov, J. L. Tholence, J. J. Capponi, O. Chmaissen, Q. Huang and M. Marezio: Physica C **217** (1993) 253.
[2]   J. L. Wagner, B. A. Hunter, D. G. Hinks and J. D. Jorgensen: Phys. Rev. B **51** (1995) 15 407.
[3]   A. Fukuoka, A. Tokiwa-Yamamoto, M.Itoh, R. Usami, S. Adachi and K. Tanabe: Phys. Rev. B **55** (1997) 6612.
[4]   C. C. Tsuei, A. Gupta, G. Trafas and D. Mitzi: Science **263** (1994) 1259.
[5]   S. H. Yun, J. Z. Wu, B. W. Kang, A. N. Ray, A. Gapud, Y. Yang, R. Farr, G. F. Sun, S. H. Yoo, Y. Xin and W. S. He: Appl Phys. Lett. **67** (1995) 2866.
[6]   L. Krushin-Elbaum, C. C. Tsuei and A. Gupta: Nature **373** (1995) 679.
[7]   S. H. Yun and J. Z. Wu: Appl. Phys. Lett. **68** (1996) 862.
[8]   Y.Yu, H. M. Shao, Z. Y. Zheng, A. M. Sun, M. J. Qin, X. N. Xu, S. Y. Ding, X. Jin, X. X. Yao, J. Zhou, Z. M. Ji, S. Z. Yang and W. L. Zhang: Physica C **289** (1997) 199.
[9]   S. L. Yan, Y. Y. Xie, J. Z. Wu, T. Aytug, A. A. Gapud, B. W. Kang, L. Fang, M. He, S. C. Tidrow, K. W. Kirchner, J. R. Liu and W. K. Chu: Appl. Phys. Lett. **73** (1998) 2989.
[10] Y. Sun, J. D. Guo, X. L. Xu, G. J. Lian, Y. Z. Wang and G. C. Xiong: Physica C **312** (1999) 197.
[11] A. Gupta, J. Z. Sun, C. C. Tsuei: Science **265** (1994) 1075.

[12] N. Khare, A. K. Gupta, S. Khare, H. K. Singh, A. K. Saxena and O. N. Srivastava: Physica C **274** (1997) 161.

[13] S. Adachi, A. Tokiwa-Yamamoto, A. Fukuoka, R. Usami, T. Tatsuki, Y. Moriwaki and K. Tanabe: *Studies of High Temperature Superconductors*, ed. A. Narlikar (Nova Science Publishers, Inc., New York, 1997) Vol. 23, Chap. 6, p. 163.

[14] Y. Moriwaki, T. Sugano, A. Tsukamoto, C. Gasser, K. Nakanishi, S. Adachi and K. Tanabe: *Advances in Superconductivity X*, ed. K. Osamura and I. Hirabayashi (Springer Verlag, Tokyo, 1998) Vol. 2, p. 1019.

[15] Y. Moriwaki, T. Sugano, A. Tsukamoto, C. Gasser, K. Nakanishi, S. Adachi and K. Tanabe: Physica C **303** (1998) 65.

[16] A. Tsukamoto, K. Takagi, Y. Moriwaki, T. Sugano, S. Adachi and K. Tanabe: Applied Physics Letters **73** (1998) 990.

[17] A. Tsukamoto, Y. Moriwaki, T. Sugano, N. Inoue, A. Kandori, S. Adachi, K. Tanabe and K. Takagi: *Advances in Superconductivity XI*, ed. N. Koshizuka and S. Tajima (Springer Verlag, Tokyo, 1999) Vol. 2, p. 1199.

[18] A. Tsukamoto, K. Takagi, Y. Moriwaki, T. Sugano, S. Adachi and K. Tanabe: IEEE Trans. Appl. Supercond. **9** (1999) 3093.

[19] Y. Moriwaki, T. Sugano, S. Adachi and K. Tanabe: IEEE Trans. Appl. Supercond. **9** (1999) 2390.

[20] Y. Moriwaki, T. Sugano, A. Tsukamoto, S. Adachi and K. Tanabe: *Advances in Superconductivity XI*, ed. N. Koshizuka and S. Tajima (Springer Verlag, Tokyo, 1999) Vol. 2, p. 1063.

[21] N. Inoue, T. Sugano, A. Tsukamoto, T. Utagawa, S. Adachi, K. Tanabe: Physica C **330** (2000) 94.

[22] X. –J. Wu, N. Inoue, T. Sugano, T. Morishita and K. Tanabe: *Advances in Superconductivity XII*, ed. T. Yamashita and K. Tanabe (Springer Verlag, Tokyo, 2000) p. 894.

[23] N. Inoue, T. Sugano, A. Tsukamoto, X. –J. Wu, T. Utagawa, S. Adachi and K. Tanabe: *Advances in Superconductivity XII*, ed. T. Yamashita and K. Tanabe (Springer Verlag, Tokyo, 2000) p. 897.

[24] N. Inoue, T. Sugano, T. Utagawa, X. –J. Wu, A. Tsukamoto, S. Adachi and K. Tanabe: Physica C **341-348** (2000) 2391.

[25] N. Inoue, T. Sugano, T. Utagawa, Y. Wu, S. Adachi and K. Tanabe: Physica C **357-360** (2001) 1444.

[26] X. –J. Wu, N. Inoue, T. Sugano and K. Tanabe: Physica C **349** (2001) 295.

[27] X. –J. Wu, N. Inoue, T. Sugano, T. Morishita and K. Tanabe: J. Mater. Res. **16** (2001) 256.

[28] A. Tsukamoto, N. Inoue, T. Sugano, S. Adachi, K. Tanabe and K. Takagi: Physica C **357-360** (2001) 1459.

[29] N. Inoue, T. Sugano, S. Adachi and K. Tanabe: Jpn. J. Appl. Phys. **41** (2002) 583.

[30] N. Inoue, A. Tsukamoto, T. Sugano, A. Ogawa, S. Adachi, K. Takagi and K. Tanabe: Jpn. J. Appl. Phys. **41** (2002) L1149.

[31] J. Shimoyama, K. Kishio, S. Hahakura, K. Kitazawa, K. Yamaura, Z. Hiroi and M. Takano: *Advances in Superconductivity VII*, ed. K. Yamafuji and T. Morishita (Springer Verlag, Tokyo, 1995) p. 287.

[32] A. Maignan, D. Pelloquin, S. Malo, C. Michel, M. Hervieu and B. Raveau: Physica C **243** (1995) 233.

[33] R. L. Meng, L. Beauvais, X. N. Zhang, Z. J. Huang, Y. Y. Sun, Y. Y. Xue and C. W. Chu: Physica C **216** (1993) 21.

[34] D. Dimos, P. Chaudhari and J. Mannhart: Phys. Rev. B **41** (1990) 4038.

[35] Z. G. Ivanov, P. A. Nilsson, D. Winkler, J. A. Alarco, T. Claeson, E. A. Stepantsov and A. Ya. Tzalenchuk: Appl. Phys. Lett. **59** (1991) 3030.

[36] R. Gross and B. Mayer: Physica C **180** (1991) 235.

[37] H. Hilgenkamp, J. Mannhart and B. Mayer: Phys. Rev. B **53** (1996) 14586.

[38] K. E. Gray, M. B. Field and D. J. Miller: Phys. Rev. B **58** (1998) 9543.

[39] T. Minotani, S. Kawakami, Y. Kuroki and K. Enpuku: Jpn. J. Appl. Phys. **37** (1998) L718.

[40] N. F. Heinig, R. D. Redwing, J. E. Nordman and D. C. Larbalestier: Phys. Rev. B **60** (1999) 1409.

[41] N. D. Browning, J. P. Buban, P. D. Nellist, D. P. Norton, M. F. Chisholm and S. J. Pennycook: Physica C **294** (1998) 183.

[42] N. D. Browning, J. P. Buban, M. Kim and S. J. Pennycook: *Extend Abstract of 2000 International Workshop on Superconductivity*, Shimane, 2000, p.47.

[43] J. Mannhart, H. Hilgenkamp, B. Mayer and Ch. Gerber: Phys. Rev. Lett. **77** (1996) 2782.

[44] L. P. Lee, J. Longo, V. Vinetskiy and R. Cantor: Appl. Phys. Lett. **66** (1995) 1539.

HIGH-T_c SUPERCONDUCTING CONDUCTORS FOR AC AND DC APPLICATIONS

Bartek A. GLOWACKI

Department of Materials Science and Metallurgy, University of Cambridge, Pembroke Street, Cambridge CB2 3QZ, England

IRC in Superconductivity, Cavendish Lab. University of Cambridge, Madingley Road, West Cambridge Site, Cambridge CB3 0HE, England

1. INTRODUCTION

Despite tremendous efforts in research and development after the discovery of high temperature perovskite-type layered superconductors, and high expectations expressed in market projection potential for electrical commercial applications of these materials are taking off very slowly. Therefore the updated the original comparative prediction of the market for the low temperature superconductors, LTS, and high temperature superconductors, HTS, is presented in Figure 1 where is no cross over between the LTS and HTS. This is because different applications of superconductivity require the simultaneous fulfilment of diverse thermal, mechanical and electromagnetic specifications; the potential to operate at higher temperatures is not the decisive factor except transmission lines and transformers. Medical care which at the mostly represented by magnetic resonance applications based on LTS will grow, and in 2010 should reach 15M€/year.
The primary focus of this chapter is on high temperature and medium temperature superconducting conductors for power engineering applications particularly the optimisation of processing routes for the production of high and medium temperature coated conductors such as $YBa_2Cu_3O_7$, $Bi_2Sr_2CaCu_2O_9$ and MgB_2 and multifilamentary superconducting conductors such as $(Bi,Pb)_2Sr_2Ca_2Cu_3O_9$ and MgB_2. Four different stages can be identified; (i) the near market design of fully engineered conductors for particular applications, (ii) incremental development of 'established' conductor processing routes, (iii) investigation of new routes for superconductor processing, and (iv) basic and enabling science to underpin materials processing. Naturally the boundaries between these types of activity are blurred, however the different activities require different levels of interaction between industry and academia. Furthermore the type of research appropriate for a particular conductor processing

240

route varies depending on the maturity of the technology and is driven by particular specifications.

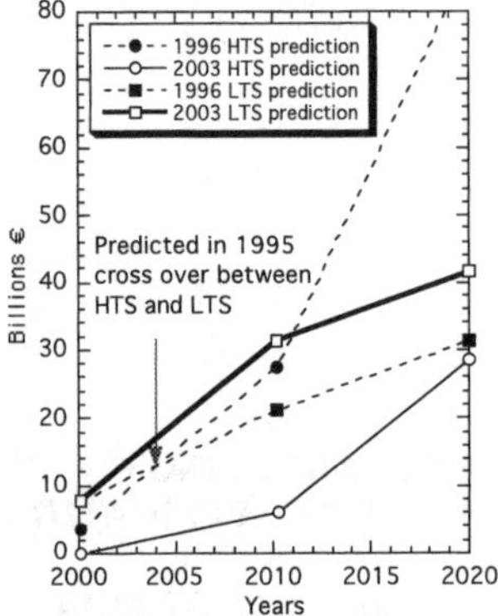

Figure 1   Projection of the LTS and HTS market which is sheared between electronics, power, industrial processing, medical care and transportation. 1995 estimation conducted at International Superconductivity Industry Summit, Yamanashi 1996, numbers corrected according to data in early 2003 (solid lines).

1.1. Material design considerations for applications

As major applications develop close interaction is necessary between design engineers and materials producers. Lists below summarises major design considerations for each of four applications in terms of eight areas of material specification and three potential design requirements. Bold typeface indicates an area of key importance, italic typeface indicates less relevance. There are distinct differences in the requirements for fully engineered conductors for each application area. Detailed design decisions must be reflected directly in the conductor specification [1]

1.1.1. Important considerations for **Magnet** design:

<u>Specification of Material:</u>
S1 **overall $J_{eng}(B,T)$**, *critical current, critical current density*
S2 **anisotropy of critical current**
S3 mechanical issues, strain response, differential contraction
S4 normal state properties, resistivity, thermal conductivity
S5 **resistive transition, n-value**
S6 *transient dissipation,* **quench stability,** *ac losses*
S7 **joints,** *contacts*
S8 **insulation issues**

<u>Design Requirements:</u>
D1 **operating temperature,** *field-temperature environment*
D2 **dimensions, composite structure, piece length, uniformity**
D3 *price target,* **pay-off against price**

1.1.2. Important considerations for **Current Lead** design:

<u>Specification of Material:</u>
S1 *overall* J_{eng} , **critical current** $I_c(B,T)$**,** *critical current density*
S2 **anisotropy of critical current**
S3 *mechanical issues, strain response, differential contraction*
S4 *normal state properties, resistivity,* **thermal conductivity**
S5 *resistive transition, n-value*
S6 *transient dissipation,* **quench stability,** *ac losses*
S7 *joints,* **contacts**
S8 *insulation issues*

<u>Design Requirements:</u>
D1 *operating temperature,* **field-temperature environment**
D2 *dimensions, composite structure, piece length, uniformity*
D3 *price target, pay-off against price*

1.1.3. Important considerations for **Cable** design:

<u>Specification of Material:</u>
S1 **overall** $J_{eng}(B,T)$**,** *critical current, critical current density*
S2 *anisotropy of critical current*
S3 **mechanical issues, strain response, differential contraction**
S4 *normal state properties, resistivity, thermal conductivity*
S5 *resistive transition, n-value*
S6 *transient dissipation, quench stability,* **ac losses**
S7 **joints,** *contacts*
S8 **insulation issues**

<u>Design Requirements:</u>
D1 **operating temperature,** *field-temperature environment*
D2 *dimensions,* **composite structure, piece length, uniformity**
D3 **price target,** *pay-off against price*

1.1.5. Important considerations for **FCL** design:

<u>Specification of Material:</u>
S1 *overall* $J_{eng}(B,T)$*, critical current,* **critical current density**
S2 *anisotropy of critical current*
S3 *mechanical issues, strain response, differential contraction*
S4 **normal state properties, resistivity,** *thermal conductivity*
S5 **resistive transition,** *n-value*
S6 **transient dissipation, quench stability,** *ac losses*

S7 **joints, contacts**
S8 *insulation issues*

<u>Design Requirements:</u>
D1 *operating temperature,* **field-temperature environment**
D2 *dimensions, composite structure, piece length,* **uniformity**
D3 **price target, pay-off against price**

In Table 1 conductors higher in the table are highly developed while those lower down are still at an early stage of development as described by Evetts and Glowacki.[1].

Table 1   Relevance of conductor families to different application areas

	Magnets	Current Leads	Cables	Fault Current Limiters	Motors, Bearings, Machines
Bi-2223 mechanically deformed	√√√	√√	√√	√	
Bi-2212 mechanically deformed	√√	√			
Bi-2212 coated conductor	√√√	√			
Bi-2212 bulk melt processed		√√√		√√	
RE-123 bulk melt processed	Future	√√		√	√√√
RE-123 coated conductor	Future	Future	Future	√√	Future
Tl, Hg based coated conductor	Future		Future		
MgB$_2$	√√	√		√√	

2. REBa$_2$Cu$_3$O$_7$ CONDUCTORS.

For more than 15 years thin films of yttrium barium copper oxide, YBa$_2$Cu$_3$O$_7$, (YBCO) have demonstrated promising superconducting properties for use at liquid nitrogen temperatures with critical current density, J_c, (at 77 K) higher than 10^6 A/cm^2, Figure 2, and high irreversibility field, H_{irr}(77 K). There are three major techniques used to manufacture so called coated conductor: (i) RABiTS (Rolling-assisted, biaxially textured substrate), (ii) Ion beam-assisted deposition (IBAD) or the inclined substrate deposition processes IDP and (iii) liquid phase epitaxy LPE and Thermo-Magnetic processing.

Recent progress in the development of biaxially oriented epitaxial buffer layers [2, 3] on biaxially oriented flexible metallic substrates, [4] has opened up possibilities for the application of YBCO in electric utility and high magnetic field dc devices such as MRI and NMR magnets. While naturally occurring defects in YBCO are capable of providing high critical current J_c values, weak links such as high angle grain boundaries (GB's) can considerably affect the overall critical current [1], [5], [6]. Several different types of deposition techniques such as Pulse Laser Deposition, Sputtering, CVD, MOCVD, Thermal Evaporation, Sol-Gel, and Liquid Phase Epitaxy have been used to manufacture short pieces of the high temperature superconducting tapes on a laboratory scale [7], [8].

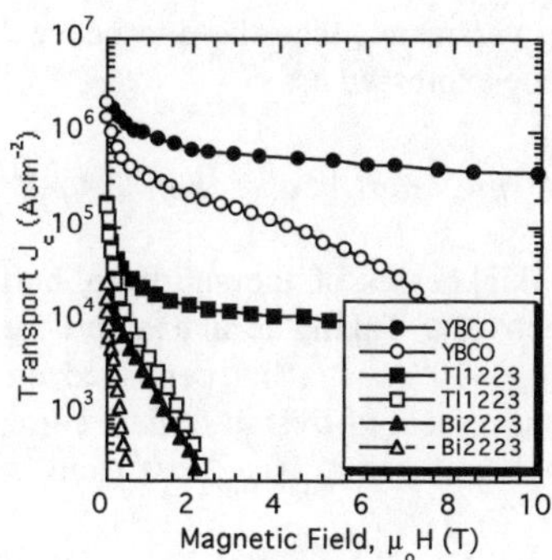

Figure 2 Critical current density vesus magnetic filed for $YBa_2Cu_3O_7$, $Bi_2Sr_2CaCu_2O_9$ and $TlBa_2Ca_2Cu_3O_9$

There is a range of Ni-based binary and ternary alloys that can be used as a substrate. Some of them are magnetic for example Ni [5], NiFe [6] and some of them are non-magnetic for example NiCu, NiV, NiCr, NiW [9], [10], [11], [12] depending on the alloy composition, and of course, the temperature of the application envisaged. As we know NiFe and also Ni is a ferromagnetic material and may not be the best for ac applications. However the recently patented method of decoupling filaments by using magnetic-superconducting heterostructures may make some ac applications economically justified [13]. For the major magnet manufactures such as Oxford Instruments plc, since their applications are DC type, the magnetism of the metallic substrates is not a critical issue.

2.1 RABiT coated conductors on Ni-Fe substrate

However, the current challenge is to find an economic route for large-scale production of the conductors including highly textured mechanically strong and thin metallic substrates in the form of a tape. The design of the final conductor under consideration consists of: (1) long lengths of well-oriented Ni-based substrate, (2) an epitaxial buffer layer grown epitaxially on the metallic substrate, (3) $REBa_2Cu_3O_7$ as the superconductor grown epitaxially on the buffer layer, (4) silver as a cryogenic and electric stabilisation layer and finally (5) polymer as the top layer for insulation and stress relief. This architecture is schematically presented in Figure 3.

The substrate must be thin because a very important consideration in most applications is a high value of engineering critical current density, J_{eng}, defined as a critical current, I_c, carried by the superconducting layer divided by the total cross section of the conductor which includes the substrate and all the functional layers. A low value of the overall critical current density J_{eng} will reduce the exploitation potential of fully engineered conductors. The thickness of non-superconducting layers and any deterioration of J_c in the superconductor with increasing thickness of the superconducting layer, t_{sup}, leads to a maximum in the dependence of J_{eng} with t_{sup} which may occur at the higher t_{sup} than usually envisaged [1]. If the $J_c(t_{sup})$ dependence on the t_{sup} is expressed as $J_c(t_{sup}) = J_c(t_{sup} \approx 0) \exp[-Ct_{sup}]$, where C is an empirical fitting parameter, then the dependence of J_{eng} on the thickness of the superconducting layer may be expressed as:

$$J_{eng}(t_{sup}) = J_c(t_{sup})\{t_{sup} \, / \, [t_{sub} + t_{buf} + t_{sup} + t_{stab} + t_{ins}]\}$$

where t_{sub}, t_{buf}, t_{stab}, t_{ins} are thicknesses of the substrate, buffer layer, stabilising (shunt) layer and insulating layer respectively. Taking as an example the $1\mu m$ thick superconducting coatings characterised by $J_c(t_{sup}) \sim 10^6 Acm^{-2}$, but deposited on the $100\mu m$ thick Ni-based substrates, this will not differ, in respect of overall critical current density from the ordinary powder-in-tube (PIT) conductor which has $J_{eng}(t_{sup}) \sim 10^4 Acm^{-2}$. Therefore for the complex

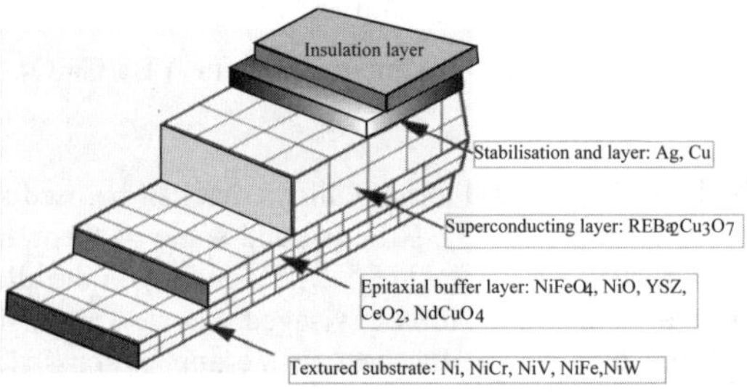

Figure 3   Schematic cross section of a fully engineered (RE)-123 coated conductor tape adopted by MUST project. Such a tape is typically 10mm wide and is required to have a uniform high critical current over long lengths. The key conductor components are the metallic or ceramic substrate material, the buffer layer and the superconducting layer; the latter has to be biaxially textured throughout so that, although granular, the misorientation from grain to grain should be less than a few degrees.

coated conductor technique to gain a significant advantage, one has to use the thinnest possible substrates but simultaneously preserve their mechanical strength. The essential condition for development of biaxially oriented epitaxial buffer layers and superconducting layers on flexible metallic substrates is the degree of texture in the metallic substrate itself. The so-called 'cube' texture, with crystallographic axes parallel to the sample axes, forms an ideal textured metallic substrate. This can be obtained in most FCC metals by cold rolling to a reduction of greater than about 95% followed by heat treatment [14], [15], [16]. Cold

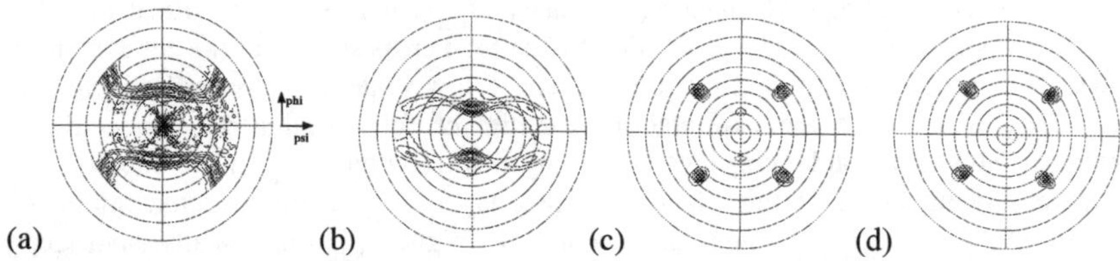

Figure 4 X-ray pole figures of NiFe tape, $25\mu m$ thick: (a) cold rolled (200) (linear scale I_{max} = 914), (b) cold rolled (111) (linear scale I_{max} = 3361) all X-ray pole figures have roll direction vertical. (c) annealed at 500°C b) annealed 800°C.

rolling gives a characteristic, but quite complex, texture described as (110)[112], (112)[111], (123)[422] and (146)[211], Figure 4(a), (b). (An ideal texture has (hkl) planes parallel to the sheet and the vector [uvw] parallel to the rolling direction.) Some of the work of deformation is stored as defects and this energy provides the driving force for recovery and recrystallisation [16], [17]. The heat treatment is required to give the optimum highly oriented grains, Figure 4(c), (d). In general a cube texture is likely to depend on the grain size and the level of impurities. For this purpose the NiFe50% tape was cast in the form of 2 tonnes ingot and deformed and cold-rolled to the 3-5mm thin stripes, followed by the final rolling to thicknesses of 50 μm, 25 μm and 13 μm and a width of 10 cm. Each final roll weighed ~ 20 kg and was approximately 1 km long. The deformation process and heat treatment in protective atmosphere was conducted in a factory environment by Carpenter Technologies (UK). The final tape was slit to 25mm and 10 mm wide tapes for our research development, see Figure 4.

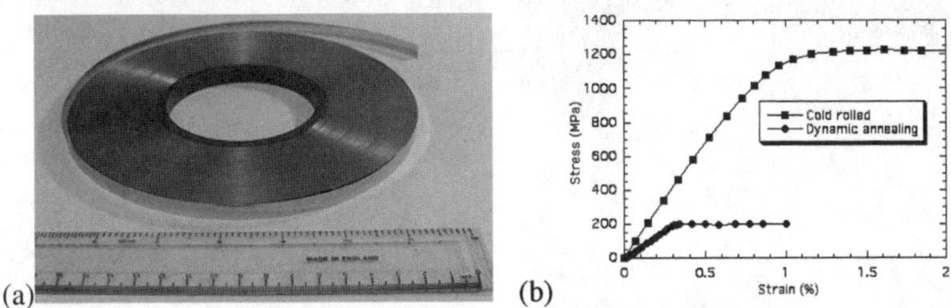

Figure 4 NiFe flexible substrates for coated conductor: (a) view of the NiFe tape, 760m long, 10mm wide and 25μm thick after continuous annealing; (b) stress - strain curves of the cold rolled and dynamically annealed in protective atmosphere NiFe tape, 25μm thick tape [18]. An important issue is the NiFe tapes have superior mechanical and structural properties compared with pure Ni substrates and some of Ni-alloys, Figure 4(b) and Figure 5.

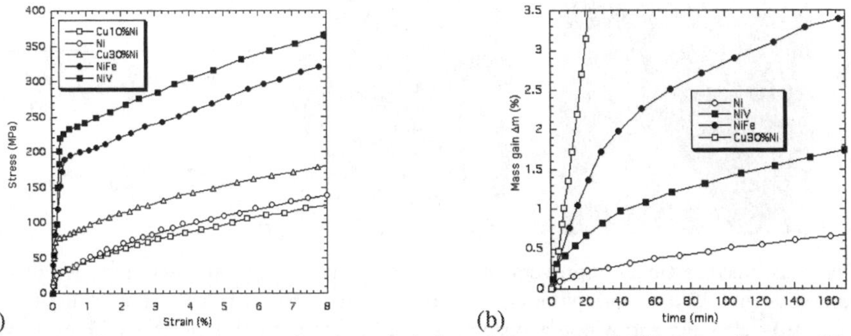

Figure 5 Ni alloys: (a) Stress-strain characteristics of the selected Ni-alloys developed in frame of European collaboration CONTEXT and MUST projects, (b) thermogravimetric data representing oxidation of the different Ni-alloys developed in frame of EU CONTEXT and MUST projects; the horizontal axis represent time during which the samples were exposed to oxygen at constant temperature 700°C.

The important issue for the uninterrupted epitaxial growth of the buffer layers and superconducting layers on a highly textured tape is the smoothness of the sample surface. Therefore continuous electropolishing technique was developed, which has greater flexibility and can be used on cube textured tapes. The electropolishing process conducted on Ni and NiFe tapes brought the roughness down from 0.20 μm to < 0.1μm and formed the base for the development of the pilot continuous electropolishing treatment unit by Europa Metalli. Trials indicated that to preserve uniform electric field distribution and to provide the optimum electropolishing, cleaning and drying conditions, the orientation of the tape should be vertical, see Fig.17. From all these observations we can argue that the most reliable technique to achieve the roughness level requested over the longer lengths without microstructure interruption is electropolishing. The profilometric measurements along the length of the NiFe(50%50%) tapes are presented in Figure 6.

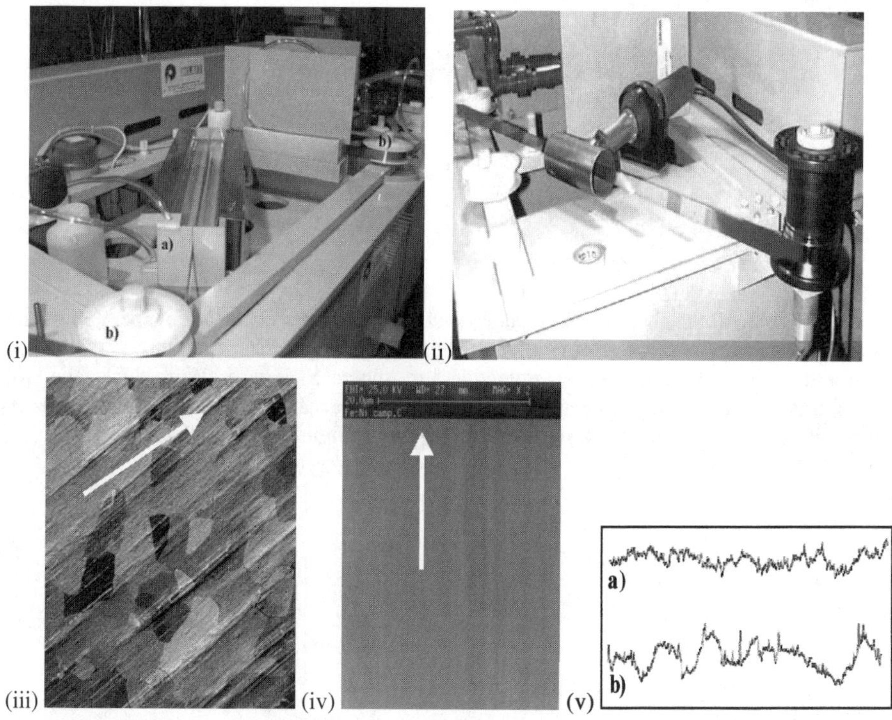

Figure 6 Surface finish improvement: (i) electropolishing unit at where long lengths of 25 mm wide and 25 μm thick NiFe tapes were electropolished: (ii) (a) - view of the treatment tank, (b) - cleaning and winding system, (ii) on-line drying unit (courtesy of Europa Metalli S.p.a. Superconductors Division), (iii) SEM picture of the NiFe tape surface before after electropolishing, (iv) SEM picture of the tape surface after electropolishing, (v) profilometric measurements of the dynamically annealed NiFe tape surface relief conducted by contact method a) - after electropolishing, b) - before electropolishing,.

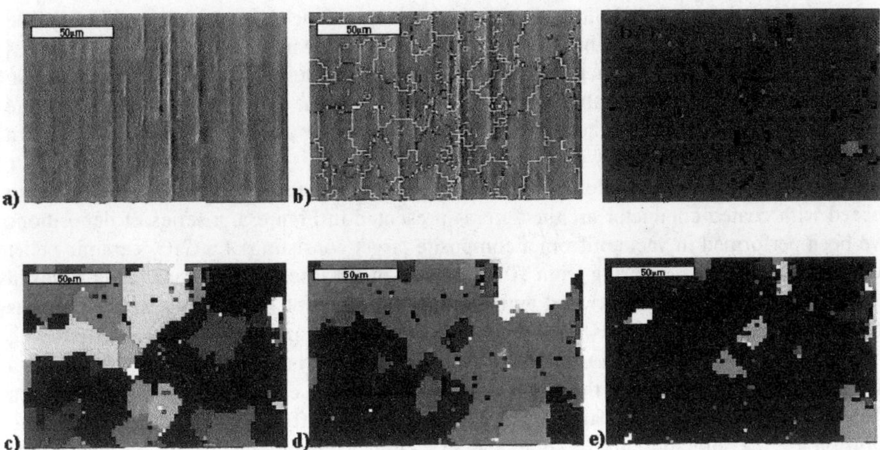

Figure 7 The NiFe 25μm thick tape dynamically anneal ed a) SEM picture of the tape surface fragment, b) SEM picture of the tape surface fragment with superimposed combined EBSP misorientation map presented in the following pictures, b') outline of the strongly out of plane oriented substrate crystals c) misorientation EBSP map for 4 degrees, d) misorientation EBSP map for 5 degrees e) misorientation EBSP map for 6 degrees; most of the grains are less than 6 degrees misaligned from normal to the surface of the substrate. *Colours used for "decoration" of the individual grains or crystallites are arbitrary for each individual picture to emphasise the boundary of the given degree of misalignment.*

The overall electron back-scattered patterns, EBSP, map of grain orientation as well as information about misorientation from grain to grain provides information about potential percolative current paths in the subsequent superconducting layer which can be derived on the base of the assumption that epitaxial growth of the subsequent buffer layers and

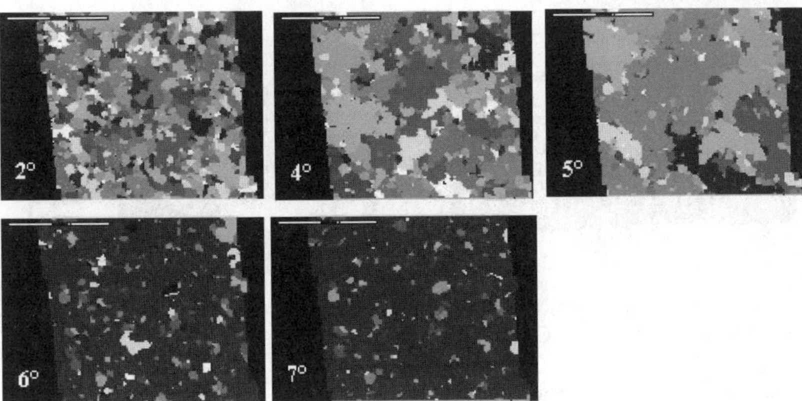

Figure 8 Large area EBSP misorientation map for 2°, 4°, 5°, 6° and 7° respectively. The width of the investigated area ~500μm. (courtesy of Dr Cecile Prouteau, EU MUST Project).

superconducting layers will follow the original texture induced by substrates [23,24] The EBSP data displayed in Figs 7 and 8 showed that there is a clear percolative path at 6 degrees but there are unresolved areas / grains which are of higher degree of misorientation and do not disappear after prolonged annealing. In Figure 7 (b') the unresolved grains which are oriented out of plane causing the strong misalignment of the subsequent epitaxially grown YBCO grains are represented by darker area.

In order to overcome the problems with the NiFe tape oxidation, Figure 5, and be able to proceed with coated conductor architecture as presented in Figure 3, a series of depositions have been performed in vacuum from a composite target consisting of a CeO_2 ceramic pellet and a piece of Pd sheet covering from 10 up to 50% of the laser ablated track. It is believed [19] that due to diffusion at elevated temperatures the effect of Pd gettering around various substrate defects (grooves, twins, surface irregularitiees) takes place similarly to the gettering of platinum by cavities in silicon recently reported by G.Mariani-Regula et al [20]. Such a gettering effect may suppress the high rate local oxidation caused by the severe oxygen diffusion through defects propagating in the overlaying buffers. Such oxidation leads to distortion of the cube-on-cube texture and cracking. Probably the deposited Pd (lattice parameter $d =3.89$ Å) also acts as an inhibitor suppressing NiO (111) formation and favoring NiO (200) ($d =4.18$ Å) formation as a result of oxygen delivered to the NiFe ($d =3.59$Å) interface by solid state oxygen diffusion through the buffer layers. In confirmation of this scenario cross-section FIB microscope images of the two samples deposited in forming gas and vacuum are presented on Figure 9 (a) and (b) respectively. This analysis reveals the much clearer features of the CeO_2:Pd/ NiO interface in comparison with those of CeO_2/ NiO. All subsequently deposited layers preserve the quality of the first interface with voids and irregularities propagating towards and within the YBCO. The region of relatively thick NiO formed during YBCO deposition is clearly seen on both samples suggesting strong oxygen diffusion through the buffer architecture. In the bottom part of the metal/oxide matrix the formation of Fe and NiFe oxide can be observed as darker regions while according to Schmalzried [21] Ni as a more noble metal oxidizes covering the outer region.

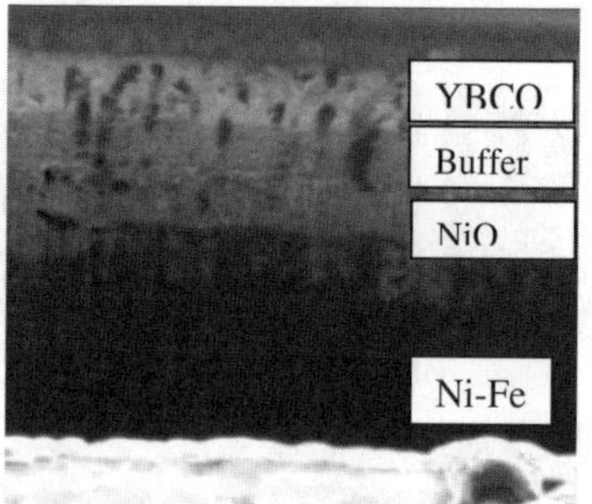

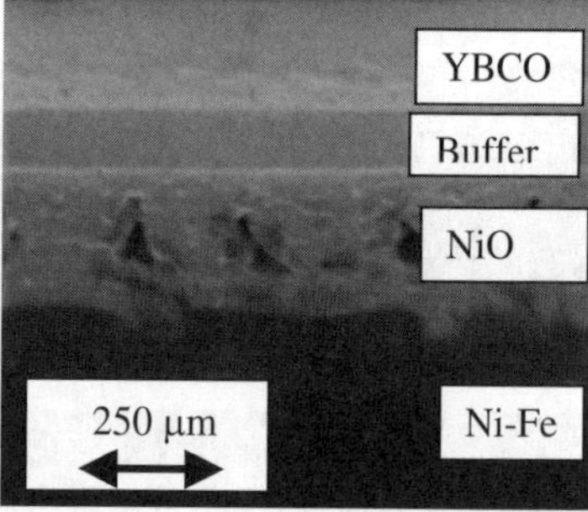

Figure 9. Cross sectional FIB microscope images of the two complete coatings (a) CeO_2 layer deposited in forming gas; (b) CeO_2:Pd layer deposited in vacuum.

The comparison of the in-plane texture quality of the architectures deposited in forming gas and vacuum is presented in Figure 10.

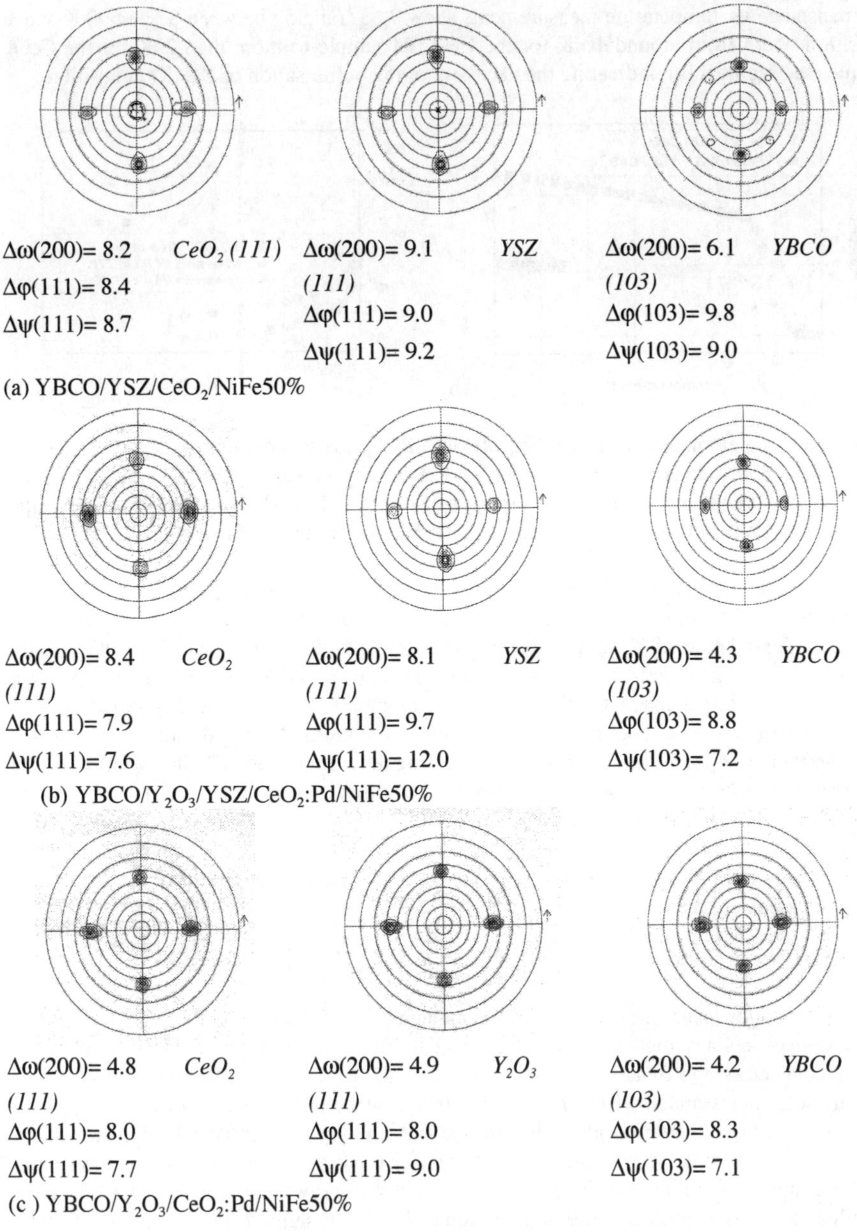

$\Delta\omega(200)= 8.2$ *CeO$_2$ (111)*
$\Delta\varphi(111)= 8.4$
$\Delta\psi(111)= 8.7$

$\Delta\omega(200)= 9.1$ *YSZ (111)*
$\Delta\varphi(111)= 9.0$
$\Delta\psi(111)= 9.2$

$\Delta\omega(200)= 6.1$ *YBCO (103)*
$\Delta\varphi(103)= 9.8$
$\Delta\psi(103)= 9.0$

(a) YBCO/YSZ/CeO$_2$/NiFe50%

$\Delta\omega(200)= 8.4$ *CeO$_2$ (111)*
$\Delta\varphi(111)= 7.9$
$\Delta\psi(111)= 7.6$

$\Delta\omega(200)= 8.1$ *YSZ (111)*
$\Delta\varphi(111)= 9.7$
$\Delta\psi(111)= 12.0$

$\Delta\omega(200)= 4.3$ *YBCO (103)*
$\Delta\varphi(103)= 8.8$
$\Delta\psi(103)= 7.2$

(b) YBCO/Y$_2$O$_3$/YSZ/CeO$_2$:Pd/NiFe50%

$\Delta\omega(200)= 4.8$ *CeO$_2$ (111)*
$\Delta\varphi(111)= 8.0$
$\Delta\psi(111)= 7.7$

$\Delta\omega(200)= 4.9$ *Y$_2$O$_3$ (111)*
$\Delta\varphi(111)= 8.0$
$\Delta\psi(111)= 9.0$

$\Delta\omega(200)= 4.2$ *YBCO (103)*
$\Delta\varphi(103)= 8.3$
$\Delta\psi(103)= 7.1$

(c) YBCO/Y$_2$O$_3$/CeO$_2$:Pd/NiFe50%

Figure 10 In-plane and out-of-plane texture quality of buffers and YBCO superconductor: (a) YBCO/Y$_2$O$_3$/YSZ/CeO$_2$:Pd/NiFe architecture deposited in forming gas, and (b) YBCO/Y$_2$O$_3$/YSZ/CeO$_2$:Pd/NiFe architecture deposited in vacuum and (c) YBCO/Y$_2$O$_3$/CeO$_2$:Pd/NiFe deposited in vacuum [19].

The resistance vs. temperature measurements show T_{conset} ranging between 87 and 90 K and a transition width from around 10 K for the CeO_2:Pd sample to more than 20K for the CeO_2 sample (see Figure 11a), indicating the necessity of the optimisation of YBCO deposition.

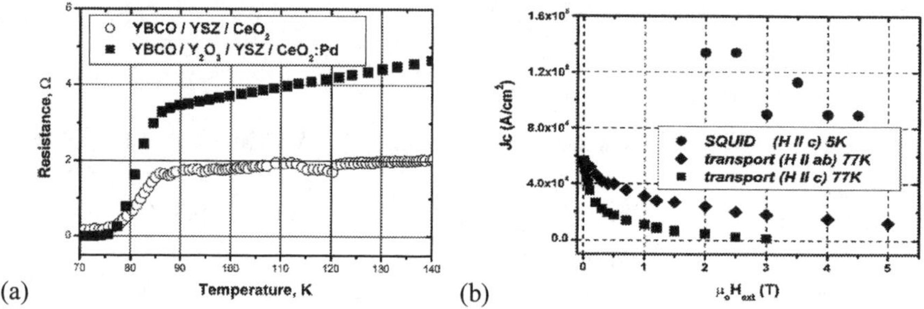

Figure 11. Superconducting properties of the YBCO coated conductors deposited on the buffered NiFe50% 25 μm thick substrates: (a) resistivity versus. T; (b) J_c versus. magnetic field; field dependence of critical current density of YBCO/Y_2O_3/YSZ/CeO_2/NiFe tape evaluated by transport (at 77 K) and SQUID (at 5 K) current measurements.

The other possible reason for a reduced T_c and wide transition could be a partial replacement of Cu atoms with Ni due to the diffusion through the grain boundaries, which automatically leads to a degradation of T_c [22] Magnetic measurements were performed in a DC SQUID magnetometer. The sample was approximately hexagonal in shape and was placed in a perpendicular external magnetic field. Magnetic moments were measured as a function of temperature in zero-field-cooling and field-cooling conditions. Magnetisation characteristics were measured at 5 K in range of applied magnetic fields ±6 T. The widths of the magnetisation loops were used to determine the critical current density J_c. A standard relation, based on the Bean critical state model [23], suitable for thin films was used: $J_c = 3\Delta M(H)/(2R)$, where ΔM is the hysteresis loop width and R is an effective radius of the sample. A plateau value of J_c (5K) = 5.5 x 10^8 A m^{-2} in the magnetic field range of 3 to 5 T was observed (see Figure 11b).

2.2 Liquid Phase Epitaxy

Among the major techniques researched for manufacture of the superconducting conductors is liquid phase epitaxy method (LPE). This method was first developed for thin film growth of semiconductors and oxides for optoelectronics [24]. LPE involves the undercooling of a molten flux supersaturated with solutes and in the presence of a suitable substrate (e.g. small lattice mismatch), the desired phase would nucleate and grow. Growing HTS materials by the one-step LPE method was pioneered by Scheel et al. [25] who surveyed the essential problems and difficulties connected with the epitaxial growth of YBCO films. LPE of (RE)$Ba_2Cu_3O_7$ is more difficult than the traditional LPE of semiconductor and garnet films since the solubility of RE elements in the melt is very low, hence nucleation and growth is more difficult. The process of LPE has four stages: (i) Producing a solution containing a quantity of solute corresponding to a saturation temperature T_s; (ii) Cooling this solution to the growth temperature T_g producing an undercooling supersaturation corresponding to

$\Delta T = (T_S - T_g)$; (iii) Bringing the solution and the substrate into contact for an appropriate period of time for the layer thickness required where solution and/or substrate rotation can be used; (iv) Terminating the growth by separating the solution from the film. The order of stages (i) and (iii) can be interchanged and the cooling can be started (or continued) after the solution and the substrate are brought into contact. Cooling of the solution to promote supersaturation may be in continuous or in a single step mode, or a combination of these two. Stages (iii) and (iv) require certain mechanical movement of the solution and/or the substrate. After growing a layer with the desired thickness, the substrate is removed and it is important to ensure that the surface is free from any drops of excess solution. Knowledge of the phase relation, especially the primary crystallisation fields, PCF, and solubility curves, in the Y-Ba-Cu-O (or other RE) system is of fundamental importance for (Y,RE)BCO crystal growth. Figure 12(a) shows a ternary phase diagram of Y211-CuO-BaCuO$_2$ system exhibiting the PCF of Y123. For LPE (and crystal growth), processing have to be performed within the PCF. Ternary phase diagrams of other REBCO systems are not fundamentally different but the areas of PCF may vary; for instance, the PCF of NdBCO is much wider than that of YBCO. Figure 12(b) illustrates the paths of liquid composition in cooling a melt of composition p. Upon cooling the liquid, solid phase Y211 crystallises out along p-q. From q to r, Y211 reacts with the liquid and forming Y123, i.e. peritectic reaction starts whereby Y211 reacts with liquid to form Y123 crystals around the Y211 particles. At r, all of Y211 has decomposed into Y123 but Y123 continues to crystallise from the liquid along r-s. Along s-t, Y123 and BaCuO$_2$ crystallise together and finally, at point t, the remaining liquid solidified via a eutectic reaction forming Y123, BaCuO$_2$ and CuO phases.

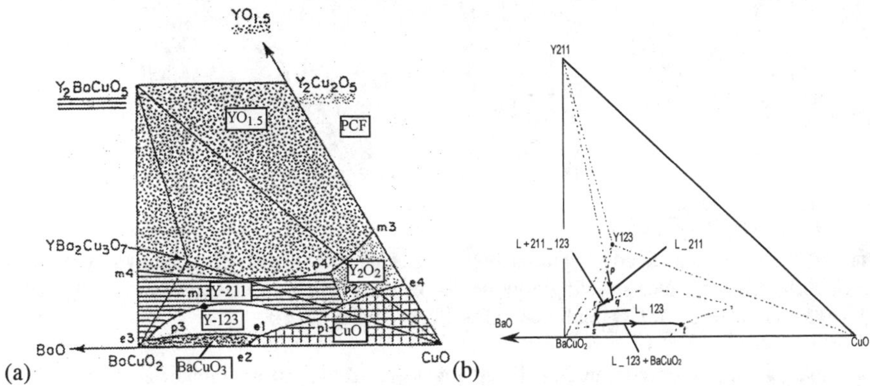

Figure 12 Ternary phase diagram of Y211-CuO-BaCuO$_2$: (a) underline of PCF of Y123 in air; after [26] (b) a schematic ternary phase diagram of Y211-CuO-BaCuO$_2$ showing the path in cooling a melt of composition p, indicating by arrows in the direction of falling temperature, adapted from [27].

The liquid phase epitaxy, LPE, growth mechanism is a complex process that is initiated by multi-nucleation, followed by spiral-mediated growth around dislocation centres beyond a certain critical film thickness according to the BCF theory (Burton, Cabrera, and Frank) [28]. The film continued to grow in a nucleation-free manner and readily developed into elliptical or polygonised (straight edge) spirals that were associated with crystallographic orientations

252

of the film as is demonstrated in Figure 13. Advantages of liquid phase epitaxy over the vacuum-base thin film techniques such as sputtering and pulse laser deposition (PLD) are evident such as process is characterised by fast growth rate, typically 1 μm/min, texture of

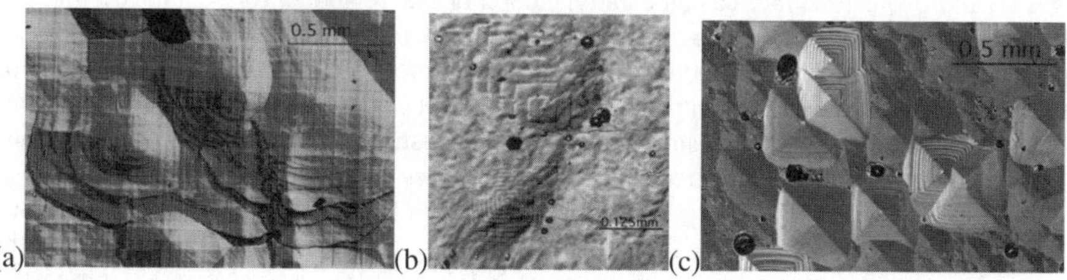

Figure 13 Optical microscopy pictures of the surface of LPE $YBa_2Cu_3O_7$ layers grown at different patterns: (a) a-axis oriented film of $YBa_2Cu_3O_7$ grown on $NdGaO_3$; (b) c-axis oriented growth of $YBa_2Cu_3O_7$ on $NdGaO_3$; (c) c-axis oriented growth of $YBa_2Cu_3O_7$ on $SrTiO_3$. After [29], courtesy of Dr Cheng.

the superconducting coating and J_c do not degrade with increasing film thickness which is unique to LPE and the process is in-situ and non-vacuum (relatively low cost), Figure 14.

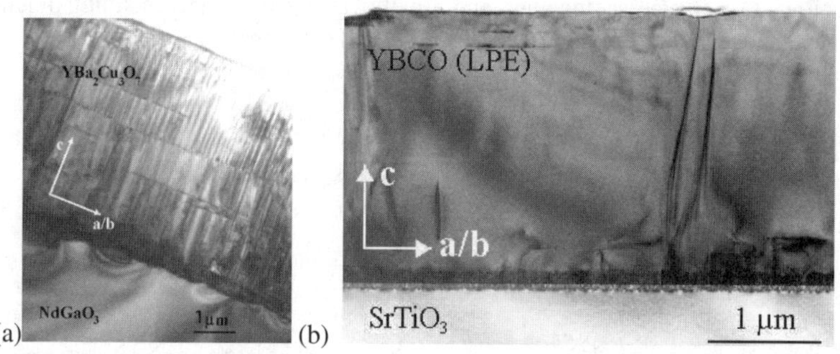

Figure 14 Transmission Electron Microscopy images across the $YBa_2Cu_3O_7$ layer grown by LPE on different substrates showing high degree of epitaxy: (a) $NdGaO_3$, Figure 13(b) after [29]; (b) MgO buffered $SrTiO_3$ substrate after [30], Figure 13(c).

Disadvantages are superconducting layer cracking above 5μm thickness dramatically degrading transport critical current. Also the process is conducted at very high temperature in a corrosive atmosphere where most of the metallic substrates will corrode and growth process is not exactly epitaxial and is sometimes called Liquid Phase Processing (LPP), therefore scalability and applicability of such LPE technique to manufacture long lengths of tape form conductor is problematic.

Although the LPE fabrication process is complex but provides excellent results in terms of production rate, texture, Figure 15 and film thickness and very high superconducting transport properties, Figure 110, when deposited on the flexible metallic tapes. To simplify the architecture of the conductor and reduce the cost of conductors, LPE fabrication of superconducting layers can be conducted directly on Ag substrates, which have a very low

Figure 15 Cross sectional Transmission Electron Microscopy observation of the LPE grown $YBa_2Cu_3O_7$ film on the $YBa_2Cu_3O_7$ seed layer on Y_2O_3-YSZ buffer layer on Ni-Cr alloy; courtesy of Dr Y.Yamada.

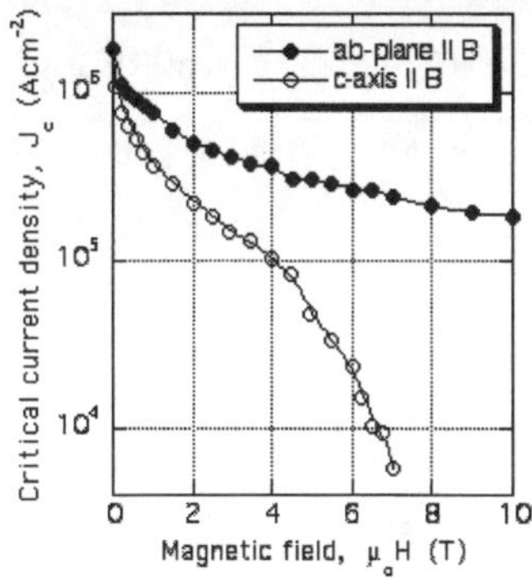

Figure 16 Critical current density vs magnetic field for the $YBa_2Cu_3O_7$ superconducting LPE film presented in Figure 15.

reactivity with $YBa_2Cu_3O_7$ and high conductivity ideal for cryomagnetic protection. Since the melting point of Ag is relatively low, the fabrication temperature of the LPE process must be lowered to around 900°C. The temperature of the LPE process can be lowered when BaF_2 is added to a Ba-Cu-O solution that is saturated with Ag. Using this procedure, Y123 films have been successfully fabricated on short Ag substrates by LPE process. Attempts are made to develop a continuous fabrication process that will allow longer tape conductors to be produced. The goal of this type of conductor is to produce long tape with a superconducting layer at least 10 μm thick using a non-vacuum procedure. The only problem with the thicker $YBa_2Cu_3O_7$ layers is the minimum bending radius of such conductors which may restrict the usage of such conductors with superconductor thickness > 5μm to specific applications e.g. Fault Current Limiters.

2.3 Magnetic Alignment and Thermo -Magnetic Growth of REBa$_2$Cu$_3$O$_7$

In traditional, thermo-mechanical processing of grain-oriented superconducting oxides for applications, thermal treatment is combined with some method of mechanical alignment of grains, such as rolling or sequential pressing. An alternative to this approach involves applying a high magnetic field before or during thermal treatment to align grains. When a grain having an anisotropic paramagnetic susceptibility in its normal state is placed in a magnetic field, the grain rotates such that the axis of the minimum susceptibility is aligned parallel to the field direction. Any anisotropic crystal placed in a uniform external magnetic field acting orientating force, F, defined by the difference of magnetic susceptibility for the perpendicular crystallographic axes $\Delta\chi$, magnitude of the applied magnetic field, H, volume of the crystal, V, and also orientation of the applied field (ϕ- angle between crystallographic axis at which c has a maximum value and the direction of external magnetic field), eq.1.

$$F = \Delta\chi \cdot V \cdot H^2 \cdot \sin(2\phi) \qquad (1)$$

Because the paramagnetic susceptibility of the rare earths is anisotropic, an alternative alignment method is to use a static magnetic field to align crystals, which precipitate from a melt. The most effective texturing effect of the $REBa_2Cu_3O_7$ compounds in the external magnetic field is for such RE ions, which have the highest magnetic moment, Table 1.

Table 1 Magnetic moments of the RE ions and axis of $REBa_2Cu_3O_7$ alignment

Atomic Number	Atomic Symbol	Ionic Radius (nm)	Magnetic Moment	Axis Parallel to H
21	Sc	0.083	-	
39	Y	0.106	0	c
57	La	0.122	0	-
58	Ce	0.118	2.56	-
59	Pr	0.116	3.62	-
60	Nd	0.115	3.68	c
61	Pm	0.114	2.83	-
62	Sm	0.113	1.55-1.65	c
63	Eu	0.113	3.40-3.50	a
64	Gd	0.111	7.94	c
65	Tb	0.109	9.7	-
66	Dy	0.107	10.6	c
67	Ho	0.105	10.6	c
68	Er	0.104	9.6	a
69	Tm	0.104	7.6	a
70	Yb	0.1	4.5	a
71	Lu	0.99	0	-

Because YBCO contains no magnetic elements, its magnetic alignment results from anisotropy in the paramagnetic susceptibility associated with the Cu-O conducting planes. Since the susceptibility parallel to the c-axis is higher than that perpendicular to the c-axis, the compound aligns with the c-axis parallel to the applied field direction. In principle there are two contributing factors to the magnetic forces tending to align superconducting cuprates: the magnetic moment tending to align the longest dimension parallel to the field and the moment which aligns the axis of largest susceptibility along the field direction for a paramagnetic material. For spherical $YBa_2Cu_3O_7$ powder, the shape factor is small, and so the c-axis aligns along the direction of the field. For rare-earth doped $(Y,RE)Ba_2Cu_3O_7$ compounds, the paramagnetic susceptibility is dominated by the R^{3+} ion, and the source of anisotropy is single-ion anisotropy associated with crystal fields at the rare-earth site. For Gd and Ho doped $(Y,RE)Ba_2Cu_3O_7$, the particles align with c-axis parallel to the magnetic field, while for Er-doped $(Y,RE)Ba_2Cu_3O_7$, the particles align perpendicular to the magnetic field. An addition of the Nd, and Eu to YBaCuO material causes density reduction of the sintered superconductor where Gd, Yb and especially Ho increase the density of the sintered $YBa_2Cu_3O_7$ coatings. There is a linear dependence between ionic radius of the rare earth elements and actual a and c lattice parameter of $REBa_2Cu_3O_7$ unit cell which is important for the magnetic alignment of the mixed rear earth, see Figure 17. The presence of a liquid

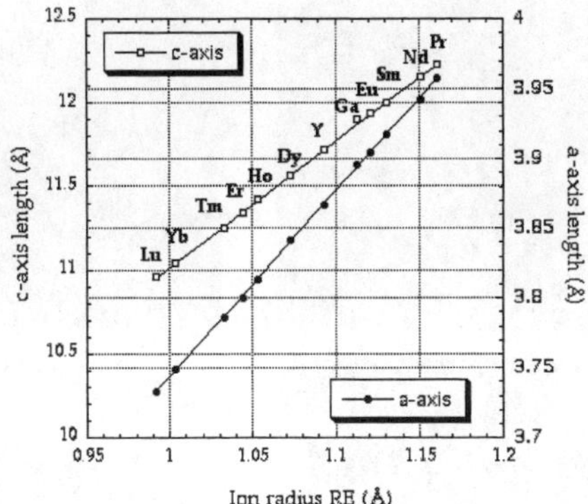

Figure 17 Lattice parameters variations of the different $REBa_2Cu_3O_{7-\delta}$ crystals versus ionic radius of rear earth element.

phase such as binders and carriers in the superconducting oxide greatly facilitates grain alignment. Such a liquid phase can be attained through high temperature treatment of the oxide during simultaneous sintering, thereby eliminating the grain interface impurities or alternatively if the liquid phase has the similar composition to the superconducting oxide may act as a interconnector between the individual particles during grain growth improving the grain interconnectivity and therefore minimising current percolation. The great advantage of self-liquid approach for high or low temperature processing over the use of an organic solvent liquid is the significant decrease in impurity phases created at the grain boundaries by reaction of the $(Y,RE)Ba_2Cu_3O_7$ compound with the organic solvents and binders and low density of the final coating. Such impurities and voids degrade connection between grains, thereby severely limiting the J_c values achievable. When these materials are textured by a high magnetic field during melt-growth, not only is a textured, larger domain structure attained, but textured small grains with better connection between grains is achieved by liquid phase binding during the high temperature processing [31] as it is described under two following headings.

Alignment During Slow Solidification of the $REBa_2Cu_3O_7$. This alignment technique essentially based on the anisotrotropic magnetic susceptibility is very similar to the second technique [32]. In this particular case slow cooling through the solidification temperature maximise $REBa_2Cu_3O_7$ grain size and simultaneously give enough time for the particles to rotate so that the direction in which the susceptibility is maximum aligns parallel with the direction of the applied magnetic field. Both calculations and experiments have shown that at temperatures which are sufficiently high to melt $REBa_2Cu_3O_7$ in preparation for resolidification under field, this magnetic force is large enough to overcome disordering forces from thermal agitations or shape anisotropy. The successful thermal procedure is: $100°C/h \Rightarrow 1085°C$ $0.4h \Rightarrow$ fast cooling to $1035°C$ cooling $2°C/h \Rightarrow 980°C$ cooling $3°C/h \Rightarrow 930°C \Rightarrow 60°C/h \Rightarrow 420°C$ $72h \Rightarrow 200°C/h \Rightarrow RT$.

Magnetically Assisted Nucleation and Growth of $YBa_2Cu_3O_7$. Such process conducted on polycrystalline untextured silver substrate by the chemical vapour deposition, CVD, appeared to possess much better J_c vs B characteristics through the improved grain connectivity [33] than that deposited on a highly textured silver substrates [34], Figure 17, Figure 18.

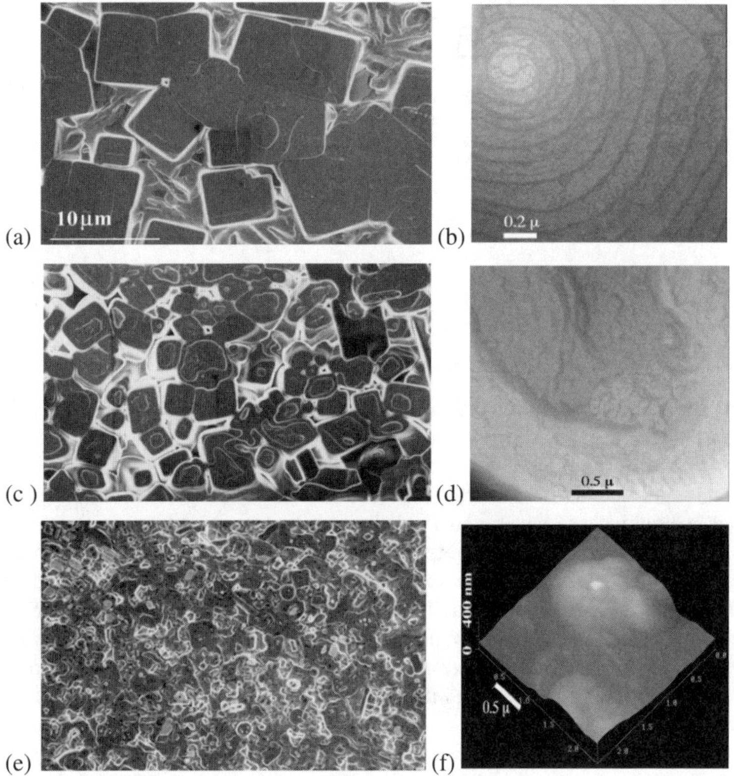

Figure 17 SEM (a), (c), e) and high-resolution AFM images of the three CVD $YBa_2Cu_3O_7$ films grown at different magnetic fields: (a) 0T, large rectangular grains; (b) 0 T, A single-core spiral; (c) 2 T, irregular grains; (d) 2 T, 2D-growth behaviour; (e) 4 T, small interconnected grains; (f) 4T, the 3D growth mode (179,215).

Images for the $YBa_2Cu_3O_7$ films deposited at different magnetic fields are presented in Figure 17. In the absence of a magnetic field, the films exhibit large rectangular grains, and large voids and poor interconnection are present. At 2 T, the grain shape seems similar to that of the 0 T films, but the grain size is reduced. Under a field of above 2 T, the crystallites of films were more homogeneous and irregular grains; 1-2 μm were well connected to each other, compared with that without the magnetic field. Clearly, the intergrain connectivity and morphology were less influenced by fields over 4 T. Thus, with increasing magnetic field, the grain size was reduced, but the grain connectivity was improved. Smoothness but not an actual texture of the silver-based substrate has a decisive role in nucleation of the preferentially align HTS coating [35]. The $YBa_2Cu_3O_7$ growth-mode change from the spiral growth mode at zero field to the two-dimensional or three dimensional island growth mode under an external magnetic field which is responsible for improvement of transport property of superconducting coatings [36], Figure 18. The thermo-magnetic process is very effective in yielding superconducting well-interconnected $YBa_2Cu_3O_7$ films on silver substrates with relatively high critical current densities, but without the use of any buffer layers. Such a novel technique may be used to produce texture in a bulk samples and in a low cost long lengths $YBa_2Cu_3O_7$ coated conductors, however further research is required to improve J_c vs B performance.

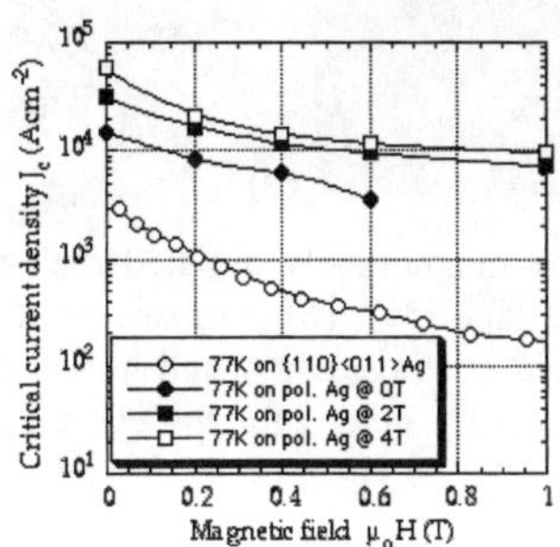

Figure 18 J_c of YBaCuO superconducting non-vacuum coatings on silver substrates. Low values of the J_c of YBCO deposited by ultrasonic spray pyrolysis [34] on highly textured silver ribbons with a single component {110}<011>, FWHM smaller than 6° in comparison with CVD deposition on highly polished polycrystalline substrates [33] prove the point that smoothness of the substrate has a decisive role in high texture alignment of YBCO on silver not actual texture of the silver [35]. Similar comparative dependence was observed in the case of Tl1223 coatings [37].

Dynamic Magnetic Alignment. As it was described earlier a static magnetic alignment by settling or immobilising monocrystal grains in a constant magnetic field is a common process to obtain uniaxially aligned $YBa_2Cu_3O_7$ [38]. In the case of the untwined $Y_2Ba_4Cu_8O_{16}$ or $Y_2Ba_4Cu_7O_{15}$ (or any rare earth-based) compound, advantage can be taken of their orthorhombic structure. The anisotropy of the structure leads to normal-state magnetic susceptibility with b < a < c for $Y_2Ba_4Cu_7O_{15}$ and $Y_2Ba_4Cu_8O_{16}$ [39]. A constant magnetic field will align freely suspended grains along the c-axis. The rotation relaxation time as the time which a particle takes to rotate to its equilibrium orientation in a static magnetic field of given magnitude can be controlled. If the field is rotated from its initial orientation by 90° for a time shorter than the c-axis rotation relaxation time so that the c-axis alignment is not disturbed, the a-axis, with intermediate susceptibility, will tend to align along the rotated direction. The b-axis of the grain, with the minimum susceptibility, will align with the zero field direction. If such an alternating or dynamic field is applied to a powder in suspension in a medium of the appropriate viscosity, biaxial alignment will be achieved. The complete process is detailed elsewhere [40]. Figure 19(a) illustrates the results of the process in the case of $Y_2Ba_4Cu_8O_{16}$. Clearly a significant fraction of the large melt-grown grains are aligned with one another and with the axes of the tape (parallel to the image's sides).

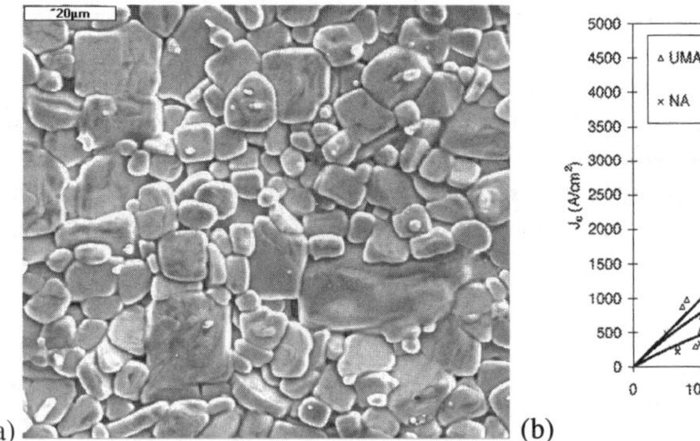
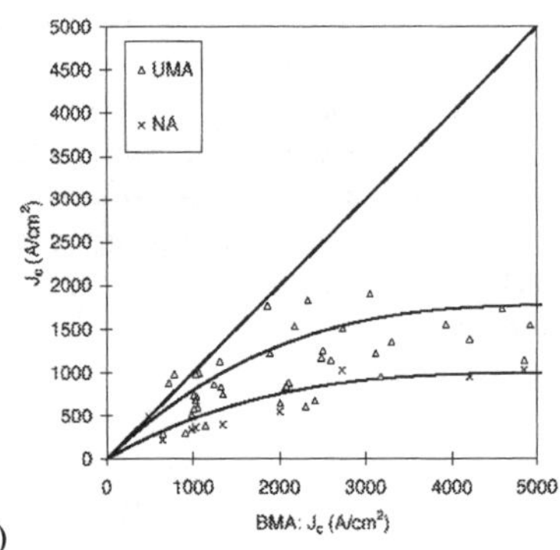

(a)   (b)

Figure 19 $Y_2Ba_4Cu_8O_{16}$ tape magnetically assisted melt-processed at 920°C, (a) SEM picture of the top surface of the tape [39]; (b) comparison of results J_c for biaxially aligned coating, BMA, uniaxially aligned coating, UMA and non-aligned coating, NA in respect to BMA coatings. However the lines are guides for the eye the benefit of biaxial alignment is obvious.

Using a dynamic alignment technique founded on similar ideas, Zhang and Budnick [41] demonstrated biaxial alignment of Sm_2Fe_{17} particles with the a-axis and b-axes distributed within about 15°. In that case, they used an oscillating field and progressively decreased the amplitude of the oscillation, so that the process finished with the magnetic field fixed in direction. It is believed that maintaining the rotating field during the entire process is more appropriate to achieve in-plane anisotropy of the final superconducting coating. This is supported by the very high degree of texture obtained. Rocking curves on samples demonstrated FWHM to be as low as 2°.

The critical current results Figure 19(b) demonstrate improvements due to the biaxial alignment process [39]. Benefit of biaxial alignment is presented in Figure 19(b). The critical current densities of the uniaxially magnetically aligned, UMA, and non-aligned, NA, tapes are plotted as a function of the critical current density of the corresponding biaxially magnetically aligned, BMA, coating. The diagonal gives the limit below which a benefit is observed. The tapes compared were in the same batch and were melt processed at the same time the only difference between them is the magnetic alignment process. All the data are situated in the bottom right part of Figure 19(b) signifying a benefit from biaxial alignment.

2.4. Current Percolation in Coated Conductors

2.4.1 RABiT conductor

In RABiT conductors, texture in the YBCO layer is developed by depositing upon a textured buffered metal tape. The preferential alignment is transferred from the substrate through the buffers into the superconductor. There is much evidence from transmission electron microscopy, Figure 20(a), and magneto-optical (MO) measurements [42] that grain assignment of a critical current to each grain boundary requires knowledge of the relationship between grain boundary angle and J_c as presented in Figure 21 and Figure 22. Despite this, all layers do not have exactly the same crystallographic alignment. The grain size of the

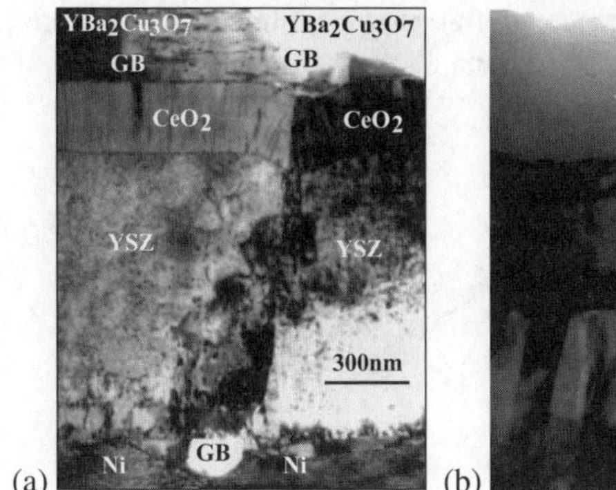
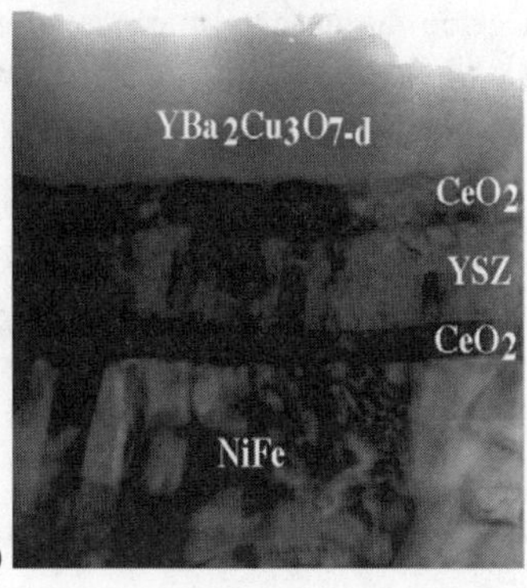

Figure 20 Transmission Electron Microscopy, TEM micrograph of the superconducting multilayer composite coated conductor cross sections manufactured by RABiT process: (a) the grain boundaries of textured buffered nickel substrate replicate the boundaries of $YBa_2Cu_3O_7$ in superconducting layer and also there is an additional grain boundary in $YBa_2Cu_3O_7$ layer probably induced by the imperfection in the CeO_2 layer induced; conductor structure: $YBa_2Cu_3O_7$/YSZ/CeO2/Ni. The average grain size of $YBa_2Cu_3O_7$ is 20μm. (Adopted after [43], courtesy of C. Prouteau,). (b) [NiFe-CeO_2-YSZ-CeO_2-$YBa_2Cu_3O_7$] the thickness of the various layers were about 50nm (1st CeO_2), 150nm (YSZ), 65nm (2nd CeO_2) and 250nm ($YBa_2Cu_3O_7$) respectively; T_c~90K. The grain boundaries are not explicitly transmitted from the buffered NiFe substrates (courtesy of the EU 'MUST' project, Birmingham University);

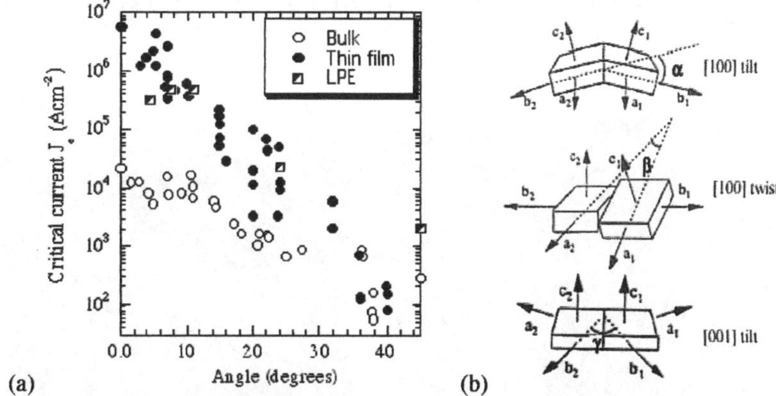

Figure 21 Transport critical current density versus grain misalignment angle: (a) J_c vs grain boundary angle data for [001] tilt boundaries in YBCO at 77K and zero external magnetic field: open symbols represent bulk samples, filled symbols thin films and crosses liquid phase epitaxy (LPE) coatings; adopted from [44], [45]. Thin film [46], [47], [48]; liquid phase epitaxy [49] and bulk [50], [51] (b) schematic representation of the misalignment angles.

deposited buffer layers is usually smaller than that of the metallic tapes. This suggests that sub-grains develop during deposition as several growing film grains nucleate in different places within a single substrate grain. In addition, the overall crystallographic texture may

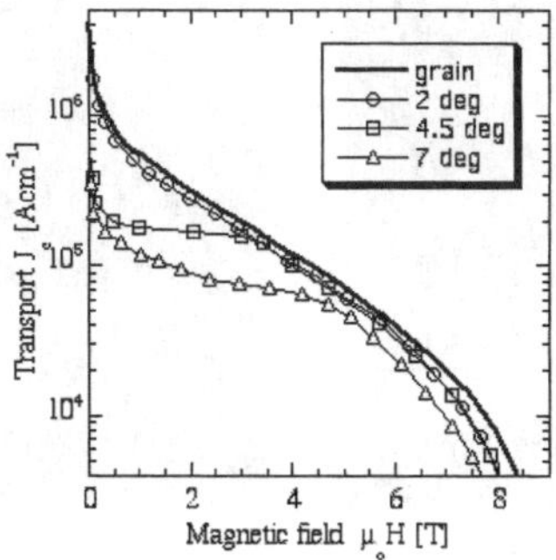

Figure 22 Transport critical current density versus magnetic field parallel to c axis of $YBa_2Cu_3O_{7-d}$ grown epitaxially with different in plane misorientation [001] tilt, see Figure 21(b); (adopted after [52], courtesy of C. Prouteau).

improve or degrade in successive layers, depending upon substrate material, adopted architecture of the buffer layers and also deposition parameters such as temperature and deposition rate, Figure 9 and Figure 20. Irrespective of the overall quality of textured metallic substrates some accidental high angle grains do exist, which may cause even growth of $YBa_2Cu_3O_7$ away from a-b plane, Figure 23 which obviously reduces the transport properties and amplifies transport current percolation. Such accidental crystals introduce a serious limitation to the overall current density of the conductors under development and should be avoided by optimisation of the cold rolling and heat treatment procedures during substrate development.

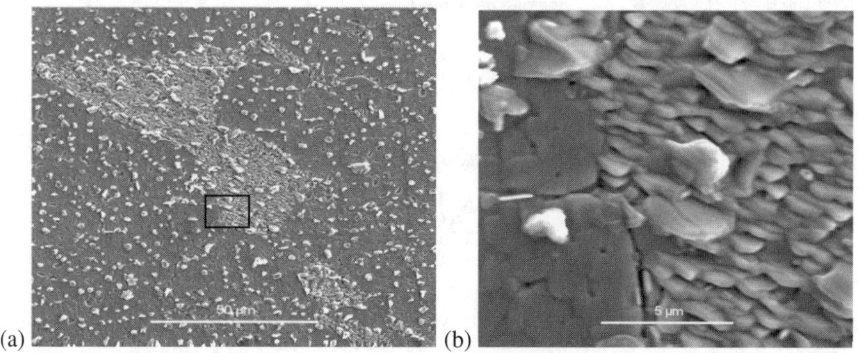

Figure 23 SEM picture of the $YBa_2Cu_2O_7$ grown on buffered misaligned Ni-based grain $Ni0.1\%Mo/NiO/Ca_{0.6}Sr_{0.4}TiO_3/YBa_2Cu_2O_7$, (a) low magnification, (b) magnification of the marked area on the (a), courtesy R.H. Huhne, University of Cambridge, compare Figure 5 (b').

2.4.2 *Current percolation improvements through grain boundary engineering - DC applications*

As recent research conducted on $YBa_2Cu_2O_7$ grown on $SrTiO_3$ bi-crystals with misalignment angle of 4 degrees shown [53], [54] an in-plane orientation of magnetic field in respect of grain
boundary has a strong influence on the actual perfomance of superconducting coating, Figure
. There is the critical angle at which the grain boundary does not degrade performance of such conductor. Such angle is a function of the temperature and field, which is in details described elsewhere [55].

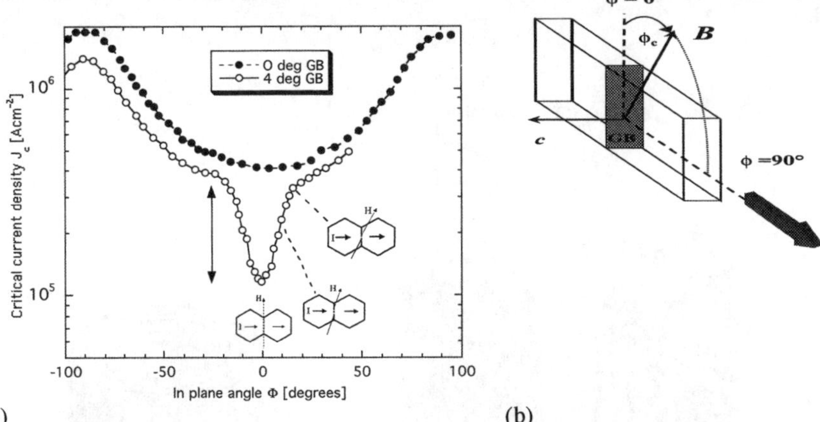

(a)   (b)

Figure 24. In plane critical current vs magnetic field measurements on YBCO thin films: (a) angular dependence of the critical current crossing low angle grain boundary at 8T. For the higher GB angle the minimum is wider and the absolute valuses are substantially lower; $\Phi=90°$ represents Lorentz force-free configuration. Hexagons represent grains whereas black outlines of hexagons represent grain boundaries; (b) schematic of the J_c vs (B,Φ) in plain measurements. [54].

To minimise the exposure of the grain boundaries to the external magnetic field hexagonal-type grains should be aligned in the conductor as demonstrated in Figure 25(a). According to percolation theory the elongated grains can be more efficient in current transfer reducing the probability of the transport current reduction. Additionally if the proposed grain elongation will coincide with grain boundary angles away from magnetic field direction as presented in Figure 25(b) such conductors will posses much better in field performance and versus temperature.
The obvious application (not only) which will benefit from this finding are self standing HTS magnets and also hybrid magnets as presented in Figure 26. Therefore an intensive research on grain boundary engineering of the metallic substrates for RABiT deposition technique is currently conducted.

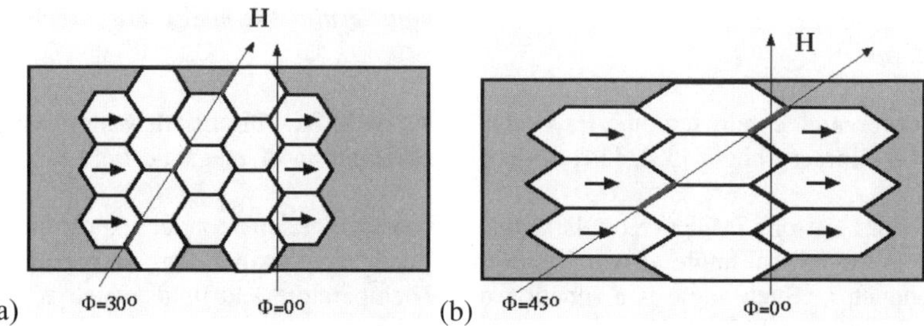

Figure 25 Elongated hexagonal grains have the better percoative properties than the simple hexagonal ones. There is a difference in the response of hexagonal grains if all of them are aligned.

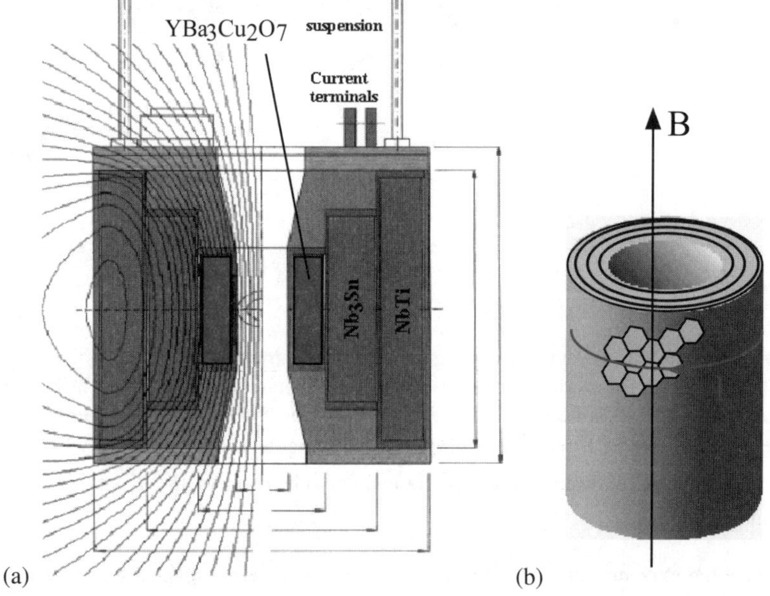

Figure 26 High magnetic field superconducting electromagnet. (a) schematic cross section of the multi-section hybrid electromagnet. The materials used are NbTi + Nb_3Sn + $NbTi_3(Sn,Ta)$ resulting in 22.59 Tesla and if additional magnetic field is generated by internal coil in the centre it would generate an additional field. The total magnetic field in such a hybrid configuration, currently exceeds 24 Tesla. (b) schematic outline of the favourable grain structure of the internal HTS coil made from the $YBa_2Cu_2O_7$ coated conductor.

There is a strong interest in chemical and structural modification of the grain boundaries to improve the electrical properties of higher angle grain boundaries. There are pronounced changes in the microstructure and improvement of the electrical properties of the small-angle GB and also high-angle GB that are induced by doping the boundaries with Ca. Ca doping provides a viable means of engineering the electrical structure of the grain boundaries themselves polycrystalline $Y_{0.8}Ca_{0.2}Ba_2Cu_3O_7$. One may accept also that J_c improvement of Ca doped $YBa_2Cu_3O_7$ coatings could change the level and shape of the J_c/J_{co}(length, width) dependence shifting J_c/J_{co} to higher values [56]. The downside of Ca doping is reduction of T_c and also necessity of avoiding oxygen overdoping, Figure 27. It is also

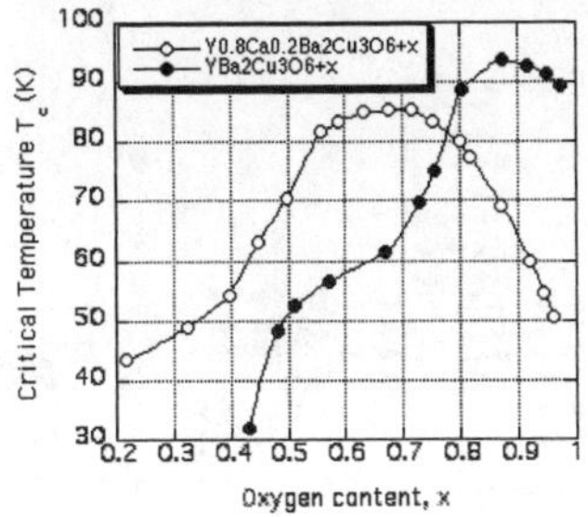

Figure 27 Critical temperature versus oxygen content x in the $Y_{0.8}Ca_{0.2}Ba_2Cu_3O_{6+x}$ and $YBa_2Cu_3O_{6+x}$ superconductor, adopted after [58].

possible to engineer the grain boundaries of the $REBa_2Cu_3O_7$ coatings with better connectivity by manipulating the composition of the GB using mixture of the rear earth elements with different ion radiuses.

Recent study of the critical current angular dependences of the polycrystalline conductors carried out by scientists in Dresden [57] conclusively proved that at low magnetic fields, up to characteristic value B^{cr}, the critical current is limited by the intergrain critical current. This is reflected in a power-law dependence of $J_c(B)$ and non-Ohmic linear differential-like signature of the V-J curves. At magnetic fields above B^{cr}, $J_c(B)$ is limited by pinning within the Y123 grain (intergrain critical current) with a typical exponential decay dependence. It was found [57] that the transition from intergrain to intragrain critical current limitation is shifted to lower magnetic fields as the temperature is increased. This crossover behavior can have important implications for coated conductor applications. For applications that require the presence of higher magnetic fields, the intergrain critical current of the Y123 coated conductor does not limit the critical current and hence J_c is completely intragrain. The irreversibility field represents in this case the upper limit for applications in high magnetic fields with the intergrain critical current having no influence. A better in-plane texture of the coated conductor with smaller GB angles does not avoid this problem since the crossover field is only shifted to lower magnetic fields, leaving the exponential J_c decrease at high fields unchanged. Only in the case of applications that require magnetic fields lower than the crossover field, an improvement of the grain boundary angles in the samples can be useful [57].

2.4.3 Conductors for AC Applications

It was established that for long tapes the average J_c increases for a smaller grain size (or larger sample) [59], [60]. Therefore in this case, both high J_c and high uniformity of J_c are favoured for a small grain size. This has significant consequences for coated conductor

264

development. Since the multifilamentary coated conductors appeared to be favourable solution for low AC loss applications, the factors which affect the percolative nature of current flow in superconducting coated conductors, such as grain size and sample dimensions, have been presented and analysed in order to predict the (grain size)/(filament-strip width) ratio influencing critical current density values in multifilamentary conductors [61], Figure 28. Also Figure 28 shows the dependence of J_c/J_{co} on sample length. As the sample becomes longer,

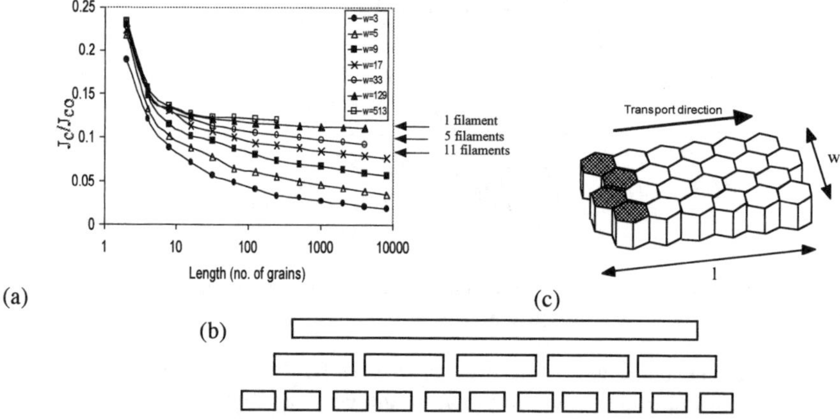

Figure 28   The variation of J_c/J_{co} with a sample length for prefixed sample width. The numbers assigned to length and width are given as a multiplication of chosen grain size. For example if the average size of the grain is of the order of ~ 20 mm, the curve 'w=129' corresponds to the conductor path width, w, equal (20 µm × 129 grains = 0.35 mm).   The shape of J_c/J_{co} vs length and width dependencies have been calculated on the basis of the existing data correlating critical current of the grain boundaries vs misorientation angles. Arrows represent the level of current reached at the length of 10000 grains for the corresponding filamentary coated conductors presented in Figure 5 (b); (b) model coated conductors cross-section (not to scale): a) monocore - rectangular cross section 5.5mmx4mm; b) 5 filaments, each of rectangular cross-section 1.1µm × 4µm, separation between the filaments-5mm; c) 11 filaments, each of rectangular cross-section 0.5µm × 4µm, separation between the filaments-5µm. J_c decreases (approximately exponentially for $l \gg w$); (c) schematic of the hexagonal grain array where I- length of the coated conductor or conductor strip and w – width of the superconducting path.

This is to be expected as a 10m long sample can be considered as ten 1m samples, with the "worst" of these small lengths determining the J_c of the long conductor.   Therefore when there is less deviation between samples (i.e. small grain size), the length dependence will be weaker. Measurements of the critical currents of long $(Bi,Pb)_2Sr_2Ca_2Cu_3O_9$ high T_c conductors show qualitatively the same exponential decrease with increasing length. Narrow coated conductors (a few grains wide) can have very low critical current densities but with increasing width, J_c increases as there are many more possible percolation paths through the sample. This effect begins to saturate at a value close to 10% of the intragrain J_c. The significance of this result is that from a percolation point of view it would be better to manufacture a single wide tape than to use several narrower ones in parallel. It will generally be favourable to have a small grain size in order to have high J_c and maintain sample consistency. If one were to produce a low loss coated conductor by division of the coating to magnetically or non-magnetically insulated ribbons a grain to width aspect ratio which cannot be too small. See for example conductor in Figure 29 where width of YBCO ribbons were design to be 10 μm similar to the grain width which will give a higher reduction of the critical current as was presented in Figure 28.

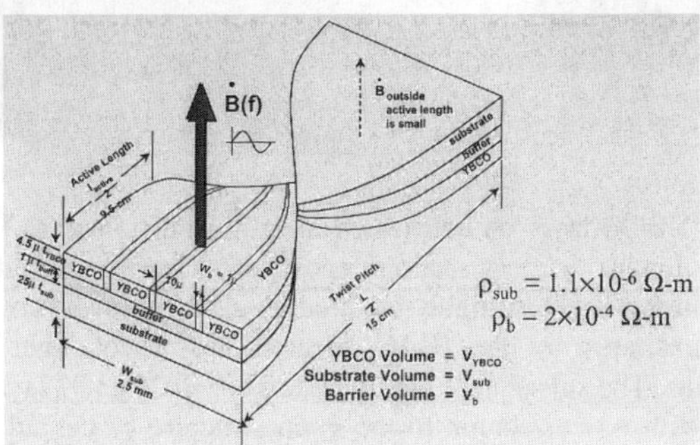

Figure 29 Schematic of the initial design of YBCO tape dimensions for minimised ac loss calculation for a 1 MW superconducting generator armature, after Oberly [62].

All the results so far are based on achieving a crystallographic texture in the superconductor, which is exactly the same as that of a given metallic substrate. The best properties have so far been obtained by pulsed laser deposition, thought there is much research into faster, more scalable techniques such as non-vacuum sol-gel technique with orientational epitactic crystallisation of the superconducting and buffer layers. One of the techniques of the coated conductor deposition which provides the finer grains than RABiT technique is ion beam-assisted deposition, IBAD, inclined substrate deposition processes, ISD. In IBAD and ISD processes Figure 30, the superconducting layer is aligned according to the biaxially aligned buffer layer. This type of conductor is usually formed on a metallic substrate that is mechanically strong, but not textured. For example biaxially aligned yttrium-stabilised zirconia (YSZ) buffer layers can be deposited on untextured metallic substrates using the IBAD and ISD process. Using this technology, Y123 tapes with a length of few meters has been fabricated in Japan and in Germany. As mentioned earlier the advantages of the YBCO layers produced by IBAD technique is the finer grain size than that of RABIT process and can be therefore formed into the finer filamentary conductors without so significant reduction of the transport current as it is for RABiT conductors as presented in Figure 28.

2.4.4 Magnetically decoupled multifilamentary YBCO coated conductors

The numerical modelling of a model $YBa_2Cu_3O_7$ coated conductor with different number of filaments of a rectangular cross-section covered with a magnetic layer schematically shown in Figure 28(b) was performed using a Finite Element Method commercial package (QuickField) supported by a specially developed AC loss evaluation software [63], [64],

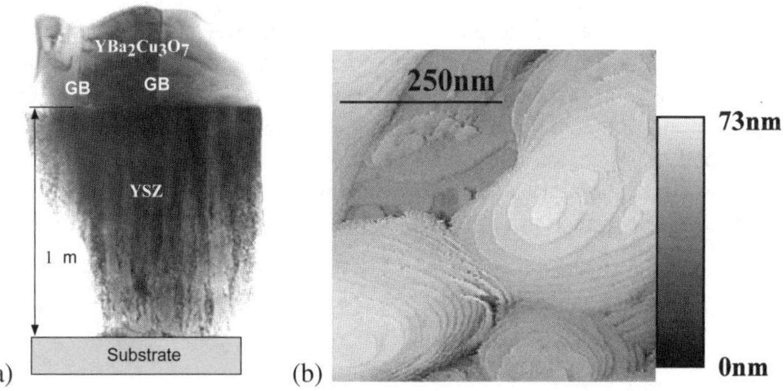

Figure 30 IBAD of YBCO layer on untextured metallic substrates. (a) TEM, micrograph of the superconducting conductor cross section deposited on flexible metallic substrates by ion beam assisted deposition, IBAD. A highly textured YSZ buffer layer formed on the randomly oriented metallic substrates by the IBAD method has a columnar structure and fine nanometre scale grains. The subsequent highly textured YBa2Cu3O7 layer, 300nm thick, has much larger grains with no correlation to the grain structure of the substrate but definitely suppressed the $YBa_2Cu_3O_7$ grain growth; courtesy of L.O. Kautschor and C. Jooss from Institute für Materialphysik, University of Goettingen; (b) the Scanning Tunnelling Microscopy of the top surface of the $YBa_2Cu_3O_7$ superconducting grains showing a-b plains formed during island growth (courtesy of Ch. Jooss and V. Born from Institute für Materialphysik, University of Goettingen). The average grain size 0.2μm.

[65], [66]. As a magnetic material a soft iron was used with B(H) characteristic and hysteretic losses. It was shown before that the best shielding effect and therefore the best decoupling of the filaments was obtained when also the buffer layer and cover layer was magnetic so that each filament was completely embedded in magnetic material. We modelled ac losses starting from a monocore conductor, Figure 28(b) which we then divided into 5 filaments and 11 filaments filaments, keeping the amount of the superconducting material unchanged. In all cases the iron layer 1 μm thick covering each filament was considered and the separation between the covered filaments was always 5 μm. The critical current density $j_c=10^{10}A/m^2$ was chosen, a typical value for YBCO films at 77 K. The part of the losses dissipated in the iron layers was less than 0.5% of the losses dissipated in the superconductor, so only the losses in the superconductor are shown. With no magnetic cover ac losses practically do not depend on the number of the filaments in the geometrical configuration under study. Covering the filaments with iron layers decreases the ac loss level significantly. With 11 filaments an ac loss decrease of about 4 times was achieved. With increasing number of the filaments ac losses decrease. The lower curve in Figure 31 represents the losses of one filament multiplied by the corresponding number of filaments, i.e. it is a theoretical limit of totally decoupled filaments and it cannot be overcome. Again the two horizontal lines in Figure 31 represent two different goals (0.45mW/A·m and 0.25mW/A·m) for high temperature superconductors (HTS) to be competitive with copper wires. Only the coated conductor with 11 filaments covered with 1mm iron layers fulfils the weaker criterion 0.45 mW/A·m. To clarify the contribution of the individual filaments to overall ac losses in 11 filamentary coated conductor further analysis was performed. AC loss

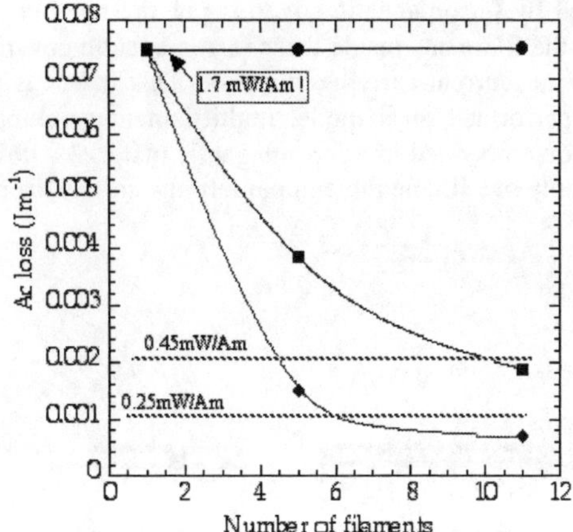

Figure 31 AC loss in superconducting material of a model coated conductor (j_c=10^6 A cm^{-2}) in dependence on number of filaments (constant amount of superconducting material, iron layer thickness 1mm, separation between covered filaments 5mm). (● - coupled filaments, i.e. no magnetic cover, ■ - magnetically screened filaments, ◆ - decoupled filaments, i.e. no magnetic cover but the filaments mutually independent). The curves are the guides for the eye. The two horizontal lines represent two different goals (0.45 mW A^{-1}·m^{-1} and 0.25 mW A^{-1}·m^{-1}) for HTS to be competitive with copper wires.

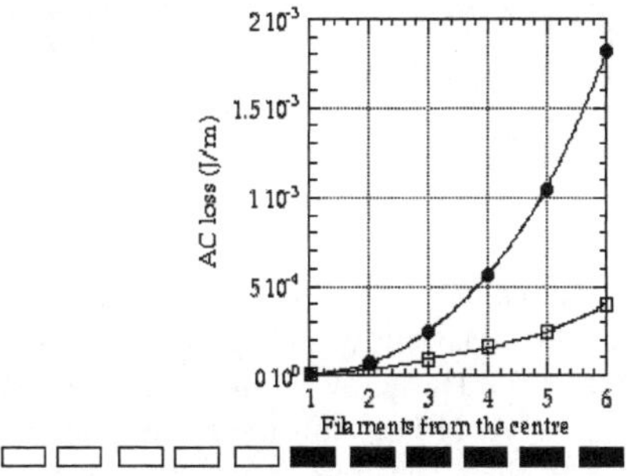

Figure 32   AC loss contribution of the individual filaments in a half of the 11 coated conductor: l - coupled filaments, o - magnetically screened filaments.

distribution among the individual filaments is very inhomogeneous as can be seen in Figure 4, central filaments tend to have the lowest losses. The highest loss level is reached in the filaments positioned at the edges of the tape due to its high aspect ratio. Covering the filaments with iron layers smoothes to some extent the ac loss gradients, but the filaments positioned at the edges of the tape still have the highest losses due to high aspect ratio of the filaments and due to the non-linear properties of the iron layers as well.

The proposed method utilises the shielding effect of magnetic material. In an idealised case magnetic field of a current-carrying filament embedded in a ferromagnetic material propagates outside and is screened by ferromagnetic coverings of the neighbouring filaments, so that the current distribution of the filaments inside these ferromagnetic coverings remains unaffected by the magnetic field of the current-carrying filament. This effect is visualised by a numerical modelling Figure 33 performed on a model multifilamentary composite with 7 filaments of rectangular cross-section embedded in a ferromagnetic material with different relative magnetic permeability m when only one filament the upper left one carries the current.

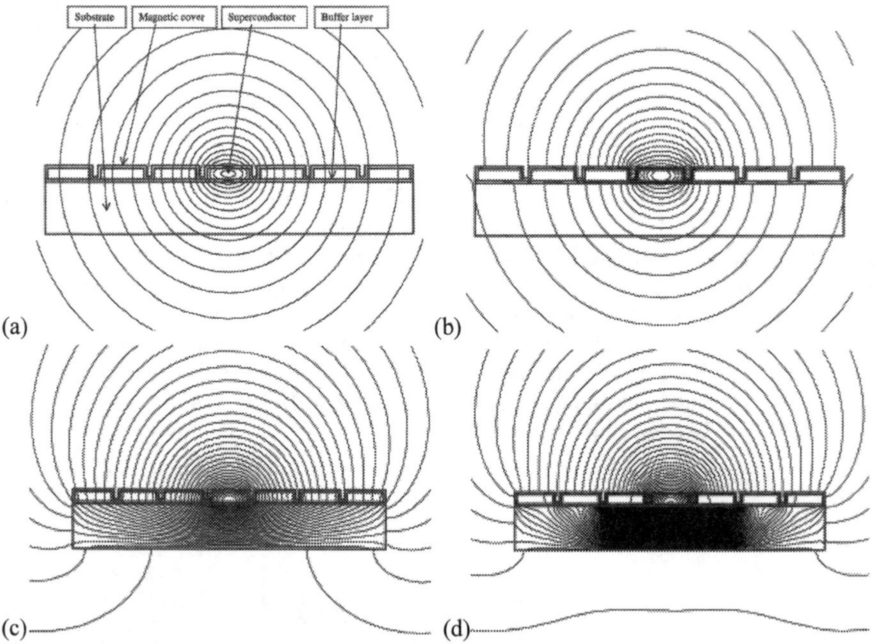

Figure 33 . Comparison of the magnetic flux profile in a superconducting multifilamentary YBCO coated conductor: (a) Hastelloy substrate no-magnetic covers around the superconducting filaments, (b) Hastelloy substrate magnetic covers around the superconducting filaments, (c) NiFe substrate no-magnetic covers around the superconducting filaments, (d) NiFe substrate magnetic covers around the filaments. The dimensions of the filaments are 8 μm in width and 2 μm in thickness. The thickness of the ferromagnetic layer in the case of (b) and (d) is 0.5 μm. The spacing between the filaments, covered with magnetic layer, in both, horizontal and vertical directions, is 1 μm. The current with a current density of 10^9 A m² passes only through the central filament [64], [65].

3. $Bi_2Sr_2CaCu_2O_{8+x}$ CONDUCTORS

3.1 Partial Melt Processing of $Bi_2Sr_2CaCu_2O_{8+x}$ Conductors

Extreme critical current anisotropy [67] and the occurrence of intergranular weak links result in current flow characteristics in polycrystals that require high $Bi_2Sr_2CaCu_2O_{8+x}$ conductors to have highly textured microstructures with an overlapping grain morphology for high current and high field applications. A variety processing techniques being developed for the production of highly textured polycrystalline $Bi_2Sr_2CaCu_2O_{8+x}$ conductors. The powder-in-tube (PIT) and melt-texture growth (MTG) processes are the most developed technologies for processing long lengths of wire or tape and bulk conductors respectively. High critical current density, J_c, values $>10^5$ Acm^{-2} (4K, 25T) have been achieved for textured Bi2212 tapes prepared by partial melting. More recently, J_c up to 1.2×10^4 Acm^{-2} (77K, 0T) have been reported for long length wires [68]. However, these conductors typically suffer from high Ag content, small cross section and thus low J_{eng}.

Also $Bi_2Sr_2CaCu_2O_{8+x}$ conductors can be manufactured in the form of multifilamentary conductors for high field applications at low temperatures, but the necessity of partial melt processing, $Bi_2Sr_2CaCu_2O_{8+x}$ compound suffers the retrograde densification and intergrowth which causes degradation of the superconducting properties more than $(Bi,Pb)_2Sr_2Ca_2Cu_3O_{10-d}$, Figure 34.

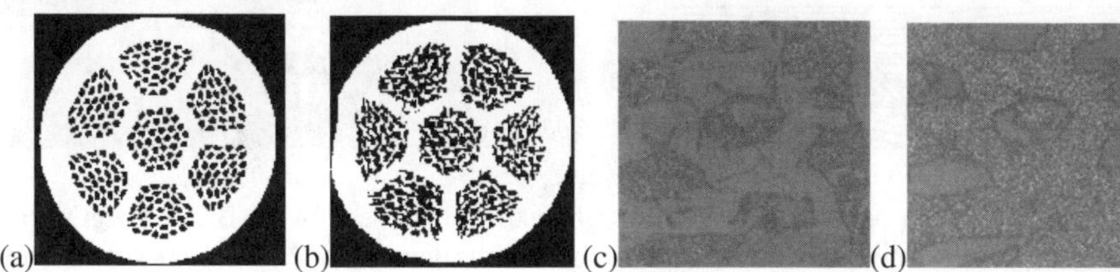

Figure 34 259-filamentary $Bi_2Sr_2CaCu_2O_{8+d}$ round wire (a) as drawn; (b) after heat treatment (c) EDAX atomic mapping of Bi (darker area) of a fragment of heat treated wire; (d) EDAX atomic mapping of Ag (darker area) of a fragment of heat treated wire.

This material is ideally suited for high field applications such an insert magnet, to increase the field strength of nuclear magnetic resonance (NMR) devices, This is due to the lack of shape anisotropy and rigorous winding procedures needed for high magnetic fields devices where Lorentz force is very high. Partial melt processing is commonly used for $Bi_2Sr_2CaCu_2O_{8+d}$ (Bi2212) ceramic superconductors [69] The process involves heating Bi2212 precursor above its peritectic point where a liquid is formed (partial melt) followed by a slow cooling process to develop a dense and uniform highly textured microstructure in Bi2212. The complicated peritectic decomposition and reformation of Bi2212 means the final properties are crucially influenced by processing parameters which include: partial melting temperature, cooling rate, final annealing temperature and time, and processing atmosphere [70].

In the crystallisation process of Bi2212 from the partial melt on cooling down, the second solid phases are not active nucleating sites and homogeneous nucleation is difficult and requires large undercooling. This fortuitously enables the control of crystal growth by the introduction of heterogeneous nucleating sites, an example of which is the substrate induced texture. The highly anisotropic growth of Bi2212 grain results in plate grains with the basal on a-b plane. Growth in a quasi-two dimensional volume leads to the strong c-axis texture observed in the microstructure of thick tape and PIT wire.

3.2 Composite Reaction Texturing

Composite Reaction Texturing (CRT) [71] is a novel process developed to process textured bulk superconductors by incorporating a chemically inert phase which acts as a nucleating seed creating a multiple-growth front, Figure 35(a), (b). A CRT process enables bulk polycrystalline Bi2212 to be fabricated with a highly textured microstructure similar to that observed in tapes and thick films. CRT has been successfully employed for routine fabrication of textured prototype artifacts, such as current leads, cylinders, coils for a wide range of applications.

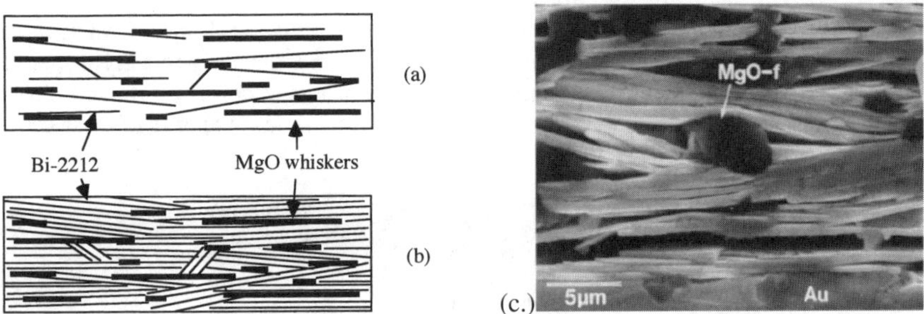

Figure 35 CRT composite: (a) and (b) a schematic cross section of Bi2212/MgO-fibre composite (a) earlier stage of nucleation during cooling down and (b) final microstructure. MgO-fibre can be envisaged to align 2D random in plane; (c) an SEM micrograph of a thermally etched cross section of CRT coating showing also the gold contact interface. (An end view of the MgO-fibre shown represent the MgO-f plane).

As presented in Figure 35(c) the textured microstructure in the bulk Bi2212 is a result of CRT [72] whereas the textured microstructure at the gold interface is the result of enhanced growth at the noble metal interface [73]. Such interfacial texturing is common for all HTS and is created by reduction of the binding energy of the superconducting ceramic at the interface by noble metals and thereby by increasing the mobility of atoms of the ceramic. This accelerates the densification and texturing of the superconducting ceramic and improve transport properties [74]. The quality and high degree of alignment away from MgO fibres which induced by it, is presented in Figure 36.

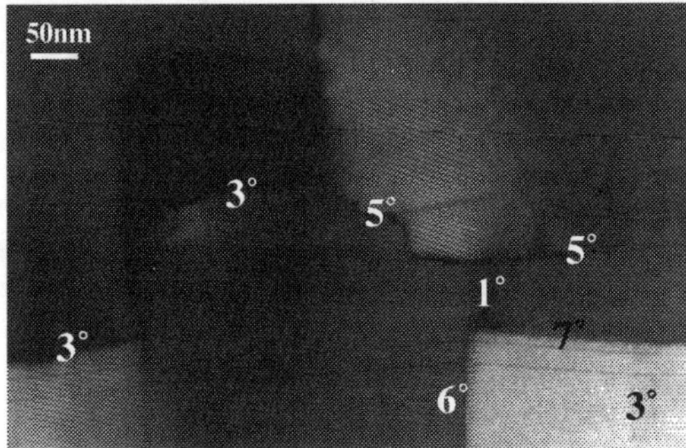

Figure 36 TEM of the melt textured Bi2212 (Composite Reaction Textured) Bi2212 Coated Conductor [75].

3.3 Bi2212 Coated Conductor

The bismuth based material, Bi2212, is being fabricated by a surface coated approach that produces several hundred meters of practically useful tape, Figure 37. This tape has already been demonstrated in useful coil configurations with high current carrying capability at temperatures below 20K and is a low cost approach for producing HTS conductor. This type of material is also ideally suited for high field applications, such as an insert magnet, to increase the field strength of nuclear magnetic resonance (NMR) devices.

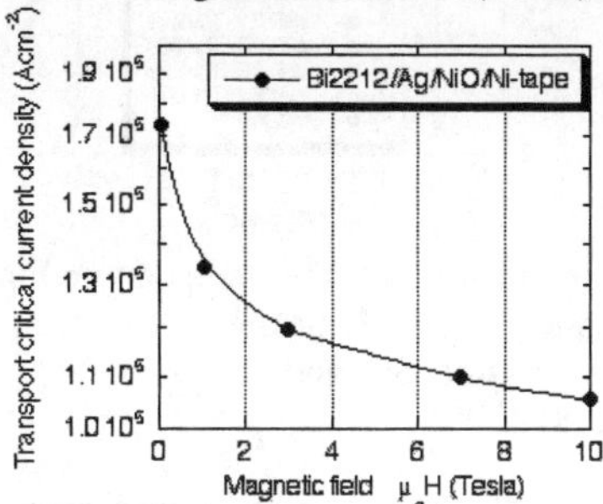

Figure 37 Transport J_c at 4.2K as a function of magnetic field B // tape surface, after [76].

The practical applications of high Tc superconductors require conductors with high Jc value under a large applied field. Melt-solidification processing of Bi2Sr2CaCu2Ox coatings on silver or silver plated Ni flexible metallic substrate (for better mechanical strength), which achieves good textured microstructure without any mechanical deformation, is a promising technique for conductors in long lengths. This value compares well with the conventional superconducting tapes, e.g., Nb_3Sn. The microstructure and superconducting properties of the tape are very sensitive to the sintering parameters during the melt-solidification process. In particular, the cooling procedure to room temperature after melt solidification and sintering is a crucial parameter in maximising tape performance [77]. To preserve the constant oxygen doping level x=8.21 and axoid overdoping or underdoping of Bi2Sr2CaCu2Ox causing degradation of the superconducting parameters, the slow cooling process in a monotonically reduced partial pressure of oxygen has been defined as a suitable cooling procedure [70]. The adjustment of x in the tapes is necessary to achieve a high Tc and Jc values. The value of x in Bi2212 phase depends on processing temperature T, and oxygen partial pressure, pO_2 [81] and is presented in the form of the diagram Figure 136. The cooling procedure of tape $SC(pO_2)$, with the exponentially decreasing cooling rate, has to be conducted under monotonically reduced oxygen atmosphere up to solid arrow 2, Figure 136, according to the eq.2, ref. [78]:

$$T = T_0 + a \cdot \log_{10}(pO_2)$$

$$(2)$$

To ensure that x=8.21 can be retain during the slow cooling procedure, the constant parameters of temperature, T_o and a, were chosen as T_o=870°C and a=60°C. For comparison, data for overdoped $Bi_2Sr_2CaCu_2O_{8.25}$ coating cooled under constant oxygen partial pressure, 0.21atm, SC, (solid arrow-2) is presented in Figures 38. The x value increases with decreasing T and with an increasing pO_2, a similar relation has been reported for $YBa_2Cu_3O_x$ [79], [80].

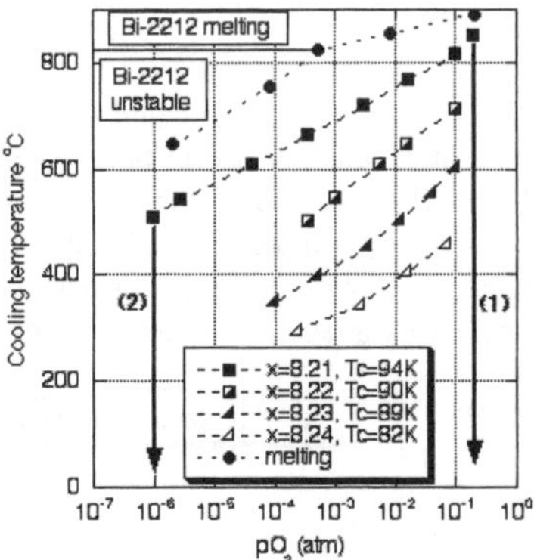

Figure 38 Schematic representation of the temperature and oxygen partial pressure dependence of the oxygen content, x, in $Bi_2Sr_2CaCu_2O_{8+x}$ phase; 1 - sample slowly cooled under 0.21atm partial oxygen pressure, SC, and 2 - sample $SC(pO_2)$. The data for different x and T_c values are taken from [81].

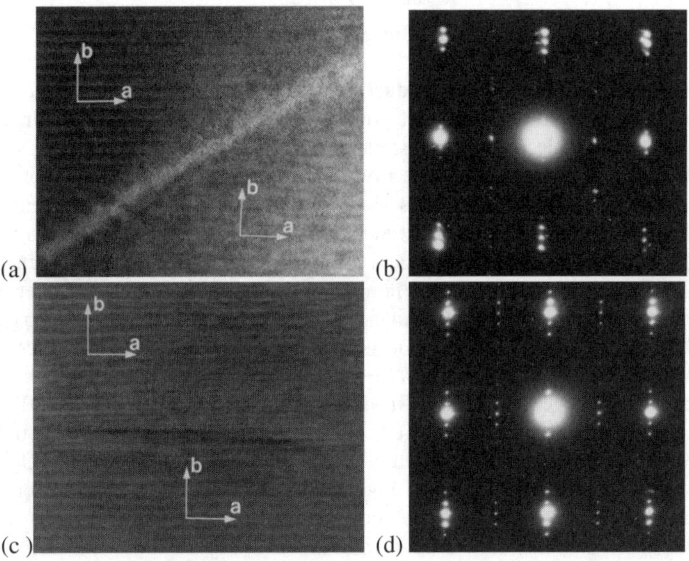

Figure 39 Representative high-resolution transmission electron microscopy (HRTEM) images of low angle grain boundaries and selected area electron diffraction (SAED) patterns recorded from individual grains of: (a) and (b) a slowly cooled conductor under constant oxygen partial pressure of 0.21 atm, SC; (c) and (d) conductor cooled under under monotonically reduced partial pressure of oxygen $SC(pO_2)$.

A typical HRTEM image and a SAED pattern taken from the tape SC are presented in Figures 39 (a) and (b). The image reveals a disordered grain boundary (approximately 20 nm wide) containing crystalline phases and a quite narrow amorphous zone (approximately 5 nm wide). This sample has many intragrain defects. In other words, most grains are not perfectly ordered. Even when an ordered area within the grain was examined, the satellite diffraction spots on the SAED pattern of sample SC were relatively weak, suggesting that order in the grains has been disturbed in the sample SC during cooling. Figures 39 (c) and (d) present a HRTEM image and a SAED pattern from the tape SC(pO$_2$). Well-ordered grains and extremely narrow grain boundaries (less than 2.5 nm wide) were observed. Grain boundaries of this width can act as superconducting transport paths, because the coherence length of the Bi2212 phase parallel to the a-b plane, $\xi\|$(a-b)(0) = 3.8 nm [82] which most strongly influences the superconducting transport properties due to the two-dimensional transport properties of the Bi2212 phase [67] is relatively higher than the width of these particular grain boundaries. Synthesised by slow cooling under controlled oxygen partial pressure, tape SC(pO$_2$) possess a very narrow grain boundaries, well ordered grains, an excellent quality of microstructure and also the best transport property at liquid nitrogen temperatures, Figure 40. The control of oxygen partial pressure during slow cooling procedure is crucial not only

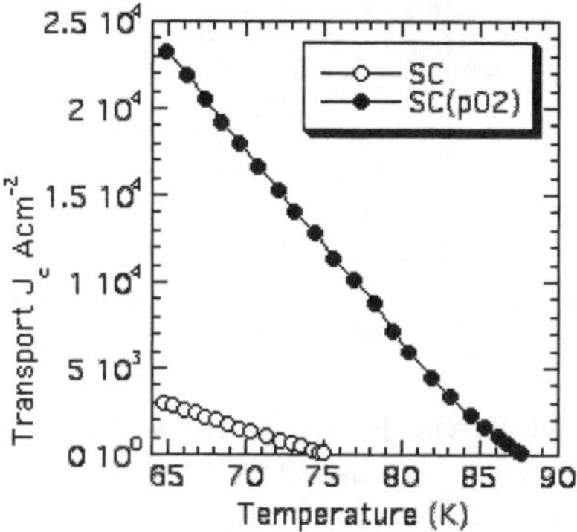

Figure 40 The temperature dependence of the transport J$_c$ characteristics for tape SC(pO$_2$) and SC.

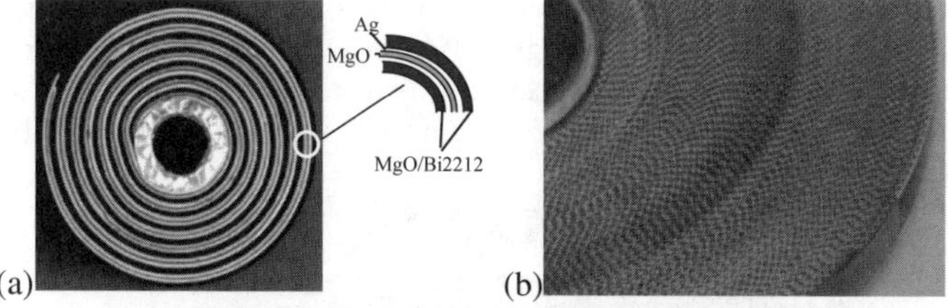

Figure 41 Top view of the Bi2212/Ag multilayer structures of the Bi2212 coils (a) monolithic 7-turn pancake coil prepared by tape casting, elastomer processing and composite reaction texturing [83], (b) the side view of the fragment of HTS high magnet component for application at liquid helium temperatures. A pancake of Bi-2212 dip coat made by wind, react and tighten process. This method allows extremely small winding radii which are hard to obtain with the typical brittle HTS. The photo shows the interium Ag corrugations after winding. This pancakes are used as high field (>21T) inserts in conventional superconducting magnets. Courtesy of H. Jones, Clarendon Lab., Oxford, UK [84].

for obtaining a high T_c value, but also for achieving a high J_c value. This oxygen optimisation process is widely used during manufacture of magnetic screens, current leads and superconducting coils world wide. Indeed, the multiturn so called pancake coils of the Bi2212 layer with $J_c > 1 \times 10^5$ A/cm^{-2} at 4.2 K in 7 T has been routinely manufactured by this process, see Figure 41. Magnetic field performance of the Bi2212 coatings is defined by the magnetic field direction in respect to a-b plane. The J_c vs (angle Θ and B) dependence can be generally presented in Figure 42 where only perpendicular magnetic field component is responsible for the dramatic reduction of the current performance.

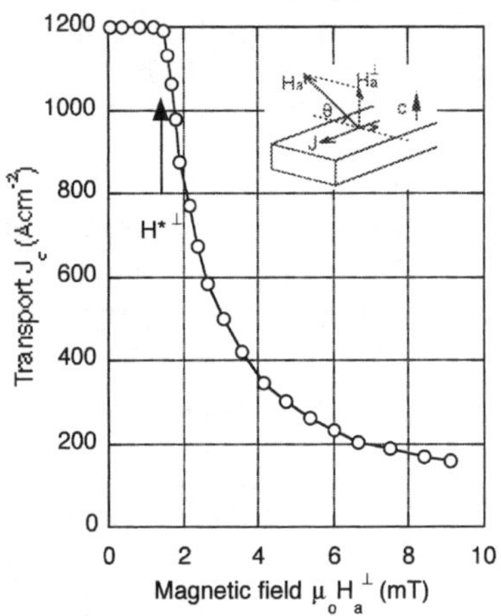

Figure 42   Universal critical current density plotted versus $H_{a,perp.}$, the field component perpendicular to the a-b plane of the Bi2212 tape or coating. H^* is a characteristic field at which J_c value is reduced $H^{*\perp}=1.3$mT [85].

3.3.1 Fault Current Limiters

The superconducting fault current limiter (FCL) is particularly attractive since no conventional limiting device exists for high voltage networks and thus corresponds to a real need. The superconducting CRT Bi2212 elements invented in Cambridge [86] and developed in collaboration with Advanced Ceramics, Figure 43, were very successfully used in the resistive FCL.

Figure 43 View of 800 A CRT Bi2212 superconducting elements for a resistive fault current limiter developed and optimised. Such elements are also used as a current leads. Courtesy of Advanced Ceramics Ltd.

Fault-current limiters help utilities deliver reliable power to their customers. HTS fault-current limiters detect abnormally high current in the utility grid (caused by lightning strikes or downed utility poles, for example, see Figure 44 (a)-blue sinusoid. They then reduce the fault current, see Figure 44(a)-violet line so the system equipment can handle it. An UK team recently produced a successful HTS fault-current limiter, Figure 44(b) is going to be produced by VA TECH Strategic Technology Group.

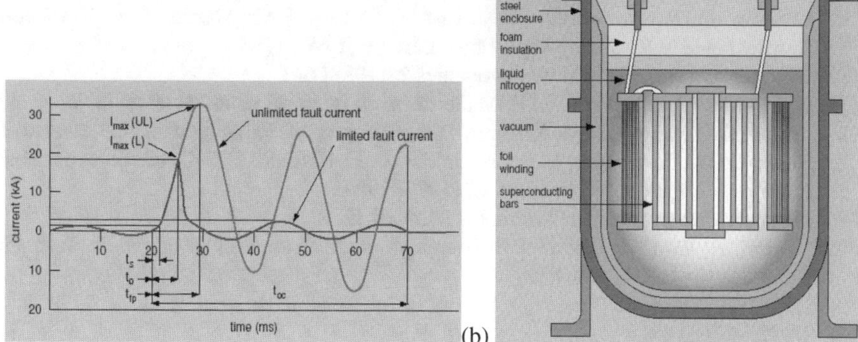

(a)   (b)

Figure 44 FCL: (a) a dynamic Fault current limiter performance parameters represented on current vs time diagram: t_s–sensing time, t_o– operating time, t; t_{fp}–time to first peak; t_{oc}–time to open circuit. (b) cross section through stainless steel vacuum flask, showing superconducting elements and field coil [87]. The resistive elements are CRT Bi2212 rods developed by Cambridg University and Advanced Ceramics.

In 2000, ABB succesfully tested a one phase 6.4 MVA prototype based on stacks of Bi2212 thick film meander composites [88]. 5 m long meanders are structured from Bi-2212 sheets (0.1 m^2) fabricated with ABB proprietary technology and are reinforced with steel bypasses and reinforced plastic composite to avoid "hot-spots" and enhance mechanical stability.

ACKNOMLEDGEMENTS

The author would like to thank many colleagues contributing to this chapter and to the script of the DVD Lectures on Superconductivity for SCENET-2. The support of the Brite-Euram MUST contract, BRPR-CT97-0331, Brite-Euram CONTEXT, AFRL/PRPS Wright-Patterson Air Force Base, Ohio, ABB, EPSRC is also gratefully acknowledged.

REFERENCES

[1] J.E. Evetts and B.A. Glowacki, Superconductor Science and Technology 13 [2000] 443.

[2] X.D. Wu, S. R. Foltyn, P. N. Arendt, W. R. Blumenthal, I. H. Campbell, J. D. Cotton, J. Y. Coulter, W. L. Hults, M. P. Maley, H. F. Safar, and J. L. Smith, Appl. Phys. Lett. 67 [1995] 2397.

[3] M. Paranthaman, A. Goyal, F. A. List, E. D. Specht, D. F. Lee, P. M. Martin, Qing He, D. K. Christen, D. P. Norton, J. D. Budai, D. M. Kroeger, Physica C 275 [1997] 266.

[4] A. Goyal, D.P. Norton, J.D. Budai, M. Paranthaman, E.D. Specht, D.M. Kroeger, D.K. Christen, Q.He, B. Saffian, F.A. List, D.F. Lee, P.M. Martin, C.E. Klabunde, E. Hartfield, and V.K. Sikka, Appl. Phys. Lett. 69 [1996] 1795.

[5] J.E. Evetts and B.A. Glowacki, Cryogenics 28(1988] 641.

[6] B.A. Glowacki, M. Vickers and E. Maher, Materials World 6,No.11(1998)683.

[7] A. Goyal, D.P. Norton, D.M. Kroeger, D.K. Christen, M. Paranthaman, E.D. Specht, J.D. Budai, Q. He, B. Saffian, F.A. List, D.F. Lee, E. Hatfield, C.E.Klabunde, P.M. Martin, J. Mathis, C. Park, J. Mater. Res. 12 [1997] 2924.

[8] Y. Yamada, Supercond. Sci. Technol. 13 [2000] 82.

[9] A. Tuissi, R. Corti, E. Villa, A.P. Bramley, M.E. Vickers and J.E. Evetts, Inst. Phys. Conf. Ser.No 167, 1[2000] 399.

[10] E. Villa, A. Tuissi, R. Tomov, J. E. Evetts, International Journal of Modern Physics B , 14: [25-27] 3145.

[11] N. Reger, B. deBoer, J. Eickemeyer, R. Optiz, B. Holzapfel, and L. Schultz, Inst. Phys. Conf. Ser. No.167 [2000] 331.

[12] J. Eickemeyer, D. Selbmann, R. Opitz, E. Maher and W. Prusseit, 'Effect of Ni purity on cube texture formation in RABiT-tapes', Proceedings of 6th International Conference on Materials and Mechanisms of Superconductivity and High-Temperature Superconductors, February 20-25, 2000, Houston, Texas, USA.

[13] M. Majoros, B.A. Glowacki and A.M. Campbell, Physica C 334 [2000]129.

[14] C.S. Barrett "The Structure of Metals" [1952] and [1966] McGraw-Hill, Inc.

[15] I.L. Dillamore Metallurgical Reviews 10 [1965] 271.

[16] K. Lucke and O. Engler, Materials Science and Technology 6 [1990] 1113.

[17] R. W. Cahn ed "Materials Science and Technology" vol.15 "Processing Metals and Alloys" (1991) VCH Verlagsgesellschaft mbH.

[18] B.A.Glowacki, M.Vickers, N.Rutter, E.Maher, F.Pasotti, A.Baldini and R.Major, Journal of Materials Science, 37 [2002] 157.

[19] R.I. Tomov, A. Kursumovic, M. Majoros, D-J. Kang, B.A. Glowacki and J.E. Evetts Superconductor Science and Technology,15 [2002] 598.

[20] G. Mariani-Regula, B. Pichaud, S. Godey, E. Ntsoenzok, O. Perner, E. Bouayadi, Materials Science and Engineering B71 [2000] 203.

[21] H. Schmalzried,Chemical Kinetics of Solids VCH [1995] 211.

[22] M. Quian, E.A. Stern, Y. Ma, R. Ingalls, M. Sarikaya, B. Thiel, R. Kurosky, C. Han, L. Hutter, I. Aksay, Phys.Rev.B 39 [1989] 9192.

[23] C.P. Bean, Rev. Mod. Phys. 36 [1964] 31.

[24] B.M. Small, E.A. Giess, and R. Ghez, in Handbook of Crystal Growth, edited by D.T. J. Hurle, Vol. 3a, North-Holland, Amsterdam, [1994].

[25] H.J. Scheel, M. Berkowski, and B. Chabot, J. Crystal Growth, 115 [1991] 19.

[26] M. Kambara, PhD Thesis, University of Tokyo, Tokyo [1999].

[27] S. Kawabata, H. Hoshizaki, N. Kawahara, H. Enami, T. Shinohara, and T. Imura, Jpn. J. Appl. Phys., Pt 2-Lett. 29 [1990] L1490.

[28] W.K. Burton, N. Cabrera, and F.C. Frank, Philos. Trans. R. Soc. London A 243[1951] 299.

[29] Y.S. Cheng, PhD Thesis, University of Cambridge, 2001.

[30] Y. Yamada, Supercond. Sci. Technol. 13 [2000] 82.

[31] J.V. Sande, MPC Industry Collegium, MIT Report. 13(3) [1996] 3.

[32] M.R. Lees, D. Bourgault, D. Braithwaite, P. De Rango, P. Lejay, A. Sulpice, and R.Tourier, Proceedings of ICMAS '91, Paris, 1991, p.209.

[33] Y. Ma, K. Watanabe, S. Aweaji, and M. Motokawa, Appl. Phys. Lett., 77 [2000] 3633.

[34] J.J. Wells, N. Male, L. Cohen, M.J. Valler-Regi and J.L. MacManus-Driscoll, Inst.Phys.Conf., Ser No.167 [1999] 319.

[35]. B.A. Glowacki, Superconductor Sci. Technol., 11 [1998] 989.

[36] Y. Ma, K. Watanabe, S. Awaji, and M. Motokawa, Phys. Rev. B, 65 [2002] 174528.

[37] T.J. Doi, T. Yuasa, T. Ozawa, and K. Higashiyama, 7th International Symposium on Superconductivity, November 8-11, 1994, Kitakyushu, Japan.

[38] D.E. Farrell, S. Chandrasekhar, M.R. DaGuire, M.M. Fang, V.G. Kogan, J.R. Clem and D. K. Finnemore, Phys. Rev. B, 36 [1987] 4025.

[39] J-Y Genoud, M. Staines, Anne Mawdsley, V. Manojlovic, and W Quinton, Supercond. Sci. Technol. 12 [1999] 663.

[40] M. Staines, J-Y. Genoud, A. Mawdsley, and V. Manojlovic, IEEE Trans. Appl. Supercond., 9 [1999] 2284.

[41] Y.D. Zhang, and J.I. Budnick, Appl. Phys. Lett. 70 [1997] 1083.

[42] D.M. Feldmann, J.L. Reeves, A.A. Polyanskii, G. Kozlowski, R.R. Biggers, R.M. Nekkanti, I. Maartense, M. Tomsic, P. Barnes, C.E. Oberly, T.L. Peterson, S.E. Babcock, and D.C. Larbalestier, Appl. Phys. Lett. 77 [2000] 18.

[43] C. Prouteau, G. Duscher, D.K. Christen, N. D. Browning, S.J. Pennycook, M. F. Chisholm, D. P. Norton, A. Goyal, and C. Park., Advances in Superconductivity X Proceedings of the 10th International Symposium on Superconductivity (ISS'97)Gifu, October 27-30 [1997] 1015.

[44] K.E. Gray, M.B. Field, and D.J. Miller, Phys. Rev. B. 58 [1998] 9543.

[45] J.G. Wen, T. Takagi, and N. Koshizuka, Supercond. Sci. & Technol. 13 [2000] 820.

[46] Z.G. Ivanov, P.A. Nilsson, D. Winkler, J.A. Alarco, T. Claeson, E.A. Stepantsov, and A.Y. Tzalenchuk, Appl. Phys. Lett. 59 [1991] 3030.

[47] N.F. Heinig, R.D. Redwing, I.F. Tsu, A. Gurevich, J.E. Nordman, S.E. Babcock, and D.C. Larbalestier, Appl. Phys. Lett. 69 [1996] 577.

[48] D.T. Verebelyi, D.K. Christen, R. Feenstra, C. Cantoni, A. Goyal, D.F. Lee, M. Paranthaman, P.N. Arendt, R.F. DePaula, J.R. Groves, and C. Prouteau, Appl. Phys. Lett. 76 [2000] 1755.

[49] N. Koshizuka, T. Takagi, J.G. Wen, K. Nakao, T. Usagawa, Y. Eltsev, and T. Machi, Physica C, 337 [2000] 1.

[50] V.R. Todt, X.F. Zhang, D.J. Miller, M. St. Louis Weber, and V.P. Dravid, Appl. Phys. Lett. 69 [1996] 3746.

[51] M.B. Field, D.C. Larbalestier, A. Parikh, and K. Salama, Physica C, 280 [1997] 221.

[52] D.T. Verebelyi, D.K. Christen, R. Feenstra, C. Cantoni, A. Goyal, D.F. Lee, M. Paranthaman, P.N. Arendt, R.F. DePaula, J.R. Groves, and C. Prouteau, Appl. Phys. Lett. 76 [2000] 1755.

[53] M.Hogg PhD Thesis, University of Cambridge, 2001.

[54] J.H. Durrel, M.J. Hogg, F. Kahlmann, B.A. Glowacki, B.P. Zeimetz, M.G. Blamire, J.E. Evetts, M.P Dlamare, R. Rossler, J.D, Pedarnig, D. Bauerle, Presented at MRS Meeting, Boston, USA, 2-6 Dec. 2002, PM S5-9.

[55] J.H. Durrel, M.J. Hogg, Z. Barber, B.A. Glowacki, M.G. Blamire, J.E. Evetts, presented at EUCAS'02 to be published.

[56] G. Hammerl, A. Schmehl, R.R. Schulz, B. Goetz, H. Bieiefedt, C.W. Schneider, H. Hilgenkamp, and J. Mannhart, Nature 407, 14 Sept [2000] 162.

[57] L. Fernandez,* B. Holzapfel, F. Schindler, B. de Boer, A. Attenberger, J. Hanisch, and L. Schultz, Physical Rev. B 67 [2003] 052503.

[58] J. Loram, K.A. Mirza, and J.R. Cooper, Research Review: High Temperature Superconductivity, IRC in Superconductivity, University of Cambridge 1998 p.77.

[59] N.A. Rutter, B.A. Glowacki and J.E. Evetts, Supercond. Sci. Technol., 13 [2000] L25.

[60] [N.A. Rutter and B.A. Glowacki, IEEE Trans. Applied Superconductivity, 11 [2001] 2730.

[61] B.A. Glowacki, M. Majoros, N.A. Rutter and A.M. Campbell, Cryogenics, 41 [2001] 103.

[62] C. E. Oberly, L. Long, G.L. Rhoads and W.J. Carr Jr., Cryogenics 41 [2001] 117.

[63] B.A.Glowacki, M.Majoros 'Transport ac losses in Bi-2223 multifilamentary tapes - conductor materials aspect' Superconductor Science and Technology, 13 [2000] 483.

[64] M.Majoros, B.A.Glowacki, in Studies of HTS superconductors (Advances in Reserach and Applications) "AC losses and Flux Pinning and Formation of Stripe Phase" ed. A.Narlikar, Nova Scienece Publishers, Inc., Huntington, New York, 32 (2000) 1-51.

[65] B.A.Glowacki, M.Majoros, N.A.Rutter and A.M.Campbell 'Superconducting-magnetic heterostructures as a new method of decreasing transport ac losses in multifilamentary and coated superconductors' Cryogenics, 41 [2001] 103.

[66] B.A.Glowacki, M.Majoros, N.A.Rutter and A.M.Campbell 'A new method for decreasing transport ac losses in multifilamentary coated superconductors' Physica C, 357-360 [2001] 1213.

[67] S. Martin, A.T. Fiory, R.M. Fleming, G.P. Espinosa, and A.S. Cooper, Appl. Phys. Lett., 54 [1989] 72.

[68] U. Balachandran, A.N. Iyer, J.Y. Huang, R. Jammy, P. Haldar, J.G. Hoehn, Jr., G. Galinski, and L.R. Motowidlo, J. of Microstructure, 46 [1994] 23.

[69] J. Kase, T. Morimoto, K. Togano, H. Kumakura, D. R. Dietderich, and H. Maeda, IEEE Trans. Magn. 27 [1991] 1254.

[70] H. Noji, W. Zhou, B.A. Glowacki, and A. Oota, Appl. Phys. Lett., 63 [1993] 833.

[71] B. Soylu, N. Adamopoulos, D.M. Glowacka, and J.E. Evetts, Appl. Phys. Lett. 25 [1992] 3183.

[72] D.R.Watson, and J.E.Evetts, Supercond. Sci. Technol. 9 [1996] 327

[73] B.A. Glowacki, W. Lo, J. Yuan, J Jackiewicz and W.Y. Liang, IEEE Trans. Appl. Superconductivity, 3 [1993] 953.

[74] B.A.Glowacki, Cryogenics, 37 [1997] 609.

[75] Y. Yan, J.E. Evetts, B. Soylu, and W.M. Stobbs, Philosophical Magazine Letters, 70 [1994] 195.

[76] Y. Nemoto, H. Miao, H. Fujii, H. Kitaguchi, H. Kumakura, K. Togano, and K. Shima, Physica C, 339 [2000] 209.

[77] A.W. Sleight, Science, 242 [1988] 1519.

[78] G. Triscone, J.-Y. Genoud, T. Graf, A. Junod, and J. Muller, Physica C 176 [1991] 247.

[79] B.A. Glowacki, R.J. Highmore, K. Peters, A.L. Greer and J.E. Evetts, Superconductor Science and Technology, 1 [1988] 7.

[80] E.D. Specht, C.J. Sparks, A.G. Dhere, J. Brynestad, O.B. Gavin, D.M. Kroeeer, and H.A. Oye, Phys. Rev. B, 37 [1988] 7426.

[81] T. Moriomoto. J. Shimoyama, J. Kase, and E. Yanagisawa, Supercond.Sci. Technol. 5 [1992] S328.

[82] B.D. Biggs, M.N. Kunchur, I.J. Lin, S. J. Poon, T.R. Askew, R.B. Flippen, M.A. Subramanian, J. Gopalakrishnan, and A.W. Sleight, Phys. Rev. B 39 [1989] 7309.

[83] M. Chen, D.R. Watson, A.J. Misson, B. Soylu, B.A. Glowacki, and J.E. Evetts, Inst. Phys. Conf. Ser., No.148 [1995] 719.

[84] D.T. Ryan, M.N. Wilson, M. Newson, H. Jones, and K. Marken, Inst. Phys. Conf. Ser. 167 [1999] 1231.

[85] N.Nakamura, G.D. Gu and N. Koshizuka, Physica C, 225 [1994] 65.

[86] J.E.Evetts and B.A.Glowacki-UK PA 8824630

[87] P. Malkin, and D. Klaus, IEE Review, 3 [2001] 41.

[88] M. Chen, W. Paul, M. Lakner, L. Donzel, M. Hoidis, P. Unternaehrer, R. Weder, and M. Mendik, Physica C, 372-376 [2002] 1677.

FABRICATION AND PROPERTIES OF AG/BI(2223) SQUARE WIRES AND ITS APPLICATION

X.D. SU[*], G. WITZ, K. KWASNITZA and R. FLUKIGER
Group of Applied Physics, University of Geneva, CH-1211 Geneva, Switzerland

1. INTRODUCTION

Extensive research has been focused on fabricating standard Bi(2223) multifilamentary tapes by the powder-in-tube PIT technique [1-3]. However, for industrial applications, in particular for the design and manufacture of coils and cables, it would be preferable to use round or square wires. This would allow to transpose square/round wires in order to produce high current conductors; but in particular also to reduce AC losses due to the combined effect of small twist pitches and smaller aspect ratios. In addition, the wire geometry would be less sensitive to the field direction.

The basic principles are the same for both, Ag/Bi(2223) tapes and square wires, in spite of their different shapes: it is essential to form a high purity Bi(2223) phase and highly textured Bi(2223) grains inside conductors. The process of fabricating multifilamentary Ag/Bi(2223) conductors comprises three steps: a) Precursor treatment and filling into Ag tube; b) Mechanical deformation; c) Reaction heat treatment including intermediate deformation.

In standard tapes, the thin and flat Bi(2223) filaments, obtained by uni-axial rolling, exhibit a large filament aspect ratio (up to 20 :1). For square wire, flat Bi(2223) monofilamentary tapes were restacked into a square Ag tube, which was then deformed to smaller sizes while the filament aspect ratio was almost kept constant as in the original tape, because the tube being simultaneously deformed in two perpendicular directions by two-axial rolling [4]. The two-axial rolling technique was used to produce the wires with square shape. However, this deformation method contains both pressures on the flat surface of the filaments (positive effect) and on the narrow surface (negative effect) during the deformation and intermediate deformation process.

Another main objective in this work was to reduce AC losses at 50Hz in the superconductor [5-7]. For this goal, some parameters for fabricating square wires have been studied, such as the reduction of the twist pitch and the reduction of the aspect ratio of the wire. Square wires have an aspect ratio close to 1, thus allowing smaller twist pitches, without J_c degradation, than in tapes. Hence, it was expected to reach a lower AC loss level in square wires as compared to tapes.

Finally, two double pancake coils, which are expected to carry a current of 200A at 0.1 T, have been made by using those novel square wires.

2. EXPERIMENTAL

Monofilamentary Ag /Bi(2223) tapes were prepared by using the powder-in-tube technique [8]. Densified precursor rods of nominal composition $Bi_{1.73}Pb_{0.35}Sr_{1.9}Ca_{2.04}Cu_3O_x$, were inserted into pure 16 / 10.5 mm Ag tubes of about 110 mm in length, which were in turn deformed by swaging and drawing to a wire of 0.9 mm diameter, which was rolled to a ~ 20 m long tape of about 350 μm in thickness and 3.5 mm in width. These monofilamentary tapes were then cut into several pieces which were stacked into 9 x 9 mm^2 square Ag tubes of 0.5 mm wall thickness with various configurations. After few passes of two-axial rolling, it is necessary to anneal tube and the tapes inside for 4 hours at 600°C in order to make them bonding well. Further, the square tube was deformed by using the two-axial rolling technique. The size in two perpendicular directions of the wire was reduced simultaneously in steps of 50 μm per pass, down to a size of 0.9 x 0.9 mm^2. Finally, the wires were reacted for 82 hours at 820°C under flowing Ar/7% O_2 with one or two intermediate deformations, performed by two-axial rolling for square wires, and the reduction is in the range of 50-200 μm. Some square wires were intermediately two-axial rolled to the desired size after 30 hours of heat treatment, and were then given twist pitches of 20, 8, 4 and 2 mm without any difficulties. Afterwards, the twisted wires were annealed again for the last 52 hours. For comparison, we also produced standard 19 filament tapes (ST19) using the same precursor powder as for the square wires.

Several short pieces of about 4 cm length were cut from fully reacted wires and tapes for measurements. The critical current was measured in 77K by the standard four-probe method with a criterion of 1μV/cm. The outer silver sheath was etched away by a solution of hydrogen peroxide and ammonium hydroxide (1:1). The etched wire was analyzed by X-ray to obtain rocking curves, using a D5000 Siemens Diffraktometer. The average misalignment degree of Bi(2223) grains was evaluated by determining the full width at half maximum (FWHM) of rocking curves of (001) peaks in the longitudinal direction of the wire. SEM images of cross and longitudinal sections of square wires and tapes were observed in polished samples.

The AC losses of wires and tapes were measured by a double Hall sensor method [5]. In order to obtain a sufficiently strong signal, 9 pieces of full-reacted wire were stacked in a 3 x 3 configuration to preserve same aspect ratio as in a single square wire, and the length of the square wires is an integer multiple of the twist pitch. For the non-twisted wire, the sample length was 20 mm each.

3. RESULTS AND DISCUSSION

3.1 Characterization of square wires

In our previous work [12, 13], we have studied a large number of filament configurations of Bi(2223) square/round wires[12]. Figure 1 gives two SEM images of square wires SQ40 and SQ36. 4 groups of filaments have a 90°-rotation symmetrical arrangement inside wires. SQ40 has 40 filaments and SQ36 has 36 filaments but main difference between two wires is precursor powder. Each filament is still keep the large aspect ratio of the original mono-core tape to provide the necessary templates for Bi2223 texturing, however, they are not flat as they are in the tape. Compared with old square wire SQ40, new square wire have high filling factor The final size of square wires is 0.8 x 0.8 mm^2 and J_c values up to 17 kA/cm^2 have been obtained.

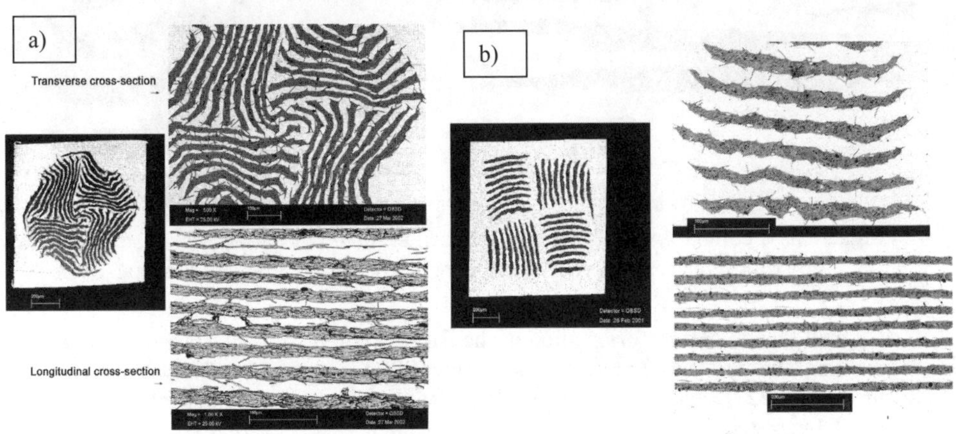

Fig. 1 Cross-section images, a) new square wire
SQ40, J_c ~17 kA/cm^2, b) old square wire SQ36, J_c ~11 kA/cm^2

Table I Comparisons between square SQ40 and ST19

Sample code	Overall size, mm^2	Filament number	Filament size, μm^2	Filling factor, %	Bi2223 content, %	Filament aspect ratio	Misalignment degree	J_c kA/cm^2
ST19	4 x 0.2	19	600 x 16.6	25	~95	~30	5°	35
SQ40	0.8 x 0.8	40	300 x 16	20	~93	~20	8°	17

The J_c value in the standard 19-filament tape made by the same precursor, following the same thermomechanical and reaction route, was about 35 kA/cm^2. The value of J_c in square/round wires is significantly lower, and the highest value being 17 kA/cm^2, as seen from table I. Table I presents the differences between SQ40 and ST19. The overall area was 0.8 mm^2 in ST19 and 0.64 mm^2 in SQ40 while the filament number is 40 in SQ40, and 19 in ST19. The filling factor was 20% in SQ40, and 25% in ST19. In term of aspect ratio, there are big differences between ST19 and SQ36 (~ 20: 1), but the individual filaments have close aspect ratios (30: 20) and similar thickness (~16 μm) in both conductors.

284

3.2 Microstructure, and texture degree development during deformation

Figure 2 shows the microstructures of uni-axially rolled ST19 and two-axially rolled square wires. One can see that filaments in ST19 have a flat shape, while they are severely bent in SQ36. XRD rocking curve measurement showed that the misalignment angle is around 8° in SQ36, compared to 5° in ST19.

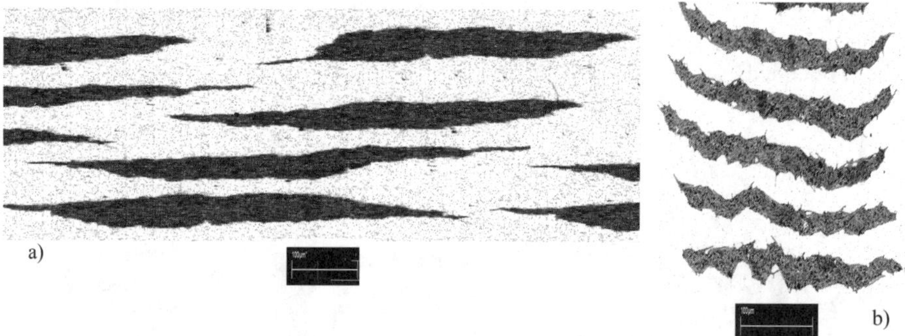

Fig. 2 SEM comparisons of filament between a) Standard flat tape ST19, deformed by rolling or pressing; b) Square wire, deformed by two-axial rolling.

It is well known that the orientation of the Bi(2212) grains in tapes is improved by cold-rolling and that this preorientation of the Bi(2212) grains will determine the texture of Bi(2223) grains after the reaction heat treatment [9]. The square wire was prepared by restacking monocore tapes in a square tube. Therefore, the rolling-induced texture in Bi(2212) grains will be decreased during the two-axial rolling deformation. Consequently, the texture degree in reacted square wires is markedly lower than that in standard flat tapes.

Because of the square shape of SQ36, the intermediate deformation method must be performed by two-axial rolling instead of rolling or pressing of the tape. We have studied the effect of intermediate deformation on the final J_c value in SQ36, as shown in Figure 3. The wire sizes were reduced in two perpendicular directions with simultaneous reduction of width and thickness, the total reductions varying from 50 to 200 μm. This total deformation was achieved by using either one or two thermomechanical sequences (see Fig. 3). The highest I_c was obtained in the wire with single intermediate deformation reductions of 100 μm. A wire treated with single deformation sequence yielded higher J_c values than after two deformation sequences, which is contrast to the standard tape, where three intermediate deformations are still effective in improving I_c. This can be explained as follows: Two-axial rolling creates a certain number of cracks in the filaments, particularly for the pressing direction parallel to the filament broad side (see Fig. 2). A certain amount of the deformation-induced degradation can be recovered after one intermediate deformation step, but this recovery is less complete for a second process.

One can conclude that, two-axial rolling is so far the most appropriate technique for square wires. However, two-axial rolling exerts force to filaments in two perpendicular directions and thus destroys a considerable part of texture in the superconductor cores. Therefore, we think that this deformation mode and correlated higher misalignment angle of Bi(2223) grains are the main reason for the low I_c in square wires. In addition, the SQ36 is four times thicker than the tape, thus causing a decreased oxygen diffusion to the most-inner Bi(2223) filaments. The Bi(2223) phase

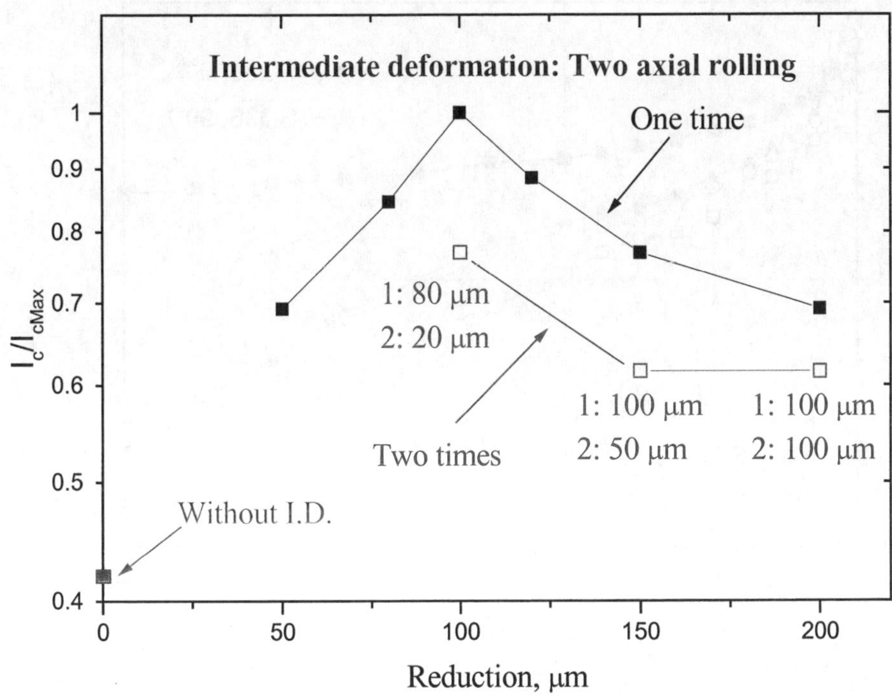

Fig. 3 Intermediate deformation and reduction dependence on I_c in SQ36.

content in final SQ36 is around ~ 5% lower than in ST19, which is another possible reason for lower J_c in square wires.

3.3 Magnetic field measurements

One crucial disadvantage of standard tapes is the anisotropy of J_c under perpendicular and parallel fields. We have found that the symmetrical configuration in the wire SQ36 leds to an isotropic behavior in magnetic field (Fig. 4). The magnetic field dependence of the normalized critical current density at 77 K for SQ36 and ST19 is shown in figure 4(a). In the low-field regime, the I_c values are controlled by the weak links between superconducting grains in both SQ36 and ST19. At high fields, the I_c behavior is governed mainly by flux pinning for the B // c direction, but also by both flux pinning and grain alignment in B // ab direction [10].

The magnetic field angle dependence of normalized I_c for SQ36 and ST19 at 77 K, 0.01T is shown in figure 4(b). When the applied field is changed from the perpendicular to the parallel direction, one can see that the normalized I_c values at B = 0.01T vary in a narrow range between 0.66 and 0.70 in the wire SQ36, compared to a larger range between 0.45 and 0.85 in the tape ST19. The results show that SQ36 is less sensitive to the magnetic field in both directions due to its particular filament configuration. Roughly, in any direction, the field direction is half way between being parallel or perpendicular to the flat surface of the filament. The relative isotropy of $I_c(B)$ in Bi(2223) wires may be very important to some field applications.

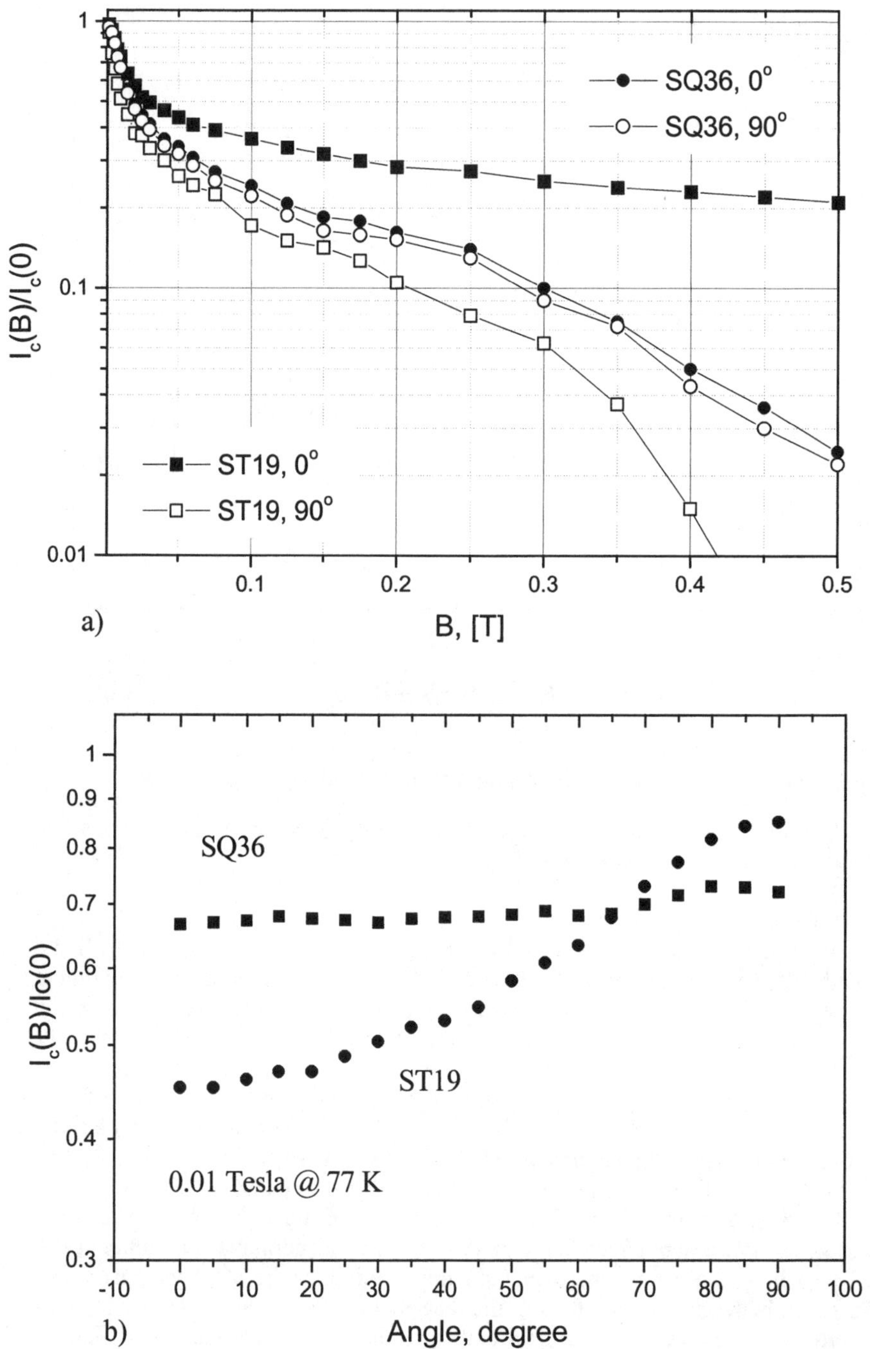

Fig. 4 Magnetic measurement for SQ36 and ST19, a) Magnetic field dependence on the normalized critical current at 77 K in both perpendicular directions; b) Magnetic field angle dependence on the normalized critical current at 77 K and 0.01T.

3.4 AC loss measurements

The square wire configuration has an aspect ratio of 1, and thus allows considerably smaller twist pitches than those in tapes (normally, 10 mm is the lower limit for tapes without breakage). Here, we must point out that the square wires were easily twisted down to very small twist lengths (l_t) without breakage of the filament even without any intermediate annealing. Figure 5 show the pictures of SQ36, in

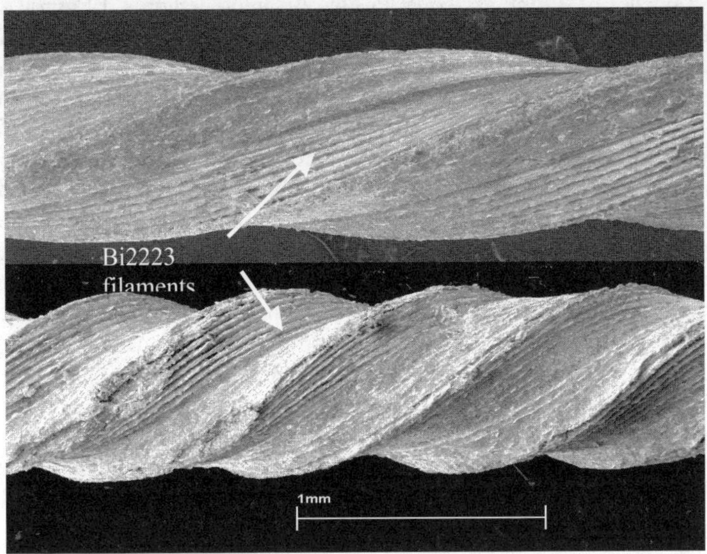

Fig. 5 Pictures of 36 filament wires with twist pitches of 4mm and 2 mm(the Ag sheath was etched away).

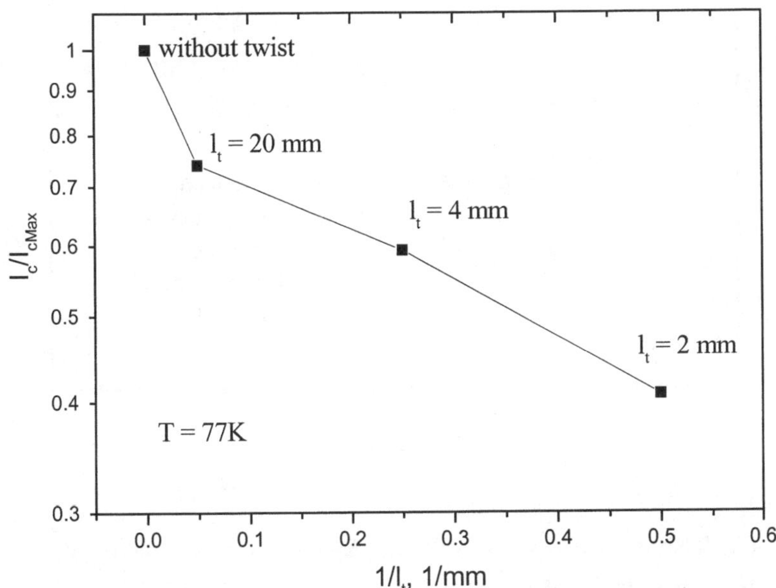

Fig. 6 Normalized I_c dependence on the inverse of twist length in SQ36.

which the outer silver sheath was etched away. Arrows indicate twisted Bi(2223) filaments in two samples, the upper one with a twist length of 4 mm, and lower one with 2 mm. The I_c dependence on the inverse of twist length $(1/l_t)$ in SQ36 is shown in figure 6. With further decrease of twist length, I_c started to decrease, in analogy to tapes.

The AC loss measurement was performed on SQ36 without twisting and with a twist pitches of 4 mm, as shown in Fig. 7 (sample length ~ 12 mm). At 77K, at a frequency of 47 Hz and an applied field $\Delta B = 63$ mT, the normalized power loss P/I_c per m of conductor length is 0.68 mW/ Am, 0.53 mW/Am for SQ36 with twist pitches of ∞ and 4 mm, respectively, while this value is 2.13 mW/Am (H // c) in a flat tape with a twist pitch of 20 mm and an aspect ratio of 14 (Tape A1 in ref. [5]). At high amplitudes ΔB, the saturation losses of square wires are significantly reduced by both, the twisting and the smaller overall width of the superconductor compared to the flat tapes. For the wire sample with $l_t = 4$ mm, the loss is proportional to $\Delta B^{1.8}$. This means that the conductor is far away from full saturation by the coupling currents. At low amplitudes, Fig. 7 shows that the losses of the flat tape are proportional to ΔB^3, indicating that this conductor acts like a large single monofilament. The losses of the untwisted and twisted square wire are proportional to ΔB^m, with m $\approx$ 2, thus indicating only incomplete screening of the applied field ΔB by the coupling currents, a behavior which is expected for the small coupling current decay time constant τ. [11].

These results indicate that even for the untwisted wire SQ36 AC losses are lower than that in standard twisted tapes, due to the appropriate small aspect ratio. In our

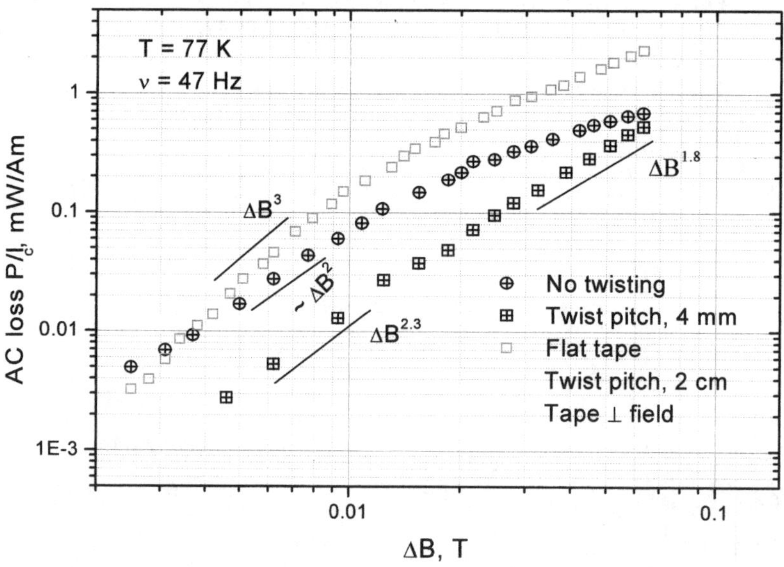

Fig. 7 Normalized power loss P/I_c in square wires and tape taken per m of conductor length

magnetization loss measurement, the loss densities (J/cm^3) were measured for all samples. To transform them to losses per m of conductor length, we must know in which area of the wires the losses occurred, and then which volume we have to use for calculation. For larger ΔB at 50 Hz, the untwisted conductors usually saturate with coupling currents, so the volume of whole filamentary zone can be used. However, the square wire with a twist pitch of 4 mm does not fully saturate, as shown in Fig. 7 (the lowest curve). Since we have used the whole filamentary zone volume instead of the actually saturated volume, this gives somewhat higher loss values than in reality. Nevertheless, the square wire has overcome the large anisotropic losses in perpendicular and parallel field.

3.5 Coils made by square wire

A prototype of double pancake coils, which will be used in a superconductor transformer, has been made in our group, as shown in Fig. 8. To obtain an ampacity of 200Apk@0.1T, 20 wires of SQ40 have been bundled into one cable. Further testing for this prototype will be performed soon.

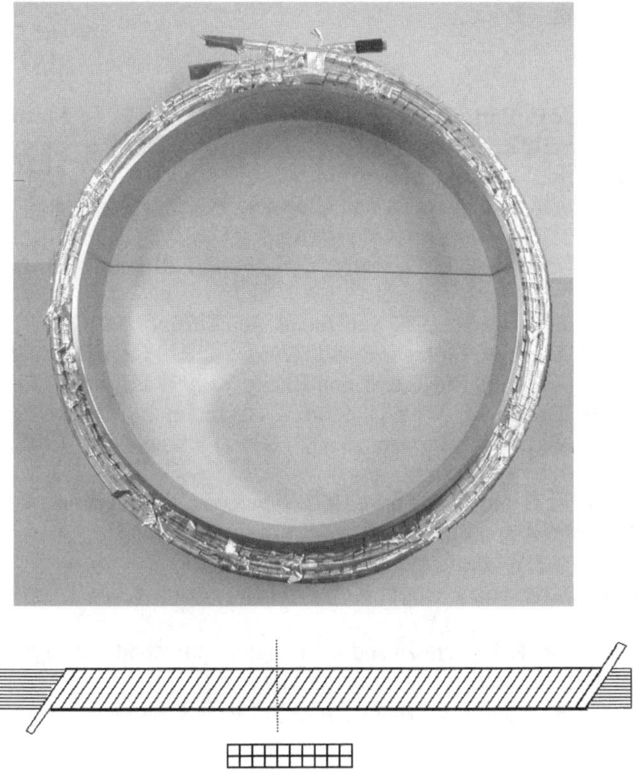

Fig. 8 Upper: Double pancake coils wound by 5m long cable. Lower: Scheme of the cable consisting of 20 square wires, which was bundled by thin Ag ribbon.

4. CONCLUSIONS

We have fabricated square wires with various filament configurations. J_c values of 17 kA/cm^2 have been obtained. Compared with standard flat tapes, J_c is lower in square wires due to the lower degree of texture in the single filaments, which were squeezed in two perpendicular directions during the deformation by two-axial rolling. The present results show that with careful arrangement of filaments, the anisotropy of critical current and the AC losses in a magnetic field can be reduced to an acceptable level for some field applications. The AC losses in the square wires are considerably smaller than in flat Ag /Bi(2223) tapes, as a consequence of the small aspect ratio(1:1) and the small twist pitches(4 mm instead of 10-20 mm). A prototype of transformer coils has been made by using square wires.

Acknowledgements
This work was supported by the European Community BIG-POWA project, project No. GRD1-1999-10461.

References
[1] Malozemoff A P, Carter W, Fleshler S, Fritzemeier L, Li Q, Masur L, Miles P, Parker D, Parrella R, Podtburg E, Riley Jr. G N, Rupich M, Scudiere J, Zhang W, 1999 IEEE Trans. on Appl. Supercond. **9** 2469
[2] Vase P, Flükiger R, Legissa M and Glowacki B, 2000 Supercond. Sci. Technol. **13** R71
[3] Cai X Y, Polyanskii A, Li Q, Riley Jr. G N and Larbalestier D C, 1998 Nature **392** 906
[4] Grasso G and Flükiger R, 1997 Supercond. Sci. Technol. **10** 223
[5] Kwasnitza K, Clerc St, Flükiger R and Huang Y, 1999 Cryogenics **39** 829
[6] Kwasnitza K, Clerc St, Flükiger R and Huang Y, 1998 Physica C **299** 113
[7] Huang Y B, Dhallé M, Witz G, Marti F, Giannini E, Walker E, Passerini R, Polcari A, Clerc S, Kwasnitza K and Flükiger R, 1998 Journal of Superconductivity **11** 495
[8] Yamada Y, Obst B and Flükiger R, 1991 Supercond. Sci. Technol. **4** 165
[9] Grasso G, Perin A and Flükiger R, 1995 Physica C **250** 43
[10] R.L. Perterson, J.W. Ekin, 1989 Physica C **157** 325
[11] Kwasnitza K, Clerc St, Flükiger R, Witz G and Huang Y, 2001 Physica C **355** 325
[12] X D Su, G Witz, K Kwasnitza and R Flükiger, Supercond. Sci. Technol. **15** (July 2002)
[13] Y. Yang, E.A. Young, C, Beduz, X.D.Su, and R Flükiger, presented in ASC 2002.

PROGRESS OF HIGH-T_c WIRES AND ITS APPLICATIONS

Yutaka YAMADA and Yuh SHIOHARA
ISTEC-SRL
2-4-1 Mutsuno, Atsuta-ku, Nagoya 456-8587, JAPAN

1. INTRODUCTION

Superconductivity has been expected as a promising application for the electric power devices; transmission cable, generator, motor, SMES, a transportation such as Maglev train and ship and medical devices such as NMR and MRI. In every application, superior materials or wires are the key points, which can enlarge the application range defined by superconducting properties of the critical temperature T_c, the critical current density J_c and the critical field, H_{c2}. When discovered in 1986, high temperature superconductor (High-T_c) was considered to bring about a revolution to the electric application and even for our daily life. Therefore, high temperature superconductors have been intensively studied and developed since then. After the detailed studies of the basics aspects and the wire applications, researchers encountered many difficulties such as the problems based on its two dimensionalities and brittleness. However, some remarkable progresses have been achieved for the wire developments. At first, Bi systems were rapidly developed by several companies and institutes. Now the state-of-the-art Bi-2223 conductor is 1000 m long and the I_c is over 100A at 77K and 0T, corresponding to J_c more than 20000 A/cm^2. Many feasibility studies of its application have been intensively proceeding as shown in this chapter. Meanwhile, YBCO is now still in the developing stage for the wire fabrication and is about 50m long as of 2003, Jan. In this chapter, we overview the basics of the fabrication method and the present status of the wire, Bi-2223,

Bi-2212 and YBCO, and then introduce the High-T_c applications which have been actively developed now; transmission cables, industrial-use-magnets such as a Si single crystal growth magnet, NMR and motors. Finally, we discuss the present significant problems for industrialization of the High-T_c superconducting wires.

2. SUPERCONDUCTING WIRES

For the wire development, now Bi-2223, Bi-2212 and YBCO materials have been intensively studied since they could be used above the liquid nitrogen temperature (77K); their T_c s are higher than 77K; they could be synthesized easily and elongated as a wire or a tape form. In principle, the fabrication methods can be scaled up. Table 1 summarizes the present status of their properties and manufacturing processes.

Table 1   Present Status of Bi and Y based Conductors. I_c is for for a single conductor.

Material	Bi-2223 [1-4]	Bi-2212 [5, 6]	YBCO [7, 8]
Typical Process	Powder-in-tube	Powder-in-tube	IBAD, RABiTS, PLD, TFA-MOD
Structure	Ag/Bi-2223 multifilament	Ag/Bi-2212 multifilament	Hastelloy(SUS)/ Oxide Buffer/ YBCO Layer, monofilament
J_c	$2\text{-}3\text{x}10^4$ A/cm^2	$>10^5$ (4.2K, 20T)	1-2MA/cm^2
I_c	>100A (77K, 0T)	>300A (4.2K, 20T)	-100A (77K, 0T)
non-SC/SC ratio	1-3	1-3	100
Length achieved	100m-1km	2-3km	46m (recorded by Fujikura, Japan. As of 2003 Jan.)
Wire Configuration	Flat Tape	Flat Tape, Round Wire	Flat Tape with thin YBCO layer
Reinforcement	Ag alloy sheath or composite with SUS	Ag alloy sheath or composite with SUS	Hastelloy, SUS Substrates
Application Tested	Cables, Large Coils	NMR, Small Coils	-
Remarks	for 77K, low field and 20K, high field	for 4.2K, high field	R&D stage for 77K, high field

Generally speaking, Bi-2223 conductors are the most advanced ones since the length has reached more than 1km and the Ic is above 100A/piece and they work at 77K. This makes it possible to design many types of superconducting devices. TheBi-2223 conductors are available from some companies such as American Superconductor Corp. (USA) and Sumitomo Electric Industries, Ltd. (Japan). On the other hand, Bi-2212 is now limited to its usage at 4.2K due to its low Tc and Birr values at the higher temperatures. Then, it should be competed with the conventional NbTi and Nb3Sn superconductors. Compared to these Bi-system conductors, YBCO wires are still amateur as a practical conductor. Although it has a high potential concerning to Jc and Birr, the length is still too short to be utilized for applications. This has been mainly due to the difficulties accompanied with the processes especially for realizing a textured structure of YBCO crystals. In the following sections, the basic understandings of the fabrication method and the current status of the respective conductor are described.

2.1 BSCCO WIRES

2.1.1 Bi-2223 WIRES

BASICS OF THE WIRE FABRICATION

Typical Bi-based superconductors are Bi-2212 andBi-2223 and fabricated by the so-called PIT (Powder-in-Tube) method. Especially Bi-2223 has the higher T_c values of 110K than the liquid nitrogen (77K). Then, Bi-2223 is now investigated to use for the cable application, in which the magnetic field is low and Bi-2223 keeps a high critical current density.

Figure 1 shows a typical processing for Bi-2223 wire. Bi-2223 precursor powders are packed into an Ag tube. Then, the composite is cold worked by drawing and rolling. Heat-treatment is carried out during the cold-rolling. When making multifilamentary wires, the single wire with one Bi-2223 core is stacked and inserted into another Ag tube again. Then, the composite is cold-worked again and heat-treated. This process is called the PIT (Powder-in-Tube) method.

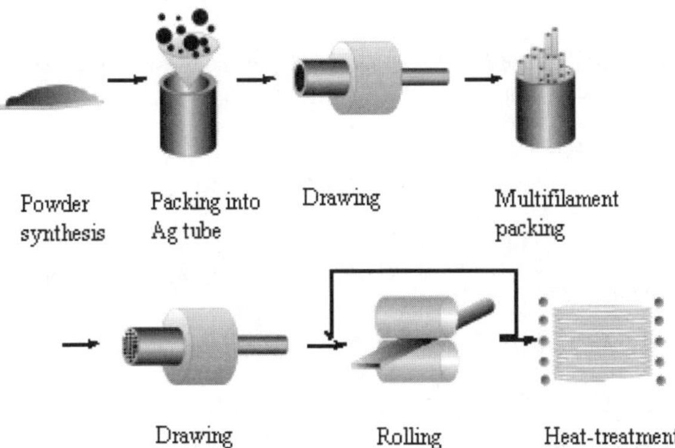

Figure 1   Schematic fabrication sequence of Ag sheathed Bi-2223 wire.
This process is called PIT (powder-in-tube) method.

The key points of this PIT process are; i) Ag sheath and ii) multiple treatments of cold-rolling and heat-treatment. Firstly, Ag sheath is indispensable because other metals react with oxygen in packed Bi-2223 powders, resulting in degradation of superconducting properties. Secondly, multiple treatments combined with rolling and heat-treatment, results in the texturing of Bi-2223 crystals in the wire. During the heat-treatment, the Bi-2223 phase forms and then the following cold-working induces the crystal grain alignment by the subsequent mechanical forces. The final heat-treatment improves the connectivity among the grains after the cold-working. The details of the wire fabrication were explained in reference [9]. A typical cross sectional area of such wires is shown in Figure 2: the width is 4mm and the thickness is 0.24mm. The dimension and capability are listed in Table 2. It should be noted that the similar specifications are also available as seen in the home-page of the American Superconductor Corp.

PRESENT STATUS OF THE WIRE

The PIT Bi-2223 wires are manufactured by a variety of companies [1-4, 10-16]. In Table 2, two major companies' wires are listed. The size is around 4×0.2-0.25mm^2 and its critical current, I_c, is over 100A. For some wires, the average engineering current density J_e for the length longer than 100 m reached 15,100 A/cm^2 at 77 K and self-field carrying 130 A [1]. The J_e value of the wire approximately doubles at 30 K under the several

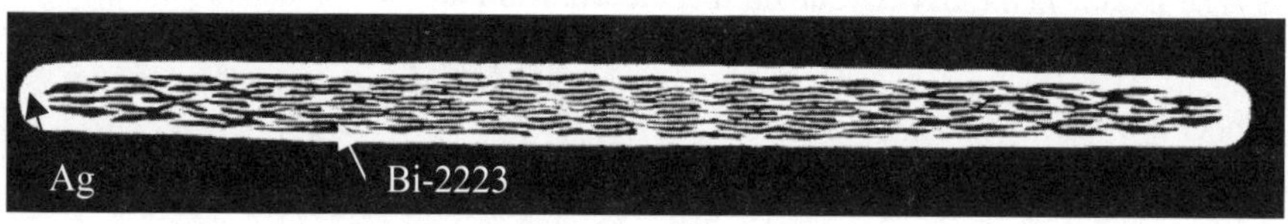

Figure 2   Typical Ag sheathed Bi-2223 conductor. Figure courtesy of S. Kobayashi, Sumitomo Electric Industries, Ltd., Japan.

Table 2   Specifications of the Bi-2223 wire.

	SEI [3, 4]	AMSC [1, 2]
J_e at 77K-self field (kA/cm^2)	12.3	15.1
Dimenion (mm)	3.9x0.24	4.1x0.2
I_c at 77K-self field (A)	110	134
Ag/Superconductor ratio	1.8	1.5
No. of filaments	85	61

Teslas of the magnetic field perpendicular to the tape surface. Furthermore, the high J_c value of the conductors longer than 1000m has been also attained as shown in Figure 3 [3, 4]. The performance of I_c, J_c, J_e and its length is adequate for commercial-scale equipments.

Mechanical properties are reasonably robust, with critical tensile stress typically 120 MPa in wires fabricated by most manufacturers, and 265 MPa in a wire reinforced with a couple of thin layers of stainless steel [2]. Then, the application is now more and more expected, especially for power cables. Table 3 shows the typical cable developed in the projects using Bi-2223 wires [2, 3, 10-16]. Mostly, the reinforced sheath was used. Then, the reliability of the conductor for handling or cabling has been enhanced. Besides those listed, IGC supplied the wire for the Southwire Cable project, 12.4kV/1.25kA 3phase 30m, [14] and NST did for the DTU-Eltek 125 MVA cable project [16].(It should be noted that NKT sold NST on Oct. 31, 2002 to America Superconductor Corporation.)

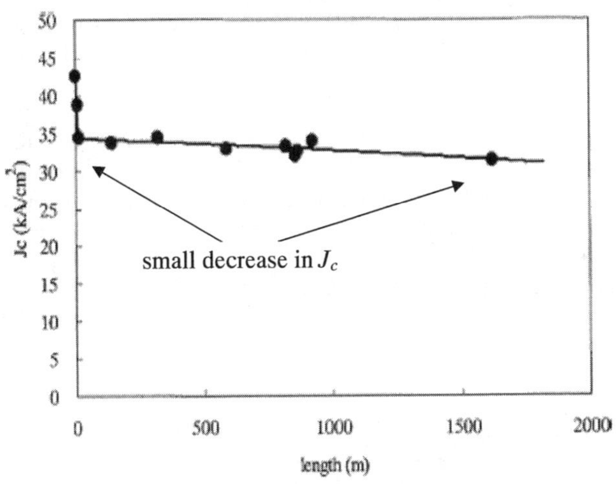

Figure 3   Critical current density of long wires of Ag sheathed Bi-2223 conductor [3, 4]. Figure courtesy of S. Kobayashi, Sumitomo Electric Industries, Ltd., Japan.

The Bi-2223 wire is now almost in the commercial stage. However, efforts for improvement of the wire characteristics, especially, for J_c is now still in progress. According to Huang et al. [17], short length performance results increased to be 170 A (77K, 0T) from 130 A. The reduction of the amount of the secondary phase of Bi-2212 resulted in increase of its J_c. As we could see in later section, the high performance Bi-2223 wire will give a confidence for use in the near-term applications such as motors, generators as well as power cables.

Table3   Typical cable projects and the wire used.

Cable Project	Wire Company	HTS/ Sheath	width mm	Thick-ness mm	J_e kA/cm^2	I_c A	Length m	Strength MPa	Ref.
TEPCO-SEI Cable	SEI	Bi2223/ AgMn	3.8	0.24	J_e=22	50	100	-	[3, 4]
TEPCO-Furukawa Cable	Furukawa	Bi2223/ AgMg	3.5	0.2	J_e=25		120	90	[12, 13]
Detroit Edison Cable	ASC	Bi2223/Ag /SUS	4.1	0.3	9.2	118-134	155	265	[2, 15]

2.1.2 Bi-2212 WIRES

BASICS OF THE WIRE FABRICATION

Bi-2212 wires show high J_c values at low temperatures under high magnetic fields. Then, the appropriate applications are considered to be a system in the liquid He (4.2K) such as NMR.

The most popular preparation method for Bi-2212 wire is the PIT, powder-in-tube, method. However, it is slightly different from the case of Bi-2223 wire: the powders inserted into an Ag tube are the precursors whose anion composition is $Bi_2Sr_2Ca_1Cu_2$; the heat-treatment makes the effective utilization of partial melting of the Bi-2212. The sample after cold-working was heat-treated according to the sequence named "partial melting method" as shown in Figures 4 and 5 [18]. The maximum temperature is above the melting temperature, typically 880ºC. Then, the sample is slowly cooled. When the Bi-2212 solidifies at the melting point, grains of Bi-2212 crystals grow large and align along the interface to the Ag sheath.

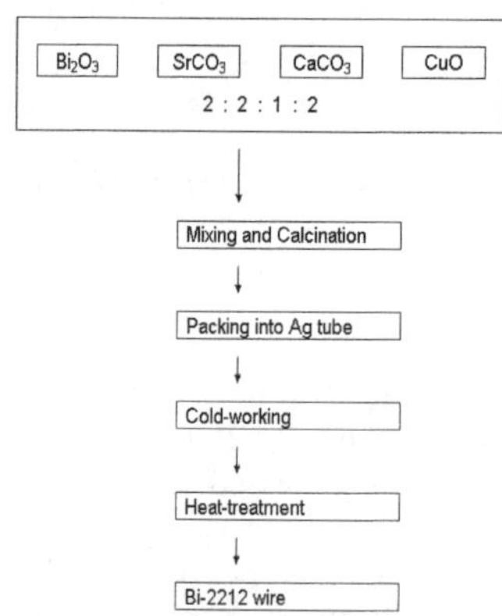

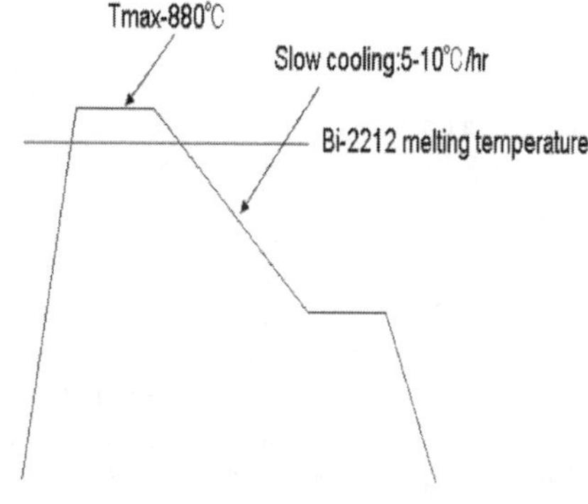

Figure 4   Typical Bi-2212 wire processing.

Figure 5   Typical Bi-2212 wire heat-treatment using partial melting of Bi-2212 phase.

In the early stage of the development, the tape was consisted of a monofilament and a thin tape configuration. However, recently, many different types of multifilamentary Bi-2212 wires have been developed for the magnet winding requirements. Now several companies are manufacturing the multifilamentary Bi-2212 wires: Hitachi Cable, Ltd. [5, 19], Showa Wire and Cable Co., Ltd. [6, 20] and Oxford Superconducting Technology [21] and so on. Here we introduce the ROSAT wire (Rotationary-Symmetric-Arranged-Tape-in-Tube Wire), which was developed to realize a kilometer long conductor by Hitachi Cable, Ltd and successfully applied to 1GHz-NMR magnet [22].

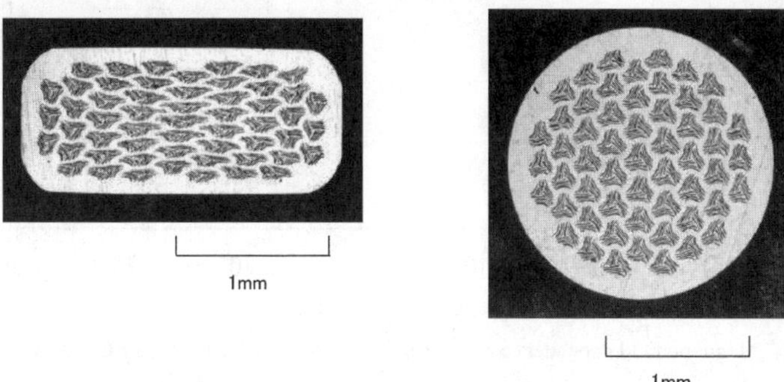

Figure 6 Various types of ROSAT Wire (Bi-2212) [5, 19]. Figure courtesy of J. Sato, Hitachi Cable, Ltd., Japan.

Table 4   Specifications of typical long Bi-2212 ROSAT Wire [5, 19]

Items	Details
Sheath material	Ag-Mg-Ni/Ag
Diameter	1.78 mm
Ag ratio	3.3
I_c at 4.2 K, 10 T	480 A (short sample)
J_e at 4.2 K, 10 T	193 A/mm^2
Filament thickness	12 μm
Filament width	110 μm
Number of filaments	660
Length	>2000 m
Weight	70 kg

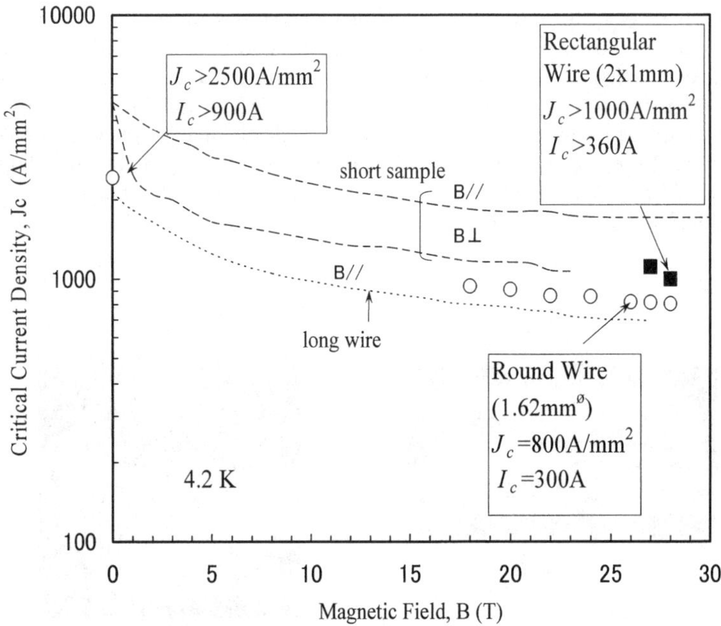

Figure 7 Magnetic field dependence of critical current density for Bi-2212 ROSAT wire [5, 19].

The ROSAT wire is fabricated by the following process;

1) Monofilamentary Bi-2212/Ag tapes were prepared by the conventional powder-in-tube method.

2) Several tapes were stacked to form a segment. Three segments are arranged in triple-rotation symmetry, and inserted into an Ag tube. Round or hexagonal wires, named sub-multi-wires, are formed by a drawing process. The ROSAT was named due to its symmetric structure.

3) The sub-multi-wires are inserted into an Ag or Ag alloy tube and are extruded or drawn into the final form.

4) Heat-treatment by a partial-melting and the solidification process at the maximum temperature of 880–890°C in a flowing oxygen atmosphere was applied.

A typical cross section of the resulting ROSAT wires and the specifications are shown in Figure 6 and Table 4. This method could make it possible to fabricate a round type and a rectangular shaped wire as shown in Figure 6. Numbers of filaments and filament size could be intentionally designed. This conductor showed superior superconducting properties as shown in Figure 7: The critical current density, J_c, exhibited high values of 800A/mm² or 1000A/mm² in a very high magnetic field above 25 T for both round and rectangular wires. Then, the wire was applied to 1GHz-NMR model coil which could generate a high magnetic field of 25T [5, 19, 22].

2.1.3 FUTURE TECHNICAL TASK FOR THE APPLICATION

As mentioned above, Bi based wires are now almost in the commercial stage. However, we need some progress to promote the applications. Such requirements for the Bi based wires are;
1. reduction of AC loss,
2. improvement of J_c especially in high temperature (Bi-2212) and high magnetic field (Bi-2223) and reduction of dead filaments in the wire,
3. cost reduction and
4. configuration appropriate for the magnet or cable fabrication.

The reduction of AC loss is significant for the electric power applications since it is used by an AC current. The useful methods to decrease the AC loss are the reduction of the filament size, twisting and a high-resistivity barrier inserted between filaments. Already some researchers succeeded in decreasing the AC loss without J_c decrease by twisting filaments with 8mm or 20mm pitches [13, 23]. Furthermore, some barrier materials such as $SrZrO_3$, $BaZrO_3$, $SrCO_3$, Sr-V-O and Bi-2201 phase are developing [24, 25]. However, these efforts are not yet available for the commercial wire. We need further studies for the AC loss reduction.

For J_c -B characteristic, we still have the problems; the J_c is still very low in high magnetic fields at 77K for Bi-2223 and in high temperatures for Bi-2212. For both cases, we need pinning centers to improve their J_c. Although the effective results have not yet been obtained, the possibility of the introduction of pinning centers have been discussed; substitution Bi by Pb [26], MgO addition [27] and so on. However, effective results could not be obtained. Related to critical current, the reduction of Ag ratio is also important, which increases engineering current density, J_e. J_e is the critical current per whole cross section of the wire and high J_e is preferred to the application because the size of the application, for example, magnet diameter, will be reduced. At present, Ag ratio is about 3. The problem with reducing the Ag ratio is the brittleness of the composite of Ag sheathed Bi-oxides. Further reduction of Ag results in the breakage of the filaments or the whole wire during fabrication and coil winding.

Other than pinning problems, Bi based conductor has the inhomogeneity problem, meaning that some part of filaments can not transport the critical current due to its low orientation degree of Bi-2212 or 2223 crystals or included impurity phases such as Bi-2212 for Bi-2223 wires. Actually, Larbalestier et al. reported much higher values than the J_c of the whole wire in some parts of a Bi-2223 filament [28]. The reduction of the dead layer will be required to draw the potential fully of the Bi based conductor.

While the cost of the wire will be discussed later from a point of economic industrialization, technological possibilities have been done to decrease the amount of expensive but indispensable Ag; Yamada et al. replaced expensive Ag sheath to Ni/NiO substrate to form Bi-2212 wires [29]. However, this is a mono-filament configuration and then further studies is needed to apply multifilament Bi based wires.

Configuration or shape of the wire is also important for the application. Some magnets like MRI need the precise magnetic field distribution to analyze the resonance signal. For this purpose, the present very thin layer type wires have a large disadvantage. For multi-strand cable type conductor, Hasegawa et al. fabricated round shaped wires of

Bi-2212 conductor [5, 20]. At present we have obtained superior properties of Bi-based conductors, which can realize a feasibility study of some application. However, the wide spread application requires the solutions for the above further problems.

2.2 YBCO COATED CONDUCTORS

As described in the former sections, the Bi-based conductor is now in the matured stage, meaning the length reached more than 1000m with I_c over 100A enough to be applied to some devices. On the contrary, YBCO coated conductor is still in the stage of research and development while the YBCO coated conductor has a potential of high J_c at high temperature and high magnetic field in contrast to the low J_c of Bi-based HTS in a high magnetic field at 77K. The length achieved so far for the YBCO is several tens meters. Then, even now, many approaches are on going around the world especially to fabricate a long conductor with a viable cost for the industrialization.

In the YBCO coated conductor processing, the most important point is the bi-axial texturing of the crystals of YBCO. This can reduce the effect of weak links at the grain boundaries, which deteriorates J_c of YBCO. To obtain the textured structure in YBCO, we have varieties of the processes, which can be classified into mainly two groups based on the texturing controlling layer:

1) Textured Buffer Layer based YBCO coated conductor
2) Textured Metallic Substrate based YBCO coated conductor

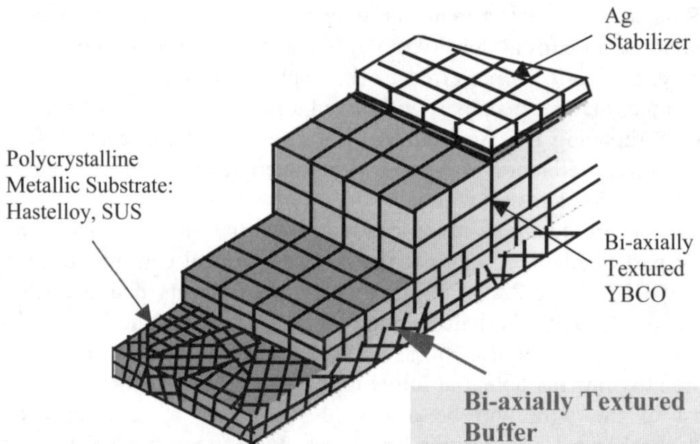

Figure 8   Architecture of the textured-buffer-based YBCO coated conductor using commercially available Ni-based metal tape or SUS without textured structure.

In the group 1, the textured microstructure is realized by the vapor deposition of an oxide material; Ion-beam-assisted deposition (IBAD) [30, 31] and Inclined Substrate Deposition [32]. As we do not need any textured metallic substrate in these methods as shown in Figure 8, high strength polycrystalline metallic substrates could be used. Both processes resulted in a textured structure irrelevant to the substrate texturing. In the group

2, various cubic metals are applied. Texturing is achieved by rolling and re-crystallization; textured Ni substrate (rolling assisted bi-axially textured substrates (RABiTS^{TM})), Ag and their alloys. The architecture is also shown in Figure9. For both cases, the two or three layers above the texture controlled layer is grown by the conventional epitaxial growth method. Sometimes the layer underneath the epitaxial layer is called "a template layer".

Following this process, various methods to synthesize the superconducting layer could be selected, which are;

    1) PLD (Pulsed Laser Deposition) process
    2) MOD (Metal Organic Deposition) process;
        Especially TFA-MOD (Trifluoroacetae MOD)
    3) Ex-situ BaF$_2$ process
    4) CVD (Chemical Vapor Deposition) process
    5) EB (Electron Beam) evaporation process
    6) LPE (Liquid Phase Epitaxy) process.

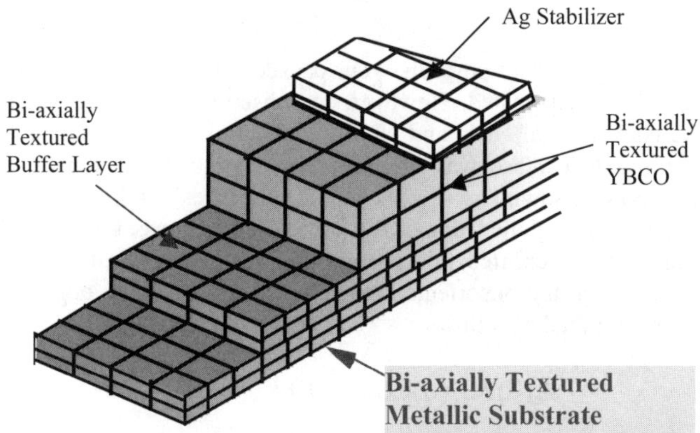

Figure 9   Architecture of the textured-metallic substrate-based YBCO coated conductor. Texturing can be obtained by a simple combination of rolling and heat-treatment for a metal substrate.

Figure 10 shows the schematic diagram of the relationship among these methods. Up to now, many methods and materials have been examined to obtain an YBCO coated conductor with high J_c at 77K. Among them, IBAD-PLD and RABiTS^{TM}-MOD routes are   the promising method from the view point of length and J_c so far achieved. In this section, we describe the basic aspects and recent achievements of these methods: RABiTS^{TM}, IBAD, PLD, MOD (TFA-MOD) and ex-situ BaF$_2$ methods.

2.2.1 EARLY WORKS AND WEAK LINK

Since the discovery of HTS or YBCO, many efforts to wire fabrication have been conducted. The first work [33] was the simple PIT YBCO which is the same as the PIT in

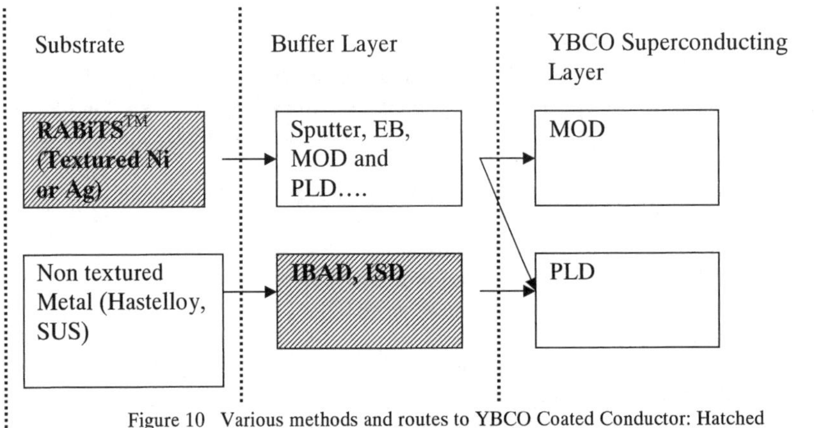

Figure 10   Various methods and routes to YBCO Coated Conductor: Hatched regions are the step controlling texturing of microstructure. In the last step, other methods of CVD, EB and LPE can be applied.

the Bi-based HTS wire. YBCO powders were packed into an Ag tube, cold worked to form a wire and then heat-treated. However, the observed J_c was desperately low, as compared to the conventional NbTi and Nb$_3$Sn conductors, in the order of 100A/cm^2 at 77K and further rapid decrease was observed when applied a magnetic field. These phenomena are now known as "weak link properties" inherent to the oxide superconductors, which degraded drastically critical current passing through the grain boundaries. Dimos et al. reported this property by using bi-crystals [34]: J_c rapidly decreases with increasing the misorientation angle especially over 5 degrees. According to the results, J_c is described as follows;

$$J_c(\theta) = J_c(o)\exp(-\theta/\theta_c) \tag{1}$$

where J_c is the ciritical current density at the grain boundaries with the misorientation angle, θ, J_c (0) is that for a single crystal YBCO without grain boundaries and θ_c is about 5 degrees. The abrupt J_c decrease at around 5 degrees is considered to be due to the order parameter reduction at the grain boundaries. The in-plane alignment in the coated conductor affects directly transport currents at the grain boundaries.

At the same time, YBCO thin films were produced on single crystalline substrates with quite high J_c values, over 10^6 A/cm^2 (77 K, 0 T), by vapor-phase deposition methods [35]. To aim at fabricating YBCO coated conductors, mainly pulsed laser disposition (PLD) and chemical vapor deposition (CVD) have been favorably used to deposit a YBCO film with high J_c values on polycrystalline flexible tape substrates [36-39]. These deposition methods have several advantages compared to conventional sputtering method: (1) good stoichiometry of Y123, (2) high oxygen gas pressures and (3) high deposition rate.

However, uni-axially textured YBCO films with c-axis perpendicular to the substrate tape surface were only grown on a polycrystalline tapes such as Hastelloy tapes and NiCr superalloy tapes. Even though they used oxide buffer layers of MgO and YSZ between

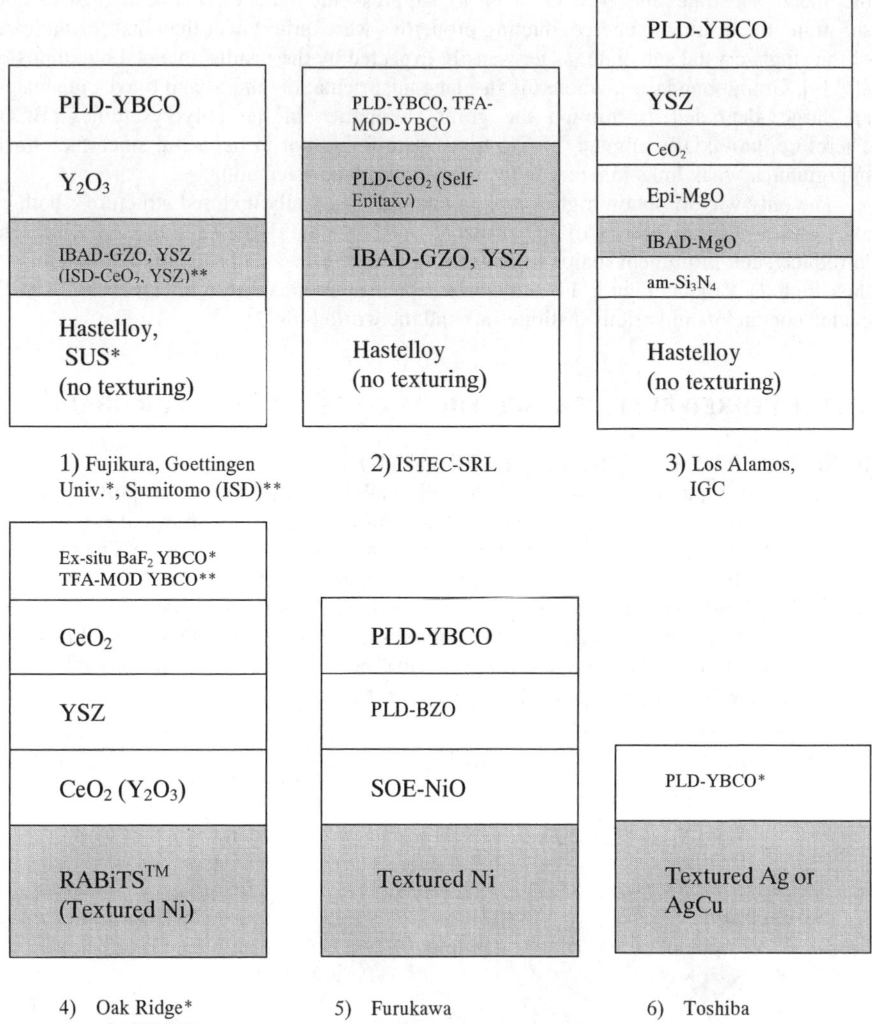

Figure 11 Various architectures of YBCO coated conductors developing in various institutes in the world [40-51]. Note that, in figure 1), Sumitomo uses ISD method. Colored regions initiate epitaxial growth and control the final YBCO layer texturing. In figure 2), PLD-CeO$_2$ also enhances the texturing degree (self-epitaxy).

the metal substrate and YBCO layer to suppress the reaction between YBCO and substrate, the resultant superconducting properties were quite lower than that for the case of the single crystal substrate, as now easily expected by the results obtained by Dimos et al. [34]. Grain boundaries, where the in-plane misorientation and a- and b-axes mismatch are large, degraded J_c through the grain boundaries in the polycrystalline YBCO. Therefore, uni-axially aligned YBCO films were found not to be useful since they have intergranular weak links inherent to the random in-plane orientation.

The only way to obtain high J_c was to attain a "bi-axially textured structure": both c-axes and a–b basal planes of all grains in YBCO should be well aligned. Here we introduce such prominent major techniques to realize a bi-axially textured structure and thus high J_c YBCO. Figure 11 summarizes present promising architectures of YBCO coated conductors in various institutes around the world [40-51].

2.2.2 TEXTURED BUFFER BASED YBCO COATED CONDUCTOR (IBAD)

BASICS OF THE CONDUCTOR FABRICATION

The most popular method to obtain a bi-axially textured buffer on polycrystalline substrate is IBAD (Ion Beam Assisted Deposition). Figure 12 shows the schematic experimental set up of the IBAD method, consisting of two Ar ion guns, YSZ target and substrate to be deposited. Ar ion sputters an oxide target such as YSZ and GZO ($Gd_2Zr_2O_7$) and sputtered species are deposited on a polycrystalline substrate with another Ar ion flux called assist ion. Then, the resulted layer on the substrate is bi-axially textured structure. This phenomenon was first observed in 1991, in the case of a polycrystalline Ni-based alloy and YSZ buffer by the IBAD method [30, 31]. They were

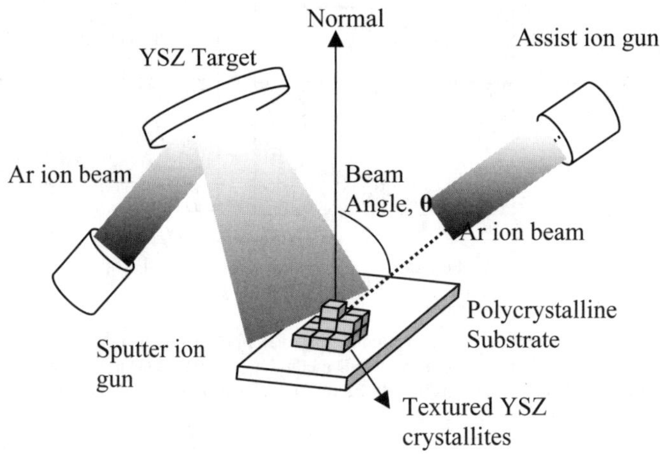

Figure 12   Schematic diagram of the ion-beam-assisted deposition system. Two sets of ion sources are used for sputtering the target material and for ion bombardment of growing films.

successfully applied to in-plane texturing of a buffer layer for further deposition of Y-123 films with J_c values over 10^5 A/cm^2. This was the first demonstration of high-J_c Y-123 superconducting tapes using a textured microstructure of a buffer layer. Subsequent detailed studies revealed that the IBAD effect is remarkable for fluorite-type oxides such as YSZ, CeO$_2$, Pr$_5$O$_{11}$ and deposited at low temperatures below 300°C [51-53]. The energy and fluency of Ar ions were 200–300eV and 100 A/cm^2, respectively. Other various types of oxides were found to be textured as well by the IBAD method. Very recently, Fujikura succeeded in increasing the deposition rate two times by using GZO instead of the conventional YSZ target [54].

The most important point in the IBAD process is the incident angle, θ, of the assisting Ar ion to a substrate as shown in Figure12. By tilting the beam axis, in-plane azimuthal ordering emerged and was optimized when the beam incident angle was 55 degrees from the normal, which corresponds to a [111] axis of the YSZ unit crystal lattice, as shown in Figure 13. Their results clearly showed the dependence of the degree of texturing of YSZ on the ion incident angle to the normal of the substrate surface. There, an abrupt change and the favorable angle for texturing existed around 55 degrees.

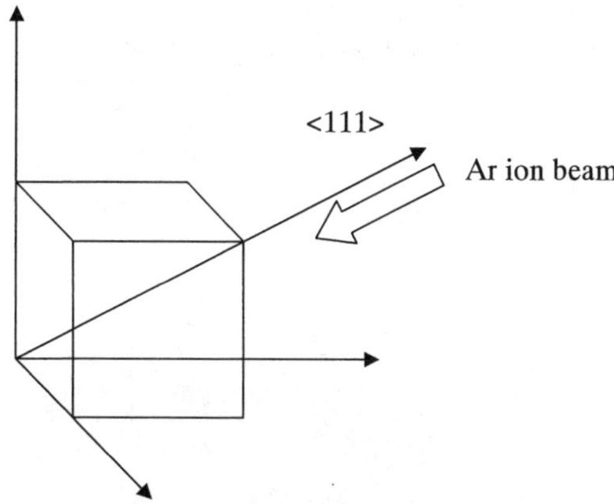

Figure 13   Incident angle, 55 degrees, of Ar ion in IBAD corresponds to <111> direction of deposited film.

PRESENT STATUS OF THE CONDUCTOR

The IBAD method provides us a good orientation of buffer layers on a polycrystalline substrate of Hastelloy as well as SUS tape. On the other hand, the disadvantage is the slow deposition rate originated from the sputtering method. This means that it takes time to fabricate a long conductor. Even recent results [55] showed it takes 4 to 6 hours to obtain a good orientation around 10 degrees in the degree of in-plane texturing, $\Delta\phi$. Then, intensive efforts have been done for solving this problem. Stanford University found another buffer material, MgO [56]. This material resulted in a fast texturing with a very thin layer of 30nm, as opposed to a thick layer of IBAD-YSZ more than 1000nm in thickness. Consequently, IBAD-MgO formed a bi-axially textured structure 100 times as fast as IBAD-YSZ. LANL group is developing this method for the long coated conductor process [57]. Similarly, Fujikura group has found another fluorite oxide, $Gd_2Z_2O_7$, buffer instead of YSZ. $Gd_2Z_2O_7$ improved the degree of texturing and texturing speed. The speed was enhanced two times, compared to the IBAD-YSZ [54].

Table 5 is the present status of IBAD processed YBCO coated conductor in the world. Fujikura installed a continuous reel-to-reel equipment as shown in Figure 14 [55] and obtained recently a 46m long conductor in 2002 [58]. The I_c value of the conductor was 74A, which can be used to some applications. They also fabricated a 100m long GZO/Hastelloy tape. Figure 15 shows typical Fujikura's IBAD YBCO coated conductor.

Goettingen university group developed also IBAD-PLD YBCO coated conductor [59, 60]. The IBAD process is similar to Fujikura's, but the PLD deposition is very fast and they use an SUS tape substrate, which is of course low cost compared to Hastelloy tape. Los Alamos National Laboratory developed the IBAD-MgO method [57]. Using this IBAD substrate, they are also developing a multi-layer YBCO coated conductor, where the other layers such as SmBaCuO were sandwiched [61]. For YBCO films, it is well known that the surface becomes rough and a-axis orientation crystal grows when increasing the thickness. This could be prevented by such a sandwiching interlayer. The sandwiched layer results in smooth surface and then YBCO second or third layers can grow with keeping high degrees of orientation. Therefore, the technique can increase the critical thickness maintaining high J_c values. Consequently, they successfully obtained high I_c of several hundreds A/cm (critical current per 1cm width).

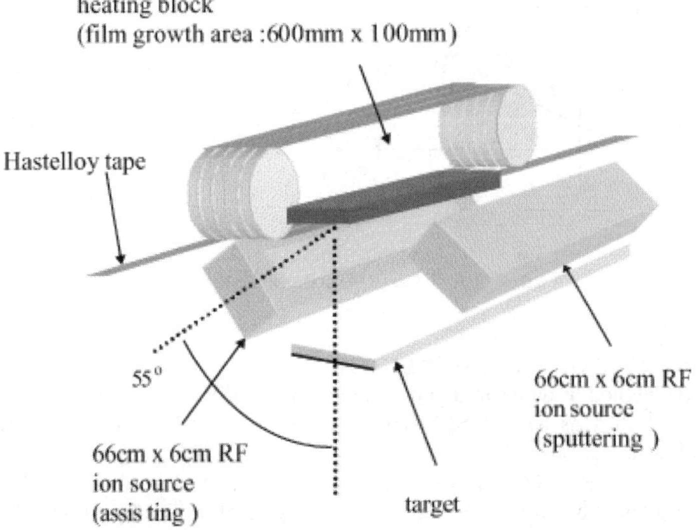

Figure 14   Reel-to-reel IBAD machine by Fujikura [55]. Figure courtesy of Y. Iijima, Fujikura, Japan.

Table 5   Current status of typical IBAD based YBCO coated conductors.

Institute	Fujikura (Japan) [54, 55, 58]	Goettingen Univ. (Germany) [59, 60]	LANL (USA) [46, 57]
Buffer	GZO/YSZ	YSZ	MgO
Buffer Thickness	1.2 µm	-1 µm	30nm
Supercond. Layer	PLD-YBCO	PLD-YBCO	PLD-YBCO
Length	46m	10m	0.8m
I_c	74A	105A	134-189A
J_c	0.6MA/cm^2	2MA/cm^2	1-2 MA/cm^2
remarks	world record (length)	low cost SUS substrate (rapid YBCO deposition)	thin MgO layer (rapid IBAD process)

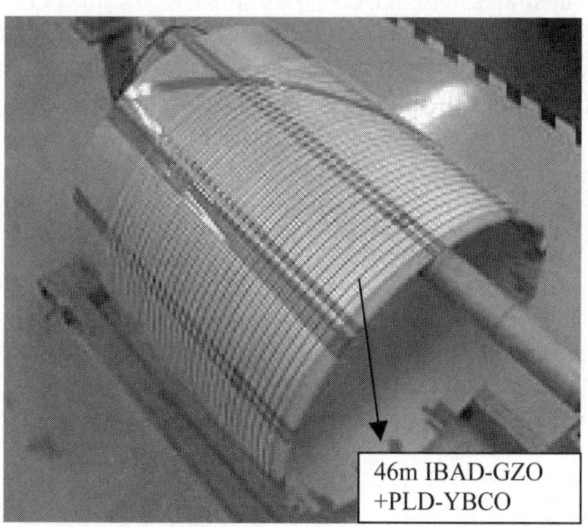

Figure 15 Typical IBAD-GZO YBCO coated conductor 46m long [58].
Figure courtesy of Y. Iijima, Fujikura, Japan.

On the other hand, Fujikura in Japan succeeded in fabricating a long YBCO coated conductor by the IBAD-PLD process as shown in Figure15 [58]. In 2002, they obtained long YBCO coated conductor, 46 m long, with an IBAD-GZO ($Gd_2Zr_2O_7$) layer deposited on a 0.1mm thick and 10mm wide Hastelloy tape. A thin Y_2O_3 layer was deposited by PLD between GZO and YBCO films in order to compensate their lattice mismatch and prevent chemical reaction. Furthermore, the IBAD tape itself has a good texturing degree of $\Delta\phi$ around 10 degrees. These recent results demonstrated that the IBAD-PLD process is the most promising in terms of the conductor length with high critical current.

2.2.3 PLD-CeO$_2$ BUFFER –SELF-EPITAXY-

The above mentioned IBAD-PLD method is the most advanced one for the industrial scale up. However, the most serious problem is its low production rate especially in the IBAD process. Typically IBAD process is so slow as 0.5 to 1m/hr to obtain a degree of texturing enough to use for the YBCO coated conductor with a high J_c. Very recently, a new method was developed as a solution for this problem [43-45]; SRL-Nagoya Laboratory developed PLD-CeO$_2$ on an IBAD seed layer, which shortens the total deposition time for the buffer oxide layer. Figure 16 is the schematic drawings of the PLD-CeO$_2$ deposition in the study compared to the conventional IBAD method. While being deposited on an IBAD layer, the PLD-CeO$_2$ layer grew epitaxially and developed its degree of texturing without any assist Ar ion beam like in the IBAD method. Hence, we call it "Self-Epitaxy". The "Self-Epitaxy" has a remarkable effect on the texturing.

For the deposition of a PLD-CeO$_2$ layer, IBAD tapes with various degrees of texturing from 10 degrees to 27 degrees, which are $\Delta\phi$ values measured by the XRD pole figure analysis, were used as substrates. The IBAD tapes consisted of a GZO (Gd$_2$Zr$_2$O$_7$) layer and a Hastelloy tape. For the CeO$_2$ buffer layer deposition, we used a KrF eximar laser. When the CeO$_2$ was deposited on the IBAD tape with the degree of an in-plane texturing, $\Delta\phi$, around 11 to 13 degrees in the Gd$_2$Zr$_2$O$_7$ layer, the degree of an in-plane texturing of the CeO$_2$ layer was improved to be the value of 2 to 3 degrees after the 600nm thick CeO$_2$ layer deposition [43]. These values of the CeO$_2$ layers are in the texturing level near a single crystalline substrate and have not been easily obtained by the conventional coated conductor processes.

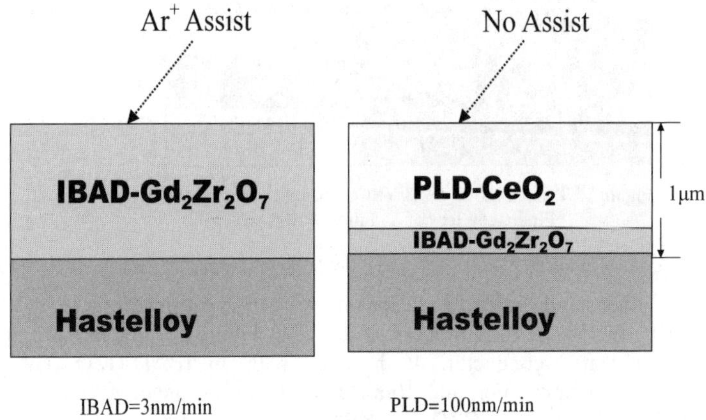

Figure 16   Normal IBAD and PLD-CeO$_2$ Self-Epitaxy Process.

However, the most striking feature is the processing time as shown in Figure17. The deposition time versus the degree of in-plane texturing is plotted for both cases of with and without a PLD-CeO$_2$ deposition. The deposition rate of PLD is as fast as 100nm/min at the laser repetition of 50Hz and 180mJ in this study, corresponding to the 30 times higher rate than that in the IBAD method. Then, when we use an IBAD substrate with a $\Delta\phi$ of 25.6 degrees and 400nm thick in the $Gd_2Zr_2O_7$ layer (open circle in Figure17), the time to obtain the degree of texturing of 10 degrees was only 1min and that for 5.6 degrees was 6min for the PLD-CeO_2 deposition. This is a remarkable effect on the processing time, compared to the conventional process of IBAD. In total, for deposition of the well aligned buffer layer, while it takes 240min by the IBAD process only, the deposition time is reduced to be 75min by this new process. The PLD-CeO_2 process, "Self-Epitaxy", drastically decreases the total processing time for a buffer layer fabrication in a coated conductor.

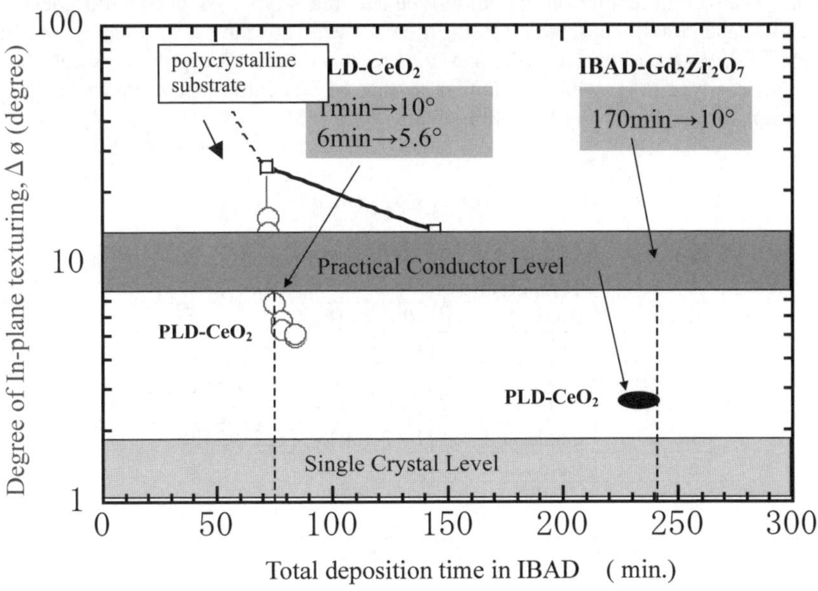

Figure 17   Comparison of the degree of bi-axially texturing, $\Delta\phi$, between normal IBAD and PLD-CeO_2 [44].

The origin of the "Self-Epitaxy" is now under investigation. According to TEM observation [62], we observed an abnormal grain growth in the initial CeO_2 layer of 100nm in thickness. The IBAD layer may interact with the CeO_2 layer and induces this phenomenon. Further studies on the mechanism to be responsible in this PLD-CeO_2 and IBAD-GZO system are in progress. Furthermore, the PLD-CeO_2 film did not show any cracks even in a thickness above 1μm. This is very attractive also from the point of simple buffer layer architecture. In Figure 11 (4), a thick CeO_2 layer on a RABiTS tape is sometimes known to result in cracking. Then, the YSZ layer was used instead. However, in the architecture using PLD-CeO_2, simple film architecture is realized, compared to the RA BiTS-CeO_2 case.

2.2.4 TEXTURED METALLIC SUBSTRATED BASED YBCO COATED CONDUCTOR (RABiTS™)

BASICS OF THE CONDUCTOR FABRICATION

Texturing microstructure of metals has been well-known in Metallurgy and utilized widely in the metal industry. Rolling and subsequent heat-treatment process result in a crystalline texture of metals. Table 6 is the summary of the textured structure for various metals obtained by this method. FCC metals such as copper and nickel show the rolling texture of "the pure-metal-type" by cold rolling. Figure 18 shows a schematic diagram of rolling with the rolling directions. This rolling texture changes into a cube texture due to re-crystallization by heat treatment [63]. The rolling plane of the cube texture is {100} and the rolling direction is <001>. The {100} plane of crystallites aligns in the rolling plane. The rolling texture of "the alloy-type (the brass-type)" is obtained in metals such as silver and brass. Those re-crystallization textures are the (110) plane in the rolling plane [63]. Warm rolling and the subsequent re-crystallization process forms the cube texture in silver [64], which is similar to that of austenitic steels, in which the cube texture is formed by raising the rolling temperature [65].

Table 6   Re-crystallization textures of pure metals and alloys.

Material	Rolling Temperature	Re-crystallization texture
Ni, Au	Cold	{100}<001>
Ag	Cold	{100}<001>
	130ºC	{100} <112>, {311} <112>
SUS 304L	800ºC	{100}<001>
Cu	Cold	{100}<001>, {122}<212>
Fe-Ni	Cold	{100}<001>

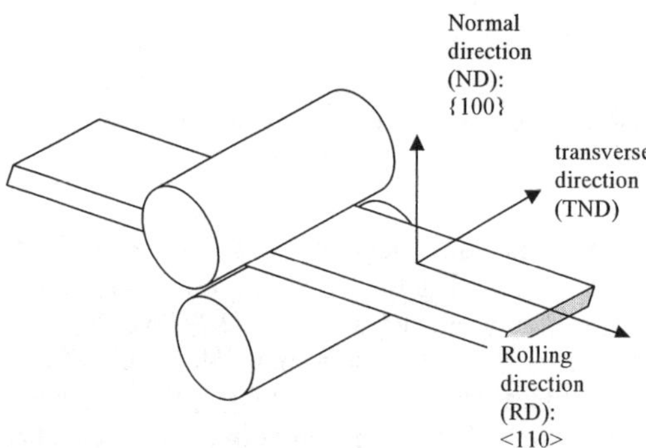

Figure18   Schematic drawing of the rolling and the directions of the texturing.

This re-crystallization texturing can be applied to YBCO coated conductor process. FCC metals such as Ni and Ag meet the requirements of bi-axially textured substrates for YBCO-coated conductors since their lattice constants are close to the a-axis length of the YBCO crystal. Furthermore, the important advantage is the simple and easy process: bi-axially texturing can be obtained only by heat-treatment after cold rolling. It is not so difficult to produce a long substrate by this method as that made by a vacuum process such as PLD and/or IBAD. This means that the cost of the wire could be reduced significantly. The rolling and re-crystallization process is a time-efficient and a cost-effective way to realize a long YBCO coated conductor.

RABiTS^{TM} -Textured Ni Substrate-

The rolling and re-crystallization process was applied to Ni metals as a substrate for YBCO coated conductor by Goyal et al. [66]. They named the resulting textured Ni substrate "Rolling assisted bi-axially textured substrates" (RABiTS^{TM}) in 1994. Ni metal bar was rolled to a tape form with a total reduction ratio more than 90% and then re-crystallized at 1000°C for 4h in a vacuum of 10^{-6} Torr. The resulted Ni tape showed a sharp {100}<100> cube texture. The degree of bi-axially texturing, $\Delta\varnothing$, was around 7 degrees and it was stable up to the melting temperature of Ni.

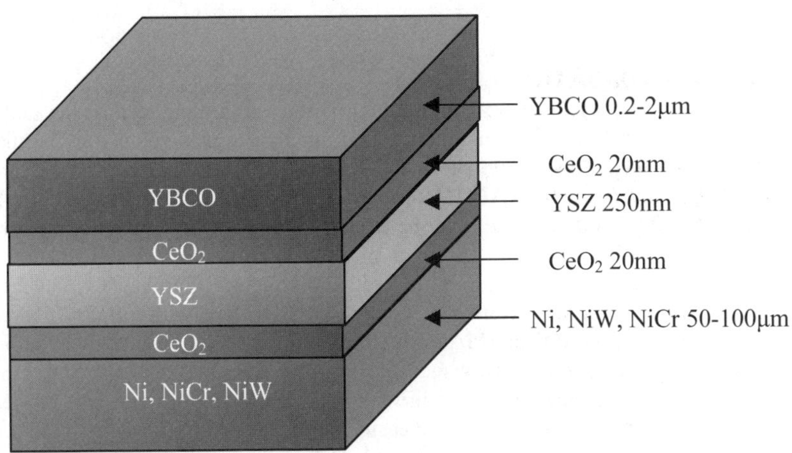

Figure 19 Typical RABiTS based YBCO coated conductor architecture [67].

Direct deposition of YBCO onto RABiTS tapes results in the chemical reaction and thus degrades superconducting properties. Then, a buffer layer between YBCO layer and RABiTS^{TM} substrate is required. Typical buffer layers are YSZ and CeO$_2$. The architecture developed to a long conductor fabrication is shown in Figure19 [67]. The metallic substrate is Ni or Ni alloys such as NiCr and NiW with 50 to 100 µm in thickness. The Ni alloys are advantageous in that they are essentially non-magnetic and have yield strengths with a factor 5 higher than that of pure Ni. The first buffer is CeO$_2$

which has an advantage to form a textured structure at a low oxygen pressure. Otherwise, a Ni based substrate will be oxidized and then can not be used for a coated conductor substrate because the degree of texturing and surface smoothness will be degraded. The second buffer is YSZ. The buffer layer should be thick enough to suppress the diffusion for the reaction between YBCO and the substrate. They also reported that a thick CeO_2 layer caused cracks. Then, YSZ is used to form a thick buffer layer. Furthermore, the top surface is again a CeO_2 layer since YSZ reacts with YBCO and forms $BaZrO_3$, which also deteriorates superconducting properties. The typical thicknesses of the first CeO_2, YSZ and the second CeO_2 are 20nm, 250nm and 20nm, respectively. The degree of bi-axially texturing, $\Delta\phi$, of typical samples obtained are 5-7 degrees for substrate to the final layer of YBCO.

Recently ORNL fabricated a 2km long RABiTS^{TM} (Ni) tape 1 cm wide and 50 µm thick [67]. They made several reel-to-reel systems to facilitate continuous fabrication of RABiTS^{TM} and YBCO films using the ex-situ BaF_2 process on the non-magnetic, Ni-13 at.%Cr alloy and obtained the J_c value of 1.4 MA/cm^2 (77 K, 0 T) [67]. Furthermore, they clarified that sulfur on the Ni metal surface controls the texturing by a c(2x2) two dimensional superstructure due to the segregation of sulfur contained in the Ni metals [68]. They apply this method to improve or keep the orientation degree constant in a continuous fabrication system and thus obtained a high J_c value of 1MA/cm^2 for a NiW alloy substrate [69].

SOE –SURFACE OXIDATION EPITAXY-

To simplify the buffer layer formation which is the following step of the textured metal substrate, the surface oxidation epitaxy, SOE, method was developed by Furukawa and SRL [49, 70, 71]. This is similar to the RABiTS method, but being different in forming an NiO layer on the RABiTS surface. The Ni rod is heat-treated at a high temperature of 800 to1100 °C as done in the RABiTS process. Subsequently, the sample is heat-treated at 1000 to 1300 °C in air or in an oxygen controlled atmosphere. The resulting NiO on the tape surface forms the epitaxial structure. Furukawa is developing this method and improves the texturing by coating the SOE tape with a $BaZrO_3$ layer by PLD. They obtained J_c more than 1MA/cm^2, which improved the SOE coated conductor characteristics [49]. Furthermore, they succeeded in preparing a 100m long SOE tape as shown in Figure 20. The tape with 5mm in width and 80µm in thickness was fabricated with the speed of 10m/hr. SOE process can easily forms the first buffer layer of NiO by a simple heat-treatment, and can protect the Ni substrate during the following treatments in an oxidized atmosphere for the deposition chamber for YBCO.

TEXTURED Ag SUBSTRATE

Other than Ni and Ni alloy, silver is also used as a substrate of the coated conductor. Ag shows an alloy-type rolling texture and {110}<112> is the most common preferred orientation for the re-crystallization texture. However, according to Yoshino et al., a good in-plane orientation of YBCO can be also obtained directly on the silver (110) plane [72]. In 1993, they first reported an example of the formation of a YBCO film by PLD on rolled silver tape with a (110) plane [72] which was reported before developments of

RABiTS. The addition of copper to silver and the use of a copper-rich target were effective for obtaining high-J_c YBCO films. This protects diffusion dissolution reactivity

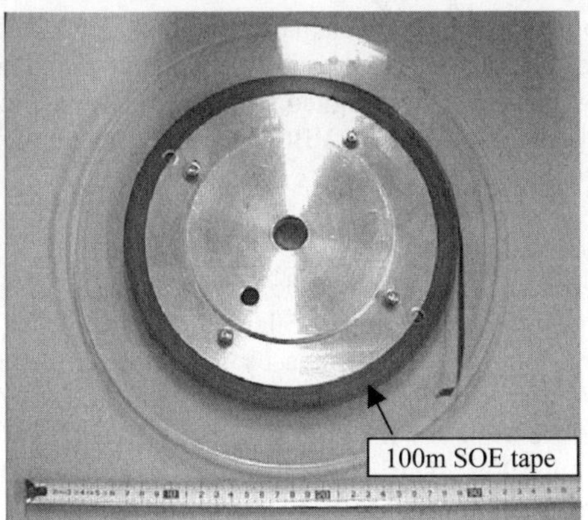

Figure 20   100m long SOE tape with the production speed of 10m/hr [49]. Figure courtesy of T. Watanabe, Furukawa and ISTEC-SRL, Japan.

of Cu at the interface from YBCO to the metal substrate [73]. Mechanically strong clad tapes such as AgCu alloy-clad AgNi alloy are also developed. Now they fabricated a 100m long bi-axially textured clad tape and a 10m long YBCO film by the PLD method. The 10m long sample showed a uniform J_c distribution, over $10^5 A/cm^2$ [74]. Similarly, Doi et al.used Ag substrate, but they applied warm-rolling instead of cold-rolling. They obtained the (100)<001> cube texture, referred to as the cube-textured silver tape (CUTE tapes), by warm-rolling and subsequent annealing at700 to 850ºC [64]. They fabricated an 80 m long Ag tape with cube texture [75]. YBCO coated conductor by CVD process is applied to the textured Ag substrate. The YBCO film showed c-axis orientation, but the crystals inclined by 45 degrees were co-existed. Then, further improvements are required for achieving higher J_c.

Although the J_c for the Ag substrate is not as high as that for the Ni one, which may be partially due to the surface roughness and the misfit between silver and YBCO, The advantage is a simple architecture without buffer oxide layer. Furthermore, this means that the conductor based on Ag textured substrates does not need any stabilizer of an Ag layer on YBCO and then skip the Ag coating process. This is a large advantage for industrialization of YBCO coated conductor.

2.2.5 NOVEL PROCESS BY F-CONTATINED REACTION: TFA-MOD AND EX-SITU BaF₂ PROCESS

As we mentioned before, mainly the PLD method has been applied to the YBCO layer deposition process so far. The achievements up to now are promising from the conductor length and I_c: Fujikura has fabricated 46m long conductor with I_c of 74A [58]. However, some disadvantages still remain: slow production rate, complicated machine system using vacuum chambers and a low efficiency of the YBCO conversion from a target to a film conductor. Essentially, the PLD method or processes using a vacuum chamber may be an expensive way.

Hence, recently two approaches have been paid attention; TFA-MOD and ex-situ BaF_2 processes, which do not need entirely or partially a vacuum chamber and moreover bring about high J_c values owing to its well-aligned structure. These methods may open a cost-effective way to the industrialization of YBCO coated conductor. Here, the two promising processes are introduced.

REACTION OF F-CONTAINED PROCESS

Both processes are essentially the same for the synthesis stage of the Y123 phase crystals. The difference is only the precursors: a calcined thin film from liquid solution coating for TFA-MOD and a thin solid film made by EB (electron beam) evaporation. Both precursors contain Y, Ba, Cu and F elements to synthesize YBCO and the subsequent heat-treatment is the same for the both cases: the precursor is heat-treated in humid oxygen, for example, at 800ºC. During the heat-treatment, the following reaction controls crucially the textured Y123 formation.

$$1/2\ Y_2Cu_2O_5 + 1/2Cu_2O \text{ (in EB process) or}$$
$$Y\text{-}Ba\text{-}Cu\text{-}O\text{-}F \text{ matrix} + CuO \text{ (in TFA-MOD process)}$$
$$+ BaF_2 \text{ (or } (Y,Ba)(O, F)_2) + H_2O \rightarrow 2HF\uparrow \tag{2}$$

This results in a so-called "conversion kinetics" control YBCO synthesis.

Although the reaction was first discussed by McIntyre and Cima et al. [76], who obtained also high J_c values in TFA-MOD samples at the same time, the detailed and extensive studies for the reaction were carried out on the sample by EB evaporation process, ex-situ BaF_2 process, by Suenaga and Solovyov et al. [77-80]. They revealed new aspects of the F-contained process, especially concerning to

1) formation kinetics and growth rate limiting process
2) nucleation and growth mechanism

According to their results, the reaction in EB samples is described as follows [79]:

$$Y + BaF_2 + Cu + O_2 + H_2O$$
$$\rightarrow Cu_2O + (Y, Ba)(O, F)_2 + O_2 + H_2O \tag{3}$$
$$\rightarrow YBa_2Cu_3O_{6.1} + HF\uparrow \tag{4}$$

The basic growth-rate-limiting process was determined to be the removal rate of the reaction product HF from the surface of the film into the reaction chamber atmosphere [79, 80]. The formation of the nuclei, c-axis-oriented YBCO, is intimately related to the growth rates. By using the resistivity measurement of ex-situ BaF_2 samples, they draw the conclusions for the growth rate, G, of YBOC in Eq. (3):

$$G=A[p(H_2O)]^{1/2}/p_t \qquad (5)$$

where p_t is total pressure; $p(H_2O)$, water vapor pressure; A, constant. The reaction, Eq. (4), is in a state of dynamic and local equilibrium at the growth front. Then, the partial pressure of HF, $p_i(HF)$, at the YBCO–precursor interface is related to the water-vapor partial pressure, $p(H_2O)$, by the equilibrium condition, $p_i(HF)/[p(H_2O)]^{1/2}=K$, the equilibrium constant of the reaction Eq.(4). Y-123 phase formation is determined by the reaction in Eq. (4) and described by each vapor pressure. At the same time, the equilibrium was established in the local film area in the non-reacted precursor region. Then the growth rate is limited by HF removal from the film surface according to the equilibrium constant, K.

The experimental results obtained by a resistivity measurement for the growth rate, G, fitted well with Eq. (5) when G is plotted as a function of the water-vapor pressure, $p(H_2O)$, for different total pressure values. The rate of HF removal is limited by gaseous diffusion in the processing atmosphere, but not by solid-state diffusion through the precursor film.

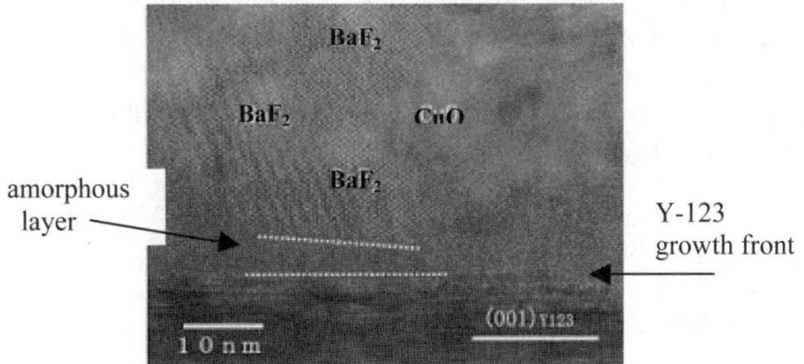

Figure 21   TEM photograph showing amorphous layer in TFA-MOD sample [82]. Figure courtesy of J. Shibata, SRL, Japan.

They also suggested a new nucleation mechanism from the detailed TEM study [80]: A precursor film consisted of Y, Cu, and BaF$_2$ quickly turns to a mixture of Cu$_2$O and a (Y, Ba)-oxy-fluoride, and then YBCO is grown on the substrate from this. According to their conclusion, first the YBCO nuclei formed from the oxy-fluoride whose (1 1 1) planes aligned with the (0 0 1) surface of the STO substrate. The oxy-fluoride is decomposed by reacting with H$_2$O. This cubic phase can order into YBCO nuclei. After nucleation, YBCO epitaxially precipitates onto the existing c-axis-oriented YBCO layer through a thin liquid layer (~7 nm) containing Y, Ba, Cu and O. The Y–Ba-oxy-fluoride decomposes through the reaction with H$_2$O, releasing F and H. This liquid layer only forms on an existing c-axis YBCO layer after a thin layer of YBCO covers the substrate surface.

This mechanism is very attractive to explain an epitaxial growth in F-contained process including TFA-MOD [81].However, according to a recent TEM study for a TFA-MOD sample [82], the observed amorphous layer in Figure 21 did not come from the melt at the firing temperature from the composition of this layer and the phase diagram of

Y-Ba-Cu-O with incongruent melting. The reaction might be different each other in ex-situ BaF_2 and TFA-MOD processes. A highly oriented structure and high J_c values are peculiar to the F-contained process and the nucleation stage affects the subsequent growth of c-axis or a-axis orientation, which determines J_c. Then, further basic studies are still required together with the scale up for industrialization.

TFA-MOD PROCESS

TFA-MOD process is a kind of the MOD method. MOD is a promising and convenient method to synthesize YBCO film. The features are as follows: (1) good homogeneity; (2) precise composition control of the final products; (4) low cost without any vacuum chambers; and (5) a potential for scaling up.

Conventionally, the MOD process used octyl salts and naphthenic acid salts as a raw material. This salt is dissolved into a solvent such as methyl alcohol. This resultant coating solution is coated on a substrate and heat-treated. However, the J_c was not as high as that by the PLD method. Then, for the time being, this simple MOD process did not attract further attention. Afterwards, McIntyre et al. reported high J_c by using metal trifluoroacetate (TFA) as a starting material which does not form barium carbonate generated during Y123 synthesis [76, 81]. $BaCO_3$ is not entirely decomposed even at the high temperature and then is considered to degrade superconducting properties. In their work, they obtained J_c values over $1MA/cm^2$ (77 K) on $LaAlO_3$ and $SrTiO_3$ substrates. Furthermore, the basic studies for TFA-MOD were also carried out [81]. The following reaction is proposed by them;

i) Calcination (room temperature to 400ºC)

$$CF_3COO\text{-}M(Y, Ba, Cu) + H_2O+O_2 \rightarrow Y\text{-}Ba\text{-}Cu\text{-}O\text{-}F + CuO \quad \text{(precursor)} \qquad (5)$$

ii) Firing (400ºC to 800ºC)

$$\rightarrow 1/2\ Y_2Cu_2O_5 + BaF_2 + 2CuO \text{ (precursor)} + 2H_2O \rightarrow YBa_2Cu_3O_{6.1} + 4HF\uparrow \qquad (6)$$

The heat-treatment consists of two steps: calcination up to 400ºC and firing at 800ºC. During calcination, trifluoroacetate of metals decomposes and forms a mixture of Y, Ba, Cu, O and F. At the same time, fine CuO particles are also formed. This precursor is fired at 800ºC in a humid oxygen atmosphere. In the firing process at 800ºC, Eq. (2) plays also an important role for the textured Y123 formation as described in the former section. H_2O reacts with F in the precursor constituent, especially BaF_2, and the reacted gas of HF is removed. During this reaction, textured YBCO develops, depending on the HF concentration.

For studying the reaction in TFA-MOD, various methods were used such as TEM and resistivity measurement during heat-treatment. Shibata et al. [82] observed the above three phases in Eq. (6) in a quenched TFA-MOD sample. Resistivity measurements also revealed the details of the reaction. Yamada et al. [83-86] measured the resistivity of TFA-MOD samples during firing as shown in Figure 22. From this measurement, we could understand the followings:

1. calcined film consisted of CuO and Y-Ba-Cu-O-F amorphous is further oxidized up to 600ºC because the film resistivity was increased from 400ºC to 600ºC (②). This

may result in the large CuO or $Y_2Cu_2O_5$ observed by McIntyre et al. [81] and by Shibata et al. [82] as shown in Figure 23.

2. the region of slowly decreasing resistivity between 600 and 670ºC (③) may be due to formation of the intermediate phases such as the amorphous YBCO, non-aligned Y123 or (Y, Ba)(O, F) [79, 80], which are discussed in the previous section.

3. textured Y123 formation, which reduces the resistivity, mostly begins at a low temperature around 670ºC (the end of the part ③ in Figure22).

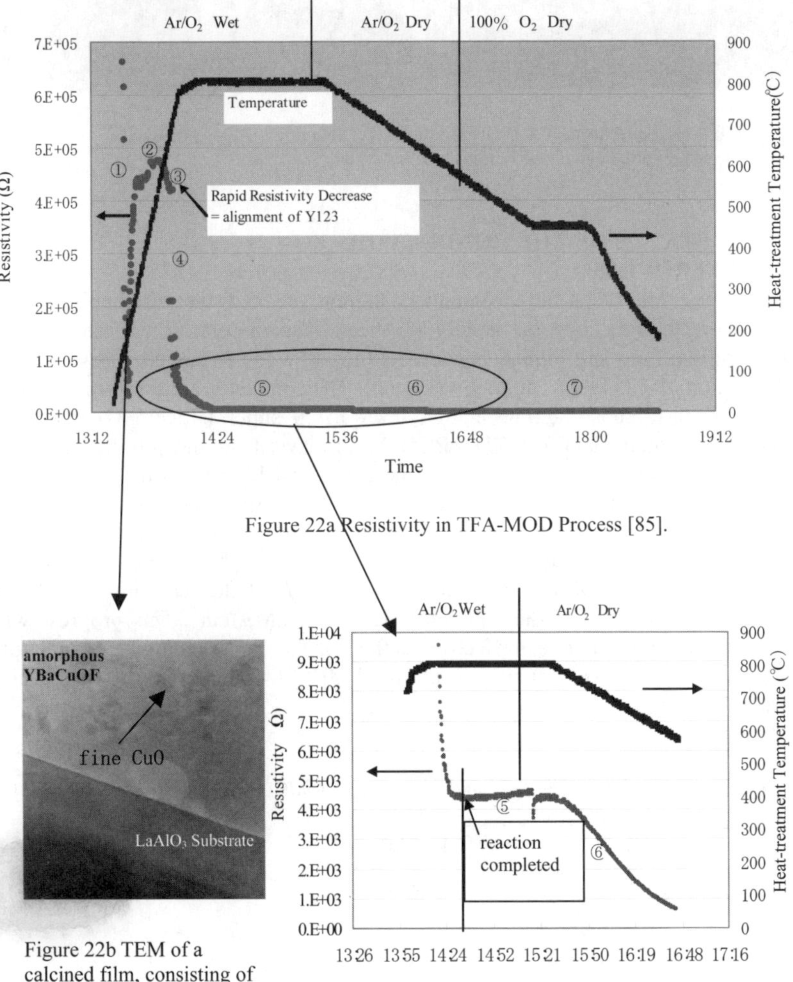

Figure 22a Resistivity in TFA-MOD Process [85].

Figure 22b TEM of a calcined film, consisting of CuO and amorphous YBaCuOF matrix.

Figure 22c Magnification of the part at 800℃ in Fig.1a.

The study using resistivity is very effective, especially being together with XRD and TEM. We can speculate the time dependent event during heat-treatment. The study of the reaction should be more meaningful for the time-dependent growth, the steady state growth and the effect of reactant partial pressure and total pressure on the growth rate are considered.

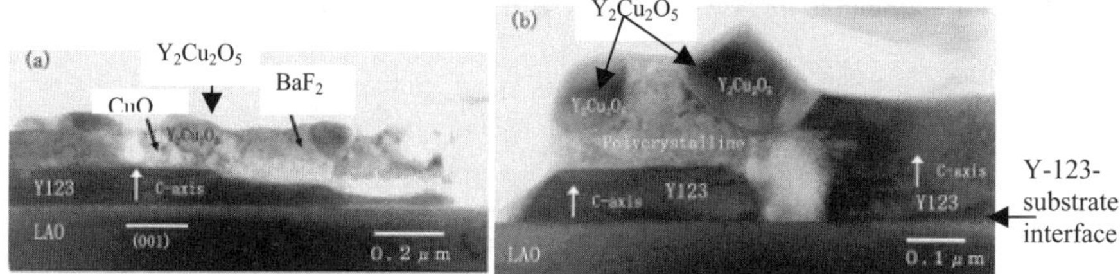

Figure 23 TEM for a quenched TFA-MOD sample [82]. Three intermediate phases, BaF_2, CuO and $Y_2Cu_2O_5$ are seen. Figure courtesy of J. Shibata, SRL, Japan.

PRESENT STATUS OF THE CONDUCTOR
TFA-MOD PROCESS

At the same time when the basic understandings proceed, the wire application of the TFA-MOD method has become rapidly advanced. Rupich et al. very recently succeeded in obtaining 1m long and high J_c conductors [46, 48]. The I_c ranged 104A to 120A and the fluctuation of 10 pieces 1m long was small. The conductor architecture is as follows: Textured Ni-alloy substrate/transient Ni layer for S super lattice/Y_2O_3 epitaxial/YSZ epitaxial/CeO_2 epitaxial/TFA-MOD YBCO/Ag cap layer from the bottom. Fabrication of the YBCO coated conductor was carried out using scalable reel-to-reel system from the substrate rolling through the Ag layer deposition. The degree of bi-axially texturing, $\Delta\o$, was 7.0 degrees or less for the NiW substrate to the final YBCO layer. The thickness of YBCO is 1μm. This wire has the J_c over $1MA/cm^2$. Then, the I_c is above 100A for 1cm wide and 1m long samples. The highest end-to-end I_c of the ten wires was 118A/cm-width. In the short sample, they obtained high J_c of $2MA/cm^2$. This progress with the low-cost substrate and the cost effective method of the non-vacuum TFA-MOD process will bring about a fast realization of industrial YBCO coated conductors replacing the present Bi-2223 wires.

Table 7 in-plane texturing degree of each layer in TFA-MOD YBCO coated conductor [48].

Substrate or Layer	Degree of in-plane texturing, $\Delta\o$ (deg.)
Ni-W	7.0
YSZ	6.5
CeO2-Y2O3	6.7
YBCO	6.4

In the above wire, the oxide buffers were deposited by a vacuum process such as EB evaporation or sputtering. For further cost reduction, MOD process is also applied to the buffers. ORNL replaced the first buffer of CeO_2 to MOD-Gd_2O_3 or $La_2Zr_2O_7$ and obtained high J_c of $1.2MA/cm^2$ and$1.9MA/cm^2$, respectively, for a short sample, $0.6MA/cm^2$ for an 80cm long sample but using ex-situ BaF_2 process [46, 69].

In Japan, TFA-MOD is also actively studied. ISTEC-SRL collaborates with Fujikura Ltd. and develops TFA-MOD YBCO on a CeO_2/IBAD-YSZ or GZO layer buffered Hastelloy substrate. The J_c value of this conductor was 1.7 to $2.5MA/cm^2$ [87, 88]. Recently, a thicker film has been obtained by optimizing the condition of the heat treatment in a multi coating process. A triple coated film of 1 µm thickness with a high J_c value of 1.6 MA/cm^2 showed a high I_c* value (I_c corresponding to a width of 1 cm) of 153 A/cm-width [88]. Recently ISTEC-SRL also obtained a higher I_c above 200A for the IBAD/CeO_2 substrate by improving the multiple coating method [89]. Showa Electric Wire & Cable Co., Ltd., has also been developing the TFA-MOD process for fabrication of a long tape by "Beed Coating" method [90]. They and SRL fabricated a 10 cm long Y-Ba-Cu-O layer on a CeO_2 capped IBAD- $Gd_2Zr_2O_7$ buffer layer on Hastelloy tape.

Thick film, hopefully more than 1µm, is a significant target to increase I_c, which limits the application. Meanwhile, TFA-MOD thick film has been well known of its degradation: the porosity increases and cracks occur. It is difficult to stably fabricate a homogeneous thick film even 1µm in thickness with high J_c of $1MA/cm^2$. To make a thick film, viscosity enhancement in coating solution or multiple coating is considered to be effective [88]. Moreover, an optimal choice of the partial pressure of H_2O and the heat-treatment time and the temperature are important factors to keep a high J_c value of a thick film because they determine the growth rate of Y123 as described before.

EX-SITU BaF$_2$ PROCESS

In this method, the precursor is fabricated by EB evaporation. The precursor film is the mixture of target materials of Y, Cu and BaF_2. The subsequent firing is the same as that in TFA-MOD process: heat-treated at 800ºC in humid oxygen. This method can make a thick and dense film more than 5 µm at a high speed [91]. As mentioned above, the reaction in this ex-situ method is similar to that in the TFA-MOD method: Y+ BaF_2+Cu $\rightarrow CuO$+(Y,Ba)(O, F)$_2$+O_2+$H_2O \rightarrow YBa_2Cu_3O_{6.1}$+HF↑. The Y123 formation is considered to be controlled by conversion kinetics of H_2O and HF.

Solovyov and Suenaga et al. obtained a high J_c value of $0.6MA/cm^2$ for a 5 µm YBCO thick film on an STO single crystalline substrate. ORNL group obtained also a high J_c value of 2 to $3MA/cm^2$ on RABiTS substrates with the buffer of CeO_2/YSZ/CeO_2 while the film thickness is 0.3 µm [46, 67, 92].

For the long tape fabrication, Solovyov proposed that Y123 formation was governed by the removal of HF from the film surface during firing and estimated the reaction rate from the equilibrium condition of HF and H_2O. He calculated the gas flow rate to be required and concluded that the practical process with a production speed enough to be applied to a realistic machine should be carried out under a reduced atmosphere, which can promote the reaction. ORNL applied a low pressure process in a batch furnace [92,

93] and also used a transverse gas flow in a reel-to-reel furnace [47]. This transverse carrier gas flow reduces a downstream HF gas stack above the tape releasing HF gas, which decreases the reaction rate. Using these techniques, they fabricated an 80cm long tape with the J_c value of 0.6MA/cm^2 [93].While the TFA-MOD process or the ex-situ process is principally simple, we need to solve a kind of a chemical engineering problem to realize a large scale and an industry-oriented reaction furnace.

2.2.6 FUTURE TECHNICAL TASK FOR THE APPLICATION

The technical task is still much remained unsolved for YBCO coated conductors. The first priority is the realization of long conductors which can be applied to an application; it is only 46m only by one company. Very soon many institutes or companies hopefully make long conductors to activate the application. The second one is the promising method from an economical point of view since the customer already learned from the results obtained by Bi based conductor application. One should fabricate YBCO coated conductor with an inexpensive way, compared to the Bi based conductors.

The other tasks peculiar to YBCO coated conductor is;

1. Reduction in substrate thickness or non-superconducting layer; this is directly significant to the application because the non superconducting area reduction increases J_e, which can decrease the size of application devices.

2. J_c improvement; as later shown in the last section, the J_c (77K) values of YBCO maintain high values even in a high magnetic field parallel to the tape surface, compared to the Bi conductors. However, in a perpendicular field, the J_c decreases by one order in 2 to 3 Tesla and by two orders in 4 to 5 Tesla. This effect, coming together with the above large non-SC/SC ratio, loses the attraction of YBCO. We need further improvement of J_c especially in a perpendicular field at 77K of liquid nitrogen temperature.

3. AC loss; we do not know the detail of AC loss in the state-of-the-art YBCO conductor since the quantity of the conductor is far small against researcher's demand. Then, of course, we do not examine the application side requirements. The feasibility study and the evaluation test of a small application will be required soon. However, generally the smaller AC loss is, the more preferred conductor is; then possible technique such as multifilamentarization or small width conductor process should be studied.

4. Coil winding or device fabrication; the present 1cm wide conductor is too wide to be utilized for some applications, from the standard of the conventional superconducting devices. As described in the Bi conductor section, although the best configuration is the round wire, the small width or isotropic configuration of the conductor is preferred. In this case, we need to study the edge-effect of the YBCO thin tape conductor, where the mechanical degradation may occurs during a fabrication process, electromagnetic influence may become large or YBCO film may be deposited with a worse orientation degree in the deposition process. Furthermore, this may be different each other between PLD and TFA-MOD processes.

Besides those, there are many tasks such as i) thick film with a large I_c value which is now intensively being studied at LANL, ii) mechanical properties and its degradation especially during cooling which is not yet studied in the new configuration with hard Hastelloy and soft Ag sheathed composite and iii) the connection; not only superconducting but also normal connection seems to be difficult because the conductor has a YBCO surface at one side, which increases the resistivity through the highly resistive Hastelloy substrates. We need to study these problems, continuing the effort to fabricate a long YBCO conductor.

3. APPLICATIONS

As shown in the former sections, since the Bi-based high temperature superconductor (HTS) is long and high J_c enough to be utilized for some applications. Meanwhile, YBCO coated conductor is still under the developing stage especially in terms of the length. Then, the feasibility study of applications is now based on the Bi-based conductor. However, the tape configuration and the current capacity may not be so different each other. Then, the present application study using Bi-based HTS will be useful in the future also for the application studies using YBCO coated conductor.

So far many application studies have been performed using Bi-based wires: cables, transformer, mortar, current limiter, current leads, high-field magnets, MRI, NMR and industrial-use magnet. Among them, the most actively developing devices, cables in Japan and US, magnet for Si single crystal growth, magnet for magnetic separation and NMR are introduced in this section.

3.1. CABLES

Superconducting cable can flow a larger current than the conventional copper conductor. Especially when using HTS, we can use low-cost liquid nitrogen for cooling instead of expensive liquid He. Then, the HTS cable may be practical from the economical point of view. For this purpose, many feasibility studies are undergoing or have been done in Japan, US and EC as listed in Table 3. All projects used Bi-2223 wires since only the Bi-2223 wire is as long as few hundreds meter to 1km and has J_c at 77K enough high to be applied. However, the J_c of Bi-2223 wire is small in a magnetic field. Then, all cables apply a special winding to reduce a magnetic field. Since the Bi-2223 wire is a flat tape and weak in J_c especially for the field perpendicular to the tape surface, in the cable, the Bi-2223 wire is wound on the former as the "spiral structure". This structure reduces the magnetic field below 1T according to the equation of $B=\mu_0 I/(2\pi r)$. Then, Bi-2223 wire has still a high carrying current capacity in the cable.

Generally the followings are the requirements for the HTS cable conductor.
1. High critical current and current density at 77K with low magnetic field.
2. Mechanical strength tolerable to the stress and strain during cabling of the wire, installation of the cable and cooling cycle.
3. Small AC loss during operation which can compensate the cooling system economically.

322

The following companies meet these demands for the cable project: Sumitomo Electric Industries, Ltd. (SEI) supplied the wires for the Tokyo Electric Power Company-SEI- Central Research Institute of Electric Power Industry (CRIEPI) cable project [94, 95] in Japan; American Superconductor, corp. (AMSC) for Detroit-Edison project with the main constructor of Pirelli Cable company [15, 96]; Intermagnetics General Corporation (IGC) and Nordic Superconductor (NST, which is now merged with AMSC in Oct., 2002) for the South wire project [14] in US and the Denmark project [16], respectively. In the following sections, we introduce two typical projects in Japan and US, which are typical in terms of the dielectric method and cable configuration.

TYPICAL PROJECT IN JAPAN

In a field test site of Central Research Institute of Electric Power Industry (CRIEPI), near Tokyo, at the Yokosuka, a 100m HTS cable project has successfully ended in June, 2002. Tokyo Electric Power Company and Sumitomo Electric Industries, Ltd. had jointly developed technologies for an HTS cable system since the starting of the project in June, 2001 [94, 95]: the technologies included are the cable conductor wound with HTSC wires, thermal insulation pipes, terminations and refrigeration system. The field test of a 100 m HTSC cable system integrating these technologies had been carried out for one year to verify the long term electric and cryogenic properties.

The cable used in the field tests was designed to be 66 kV/1 kA/114 MV. Figures 24 and 25 show the structure of the 3-core HTS cable. As the targeted practical HTS cable is to be installed in an existing duct 150 mm in diameter, the outer diameter of the cable was restricted below135 mm. The HTS conductor in the cable was composed of 4 layers of Bi-2223 wires wound spirally around a former. Polypropylene laminated papers were adopted as a cable insulation for the properties of high insulation strength and low dielectric loss. Bi-2223 wires were also wound around the PPLP electrical insulation to form an electrical and magnetic shield. To reduce heat invasion, a multi-layer insulation was wound between the co-axial stainless corrugated pipes for vacuum. The total number of the HTS wires of the conductor and the electrical shield were 52 and 53 by a cable core, respectively. Their critical currents are estimated to be 2.7 kA DC at 77K.

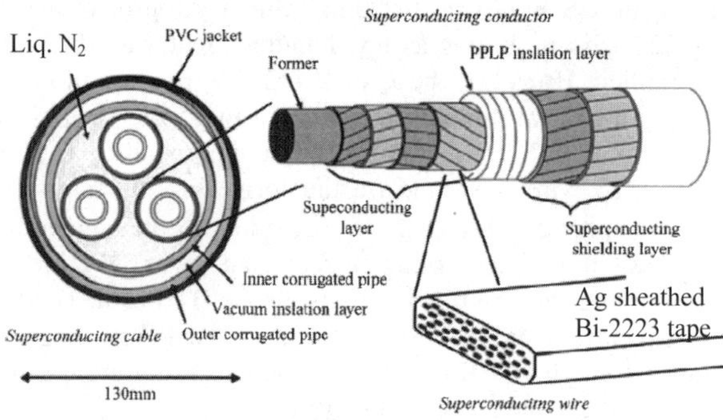

Figure 24   Schematic cross-section of the 3-phae HTS cable (TEPCO/Sumitomo) [94]. Figure courtesy of S. Honjo, Tokyo Electric Power Company, Japan.

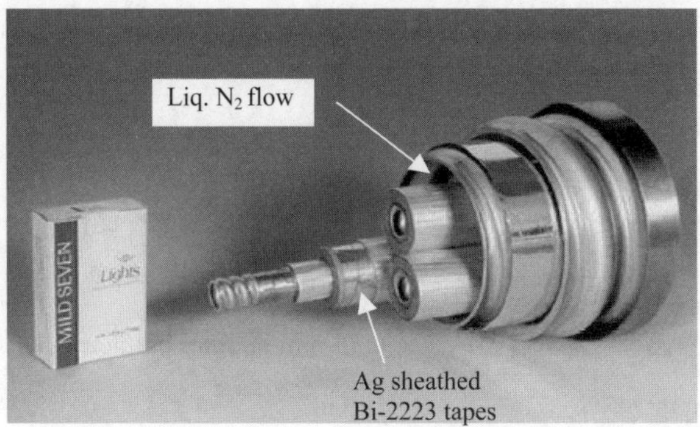

Figure 25   Cross section of three-phase cable model (TEPCO/Sumitomo) [94].Figure courtesy of S. Honjo, Tokyo Electric Power Company, Japan.

The cable was partially installed into150 mm ø duct and formed in a U-shape as shown in Figure 26. Each end has a splitter box and three terminations. The cable and the terminations were cooled using two separate sets of a pressurized and sub-cooled liquid nitrogen cooling system. The specifications are in Table 8. The cable system was composed of a 100 m HTS cable, two splitter boxes, six terminations and cooling systems. The cable was bent at its center and forms a U-shape. Single phase voltage was applied to the terminations through a transformer. The cable and the terminations were cooled by two sets of a pressurized and sub-cooled liquid nitrogen cooling system so as to evaluate the heat loss of the cable and termination, separately.

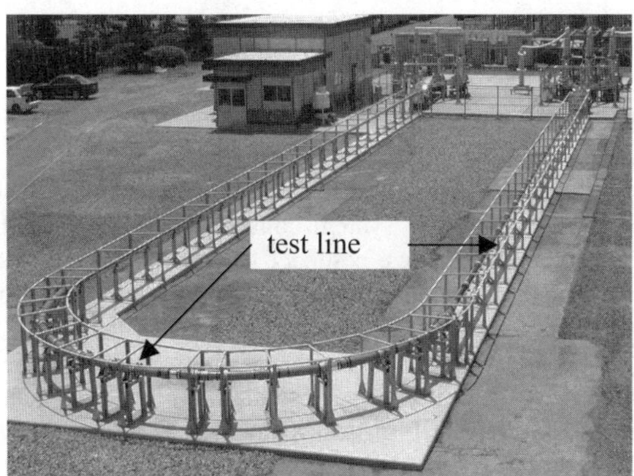

Figure26   100 m, 3 core 66 kV-1 kA HTS cable system [4] (TEPCO/Sumitomo). Figure courtesy of K. Sato, Sumitomo Electric Industries, Ltd., Japan.

Table 8   Specifications of the HTS cable test system [95].

Rated Voltage	66 kVrms
Nominal current	1 kArms
Nominal capacity	114 MVA

HTS cable

Shape	3 core in a cryostat (136 mm in outer diameter)
Length	100 m
Electrical insulation type	Cold electrical insulation
Cooling system	Sub-cooled Liquid N_2 circulation

In this test line with a 100 m 66 kV/1 kA/114 MV, the first cooling test was carried out in 2001. The measured critical current and AC loss prove that the cable did not suffer any serious damage during manufacturing and cooling. The rated voltage, 40 kV, and the nominal current, 1 kA, were successfully loaded to the cable for 350 h. The obtained I_c value of the cable was 2.7kA, which was the same as the expected one from the single Bi-2223 wires. This means that superconducting properties including AC losses were not degraded by the cooling cycle in the field test. The total loading time of the current was more than 2000 hours. Although we still need the requirements for the longer term reliability and an accident, the success of this feasibility test will enhance the possibility of HTS cable application.

TYPICAL PROJECT IN USA

Also in USA, a similar project is under way, the Detroit Edison HTS Cable Project in Detroit, Michigan [96]. The project members are Pirelli cable, AMSC, DOE, EPRI, LANL and Lotepro Corporation. The system consists of 24 kV/100 MVA three phase superconducting warm dielectric cables, compact terminations and refrigeration system. Nine conventional cables in a duct have been substituted with only three HTS cables operating at the same voltage and overall current.

The most important purpose of this project is to survey and solve the problems in the real site. Then, they used the Detroit Edison's Frisbie Station for the construction of HTS cable while the above TEPCO-SEI project constructed a new line for HTS cable construction. This represents the conditions which will be encountered for retrofit/upgrade projects that will be found in many urban areas. The demonstration system will include outdoor and indoor terminations, a field splice, a repair splice, and a fully engineered refrigerator suitable for unmanned operation.

The installation requirements in this line for the cable design are;

1. 24 kV/100 MVA power transfer,
2. Installation into a 102 mm duct,

3. 120 m continuous length with a minimum bending radius of 0.94 m.

The application requires the compact size and high flexibility of a warm dielectric cable. Table 1 resumes the main characteristics of the cable and system for Detroit Edison cable project.

Table 9   Specifications of Detroit Edison Cable Project [2, 96].

Tape	Bi-2223/SUS laminated type
I_c of each tape	110 A
Conductor	Four layer and winding for uniform current distribution and reduced AC losses
Cable dimension	Retrofitting of 4 inch (102 mm) duct
Installation	Retrofit underground
Operation	Real network
Refrigeration	Closed cycle + backup
Length	120 m
Electrical insulation type	Warm dielectric insulation
Rated Voltage /Nominal current	24 kVA/2.5 kArms
Nominal capacity	104 MVA

The Bi-2223 tapes were sandwiched with two thin stainless steel stripes, for improving robustness and reliability [2]. This resulted in the margin for cabling and installation against the conductor pitch angle, bending radius and pulling load. The average DC critical current of these tapes was 118 A. Furthermore, this laminated configuration avoided the so-called "ballooning phenomena". This phenomenon appears frequently in Ag sheathed Bi-based conductors: when the tape is kept in pressurized liquid nitrogen for a long time, liquid nitrogen penetrates and diffuses into the HTS tape through the surface defects in the Ag sheath. Then, a large volume expansion will occur during the warming up. This, of course, induces the considerable degradation in J_c of the Bi-2223 tapes.

The cable conductor was manufactured in a single length, more then 400-m long. The conductor consisted of 4 layers with variable angle configuration designed to uniformly distribute the current among the layers and reduces the electrical AC losses. The thermal barrier cryostat was realized directly over the conductor, composed of two concentric stainless steel corrugated tubes separated by multi-layer reflective sheets and spacers. Design of the corrugated tubes was mechanically optimized in order to have enough flexible characteristics and a good axial tension behavior for the distribution of the pulling load during the cable installation.

The cable has an outer dimension of 89 mm and a weight of about 6 kg/m. The 400-m long cable section was finally cut into the three pieces and delivered to the installation site in Detroit. This cable could carry a large critical current of 6000A. However, reportedly, the problem concerning to the vacuum and thermal insulation has been found recently [97]. Then, Pirelli is now continuing the repairing.

The HTS cable technology is a promising technology for the future power system design when traditional solutions do not meet the requirements economically or socially. For the wide incorporation into the utility network, a large amount of field studies and experience must be carried out.

3.2. HTS MAGNET FOR INDUSTRIAL USE: Si SINGLE CRYSTAL GRWOTH

In this section, we introduce a typical industrial use application, an HTS split magnet for Si single crystal growth. This project was done jointly by Toshiba, SEI and Shin-Etsu Handotai Co., Ltd. [98, 99]. Application of the magnetic field to molten silicon during crystal growth is effective for fabricating single crystal silicon with high purity. A magnetic field prevents natural convection of molten silicon causing dissolution of oxygen from the SiO_2 crucible. This degrades the purity of the grown single crystal. Conventional NbTi superconducting magnets for Si single crystal growth usage are already commercialized and widely used. However, HTS magnet is expected to work efficiently and reliably for magnet quenching because of high T_c and high heat capacity of HTS, resulting in a large temperature margin.

Table 10   Specifications of the conductor for Si single crystal growth magnet [99].

Conductor	3xBi-2223 tapes and stainless steel tape
Width	4.5mm
Thickness	1.1mm

Wire	Ag or Ag-Mn alloy sheathed Bi-2223 by PIT method
Width	3.8mm
Thickness	0.24mm
No. of filaments	61
Ag ratio	3
Ic at 77K and 0T	30 to 70A

The wire used is an Ag sheathed Bi-2223 conductor. The total length of the wire used was approximately 80 km. The specifications are shown in Table10. The magnet required 108 pieces of the wire. The average I_c of the wires was 49 A. Three Bi-2223 wires and one SUS tape were stacked and wound for the magnet. The magnet is composed of two coils, each consisting of 18 pancakes. The following points were taken into consideration for the magnet design.

1. Cross sectional shape of the magnet should be designed to reduce the effect of the anisotropy in I_c or J_c.

2. Mechanical strain and stress to the tape wires should be considered no to damage the Bi-2223 wire during the manufacturing process, the cooling process and the excitation of the magnet generating electromagnetic force.

J_c of the Bi-2223 tape decreases much more for the magnetic field perpendicular to the tape surface than for the parallel field. Then, the designed HTS magnet had a high aspect ratio in the coil cross-section for reduction of the magnetic field perpendicular to the tape surface. Figure 27 shows the whole magnet system with two splits coils conductively cooled to 20K by a GM cryo-cooler. In the figure, a crucible for Si melt will be placed at the center between two splits coil and the operation will be carried out in a magnetic field of 0.3 to 0.5 T. Table 11 is the specifications of the magnet. The size and the stored energy are the largest in the world among the HTS magnets so far made.

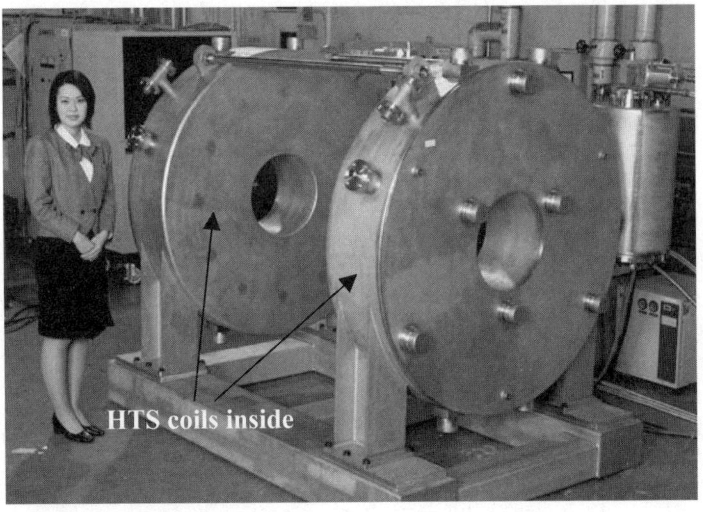

Figure 27   Whole view of Si single crystal growth HTS magnet system [98].
Figure courtesy of M. Ono et al., Toshiba Corp., Japan.

328

Table 11   Specifications of the Si crystal growth magnet [98].

Magnet system	Split type
Outer diameter	1200mm
Inner diameter	600mm
No. of pancake coils	36
Operating current	208A
Operating temperature	Below 20K
Overall current density	33A/mm2
Stored energy	1MJ
Maximum magnetic field	1.8T
Excitation speed	3.3A/second
Total length of Bi-2223 wire	80km

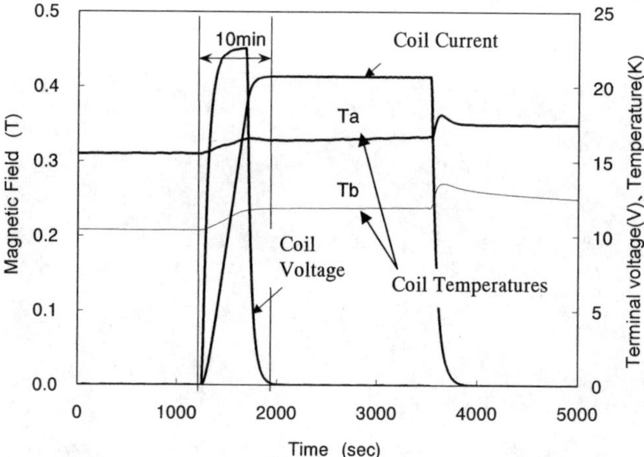

Figure 28   Excitation test result of Si single crystal growth magnet [98]. Operating current of the magnet, voltage during excitation and two split coils' temperature are shown.

Figure 28 is the excitation results of this magnet. The excitation time to the maximum operating current of 215A was only 10min and the temperatures of the magnets were 11K and 16K for two coils. During the excitation, the increase in temperature was only1K. The temperature of the magnet was stable during the steady state operation while generating the constant magnetic field of 0.4T. The stored energy is 1.1MJ. This is the largest in the world. The magnet could be also excited with a fast operation of 1min to the rating current. Furthermore, the magnet worked for 8hours stably. This experiment demonstrated the following merits of HTS systems in this kind of industrial use systems.

1. Fast and stable excitation of the magnet; the specific heat of the conductor is increased by 100 times compared to the value at 4.2K and HTS has high T_c. Then, quenching of the coil due to the wire motion hardly occurs even though the magnet is exited fast.

2. The transient from the superconducting state to the normal state is so slow that quenching could be avoided.

These merits realize the safe and reliable superconducting magnet systems.

3.3 MAGNETIC SEPARATION MAGNET

The magnetic separation system was fabricated by National Institute of Metals and Science, NIMS, Japan, using Bi-2223 wires [100]. The whole view is shown in Figure 29 and the specifications are shown in Table12.

Table 12   Specifications of HTS magnet for magnetic separation using Bi-2223 wire [100].

Coil inner diameter	240mm
Coil outer diameter	306mm
Coil height	352mm
Inductance	1.6H
Maximum exciting rate	1.7T/min
Maximum field	1.82T at 20K
Maximum Temperature during operation	38K at 1.7T
Operating current	180A at 1.7T
Conductor	Bi-2223

The magnet cooled by a GM cryo-cooler, generates 1.7 T at the center and more than 1 T in the 200mm room temperature bore. The magnet could be excited up to the maximum magnetic field in 1 min. During 20 times cyclic excitations to the maximum magnetic field with the cycle length of nine minutes, the magnet temperature stabilized at 35 and 38 K.

Using the Bi-2223 magnet system with a 200 mm room temperature bore, the magnetic separation test was done for separation of particles from slurries containing fine α-hematite paramagnetic particles [100]. To realize efficient magnetic separation, rapid energizing and de-energizing are important. The separation test showed that almost all hematite particles were separated from the slurry. During the cyclic excitation test, the Bi-2223 magnet was thermally stable. Therefore, the HTS magnet system is suitable for

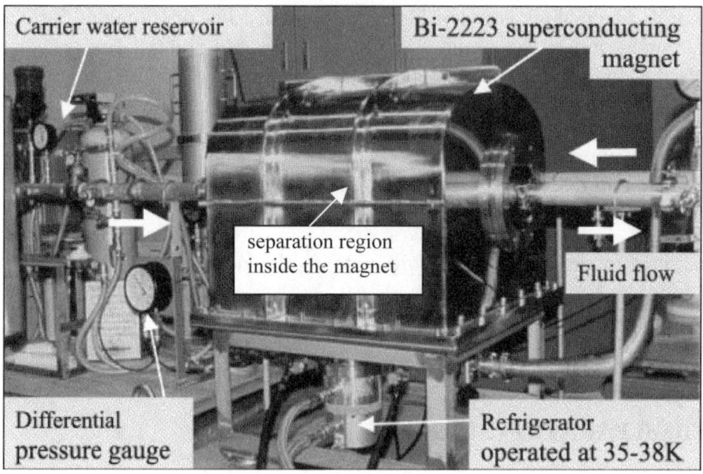

Figure 29   Bi-2223 HTS magnet and magnetic separation system developed by the National Institute for Metals and Science (NIMS, Sumitomo) [100]. Figure courtesy of T. Ohara, NIMS, Japan.

the magnetic separation applications. It is difficult to realize such a rapid cyclic excitation without quenching when we use a conventional NbTi conductor or Nb$_3$Sn magnet.

3.4 OTHER APPLICATIONS: MOTOR and NMR

The mortar application has been done mainly in USA. AMSC company succeeded in developing recently a 5000HP motor [101]. The motor has an HTS field winding and a water-cooled stator. The size reduction and energy efficiency improvement will be expected. This development initiated a ship propulsion system using HTS motor or rotating machine. The HTS motor is considered to have weight, size and power savings advantages [102].

Being different from the above all applications, the application of NMR has been investigated by using Bi-2212 conductor. This system requires a very high magnetic field, corresponding to the resonance frequency. NIMS and Hitachi, Ltd. fabricated two double pan-cake coils using Bi-2212 wire, inserted them into 18T of a conventional superconducting magnet and then succeeded in generating a high magnetic field of 23.4T [103]. It was the first time to generate a magnetic field over 23T only by a superconducting magnet. This HTS magnets consist of two coils: the outer coils has the outer diameter of 147mm and the inner diameter of 64mm, which used 1000m of Bi2212 wires with 5mm in width and 0.34mm in thickness; the coil diameter is 48mm in outer diameter and 16.5mm in inner diameter. They generated 5.4T under 18T by a backup magnetic field of the conventional superconductor. This value of 23.4T is almost the same as the magnetic field of 1GHz NMR, 23.5T, which can analyze the protein structure, then this study opens up a new promising application. Other institutes such as NHMFL

(National High Magnetic Field Laboratory, USA) and OST Company also have been studying Bi-2212 wires for NMR application [104]. For the high magnetic field coil, especially the inner HTS coil has a small diameter and is under a high magnetic field. This means that the HTS magnet receives a strong stress and a large strain. Therefore, in a high magnetic field use, the conductor should be mechanically strong and strain-tolerant in addition to high J_c property at a high magnetic field.

4. TASK FOR INDUSTRIALIZATION

As mentioned above, we looked into the present status of the HTS wire development and of the on-going applications. While the application was proceeding mainly with BSCCO wire, the requirements for industrialization have been argued intensively. Here we discuss the possibility of industrialization of HTS conductors from the points of its cost performance and superconducting properties, especially J_c and I_c.

COST PERFORMANCE

Recent paper of AMSC says that Ag/Bi-2223 wires sell about \$200/kAm as of 2001 [1]. It reduced from the price of \$300/kAm one year ago. Then, they speculate the price in the future, where the price will go down and reach \$50/kAm in 2004. They said that mortars, generators and power cables are economically viable at the price of \$50/kAm.

Similarly, the cost/performance ratio has been analyzed on the industrial use magnet in Japan in detail [105].They calculated the limiting cost which can compete with the conventional devices and equals to the conventional cost from the savings of the energy. For example, they considered the cost reduction due to the scale merit by HTS, electric power saving merit, refrigerator cost and so on. According to the analysis, the limiting costs of HTS are \$55-96/kAm for mortars such as 5000HP or ship-use mortars, \$30-60/kAm for Si single crystal growth magnet, \$1950/kAm for the magnetic separation system and \$10-20/kAm for maglev train. Particularly, the cost performance is severe for the cases already employed the conventional superconductor of low cost NbTi such as the magnet for Si single crystal growth and maglev trains.

Furthermore, we also take into account for the reliability of the system as all the HTS system uses a refrigerator or a liquid coolant. We also need to enhance the reliability to be accepted in the conventional system and to replace them.

J_c-B CHARACTERISTIC

The above cost performance is, of course, related, to the superconducting properties, especially critical current density J_c. However, essentially J_c affects and limits the utilization of the superconductor. Figure 30 [83, 85] illustrates the J_c-B characteristic for typical superconductors: YBCO coated conductor, Bi-2223 and NbTi. For YBCO conductor's sample, we choose the TFA-MOD sample on an IBAD tape, which is not so different from other types of YBCO coated conductor such as PLD and ex-situBaF$_2$ methods. In this figure, YBCO coated conductor is found to show a higher J_c compared to Bi-2223 and NbTi. However, the J_c value of YBCO decreases much especially for a magnetic field perpendicular to the tape surface. J_c is 1.7MA/cm^2 at 0T while degrades by one tenth in a few tesla and one hundredth in a 5T. Compared to the NbTi and Bi-

2223, the non-SC/SC ratio is 100 for YBCO conductors and 1 to 3 for NbTi and Bi-2223 conductors. Then, the engineering critical current density, J_e, for YBCO almost equivalent to those is much lower for YBCO than that for NbTi and Bi-2223, which determines the coil design: the smaller J_e is, the larger coil is. Of course, we can design the magnet to reduce this effect as done in the coils for Si single crystal growth. However, principally we still need to increase J_c or J_e by the material improvement and the combination of thinner substrates and a thicker superconducting layer.

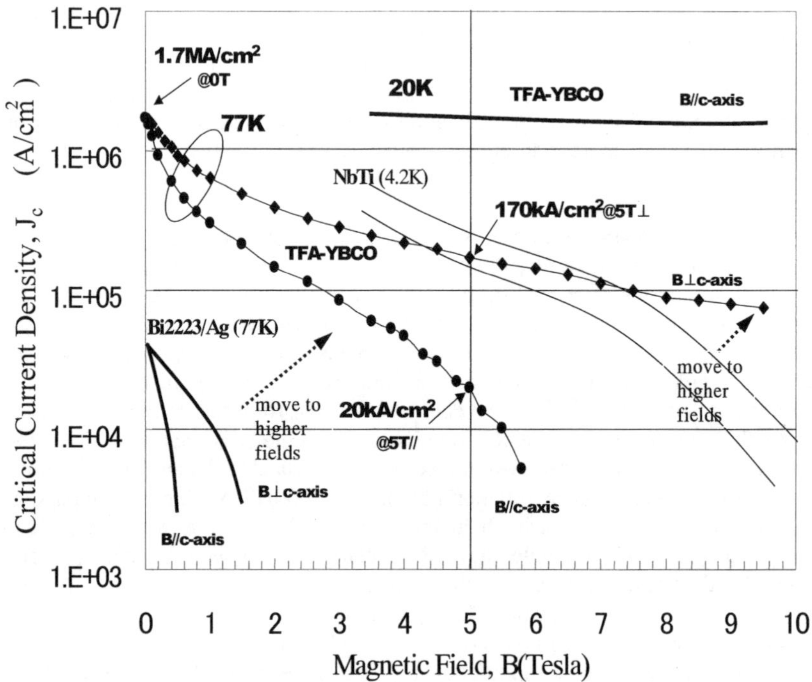

Figure 30   Typical J_c-B characteristic of YBCO coated conductor, Bi-2223 and NbTi conductors [83, 85].

5. CONCLUSIONS

Since the discovery of High-T_c materials in 1986, a great deal of studies has been done around the world. Then, some problems have been solved and others are still remained unsolved in the basic fields as well as the application field. However, now many companies are selling HTS wires of Bi-2223 and Bi-2212. Furthermore, many national projects around the world are actively undergoing. Especially, the cable projects attract much attention in the USA. For the time being, the companies can supply the wire and the researchers should solve the problems encountered in the larger application than before. On the other hand, it is certain that the Bi conductor has a limitation for some application as we described in this chapter. Therefore, the development of the YBCO coated conductor should be accelerated, together with considering significantly the results of the feasibility studies using Bi based conductors. Very recently we have heard of two significant fruitful achievements of a 46m long IBAD-PLD YBCO conductor in Japan and a 10m long and cost-effective TFA-MOD YBCO conductor in the USA. As we have learned historically the rapid growth of Bi-2223 production ability for several years, it is strongly expected that YBCO will follow the same trend. From now on, the collaborative work should become significant among the researchers in materials science and in the application side as well.

ACKNOWLEDGEMENTS

The authors wished to thank all those who have contributed to this area, especially, YBCO coated conductor, the members of ISTEC-SRL Nagoya and Tokyo, editors of this book and the original authors who has been cited. An apology is offered to those whose pertinent work has not been cited here. Some part of the work cited here was supported by the New Energy and Industrial Technology Development Organization (NEDO) as Collaborative Research and Development of Fundamental Technologies for Superconductivity Applications.

RERFERENCES

1. L. Masur, J. Keller, F. Li, S. Fleshler and E. Podtburg, Submitted to Magnet Technology Conference, Geneva, Switzerland, Sept. 24-28, 2001.
2. L. Masur, D. Parker, M. Tanner, E. Podtburg, D. Buczek, J. Scudiere, P. Caracino, S. Spreafico, P. Corsaro and M. Nassi, *IEEE Trans. Appl. Supercond.* **11**(2001), 3256.
3. K. Hayashi, T. Hikata, T. Kaneko, M. Ueyama, A. Mikumo, N. Ayai, S. Kobayashi, H. Takei, K. Sato, *IEEE Trans. Appl. Supercond.* **11** (2001), 3281.
4. S. Kobayashi, T.Kaneko, N. Ayai, K. Hayashi, H. Takei and R. Hata, *Physica C* **357-360** (2001), 1115.K. Sato and K. Hayashi, *OYO BUTURI* **71**, 66(2002).
5. J. Sato, K. Ohata, M. Okada, K. Tanaka, H. Kitaguchi, H. Kumakura, T. Kiyoshi, H. Wada and K. Togano, *Physica C* **357-360** (2001), 1111.
6. T. Hasegawa, N. Ohtani, T. Koizumi, Y. Aoki, S. Nagaya, N. Hirano, L. Motowidlo, R. S. Sokolowski, R. M. Scanlan, D. R. Dietderich and S. Hanai , *IEEE Traans. Appl. Supercond.* 11(2001), 3034.
7. DOE Annual Peer Review, July 17-19, 2002, Superconductivity for Electric Systems Program Review.
8. T. Watanabe, Y. Shiohara and T. Izumi, presented at the Applied Superconductivity Conference, Houston, USA, August 5, 2002. To be published.
9. Y. Yamada, in *BISMUTH-BASED HIGH-TEMPERATURE SUPERCONDUCTORS*, Edited by H. Maeda and K. Togano, Marcel Decker, Inc. (New York), 1996, p289, chapter 14, "Powder-in-tube Bi-2223 Ag-sheathed tape, Basic Process, Microstructure and Critical Current Density".
10. B. Fischer, T. Arndt, J. Gierl, H. Krauth, M. Munz, A. Szulczyk, M. Leghissa, H.-W. Neumueller, *IEEE Trans. Appl. Supercond.* **11** (2001), 3261.
11. W.G. Wang, B. Seifi, Y.-L. Liu, M. Eriksen, P. Skov-Hansen, J.-C. Grivel and P. Vase, *IEEE Trans. Appl. Supercond.*, 11 (2001), 2983.
12. K. Miyoshi, S. Mukoyama, H. Tsubouchi A. Takagi, S. Meguro, K. Matsuo, S. Honjo, T. Mimura and Y. Takahashi, *Physica C* **357-360** (2001), 1259.
13. S. Mukoyama et al., *IEEE Trans. Appl. Supercond.* **11** (2001), 2192.

14. J.P.Stovall, J.A. Demko, P.W. Fisher, M. J. Gouge, J. W. Lue, U. K. Sinha, J. W. Armstrong, R. L. Hughey, D. Lindsay and J. C. Tolbert, *IEEE Trans. Appl. Supercond.* **11** (2001), 2467.
15. N. J. Kelley, C. Wakefield, M. Nassi, P. Corsaro, S. Spreafico, Don W. Von Dollen, J. Jiping,, *IEEE Trans. Appl. Supercond.* **11** (2001), 2461.
16. M. Däumling, C. N. Rasmussen, F. Hansen, D. W. A. Willén, O. E. Schuppach, B. S. Hansen, C. Træholt, K. H. Jensen, S. K. Olsen, C. Rasmussen, E. Veje, O. Tønnesen and J. Østergaard, *Physica C* **357-360** (2001), 1241.
17. Y. Huang, X. Y. Cai, G. N. Riley, Jr., D.Larbalestier, D. Yu, M. Teplitsky, A. Otto, S. Fleshler and R. D. Parrella, *Adv. in Cryogenic Engineering* **48B** (2002), 717.
18. H. Kumakura, in *BISMUTH-BASED HIGH-TEMPERATURE SUPERCONDUCTORS*, Edited by H. Maeda and K. Togano, Marcel Decker, Inc. (New York), 1996, p451, chapter 21, "Bi-2212/Ag Composite Tapes Processed by Doctor-Blade or Dip-Coating Process".
19. J. Sato, *OYO BUTURI* **71** (2002), 69.
20. Y. Aoki, et al., *Physica C* 335(2000), 1.
21. H. Miao, K.R. Maraken, J. Sowa, J. A. Parrell and S. Hong, *Adv. in Cryogenic Engineering* **48B** (2002), 709.
22. M. Okada, H. Morita, J. Sato, T. Kiyoshi, H. Kitaguchi, H. Kumakura, K. Togano and H. Wada, *Adv. in Cryogenic Engineering* **48B** (2002), 703.
23. T. Mimura and Y. Takahashi, *Advances in Superconductivity* **XII** (2000), 661.
24. Y.B. Huang and R. Flükiger, *Physica C* **11** (1998), 495.
25. H. Maeda, T. Inaba, M. Sato, P. X. Zhang and H. Kumakura, *IEEE Trans. Appl. Supercond.* **11** (2001), 2959, and other papers in this proceedings.
26. J. Shimoyama, K. Murakami, K. Shimizu, Y. Nakayama and K. Kishio *Physica C* **357-360** (2001), 1091.
27. S. X. Dou, X. L. Wang, Y. C. Guo, Q. Y. Hu, P. Mikheenko, J. Horvat, M. Ionescu and H. K. Liu, *Supecond. Sci. Technolo.* **10** (1997), A52.
28. D. Larbalestier, X. Y. Cai, Y. Feng, H. Edelman, A. Umezawa, G. N. Riley and W. L. Carter, *Physica C* **221** (1994), 299.
29. Y. Yamada and T. Hattori, *Physica C* **335** (2000), 78.
30. Y. Iijima, N. Tanabe, O. Kohno and Y. Ikeno, *Appl. Phys.Lett.* **60** (1992), 769.
31. Y. Iijima, K. Onabe, N. Futaki, N. Tanabe, N. Sadakata, O. Kohno and Y. Ikeno, *IEEE Trans. Appl. Supercond.* **3** (1993), 1510.
32. K. Hasegawa, N. Yoshida, K. Fujino, H. Mukai, K. Hayashi, K. Sato, T. Ohkuma, S. Honjo, H. Ishii and T. Hara, in *Proc. ICEC16/ICMC (Kitakyushu, Japan)* ed T. Haruyama, T. Mitsui and K. Yamafuji (Tokyo: Elsevier) (1997), p 1413.
33. Y. Yamada, N. Fukushima, S. Nakayama, H. Yoshino and S. Murase, *Jpn. J. Appl. Phys.* **26**(1987), L865.
34. D. Dimos, P. Chaudhari and J. Mannhart, *Phys. Rev. B* **41** (1990), 4038.
35. B. Roas, L. Schultz and G. Saemann-Ischenko, *Phys. Rev. Lett.* **64**(1990), 479.
36. D. T. Shaw, *MRS Bulletin* **XVII** (1992), 33.
37. N. Yoshida, M. Kubota, S. Okuda, S. Takano, M. Nagata, T. Hara, K. Okaniwa and T. Yamamoto, in *Advances in Superconductivity* vol 3 (1991), ed K Kajimura and H Hayakawa (Springer: Tokyo), p 901.
38. Y. Iijima, H. Hayakawa, N. Sadakata and O. Kohno, in *Proc. MT-11 (Tsukuba, Japan 1988)* (Elsevier: London) (1989), p 1442.
39. J. Saitoh, M. Fukutomi, K. Komari, Y. Tanaka, T. Asano, H. Maeda and H. Takahara, *Japan. J. Appl. Phys.* **30**(1991), L898.
40. K. Kakimoto, Y. Iijima and T. Saitoh, *Physica C* **378-381** (2002), 937.
41. A. Usoskin, J. DZick, A. Issaev, J. Knoke, F. Garcia-Moreno, K. Sturm, H. C. Freyhardt, *Supercond. Sci. Technol.* **14** (2001), 676.
42. K. Ohmatsu, K. Muranaka, S. Hahakura, T. Taneda, K. Fujino, H. Takei, Y. Sato, K. Matsuo and Y. Takahashi, *Physica C* **357-360** (2001), 946.
43. T. Muroga, T. Araki, T. Niwa, Y. Yamada, I. Hirabayashi, Y. Iijima, T. Saitoh, preprint of the Applied Superconductivity Conference in Houston, Aug.4, 2002, program No. 4MH01.
44. Y. Yamada, T. Muroga, H. Iwai, T. Izumi and Y. Shiohara, WS-4 / Presented at ISS 2002, Submitted 13 November 2002, to be published in *Physica C* (2003).

45. T. Muroga, H. Iwai, T. Niwa, T. Araki, Y. Yamada, T. Izumi, Y. Shiohara, Y. Iijima, T. Saito, T. Kato, Y. Sugawara, T. Hirayama, Presented at ISS 2002, Submitted 13 November 2002. To be published in *Physica C* (2003).
46. in the part of LANL, ORNL, IGC and ASC, DOE Annual Peer Review, July 17-19, 2002, Superconductivity for Electric Systems Program Review.
47. DOE Annual Peer Review, Aug.1-3, 2001, Superconductivity Program for Electric Systems.
48. M. W. Rupich, U. Schoop, D. T. Verebelyi, C. Thieme, W. Zhang, X. Li, T. Kodenkandath, N. Nguyen, E. Siegal, D. Buczek, J. Lynch, M. Jowett, E. Thompson, J-S. Wang, J. Scudiere, A. P. Malozemoff, Q. Li, S. Annavarapu, S. Cui, L. Fritzemeier, B. Aldrich, C. Craven, F. Niu, A. Goyal, and M. Paranthaman, Applied Superconductivity Conference August 4 – 9, 2002 Paper 3MA03.
49. T. Watanabe, K. Wada, Y. Ohashi, M. Ozaki, K. Yamamoto, T. Maeda and I. Hirabayashi, *Physica C* **378-381** (2002), 911.
50. H. Yoshino, M. Yamazaki, T. D. Thanh, Y. Kudo and H. Kubota, *Physica C* **357-360** (2001), 923.
51. Y. Iijima, M. Hosaka, N.Tanabe, N. Sadakata, T. Saitoh, O. Kohno and K. Takeda, *J. Mater. Res.* **12** (1997), 2913.
52. S. Zhu, D. H. Lowndes, J. D. Budai and D. P. Norton, *Appl.Phys. Lett.* **65** (1994), 2012.
53. V. Betz, B. Holzapfel, D. Raouser and L. Schultz, *Appl.Phys. Lett.* **71** (1997), 2952.
54. Y. Iijima, K. Kakimoto, T. Saitoh, T. Kato and T. Hirayama, *Physica C* **378-381** (2002), 960.
55. K. Kakimoto, Y. Iijima and T. Saitoh, *Physica C* **378-381** (2002), 937.
56. C. P. Wang, K. B. Do, M. R. Beasley, T. H. Geballe and R. H. Hammond, *Appl. Phys. Lett.* **71** (1997), 2955.
57. P. N. Arendt, in the part of LANL, DOE Annual Peer Review, July 17-19, 2002, Superconductivity for Electric Systems Program Review
58. K. Kakimoto, Y. Iijima and T. Saitoh, Presented at ISS 2002, Submitted 13 November 2002. To be published in *Physica C* (2003).
59. A. Usoskin, A. Issaev, H. C. Freyhardt, M. Leghissa, M. P. Oomen and N.-W. Neumueller, *Physica C* **372-376** (2002), 857.
60. A. Usoskin, J. Knoke, F. García-Moreno, A. Issaev, J. Dzick, S. Sievers and H.C. Freyhardt , *IEEE Trans. Appl. Supercond.* **11** (2001), 3385.
61. S. R. Foltyn, in the part of LANL, DOE Annual Peer Review, Aug.1-3, 2001, Superconductivity Program for Electric Systems.
62. T. Kato, Y. Iijima, T. Muroga, T. Saito, T. Hirayama, I. Hirabayashi, Y. Yamada, T. Izumi, Y. Shiohara, Presented at ISS 2002 in 13 November 2002. To be published in *Physica C* (2003).
63. F. A. Underwood 1961 Textures in Metal Sheets (London: Macdonald)
64. T. Doi, N. Sugiyama, T. Yuasa, T. Ozawa, K. Higashiyama, S. Kikuchi and K. Osamura, *Advances in Superconductivity* vol. 8 (1996), ed H, Hayakawa and Y. Enomoto (Tokyo: Springer), p 903.
65. S. R. Goodman and H. Hu, *Trans. Met. Soc. AIME* **233**(1965) 105
66. A. Goyal, D. P. Norton, D. M. Kroeger, D. K. Christen, M. Paranthaman, E. D. Specht, J. D. Budai, Q. He, B. Saffian, F. A. List, D. F. Lee, E. Hatfield, P. M. Martin, C. E. Klabunde, J. Mathis and C. Park, *J. Mater. Res.* **12** (1997), 2924.
67. A. Goyal, D. F. Leea, F. A. Lista, E. D. Spechta, R. Feenstraa, M. Paranthaman, X. Cuib, S. W. Luc, P. M. Martina, D. M. Kroeger, D. K. Christen, B. W. Kang, D. P. Norton, C. Park, D. T. Verebelyi, J. R. Thompson, R. K. Williams, T. Aytug and C. Cantoni, *Phsyica C* **357-360** (2001), 903.
68. C. Cantoni, D. K. Christen, R. Feenstra, A. Goyal, G. W. Ownby, D. M. Zehner, and D. P. Norton, *Appl. Phys. Lett.* **79** (2001), 3077.
69. M. Paranthaman, T. Aytug, S. Sathyamurthy, D. B. Beach, A. Goyal, D. F. Lee, B. W. Kang, L. Heatherly, E. D. Specht, K. J. Leonard, D. K. Christen and D. M. Kroeger, *Physica C* **378-381** (2002), 1009.
70. K. Matsumoto, Y. Niiori, I. Hirabayashi, N. Koshizuka, T. Watanabe, Y. Tanaka and M. Ikeda, in *Advances in Superconductivity* **vol.10** (1998), ed K Osamura and I Hirabayashi (Tokyo: Springer), p 611.
71. T. Watanabe, K. Matsumoto, T. Maeda, T. Tanigawa and I. Hirabayashi, *Physica C* 357-360 (2001), 914.
72. H. Yoshino, M. Yamazaki, H Fuke, T. D. Thanh, Y. Kudo and S. Oshima, in Advances in Superconductivity vol. 6 (1994), ed T Fujita and Y Shiohara (Springer: Tokyo), p 759.

73. H. Yoshino, H. Kubota, M. Yamazaki, T. D. Thanh and Y. Kudo, *IEEE Trans. Appl. Supercond.*, **11** (2001), 3142.

74. H. Yoshino, M. Yamazaki, T. D. Thanh, Y. Kudo and H. Kubota, *Physica C* **357-360** (2001), 923.

75. K. Higashiyama, T Yuasa, T Fujiwara, H Akata, R Takahashi and K Osamura, *Extended Abstract of Int. Workshop on Superconductivity (Tokyo, ISTEC) (1998)*, p 65.

76. P.C. McIntyre, M.J. Cima and Fai Ng Man, *J. Appl. Phys.* **68** (1990), 4183.

77. V.F.Solovyov, H.J.Wiesmann, Li-jun Wu, M.Suenaga and R.Feenstra, *IEEE Trans. Applied. Supercond.* **9** (1999), 1467.

78. V.F. Solovyov, H.J. Wiesmann, L.-J. Wu, Y. Zhu and M. Suenaga, *Appl. Phys. Lett.* **76** (2000), 1911.

79. M. Suenaga, *Physica C* **378-381**(2002), 1045.

80. L. Wu, V.F. Solovyov, H.J. Wiesmann, Y. Zhu and M. Suenaga, *J. Mater. Res.* **16** (2001), 2869.

81. P.C. McIntyre and M.J. Cima, *J. Mater. Res.* **9** (1994), 2219.

82. J. Shibata, T. Honjo, H. Fuji, R. Teranishi, T. Izumi, Y. Shiohara, T. Yamamoto and Y. Ikuhara , *Physica C* **378-381** (2002), 1039.

83. Y. Yamada, S. B. Kim, T. Araki, Y. Takahashi, T. Yuasa, H. Kurosaki, I. Hirabayashi, Y. Iijima and K. Takeda, *Physica C* 357-360 (2001), 1007.

84. Y. Yamada, T. Araki, H. Kurosaki, S.B. Kim, T. Yuasa, I. Hirabayashi, Y. Shiohara, Y. Iijima, T. Saitoh,, J. Shibata, Y. Ikuhara, T. Katoh and T. Hirayama, in Extended Abstracts of 2001 International Workshop on Superconductivity (2001), p.229.

85. Y. Yamada, T. Araki, H. Kurosaki, S.B. Kim, T. Yuasa, I. Hirabayashi, Y. Shiohara, Y. Iijima, T. Saitoh,, J. Shibata, Y. Ikuhara, T. Katoh and T. Hirayama, *Adv. in Cryogenic Engineering* **48B** (2002), p631.

86. Y. Yamada, S.B. Kim, T.Araki, T. Yuasa, Y. Takahashi, H. Kurosaki, Y. Shiohara, , I. Hirabayashi, Y. Iijima and K. Takeda in Proceedings of the MRS 2000 Fall Meeting, in *Materials Research Society Symposium Proceedings* **Vol. 659** (2001), edited by U. Balachandran, H. C. Freyhardt, T. Izumi and D. C. Larbalestier, 2001, Materials Research Society (Warrendale, Pennsylvania, USA), p II4.7.

87. T. Araki, Y. Takahashi, K. Yamagiwa, Y. Iijima, K. Takeda, Y. Yamada, J. Shibata, T. Hirayama and I. Hirabayashi, *Physica C* **357-360** (2001), 991.

88. H. Fuji, T. Honjo, Y. Nakamura, T. Izumi, Y. Shiohara, R. Teranishi, M. Yoshimura, Y. Iijima and T. Saito, *Physica C* **378-381**(2002), 1014.

89. T. Honjo et al., Presented at ISS 2002 in 13 November 2002. To be published in *Physica C* (2003).

90. Y. Takahashi, Y. Aoki, T. Hasegawa, Y. Iijima, T. Saitoh, I. Hirabayashi, T. Honjo and Y. Shiohara, *Physica C* **378-381**(2002), 1024.

91. V.F. Solovyov, H.J. Wiesmann, L. J. Wu, Y. Zhu, M. Suenaga and R. Feenstra, *Physica C* **309** (1997), 269.

92. R. Feenstra, in the part of ORNL, DOE Annual Peer Review, July 17-19, 2002, Superconductivity for Electric Systems Program Review.

93. M. Paranthaman , in the part of ORNL, DOE Annual Peer Review, Aug.1-3, 2001, Superconductivity Program for Electric Systems.

94. S. Honjo, K. Matsuo, T. Mimura and Y. Takahashi, *Physica C***357-360** (2001), 1234.

95. T. Masuda, T. Kato, H. Yumura, M. Watanabe, Y. Ashibe, K. Ohkura, C. Suzawa, M. Hirose, S. Isojima, K. Matsuo, S. Honjo, T. Mimura, T. Kuramochi, Y. Takahashi, H. Suzuki and T. Okamoto, *Physica C* **378-381** (2002), 1174.

96. P.Corsaroa, M. Bechisa, P. Caracinoa, W. Castiglionia, G. Cavalleria, G. Colettaa, G. Colomboa, P. Ladièa, A. Mansoldoa, R. Melea, S. Montagnera, C. Moroa, M. Nassia, S. Spreafico, N. Kelley and C. Wakefield, *Physica C* **378-381** (2002), 1168.

97. News release in Jul.19, 2002 at the America Supserconductor's home page, http://www.amsuper.com/html/newsEvents/news/ (as of Feb., 2003)

98. M. Ono, K. Tasaki, Y. Otani, T. Kuriyama, M. Kyoto, T. Shimonosono, S. Hanai, Y. Sumiyoshi, S. Nomura, M. Shojiyu, N. Ayai, T. Kaneko, S. Kobayashi, , K. Hayashi, H. Takei, K. Sato, T. Mizuishi, M. Kimura and T. Masui, in *Japanese Proceedings of the 64th meeting on Cryogenics and Superconductivity* (2001), P73.

99. M. Ono, K. Tasaki, T. Kuriyama, M. Kyoto, T. Shimonosono, S. Hanai, Y. Sumiyoshi, S. Nomura, M. Shojiyu, N. Ayai, T. Kaneko, S. Kobayashi, K. Hayashi, H. Takei, K. Sato, M. Kimura and T. Masui, *Physica C* **357-460**(2001), 1281 and M. Ono, S. Hanai, K. Tasaki, M. Hiragishi, K. Koyanagi, C. Nomura, T. Yazawa, Y. Otani, T. Kuriyama, Y. Sumiyoshi, S. Nomura, Y. Dozono, H. Maeda, T.

Hikata, K. Hayashi, H. Takei, K. Sato, M. Kimura and T. Masui, *IEEE Trans. Appl. Supercond.* **10** (2000), 499.
100. T. Ohara, Hiroaki Kumakura and Hitoshi Wada, *Physica C* **357-360** (2001), 1272.
101. D. Madura, M.Richardson, D. Bushko, G. Snitchler, P. Winn, S. Kalsi and B. Gamble, Proceedings of Applied Superconductivity Conference, August 2002, Houston, USA, to be published.
102. S. C. Karon, Maritime Reporter & Engineering News, September, 2002.
103. T. Kiyoshi et al., *IEEE Trans. Appl. Supercond.* **10** (2000), 472.
104. D. Markiewicz et al., IEEE Trans. Appl. Supercond. 10(2000), 728.
105. ISTEC-SRL Report on "Survey of the applications of high temperature superconductor to the industrial use devices", March, 2002.

$$\textbf{ELECTRICAL INSULATION}$$
$$\textbf{FOR SUPERCONDUCTING POWER APPARATUS}$$

Naoki HAYAKAWA and Hitoshi OKUBO
Department of Electrical Engineering, Nagoya University
Furo-cho, Chikusa-ku, Nagoya 464-8603, JAPAN

1. INTRODUCTION

Application of superconducting power technology to electric power apparatus such as generators, transformers, fault current limiters, power cables, energy storage devices and even to their integrated systems, will give rise to enhanced power supply efficiency and capacity. Research and development of superconducting power technology have led to rapid progress in a variety of projects, whose principal focus has been the large current performance of superconducting materials and superconducting power apparatus fabricated on a trial basis. However, in order to incorporate superconducting power apparatus into power systems for the next generation, the high voltage performance of cryogenic liquids and composite electrical insulation systems for superconducting power apparatus must be established.

Electrical insulation in the cryogenic environment is an essential and common technology for the practical design and development of superconducting power apparatus. The importance of electrical insulation technology at cryogenic temperature has been extensively recognized; e.g., a "Special Issue on Electrical Insulation in Superconducting Power Apparatus" was published by *Cryogenics* in 1998 [1]. Although there have been fundamental and applied studies on the electrical insulation performance of cryogenic liquids such as liquid nitrogen (LN_2) and liquid helium (LHe), useful data applicable to practical and efficient insulation design are less satisfactory than those for conventional SF_6 gas, insulating oil etc. For the practical design and development of superconducting power apparatus, not only basic, but also applied-level research on the electrical insulation performance of cryogenic liquids have been highly desired.

Since transformers include many types of electrical insulation, let us take the superconducting transformer as an example. Figure 1 shows the electrical insulation components to be considered in the insulation design of a superconducting transformer

340

connected with superconducting cable. Although Fig.1 basically resembles the conventional high voltage oil-filled transformer, it also contains some elements peculiar to the superconducting transformer and electrical insulation at cryogenic temperature. To clarify the electrical insulation characteristics of each part ①-⑨ in Fig.1, many electrical insulation characteristics of cryogenic liquids must be experimentally investigated and systematized.

In the following sections, as common and fundamental electrical insulation characteristics corresponding to Fig.1 ①-⑨, we show the results of measurement and analysis on the area and volume effects on dielectric strength, V-t characteristics and electrical insulation under the thermal and electrical combined stress peculiar to the cryogenic environment. These data, obtained in terms of partial discharge (PD) inception and breakdown, will be applicable to the practical insulation design of superconducting transformers, fault current limiters, power cables and so on. In closing, a flow chart for electrical insulation design of superconducting power apparatus is summarized.

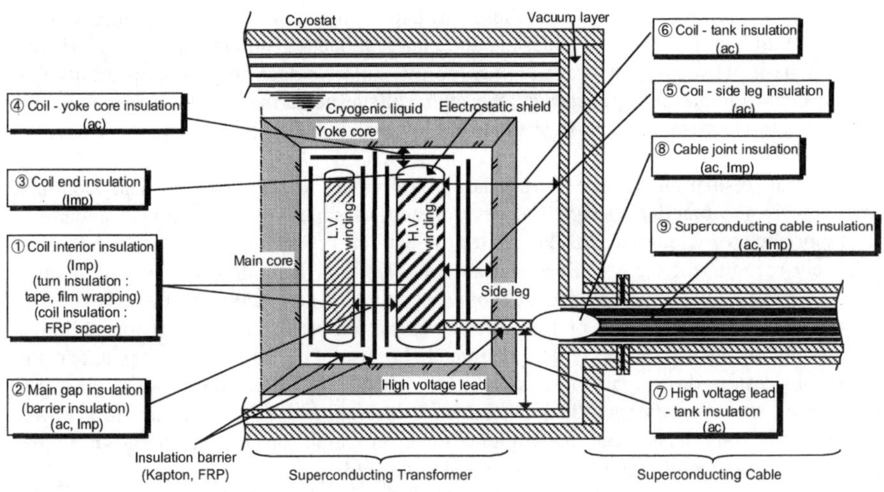

Figure 1   Electrical insulation components for superconducting transformer

2. FUNDAMENTALS FOR ELECTRICAL INSULATION TEST AND ANALYSIS AT CRYOGENIC TEMPERATURE

Before discussing the electrical insulation characteristics of cryogenic liquids, we will begin with 2 fundamental factors for electrical insulation test and analysis at cryogenic temperature: cryostat for electrical insulation test and Weibull distribution for statistical analysis of discharges. These factors are crucial for each experiment and analysis of discharges at cryogenic temperature to be described in this article.

2.1 Cryostat for electrical insulation test

A cryostat that can withstand the application of high voltage and large current in a cryogenic environment is essential for electrical insulation tests of cryogenic liquids and superconducting power apparatus. It is particularly necessary that the cryostat be equipped with a high voltage bushing connecting the cryogenic environment with the room-temperature environment.

Figure 2 shows the cryostat developed for electrical insulation tests in a cryogenic environment. The cryostat, made of stainless steel, consists of an outer vessel for cryogenic liquids at atmospheric pressure and an inner vessel for cryogenic liquids at atmospheric or high pressure. The inner vessel can be pressurized up to an absolute pressure of 0.5 MPa for various conditions of cryogenic liquids; e.g. sub-cooled LN_2 and supercritical helium. The capacitor core of the bushing is mainly composed of FRP, which has been verified corona free at 100 kV_{rms} in LN_2 and at 75 kV_{rms} in LHe. The cryostat has the added benefit of 4 observation windows, each with an effective diameter of 50 mm. The behavior of thermal bubbles and discharge phenomena in cryogenic liquids can be directly observed through the windows and recorded by a high-speed video camera.

The different kinds of experimental models described in the following sections can be arranged in the inner vessel of the cryostat and cooled to cryogenic temperature using LN_2 or LHe. High voltage with ac, dc or impulse waveform can be applied to the samples through the bushing, inducing PD and breakdown. In the case of ac high voltage application, the PD signal can be detected by a CR circuit and fed into 2 digital oscilloscopes with different time scales of 2 ms/div and 0.5 μs/div, to measure PD phase characteristics and PD charge, respectively. PD detection sensitivity is less than 1 pC.

2.2 Weibull distribution for statistical analysis of discharges

Discharge characteristics of gases and liquids should be measured and analyzed using a statistical approach, since the experimental data usually have considerable scattering associated with the probabilistic aspect of discharges, even under fixed condition. Weibull distribution based on the weakest-link theory is generally used for such statistical problems [2].

Suppose a discharge in a liquid divided into n pieces of small volume units. The discharge probability of unit i is pi, where each unit is small enough for the electric field strength to be regarded as constant. The weakest-link theory gives the entire discharge probability p of the liquid gap as expressed by

$$\ln(1 - p) = \ln \prod_{i=1}^{n} (1 - p_i) = \sum_{i=1}^{n} \ln(1 - p_i) \qquad (1)$$

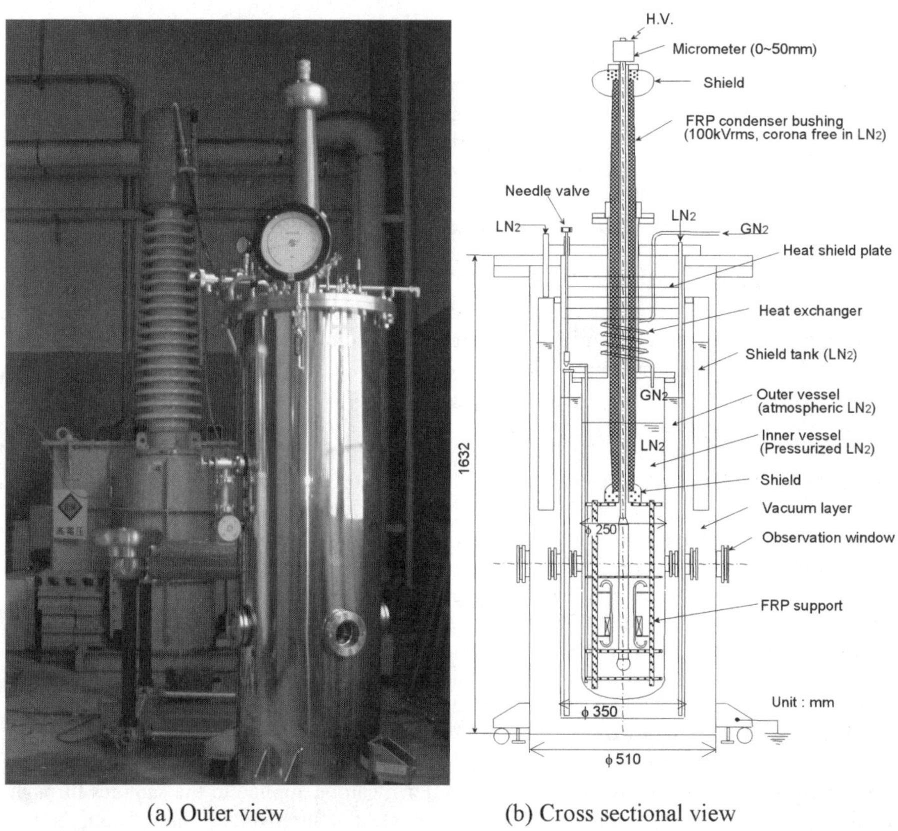

(a) Outer view   (b) Cross sectional view

Figure 2   Cryostat for electrical insulation tests in cryogenic environment

When E_i and p_i are the electric field strength and the discharge probability at unit i, respectively, Weibull statistics permit p_i to be given by

$$\ln(1 - p_i) = -\left(\frac{E_i}{E_0}\right)^m \qquad (2)$$

where m is the shape parameter and E_0 is the scale parameter in Weibull statistics. Shape parameter m corresponds to the magnitude of statistical scattering in the dielectric strength; scale parameter E_0 corresponds to the intrinsic dielectric strength. Substitution of Equation (2) into Equation (1) yields

$$\ln(1 - p) = -\sum_{i=1}^{n}\left(\frac{E_i}{E_0}\right)^m = -\frac{1}{v_0}\iiint_v\left(\frac{E_i}{E_0}\right)^m dv = -\frac{1}{v_0}\iiint_v\left(\frac{E_i}{E_m}\right)^m dv\left(\frac{E_m}{E_0}\right)^m = -\frac{v}{v_0}\left(\frac{E_m}{E_0}\right)^m \qquad (3)$$

where E_m is the maximum electric field strength and v_0 represents the reference value of the liquid volume. v is defined as the statistical stressed liquid volume ($SSLV$) as given by

$$v = \iiint_v\left(\frac{E_i}{E_m}\right)^m dv = SSLV \qquad (4)$$

where $(E_i/E_m)^m$ represents the relative discharge probability at unit i, and v/v_0 in Equation (3) corresponds to the number of weak points in the liquid gap. Thus, the entire discharge probability p, i.e. the Weibull distribution function for discharges, can be obtained as

$$p = 1 - \exp\left[-\frac{v}{v_0}\left(\frac{E_m}{E_0}\right)^m\right] \qquad (5)$$

Similarly, when the discharge is attributed to the electrode surface condition, the statistical stressed electrode area ($SSEA$) is defined by

$$S = \iint_s\left(\frac{E_i}{E_m}\right)^m dS = SSEA \qquad (6)$$

When dielectric strength is iteratively measured N times under fixed condition, the entire or cumulative discharge probability p at j-th ($j < N$) magnitude E in a series of dielectric strength can be approximately expressed by $j/(N+1)$. Equation (5) can then be modified into

$$\ln\ln\frac{1}{1-p} = \ln\ln\frac{1}{1-\dfrac{j}{N+1}} = \ln\left[\frac{v}{v_0}\left(\frac{E}{E_0}\right)^m\right] = m(\ln E - \ln E_0) + \ln\frac{v}{v_0} \qquad (7)$$

The values of $\ln\ln[1/\{1\text{-}j/(N+1)\}]$ can be plotted as a function of $\ln E$ for the experimental data ($j = 1$ to N) sorted by magnitude; this is called the "Weibull plot". If the Weibull plot yields a straight line, we can confirm that the Weibull distribution is applicable to the statistical problem. In addition, the Weibull plot also yields shape parameter m as the gradient of the straight line, and scale parameter E_0 as the electric field strength with $p = 0.632$.

Let us draw and understand the Weibull plot, using some examples. Figure 3 shows basic experimental models for electrical insulation tests: (a) sphere-to-plane electrode model and (b) coaxial cylindrical electrode model. Each electrode model was immersed in LN_2 or LHe in the inner vessel of the cryostat in Fig.2, and stressed under 60 Hz ac high voltage until breakdown was induced. Figure 4 represents the sequential data ($N = 50$) of ac breakdown strength for both electrode models in LN_2. The Weibull plots of breakdown strength in LN_2 and LHe are shown in Fig.5, where the vertical axis designates the cumulative breakdown probability. Each Weibull plot yielded a straight line, respectively, verifying that the statistical scattering in breakdown strength of cryogenic liquids could be evaluated by the Weibull distribution. Note that the smaller value of shape parameter m corresponds to larger scattering in the breakdown strength.

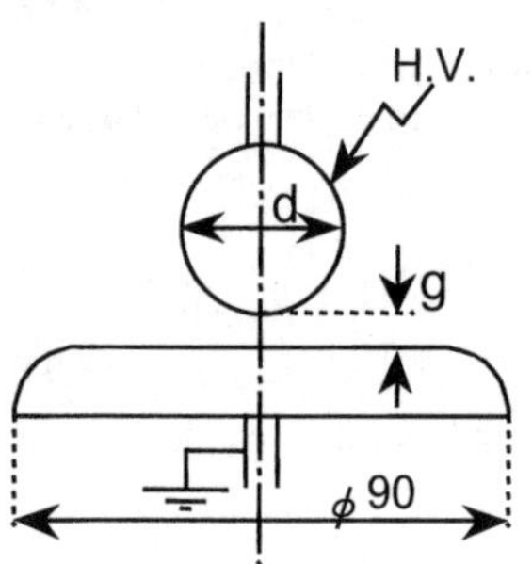

Sphere diameter : d
6, 10, 20, 50mm
Gap length : g
0.5, 0.75, 1.0mm

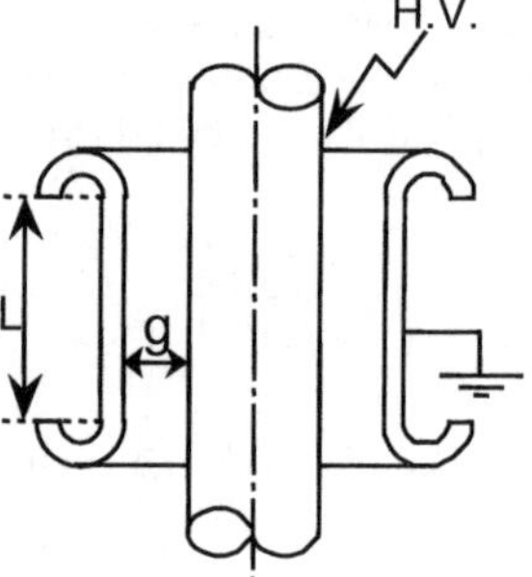

Gap length : g
2.3, 4.5mm
Electrode length : L
30, 100, 350mm

(a) Sphere-to-plane electrode model

(b) Coaxial cylindrical electrode model

Figure 3   Basic experimental model for electrical insulation tests

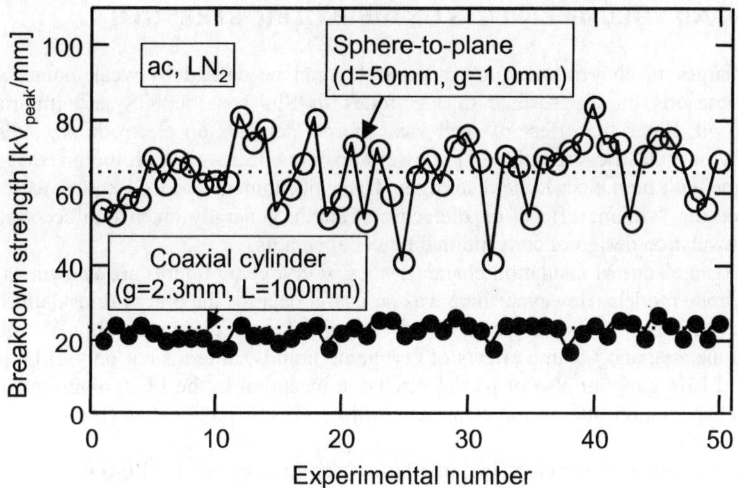

Figure 4   Sequential data of ac breakdown strength in LN2
(Sphere-to-plane electrode model: $d = 50$ mm, $g = 1$ mm
Coaxial cylindrical electrode model: $g = 2.3$ mm, $L = 100$ mm)

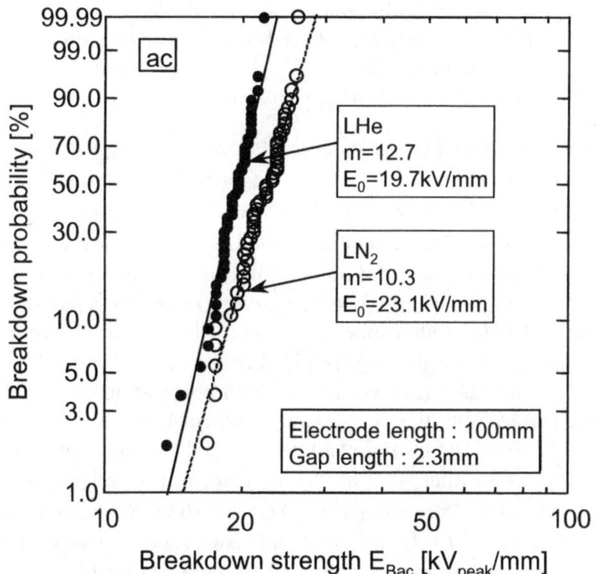

Figure 5   Weibull plots of ac breakdown strength in LN2 and LHe
(Coaxial cylindrical electrode model: $g = 2.3$ mm, $L = 100$ mm)

3. AREA AND VOLUME EFFECTS ON DIELECTRIC STRENGTH

Discharges in conventional power apparatus can be caused by weak points such as micro-protrusions on the surface of electrodes in SF$_6$ gas, bubbles and impurities in insulating oil. Since the effect of such weak points depends on electrode size, discharge probability or dielectric strength of the power apparatus decreases with the increase in area and volume with high electric field strength. These phenomena are well known as the "area effect" and the "volume effect" on dielectric strength, generally taken into account in the practical insulation design of conventional power apparatus.

The basic electrical insulation characteristics of cryogenic liquids are known for small-scale electrode models. However, there was no useful data for the practical insulation design of superconducting power apparatus with large-scale electrode models. We therefore examined the area and volume effects of cryogenic liquids, in terms not only of breakdown in LN$_2$ and LHe gap, but also of partial discharge inception in the LN$_2$/polypropylene (PP) laminated paper composite insulation system for high-T$_c$ superconducting (HTS) cables.

3.1 Critical stress level for electrical insulation design in cryogenic liquids

SSEA and *SSLV* in Equations (4) and (6) are useful in discussing the area and volume effects on discharge characteristics in dielectric materials. In general, however, *SSEA* and *SSLV* are difficult to calculate, because they depend on the experimental measurement of Weibull shape parameter *m*. Therefore, in the case of area and volume effects on discharge characteristics in SF$_6$ gas and insulating oil for conventional power apparatus, stressed electrode area (*SEA*) and stressed liquid volume (*SLV*) with electric field strength exceeding 90 % of maximum electric field strength *Em* have been taken as the critical stress level to be considered in practical insulation design [3]. In the practical insulation design of superconducting power apparatus, therefore, the critical stress level for cryogenic liquids should be evaluated.

SEA and *SLV* were calculated by electric field analysis using the charge simulation method. Figure 6 shows calculated *SEA* and *SLV* as a function of sphere diameter for the sphere-to-plane electrode model in Fig.3 (a) [4]. The dotted lines represent η % *SEA* and η % *SLV* for different values of η, which can be calculated as *SEA* and *SLV* with the electric field strength higher than η % of *Em*. The solid lines, on the other hand, represent *SSEA* and *SSLV* calculated taking account of Weibull shape parameter *m* as measured by ac breakdown experiments in LN$_2$ and LHe at atmospheric pressure. The best fit of experimental results (solid lines) to calculated results (dotted lines) is obtained at η = 75-85 % in LN$_2$ and η = 65-80 % in LHe. These results can be verified through observation of breakdown traces on the sphere electrode surface after the experiment, as schematically shown in Fig.7, with the sphere diameter *d* = 50 mm. The breakdown traces were distributed within the limited area beyond η = 85 % for LN$_2$ and η = 80 % for LHe, respectively, which was in good agreement with the plots at *d* = 50 mm in Fig.6. Figure 8 shows the relation between η and the corresponding *SSEA* and *SSLV* for different parameter combinations of *d* and *g* under ac and dc voltage applications. The average values of η are 82 % for LN$_2$ and 75 % for LHe, irrespective of *SSEA* and *SSLV*, which can be regarded as the critical stress level for cryogenic liquids. The value of η for LHe, lower than that for LN$_2$, corresponds to the latent

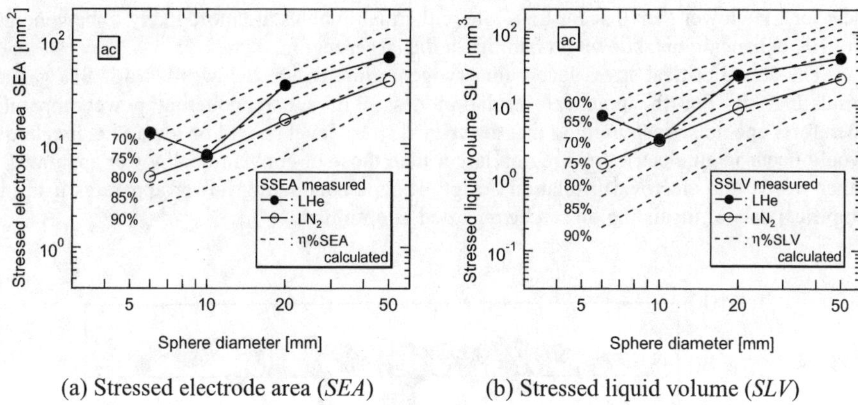

(a) Stressed electrode area (*SEA*) (b) Stressed liquid volume (*SLV*)

Figure 6 Calculated *SEA* and *SLV* as a function of sphere diameter
(Sphere-to-plane electrode model: $g = 1$ mm)

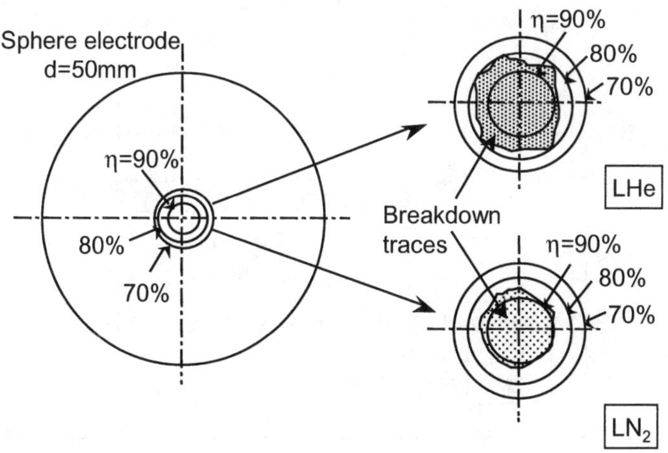

Figure 7 Breakdown traces on sphere electrode surface
(Sphere-to-plane electrode model: $d = 50$ mm, $g = 1$ mm)

348

heat for LHe lower than that for LN2, where thermal bubbles are more likely to be generated in LHe, enhancing breakdown probability in the larger region.

The above critical stress levels for cryogenic liquids are below 90 % for SF6 gas and insulating oil. For the practical insulation design of superconducting power apparatus, therefore, one must keep in mind that the critical stress level should be lower, i.e. breakdown would occur in an electric field region larger than those of conventional power apparatus. In other words, the electrical insulation design of superconducting power apparatus using the empirical data of insulating oil can be regarded as optimistic.

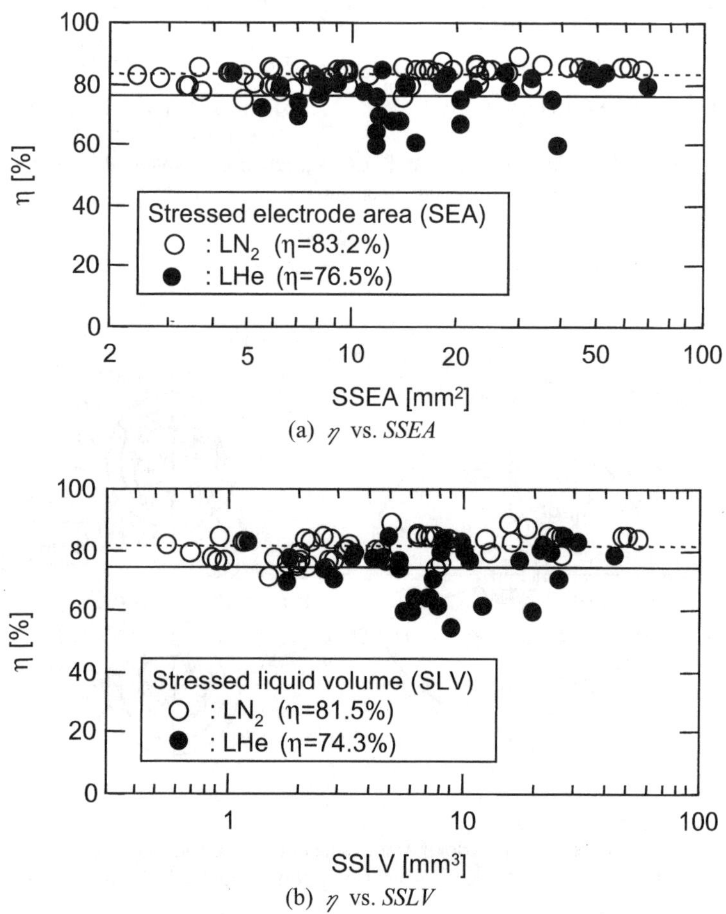

(a) η vs. *SSEA*

(b) η vs. *SSLV*

Figure 8   Relationship between η and *SSEA* and *SSLV*
(Sphere-to-plane electrode model, ac and dc voltage application)

3.2 Area and volume effects on breakdown strength of LN₂ and LHe gap

Figure 9 shows an example of ac breakdown voltage Vb of LN₂ and LHe as a function of sphere diameter for the sphere-to-plane electrode model in Fig.3 (a) [4]. Vb in LN₂ and LHe decreased at the larger sphere diameter. In general, electric field distribution becomes more uniform with the increase in sphere diameter. Therefore, if breakdown were determined only by maximum electric field strength, Vb would increase with the increase in sphere diameter. Figure 9 therefore indicates that area and volume effects would appear in the breakdown strength of LN₂ and LHe.

Area and volume effects on the breakdown strength of LN₂ and LHe were systematized by compiling data from different bibliographic sources [5]. Figure 10 shows ac breakdown strength $E_{B(LN2)}$ of LN₂ at atmospheric pressure as a function of (a) SEA, i.e. area effect, and (b) SLV, i.e. volume effect. Figure 11 shows area and volume effects on the ac breakdown strength $E_{B(LHe)}$ of LHe. Note that the horizontal axis in Fig.10 represents 82 % SEA and 82 % SLV for LN₂, and that in Fig.11 represents 75 % SEA and 75 % SLV for LHe, as described in the previous sub-section. Breakdown strength in LN₂ and LHe decreased linearly in the log-log scale with the increase in SEA and SLV over extensive ranges (SEA: 10^{-1}-10^{5} mm², SLV: 10^{-2}-10^{5} mm³). Area and volume effects on the ac breakdown strength of LN₂ and LHe can be expressed by

$$E_{B(LN2)} = 105 \times SEA^{-\frac{1}{6.23}} = 83.1 \times SLV^{-\frac{1}{7.30}} \, [\text{kV}_{\text{peak}}/\text{mm}] \qquad (8)$$

$$E_{B(LHe)} = 53.0 \times SEA^{-\frac{1}{8.48}} = 44.2 \times SLV^{-\frac{1}{11.1}} \, [\text{kV}_{\text{peak}}/\text{mm}] \qquad (9)$$

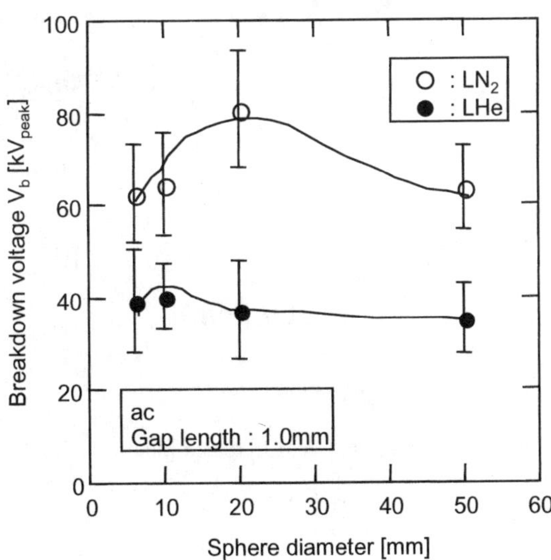

Figure 9   ac breakdown voltage of LN₂ and LHe as a function of sphere diameter (Sphere-to-plane electrode model: g = 1 mm)

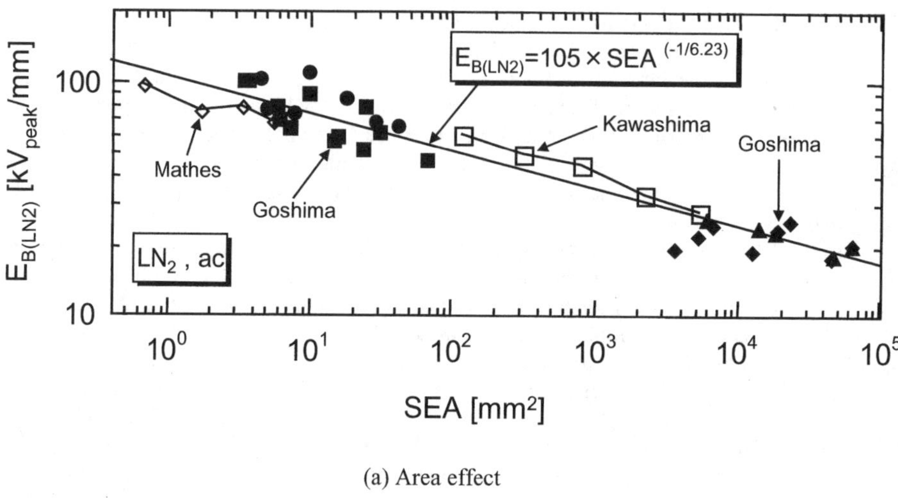

(a) Area effect

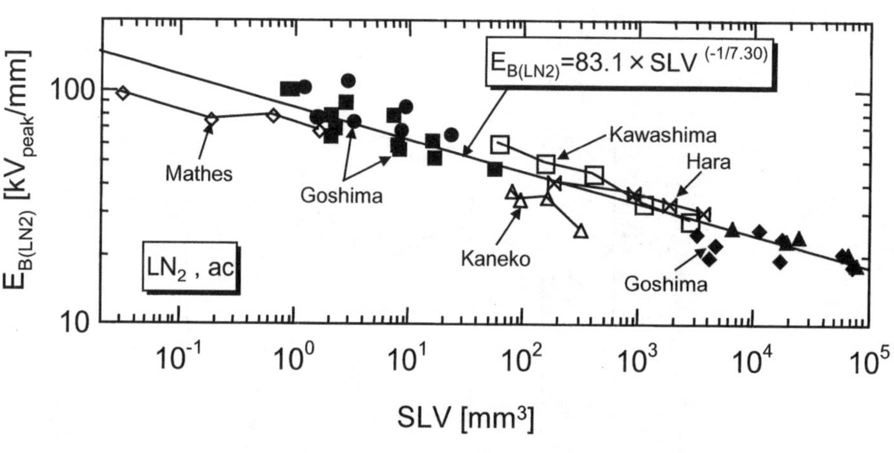

(b) Volume effect

Figure 10   ac breakdown strength of LN$_2$ as a function of *SEA* and *SLV*

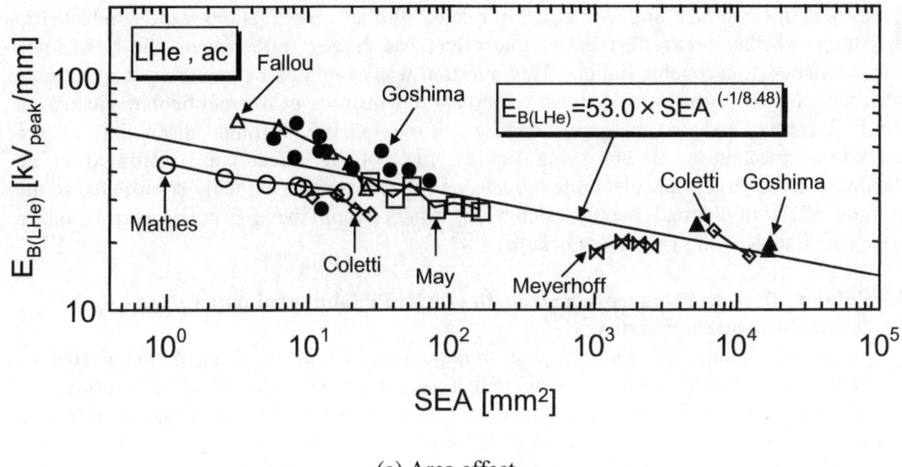

(a) Area effect

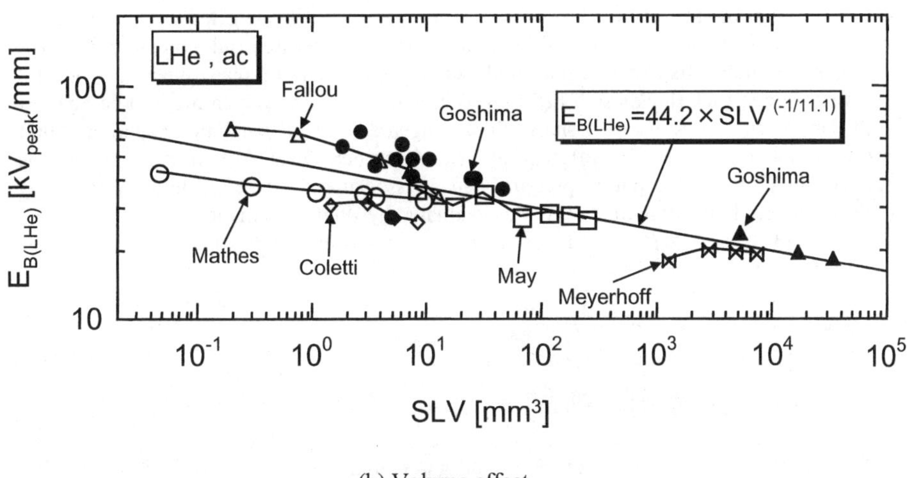

(b) Volume effect

Figure 11   ac breakdown strength of LHe as a function of *SEA* and *SLV*

Since *SEA* and *SLV* are geometrically related, further investigations were conducted to determine whether area effect or volume effect has greater influence on the breakdown characteristics of cryogenic liquids. This question was examined by changing the electrode surface treatment (mirror finish, rough finish) and the quantity of thermal bubbles (increased using a heater, reduced by pressurization). Consequently, through discussion of the breakdown mechanism in cryogenic liquids, the volume effect was confirmed to be dominant in the large-scale electrode models, and the area effect mutually contributed to the volume effect in thermal bubble conditions, which would be a peculiar correlation to cryogenic liquids with lower latent heat [6].

3.3 Volume effect on PD inception strength of LN$_2$/PP laminated paper composite insulation system

Area and volume effects are quite important in the practical insulation design of superconducting power cables of long length, i.e. large *SEA* and *SLV*. In addition, the LN$_2$/PP laminated paper composite insulation system is expected to be the most promising electrical insulation system for the cold dielectric type HTS cables, where butt gaps are produced between PP laminated papers in the winding process. The butt gaps are filled with LN$_2$ of a dielectric strength lower than that of PP laminated papers, so that partial discharge (PD) may occur in the butt gaps as the precursor of breakdown. Therefore, PD inception characteristics, rather than breakdown characteristics, should be understood in the reliable insulation design of HTS cables. From these backgrounds, we investigated volume effect on the PD inception strength of the LN$_2$/PP laminated paper composite insulation system [7, 8].

Figure 12 shows the experimental model composed of parallel plane electrodes and PP laminated papers, each thickness 0.125 mm. Butt gaps were arranged in the middle layer of the PP laminated papers. Table 1 shows the arrangements of PP laminated papers and butt gaps, each diameter 5 mm, for different layers and numbers. Each experimental model was immersed in LN$_2$ at atmospheric pressure and stressed under 60 Hz ac high voltage. PD inception electric field strength (*PDIE*) was evaluated by Weibull statistics.

Figure 13 shows *PDIE* as a function of butt gap volume (*BGV*) for 3-layer models (Case-B, C, D in Table 1). *PDIE* decreased linearly in the log-log scale with the increase in *BGV* for 3-layer models. This result suggests the volume effect on *PDIE* against *BGV* for 3-layer models, which can be expressed by

$$PDIE = 60.9 \times BGV^{-\frac{1}{9.0}} \ [\text{kV}_{\text{rms}}/\text{mm}] \qquad (10)$$

On the other hand, *PDIE* for different butt gap conditions (Case-A, B, E, F, G in Table 1) is shown in Fig.14. Experimental results revealed the followings:

(a) For 3-layer models, *PDIE* without butt gaps (case-A) is 9 % higher than that with a butt gap of 0.125 mm thickness (case-B).
(b) For 5-layer models, *PDIE* without butt gaps (case-E) is almost the same as that with a butt gap of 0.125 mm thickness (case-F), and 14 % higher than that with butt gaps of 0.375 mm thickness (case-G).
(c) For experimental models without butt gaps, *PDIE* for 3-layer model (case-A) is 20 % higher than that for 5-layer model (case-E).

(d) For experimental models with a butt gap of 0.125 mm thickness, *PDIE* for 3-layer model (case-B) is 10 % higher than that for 5-layer model (case-F).

The above results suggest that PD inception may be attributed not only to the number of butt gaps or *BGV*, but also to the number of PP laminated papers in the LN$_2$/PP laminated paper composite insulation system. In other words, there could be LN$_2$-filled thin layers between PP laminated papers, where total stressed volume might become large and contribute as weak points sufficient to induce PD.

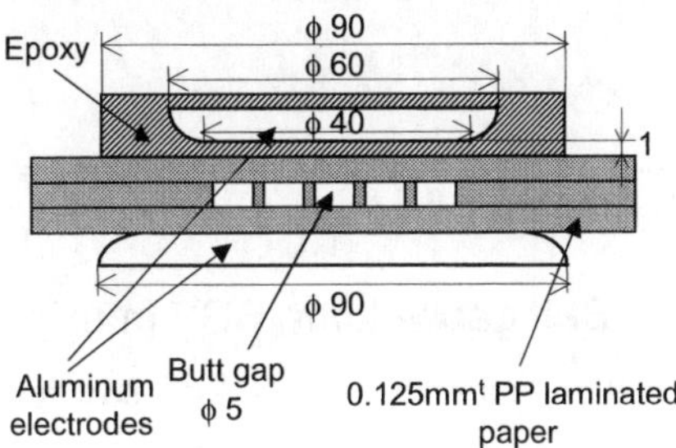

Figure 12   Experimental model of LN$_2$/PP laminated paper composite insulation system

Table 1   Arrangement of PP laminated papers and butt gaps

	Experimental model	Number of PP laminated papers	Number of butt gaps	Butt gap volume (*BGV*) [mm^3]
A		3	0	---
B			1	2.45
C			5	12.25
D			13	31.85
E		5	0	---
F			1	2.45
G			3	7.35

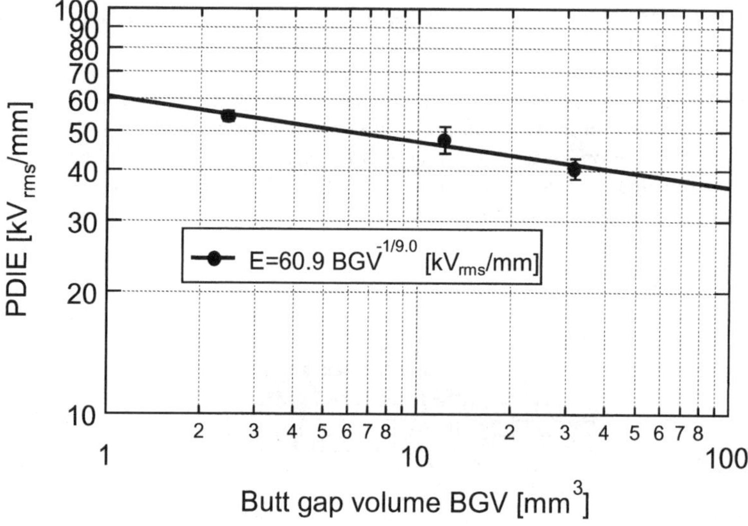

Figure 13 PDIE as a function of *BGV* for 3-layer models

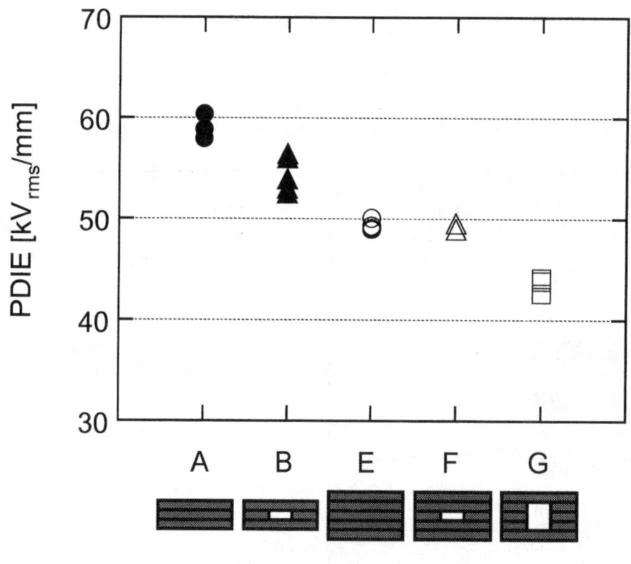

Figure 14 PDIE for different butt gap conditions

4. V-t CHARACTERISTICS

V-t characteristics play an important role in determining testing voltage level and estimating the lifetime of conventional power apparatus with a certain rated voltage. In practical and reliable insulation design of superconducting power apparatus, V-t characteristics should be obtained in a cryogenic environment. Furthermore, in the case of HTS cables, PD inception in the LN_2/PP laminated paper composite insulation system may diminish electrical insulation performance, ultimately resulting in breakdown. Therefore, V-t characteristics at PD inception should also be obtained in the LN_2/PP laminated paper composite insulation system. From these backgrounds, we investigated the V-t characteristics for ac voltage application not only at breakdown in the LN_2 and LHe gap, but also at PD inception in the LN_2/PP laminated paper composite insulation system for HTS cables.

4.1 V-t characteristics at breakdown in LN_2 and LHe gap

V-t characteristics can be analyzed by Weibull statistics as a function of time to discharges and electric field strength. The Weibull distribution function with such double variables is obtained as

$$p = 1 - \exp\left[-\left(\frac{t}{t_0}\right)^a \left(\frac{E}{E_0}\right)^m\right] \qquad (11)$$

where t is the time to discharges under a fixed applied voltage, t_0 and a are the scale parameter and the shape parameter, respectively, on the time to discharges. The Weibull plot can also be drawn for cumulative discharge probability as a function of time to discharges. Figure 15 shows the Weibull plot of the time to ac breakdown in LN_2 and LHe at atmospheric pressure for the coaxial cylindrical electrode model in Fig.3 (b). Each Weibull plot yielded a straight line, verifying that the statistical scattering in time to breakdown could be evaluated by the Weibull distribution. Breakdown time t_{50} with a breakdown probability of 50 % can be derived from the Weibull plot. Figure 16 shows the time to breakdown in LHe for different applied ac voltages. The total number of breakdowns at each applied voltage is 20, where the number in parentheses indicates the number of breakdowns before the target voltage is reached. The 50 % breakdown time t_{50} is designated by a closed symbol at each applied voltage; t_{50} increased with the decrease in applied voltage; this is called the "V-t characteristics". When 50 % breakdown strength E_{50} is used in the vertical axis, instead of breakdown voltage, the V-t characteristics can be obtained with $p = 0.5$ in Equation (11) and expressed by

$$E_{50} = 0.693^{\frac{1}{m}} E_0 \left(\frac{t}{t_0}\right)^{-\frac{1}{n}} \qquad (12)$$

where $n = m/a$ and is called the "lifetime index".

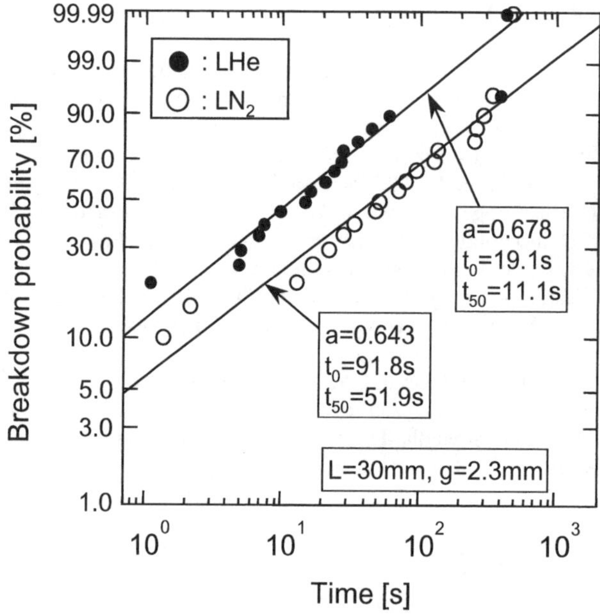

Figure 15 Weibull plot of the time to ac breakdown in LN2 and LHe
(Coaxial cylindrical electrode model: g = 2.3 mm, L = 30 mm)

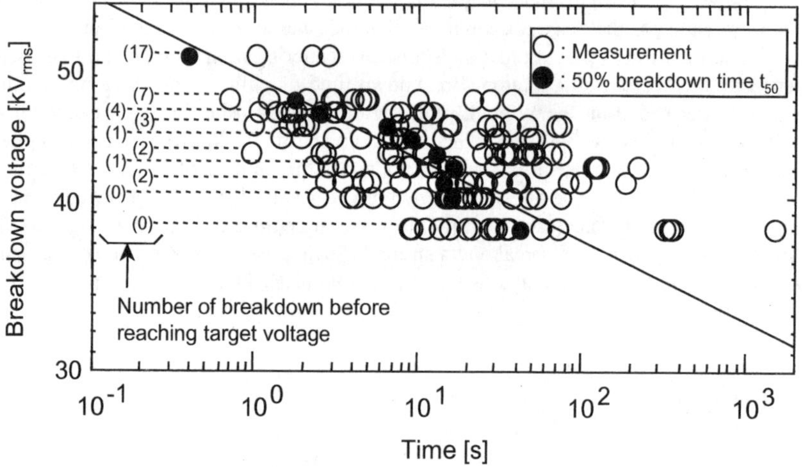

Figure 16 Time to breakdown in LHe for different applied ac voltage
(Coaxial cylindrical electrode model: g = 4.5 mm, L = 100 mm)

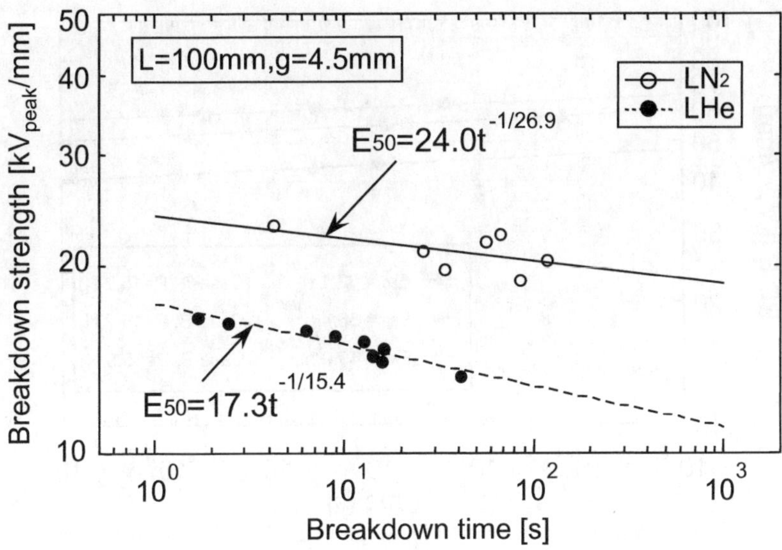

Figure 17   V-t characteristics at breakdown in LN2 and LHe
(Coaxial cylindrical electrode model: g = 4.5 mm, L = 100 mm)

Figure 17 shows V-t characteristics at breakdown in LN2 and LHe at atmospheric pressure for a coaxial cylindrical electrode [5]. Lifetime indices n of V-t characteristics at breakdown in LN2 and LHe were 26.9 and 15.4, respectively. The difference in n values between LN2 and LHe may be attributed to the difference in latent heat; LN2, with the latent heat higher than LHe, has the longer lifetime because of the lower generation of thermal bubbles. In addition, n values for cryogenic liquids are smaller than n = 38-58 for insulating oil [9]. This means that the lifetime of superconducting power apparatus may be shorter than that of conventional power apparatus, from the viewpoint of n values at breakdown.

4.2 V-t characteristics at PD inception in LN2/PP laminated paper composite insulation system

V-t characteristics at PD inception in the LN2/PP laminated paper composite insulation system are shown in Fig.18 for different experimental models (Case-A, B, F in Table 1), under the similar surface pressure on PP laminated paper, which was taken as a parameter [8]. Figure 18 indicates that the lifetime indices n of V-t characteristics at PD inception were 86.8-111.5, irrespective of butt gap condition. V-t equations with n values for different butt gap conditions and surface pressures are summarized in Table 2. The difference in magnitude of V-t characteristics can be understood by the volume effect on PD inception strength, described in the previous section.

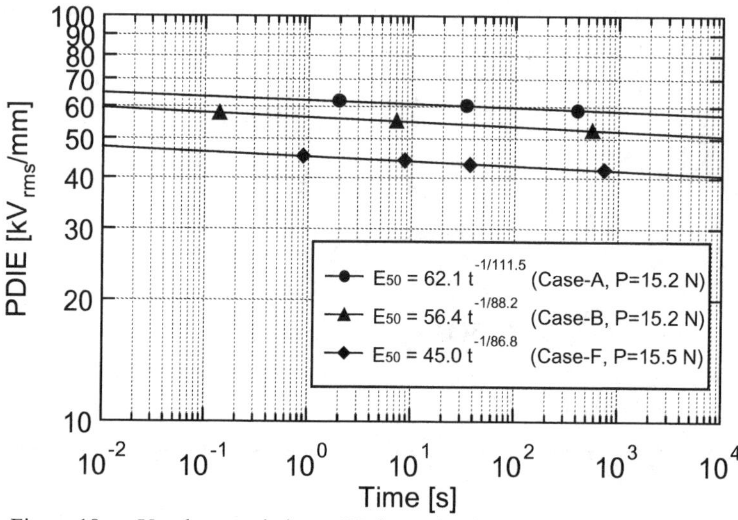

Figure 18   V-t characteristics at PD inception in LN$_2$/PP laminated paper composite insulation system

Table 2   V-t equations at PD inception in LN$_2$/PP laminated paper composite insulation system

	Experimental model	Number of PP laminated papers	Number of butt gaps	Surface pressure P [N]	V-t equation
A			0	15.2	$E_{50}=62.1\,t^{-1/111.5}$
B		3		9.4	$E_{50}=52.7\,t^{-1/33.7}$
			1	15.2	$E_{50}=56.4\,t^{-1/88.2}$
				21.0	$E_{50}=57.7\,t^{-1/108.5}$
E			0	4.9	$E_{50}=53.6\,t^{-1/117.6}$
F			1	9.7	$E_{50}=50.3\,t^{-1/117.6}$
		5		4.9	$E_{50}=42.7\,t^{-1/192.2}$
G			3	9.7	$E_{50}=44.1\,t^{-1/119.7}$
				15.5	$E_{50}=45.0\,t^{-1/86.8}$

The n values of V-t characteristics in the LN$_2$/PP laminated paper composite insulation system were compared between at PD inception and at breakdown, where $n = 15$-89 at breakdown [10-12] was lower than $n = 68$-192 at PD inception in most cases in Table 2. The difference in n values may be attributed to the difference in discharge mechanism. PD inception could occur not only in the butt gaps, but also in the thin layers between PP laminated papers, as was discussed in Section 3.3. On the other hand, breakdown would occur via the butt gap filled with bubbles. The lower n values at breakdown can therefore be interpreted by the following discharge mechanism: PD in the butt gap generated bubbles with dielectric strength lower than LN$_2$, which could accelerate PD development in the butt gap, easily resulting in breakdown.

5. ELECTRICAL INSULATION UNDER THERMAL AND ELECTRICAL COMBINED STRESS

As described in Section 3.1, thermal bubbles are likely to be generated in cryogenic liquids due to the lower latent heat. Since the dielectric strength of bubbles is much lower than that of liquids, the possible generation of thermal bubbles in cryogenic liquids should be understood and taken into account in the practical insulation design of superconducting power apparatus. Thermal bubbles may be generated transiently by quench of superconductors and continuously by ac loss, heat in-leak and so on. We therefore discussed the electrical insulation characteristics of cryogenic liquids under thermal and electrical combined stress, assuming the cooling channel of cryogenic liquid gap for superconducting magnets and coils, superconductors molded with epoxy resin for bushings, terminals and resistive fault current limiters.

5.1 Quench-induced dynamic insulation performance of LHe

Quench of superconducting power apparatus can occur in such cases as short-circuit fault in a future electric power network, suddenly generating a large Joule heat, resulting in violent disturbance of thermal bubbles in cryogenic liquids. Thermal bubble disturbance under high electric field stress diminishes the electrical insulation performance of cryogenic liquids and may induce PD and breakdown. Such electrical insulation performance of cryogenic liquids under thermal bubble conditions is referred to as "quench-induced dynamic insulation performance", which should be distinguished from "static insulation performance" without consideration of the quench phenomena. Quench-induced dynamic insulation performance is quite important, especially for resistive superconducting fault current limiters utilizing the quench phenomena. The dynamic insulation performance of LHe was investigated and compared with the static insulation performance [13, 14].

Figure 19 shows the experimental setup for quench-induced dynamic insulation performance of LHe. A superconducting wire-to-plane electrode model (Fig.20 (a)) or a superconducting coil-to-plane electrode model (Fig.20 (b)) with gap length g was placed in the inner vessel of the cryostat filled with LHe at different pressure P. 60 Hz ac high voltage V_a was applied to the plane electrode, whereas the superconducting wire or coil was grounded. With the superconducting wire or coil exposed to high electric field stress, a prospective 60 Hz ac large current I_{pro} was supplied for several cycles to the superconducting wire or coil, inducing the quench. Such combined stress of high voltage and large current could induce breakdown in the LHe gap, which corresponds to the dynamic breakdown.

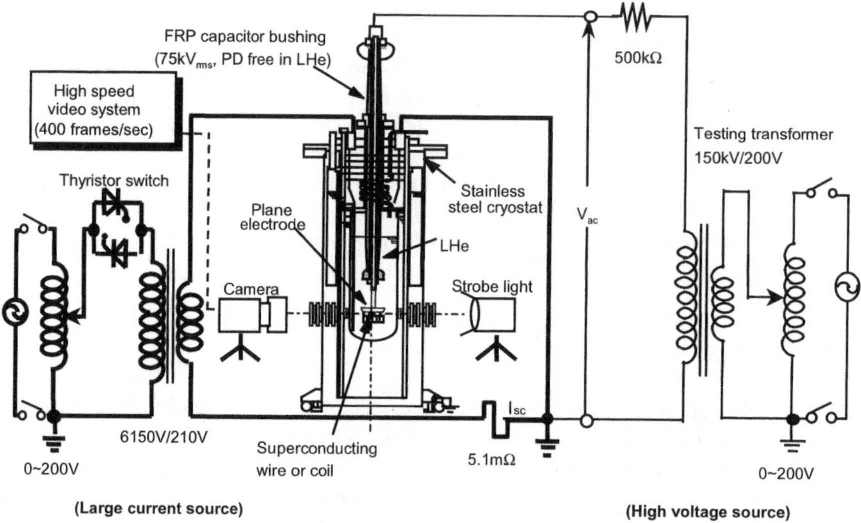

Figure 19   Experimental setup for quench-induced dynamic insulation performance of LHe

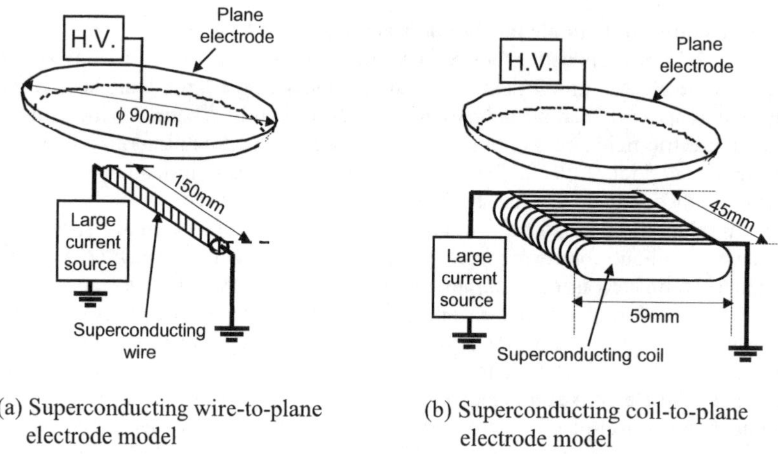

(a) Superconducting wire-to-plane
electrode model

(b) Superconducting coil-to-plane
electrode model

Figure 20   Experimental model of superconducting wire (coil)-to-plane electrode

Figure 21 shows an example of thermal bubble behavior and dynamic breakdown of LHe at $P = 0.1$ MPa, $g = 9$ mm, $V_a = 31$ kV$_{peak}$, $I_{pro} = 190$ A$_{peak}$ for a superconducting coil made of NbTi with quench current $I_q = 140$ A$_{peak}$. A thermal bubble cluster, generated due to quench of the superconducting coil, developed into the gap space. Consequently, at 40 ms after quench, a bright discharge path of dynamic breakdown was observed in the thermal bubble disturbance.

Through the above experiments, dynamic breakdown voltage V_d and strength E_d were defined as the minimum values of applied stress below which dynamic breakdown was no longer induced. On the other hand, static breakdown voltage V_s and strength E_s were measured by gradually increasing the applied voltage without current flow into the superconductors. Figure 22 shows E_s and E_d of LHe at $P = 0.1$ MPa as a function of gap length under a uniform electric field composed of superconducting coil-to-plane electrode models for different effective post-quench thermal energy density J_{eff}. J_{eff} is the accumulated thermal energy per unit length of the superconducting wire in quench condition, which can be controlled by using different superconducting wires and prospective currents. E_s was slightly reduced with the increase in gap length, due to the volume effect as described in Section 3.2. E_d was much lower than E_s, and improved slightly with the increase in gap length, where the quench-induced thermal bubble disturbance was suppressed in the larger gap space. Figure 23 shows E_s and E_d as a function of LHe pressure at $g = 7$ mm. E_s and E_d were improved in pressurized LHe, where thermal bubble disturbance was suppressed and gas density in the thermal bubbles was enhanced. However, E_d decreased with the increase in J_{eff}, due to activated thermal bubble disturbance.

The quench-induced dynamic insulation performance of LHe was obtained for different parameter combinations of P, g and J_{eff}. Figure 24 shows a 3-dimensional diagram of V_d as a function of g and J_{eff} at (a) $P = 0.1$ MPa, (b) $P = 0.15$ MPa and (c) $P = 0.2$ MPa, where V_d at $J_{eff} = 0$ J/cm corresponds to V_s. These diagrams make clear the drastic reduction of V_d at higher J_{eff} and lower g. Figure 25 was derived from these diagrams as an equi-breakdown voltage surface for (a) $V_d = 30$ kV$_{peak}$, (b) $V_d = 40$ kV$_{peak}$ and (c) $V_d = 50$ kV$_{peak}$, as a function of P, g and J_{eff}. These diagrams can be used as follows: if J_{eff} is expected to be 0.05 J/cm, $g = 10$ mm is at least required at $P = 0.1$ MPa (point A) to prevent the dynamic breakdown in quench condition. However, when the pressure of LHe is raised to 0.2 MPa, the gap length can be reduced to $g = 5$ mm (point B). Furthermore, in the case of J_{eff} as low as 0.01 J/cm, $g = 5$ mm can be applicable even at $P = 0.1$ MPa (point C).

Similarly, the quench-induced dynamic insulation performance of LHe was also obtained under non-uniform electric field composed of superconducting wire-to-plane electrode models. Figures 26 and 27 show 3-dimensional diagrams of V_d and equi-breakdown voltage surfaces, respectively, under non-uniform electric field.

The parameter combination of P, g and J_{eff} can be optimized by using the equi-breakdown voltage surface in Figs.25 and 27; this will contribute to the practical and efficient insulation design of superconducting power apparatus taking into account the quench condition.

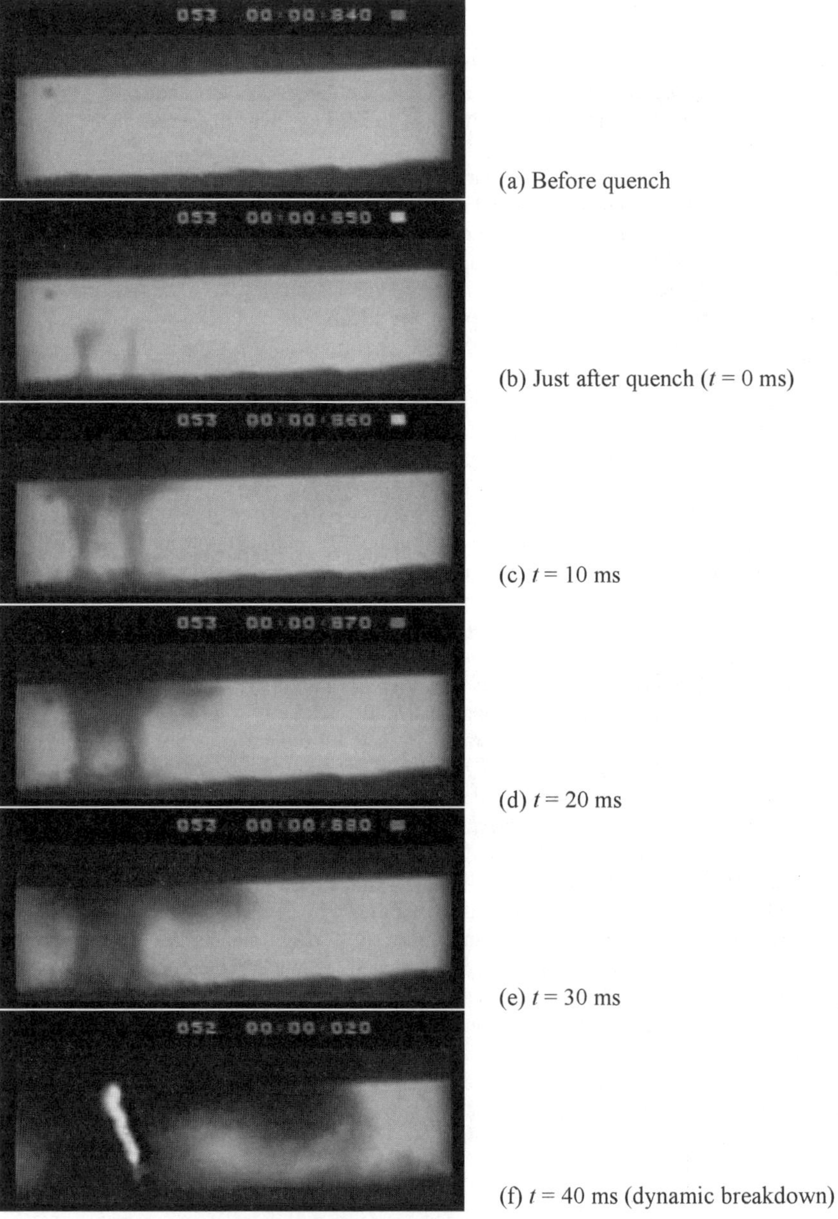

(a) Before quench

(b) Just after quench (t = 0 ms)

(c) t = 10 ms

(d) t = 20 ms

(e) t = 30 ms

(f) t = 40 ms (dynamic breakdown)

Figure 21 Thermal bubble behavior and dynamic breakdown of LHe
(Superconducting coil-to-plane electrode model,
P = 0.1 MPa, g = 9 mm, V_a = 31 kV$_{peak}$, I_{pro} = 190 A$_{peak}$, I_q = 140 A$_{peak}$)

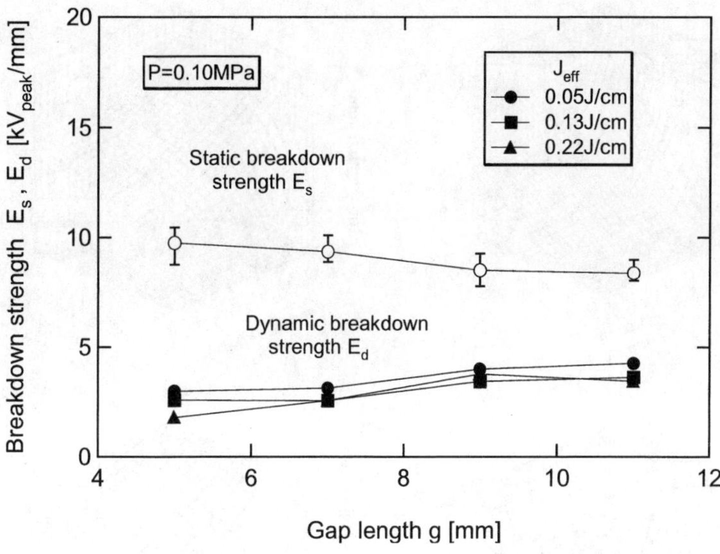

Figure 22   Static and dynamic breakdown strength of LHe as a function of gap length
(Superconducting coil-to-plane electrode model, $P = 0.1$ MPa)

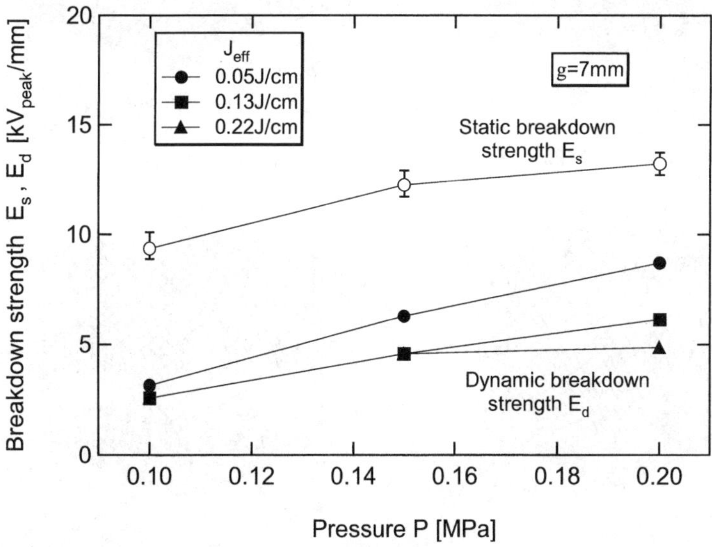

Figure 23   Static and dynamic breakdown strength as a function of LHe pressure
(Superconducting coil-to-plane electrode model, $g = 7$ mm)

364

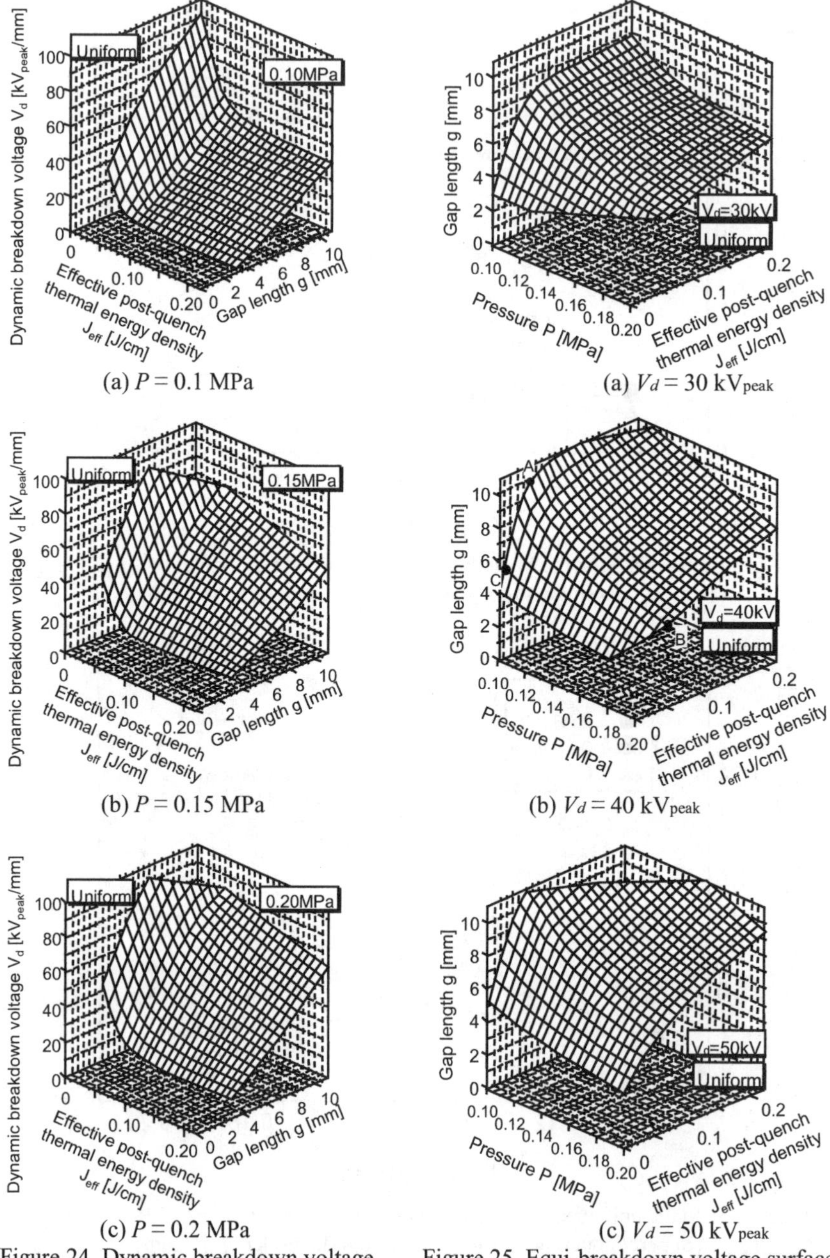

(a) $P = 0.1$ MPa

(a) $V_d = 30$ kV$_{peak}$

(b) $P = 0.15$ MPa

(b) $V_d = 40$ kV$_{peak}$

(c) $P = 0.2$ MPa

(c) $V_d = 50$ kV$_{peak}$

Figure 24 Dynamic breakdown voltage under uniform electric field

Figure 25 Equi-breakdown voltage surface under uniform electric field

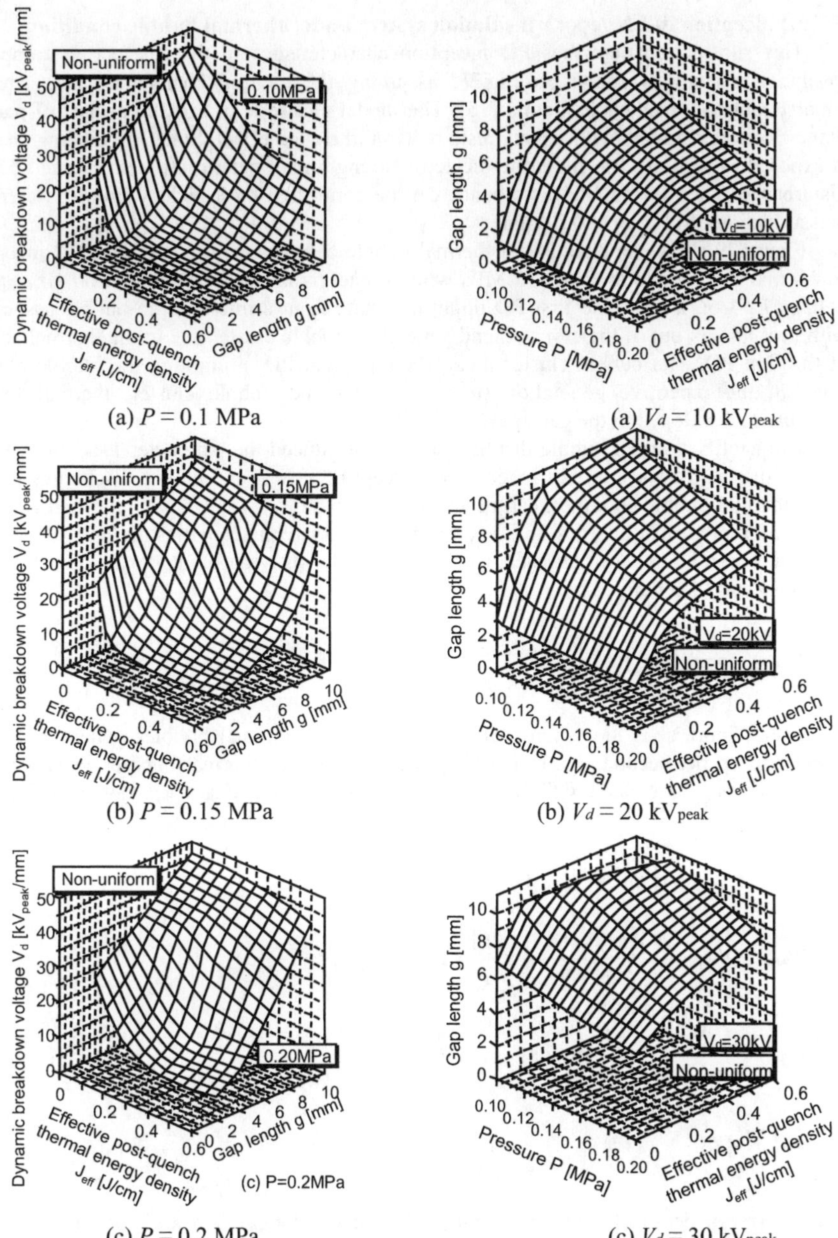

(a) P = 0.1 MPa

(a) V_d = 10 kV$_{peak}$

(b) P = 0.15 MPa

(b) V_d = 20 kV$_{peak}$

(c) P = 0.2 MPa

(c) V_d = 30 kV$_{peak}$

Figure 26   Dynamic breakdown voltage under non-uniform electric field

Figure 27   Equi-breakdown voltage surface under non-uniform electric field

5.2 PD inception in LN2/epoxy insulation system under thermal bubble condition

This sub-section describes PD inception characteristics in the LN2/epoxy composite insulation system as shown in Fig.28, assuming the cooling channel of LN2 gap for superconducting magnets and coils [15]. The model electrode in Fig.28 consists of parallel plane electrodes molded with epoxy resin to avoid the edge effect. Since the cooling channel is exposed to high electric field, the dielectric strength may deteriorate with thermal bubble disturbance. Thermal bubble disturbance can be controlled by a nickel-chrome heater and pressurization of LN2 up to 0.2 MPa.

Figure 29 shows an example of thermal bubble behavior before and after PD inception ($t = 0$ ms) in LN2 gap at $P = 0.15$ MPa without energizing the heater under 60 Hz ac high voltage $V_a = 56$ kV$_{rms}$. The first PD might originate from a microscopic bubble associated with heat in-leak out of the cryostat, and generate a bubble cluster due to the thermal energy of the first PD. The bubble cluster induced subsequent PD, promoting the bubble cluster. Through the repetitive generation process of PD and bubble cluster, thermal bubble disturbance developed in the gap space.

To quantify thermal bubble disturbance, we introduced the parameter V_{rate}: the volume ratio of thermal bubbles to gap space. The concept of V_{rate} is as follows: discharges in LN2 under thermal bubble condition can occur not only in an isolated large bubble, but also in a cluster of microscopic bubbles. In other words, discharge characteristics mutually depend on the diameter and number of bubbles, which are regarded as weak points for PD generation in LN2. V_{rate} was then defined by

$$V_{rate} = \frac{V_{bubble}}{V_{gap}} \times 100 \ [\%] \qquad (13)$$

where V_{gap} [mm^3/s] is the volume of gap space, where thermal bubbles generated by the heater rise up per second. V_{bubble} [mm^3/s] is the volume of thermal bubbles generated per second, which is associated with the temperature rise of heater. V_{gap} and V_{bubble} are calculated by [16]

$$V_{gap} = b \times d \times v \, [\text{mm}^3/\text{s}] \qquad (14)$$

$$V_{bubble} = V_{mol} \times \frac{22.4 \times (T_b + \Delta T)}{273} \times \frac{0.1}{P} \times 10^6 \ [\text{mm}^3/\text{s}] \qquad (15)$$

$$V_{mol} = \frac{R I_h^2}{L + C_{PG}\Delta T} \times \frac{1}{M} \ [\text{mol/s}] \qquad (16)$$

$$\Delta T = \frac{q}{h} \ [\text{K}] \qquad (17)$$

where b [mm]: depth of heater (40 mm), d [mm]: gap length (2.5 mm), v [mm/s]: rising velocity of bubble to be measured, V_{mol} [mol/s]: mol number of bubbles generated by the heater per second, T_b [K]: boiling temperature of nitrogen, ΔT [K]: temperature rise from T_b, P [MPa]: pressure of LN2, R: resistance of heater (13.3 Ω), I_h [A$_{rms}$]: heater current, L [J/kg]:

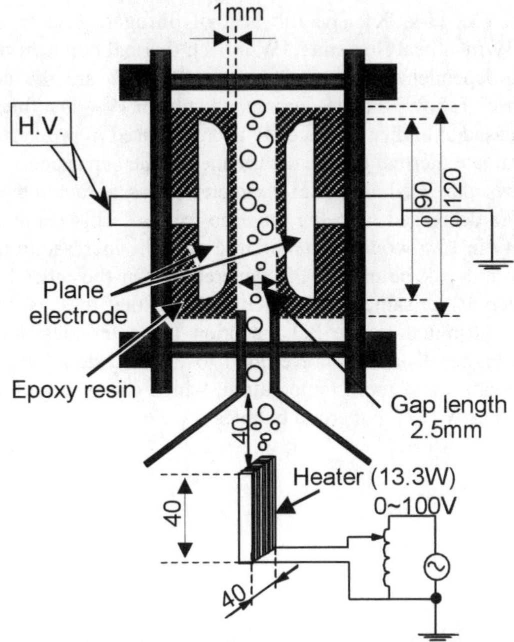

Figure 28   Experimental model of LN₂/epoxy composite insulation system

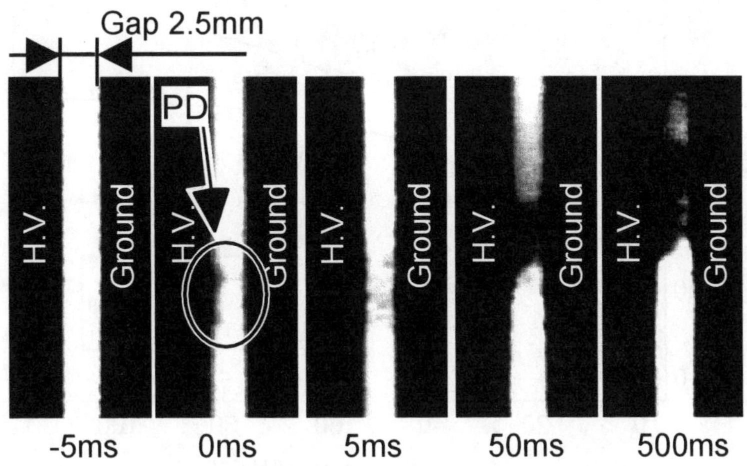

Figure 29   Thermal bubble behavior before and after PD inception in LN₂
($P = 0.15$ MPa, $V_{rate} = 0$ %, $V_a = 56$ kV$_{rms}$)

368

latent heat of LN2, C_{PG} [J/kg/K]: specific heat of nitrogen gas, M: molecular weight of nitrogen (28.0), q [W/m^2]: heat flux and h [W/m^2/K]: thermal conductivity of LN2. T_b, L, C_{PG} and h are pressure-dependent variables, whereas P and I_h are the control parameters at measurement. Figure 30 shows the measurement result for V_{rate} as a function of heater power $W = RI_h^2$. V_{rate} increased at higher values of W and decreased in pressurized LN2, enabling us to quantitatively evaluate thermal bubble disturbance in the gap space.

Figure 31 shows the Weibull scale parameter E_0 as a function of V_{rate} for different pressures of LN2. E_0 decreased with the increase in V_{rate}. This result makes clear that the discharge probability in LN2 would become large with the increase in diameter and number of thermal bubbles as weak points for PD generation. On the other hand, E_0 increased in pressurized LN2 even at the same V_{rate}, which suggests that the gas density in the thermal bubbles mutually contributed to the PD inception characteristics, together with V_{rate}. In addition, E_0 at the higher V_{rate} may correspond to the withstand strength of nitrogen gas between LN2 temperature and room temperature, which can be affected by the gas density and the vapor-mist density [17] in thermal bubbles.

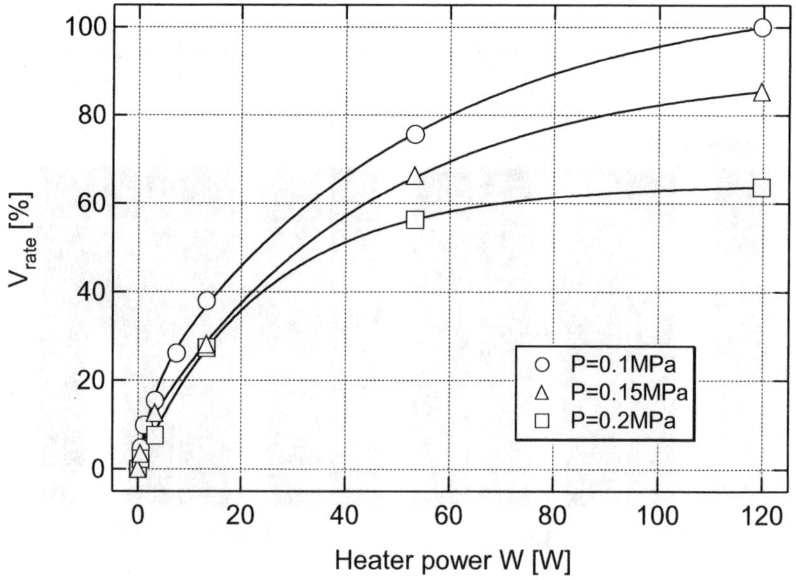

Figure 30 V_{rate} as a function of heater power

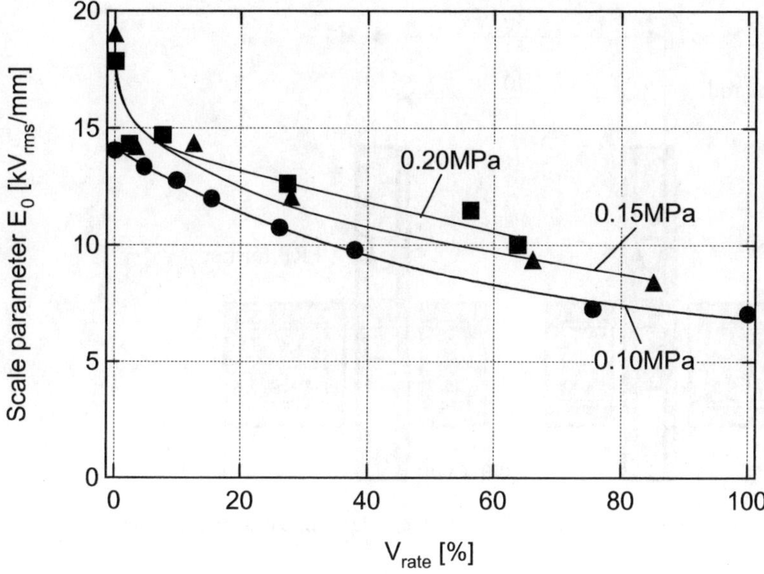

Figure 31   Scale parameter E_0 as a function of V_{rate}

5.3 LN₂/epoxy insulation system under thermal and electrical combined stress

Resistive superconducting fault current limiters (SFCL) featuring compactness and low cost are expected to be new power devices in the future electric network. As one promising countermeasure for enhancing the mechanical and dielectric strength of the resistive SFCL, such as a Melt Cast Processed (MCP)-BSCCO 2212 bifilar coil [18], the superconductors can be molded with epoxy resin, i.e. an LN₂/epoxy composite insulation system may also be used. However, since the resistive SFCL utilizes the quench of superconductors, thermal stress during the fault current limiting operation will be superposed on the electric field stress in the epoxy resin. We therefore investigated the electrical insulation performance of the LN₂/epoxy composite insulation system exposed to thermal and electrical combined stress, which may be regarded as the quench-induced dynamic insulation performance of the resistive SFCL [19].

Figure 32 shows the model electrode composed of a parallel pair of ring electrodes simulating the MCP-BSCCO 2212 bifilar coil molded with epoxy resin. The outer coil serves to induce electromagnetic current and temperature rise in the ring electrodes, whereas the inner coil generates thermal bubbles in the inner space of the model electrode. The materials of the upper and lower ring electrodes are aluminum and SUS304, respectively; the outer and inner coils consist of copper wire. Aluminum was selected as the material of the upper ring electrode in order to induce a temperature rise only in the upper ring electrode. Note that the temperature of the lower ring electrode does not increase due to the high resistivity of SUS304. Such a composite material for the ring electrodes brings high thermal stress, superposed on high electric field stress, to the upper ring electrode.

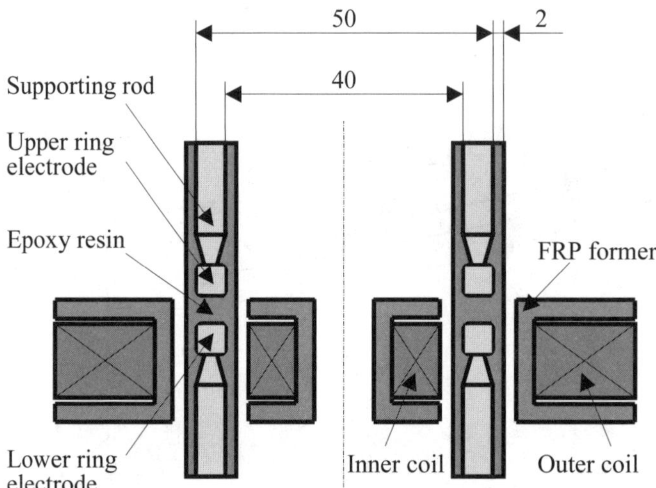

Figure 32 Parallel pair of ring electrodes molded with epoxy resin

The model electrode was immersed in LN2 at atmospheric pressure and exposed to 50 Hz ac high voltage V_a. PD inception voltage of the model electrode was confirmed higher than 30 kVrms and the breakdown (BD) voltage to be 60 kVrms, which corresponded to the intrinsic or static insulation level. Under electric field stress below the static insulation level, the outer and inner coils connected in series were energized by 50 Hz ac large current I_m for 1 second. V_a and I_m were enhanced in steps by 5 kVrms and about 100 Arms, respectively, until PD or BD was induced in the model electrode. The temperature rise of the ring electrodes was measured in advance by embedded Chromel-Alumel thermocouples, and calibrated in consideration of thermal response time of the thermocouples.

Figure 33 shows the PD and BD characteristics, where the vertical axis represents maximum electric field strength on (a) the upper ring electrode and (b) the epoxy surface, and the horizontal axis represents maximum temperature of the upper ring electrode. Both figures have a criterion designated by a solid curve to discriminate PD and/or BD region from no PD&BD region. The electric field strength and temperature rise of the superconductors during fault current limiting operation should therefore be reduced below the criteria in Figs.33 (a) and (b), which can be used as a guideline in electrical insulation design of the resistive SFCL.

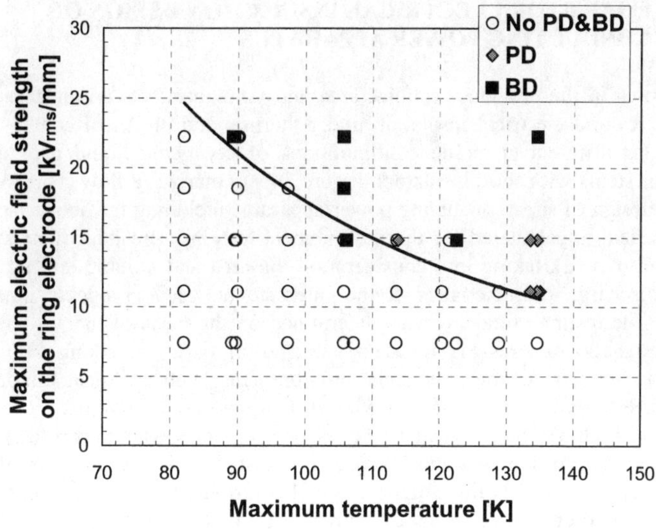

(a) Maximum electric field strength on upper ring electrode vs. temperature

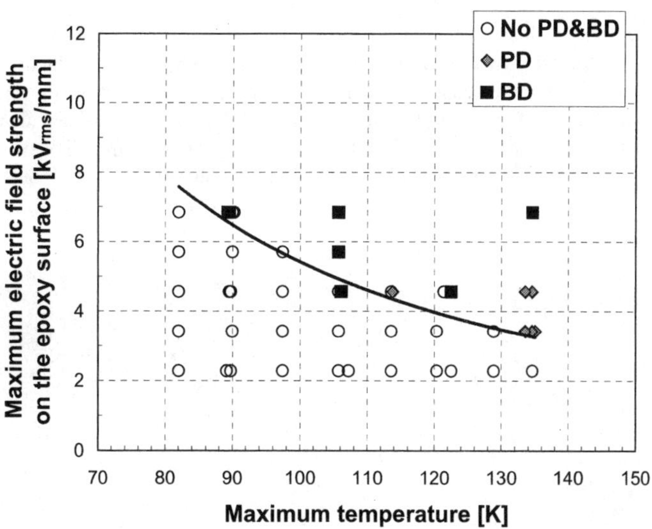

(b) Maximum electric field strength on epoxy surface vs. temperature

Figure 33   PD and breakdown characteristics under thermal and electrical combined stress

6. FLOW CHART FOR ELECTRICAL INSULATION DESIGN OF SUPERCONDUCTING POWER APPARATUS

As shown in the previous sections, area and volume effects on dielectric strength, V-t characteristics and electrical insulation under thermal and electrical combined stress were discussed for different electrode configurations of cryogenic liquid gaps and composite insulation systems with solid insulators. Figure 34 summarizes a flow chart for the electrical insulation design of superconducting power apparatus including the above factors [5]. First, an electric field is calculated for the design specifications. Breakdown stress level is then evaluated in part (a), taking into consideration the area and volume effects, as well as the creepage discharge characteristics at the interface between cryogenic liquids and solid insulators. The testing voltage can be determined on the basis of the V-t characteristics in part (b). Breakdown stress is then re-evaluated in part (c), taking into consideration phenomena peculiar to the cryogenic environment, such as quench-induced dynamic breakdown. In addition, safety factors (design margins) are determined, in view of system coordination and the statistical scattering of discharge phenomena, in part (d).

The procedure in Fig.34 should be greatly assisted by systematization of the electrical insulation data described in this article, and will contribute to the establishment of electrical insulation technology at cryogenic temperature applicable to practical and efficient insulation design and the development of superconducting power apparatus.

7. ACKNOWLEDGEMENTS

The research in this article was supported in part by a Grant-in-Aid for Scientific Research of the Ministry of Education, Science, Sports and Culture of Japan, and by Super-ACE (R&D of fundamental technologies for superconducting AC power equipment) project of the Ministry of Economy, Trade and Industry of Japan. The authors also express their sincere thanks to the collaborators of Chubu Electric Power Co. Inc., Tokyo Electric Power Company, Super-GM, Central Research Institute of Electric Power Industry, Sumitomo Electric Industries Ltd. in Japan, and Forschungszentrum Karlsruhe in Germany.

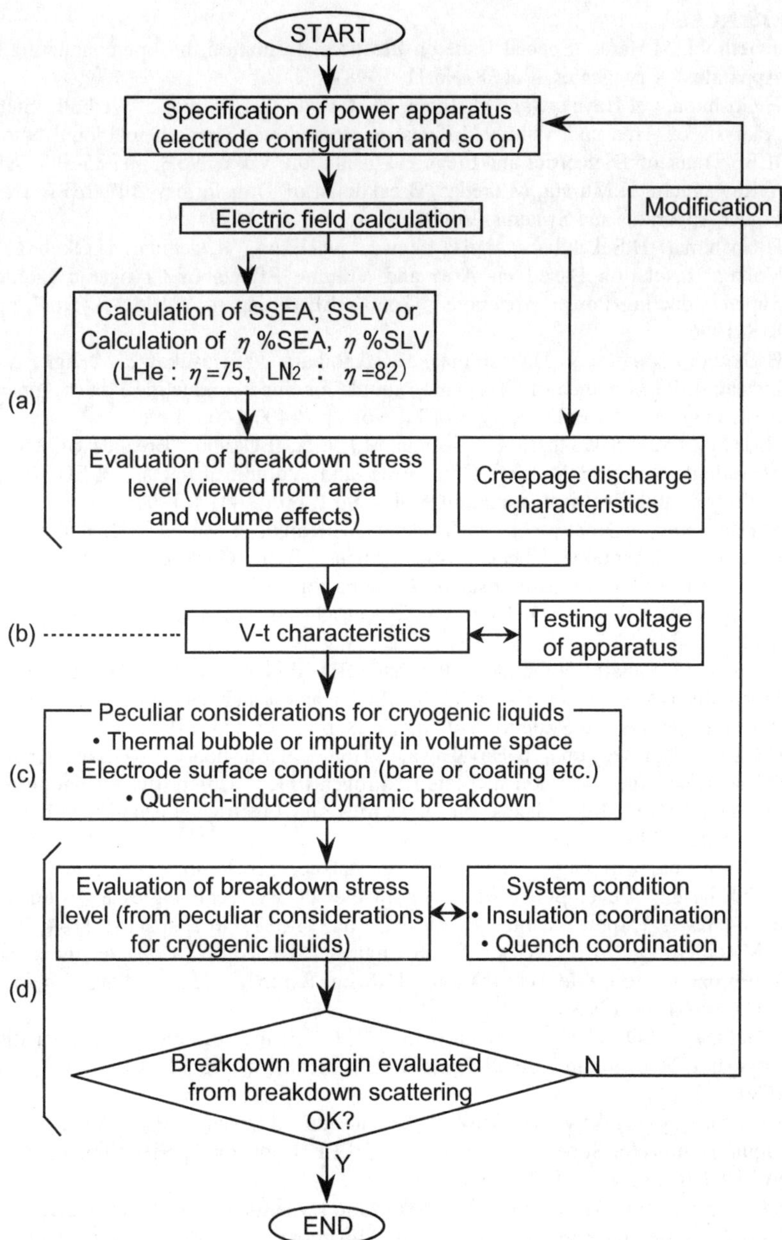

Figure 34   Flow chart for electrical insulation design of superconducting power apparatus

REFERENCES

[1] J.Gerhold, M.Hara: "Special Issue on Electrical Insulation in Superconducting Power Apparatus", Cryogenics, Vol.38, No.11, 1998

[2] H.Goshima, N.Hayakawa, M.Hikita, K.Uchida, H.Okubo: "Weibull Statistical Analysis of Area and Volume Effects on Breakdown Strength in Liquid Nitrogen", IEEE Trans. on Dielectrics and Electrical Insulation, Vol.2, No.3, pp.385-393, 1995

[3] Y.Kawaguchi, H.Murata, M.Ikeda: "Breakdown of Transformer Oil", IEEE Trans. on Power Apparatus and Systems, Vol.91, No.1, pp.9-23, 1972

[4] H.Goshima, H.Sakakibara, N.Hayakawa, M.Hikita, K.Uchida, H.Okubo: "High Voltage Insulation Based on Area and Volume Effects of Cryogenic Liquids for Superconducting Power Apparatus", Trans. IEE of Japan, Vol.116-B, No.7, pp.812-818, 1996

[5] H.Okubo, M.Hikita, H.Goshima, H.Sakakibara, N.Hayakawa: "High Voltage Insulation Performance of Cryogenic Liquids for Superconducting Power Apparatus", IEEE Trans. on Power Delivery, Vol.11, No.3, pp.1400-1406, 1996

[6] N.Hayakawa, H.Sakakibara, H.Goshima, M.Hikita, H.Okubo: "Breakdown Mechanism of Liquid Nitrogen Viewed from Area and Volume Effects", IEEE Trans. on Dielectrics and Electrical Insulation, Vol.4, No.1, pp.127-134, 1997

[7] M.Hazeyama, T.Kobayashi, N.Hayakawa, S.Honjo, T.Masuda, H.Okubo: "Partial Discharge Inception Characteristics under Butt Gap Condition in Liquid Nitrogen/PPLP® Composite Insulation System for High Temperature Superconducting Cable", IEEE Trans. on Dielectrics and Electrical Insulation, Vol.9, No.6, pp.939-944, 2002

[8] M.Ikeda, T.Yanari, H.Okubo: "PD and BD Probability Distribution and Equi-Probabilistic V-t Characteristics of Oil-filled Transformer Insulation", IEEE Trans. on Power Apparatus and Systems, Vol.101, No.8, pp.2728-2735, 1982

[9] H.Okubo, M.Hazeyama, N.Hayakawa, S.Honjo, T.Masuda: "V-t Characteristics of Partial Discharge Inception in Liquid Nitrogen/PPLP® Composite Insulation System for HTS Cable", IEEE Trans. on Dielectrics and Electrical Insulation, Vol.9, No.6, pp.945-951, 2002

[10] S.Mukoyama, M.Yagi, S.Tanaka, K.Matsuo, S.Honjo, T.Mimura, T.Aiba, Y.Takahashi: "Development of important elementary technologies for a 66kV-class three-phase HTS power cable", Physica C, Vol.378-381, No.2, pp.1181-1184, 2002

[11] G.M.Hathaway, A.E.Davies, S.G.Swingler: "Dielectric Considerations for a Superconducting Cable Termination", 11th Int. Symp. on High Voltage Engineering, Vol.4, pp.84-87, 1999

[12] A.Bulinski, J.Densley, T.S.Sudarshan: "The Ageing of Electrical Insulation at Cryogenic Temperature", IEEE Trans. on Electrical Insulation, Vol.15, No.2, pp.83-88, 1980

[13] S.Chigusa, N.Hayakawa, H.Okubo: "Quench-induced Dynamic Breakdown Strength of Liquid Helium for Superconducting Coils", IEEE Trans. on Applied Superconductivity, Vol.10, No.1, pp.1518-1521, 2000

[14] S.Chigusa, N.Hayakawa, H.Okubo: "Static and Dynamic Breakdown Characteristics of Liquid Helium for Insulation Design of Superconducting Power Equipment", IEEE Trans. on Dielectrics and Electrical Insulation, Vol.7, No.2, pp.290-295, 2000

[15] N.Hayakawa, H.Maeda, S.Chigusa, H.Okubo: "Partial Discharge Inception Characteristics of LN_2 / Epoxy Composite Insulation System under Thermal Bubble Condition", Cryogenics, Vol.40, pp.167-171, 2000

[16] A.Kamitani, T.Amano, S.Sekiya, A.Ohara: "Stability Analysis of a Forced-Flow Cooled Superconducting Coil: Numerical Simulation of Multiple Stability", Cryogenics, Vol.31, pp.110-118, 1991

[17] H.Goshima, T.Suzuki, N.Hayakawa, M.Hikita, H.Okubo: "Dielectric Breakdown Characteristics of Cryogenic Nitrogen Gas above Liquid Nitrogen", IEEE Trans. on Dielectrics and Electrical Insulation, Vol.1, No.3, pp.538-543, 1994

[18] M.Noe, K.-P.Juengst, F.N.Werfel, S.Elschner, J.Bock, A.Wolf, F.Breuer, "Measurements and tests of HTS bulk material in resistive fault current limiters", Physica C, Vol.372-376, Part III, pp.1626-1630, 2002

[19] N.Hayakawa, M.Noe, K.-P.Juengst, H.Okubo: "Electrical Insulation Performance under Thermal and Electric Combined Stress for Resistive Superconducting Fault Current Limiters", Applied Superconductivity Conference, 2LE02, 2002

MODELING CURRENT FLOW IN GRANULAR SUPERCONDUCTORS AND IMPLICATIONS FOR POTENTIAL APPLICATIONS

N. A. RUTTER and A. GOYAL
Metals & Ceramics Division, Oak Ridge National Laboratory,
Oak Ridge, TN 37831, USA

1. INTRODUCTION

A coated conductor, fabricated by the Rolling Assisted Biaxially Textured Substrates (RABiTS) route consists of a metallic substrate, one or more buffer layers and a high temperature superconducting layer such as $YBa_2Cu_3O_{7-\Gamma}$ (YBCO) [1, 2]. The metallic tape, which is usually Ni-based, is cold-rolled and recrystallized in order to induce the cube texture $\{001\}<100>$ [3]. If the tape is a Ni-alloy, it may be coated with a layer of pure Ni [4]. The substrate may also be annealed in a sulfur-containing environment in order to produce a surface onto which oxide layers may be grown epitaxially [5]. Buffer layers such as Y_2O_3, YSZ and CeO_2 are deposited, maintaining the cube texture of the underlying grains [6]. The superconductor is then deposited such that it grows biaxially, with the crystallographic c-axis perpendicular to the tape and with the a and b axes aligned with the rolling and transverse directions of the tape. Electron backscatter measurements have demonstrated that the crystallographic orientations of the superconducting grains are indeed determined by those in the underlying layers [7]. Hence the superconducting film is made up of a network of grains which have crystallographic orientations that are within a few degrees of $\{001\}<100>$.

The critical current density (J_c)[#] of such a superconducting grain network is likely to be limited by dissipation at the low-angle grain boundaries, which generally have less current-carrying capability than the grains [8]. Current flow in polycrystalline coated conductors therefore relies on the transfer of current across the grain boundaries. Thus when attempting to model current flow, the physical structure of the superconducting film may be represented as a 2-D network of grains, each being oriented close to {001}<100>. The current which is able to flow between any two neighboring grains depends upon the relative misorientation of the grains and the surface area of the common interface.

2. COATED CONDUCTOR MODELS

A number of models which study current percolation in networks of low-angle grain boundaries have been developed [9-14]. The factors studied in such models include grain size, sample dimensions, extent of grain misorientation and grain shape.

2.1 Grain Shape

In most models the conductor consists of a regular array of equally sized, identically shaped grains. The models of Specht et al. [10] and Rutter et al. [11,12] use an array of hexagons, whilst Nakamura et al. [13] uses squares. The only model in which the grains are not identically shaped is that of Holzapfel et al. [14]. This model is unlike the others in that the grain structure used is one that has been measured by Electron BackScatter Diffraction (EBSD) rather than being simulated.

Each of these methods of modeling the structure has various advantages and disadvantages. The square-grain structure is the simplest to model, but its main limitation is that each grain has only four neighbors whereas in real tapes the number of neighbors is around six. The other problem with the square-based model is that the length of the shortest grain boundary path across the width of the sample is exactly equal to the tape width. In a real tape, this is unlikely to be true, thus the critical current (which is proportional to the length of the grain boundary path across which dissipation occurs) is likely to be underestimated by the square model.

The use of a hexagonal model addresses these issues. The number of nearest neighbors is six, and the shortest boundary path length is increased in excess of the tape width. One of the main problems of the hexagonal grain model is that one of two distinctive orientations, which are shown in figure 1, must be chosen.

[#] The uppercase symbol J will be used to represent macroscopic current densities of samples and the lowercase symbol j to represent the microscopic current densities in individual grains and boundaries.

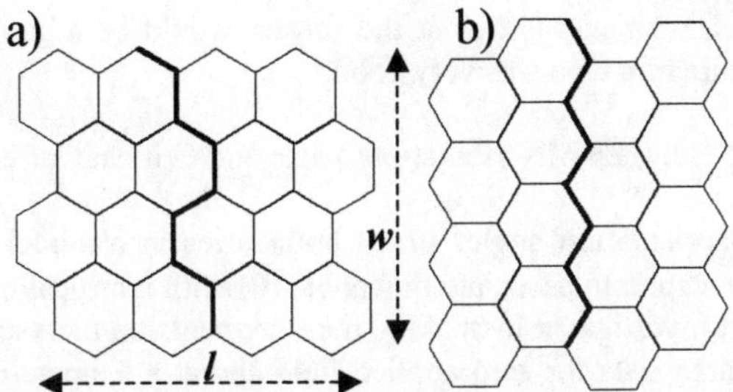

Figure 1. Different ways in which a 4x4 array of hexagonal grains can be arranged. In a) the minimum grain boundary path across the conductor width is $4w/3$, whereas in b), the value is $2w/\sqrt{3}$.

Both existing models [10, 11] placed the hexagons in the orientation shown in figure 1a), relative to the tape length and width directions. The minimum boundary path length (marked in bold) is now approximately a factor of 4/3 greater than the tape width. In the arrangement shown in figure 1b) this minimum length is just $2/\sqrt{3}$ times the width. Also it is clear from the figure that the number of grains along the length of the tape and the number across the width are not as easily defined as for the square grains. Whilst both cases would be well described as a 4 by 4 array of hexagons, the aspect ratios of the samples are quite different.

The Holzapfel model [14] has grains with more irregular shapes. In this case the structure is that of a real tape measured by EBSD. The resolution of the grain edges is determined by the size of the square pixels in the array. This model is more realistic in that the grains have a range of shapes and sizes and have differing numbers of nearest neighbors.

2.2 Grain Boundary Misorientation

The first attempt to study critical currents in coated conductors using percolation modeling [10] used a bimodal distribution of boundary currents, assigning values of 0 or 1 randomly to the grain boundaries, rendering a given fraction, f, conducting and the remainder, $(1-f)$, non-conducting. Subsequent models have attempted to assign critical currents to the boundaries based on grain misorientation. Holzapfel was able to use the true orientations of the grain as measured by EBSD to determine the boundary misorientation. An alternative is to use a gaussian distribution to assign grain orientations [13] before calculating the boundary misorientations. This effectively simulates a conductor for which the grains are misoriented about one axis (for example only in-plane rotations). However the result is a grain boundary misorientation distribution (GBMD) which is not quite accurate. The GBMD produced by this approach will always give a peak at 0°, indicating that that is the most common grain boundary angle. If a 3-D orientation is assigned to each of the grains [11], the GBMD instead indicates no 0° boundaries and a peak at an angle of a few degrees. The latter approach produces results similar to EBSD measurements [12], but the former would be a good approximation if the out-of-plane texture of a tape was very good.

2.3 The Relationship Between Misorientation Angle and Critical Current Density

After the misorientation angles of all boundaries in a model sample have been determined, the next step is to associate this angle (θ) with a critical current density. The relationship was first investigated by making measurements on bicrystal grain boundaries [15, 16]. As most data exists for zero applied field and at a temperature of 77 K, this is the situation that is usually considered in modeling. This relationship between j_c and θ [17-20] is not precisely known and thus there have been various different attempts at quantifying it. It is most generally expressed as

$$j_c(\theta) = j_c' \exp\left(\frac{-(\theta - \beta)}{\alpha}\right) \quad \text{for } \theta > \beta \tag{1}$$
$$j_c(\theta) = j_c' \quad \text{for } \theta < \beta$$

where j_c' is the value measured for single crystals.

Equation 1 therefore suggests that there is a plateau region for very small angles where there is no decrease in j_c, followed by an exponential decrease with increasing angle. It is believed that this "plateau region" is simply a region in which the measured properties are determined by the properties of the grain [12, 19] and that the grain boundary properties probably continue to improve all the way to 0°. It has been noted that twin boundaries, which are present within the granular regions, are effectively 1.8° boundaries [21], and thus it is assumed that when a measurement is made on a bicrystal boundary which has a misorientation angle less than 1.8°, the critical current will be limited by the grains.

Table 1. A summary of the values of α for measurements on [001] tilt boundaries in thin film YBCO in self field at 77 K

Author	α (°)
Ivanov [17]	4.4
Heinig [18]	3.8
Verebelyi [19]	3.2
Holzapfel [14]	2.4

Table 2. Values of α and β used in percolation models

Author	α (°)	β (°)
Rutter [11,12]	3.4	0
Holzapfel [14]	2.4	2.4
Nakamura* [13]	3.0	3.0

not stated explicitly – estimated from text and figure 3

Whether or not the plateau is included, there is still some variation in the values quoted for the factor α. Tables 1 and 2 show some values estimated from bicrystal measurements and the values used in percolation models. Figure 2 shows this relationship schematically. The straight line portion represents the properties of grain boundaries with misorientation angle (θ), whilst a plateau is shown representing cases where the grains have the same j_c as a) 4° boundaries [14] or b) 1.8° boundaries [21].

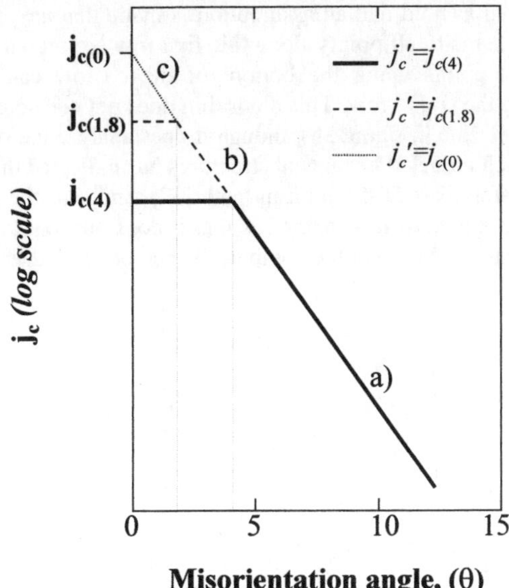

Figure 2. Schematic showing the relationship between critical current and boundary angle, with plateaus at a) 4° and b) 1.8° due to granular dissipation.

2.4 Algorithms To Calculate The Critical Current

When the grain structure has been determined and the critical currents of the grain boundaries assigned, the calculation of the critical current depends upon the solution of a network problem. Consider a path along the grain boundaries, which traverses the width of the sample. The maximum current which can cross this path is equal to the sum of the critical currents of the grain boundary segments which comprise the path. When all such paths which traverse the sample width are considered, the 'weakest' of these is the one that limits the amount of current which can flow along the sample length. If this path can be found and the sum of the boundary currents comprising the path is calculated, the critical current of the sample is known. This method is based on a property of networks known as the maximum-flow/minimum-cut theorem, proved by Ford & Fulkerson [22]. In physical terms, when the critical current of a sample is reached this limiting path of grain boundaries will be where dissipation occurs and in the presence of a magnetic field, the path will act as a flux flow channel [23, 24].

One inexact algorithm which has been utilized to calculate the minimum boundary sum for hexagonal grains [11, 12] is the following. The minimum path across

382

the conductor width along the grain boundaries is sought. The width is comprised of a number of rows of grains as shown in figure 3a). Several possible routes from the top of the sample to the point at the bottom of row 1 are indicated in bold. The route involving the lowest critical current sum from the top of the conductor to this point is likely to be one of the paths marked in bold and an assumption is made that any other paths will be longer. The minimum sums to all points along this first row are found. Subsequently, the minimum sums to all points along the bottom of the rth row can be calculated by considering paths from the $(r-1)^{th}$ row. This algorithm does not consider paths such as the one shown by the dotted line in figure 3b), though it does analyze the type of path shown in bold. Analysis of such modeled hexagonal structures has indicated that the paths which are not analyzed are rarely part of the limiting path [12], and thus the algorithm is often accurate. The main advantage of this method is that it does not require that all the grain boundary critical currents be stored in the computer's memory at the same time.

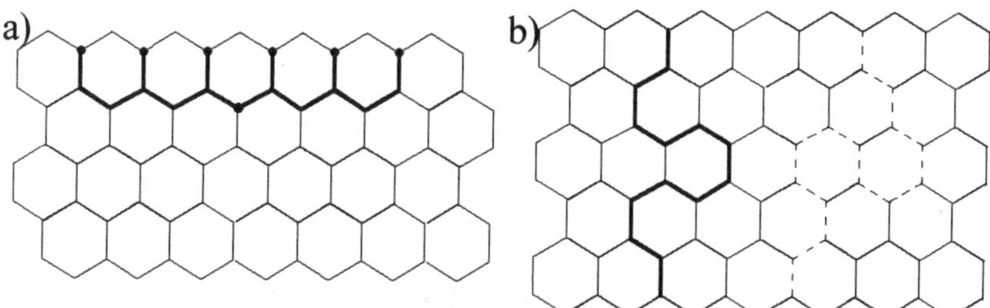

Figure 3. a) Illustration of the algorithm for finding the minimum current sum across the tape width. 6 alternative paths from the top of the conductor to a point at the bottom of the top row are shown in bold. b) shows a path across the conductor which will be considered (bold) and one which will not (dotted line).

The method used by Holzapfel et al. [14] does produce an exact solution. It describes each square in the array as a 'room' and each boundary as a 'door'. A unit of current is represented by a 'mouse', which attempts to walk from one end of the conductor to the other by passing through the doors using a rule such as always keeping the walls on the right hand side. When a mouse gets to the end, the capacity of each door for carrying further mice is reduced by one. When a mouse fails to reach the other end of the tape, the critical current has been reached. Using this method the current is quantized and if the ratio of the highest possible boundary critical current to the lowest is to be sufficiently high ($\sim 10^4$), then the number of "mice" which must travel through the sample will be very high and the calculation will be time consuming for large networks.

There are a number of ways in which this method may be made more efficient. The first is to close permanently any door which a single mouse goes through in both directions in order to prevent further mice going down dead-ends. An alternative is to use mice which can carry different amounts of current, with the large capacity mice going through the network first, followed by smaller capacity mice.

The mouse-walk method is based upon the Ford-Fulkerson algorithm of flow-augmenting paths [25-27]. This algorithm finds a path through the network along which the capacity has not yet been reached (a flow-augmenting path). The flow (current) is then increased by the maximum amount which may flow along that path (δ), and the remaining capacity of each boundary along the path is reduced by δ. When no more flow-augmenting paths remain, the maximum flow has been reached and the critical current has been calculated.

3. DEVELOPMENT OF A NEW MODEL

The main limitation of most existing models is that they use grain shapes which are unrealistic (i.e. hexagons or squares). The Holzapfel model [14] calculates a critical current in a more realistic grain network, but is limited by the need to perform the EBSD measurement. The aim is therefore to develop a model which combines the type of grain structure in the Holzapfel model with flexibility of conductor scale and crystallographic orientation.

3.1 Simulation of a 2-D Grain Network

Grain structure simulation is carried out using the Monte-Carlo Potts method [28] which was developed as a grain growth model by Anderson et al. [29-31]. The method has been used recently by a number of authors [32-35] to generate both 2-D and 3-D grain structures. For the purposes of a coated conductor model, we use a 2-D array of square pixels. Initially each pixel is attributed a unique index number or letter, which identifies it as a grain and hence at the start of the simulation, the sample consists of a large number of small, square grains. The algorithm used to simulate grain growth begins by randomly selecting an individual pixel. The overall "energy" associated with that pixel is determined based on the number of near neighbors (of the 8 linked by edges or vertices) which are in different grains. The algorithm then evaluates the effect on the energy of changing the index number of the pixel such that it is in the same grain as one of its 4 nearest neighbors (linked by edges). If this energy change is favorable (i.e. negative) or neutral (zero), then the change is made with probability (P) equal to 1. If however, the energy change would be positive, then the change is made with a probability given by equation 2.

$$P = \exp\left(\frac{-\Delta G}{kT}\right) \tag{2}$$

ΔG is the difference between the number of neighboring pixels in different grains before and after the change is made. The factor kT is a thermodynamic variable and has been set equal to 1 for the purposes of this model.

An example of this process is shown in figure 4. The grain structure starts out as shown in a) with the center pixel selected. It has an "energy" of 5 (5 out of 8 neighbors which share an edge or vertex are in other grains). It will then be provisionally assigned the index of one of the four neighbors with which it shares an edge and the consequences of making the change are evaluated. The trivial case is that it 'changes' to grain A, and the structure is maintained with no effect on the overall energy. Alternatively, it could attempt to change to B which would result in the situation shown in figure 4b). The new energy for this pixel would be 4 units, and as this is favorable, the change would be confirmed. The remaining alternative is to attempt to become a C grain, resulting in the structure shown in c). As this energy change is positive (+2 units), the change will occur with a probability of exp(-2).

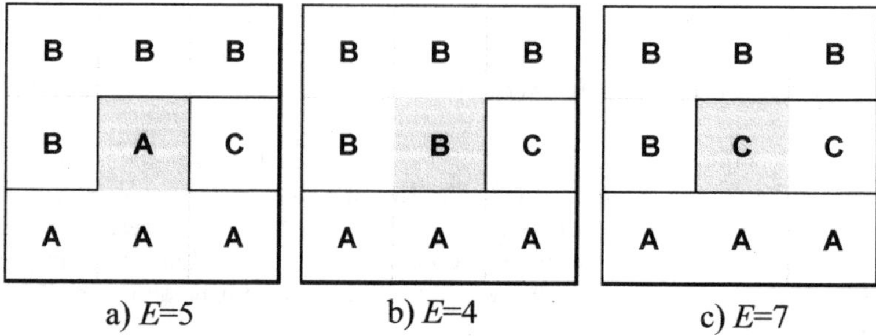

Figure 4. The center pixel has an energy of either a) 5, b) 4 or c) 7 depending on which grain it is in.

When the number of pixel index changes attempted is equal to the total number of pixels in the lattice, one time period of grain growth, known as a Monte Carlo Step (MCS) has occurred. Grain growth progresses to produce structures such as those shown in figure 5 for a 300 x 300 pixel lattice.

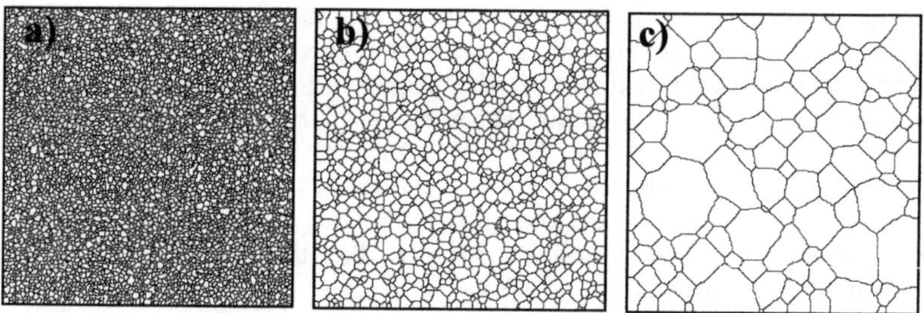

Figure 5. Grain structures generated by the model on a 300x300 pixel grid with MCS numbers of a) 10, b) 100 and c) 1000

Once a physical grain structure has been established, the crystallographic orientation of each grain is assigned. This assignment is carried out randomly based on a typical X-Ray gaussian distribution as is the case in other simulations. In order to simplify the model slightly, only a single grain misorientation (i.e. in-plane) is considered here, rather than the 3-D misorientation used elsewhere [11, 12]. Except where stated otherwise a FWHM equal to 10° is used.

3.2. Defining Sample Parameters

A modeled sample, which carries current along its length, is composed of P_l pixels along its length and P_w pixels across its width. There are an average of N_l grains in the length direction and N_w grains across the width. Each grain is, on average composed of p_l pixels in the "length" direction and p_w in the "width" direction. This is shown in figure 6.

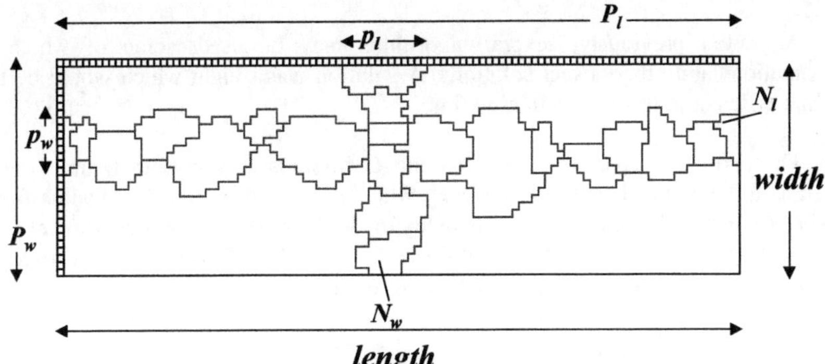

Figure 6. Definition of the total number of pixels (P), the average number of pixels per grain (p) and the number of grains in the width (N_w) and length (N_l) directions.

Hence, $N_l \times p_l = P_l$ and $N_w \times p_w = P_w$, thus so for equiaxed grains, $p_l = p_w = p$.

The average grain size is determined by calculating the total numbers of "vertical" grain boundary segments, v, and "horizontal" boundary segments, h, in the sample.

$$p_l = \frac{(P_l - 1)(P_w - 1)}{v} \tag{3}$$

$$p_w = \frac{(P_l - 1)(P_w - 1)}{h} \tag{4}$$

Note that the (P-1) term arises from the fact that the sample edge boundaries are not counted when summing the grain boundaries. This method is essentially equivalent to using a linear intercept method in the two perpendicular sample directions.

3.3 Calculating the Critical Current

The relationship between grain boundary angle and j_c was discussed earlier (see equation 1). Values of $\alpha = 2.4°$ and $\beta = 1.8°$ will be used in the initial calculations. Hence all grain boundaries for which $\theta \leq 1.8°$ will be indistinguishable from the grains. Results will be normalized by dividing the calculated critical current density (J) by the single crystal value j_c'.

Once the network of grain boundaries, each with an assigned critical current, has been established one needs to calculate the maximum current that may flow through such a network. One requires an algorithm which determines the path of grain boundaries across the width of the conductor whose critical current sum is the minimum of all such possible paths.

$$J_c = \frac{1}{w} \sum_{min} (j_c \cdot l) \tag{5}$$

As noted previously, several algorithms may be used, some of which are approximations and others exact solutions. A solution was sought which would be both exact and efficient to reduce calculation time.

The pixel grid consists of R rows and C columns as shown in figure 7 and is indexed as follows. Pixels are indexed $[r,c]_p$ with r running from 0 to $(R-1)$ and c from 0 to $(C-1)$. A horizontal boundary with the index $[r,c]_h$ forms the upper edge of pixel $[r,c]_p$. A vertical boundary with the index $[r,c]_v$ forms the left edge of pixel $[r,c]$. A vertex with the index $[r,c]_x$ is located at the upper left corner of pixel $[r,c]_p$.

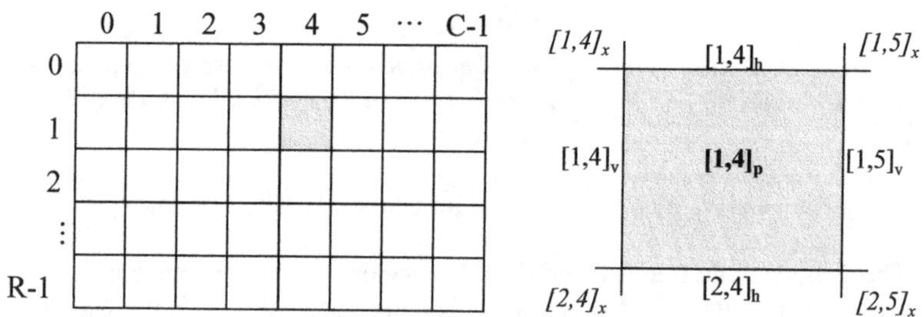

Figure 7. Indexing of cells, edges and vertices in the square pixel model for a conductor C columns long and R rows wide.

Each pixel boundary has already been assigned a critical current. Hence values of $[r,c]_h$ are filled from $r=1$ to $R-1$ and $c=0$ to $C-1$ and values of $[r,c]_v$ are filled from $r=0$ to $R-1$ and $c=1$ to $C-1$. The calculation is performed by finding minimum values of paths to the vertices and thus initial values of the vertices must be assigned as follows :

$$\begin{aligned} [r,c]_x &= 0 && \text{for } r=0 \\ [r,c]_x &= \infty && \text{for all other } r \end{aligned} \tag{6}$$

The problem is to find the 'shortest' route from the vertices in row 0 to those in row R, where shortest means having the lowest critical current sum. The calculation is carried out in 2 stages as follows :

Consider a vertex $[1,c]_x$. One route from the top of the sample to it is along vertical boundary $[0,c]_v$. The critical current sum along this path is equal to the sum of the value stored at $[0,c]_x$ and the value of $[0,c]_v$. If this sum is lower (which must be the case here since $[1,c]_x=\infty$) then the new value replaces the old one,

Step 1: $\qquad [r,c]_{x(new)} = \min \{ [r,c]_{x(old)} , [r-1,c]_x + [r-1,c-1]_v \} \qquad (c=1 \to C\text{-}1) \qquad$ (7)

where $\min \{ a , b \}$ represents the lower of values a and b.

When this calculation is performed with c ranging from 1 to C, the vertex values in row 1 are now critical current sums between there and the top of the conductor. However they may not be the minimum values, a lower-sum paths may exist which include the horizontal boundaries in row 1. Thus the following calculations are performed for the vertices in row 1 :

Step 2: $\qquad [r,c]_{x(new)} = \min \{ [r,c]_{x(old)} , [r,c-1]_x + [r,c+1]_h \} \qquad (c=1 \to C\text{-}1) \qquad$ (8)

Step 3: $\qquad [r,c]_{x(new)} = \min \{ [r,c]_{x(old)} , [r,c+1]_x + [r,c]_h \} \qquad (c=C\text{-}1 \to 1) \qquad$ (9)

The 3 steps are carried out in sequence along each row of the array from top to bottom, and when this has been completed, the values of the vertices along row R represent critical current sums for possible routes across the sample width, but these are not necessarily the lowest J_c paths. The algorithm can be made to produce an exact solution by continuing as follows :

Step 4: $\qquad [r,c]_{x(new)} = \min \{ [r,c]_{x(old)} , [r+1,c]_x + [r,c]_v \} \qquad (c=1 \to C\text{-}1) \qquad$ (10)

followed by steps 2 and 3, first in the $(R\text{-}1)^{th}$ row, then in each subsequent row up to row 1. Steps 1-3 are then repeated from the top of the sample to the bottom.

If these steps leave the values at all the vertices in the grid unchanged, then the solution found in the initial stage (steps 1-3 only) was exact and the lowest value along row R is the critical current. Otherwise, this scan along the rows of vertices is repeated up to the top of the sample (steps 4,2,3) and back to the bottom (steps 1,2,3) until no changes are made to the vertex values.

This method was used for a number of model structures alongside the "mouse-walk" algorithm [14] and the Ford-Fulkerson algorithm [25-27] to ensure that the results were identical. The time taken to solve a 600 x 600 pixel array with grain sizes of 5 x 5 pixels completely is approximately 80 seconds, of which the initial scan of the grid (steps 1,2,3) is completed within the first 10 seconds. The answer produced by just the single downward scan is on average an overestimate of approximately 10-15%. This algorithm as coded is many times faster than the Ford-Fulkerson method. As the values of all grain boundaries are held in memory at the same time, the limitations of the hardware used to carry out the calculation restrict the size of network used to around 400,000 pixels.

4. MODELING RESULTS

4.1 Features of the Model

First we will examine the nature of the grain structures produced by the Anderson grain growth model. One Monte-Carlo Step (MCS) is defined as attempting $P_l \, x \, P_w$ grain transitions, and this is the basic time unit for growth. Growth is symmetrical in the vertical and horizontal directions and so p_l and p_w will be equal. When MCS=0, the grains are all composed of justs a single pixel ($p = p_l = p_w = 1$). The dependence of p, the average grain size expressed in pixels, as MCS is increased is shown in figure 8. The structures from which the grain sizes were calculated were generated on a 600x600 square pixel lattice. The model replicates the result found by Burke [36] such that the grain size is proportional to the square root of the growth time, which is represented by the MCS number. Note that the resulting grain size for a given MCS value is not dependent upon the sample size or shape (as long as there remain several grains across both width and length).

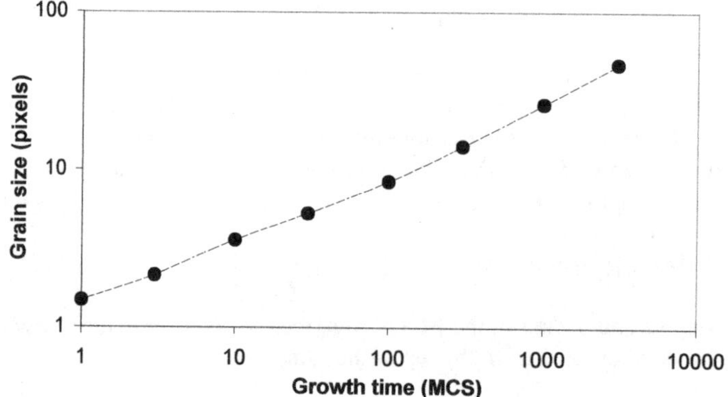

Figure 8. Plot showing grain growth rate in the Anderson model. Grain size is the average linear dimension.

The fact that there are grains of different sizes for cases other than p=1 is an important feature of the model. Figure 9a) shows the linear grain size distribution functions for average grain sizes of 1.9 and 4.7 pixels and 9b) shows a normalized plot, in which the grain size is divided by the average. These plots demonstrate that the peak in the distribution is just below the mean grain size, and only approximately 1% of grains have a linear dimension which is greater than 3 times the average. This distribution has been noted to correspond closely with that measured in real materials [33].

In order to discover the essential differences between this percolation model and previous models, samples containing N_l grains along the length and N_w grains across the width are considered. In the simplest case, no MCS steps are carried out and the grain structure consists of square grains, each 1 pixel by 1 pixel in size. This situation is then identical to the Nakamura model [13]. We compare this with samples which still have N_l x N_w grains, but the grains are irregularly shaped and have $p>1$.

Figure 10 shows three such modeled samples, for which $N_l=25$ and $N_w=10$. The simplest case is that where the grains are regular, identical squares, $p=1$ (Fig. 10a). Figure 10b shows a structure for which $p=2$. Whilst the grain structure is still not realistic, it does at least have a range of grain sizes and there are two other differences which may be important for percolation. The average number of neighbors with which each grain shares a boundary is greater than 4 and also the minimum grain boundary path length connecting the upper and lower edges of the sample is now somewhat greater than the sample width. In figure 10c), for which $p=5$, it is evident that a much more realistic grain structure exists than in the previous cases, even though the pixel resolution is still apparent.

In order to understand the differences between the square-grain model and one with more realistic grains, the critical current densities of samples such as those illustrated in figure 10 have been calculated. Samples with $N_l=200$ and $N_w=20$ have been considered with the value of p varying from 1 to 10 in order to study the effect upon J_c. Additionally, the $p=1$ and $p=5$ models are compared for samples of fixed width over a range of lengths.

Figure 11a shows results of simulations of 100 samples for each value of p. It demonstrates that the square pixel model ($p=1$) produces a significantly different result, with a much lower average critical current. The reason for this is associated with the fact that the value of J_c is directly proportional to the length of the path which limits the current. In the case of square grains, this path is likely to be a straight line across the sample width. However, when the grains are irregular, this path is likely to be more circuitous, leading to higher values for J_c. It is interesting to note that increasing p above around 5 has little effect. The $p=1$ and $p=5$ models are compared over a range of sample lengths in figure 11b. This illustrates again the lower values of J_c produced by the simple square model.

4.2 Sample Dimensions

Whilst the three parameters of tape length, tape width and average grain size may each be independently varied, the grain network is best described in a dimensionless manner, describing the sample in terms of the number of grains along the length and the number of grain across the width. In terms of the critical current density of a network, the solutions to a 10 cm long, 1 cm wide tape with 100 μm grains is identical to a 1 cm x 1 mm tape with 10 μm grains. We can vary the grain size for a sample of fixed dimensions, which is equivalent to varying the sample area (maintaining the aspect ratio) whilst holding the grain size constant.

Firstly, we will consider a square sample of dimensions $(10 \text{ mm})^2$. The grains are isotropic and the average linear grain size is 5 pixels. The average number of   grains   in

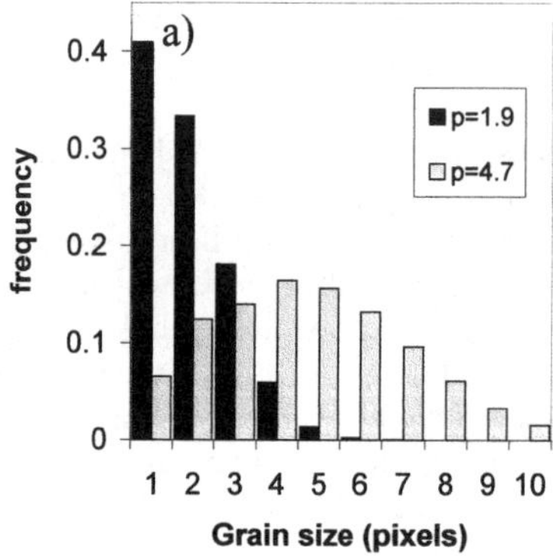
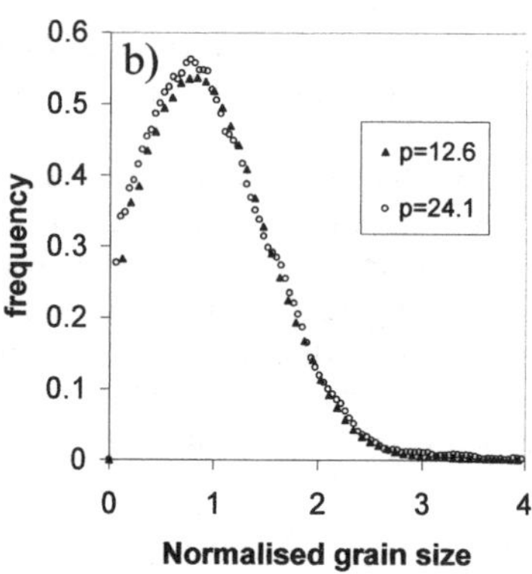

Figure 9. a) Histogram showing the distribution of absolute grain sizes for MCS=3 (p=1.9) and MCS=30 (p=4.7) and b) the normalized distribution which for high values of p lie on a single curve.

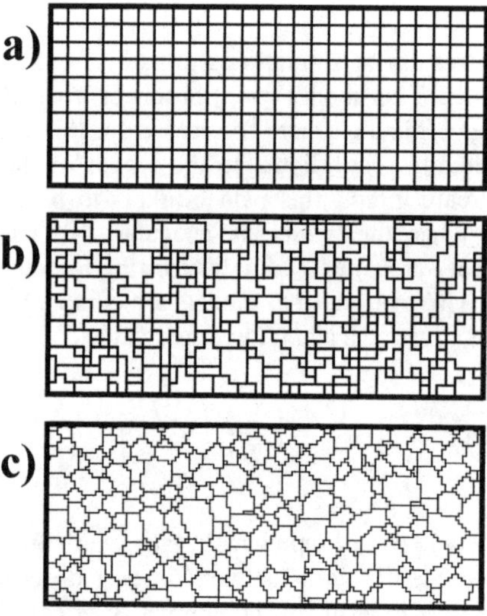

Figure 10. Modeled samples for which $N_l=25$ and $N_w=10$. a) p=1 generates the simple square model, b) p=2 produces irregular grains, but which are highly pixelated and c) p=5, with more realistic grain shapes.

each dimension is varied from 5 up to 120. This is compared to a sample that is 30 mm long and 3 mm wide in which N_w ranges from 4 to 60. Thus in both cases, the range of grain sizes examined is approximately 25-250 µm. The calculated critical current

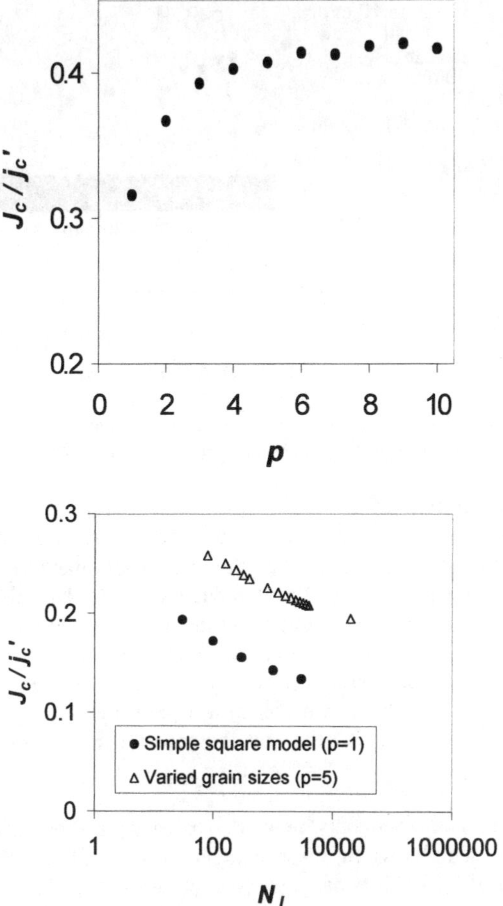

Figure 11. a) the dependence of the calculated sample J_c on the value of p (N_l=200, N_w=20) and b) comparison of p=1 and p=5 for various conductor lengths (N_w=40)

densities are shown in figure 12, demonstrating that grain size has little or no effect for the square shaped conductor, but is important for tapes which are longer than they are wide.

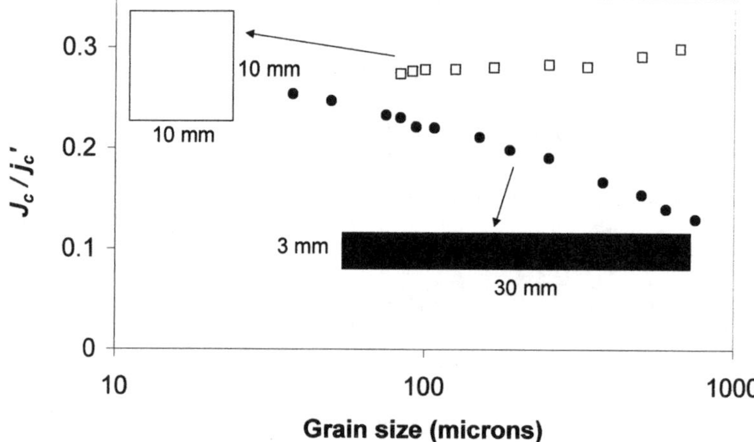

Figure 12. Dependence of calculated J_c on grain size for two different conductor shapes, a square 10 mm x 10 mm and a rectangle 3 mm x 30 mm (p=5)

The critical current of the tape which has an aspect ratio of 10 falls rather significantly as the size of the grains become a significant fraction of the tape width. For a conductor with this shape and size, the grain size would have to be around 50 µm or smaller in order to avoid any significant reduction in J_c. The difference in behavior between the two conductor shapes is important to consider when designing an experiment to verify the proposition that smaller grains are favorable for coated conductors. In order to see a reduction in J_c, one would need to ensure that the length (which corresponds to the distance between the voltage contacts in a 4-point measurement) was significantly greater than the tape width. This often not the case for measurements made on small-scale samples.

Previous models of grain boundary networks using square or hexagonal grains have predicted that the sample width must contain at least 50 to 100 grains to avoid "percolative pinch-off" [6,10]. It has also been observed that there is an approximately logarithmic decrease in J_c as the sample length is increased [13]. These ideas have been investigated here using the more realistic grain shape model.

It has been proposed [12] that for a given sample width, there is the following relationship between J_c and conductor length.

$$J_c \propto \left(\frac{1}{l}\right)^{\frac{1}{m}}$$

(11)

This expression arises from a Weibull analysis [37], and *m* represents the Weibull modulus amongst a set of sections of the tape.

Calculations have been carried out over a range of widths and lengths for *p*=1 (the square-grain model) and *p*=5 (irregular grain model). A plot of log J_c vs. log (1/N_l) for both cases is shown in figure 13. It turns out that the gradients of the lines in these plots varies with and is approximately equal to the conductor width.

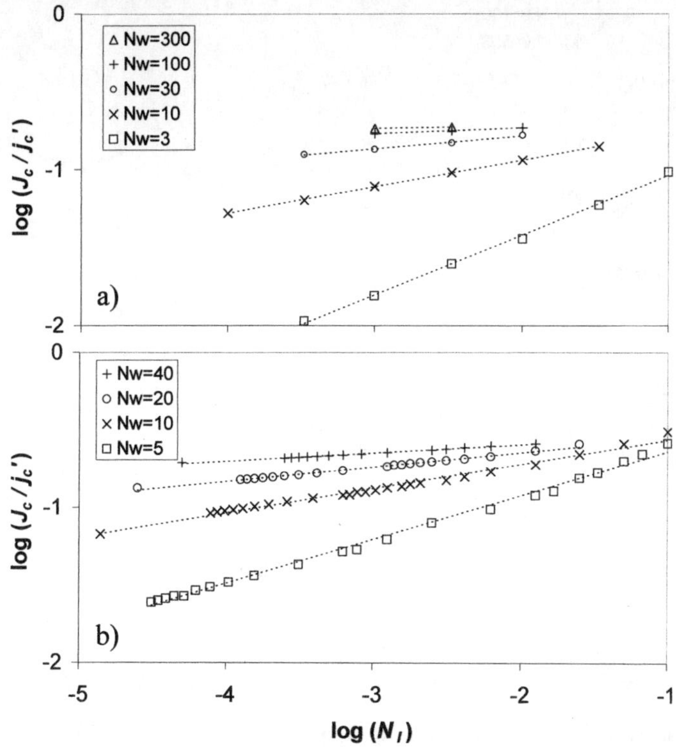

Figure 13. Log-log plots of J_c vs. length for various conductor widths for a FWHM of 10° with a) p=1 and b) p=5.

Hence the following relationship between tape width and length is proposed to approximate J_c over limited ranges :

$$J_c \approx \left(\frac{1}{N_l}\right)^{\frac{1}{N_w}}$$

(12)

Figure 14 shows the data from figure 13 replotted in terms of equation 12 for *p*=1 and *p*=5. The plot indicates a series of lines for different widths rather than a single

394

straight line, though for tapes which are highly aspected ($l>>w$), the values do lie along a single line.

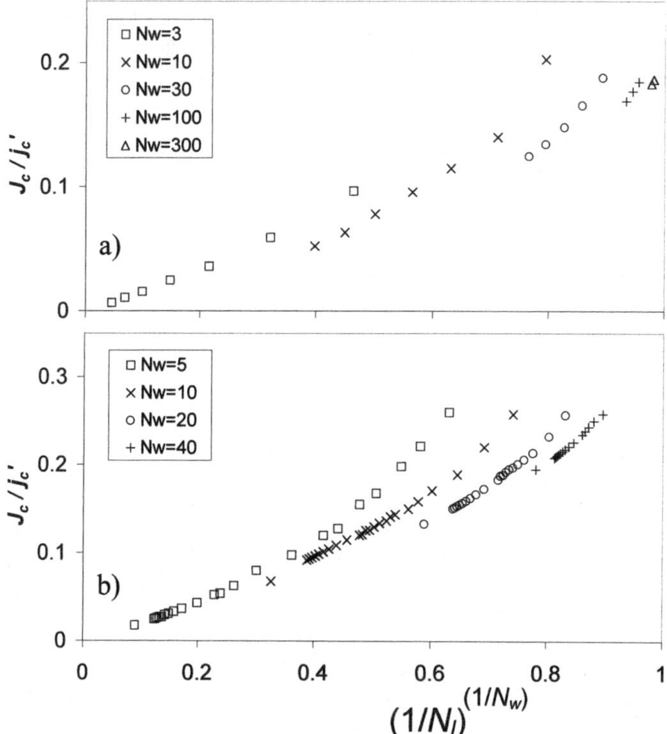

Figure 14. Plots demonstrating the relationship between critical current density and sample width and length for a FWHM of 10° with a) p=1 and b) p=5.

These two figures demonstrate how the critical current is reduced as the tape length increases and that this dependence becomes stronger as the tape width narrows. An alternative way of looking at it is that the current is reduced as the width decreases and that this dependence becomes more significant for long tapes. Figure 14 demonstrates that equation 12 is most accurate in predicting the width and length dependence of tapes when the grain size is a small fraction of the sample dimensions and the tape is highly aspected. Fortunately, this is the case for most practical coated conductors.

4.3 Crystallographic Texture and Intra-Granular Dissipation

Another advantage of the model is that it can be adapted to study the effect of intra-granular dissipation. As each grain in the model is made up of a number of small square pixels it is possible to assign critical current values to the boundaries between the intra-granular regions in order to simulate the intra-granular J_c. The effect of the intra-granular regions will be considered together with the effect of changing the FWHM of the grain orientations as the effects of the two parameters are closely inter-related. Consider for example the situation if the grains were able to carry the same amount of current as 1.8° grain boundaries, which may occur due to the existence of twins. In such a case, it is evident that intra-granular dissipation in a sample would be significant if all grains are aligned to within a few degrees. If however the sample has a large FWHM the effect of the grains is likely to be less significant, with most dissipation occurring at the grain boundaries, most of which will have a lower j_c.

It is useful to define a parameter which describes the current-carrying capability of the granular regions in a similar way to that of the boundaries. Instead of using absolute values of j_c, we will instead define the "equivalent angle" of the grains, ε, such that grains which carry the same amount of current as a 2° grain boundaries are described as $\varepsilon=2°$ grains. We will assume that all the grains within a sample have the same j_c and that this value is independent of the orientations of the grains.

The effect of varying the FWHM of the distribution of grain orientations has been modeled for samples with different intra-granular critical currents. Figure 15 shows how J_c varies with FWHM when the grains carry the same j_c as 0°, 1.8° and 4° boundaries. Also shown by the dotted line is the situation where the grains are not considered, and hence effectively can carry an infinite amount of current. In the case where the grains

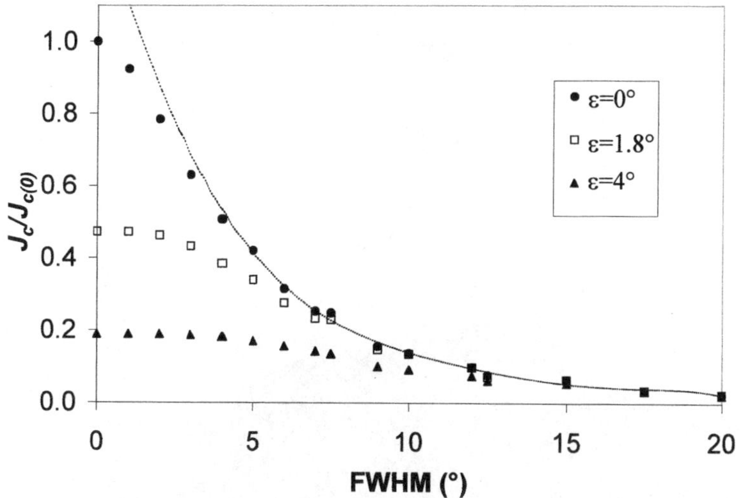

Figure 15. Graph showing how the critical current is reduced as the FWHM increases, and how this depends on the current carried by the grains. The broken line shows the values of J_c if intra-granular dissipation is not considered.

have $\varepsilon=0°$ and the FWHM is very low we can draw two conclusions. As the critical current decreases with increasing FWHM this must be a regime in which grain boundaries affect J_c. Also as J_c is lower than would be the case if the grains could carry

396

infinite current (the dotted line), then there must also be some intra-granular dissipation. Hence for high j_c grains, when the FWHM is low, there is both intra- and inter-granular dissipation. If the FWHM is above 5° however, this sample has the same critical current as it would if the grain j_c were infinite. Therefore there must be no intra-granular dissipation for a FWHM above 5° and sample J_c is only limited by the grain boundaries.

Now consider the situation when ε=4°, that is that the grains in the sample can only carry the same amount of current as a 4° grain boundary. Figure 15 shows that up until approximately 3°, J_c does not fall with increasing FWHM, hence in this regime, the critical current of the sample must be limited only by dissipation within the grains. At higher FWHM values (between 5° up to around 20°), J_c changes with FWHM and is below the value which would be achieved with perfect grains and hence is in the mixed regime with some dissipation within grains and some at grain boundaries. Above around 20°, J_c is determined only by the grain boundaries.

Figure 16 shows in a 3-D plot how J_c varies with both FWHM (a factor which affects the current-carrying ability of the grain boundaries) and ε (a measure of the current-carrying capability of the grains). Depending upon where samples produced by a particular coated conductor process lie on the surface plotted, one may determine whether it is more important to concentrate on the quality of the grains or the boundaries, or both,

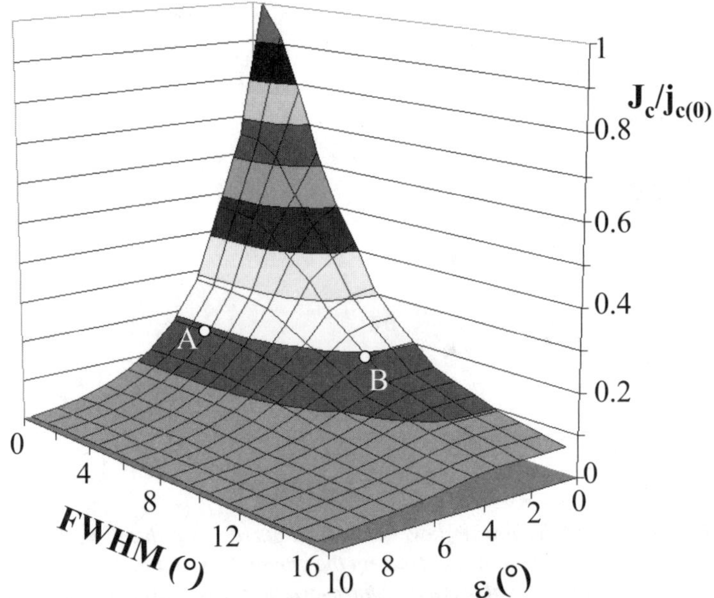

Figure 16. The effect of the FWHM and the intra-granular j_c (represented by ε, the equivalent angle) on J_c as predicted by the model. Points A and B are possible samples discussed in the text.

in order to best improve the critical current. Consider the following examples; process A produces a sample which is very well textured with a FWHM of just 2° but has rather poor grains, which behave like 4° grain boundaries (ε=4°). Process B gives a FWHM of 8°, but has better grains, which are able to carry the same currents as 2° boundaries. Figure 16 indicates that these samples will carry similar critical currents. It also shows that in order to improve sample A, it is of little use trying to improve the texture further, or to dope the boundaries with calcium as this will have little effect on the overall J_c. Instead, one needs to improve the quality of the grains. Conversely for sample B, concentrating on the grain boundaries is more important.

From the data plotted in figures 15 and 16, it is possible to produce a schematic phase diagram indicating the regions in which dissipation is entirely intra-granular, or entirely inter-granular or a combination of both. To calculate the points comprising the boundaries in figure 17, one considers in turn given values of ε. The transition between the intra-granular region (IG) and the mixed region is determined by increasing the FWHM from zero until a fall in J_c is observed. The FWHM is then increased further until the point at which J_c is indistinguishable from that in a sample with perfect grains. This marks the transition to the regime of grain-boundary (GB) dissipation only. This has been done for values of ε up to 10°.

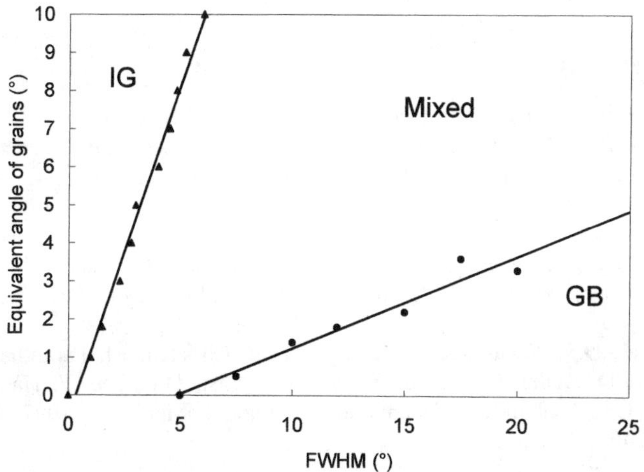

Figure 17. Graph showing regimes of dominance of grain boundary dissipation (GB), intra-granular dissipation (IG) and the combination of the two (mixed). Lines are plotted for changes of 1% in J_c.

Figure 17 indicates that most coated conductors are probably in the mixed zone, but closer to the GB regime (i.e. most dissipation occurs along grain boundaries, but grains also have some effect). For example, if we assume that the quality of grains is limited to ε=1.8° by the presence of twin boundaries, then only conductors in which the FWHM is higher than 12° have no significant granular dissipation. The figure also shows that very sharply aligned (almost single-crystalline) samples have to be achieved before the regime of exclusively intra-granular dissipation is reached. Textures with FWHM values of 5-6° are currently being reported in RABiTS tapes [38].

It is worth noting that the mechanism of dissipation will have important consequences for the effect of the sample dimensions, which was discussed in section 4.2. Percolation modeling generally considers samples in which dissipation occurs only at grain boundaries (i.e. in the GB zone). If a sample were in the mixed zone, it is likely that the length and width dependences would be lessened, and ultimately for samples in the IG zone, J_c should be totally independent of the sample dimensions.

ACKNOWLEDGEMENTS

This work is sponsored by the United States Department of Energy, Office of Energy Efficiency and Renewable Energy, Office of Distributed Energy and Electric Reliability- Superconductivity program. This research was performed at the Oak Ridge National Laboratory, managed by UT-Battelle, LLC for the United States Department of Energy under contract No. DE-AC05-00OR22725. The support of Oak Ridge Associated Universities (ORAU) is also acknowledged.

REFERENCES

1. D.P. Norton, A. Goyal, J.D. Budai, D.K. Christen, D.M. Kroeger, E.D. Specht, Q. He, B. Saffian, M. Paranthaman, C.E. Klabunde, D.F. Lee, B.C. Sales, and F.A. List, Science, 1996. **274** (5288): p. 755-757.

2. A. Goyal, D. P. Norton, J. D. Budai, M. Paranthaman, E. D. Specht, D. M. Kroeger, D. K. Christen, Q. He, B. Saffian, F. A. List, D. F. Lee, P. M. Martin, C. E. Klabunde, E. Hatfield, V. K. Sikka, Appl. Phys. Lett., **69** (1996) 1795.

3. A. Goyal, D.P. Norton, D.M. Kroeger, D.K. Christen, M. Paranthaman, E.D. Specht, J.D. Budai, Q. He, B. Saffian, F.A. List, D.F. Lee, E. Hatfield, P.M. Martin, C.E. Klabunde, J. Mathis, and C. Park, J. Mat. Res., 1997. **12** (11): p. 2924-2940.

4. N.A. Rutter, A. Goyal, C.E. Vallet, F.A. List, D.F. Lee, L. Heatherly, D.M. Kroeger, *submitted to Physica C.*

5. C. Cantoni, D.K. Christen, L. Heatherly, M.M. Kowalewski, F.A. List, A. Goyal, G.W. Ownby, D.M. Zehner, B.W. Kang, D.M. Kroeger, J. Mater. Res., 2002. **17** (10): p. 2549-2554.

6. A. Goyal, D.F. Lee, F.A. List, E.D. Specht, R. Feenstra, M. Paranthaman, X. Cui, S.W. Lu, P.M. Martin, D.M. Kroeger, D.K. Christen, B.W. Kang, D.P. Norton, C.

Park, D.T. Verebelyi, J.R. Thompson, R.K. Williams, T. Aytug, C. Cantoni, Physica C, 2001. **357** (2): p. 903-913.

7. A. Goyal, S.X. Ren, E.D. Specht, D.M. Kroeger, R. Feenstra, D. Norton, M. Paranthaman, D.F. Lee, D.K. Christen, Micron, 1999. **30**: p. 463-478.

8. D.M. Feldmann, J.L. Reeves, A.A. Polyanskii, G. Kozlowski, R.R. Biggers, R.M. Nekkanti, I. Maartense, M. Tomsic, P. Barnes, C.E. Oberly, T.L. Peterson, S.E. Babcock, D.C. Larbalestier, Appl. Phys. Lett., 2000. **77** (18): p. 2906-2908.

9. B. Zeimetz, B.A. Glowacki, J.E. Evetts, *submitted to European Physical Journal B.*

10. E.D. Specht, A. Goyal, D.M. Kroeger, Supercond. Sci. Technol., 2000. **13**: p. 592-597.

11. N.A Rutter, B.A. Glowacki and J.E. Evetts Supercond. Sci. & Technol., 2000. **13**: p. L25-L30.

12. N.A. Rutter, Ph.D. Thesis, University of Cambridge, Cambridge, UK, 2001. *ftp://dmg-stairs.msm.cam.ac.uk/pub/theses/*

13. Y. Nakamura, T. Izumi, Y. Shiohara, Phys. C., 2002. **371**: p. 275-284.

14. B. Holzapfel, L. Fernandez, F. Schindler, B. de Boer, N. Reger, J. Eickemeyer, P. Berberich, W. Prusseit, IEEE. Trans. Appl. Supercond., 2001. **11** (1): p. 3872-3875.

15. D. Dimos, P. Chaudhari, J. Mannhart, and F.K. LeGoues, Phys. Rev. Lett., 1988. **61**(2): p. 219-222.

16. D. Dimos, P. Chaudhari, and J. Mannhart, Phys. Rev. B, 1990. **41** (7): p. 4038-4049.

17. Z.G. Ivanov, P.A. Nilsson, D. Winkler, J.A. Alarco, T. Claeson, E.A. Stepantsov, and A.Y. Tzalenchuk, Appl. Phys. Lett., 1991. **59** (23): p. 3030-3032.

18. N.F. Heinig, R.D. Redwing, I.F. Tsu, A. Gurevich, J.E. Nordman, S.E. Babcock, and D.C. Larbalestier, Appl. Phys. Lett., 1996. **69** (4): p. 577-579.

19. D.T. Verebelyi, D.K. Christen, R. Feenstra, C. Cantoni, A. Goyal, D.F. Lee, M. Paranthaman, P.N. Arendt, R.F. DePaula, J.R. Groves, and C. Prouteau, Appl. Phys. Lett., 2000. **76** (13): p. 1755-1757.

20. K.E. Gray, M.B. Field, and D.J. Miller, Phys. Rev. B., 1998. **58**(14): p. 9543-9548.

21. D.T. Verebelyi, C. Cantoni, J.D. Budai, D.K. Christen, H.J. Kim, J.R. Thompson, Appl. Phys. Lett., 2001. **78** (14): p. 2031-2034.

22. L.R. Ford, and D.R. Fulkerson, Naval Research Logistics Quarterly, 1957. **4** (1): p. 47-54.

23. N.A. Rutter, B.A. Glowacki, Supercond. Sci. Technol., 2001. **14** (9): p.680-684.

24. J.E. Evetts, M.J. Hogg, B.A. Glowacki, N.A. Rutter, V.N. Tsaneva, Supercond. Sci. Technol., 1999. **12** (12) : p.1050-1053.

25. L.R. Ford, Jr. and D.R. Fulkerson, Canadian Journal of Mathematics, 1956. **8**: p399-404.

26. L.R. Ford, Jr. and D.R. Fulkerson, Canadian Journal of Mathematics, 1957. **9**: p.210-218.

27. L.R. Ford, Jr. and D.R. Fulkerson. Flows in Networks. Princeton University Press, Princeton, NJ, 1962.

28. R B Potts, Proceedings of the Cambridge Philosophical Society, 1952. **48** (106) 1952.

29. M.P. Anderson (1988). Simulation of grain growth in two and three dimensions. Roskilde, Denmark, Risø National Laboratory.

30. M.P. Anderson, D. J. Srolovitz, G. S. Grest and P. S. Sahni 1984. Act. Metall., **32** (5): p.783-791.

31. D.J. Srolovitz, M. P. Anderson, P. S. Sahni and G. S. Grest 1984. Act. Metall., **32** (5): p.793-802.

32. E.A. Holm, C.C. Battaile, Journal of Materials, 2001. **53** (9): p. 20-23.

33. K. Mehnert, P. Klimanek, Scripta Materialia, 1996. **35** (6): p.699-704.

34. M. Morhac, E. Morhacova, Cryst. Res. Technol., 2000. **35** (1): p.117-128.

35. P. Blikstein, A.P. Tschiptschin, Materials Research, 1999. **2** (3): p.133-137.

36. J.E. Burke, Trans. Am. Inst. Min. Engrs, 1949. **180** : p. 73.

37. W. Weibull, J. Appl. Mech., 1951. **18**: p. 293.

38. A. Goyal et al., Florida Wire Workshop, January 2003.

CRITICAL CURRENT DENSITY, FLUX PINNING AND MICROSTRUCTURE IN MgB$_2$ SUPERCONDUCTORS

Y. FENG[1,2*] , Y. ZHAO[3,2], G. YAN [1], A. K PRADHAN[2], L. ZHOU[1], N. KOSHIZUKA[2] and M. MURAKAMI[2]

[1]Northwest Institute for Nonferrous Metal Research, P.O.Box 51, Xi'an, 710016 CHINA

[2] SRL-ISTEC, 1-10-13 Shinonome, 1-chome, Koto-ku, Tokyo 135 JAPAN

[3] School of Materials Sciences and Engineering, University of New South Wales, Sydney, 2052, NSW, AUSTRALIA

* Corresponding address: yfeng@c-nin.com

I. INTRODUTION

The discovery of superconductivity at 40 K in MgB$_2$ by Nagamatsu et al [1] has generated a great deal of excitement in both the fundamental and practical investigations of this material. The transition temperature T_c of MgB$_2$ is extremely higher than Nb$_3$Ge (by almost a factor of 2) having the highest T_c in the conventional superconductors. The advantage of MgB$_2$ is its applications at higher temperature (20-30 K) region, where the conventional superconductors can not play any role due to its low T_c. Also, the progress in cryogen free cooling technique at temperature region of 20-30 K will promote the development and applications of MgB$_2$. Some investigations including the observation of the isotope effect, band structure and tunneling measurements of superconducting gap suggest that MgB$_2$ is a conventional phonon-mediated BCS superconductor [2-4]. In contrast to the BCS theory, another model was proposed to explain superconductivity in MgB$_2$ by the pairing of dressed holes [5].

On the other hand, it is found that supercurrent flow in MgB$_2$ bulk material is not reduced by grainboundaries [6] which are well known in high-T_c ceramic superconductors. As revealed by the irreversible magnetization experiment [7], the behavior of the magnetic hysteresis loops of MgB_2 is dominated by bulk pinning. Furthermore, measurements of $H_{c2}(T)$, dynamics critical field and J_c suggest that MgB_2 is a typical type-II superconductor, similar to Nb_3Sn except for the extremely high T_c [8,9]. In addition, it is reported that the grain boundaries may be the main source of vortex pinning in MgB_2 like in Nb_3Sn [10]. The recent investigation of MgB_2 wire under different heat treatments indicates that MgB_2 bulk samples fabricated by solid-state reaction method contain many small $Mg(B,O)_2$ precipitates within the MgB_2 matrix, which are suitable to be pinning centers [11]. For the electrical applications of superconductors, high critical current density is required. Therefore, much work has been carried out to improve the superconducting characteristics of MgB_2 in its various shapes, either bulk, or thin film or wire. Just recently, a very high J_c of $1.2 \times 10^7 A/cm^2$ at 4.2 K in zero-field was obtained in the in-situ epitaxial MgB_2 thin film, suggesting that MgB_2 can reach extremely high intrinsic J_c [12]. Unfortunately, the bulk materials prepared at ambient pressure always show lower J_c due to the bad grain connectivity, low density and poor flux pinning (the highest J_c is around $2 \times 10^5 A/cm^2$ at 10 K in zero field) [13]. It is found that the most efficient techniques used to densify the sample and to achieve high J_c are hot isostatic pressing and uniaxial hot pressing at high pressure [14-16]. Also, proton and neutron irradiation are applied to enhance J_c properties in MgB_2 superconductors [9,17]. However, these techniques are not available for the preparation of MgB_2 wires, tapes and large bulks because of its complicated system and the practical difficulty in use of very large vessels and dyes in the case of high temperature and high pressure. On the other hand, it is reported that chemical doping is effective to increase J_c in high-T_c superconductors. The similar technique has also been utilized to the MgB_2 materials. Although some results show positive effects, they might introduce a big reduction of T_c. It was observed that the partial substitution of Zn or Cu for Mg resulted in a decrease in T_c [18]. Meanwhile, most substitutions of different elements such as Al, Si, Co, Fe, Be etc for Mg also lead to a reduction of T_c [19]. In addition, contrast to a paper describing a deleterious effect of Ti-doping[20], previous study shows a suitable amount of Ti doping will increase J_c in MgB_2 [21].

On the other hand, many groups are fabricating MgB_2 wires and tapes with high performance by using different metal sheaths. Canfield *et al* fabricated high-density MgB_2 wires ($160\mu m$ in diameter) through the exposure of boron filaments to Mg vapor [22]. Also, the recent reports on preparation of MgB_2 wires by PIT using either Ag or Cu sheath and MgB_2 strands by filling Nb-lined monel tubes with commercial MgB_2 powders were the first steps to put MgB_2 superconductors into applications [23]. By using Cu as a sheath, transport Jc of $50000A/cm^2$ at 15 K in self-field was obtained in MgB_2 wires. For the non-sintered MgB_2/Ni tape, J_c reached around $10^5 A/cm^2$ at 4.2 K in self-field [24]. Recently, Wang et al reported their results on a Fe-clad MgB_2 wire, in which Jc achieved $4.2 \times 10^5 A/cm^2$ at 4.2 K in self-field [25]. J_c is further improved to $1.7 \times 10^4 A/cm^2$ in 1 T at 29.5 K and at 33K in self-field [26]. Also, the transport Jc of $8700A/cm^2$ and $55830A/cm^2$ at 4.2 K in self-field were measured for Cu and Fe/Cu sheathed MgB_2 square wires by using commercial MgB_2 powder [27]. To date, two important factors of low density and poor flux pinning are the main obstacles to obtain J_c in MgB_2 samples.

In this chapter, we will review our recent progresses on thermodynamic behavior, flux pinning characteristics, transport properties and microstructure of MgB_2 bulks, wires and tapes in the past two years. We developed several new, simple and feasible methods to fabricate high performance MgB_2 bulks and wires with different sheaths at ambient pressure. By Zr doping, critical current density in MgB_2 bulks or tapes can be significantly improved. Among MgB_2 wires and tapes with different metal sheaths, the Fe-clad MgB_2 wires show the highest irreversibility fields and excellent performance in high fields and high temperatures. The mechanism of high critical current density in Zr-doped bulks and MgB_2/Fe wires will be discussed in this chapter.

II. Thermodynamic behavior and the phase formation in Mg-B system

Unfortunately, because of the high volatility of Mg at elevated temperature, it is very difficult to fabricate MgB_2 with high density at ambient pressure. The MgB_2 wires prepared by powder in tube (PIT) method at ambient pressure have 25% lower density than dense bulk samples fabricated under high pressure [28]. Therefore, it is still a big challenge to synthesis MgB_2 superconductor with high density at ambient pressure in order to remarkably improve superconducting properties. One of the key problems is to obtain the clearly thermodynamic characteristics of Mg-B system. Although the thermodynamics of Mg-B system were calculated by the phase diagram modeling technique [29], no detail of experimental thermodynamic information is available in the literature until now. However, it is very necessary and important to understand the thermodynamic behavior and the phase formation of Mg-B system for preparation of high quality MgB_2 bulks, thin films, wires and tapes.

All the MgB_2 bulk samples were fabricated by a standard solid state reaction technique at ambient pressure. The stoichiometric mixture of Mg powder and submicron amorphous B powder was well ground in an agate mortar for 1h and then pressed into cylindrical pellets with a diameter of 12 mm. Five pellets settled in Al_2O_3 crucibles were put into a quartz tube. To prevent the oxygenation of the Mg powder and B powder, the quartz tube was vacuumized until 3×10^{-3}Pa and then filled with pure argon. The samples were sintered at 600°C, 650°C, 700°C, 750°C, 850°C for 2h respectively. Then all the samples were quenched to room temperature under pure argon atmosphere in 20min. In addition, in order to investigate the effects of sintering time and the increasing rate in temperature on the phase formation of MgB_2 phase, a series of samples were fabricated under different conditions. Two samples were sintered at 650°C for 2h at the temperature increasing rate of 60°C/min and 5°C/min. Also, several samples were held at 650°C for 3 min, 60 min and 120 min at the increasing rate of 60°C/min.

In order to get the thermodynamic information of MgB_2 synthesis process, the diffraction thermal analysis (DTA) experiment was performed for mixture powder with the stoichiometric Mg:B=1:2. Powder X-ray diffraction (XRD) with Cu Kα radiation was employed to investigate the phase composition of the samples.

It is well known that MgB_2 becomes very porous after high temperature sintering, which can be explained by the evaporation of Mg particles. Vapor pressure is a property constant of a material, which denotes the gasification uptrend of solid materials. As for the phase transformation of solid-liquid and liquid-gas, the increase of vapor pressure means the much evaporation of solid materials. According to the Clapeyron equation, the vapor pressure of a material (P) can be expressed as a function of temperature (T): $logP=-A/T+BlogT+CT+DT^2+E$, in which A,B,C,D,E constants are different in different materials. Figure 1 shows the vapor pressure of Mg as a function of temperature in ambient pressure.

404

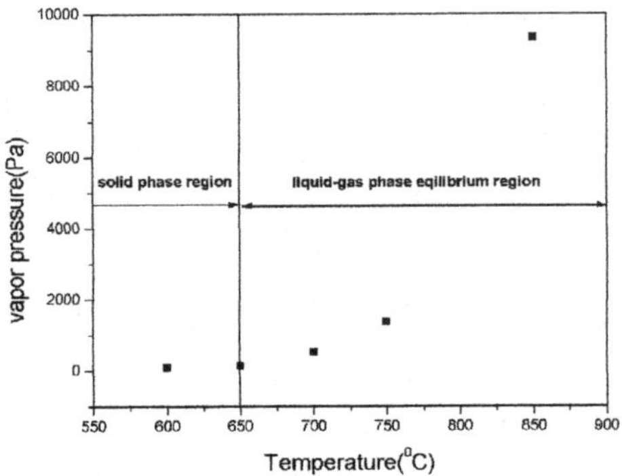

Figure.1. Vapor pressure vs. temperature of Mg at ambient pressure

There are solid-solid equilibrium region under 650°C and liquid-gas-solid equilibrium region above 650°C in Mg-B system. The vapor pressure of Mg reaches 1383Pa at 750°C, indicating even at this temperature a lot of liquid Mg will turn into vapor. For the *in-situ* fabrication of MgB_2 wires and tapes, too much gasification of Mg will induce the loss of initial stoichiometric elements and more voids in the samples.

 Figure 2 illustrates the DTA curve for the mixture of Mg:B=1:2. The two peaks in the DTA curve represent the two reactions as following:

$Mg_{(s)} \rightarrow Mg_{(l)}$ (a)

$Mg_{(l)} + 2B_{(s)} \rightarrow MgB_{2(s)}$ (b)

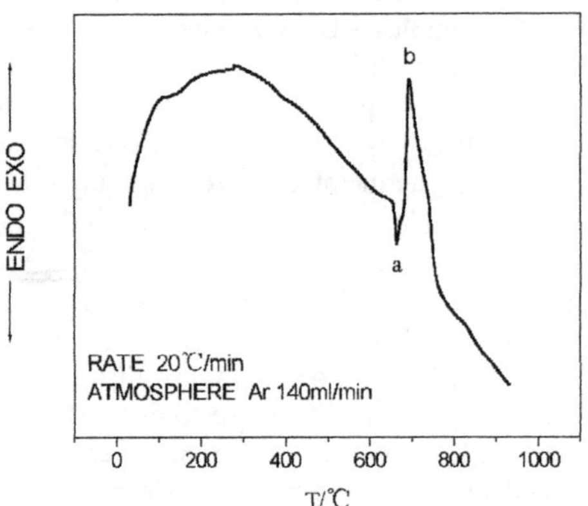

Figure.2. The DTA result of the mixture of Mg:B=1:2

Here *s* denotes the solid phase and *l* is the liquid phase. Reaction (a) suggests that the solid state Mg becomes liquid state, while Reaction (b) shows the formation of MgB_2 phase. It is clear that the reaction of MgB_2(b) appears following the Mg melting reaction (a). The vapor pressure at Mg melting point 650°C is only 135Pa, which implies that the evaporation of Mg at this temperature is much less. The liquid-solid reaction between Mg and B will improve the density of MgB_2.

On the other hand, there is another reaction of Mg+4B→MgB$_4$ in the Mg-B system during the reaction process between Mg and B. MgB$_4$ is of an orthorhombic structure with a=0.5464 nm, b=0.7472 nm and c=0.4428 nm. The X-ray diffraction patterns of five MgB$_2$ pellets prepared at different temperatures were presented in Fig.3. It is clearly observed that the sample sintered at 600°C has MgB$_4$ and MgO impurity phases and the samples sintered above 650°C have more MgO impurities as temperature increases. The sample fabricated at 650°C is single phase with few MgO impurities. The results indicate that the reaction between Mg and B is not sufficient when the sintering temperature is under 650°C. On the other hand, the evaporation of Mg results in more voids in the MgB$_2$ samples prepared at temperatures above 650°C.

Fig.4 displays XRD results for the samples prepared at 650°C for 2 h with different increasing temperature rates. It can be observed that more MgB$_4$ phases appear in the sample with the lower rate of 5°C/min. MgB$_4$ phase has a deteriorate effect on superconducting properties because it usually stays in the grain boundary and causes a loose connection between grains due to the larger mismatch in the crystal structure and lattice parameters with MgB$_2$ [30]. As for the sample prepared at high increasing rate of 60°C/min, the sample has a better phase composition. Therefore, large increasing rate is required in order to obtain high quality MgB$_2$ superconductors. In addition, X-ray diffraction patterns of the samples sintered at 650°C for different time with the increasing rate of 60°C /min are given in Fig. 5. The amount of MgB$_2$ phase does not change much with the sintering time, suggesting that the formation rate of MgB$_2$ phase is very fast. However, it is necessary to prolong the sintering time to get better microstructure.

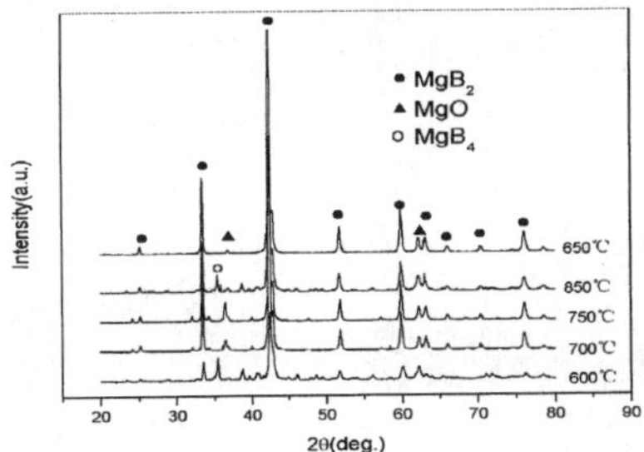

Fig.3. The XRD patterns of MgB$_2$ sintered at different temperatures

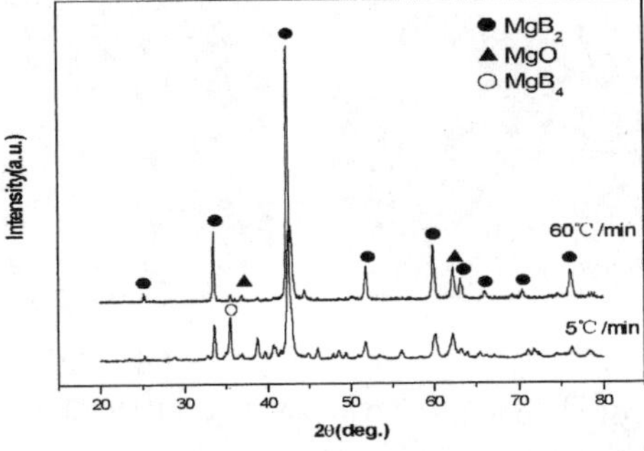

Fig.4. The XRD patterns of MgB$_2$ bulks sintered at 650°C for 2 h with different increasing temperature rates

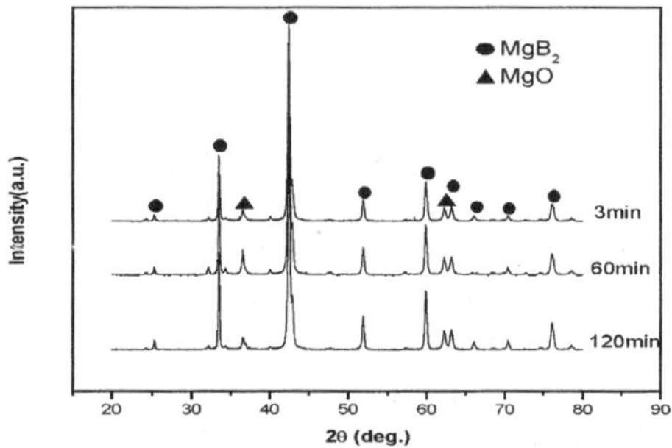

Fig.5. XRD patterns of MgB_2 sintered at 650°C for 2 h with different holding time

As expected, the processing conditions have a significant influence on microstructure features in superconductors. Typical SEM photos of the samples treated at different temperatures were shown in Fig. 6. One can see that the grain size, grain connectivity and morphology change with the processing temperature. For the sample prepared at lower temperature of 650°C and higher temperature of 800°C, there are more voids and the shape of grains is not regular. Also, the grain size is not uniform. The MgB_2 grains of the sample at 750°C are much small, well compacted, and more homogenous. The MgB_2 grain is of round shape and this sample has a higher density than other samples. These microstructure features are beneficial for obtaining high J_c in MgB_2 samples.

(a). 650°C

(b). 700°C

(c). 750°C

(d). 800°C

Fig. 6. SEM photos for the samples prepared under different temperatures (a). 650°C, (b). 700°C, (c). 750°C and (d). 800°C

III. Effects of Zr-doping on superconducting properties and microstructure in MgB_2 superconductors

Recently, a lot of efforts have been carried out to improve critical current density in MgB_2 superconductors. Although several methods including high pressure processing and proton irradiation are found to be effective in increasing J_c, they are not available to prepare MgB_2 wires and tapes for applications. Here we developed a simple and feasible way to significantly enhance the flux pinning in MgB_2 superconductors at ambient pressure [31,32]. We prepared a series of $Mg_{1-x}Zr_xB_2$ bulk samples with different x values by the solid state reaction at ambient pressure and investigated the effects of the Zr-doping on phase compositions, microstructure and flux pinning behavior. Microstructure analysis indicates that the Zr-doping results in an improved intergrain connectivity and significantly refines the MgB_2 grains. The MgB_2 nano-particle (around 20 nm in diameter) structure with nanometer and clean grain-boundaries (< 1nm thick) is formed in the Zr-doped MgB_2 samples. These microstructural features are proposed to be responsible for the increase of flux pinning over a large range of temperatures and fields. At 20 K in self-field and 1 T, critical current density is as high as 1.86 MA/cm^2 and 5.51×10^5 A/cm^2 in the Zr-doped sample.

A series of $Mg_{1-x}Zr_xB_2$ (x=0, 0.05, 0.1 and 0.2) were prepared by the solid state reaction method at ambient pressure. The high purity powders of Mg (99%), Zr (99%) and B(99%) were well ground and cold pressed into small cylindrical pellets. 5% extra Mg powder was added in order to compensate the loss of Mg in high temperature. Then, the pellets were put on a Ta plate and sintered at the temperatures between 600 °C to 900 °C for 3h in flowing argon without high pressure. Finally, a furnace cooling to room temperature followed the pellets. The crystal structure of all samples was performed by an AC 10000 x-ray diffractometer with a Cu Kα irradiation. The DC magnetization was carried out by a SQUID magnetometer (Quantum Design, MPMSR2) at different temperatures in the magnetic field up to 7T. J_c values were calculated from the magnetization curves using the Bean critical model of $J_c=20(M^+-M^-)/a(1-a/3b)$, M^+ and M^- are magnetization moments at increasing and decreasing field, respectively. a and b are sample dimensions perpendicular to the field in cm with a<b [33]. All the samples for the magnetic measurements have the same dimensions of $0.7 \times 0.9 \times 1.4 mm^3$.

Fig. 7 shows x-ray diffraction patterns for all samples with different Zr content. The line widths of the peaks are sharp and are independent of Zr content in $Mg_{1-x}Zr_xB_2$ up to x=0.2, indicating all samples have good crystallinity. Also, no MgB_4 phase was found in all samples. It can be seen that the ZrB_2 phase is presented in addition to the MgB_2 phase and the ratio of the strongest peak intensities of MgB_2 and ZrB_2 decreases with the increase of Zr content in Zr-doped samples. The molar percentage of ZrB_2 in the samples increases from 3.63% in $Mg_{0.95}Zr_{0.05}B_2$ through 7% in $Mg_{0.9}Zr_{0.1}B_2$ to 14% in $Mg_{0.8}Zr_{0.2}B_2$.

The lattice parameters of all samples calculated by the least-square fitting are displayed in Fig. 8. The a-axis and c-axis show completely different behaviors, in which the a-axis increases with the increase of the doping level and the c-axis remains almost unchanged. The a-axis is increased by 1.36% through a small amount of Zr atoms in the lattice of MgB_2. The similar result was observed in Zn-doped MgB_2 bulks [34], while it is much different from the report on Ti-doped MgB_2 [21]. However, the lattice constants in the sample with x=0.2 do not change as compared with those in $Mg_{0.9}Zr_{0.1}B_2$, implying that the solubility limit is reached in the sample with x=0.1.

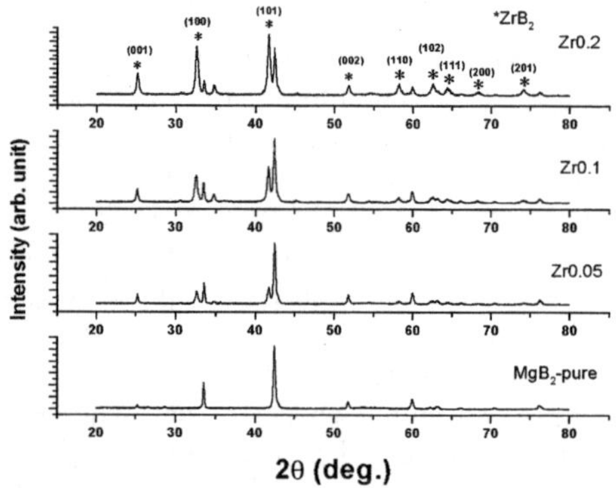

Fig. 7. X-ray diffraction patterns for the $Mg_{1-x}Zr_xB_2$ with x=0, 0.05, 0.1 and 0.2. The ZrB_2 phase is observed in Zr-doped specimens and the amount of ZrB_2 increases with the increase of the Zr-doping.

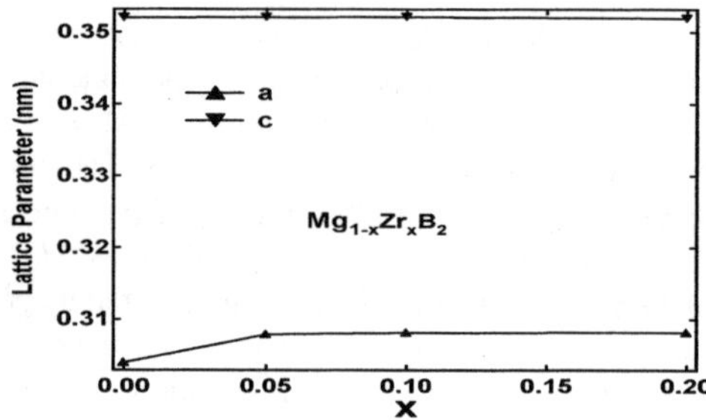

Fig. 8. Lattice parameters a and c as a function of the Zr content(x) for all samples.

Figure 9 displays the temperature dependence of magnetization for all the samples at 20 Oe. It can be observed that the diamagnetic signal in ZFC condition changes non-monotonously with the doping level and is increased by the Zr-doping. The maximum diamagnetic signal is achieved in the $Mg_{0.9}Zr_{0.1}B_2$. As shown in Fig.10, $Mg_{0.9}Zr_{0.1}B_2$ has the largest difference between ZFC and FC curves, indicating the very strong flux trapping in the sample with x=0.1. In addition, critical temperature varies with the Zr content up to x=0.1 and slightly decreases from 38.4 K in pure MgB_2 to 37.3 K in $Mg_{0.8}Zr_{0.2}B_2$ as illustrated in Fig.11. The decreasing rate of T_c is around dT_c/dx=-11K in the Zr-doped samples up to x=0.1, which is much smaller than those in $MgB_{2-x}C_x$ and $Mg_{1-x}Al_xB_2$ [34,35]. The slight variation of T_c in Zr-doped samples may be related to the partial substitution of Zr for Mg. The expansion of the a-axis leads to the decrease of T_c in Zr-doped specimens, which may support the model of hole superconductivity in MgB_2 [5]. In this model, the increase of the a-axis will reduce T_c. Furthermore, it is noted that the $Mg_{0.9}Zr_{0.1}B_2$ sample shows a superconducting transition at 37.3 K with a sharp transition width of 0.7 K, indicating the good phase homogeneity and better connection between grains. Interestingly,

$Mg_{0.8}Zr_{0.2}B_2$ has the same T_c as $Mg_{0.9}Zr_{0.1}B_2$, but the transition width becomes larger. This result may be due to the solubility limit and more second phases in the sample. The non-monotonous change of T_c with the doping level is also found in Zn-doped MgB_2 samples [34], while it is much different from the results in Ti-doped samples [21].

Figure 12 shows the irreversibility field H_{irr} as a function of temperature for pure MgB_2 and all Zr-doped samples. The irreversibility field was determined from the closet point of the hysteresis loops at a given temperature using a criterion of $10A/cm^2$. H_{irr} is found to increase as the Zr content from x=0 to 0.1 in $Mg_{1-x}Zr_xB_2$. Then, H_{irr} decreases in the sample with x=0.2. However, all the Zr-doped samples

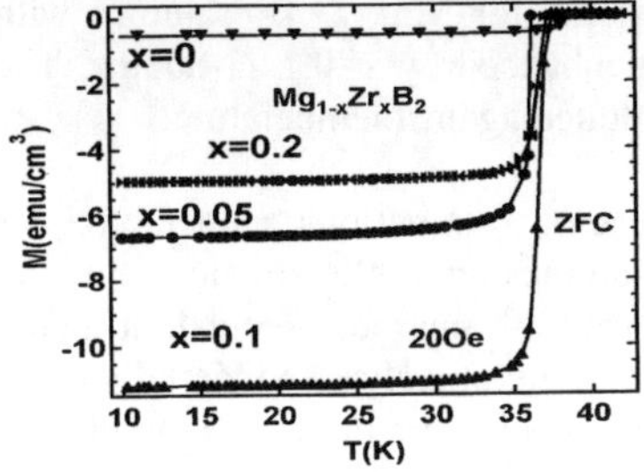

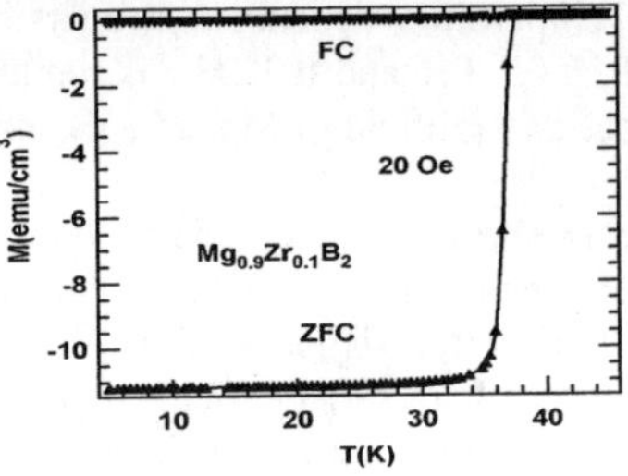

Fig. 9. M-T curves measured in ZFC condition at 20 Oe for all samples.

Fig. 10. M-T curves in ZFC and FC for $Mg_{0.9}Zr_{0.1}B_2$.

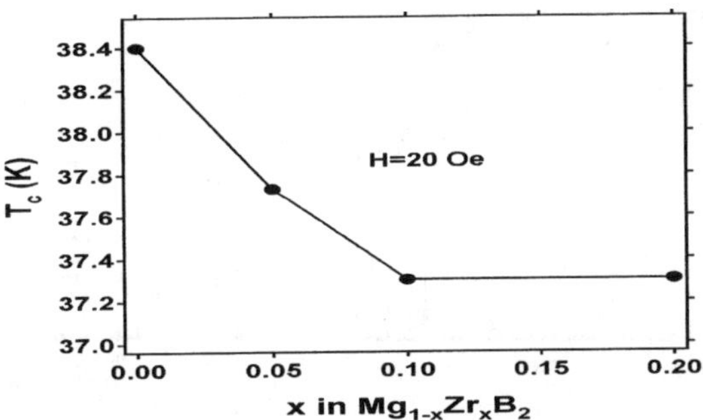

Fig. 11. Critical temperature versus the Zr content in the samples.

have a higher H_{irr} than pure MgB_2, indicating that the flux pinning is enhanced by the Zr-doping. Surprisingly, H_{irr} for the sample with x=0.1 remains the highest among all the samples. At 20 K, H_{irr} is increased from 3.6 T in pure MgB_2 to 4.3 T in $Mg_{0.9}Zr_{0.1}B_2$. The irreversibility field in the sample with x=0.1 is much higher than those in the dense MgB_2 samples fabricated by high pressure sintering. In addition, the upper critical field H_{c2} for $Mg_{0.9}Zr_{0.1}B_2$ is also displayed in Fig.12. A linear relation between H_{c2} and temperature is found, which yields the slope of dH_{c2}/dT=-0.56 T/K. The higher upper critical field $H_{c2}(0)$=20 T in $Mg_{0.9}Zr_{0.1}B_2$ is estimated by the extrapolation of $H_{c2}(T)$. Using the above estimation of $H_{c2}(0)$=20 T, the coherence length $\xi_0=[\phi_0/2\pi H_{c2}(0)]^{1/2}$ is found to be around 3.9 nm, which is smaller than that of pure MgB_2 bulk samples [36] and consistent with that in high quality MgB_2 films[37].

410

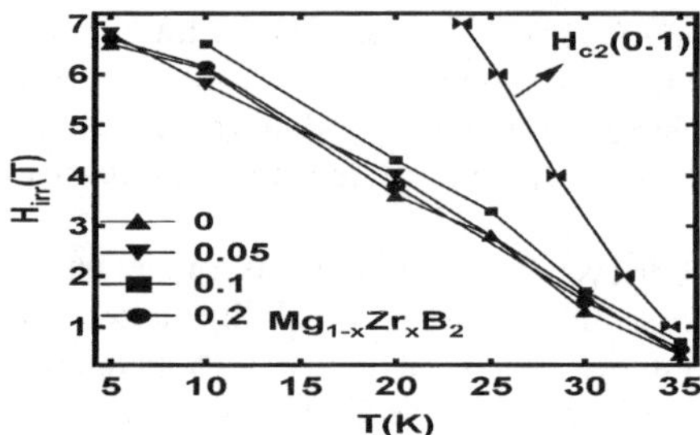

Fig. 12. Temperature dependent irreversibility fields for $Mg_{1-x}Zr_xB_2$ samples with x=0, 0.05, 0.1 and 0.2. H_{irr} is remarkably enhanced by x=0.1 Zr-doping. The upper critical field of $Mg_{0.9}Zr_{0.1}B_2$ is also plotted against temperature.

Figure 13 shows the magnetic field dependence of J_c at various temperatures for all the specimens. The similar features of these samples are outlined below. First, at low fields, a plateau region of J_c is observed for all the samples. Second, at higher fields above a characteristic field, J_c decreases approximately exponentially. It is found that the characteristic field drops with increasing temperature for all samples. In addition, the Zr-doping leads to a higher characteristic field at the same temperature, suggesting a stronger pinning force in the Zr-doped sample. The J_c values at low temperatures (from 5 K to 15 K) and low fields are very close. The similar observation is also found in MgB_2 bulk sample prepared under high pressure.

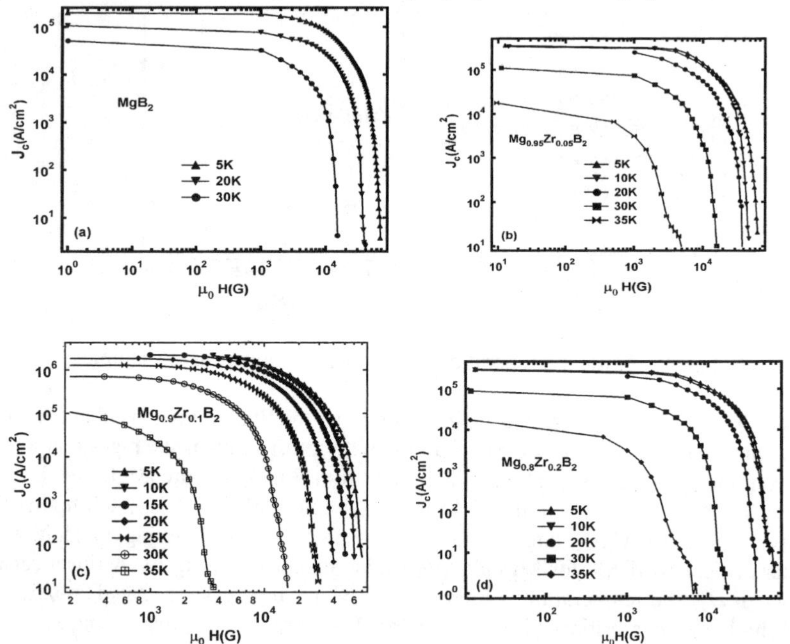

Fig. 13. Magnetic field dependence of J_c at different temperatures for the samples with x=0 (a), 0.05 (b), 0.1 (c) and 0.2 (d). Surprisingly, J_c in $Mg_{0.9}Zr_{0.1}B_2$ is very high about 1.83×10^6 A/cm^2 in self-field at 20 K.

J_c values at several selected temperatures and fields as a function of the Zr content are illustrated in Fig. 14. It is noted from Fig. 14 that J_c increases remarkably with increasing the Zr content, reaches the maximum for x=0.1 Zr doping and then decreases for the further increase of the Zr-doping at x=0.2. J_c in all the Zr-doped samples is always higher than that in pure MgB_2. Within all temperatures and the whole field region up to 7 T, the J_c value of the sample with x=0.1 is the highest among these sample (as shown in Fig. 13 and 14). A high J_c of $2.1x10^6$ A/cm^2 in 0.56 T and $5.7x10^5$ A/cm^2 in 2 T at 5 K is obtained in the $Mg_{0.9}Zr_{0.1}B_2$ sample. At 20 K, J_c reaches $1.83x10^6$ A/cm^2 in self-field, $5.51x10^5$ A/cm^2 in 1 T and $1.22x10^5$ A/cm^2 in 2 T. Surprisingly, J_c is as high as $7.2 x10^5$ A/cm^2 in self-field and $1.2x10^4$ A/cm^2 in 1 T even at 30 K. These J_c values are much larger than the best data on the MgB_2 bulk samples including hot isostatic pressed sample, MgB_2 specimen prepared under high pressure and proton-irradiated fragments.

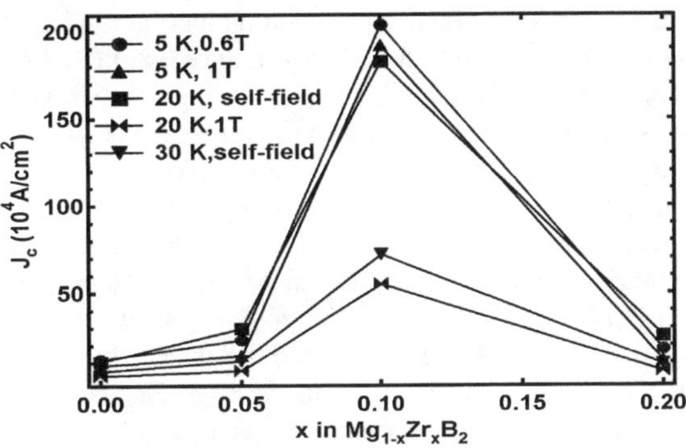

Fig. 14. J_c as a function of the Zr content at several temperatures and fields, showing that the highest J_c is achieved in $Mg_{0.9}Zr_{0.1}B_2$.

The above results clearly demonstrate that the Zr doping has a tremendous effect on the flux pinning behavior of MgB_2 samples. There is an optimum amount of the Zr doping which displays the highest J_c in MgB_2 bulk superconductors. The nature of flux pinning is different in different regions of the field. Furthermore, the Zr-doping in MgB_2 opens up a new effective and easily controlled method to improve J_c. It should be noted that this technique is very suitable for industrial scale fabrication of MgB_2 bulks and wires since the ambient pressure is utilized to fabricate the Zr-doped samples with high J_c.

The bulk pinning force F_p is calculated by $F_p=\mu_0 H J_c(H)$ and Fig.15 shows the field dependence of the pinning force at various temperatures for $Mg_{0.9}Zr_{0.1}B_2$ sample. The maximum force $F_{p.max}$ shifts towards higher field with decreasing temperature. The magnetic field at the peak of F_p is below 1.5 T, in consistent with those values of the MgB_2 wires [38]. At 5 K, $F_{p,max}$ achieves a very high value of $1.3x10^{10}$N/m^3, which is close to that of Nb47wt%Ti and Nb$_3$Sn with the range of 15-30GN/m^3. As seen in the inset of Fig. 15, the normalized pinning force, $F_p/F_{p,max}$ at various temperatures can be scaled into a single curve, which suggests that the flux pinning mechanism is the same at these temperatures. The F_p plots also show a peak at 0.18, followed a long tail towards high H/H_{irr}.

412

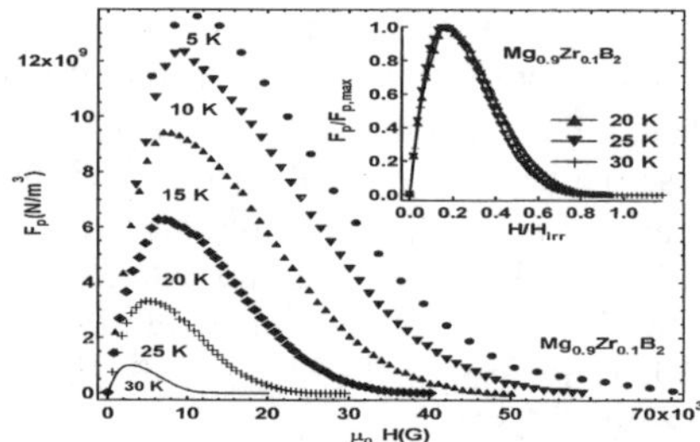

Fig. 15. Bulk flux pinning density as a function of field for the Zr-doped sample at various temperatures. Inset: Reduced field dependence of normalized flux pinning density at 20, 25 and 30 K, showing a single peak around 0.18.

It is well established that the superconducting and physical properties of superconductors are sensitive to the microstructure features. As mentioned above, the Zr-doped sample shows the excellent flux pinning , which should be related to the modification of microstructure. The significant increase in J_c by Zr-doping may be due to the higher density, excellent grain connectivity and very strong pinning force. With Zr-doping, a much better connection between the grains is obtained. The density and J_c value increase with the increase of the doping level. The Zr-doped sample is even denser than the MgB_2 sample fabricated at high pressure. However, the $Mg_{0.9}Zr_{0.1}B_2$ sample has the optimum microstructure in all specimens. The grains in $Mg_{0.9}Zr_{0.1}B_2$ sample are well connected and the sample has a high density. Almost no poles , no MgO particles are observed in the $Mg_{0.9}Zr_{0.1}B_2$ sample. Furthermore, the smaller grain size of MgB_2 is observed in the samples with the high doping level. The average grain size in $Mg_{0.95}Zr_{0.05}B_2$ is around 90nm, almost the same as that in pure MgB_2. As the Zr content is increased to x=0.1, the grain size is much reduced, which results in a significant improvement of the flux pinning. A typical TEM image of $Mg_{0.9}Zr_{0.1}B_2$ sample is illustrated in Fig 16. Most of the MgB_2 grains are just around 20 nm in diameter, which is much smaller than those about 0.4-8μm in the samples fabricated by high pressure, hot isostatic pressing and in the nanoparticle doped MgB_2 samples [39,40]. Some ultrafine MgB_2 grains around 5 nm can be also found in the Zr-doped sample. Figure 17 shows the typical grain boundary (GB) of MgB_2 grains for the $Mg_{0.9}Zr_{0.1}B_2$ sample. Grain boundaries are found to be very clean and the thickness of GB is very small around 0.8 nm, which is much different from other reports. The earlier research work revealed that the amorphous GB with 10 nm or more thick are seen and some element-doped MgB_2 samples have additional phases at the GB [41]. These microstructure characteristic may deteriorate the critical current. The very thin GB in the Zr-doped sample makes the nano-MgB_2 particles tightly bonded, inducing the good connectivity of MgB_2 grains and the high density in the Zr-doped samples. Also, a small orientation difference between two adjacent grains can be seen in Fig. 17. As reported early, GB pinning is effective in MgB_2 superconductors. Therefore, the nano-MgB_2 particles and ultra-thin GB in the Zr-doped sample can greatly improve the flux pinning. In addition to the reduction of the grain size, very small ZrB_2 particles form in the MgB_2 matrix similar to the results in Ti-doped MgB_2, which also contributes to the strong flux pinning in Zr-doped samples.

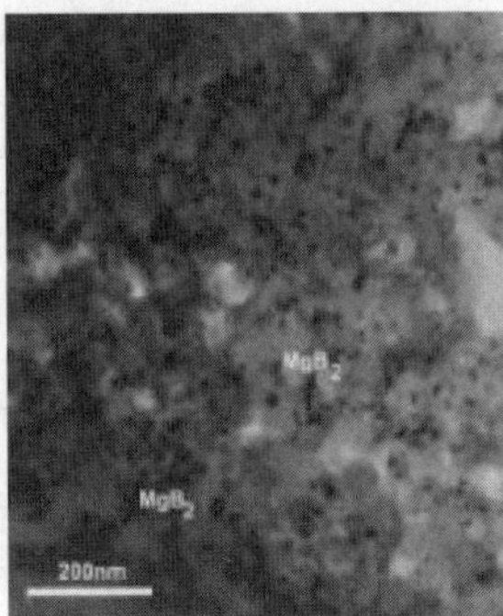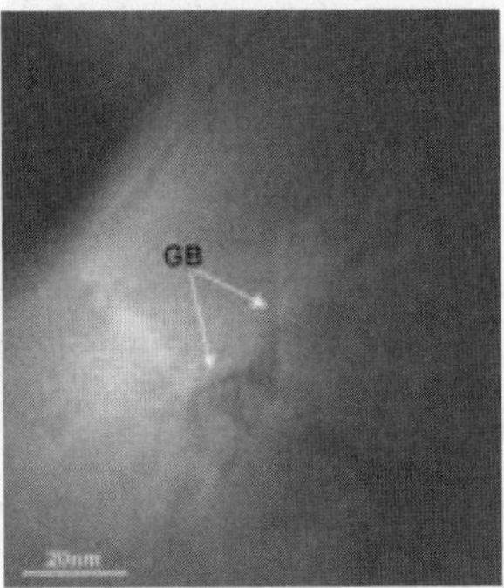

Fig. 16. Typical TEM images for the $Mg_{0.9}Zr_{0.1}B_2$ sample, showing good grain connectivity and MgB_2 nano-particles (around 20 nm).

Fig. 17. Typical ultra-thin and clean grain boundaries with a thickness about 0.8nm in $Mg_{0.9}Zr_{0.1}B_2$ sample.

At the low doping level of x=0.05, the density is higher than pure MgB_2, but not as high as in $Mg_{0.9}Zr_{0.1}B_2$. Thus, the J_c value in this sample is larger than that in pure MgB_2, but much lower than that in $Mg_{0.9}Zr_{0.1}B_2$. As for the sample with x=0.2, it has higher density, while J_c is still smaller than that of $Mg_{0.9}Zr_{0.1}B_2$. This is due to the decrease of the volume fraction of MgB_2 phase (79 mol%) and the increase of the ZrB_2 phase (14 mol%).

IV. Transport behavior , flux pinning and microstructure in $MgB_2/Ta/Cu$ wires

For the engineering applications, it is necessary to make MgB_2 superconductors into wires or tapes. However, unlike conventional superconductors, the fabrication of an actual composite conductor of MgB_2, that shows uniform properties over a large length of a wire form remains a challenge, mainly due to the ceramic and brittle nature such as high-T_c superconductors, and high pressure synthesis method due to high reactivity with O_2. This suggests that embedding MgB_2 in a metal sheath by a powder-in-tube (PIT) technique, generally used for fabrication of Ag sheathed Bi-2223 superconducting tapes, could be a potential solution for manufacturing long length wires for high current applications. Here, we present a very similar technique used for fabrication of Bi-2223/Ag-sheathed tapes by the PIT method to manufacture MgB_2 round wires that are generally demanded for magnet fabrications. We show that these round wires demonstrate remarkably low resistance, moderately high critical current density J_c , and higher critical fields H_{c2} without optimization. Also, the relaxation behavior and TEM observations in the sample were reported in this section. Our results strongly suggest that high quality MgB_2 wires can be successfully fabricated at high T [42-44].

The MgB_2 composite wires with and without Ta buffer layer were prepared by the in-situ powder-in-tube method. Mg(99%) and B(99%) powders were well mixed in air for a short time. 5wt% extra Mg powder was added in the precursor in order to compensate the loss of Mg at high temperature. Then, the mixture was filled into a tantalum tube of 7mm in outside diameter and 1.5mm wall thickness and sealed in a copper tube of 9mm in outside diameter and 1mm wall thickness. The tantalum tube was used to prevent the reaction between MgB_2 and Cu effectively. Then, the tubes were swaged and drawn to a wire of 2mm in diameter. Finally the wires were annealed at 700-900°C for 2 h.

High-resolution transport measurements were carried out using standard four-probe ac technique (I=1 mA and f=17 Hz). The transport J_c was measured using a current reversal dc technique. The magnetization measurements were performed in a superconducting quantum interference device magnetometer.

Figure 18 shows the magnetic field dependence of resistivity of MgB_2/Ta/Cu wire. The midpoint value of normal-state resistivity of the superconducting transition at H=0 T is 38.3K. A similar T_c value of 38.4 K with transition width of 0.6K was determined from the magnetization measurements as shown in the lower inset of Fig. 18. The dense bulk extracted from the core of the wire shows sharp transition with significantly low transition width (shown as dashed lines in the inset). The normal-state resistivity curve shown in the top inset of Fig. 18 follows roughly usual T^2 temperature dependence over the entire temperature region between 300 K and the onset of T_c as obtained for dense polycrystalline samples [16, 45]. We mention that the resistance due to the metallic sheath is subtracted from the wire resistance and the actual conductor resistance is normalized to the bulk sample extracted from inside the wire. No magnetoresistance was observed from 300 K to T_c in these wires and this is consistent with previous results on polycrystalline samples [16,45], but different from the another report [8]. Although no significant broadening of the transition was observed at low fields, a considerable resistive broadening was seen at low temperatures at higher fields, which may be caused by the flux–flow effects at high magnetic fields. This becomes very clear in the Arrhenius plot of resistivity shown in Fig. 19 that shows large slopes at higher field values indicating the thermal activation of the dissipation. However, the role of layered structure in MgB_2 in causing such dissipation is not known yet. The other option is to consider the impurity effects due to small amount of MgO in the interfaces. The resistive behavior of MgB_2 is significantly similar to that of intermetallic Borocarbide and A15 compounds.

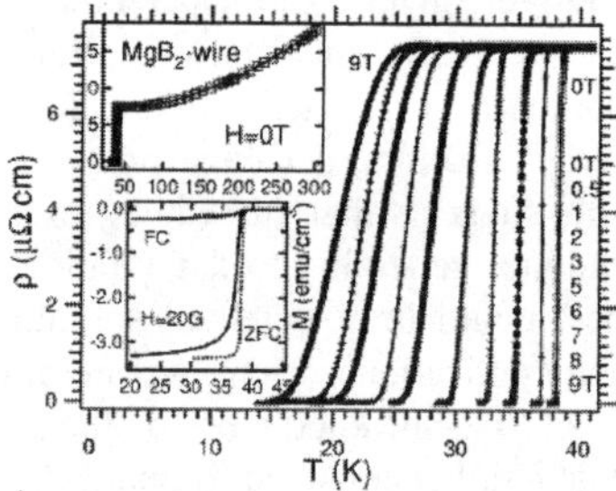

Fig. 18. Temperature-dependent electrical resistivity of MgB_2–Ta wire in applied fields from 0 T to 9 T (right-hand side to left-hand side) for H parallel to the wire length is shown. The top inset shows the T dependence of the normal-state resistivity of the same wire for H=0 T. The bottom inset shows the T dependence of the field-cooled (FC)and zero-FC magnetization of the same wire (solid lines) and the dense bulk (dashed lines) extracted from the core of the wire.

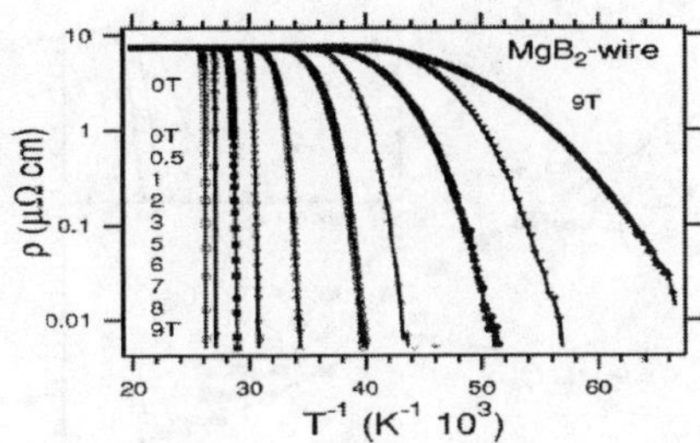

Fig. 19. The Arrhenius plot of resistivity curves at various fields 0 to 9 T (left-hand side to right-hand side) that show large broadening at higher fields is shown.

In order to compare the resistive broadening in fields, we plot the upper critical field H_{c2} as function of reduced temperature (T/T_c) in Fig. 20 for MgB$_2$/Ta/Cu and MgB$_2$/Cu wires. The slope of the H$_{c2}$– T curve yields dH_{c2}/dT=0.39, which gives $H_{c2}(0)$=15 T and $H_{c2}(4.2$ K)=13.2 T for MgB$_2$/Ta/Cu wire. These values are almost half the value of $H_{c2}(0)$ for Nb$_3$Sn. In this extrapolation method, we have considered only the linear region above 2 T in the phase diagram, because a small but definite positive curvature in H_{c2} was observed in the low-field region in both samples, similar to the borocarbides. Furthermore, it is remarkable to note that the Tc is found to be very much current-induced due to the self-field effect as shown in the inset of Fig. 20. The pronounced leg feature indicates a large dissipation that is similar to the application of a magnetic field. Hence, H_{c2} measurements were performed using a current as low as 1 mA. Using the estimation of $H_{c2}(0)$ =15 T, the coherence length $\xi_0=[\phi_0/2\pi Hc2(0)]^{0.5}$ is found to be around 4.5 nm and consistent with the bulk samples [46]. In order to evaluate the electronic mass anisotropy, $\gamma= H_{c2}^{\parallel}/H_{c2}^{\perp}$, we measured both $H_{c2}^{\perp}$ (H perpendicular the wire length! and $H_{c2}^{\parallel}$ (H parallel to the wire length) . In MgB$_2$–Ta for both perpendicular and parallel fields, respectively, and shown in Fig. 20. In contrast to high-T_C layered Bi-2223/Ag multifilamentay tapes, γ is about 1.3±0.15 in MgB$_2$–Ta wire, which is almost half of the value of Bi tapes [47] and less than the c-axis-oriented thin film [48], aligned MgB$_2$ bulk [49], and much less than Mg^{11}B$_2$ polycrystalline sample [50]. However, this value does not reflect the actual value of γ , rather indicates that there remain some misoriented grains whose field component is closer to $H_{c2}^{\perp}$ (or $H_{c2}^{\perp c}$). This shows that the wires are textured to some extent and further optimization can yield highly textured wires for real applications, although enhancing the density of MgB$_2$ in the wire core remains the best option to achieve the largest Jc . However, such a low value of γ can have a serious consequence on reducing transport Jc .

Figure 21 presents data on both magnetic and transport Jc of MgB$_2$/Ta/Cu wire for magnetic fields applied parallel to the wire length. The dashed lines with symbols are Jc values extracted from the direct transport measurements of the current–voltage curves at a fixed T and H. The solid lines are Jc values deduced from the magnetization loops using Bean's model. The transport Jc values (limited to only 320 A/cm ^{2} due to local heating effects)are consistent with that of magnetization if extrapolated to the low-field region. Two significant observations are recorded. First, magnetic Jc decays approximately exponentially above a characteristic plateau in field dependent Jc curve of the self-field region and may be related to the penetration

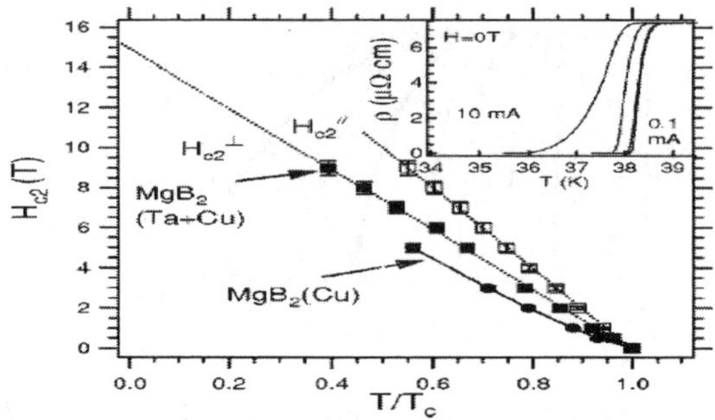

Fig. 20. The upper critical fields H_{c2} as a function of reduced temperature T/T_C for MgB$_2$–Ta and MgB$_2$–Cu wires are shown. The inset shows the resistive transition of MgB$_2$–Ta wire for different excitation currents I =0.1, 1, 3, and 10 mA in zero field.

field, especially at a low temperature. Second, the transport Jc is higher than the magnetic one at a fixed H and T in the high-field region and can be understood in terms of criteria fixed for two independent experiments. At 5K and in self-field, our wire have a high Jc value above 10^5A/cm^2. The J_c also reaches 10^4A/cm^2 at 20 K and 1 T, which is comparable to the high-pressure sample [51]. Figure 22 illustrates J_c values as a function of field at various temperatures for MgB$_2$/Cu wire. It can be seen that J_c is much lower in MgB$_2$/Cu wire than that in MgB$_2$ /Ta/ Cu wire, which should be related to the serious reaction between MgB$_2$ and Cu. Also, as compared to MgB$_2$/Ta/Cu wire, MgB$_2$/Cu wire has a lower T_c of 37.6K with broader transition width (ΔTc=1.2 K). In addition, a much lower irreversibility field is observed in MgB$_2$/Cu wire as compared to MgB$_2$/TaCu. However, the overall Jc values in MgB$_2$ /Ta/Cu both from transport and magnetization measurements are lower than that of MgB$_2$ strands. Moreover, the wires fabricated through *in situ* reaction by a PIT technique have 25% lower density than the dense bulk prepared by high pressure synthesis. Hence, the effective radius of the wire R_{eff} is less than the actual radius R while calculating using Bean's model defined for magnetic J_c as J_c (H)=3/2($\Delta M/R$). Consequently, J_c is reduced by 25% of the actual value for the wire. This is mainly due to lower density arising from some scattered porous MgB$_2$ matrix in the wire. We note that the extracted dense MgB$_2$ bulk from the core the wire produces comparable Jc values to strands. This suggests many scopes to enhance the flux pinning in the wire by optimizing the fabrication process. The misorientation and weak connectivity between domains seem to limit Jc although not as severe as high-T_C superconductors. We mention here that although weak-link phenomenon may not be a reason for low Jc in MgB$_2$ as reported [6], the link between the domains due to poor density inside the wire remains a primary concern and needs to be addressed further.

The time decay of magnetization was measured for the MgB$_2$/Ta/Cu wires by SQUID and Fig.23 shows the results. It can be observed that the magnetization exhibits the linear dependence of time, indicating that this behavior follows the Anderson-Kim model. The relaxation rate S was calculated by the equation of S=-dlnM/lnt. Fig 24 presents the relaxation rate as a function of field at 20 K and 10 K and as a function of temperature at 0 T and 0.5 T. The relaxation rate is very small, just around 0.086% at 5 K in 0 T. Also, S is 0.003 at 0.5 T , being the same as the MgB$_2$ bulk sample [52,53]. This value is an order of magnitude lower than that of HTSC.

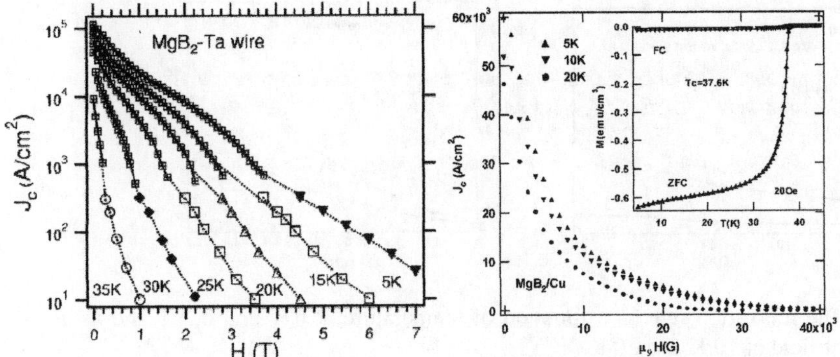

Fig. 21.Field dependence of J_c at various temperatures of MgB_2–Ta wire.The symbols with solid lines are the magnetic J_c. The symbols with dashed lines are the transport J_c.

Fig.22. Field dependence of J_c for MgB_2/Cu wire. Inset: Temperature dependence of magnetization.

Furthermore, the relaxation rate shows a very weak temperature dependence in the measured temperature range. At 0.5 T, S increases from 0.003 at 5 K to 0.008 at 25 K. As described by Thompson et al, the exponent n of electric field $E-E_0(J/J_0)^n$ is simply estimated by $(n-1)=1/S$, where J is current density. This will lead to a n value around 300, corresponding to very steep curves in transport I-V characteristic. On the other hand, the relaxation rate increases slowly with field and may be high at a higher field as observed in Fig.24. For example, at 20 K S increases from 0.0046 at 0.2 T to 0.062 at 2.5 T. The very small relaxation rate and weak temperature dependence may be related to the strong pinning barrier. It is found that the intrinsic pinning energy is very high and the pinning well of MgB_2 superconductors is so deep that the thermal activation and fluctuation has a trivial effect [54, 55].

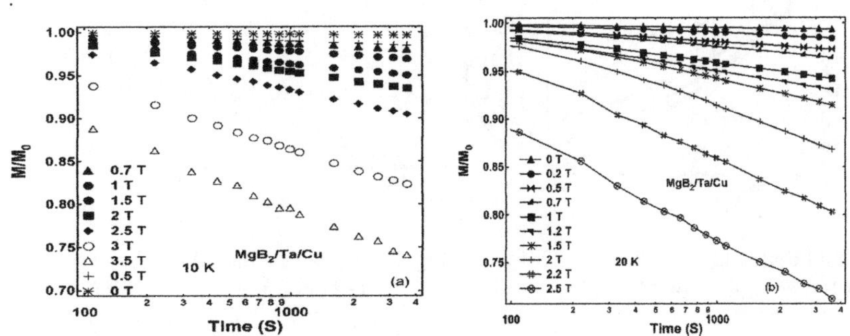

Fig. 23. Time dependence of magnetization at 10 K (a) and 20 K (b) for MgB_2/Ta/Cu wires, showing linear time decay.

The dc magnetization was performed by a SQUID magnetometer. All the magnetization measurements were measured by first cooling the sample in zero field and then applying a field to begin the measurement. After it finished at a given temperature, the field was set to zero and the temperature was warmed to 50K in order to completely remove the trapped field inside the sample.

418

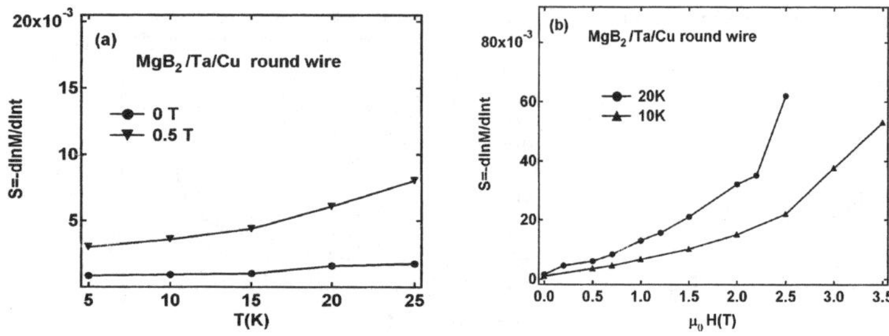

Fig. 24. The relaxation rate as a function of temperature at 0 T and 0.5 T and as a function of field at 10 K and 20 K.

Fig 25 shows the field dependence of magnetization at various temperatures for MgB_2/Ta/Cu wires . It is evident that all the curves exhibit the common linear field dependence of magnetization induced by the Meissner effect at low fields. The linear field dependence on magnetization is observed in the fields below 400Oe at 5 K. The lower critical field H_{c1} is determined by the departure point from the linearity on the slope of the magnetization curve and the results are displayed in Fig.26. At 5 K, H_{c1} is around 39 mT, being higher that MgB_2 bulk samples and wires [56]. Also, the lower critical field shows the linear dependence of temperature , which is similar to that in the YNi_2B_2C system and in high T_c superconductors . It is believed that this behavior may be related with the linear dependence of upper critical field in MgB_2 as reported by some authors [57].

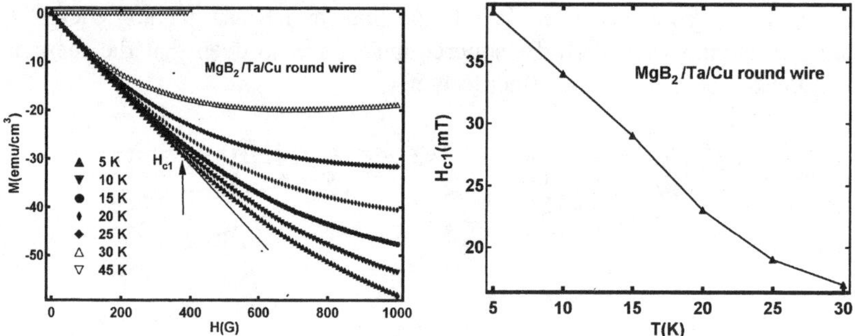

Fig. 25. Field dependence of magnetization at various temperatures for MgB_2/Ta/Cu wire.

Fig. 26. H_{c1} as a function of temperature.

As reported by our previous work , Ta is a good candidate to be the buffer layer for preparing MgB_2 wires [42]. Fig. 27 illustrates TEM image of the interfaces between MgB_2 grains and Ta. There is no reaction between them. By EDAX analysis, some MgO particles are found as marked as 1,2,3,4 in this sample. Also, the size of these particles is very small less than 50 nm. However, a serious reaction of Mg and Cu is observed in the MgB_2/Cu wires without Ta buffer layer. The $MgCu_2$ compound is formed within the MgB_2 core/Cu-sheath interface, which leads to a big reduction in J_c as shown in Fig.22. The similar result was also reported in Ref. [23]. Furthermore, typical TEM images for the MgB_2/Cu wires with and without Ta buffer layer are shown in Fig. 28. The MgB_2/Cu sample has the smaller and more homogenous MgB_2

grains, while the superconducting grains are larger in MgB_2/Ta/Cu wire. As observed by A.Gumbel et al [10], the grain boundary pinning is very effective in MgB_2 superconductors. The smaller grains should result in the enhancement of flux pinning. However, the serious reaction between Cu sheath and MgB_2 and the lower T_c with broader transition lead to the decrease in J_c for MgB_2/Cu wires.

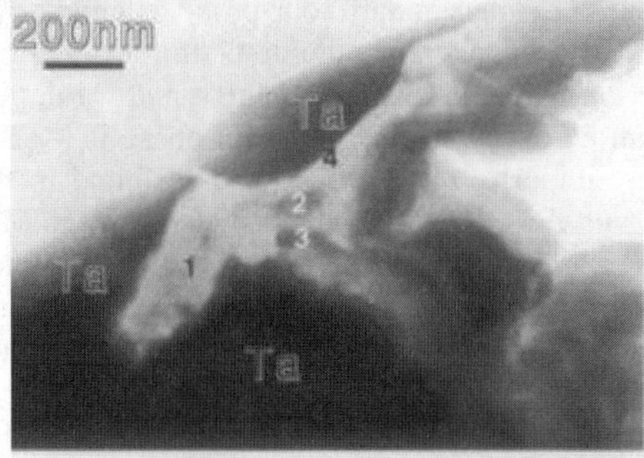

Fig. 27. TEM image of interface between MgB_2 grains and Ta for MgB_2 /Ta/Cu.

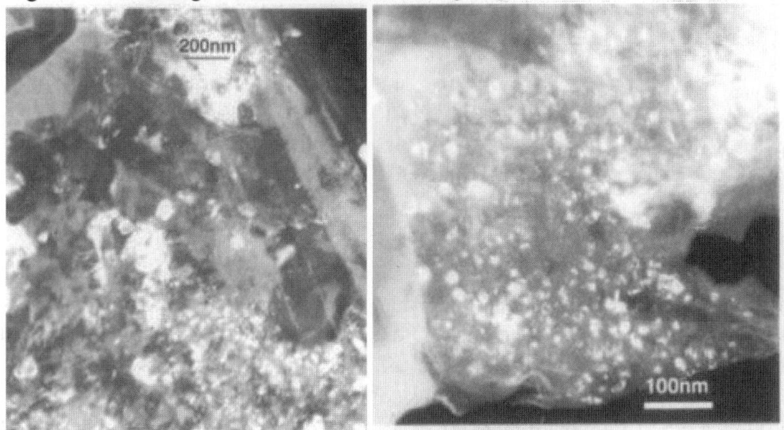

(a). MgB_2 /Ta/Cu wires   (b). MgB_2 /Cu wires

Fig. 28 Typical TEM images for MgB_2/Cu wires with (a) and without Ta (b).

V.   Improvement of J_c in MgB_2 tapes by Ti-doping

In this section, we presented the superconducting properties, structure and microstructure results of Ti-doped MgB_2/Cu tapes with Ta as a buffer layer by the in-situ powder-in-tube method at ambient pressure. It is found that the Ti-doping can significantly enhance the flux pinning. The high critical current density of $1.5 \times 10^6 A/cm^2$ (10 K, self-field) and $9.3 \times 10^5 A/cm^2$ (20 K, self-field) were obtained in $Mg_{0.9}Ti_{0.1}B_2$ tape. Also, the reasons of high J_c in this sample were briefly discussed [58]

The MgB_2 composite tapes with and without Ti doping were prepared by the in-situ powder-in-tube method. Mg(99%), Ti(99%) and B(99%) powders were well mixed in air for a short time. The detailed description in preparation of $Mg_{0.9}Ti_{0.1}B_2$/Ta/Cu wires can be found in Section IV. Then, the wires were subsequently rolled to tapes of 3.4×0.25mm . Finally, the tapes were sintered at 900~950ºC for 2~3 hours in argon at

420

ambient pressure and followed by furnace cooling to room temperature. The phase analysis was carried out using x-ray diffractiometer. The magnetization measurements were performed on a commercial superconducting quantum interference device (SQUID) magnetometer at different temperatures in the magnetic field up to 7T and the microstructure was examined by scanning electron microscope (SEM). The transport currents were measured by the standard four-probe technique. The available cross section areas of $Mg_{0.9}Ti_{0.1}B_2$ tape were about 0.6~0.7mm^2.

The typical x-ray diffraction patterns of pure MgB_2 and Ti-doped tapes are illustrated in Figure 29. It can be observed that the main phase is MgB_2 in the pure MgB_2 sample while only few impurity phases of MgO are found in the spectrum. However, some TiB_2 phases and fewer MgO can be detected besides MgB_2 phase in the Ti-doped sample. Moreover, no MgB_4 is observed in these two samples. The XRD analysis shows that the lattice parameters are not changed by Ti-doping, suggesting that Ti atoms do not enter the structure of MgB_2. The critical temperature (38 K) of the Ti doped sample is slightly lower than that of pure MgB_2 (38.4 K).

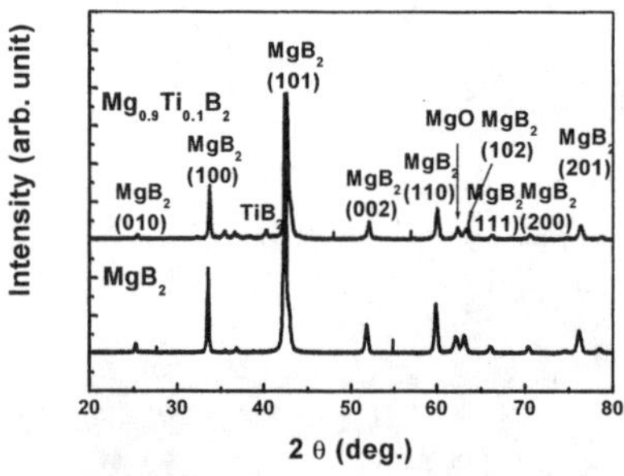

Fig. 29 X-ray diffraction patterns of MgB_2 tape with and without Ti doping. Few impurity phases of MgO are found in both tapes. TiB_2 phases are found in Ti-doped tape.

Figure 30 shows the magnetic field dependence of J_c for the MgB_2/Ta/Cu tapes with and without Ti doping at different temperatures. The J_c values were deduced from the hysteresis loops by using the Bean model of $J_c=30\Delta M/d$ (where d is the diameter of the sample). In this figure, the symbols presented the experiment data and the lines are the fitting curves of the equation $J_c(B)=Jc(0)\exp(-(B/B_0)^{0.65})$. It can be observed that the J_c value is significantly improved by the Ti-doping in MgB_2 tape. At 10 K and self-field, the J_c of the Ti-doped tape reaches a high value above $1.5\times10^6 A/cm^2$. As the magnetic field is increased to 1 T, J_c is as high as 2.7×10^5 A/cm^2 . However, J_c is only around of 1.9×10^4 A/cm^2 at 10 K in 1 T for pure MgB_2 tape. Interesting, at 20 K, J_c achieves 9.5×10^5 A/cm^2 in self-field. This result means that MgB_2 can be applied in temperatures around 20 K and becomes a possible candidate to substitute the low temperature superconducting materials possibly. Because the limit of measure conditions and heat effects of samples, the maximum measure current is only 80A. In 20K, the critical transport current reaches 71A in 1T field. Figure 31 shows the results of magnetic field dependence of magnetic J_c and transport J_c for $Mg_{0.9}Ti_{0.1}B_2$/Ta/Cu tape at 20K and 30K. They are well fit with the curves of magnetic current density. In addition, a plateau region of J_c can be observed from Figure 30 at low magnetic fields. At this stage, J_c has a weak dependence on the field. But when the magnetic filed is increased above a crossover field B_{sb}, J_c begins to decrease quickly. Also, the crossover field decreases with increasing temperature. The crossover field of Ti doping MgB_2 tapes is relatively high compared to pure MgB_2, which indicates the $Mg_{0.9}Ti_{0.1}B_2$ sample has a very strong flux pinning ability. Furthermore, when the field is lower than crossover field B_{sb}, it is found that the experimental data are well fit by $J_c(B)=Jc(0)\exp(-(B/B_0)^{0.65})$ at temperatures below 35 K in the $Mg_{0.9}Ti_{0.1}B_2$ tape. This field dependence of J_c suggests that the dominant pinning mechanism in the $Mg_{0.9}Ti_{0.1}B_2$ tape may be explained by the core pinning model [59] when the applied field is lower than B_{sb}. But as the applied magnetic field is larger than B_{sb}, the law of this curves submits $J_c(B)\sim B^{-3}$. This result indicates that three-dimensional pinning model is taken effect. However, a previous result showed that Ti-doping decreased the critical current density in MgB_2 tapes. The reason is that the Ti-doping level is only 5 mol% In our work, the tape with 5 mol% Ti-doping also deteriorates the critical current density. When the Ti-doping level reaches 10 mol%, the J_c value increases significantly. Figure 32 illustrates the temperature dependence of the irreversibility field (H_{irr}) for $Mg_{0.9}Ti_{0.1}B_2$ tape and MgB_2 tape. The H_{irr} value was determined from the closure of the magnetization curves. The H_{irr} achieves 7.4 T at 10 K and 1.4 T at 30 K in $Mg_{0.9}Ti_{0.1}B_2$, which is similar to the result in hot pressed MgB_2 bulks. At 10 K, H_{irr} is increased from 6.1 T in pure MgB_2 to 7.4 T in $Mg_{0.9}Ti_{0.1}B_2$. But the H_{irr} value merges at high temperature, suggesting that the flux pinning ability in Ti-doped sample at high temperature is not much improved in high fields.

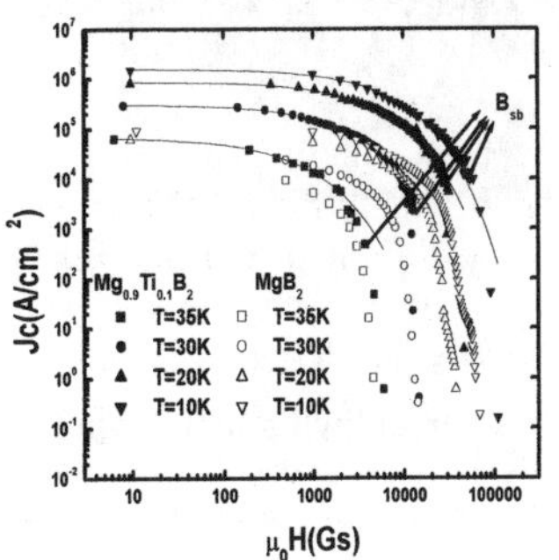

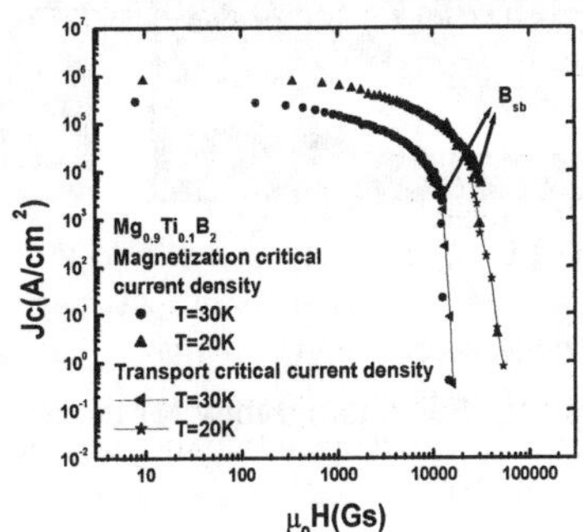

Fig.30. Magnetic field dependence of J_c for tape with and without Ti doping at different temperatures. Fitting curves of the equation $J_c(B)=Jc(0)\exp(-(B/B_0)^{0.65})$ are also shown.

Fig. 31. Magnetic field dependence of $MgB_2/Ta/Cu$ magnetization J_c and transport J_c for $Mg_{0.9}Ti_{0.1}B_2/Ta/Cu$ tape at 20K and 30K.

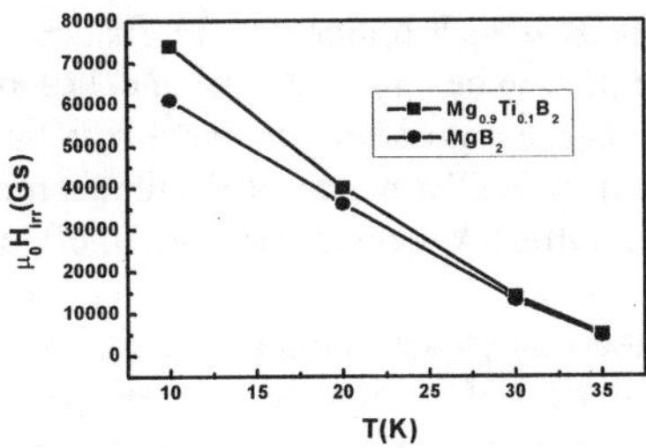

Fig. 32. Temperature dependence of irreversibility field for $MgB_2/Ta/Cu$ tape with and without Ti doping

The typical SEM photograph of the cross section in pure MgB_2 and $Mg_{0.9}Ti_{0.1}B_2$ tapes are given in Figure 33. More voids formed by the evaporation of Mg can be

422

observed in the pure MgB_2. However, the $Mg_{0.9}Ti_{0.1}B_2$ sample has a much higher density with few voids. In fact, with Ti doping, the temperature of forming MgB_2 phase is higher than pure MgB_2. Also, the connections between grains are much improved and the fine grains of MgB_2 are found in the Ti-doped tape. A very thin layer of TiB_2 forms around the MgB_2 particles and MgO nano-particles are observed in Ti-doped MgB_2 samples[21]. Therefore, it can be concluded that the TiB_2 phases in the tapes may prevent the growth of grain size of MgB_2 and lead to the very fine MgB_2 particles. In the mean while, the fine grain size creates many grain boundaries which may act as the important pinning centers in MgB_2 material and enhances the critical current density of MgB_2.

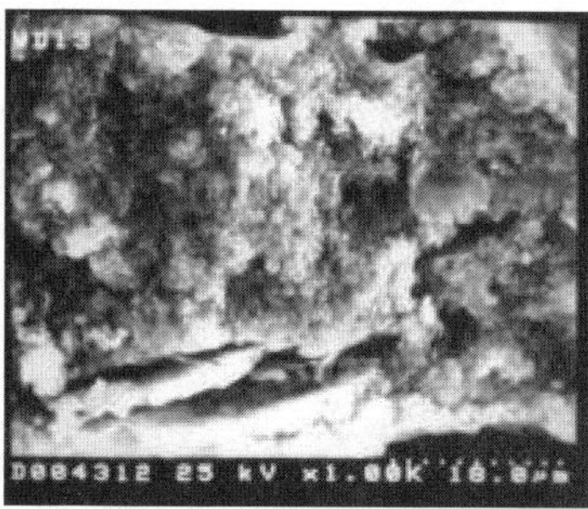
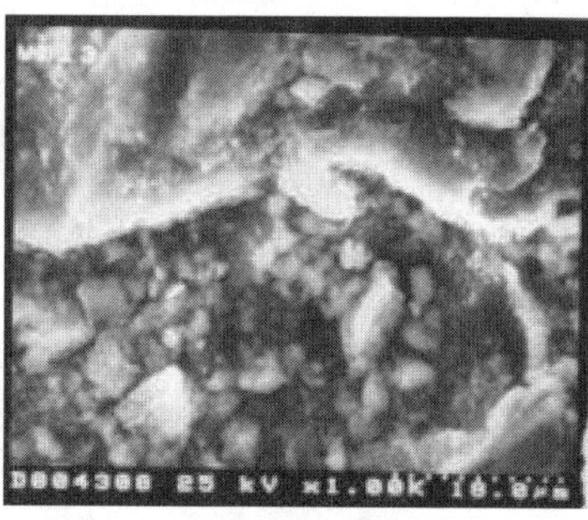

(a) Pure MgB_2 (b) $Mg_{0.9}Ti_{0.1}B_2$

Fig.33. Typical SEM photography for the MgB_2 tape with and without Ti doping.

VI. High critical current density and microstructure in MgB_2/Fe wires

The most promising applications for MgB_2 superconductors are transformer, current limiter, MRI magnet, etc. High critical current density in high fields and high temperatures (10-30 K) is necessary for these application. However, the recent results on MgB_2 wires and tapes are not very exciting. In high fields or at high temperatures, J_c of MgB_2 superconductors is low. We prepared 18 filament Cu/NbZr/MgB_2 tapes through in-situ PIT method. J_c reaches $80000 A/cm^2$ and $14000 A/cm^2$ respectively in self-field and 1 T at 10 K. These values are comparable with those reported for the wires with different sheaths [23,24]. However, J_c decreases rapidly with temperature and field, which may be due to the low density and lack of effective pinning centers in this tape.

Here, we developed a good method to prepare high quality MgB_2 wires with Fe sheath by the powder-in-tube (PIT) method at ambient pressure. The phase compositions, microstructure features and flux pinning properties are investigated by using x-ray diffractometer, SEM , TEM and standard four-probe technique. The results indicate that MgB_2 wires are dense and only MgB_2 and few MgO phases exist in the core area. MgB_2/Fe wires prepared at 750°C exhibit very high transport critical current densities of $1.7 \times 10^5 A/cm^2$ (4.2 K, 2 T), $4.4 \times 10^4 A/cm^2$ (4.2 K, 6 T), $3.72 \times 10^4 A/cm^2$ (15 K, 4T) and $2.34 \times 10^4 A/cm^2$ (25 K, 3T). The maximum flux pinning force shows a peak at 4 T, suggesting the strong pinning at high fields [60].

Single filamentary MgB$_2$ composite wires were fabricated by the *in-situ* powder-in-tube process. Mg powder and amorphous B powder were used as starting materials in an atomic ratio of Mg:B=1:2. 5% extra Mg was added to compensate the loss of magnesium in high temperature. The mixture of these powders were well ground and filled into an iron tube of 6 mm in diameter. The composite tube was swaged and drawn down to a wire of 2.0 mm in diameter with an intermediate annealing. Finally, the wires were sintered at 600-900 °C for 2 hours in argon at ambient pressure. In order to improve superconducting properties, a special rolling was developed to obtain the wires with high density.

The phase composition was analyzed by x-ray diffraction measurements. The microstructure features of these wires were observed by scanning electron microscopy (SEM) and transmission electron microscopy (TEM). TEM sample was obtained by mechanically grinding the MgB$_2$ samples to a thickness of about 40 µm and further thinning them using the polishing system. The current-voltage characteristics of the wires were measured at various fields and temperatures by the standard four probe method.

Figure 34 shows x-ray diffraction patterns for the MgB$_2$/Fe wires. The line widths of the peaks are sharp, indicating that the sample have good crystallinity. As seen in Fig.34, the MgB$_2$ grains are not well textured. Also, no MgB$_4$ phase is found in the sample, while only few MgO phase is observed.

Figure 35 shows the J_c values as a function of field at 4.2 K for the MgB$_2$/Fe wires prepared at various temperatures. It is noticed that J_c is very high especially in high fields and J_c exhibits different behaviors in field for these samples. The sample fabricated at 850°C has the maximum J_c at fields below 4 T, while J_c of the sample prepared at 750°C is the highest among them in the fields above 4 T. For the sample at 750°C, J_c is as high as 4.4×10^4A/cm^2 and 6560A/cm^2 evan at 6 T and 10 T, which is the best data reported for PIT wires and tapes. At 1 T, all the samples have large J_c values above 2.0×10^5A/cm^2 with the highest J_c around 3.2×10^5A/cm^2. Also, in the field range of 4-7 T, the J_c values for all samples are higher than 1.6×10^4A/cm^2. The high J_c in our samples may be due to the good grain connectivity and strong flux pinning force. By computer fitting, it is found that J_c decreases exponentially with the field as shown in the inset of Fig. 35.

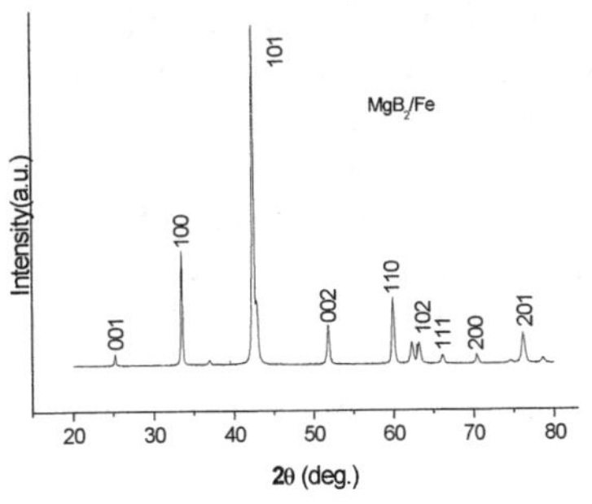

Fig.34. X-ray diffraction pattern of the MgB$_2$/Fe wire. No MgB$_4$ is observed and only few MgO phase exists.

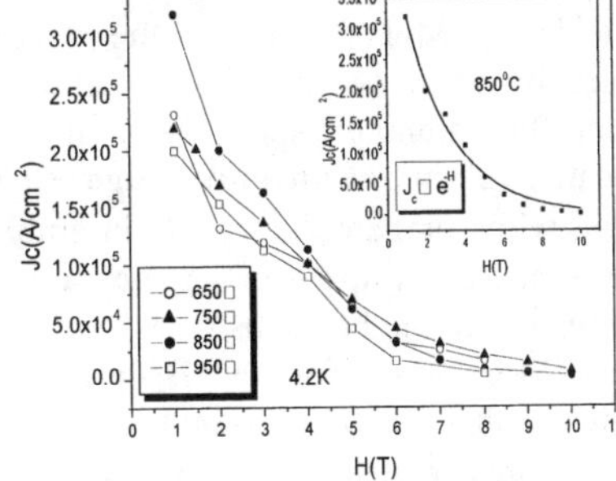

Fig. 35. Jc as a function of field at 4.2K for MgB$_2$/Fe wires prepared at different temperatures.

Furthermore, the MgB$_2$/Fe wire shows the high irreversibility field and excellent performance in high field and high temperature. Fig.36 illustrates the field dependence of transport J_c values at 15 K and 25 K for the MgB_2/Fe wires prepared at 750°C. It can be seen that high J_c values of $5.6 \times 10^4 A/cm^2$ and $3.72 \times 10^4 A/cm^2$ are obtained in the MgB_2/Fe wires at 15 K in 2 T and 4 T. Also, it should be noted that J_c reaches $2.34 \times 10^4 A/cm^2$ even at 25 K in 3 T for the sample. These J_c data are much higher than the best results on the MgB_2 wires and tapes and even on the bulk samples [61] . In addition, J_c curves at different temperatures exhibit an exponential behavior as a function of the magnetic field.

The bulk pinning force F_p is calculated by $F_p = \mu_0 H J_c(H)$ and Fig.37 shows the field dependence of the pinning force at 4.2 K for the MgB_2/Fe wire fabricated at 750°C. The maximum force $F_{p.max}$ displays a peak at 4 T, which is much higher than those values of the MgB_2 wires [38]. This result clearly indicates that the flux pinning is much stronger in our sample in high field. At 4.2 K, $F_{p,max}$ achieves a high value of $4.2 \times 10^9 N/m^3$.

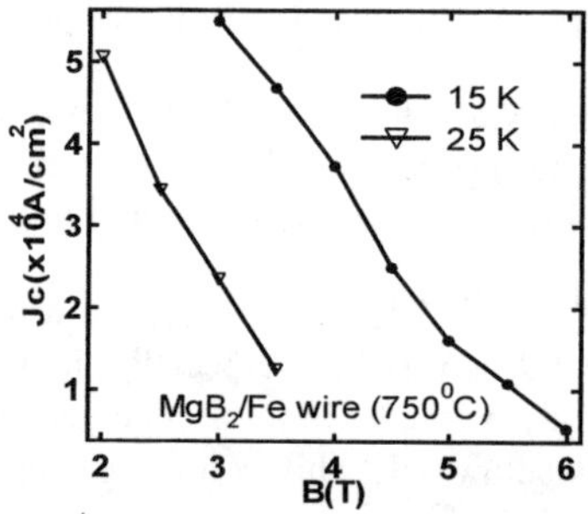

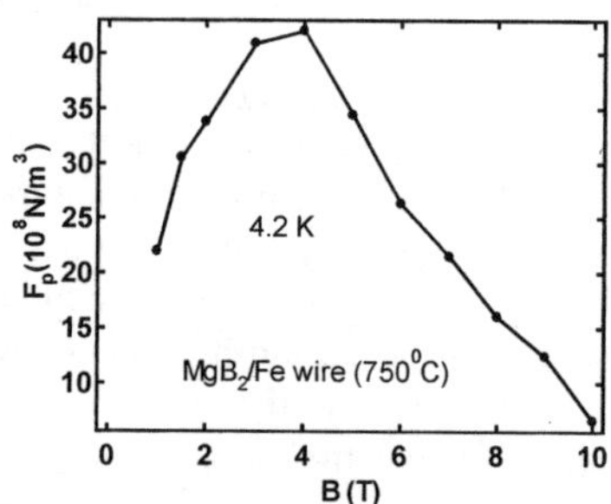

Fig.36. Field dependence of J_c at 15 K and 25 K. Note that J_c is very high around $1.43 \times 10^5 A/cm^2$ at 4.2 K in 4 T and $2.34 \times 10^4 A/cm^2$ at 25 K in 3 T.

Fig. 37. Field dependence of bulk flux pinning force density at 4.2 K, showing a maximum value at 4 T.

It is well known that critical current density is controlled by the flux pinning characteristics of superconductors. A lot of crystal defects with suitable dimensions are the candidates for the flux pinning centers. To date, the grain boundary, fine second-phase particles, nano MgO particles and dislocations are proposed to be effective pinning centers in MgB_2. In order to explore the mechanism of high J_c in high fields for MgB_2/Fe wires, the microstructure features of the samples were examined by SEM and TEM. As mentioned above, the MgB_2/Fe wire shows the excellent flux pinning especially in high fields, which should be related to the modification of microstructure. Figure 38 shows the typical SEM photo for the MgB_2/Fe wires prepared at 750°C in ambient pressure. It can be observed that the sample is dense and the MgB_2 grains are small, well connected and more uniform. The sample has higher density as compared to other MgB_2 wires with Cu or Ni sheaths. Also, most of the MgB_2 grains are found to be round shape , which will result in an enhancement of grain boundary pinning. Fig. 39 shows TEM image of the MgB_2/Fe wires treated at 750°C. Note that the grain size in the sample is small around 200-500nm, which is beneficial for the enhancement of the grain boundary pinning. In addition, neither voids or MgO are found at MgB_2 grain boundaries. However, it is interesting to note that some precipitates with size less than 60nm can be seen in MgB_2 matrix. These precipitates appear darker than the MgB_2 matrix. By using X-ray energy dispersive spectroscopy, these particles are identified to be iron. As reported by other authors, the precipitates can be the candidates for the flux pinning centers [62]. Furthermore, the electron diffraction pattern for MgB_2/Fe is shown in Fig.40. The spots are consistent with MgB_2 and the shape of the spots implies that many crystal defects exist in this sample. Fig. 41 illustrates high resolution TEM image for the MgB_2/Fe wire prepared at 750°C.. A high density of stacking faults are found, which will contribute to the flux pinning significantly. In other reports, a number of dislocations are observed in hot isostatic pressed samples and SiC-doped samples, leading to an improvement of J_c in MgB_2 bulk samples. As discussed above, the combination effects of high density, small grains, precipitates and high density of stacking faults may be responsible for the high J_c and large irreversibility field. However, the MgB_2/Fe wire has a low density as compared to the high quality bulk samples. Therefore, we can expect that J_c can be further improved by increasing the density of the sample and introducing strong pinning centers. The preliminary results shown above suggest that the process used here is simple, and promising to produce high quality MgB_2 superconducting wires for practical applications.

Fig. 38. SEM photograph of MgB_2/Fe wire prepared at 750°C

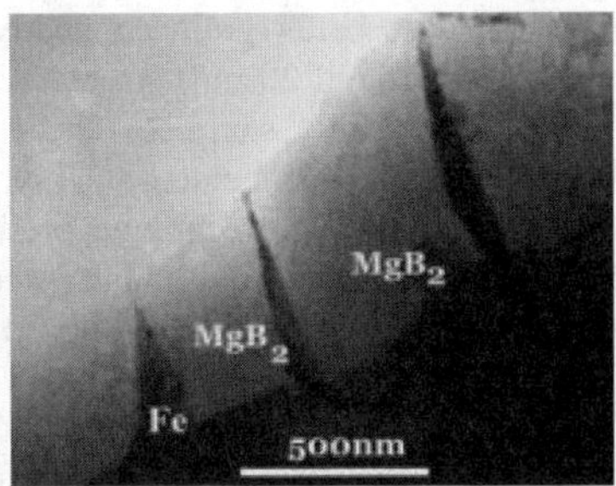

Fig. 39. TEM image of MgB_2/Fe wires

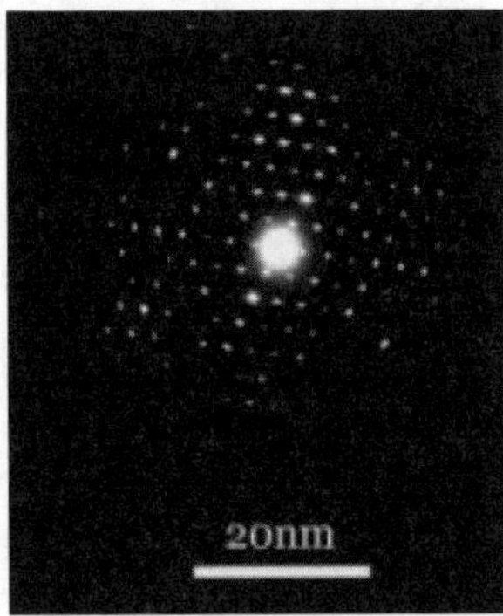

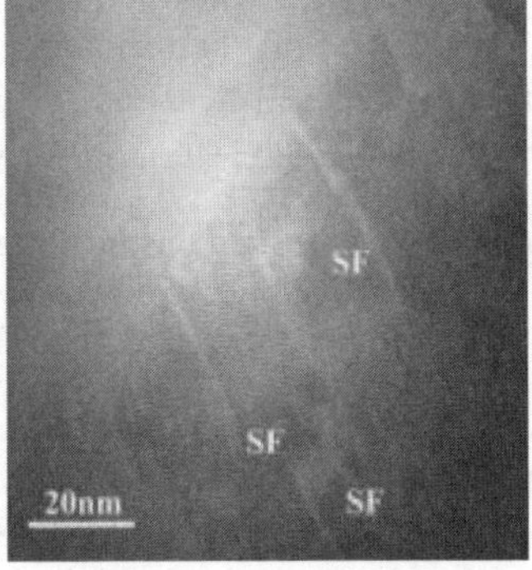

Fig.40. Electron diffraction pattern for MgB_2/Fe wire.

Fig. 41. HRTEM image for MgB_2/Fe wires. High density of stacking faults is observed.

VI.   Conclusion

In summary, we investigated the thermodynamics behavior and phase formation in Mg-B system. It is found that the formation of MgB_2 phase is very fast and the high increasing rate in temperature is necessary to prepare high quality MgB_2 superconductors. The MgB_2 phase can form in a wide temperature range , while more voids may be created by the serious evaporation of Mg at high temperature. Moreover, the MgB_2 superconductors prepared around 750°C has the better microstructure features, beneficial to obtain high J_c samples.

We developed a new, feasible and simple method to significantly improve flux pinning characteristics by Zr-doping in MgB_2 superconductors at ambient pressure. The Zr-doping results in the formation of ZrB_2 and a small substitution of Zr for Mg. T_c decreases with the increase of the Zr content up to x=0.1 in $Mg_{1-x}Zr_xB_2$. The $Mg_{0.9}Zr_{0.1}B_2$ sample exhibits the highest J_c and irreversibility field. At 5 K, J_c in this sample is enhanced to 2.1×10^6 A/cm^2 in 0.56 T and 5.7×10^5 A/cm^2 in 2 T. At 20 K, J_c is as high as 1.83×10^6 A/cm^2 in self-field and 5.51×10^5 A/cm^2 in 1 T. The Zr-doping can lead to the better grain connectivity, high density and a formation of MgB_2 nanoparticle (around 20 nm) structure with ultra-thin, clean grain boundaries (<1 nm) in the MgB_2 bulk samples. These microstructure modifications may be responsible for the huge enhancement of flux pinning over a large temperature range in the Zr-doped sample. The technique presented here shows a great potential to fabricate MgB_2 bulks and wires with high performance on an industrial scale.

Also, We have successfully prepared dense single filamentary $MgB_2/Ta/Cu$ wires and tapes with and without Ti-doping, MgB_2/Fe wires and 18 filament $MgB_2/NbZr/Cu$ tapes by powder-in-tube technique. It is found that Cu has a serious reaction with MgB_2. By using Ta as a buffer layer, this reaction can be greatly suppressed. Temperature and magnetic-field-dependent resistivity in $MgB_2/Ta/Cu$ wires displays a high conductivity and upper critical field H_{c2} generally observed in dense samples. The electronic mass anisotropy $\gamma \approx 1.3 \pm 0.15$ predicts some texturing in the wire. J_c in $MgB_2/Ta/Cu$ wires is higher than $10^5 A/cm^2$ at 5 K in self-field and $10^4 A/cm^2$ at 20 K in 1 T. Furthermore, J_c value is significantly increased to $1.5 \times 10^6 A/cm^2$ at 10 K in self-field and 2.7×10^5 A/cm^2 at 10 K in 1 T by Ti-doping, which may be due to the high density , fine MgB_2 particles and thin layer of TiB_2 in Ti-doped $MgB_2/Ta/Cu$ tapes.

We found that Fe has the high chemical stability against MgB_2 and is the best one among these metal sheaths. MgB_2/Fe wires exhibit very high transport critical current densities at high temperatures and high fields. J_c values achieve as high as $1.43 \times 10^5 A/cm^2$ (4.2 K, 4 T), $3.72 \times 10^4 A/cm^2$ (15 K, 4T) and $2.34 \times 10^4 A/cm^2$ (25 K, 3T). Also, MgB_2/Fe wires have high J_c of $4.4 \times 10^4 A/cm^2$ and $6560 A/cm^2$ evan in 6 T and 10 T at 4.2 K . The maximum flux pinning force shows a peak at 4 T, suggesting the strong pinning at high fields. The high density, small grains, fine precipitates and high density of stacking faults in the MgB_2/Fe wires may be responsible for the high J_c and large irreversibility field.

ACKNOWLEDGMENTS

The authors would like to thank C. F. Liu, B. Q. Fu, P. Ji at NIN, Xi'an; L. Z. Cao, K. Q. Ruan, X. G. Li and Y. H. Zhang at USTC in Hefei; X. J. Wu in Shanghai and E. Mossang, A. Sulpice and E. Hebral in Grenoble for their useful discussion and kind assistance in a part of experiments. This work was supported by National "863" project, National Natural Science Foundation of China under Contract No. 50172040 and NEDO of Japan.

Reference:

1. J. Nagamatsu, N. Nakagawa, T. Muranaka, Y. Zenitani, and J. Akimitsu, Nature **410** (2001). 63
2. S. L. Bud'ko, G. Lapertot, C. Petrovic, C. E. Cunningham, N. Anderson, and P. C. Field, Phys. Rev. Lett. **86** (2001) 1877
3. J. Kortus, I. I. Mazin, K. D. Belashchenko, V. P. Antropov, and L.L. Boyer, Phys. Rev, Lett. **86** (2001) 4656
4. A. Sharoni, I. Felner, and O. Millo, Phys. Rev. B **63** 220508 (2001)
5. J. E. Hirsch, Phys. Lett. A **282** (2001) 392
6. D.C .Larbarlestier, M. O. Rikel, L. D. Cooley, A. A. Polynaskil, J. Y. Jiang, S. Patniak, X. Y. Cai, D. M. Feldman, A. Gurevich, A. A. Squitieri, M. T. Naus, C. B. Eom, E. E. Hellstrom, R. J. Cava, K. A. Regan, N. Rogado, M. A. Hayward, T . He, J. S. Slisky, P. Khalifah, K. Inumara, and M. Haas, Nature **410** (2001) 186
7. M. S. Kim, C. U. Jung, M. S. Park, S. Y. Lee, H. P. Kim, W. N. Kang, and S. I. Lee, Phys. Rev. B **64** (2001) 012511
8. D. K. Finnemore, J. E. Ostenson, S. L. Bud'ko, G. Lapertot, and P. C. Canfield, Phys. Rev. Lett. , **86** (2001) 2420
9. Y. Bugoslavsky, G. K. Perkins, X. Qi, L.F. Cohen, and A. D. Caplin, Nature **411** (2001) 561
10. A.Gumbel, J. Eckert, G. Fuchs, K. Nenkov, K. H.Muller, and L. Schultz, Appl. Phys. Lett ., **80** (2002). 2725
11. A. Serquis, L. Civale, D. L. Hammon, J. Y. Coulter, X. Z. Liao, Y. T. Zhu, D. G.. Peterson, and F. M. Mueller, unpublished; X..Z. Liao, A. Serquis, Y. T.. Zhu, J..Y. Huang, D. E. Peterson, F. M. Mueller, and H. F. Xu, Appl. Phys. Lett. **80** (2002) 4398.
12 X. H. Zeng, A. L. Pogrebnyakov, A. kotcharov, J. E. Jones, X. X. Xi, E. M. Lysczek, J. M. redwing, S. Y. Xu, Q. Li, J. lettieri, D. G. Schlom, W. Tian, X. Q. Pan, A. K. Liu, Nature Materials **1** (2002) 1.
13 M. Kambara, N. H. Babu, E. S. Sadki, J. R. Cooper, H. Minami, D. A. Cardwell, A. M. Campbell, and I. H. Inoue, Supercond. Sci. Technol. **14** (2001) L5
14. A. Serquis, X. Z. Liao, Y. T. Zhu, J. Y. Coulter, J. Y. Huang, J. O. Willis, D. E. Peterson, F. M. Mueller, N. O. Moreno, J.D. Thompson, V. F. Nesterenko, and S. S. Indrakanti, J. Appl. Phys. **92** (2002), 351
15. P. Toulemonde, N. Musolino, and R. Flukiger, Unpublshed,
16. Y. Takano, H. Takeya, H. Fuji, H. Kumakura, T. Hatano, K. Togano, H. Kito, and H. Ihara, Appl. Phys. Lett. **78** (2001) 2914
17. M Eisterer, M Zehetmayer, S Tonies, H W Weber, M Kambara, N H Babu, D A Cardwell and L R Greenwood, Supercond. Sci. Technol. **15** (2002) L9
18. S. M. Kazakov, M. Angst and J. Karinski, Cond-mat/0103350
19. C. Buzea and T. Yamashita, Supercod. Sci. Technol. **14** (2001). 1215
20 .S. Jin, H. Mavoori, and R. B. Vandover, Nature **411** (2001) 563
21. Y. Zhao, Y. Feng, C. H. Cheng, L. Zhou, Y. Wu, T. Machi, T. Fudamoto, N. Koshizuka, and M. Murakami, Appl. Phys. Lett. **79** 〔2001〕1154
22. P. C. Canfield, D. K. Finnemore, S. L.Bud'ko, J. E. Ostenson, G. Lapertot, C. E. Cunningham, and C. Petrovic, Phys. Rev. Lett. **86** (2001) 2423
23. E. Martinez, L. A. Angurel, and R. Navarro, Supercond. Sci. Technol. 15 (2002)

1043
24. G.. Grasso, A. Malagoli, C. Ferdeghini, S. Roncallo, V. Braccini, M. R. Cimberle, and A. S. Siri, Appl. Phys. Lett. 79 (2002) 230.
25. X. L. Wang, S. Soltanian, J. Horvat, A. H. Liu, M. J. Qin, H. K. Liu and S. X Dou, Physica C 361 (2001) 149
26. S. Soltanian, X. L. Wang, I. Kusevic, E. Babic, A. H. Liu, M. J. Qin, J. Horvat, H. K. Liu, E. W. Collings, E. Lee, M. D. Sumption and S. X. Dou, Physica C 361 (2001) 84
27. P. Kovac, I. Husek, W. Pachita, T. Melisek, R. Diduszko, K. Frohlich, A. Morawski, A. Presz, and D. Machajdik, Supercond. Sci. Technol. 15 (2002) 1127.
28. A Glowacki et al. Supercond.Sci.Technol.78 (2001) 2914
29. Zi-Kui Liu.et al. Appl.Phys.Lett.78 (2001) 3678
30. Y. Zhu, L. Wu, V. Volkov et al, Physica C 356 (2001)239
31. Y. Feng, Y. Zhao, A. K. Pradhan, C. H. Cheng , L. Zhou. J. K. F. Yau , N. Koshizuka and M. Murakami, **J. Appl. Phys. 92** (2002) 2614
32. Y. Feng□ Y. Zhao, Y. P. Sun , F. C. Liu, B. Q. Fu, L. Zhou, C. H. Cheng, N. Koshizuka and M. Murakami M., Appl. Phys. Lett. **79** (2001) 3983
33. C. P. Bean, Rev. Mod. Phys. **36** (1964) 31
34.. Takenobu, T. Itoh, D. H. Chi, K. Prassides and Y. Iwasa, cond-mat/0103241
35. J.. S. Slusky, N. Rogado, K. A. Regan, M. A. Hayward, P. Khalifah, T. He, K. Inumaru, S. M. Loureiro, M. K. Haas, H. W. Zendbergen, and R. J. Cava, Nature **410** 343 (2001)
36. A. Handstein, D. Hinz, G. Fuchs, K. H. Muller, K. Nenkov, O. Gutfleisch, V. N. Narozhnyi, and L. Schultz, cond-mat/0103408
37. M. H. Jung, M. Jaime, A. H. Lacerda, G. S. Boebinger, W. N. Kang, H. J. Kim, E. M. Choi and S. J.Lee, cond-mat/0106146
38. G.. W. Goldacker, S. I. Shachter, S. Zimmers, and H. Reiner, 2001 Supercond. Sci. Technol. **14** (2001) 787.
39. J. Q. Li, L. Li, Y. Q. Zhou, Z. A. Ren, G. C. Che, and Z. X. Zhao, Chin. Phys. Lett. **18** (2001) 680
40. J. Wang, Y. Bugoslavsky, A. Berenov, L. Cowey, A. D. Caplin, L. F. Cohen, J. L. MacManus Driscoll, L. D. Cooley, X.. Song, D. C. Larbalestier, Appl. Phys. Lett., **81** (2002) 2026.
41. Y. Zhu, L. Wu, V. Volkov, Q. Li, G. Gu, A. R. Moodenbaugh, M. Malac, M.. Suenaga, and J. Tranquada, Physica C **356** (2001). 239
42. Y. Feng , Y. Zhao, A. K. Pradhan, L. Zhou , P.X. Zhang, X. H. Liu, P. Ji , S. J. Du , C. F. Liu, Y. Wu , and N. Koshizuka N.,: Supercond. Sci. Technol. 15 (2002) 12
43. A. K. Pradhan, Y Feng., Y. Zhao, N. Koshizuka, l. Zhou, P. X. Zhang, X. H. Liu, P. Ji, S. J. Du and C. F. Liu, Appl. Phys. Lett. 79 (2001)1649
44. C. F. Liu, Advances in Cryogenic Engineering, 48 (2002) 832
45. W. N. Kang, C. U. Jung, H. P. Kim, M. Park, S. Y. Lee, H. Kim, E. Choi,K. H. Kim, M. S. Kim, and S. I. Lee, cond-mat/0102313 (2001).
46. K.-H. Muller, G. Fuchs, A. Handstein, K. Nenkov, and D. Eckert (unpublished).
47. A. K. Pradhan, Y. Feng, Y. Wu, K. Nakao, N. Koshizuka, P. X. Zhang, L. Zhou, and C. S. Li, J. Appl. Phys. **89**, (2001) 3861.
48. S. Patnaik, L. D. Cooley, and D. C. Larbalestier, cond-mat/0104562
49. O. F. de Lima, R. A. Riberio, M. A. Avila, C. A. Cardoso, C. U. Jung, and S. I. Lee, cond-mat/0103179 .
50. F. Simon, J. Janossy, T. Feher, F. Muranyi, S. Garaj, L. Forro, C. Petrov, S.

L. Bud'ko, G. Lapertot, V. G. Kogan, and P. C. Canfield, cond-mat/0104557 (2001).

51. Y. Takano, H. Takeya, H. Fuji, H. Kumakura, T. Hatano, K. Togano, H. Kito, and H. Ihara, Cond-mat./0102167

52. J. R. Thompson, M. Paranthaman, D. K. Christen, K. D. Sorge, H. J. Kim, and J. G. Ossandon, Supercond. Sci. Technol. 14 (2001)L17

53. H. H. Wen, S. L. Li, Z. W. Zhao, H. Jin, Y. M. Ni, Z. A. Ren, G. C. Che, and Z. X. Zhao, Physica C 363 (2001) 170

54. Z.W. Zhao, H. H. Wen, S. L.Li, Y. M. Ni, Z. A. Ren, G. C. Che, H. P. Yang, Z. Y. Liu, Z. X. Zhao, Chin. Phys. Lett., 10 (2001) 340

55. H. Jin, and H. H. Wen, Chin. Phy. Lett. 18 (2001) 823

56. S.L.Li, H. H. Wen, Z. W. Zhao, Y. M. Ni, Z. A. Ren, G. C. Che, H. P. Yang, Z. Y. Liu and Z. X. Zhao, Phys. Rev. B 64 (2001) 094522

57. S. L. Bud'ko, C. Petrovic, G. Lapertot, C. E. Cunningham, P. C. Canfield, M. H. Jung, and A. H. Lacerda, .Phys. Rev. B 63 (2001) 220503

58. B. Q. Fu, Y. Feng, G. Yan, Y. Zhao, A. K. Pradhan, C. H. Cheng, P. Ji, X .H. Liu, C. F. Liu, L. Zhou and K. F. Yau, J. Appl. Phys. 92 (2002) 7341

59. M.J.Qin, X.L.Wang, H.K.Liu, S.X.Dou, cond-mat/0108172

60. Y. Feng, G..Yan, Y. Zhao, et al Phyica C (2003) in press.

61. C. Beneduce, et al, Preprint cond-mat/0203551.

62. S. X. Dou, A. V. Pan, S. Zhou, M. Iouescu, H. K. Liu, and P. R. Munroe, Supercond. Sci. Technol. 15 (2002) 1587

AC LOSSES UNDER SELF-FIELD IN A SUPERCONDUCTING TUBE

Jean LEVEQUE, Bruno DOUINE, Denis NETTER
GREEN Faculté des Sciences
BP 239 F-54506 Vandoeuvre Cedex France

1 INTRODUCTION

About one century after the discovery of superconductivity by K. Onnes, several applications using these materials have become usual. Until the sixties the phenomenon of superconductivity was essentially studied in laboratories. It is the implementation of superconducting wire that allows its use in electromechanical machinery. The main hope is to replace copper by a resistanceless superconductor in electrical devices. Therefore, there are no losses. Since these years, the field of application of the superconductivity did not stop growing from motors to filters for mobile telephony, transformers and electronic microscopy.

Today, applications of the superconductors spread in many fields in the electrical engineering. The main applications are the large electromagnet for the great physics instruments and for medical applications. The development of several kinds of electric motors is very important. There are many realisations, either with low critical temperature superconductors, or with high critical temperature superconductors. The power of some motors is about 20 MW; the power of the superconducting generator is greater than 300 MVA. The superconducting transformer is now in a phase of industrial development; they are built and tested on electrical network. Another great application is the superconducting current limiter, we can find several devices tested on electrical networks. Now, electric companies are testing transmission lines for large electrical networks. There are a lot of plans about nuclear fusion, or vehicle levitation. This list of applications is not exhaustive, but puts forward some domains in which the study of the superconductors is particulary important.

The superconducting materials are characterised, to put it simply, by two essential properties: zero resistivity and diamagnetism. These properties are revealed under some conditions of field, density of current and temperature. Therefore, when these materials lead a direct current, there are not Joule losses. However, if they are fed by variable current, it occurs losses. These are small losses compared to the losses in a copper wire, but unfortunately these losses occur at low temperature. Heating at this temperature requires a large amount of refrigeration power. Therefore, these losses are very costly. For all applications, their knowledge is part of criteria used to elecrical design some machines using superconductors.

Unless a specific superconducting wire is made, these losses prevent the low critical temperature superconductors from being used for the leading of alternating current in the industrial frequencies. Either they were too important to succeed in maintaining the superconducting material at low temperature, or the consumption of cryogenic fluid was too important. This problem has limited the applications of the superconductors in electrical engineerin for a long time.

The high critical temperature superconductors are also the seats of losses when they lead alternating current. These are less critical on the cryogenic plan than for the low critical temperature superconductor. Nevertheless, the realisation of superconducting wire for a commercial use, in alternating current, requires the restraint of these losses.

All this explains that AC losses in superconductor are subject of many researches. In this chapter, we bring our contribution to this important work.

There are many constitutive laws, like Ohm's law for classical conductors, linking electrical field and current density in superconductors. We shall present two laws: the Bean's model and a J^n law.

They are many interests in a constitutive J^n law. On the one hand, we are not limited to the study of lower current to the critical current whose definition is not clearly established. Indeed, the definition of the critical current is not obvious in superconductors, especially in high critical temperature superconductors. On the other hand, we have a better approach of the real measured characteristics $E(J)$ of the superconductors using a power J^n law.

In the second part, we shall calculate some typical cases thanks to the Bean's model. We calculate losses for a superconducting plate, for a superconducting barrel submitted to an external magnetic field and for a hollow tube submitted to its magnetic self-field. These cases are going to serve us as a reference for the rest of our work.

In the third part, we shall present the sample that we have characterised and studied.

The two following parts are devoted to the calculation of losses in incomplete and complete penetration of the current in the superconductor. In the first case, we use a dimensional analysis and numerical results to obtain analytical formulas. In the second case, we base this study on the previous work of I. Mayergoyz, as in the previous case we deduce analytical formulas.

In the last part, we shall present an experimental study of losses. Two types of measuring are used. The first one uses a differential amplifier associated with a computer, which calculate the instantaneous power by numerical integration. The second series of experiments, only valid with sinusoidal current, are made using a lock-in amplifier measuring the RMS voltage. The value of transport current is high enough to be compatible with applications. Both methods give nearly the same results for sinusoidal signal.

Measurements are made with non-sinusoidal current using a dimmer switch to feed superconductor sample. The experiments show that losses essentially depend on the maximum of the current, not on the firing angle. We conclude that the shape of the current is less important than its peak value. We compare results of measures to our results of calculations.

2 MODELLING OF THE PROBLEM

2.1 Electrical field and current density

At low frequencies, the electric and magnetic fields obey the simplified Maxwell's equations:

$$\overrightarrow{\text{curl}}\,\vec{E} = -\frac{\partial \vec{B}}{\partial t} \tag{II.1}$$

$$\overrightarrow{\text{curl}}\,\vec{B} = \mu_0 \vec{J} \tag{II.2}$$

$$\text{div}\vec{B} = 0 \tag{II.3}$$

$$\text{div}\vec{E} = 0 \tag{II.4}$$

Many constitutive laws can be used. For a classical conductor, it exists a simple relation between the current density $\vec{J}$ and the electrical field $\vec{E}$: $\vec{J} = \sigma\vec{E}$.

For the superconductors, they are mainly two constitutive laws. The oldest and simplest one is the Bean's model. In this case, we supposed that J could take three values: $\pm J_c$ or 0. Another model has the following form:

$$\vec{E} = E_C \left(\frac{J}{J_c} \right)^n \frac{\vec{J}}{J} \tag{II.5}$$

For low critical temperature superconductors, n is great and worth several tens. But, for the high critical temperature superconductors, n varies between 5 and 20. This value of n is determined by measurements. If n tends to infinity, we obtain the Bean's model. Bean's model and J^n law are represented in figure 1.

The combination of (II.1), (II.2) and (II.5) give the following results:

$$\Delta E = \mu_0 k \frac{\partial E^{\frac{1}{n}}}{\partial t} \tag{II.6}$$

or

$$\Delta J^n = \mu_0 k^n \frac{\partial J}{\partial t} \tag{II.7}$$

with $\quad k = \dfrac{J_c}{E_c^{\frac{1}{n}}} \quad$ and $\quad J_c = kE_c^{\frac{1}{n}}$

These equations govern the distribution of electrical field and current density. We intend to solve them.

In this chapter, we shall use cylindrical co-ordinates, as we can see in figure 2. The current i(t) circulate along z-axis.

The calculs aim is to determine J and E with $r \in [R_{in}, R_e]$. Current density and electrical field depend only on r and t. We note them J(r,t) and E(r,t). In order to calculate them, we need to solve one of the equations (II.6) or (II.7). We will solve the equation with electric field and we will deduce the density of current.

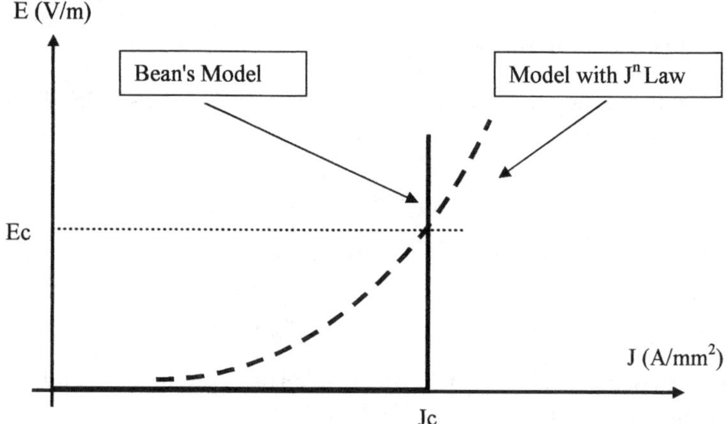

Figure 1: comparison between constitutive law

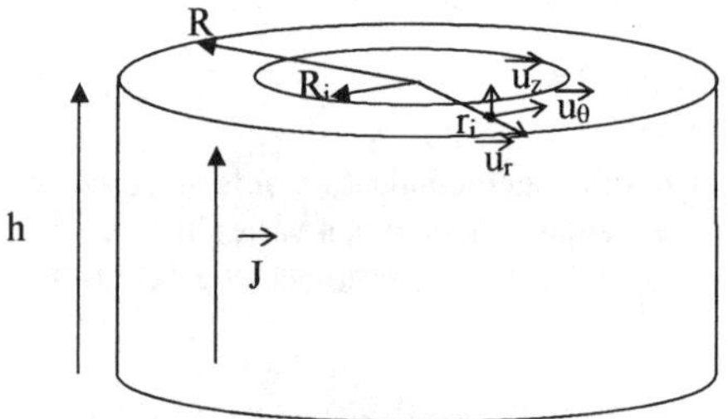

Figure 2: Base used

The equation in cylindrical co-ordinates is as follows.

$$\frac{1}{r}.\frac{\partial E}{\partial r} + \frac{\partial^2 E}{\partial r^2} = \mu_0.k.\frac{\partial E^{\frac{1}{n}}}{\partial t} \qquad (II.8)$$

Once we have expressed the equation to solve, we clarify boundary conditions. On the external radius of the tube, we know the magnetic field B(Re,t) thanks to the ampere's law:

$$B(R_e,t) = \frac{\mu_0.i(t)}{2.\pi.R_e} \qquad (II.9)$$

From the equation (I.1), we deduce the relation between E and B:

$$\frac{\partial E}{\partial r}(r = R_e) = \frac{\partial B}{\partial t} = \frac{\mu_0}{2.\pi.R_e}\frac{\partial i(t)}{\partial t} \qquad (II.10)$$

We take $\dfrac{\partial E}{\partial r} = 0$ on the internal radius of the tube.

At t=0, electrical field E is equal to 0.

2.2 Losses

AC losses in superconducting materials are hysteresic losses. At microscopic level, they are due to the movement of vortices, linked to the variation of the local magnetic field. With current density, this movement creates an electric field that leads losses. At macroscopic level, the variation of magnetic induction produces a variation of electric field in superconductor. In every point of the material where electric field and current density exist simultaneously, losses appear. We have two ways to calculate the distribution of current density and electric field. Either, we use the Bean's model, or we use the power law $E=f(J^n)$.

Two methods of calculation can be used to evaluate losses. Either we integrate the M(H) cycle. Therefore, we have the dissipated energy:

$$Q = \int_{cycle} MdH$$

Or, we integrate the volume V of the material and on a period, the scalar product of the electric field and the current density:

$$Q = \int_T \int_V \vec{E}\,\vec{J}dv.dt \qquad (II.11)$$

We shall use the latter to calculate losses.

2.3 Some classical results

We use a model wich was presented for the first time by C. P. Bean in 1964 [1] and the relation (II.11) to calculate losses.

Bean's model imposes that the local current density in a superconducting material is either equal to 0 or to its critical value J_C:

$$J=\pm Jc \text{ or } J=0 \qquad (II.12)$$

It is a local macroscopic law. This hypothesis may seem strong, but it permits to give the analytic formulas of losses for many simple cases.

The critical current density J_C is considered constant in the Bean's model Nevertheless, it is possible to take into account of the variation of J_C versus to the magnetic induction.

The calculation of losses, using this method, in a superconducting cylinder supplied by a current is well known [1-4]. We use these results to calculate losses in incomplete penetration of the current in a hollow cylinder. These results are classic, therefore, they are not going to be the subject of an important development. Nevertheless, it is necessary to remind them because they will be used to establish the new relations and to validate them.

2.3.1 Calculation of losses in an infinitely long plate submitted to a external longitudinal magnetic field

We are going to calculate losses in an infinitely long superconducting plate of width 2a, submitted to an external magnetic field Ha parallel to its surface and directed according to y-axis. This problem is independent of z and y, H, E and J only depends on x. We will consider that the induction B in the material is superior to $\mu_0 H_{C1}$. Therefore, the superconductor is the seat of losses.

As the field magnetic Ha increases, the magnetic induction B is going to penetrate progressively inside the material. We are going to use the Bean's model that permits to determine easily the magnetic induction B and the distribution of current inside the plate. In this considered case, electric and magnetic fields obey the simplified Maxwell's equations.

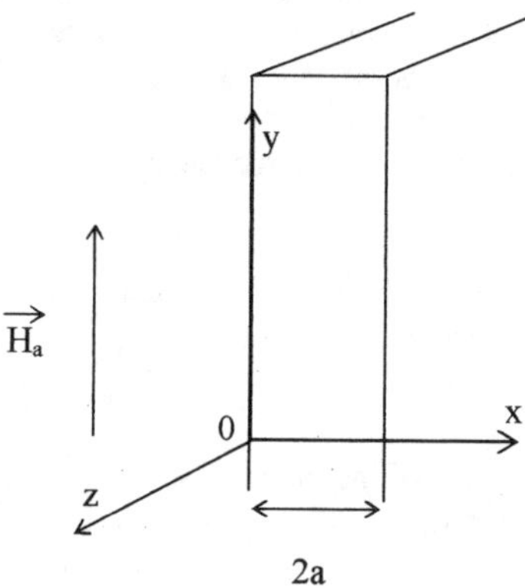

Figure 3: Superconducting plate submitted to an external magnetic field

The problem that we have to solve is defined by the following equation and boundary conditions:

The equation (II.2), combined with the Bean's model (II.12), give:

$$\mathrm{curl}\,\vec{B} = \pm\mu_0.\vec{J}_C \ \ \text{or} \ \ \mathrm{curl}\,\vec{B} = 0 \tag{II.13}$$

The induction B(x,t) is directed according to y-axis:

$$\vec{B} = B(x,t).\vec{u}_y$$

According to boundary conditions, the induction on the surface of plate is worth:

$$B_0(0,t) = B(2a,t) = \mu_0.H_a(t)$$

H_a and B_0 are alternating with an amplitude, H_{max} and B_{max}. First, we consider that $B_0(t)$ increases from 0 to B_{max}.

The current density is distributed in the plate to be opposed to the penetration of the magnetic induction. Therefore, J is oriented along z-axis, with J = -J_C or 0 for 0<x<a and J = +J_C or 0 for a<x<2a.

With equations (II.13) and (II.1) we obtain:

$$\text{Along z-axis} \quad \frac{\partial B}{\partial x} = \pm\mu_0.J_C \tag{II.14}$$

Along y-axis $\quad \dfrac{\partial E}{\partial x} = \dfrac{\partial B}{\partial t}$ (II.15)

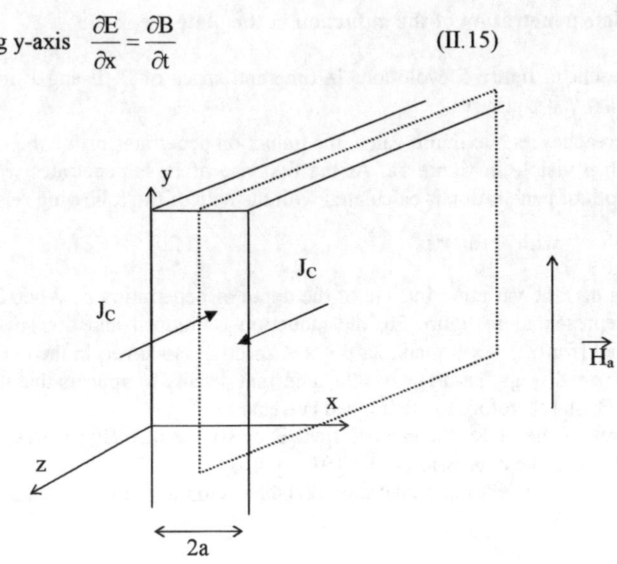

Figure 4: Induced currents in the plate submitted to an outside magnetic field.

The electrical field E is present where the magnetic field $\partial B/\partial t$ exists, and E and J are directed according to z-axis:

$$\vec{E} = E(x,t).\vec{u}_z$$

$$\vec{J} = J(x,t).\vec{u}_z$$

Losses are given by the following expression:

$$\vec{E}.\vec{J} = |E|.J_c \qquad (\text{II}.16)$$

The integration of the equation (II.14) allows us to calculate the magnetic induction for $0 < x < d(t)$:

$$B(x,t) = -\mu_0 J_c.x + B_0(t) \qquad (\text{II}.17)$$

We deduce the variation of the induction versus time

$$\dfrac{\partial B}{\partial t} = \dfrac{\partial B_0}{\partial t} \qquad (\text{II}.18)$$

The depth of penetration of the magnetic field in the material varies with $B_0(t)$, we note it $d(t)$. Using the equations (II.15) and (II.18), we calculate the electric field:

$$E(x,t) = \dfrac{\partial B(x,t)}{\partial t}.(x - d(t)) \qquad (\text{II}.19)$$

The maximal value of the penetration depth changes according to B_{max}. If B_{max} is low, there is incomplete penetration of the magnetic field in the plate. If B_{max} is high, the magnetic field penetrates the plate completely.

We are going to consider two cases: a complete penetration and an incomplete penetration of the magnetic field in the plate.

438

2.3.1.1 Incomplete penetration of the induction in the plate

We represent in figure 5 evolutions in time and space of B, E and J, in incomplete penetration, for $|\partial H_a/\partial t|$ constant.

When B reaches its maximum value, the induction penetrates inside the material with a maximal depth p visible in figure 5a. At the first rise of B, E penetrates with the same depth p. This depth of penetration is calculated with the help of the following relation:

$$p = \frac{B_{max}}{\mu_0 J_C} \qquad \text{with } 0<d(t)<p \qquad (II.20)$$

One calls $d(t)$ the variation in time of the depth of penetration p. When B decreases, from B_{max}, as represented in figure 5b, the induction is trapped and don't vary anymore inside the material (for $d(t) < x < p$ and $2a-p < x < 2a-d(t)$). However, in the external part of the plate, for $0 < x < d(t)$ and $2a-d(t) < x < 2a$, a current density J_C appears that is opposed to the reduction of B, and therefore to the trapped currents.

In this part of the plate, the electric field E exists because $B(t)$ varies according to time, this one has the same expression as (II.19).

One calculates the depth of penetration $d(t)$ thanks to the following relation:

$$d(t) = \frac{B_{max} - B_0(t)}{2.\mu_0.J_C} \qquad (II.21)$$

One deduces the instantaneous dissipated electromagnetic power per unit of volume:

$$p_V(t) = \frac{2}{2a}.\int_0^{d(t)} E.J_C.dx = \frac{\partial B_0}{\partial t}.\frac{J_C}{a}.\left[\frac{x^2}{2} - d(t).x\right]_0^{d(t)} = -\frac{1}{a.\mu_0^2.8.J_C}.\frac{\partial B_0}{\partial t}.\left[B_{max} - B_0(t)\right]^2$$

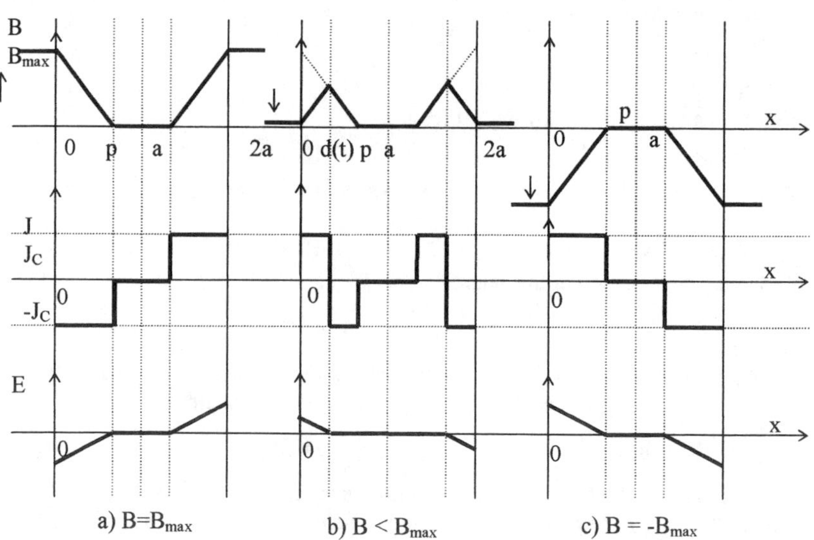

Figure 5: Distributions of B, J and E in incomplete penetration.

One deduces the instantaneous dissipated electromagnetic power per unit of volume:

$$p_V(t) = \frac{2}{2a}.\int_0^{d(t)} E.J_C.dx = \frac{\partial B_0}{\partial t}.\frac{J_C}{a}.\left[\frac{x^2}{2} - d(t).x\right]_0^{d(t)} = -\frac{1}{a.\mu_0^2.8.J_C}.\frac{\partial B_0}{\partial t}.\left[B_{max} - B_0(t)\right]^2$$

Therefore:

$$p_V(t) = \frac{1}{24.a.\mu_0^2.J_C}\frac{\partial\left[(B_{max} - B_0(t))^3\right]}{\partial t} \qquad (II.22)$$

$B_0(t)$ varies from B_{max} to $-B_{max}$, during half a period. One can calculate volumic losses on this half period and can deduce the total volumic losses P_V:

$$P_V = \frac{2}{T}.\int_0^{\frac{T}{2}} p_V(t).dt$$

$$= \frac{f}{12.a.\mu_0^2.J_C}.\int_0^{\frac{T}{2}} \frac{\partial\left[(B_{max} - B_0(t))^3\right]}{\partial t}dt \qquad (II.23)$$

Therefore:

$$Pv = \frac{f}{12.a.\mu_0^2.J_C}.\int_{B_{max}}^{-B_{max}} d\left[(B_{max} - B_0(t))^3\right]$$

$$= \frac{f}{12.a.\mu_0^2.J_C}.\left[(B_{max} - B_0(t))^3\right]_{B_{max}}^{-B_{max}}$$

We finally obtained:

$$P_V = \frac{2.f.B_{max}^3}{3.a.\mu_0^2.J_C} \qquad (II.24)$$

Now, we are going to study the case of the complete penetration.

2.3.1.2 Complete penetration of the induction in the plate

In this case, B penetrates entirely in the plate from a value B_p:

$$B_p = a.\mu_0.J_C. \qquad (II.23)$$

When B varies from B_{max} to $-B_{max}$, it exists two stages. First, the induction only varies in a thickness $d(t)$ wich is (figure 6b), between 0 and $d(t)$ and between $2a-d(t)$ and $2a$. It lasts until $d(t)$ reaches "a" and therefore at the instant t_1 where $B_0(t1) = B_{max}-2B_p$. One has the same instantaneous dissipated power $p_V(t)$ as previously. Secondly, the induction having penetrated all the plate, increases progressively in every point as indicated in the figure 6c. It occurs for $B_{max}-2B_p > B_0(t) > -B_{max}$. In this case, losses by unit of volume are independent of time.

$$p_V(t) = \frac{\partial B}{\partial t}.J_C.\frac{a}{2} \qquad (II.24)$$

So, we deduce:

440

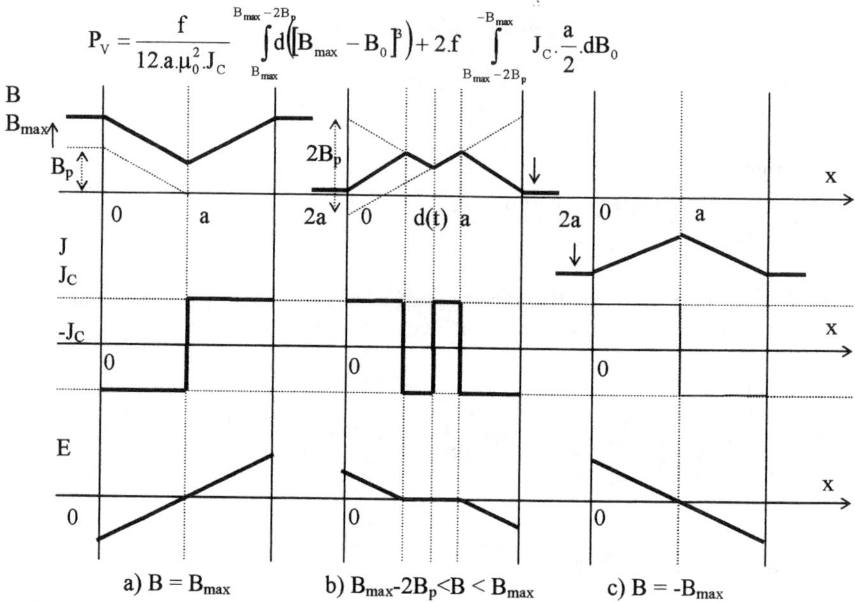

$$P_V = \frac{f}{12.a.\mu_0^2.J_C} \int\limits_{B_{max}}^{B_{max}-2B_p} d\left(\left[B_{max}-B_0\right]^3\right) + 2.f \int\limits_{B_{max}-2B_p}^{-B_{max}} J_C.\frac{a}{2}.dB_0$$

Figure 6: Distributions of B, J and E inside a superconducting plate submitted to an external magnetic field.

$$p_V(t) = \frac{\partial B}{\partial t}.J_C.\frac{a}{2} \tag{II.24}$$

So, we deduct:

$$P_V = \frac{f}{12.a.\mu_0^2.J_C} \int\limits_{B_{max}}^{B_{max}-2B_p} d\left(\left[B_{max}-B_0\right]^3\right) + 2.f \int\limits_{B_{max}-2B_p}^{-B_{max}} J_C.\frac{a}{2}.dB_0$$

Therefore, after integration, we obtain:

$$P_V = 2.f.J_C.a.\left[B_{max} - \frac{2.\mu_0.J_C.a.}{3}\right] \tag{II.25}$$

2.3.1.3 Factor of losses

In order to compare losses in the different studied cases, we express the losses P_V according to the factor of losses $\Gamma(\beta)$ where $\beta = B_{max}/B_p$, with $B_p = a.\mu_0.J_C$.

So, we obtain

$$P_V = 2.f.\frac{B_{max}^2}{\mu_0}.\Gamma(\beta) \tag{II.26}$$

With

$\Gamma(\beta) = \beta / 3$ in incomplete penetration ($\beta<1$).

$$\Gamma(\beta) = \frac{1}{\beta} - \frac{2}{3.\beta^2}$$ in complete penetration ($\beta>1$)

2.3.2 Calculation of losses in a superconducting cylinder submitted to an external periodic magnetic field

We are going to calculate losses in a cylinder, submitted to an external magnetic field $H_a(t)$, parallel to its axis. Let us note R its radius. The maximal values of this applied magnetic field and of the magnetic induction $B_0(t)$ at the surface of the cylinder, are, as previously, equal to H_{max} and B_{max}.

The physical phenomenon, in this case, is the same as the one we had for the plate, only equations change. One replaces cartesian co-ordinates by cylindrical co-ordinates.

We represent in the figure 7, the distribution of the electric field and the current density in the case of a complete and incomplete penetration of the field.

The cylinder and the applied magnetic field $H_a(t)$ are z-axis. Therefore, J and E are oriented according to θ and vary according to r.

So, we have the following equations to solve:

$$-\frac{\partial B}{\partial r} = \pm\mu_0.J_c \qquad (II.27)$$

and

$$\frac{1}{r}.\frac{\partial(r.E)}{\partial r} = -\frac{\partial B}{\partial t} \qquad (II.28)$$

Boundary conditions are of the very same kind as the latter.

As previously, we are going to study losses in incomplete and complete penetration.

2.3.2.1 Incomplete penetration of the induction

We shall consider a stage of decrease of $H_a(t)$. The magnetic field penetrates until R_p (figure 7) that one can calculate:

$$R_p = R - \frac{B_{max}}{\mu_0.J_c} \qquad (II.29)$$

The electric field is different from zero between $R_1(t)$ and R. One calculates it with the help of the relation (II.28):

$$E(r,t) = -\frac{\partial B}{\partial t}.\frac{\left(r^2 - R_1^2(t)\right)}{2r} \qquad (II.30)$$

The value of $R_1(t)$ is the following:

$$R_1(t) = R - \frac{B_{max} - B_0(t)}{2.\mu_0.J_c} \qquad (II.31)$$

So, we deduce the value of losses by unit of volume:

442

$$p_V(t) = -\frac{J_C \cdot \frac{\partial B_0}{\partial t}}{R^2}\left[\frac{R^3}{3} + \frac{2.R_1^3(t)}{3} - R_1^2.R\right] \qquad (II.32)$$

We calculate losses per unit of P_V volume integrating $p_V(t)$ on one period.

$$P_V = \frac{2}{3}f\frac{B_{max}^3}{\mu_0 R J_c}(2 - B_{max}) \qquad (II.33)$$

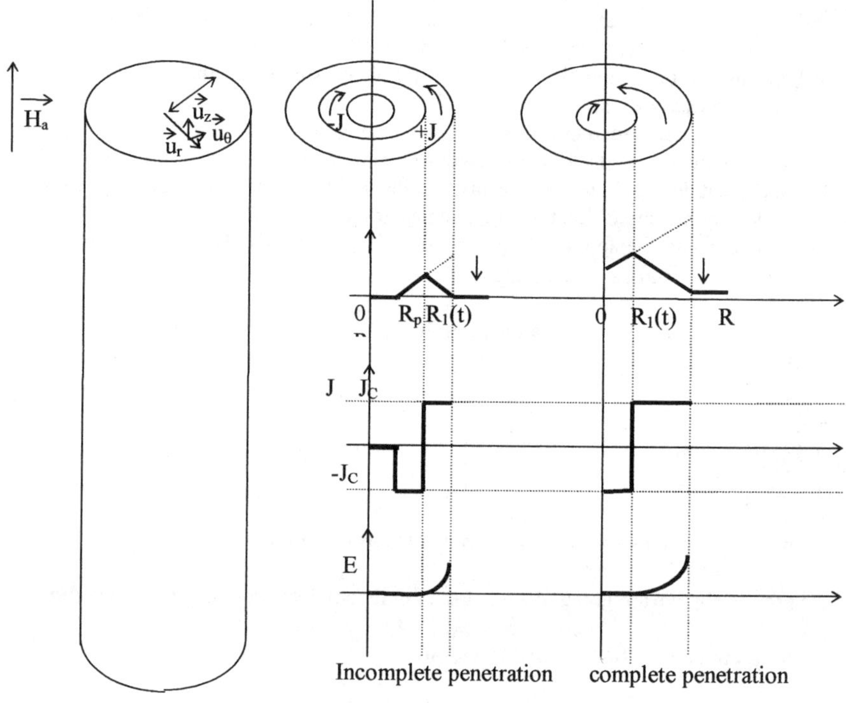

Figure 7: Distribution of B, J and E in a superconducting cylinder submitted to a longitudinal external magnetic field.

2.3.2.2 Complete penetration of the induction

The induction of complete penetration B_p is worth:

$$B_p = \mu_0 J_C.R \qquad (II.34)$$

In complete penetration R_p is equal to 0 (figure 7). When $R_1(t)$ is equal to zero, $E(r,t)$ remains constant in time and is equal to:

$$E(r,t) = -\frac{\partial B_0}{\partial t}\cdot\frac{r}{2} \qquad (II.35)$$

We have an instantaneous power of losses then per unit of volume, equals:

$$p_V(t) = -\frac{\partial B_0}{\partial t} \cdot \frac{J_C.R}{3} \qquad (II.36)$$

We calculate losses per unit of volume P_V integrating $p_V(t)$ on one period.

$$P_V = \frac{2}{3} f \, R \, J_C \left(2B_{max} - R\mu_o J_C\right) \qquad (II.37)$$

2.3.2.3 Factor of losses

In this case, we obtain, as previously:

$$P_V = 2.f. \frac{B_{max}^2}{\mu_0} \Gamma(\beta) \quad \text{with} \quad \beta = B_{max}/B_p . \qquad (II.38)$$

With:

$$\Gamma(\beta) = \frac{2\beta}{3} - \frac{\beta^2}{3} \text{ in incomplete penetration } (\beta<1).$$

$$\Gamma(\beta) = \frac{2}{3\beta} - \frac{1}{3.\beta^2} \text{ in complete penetration } (\beta>1)$$

2.3.3 Comparison of loss factors

We represent in figure 8 the factor of losses $\Gamma(\beta)$ for both previous studied cases.

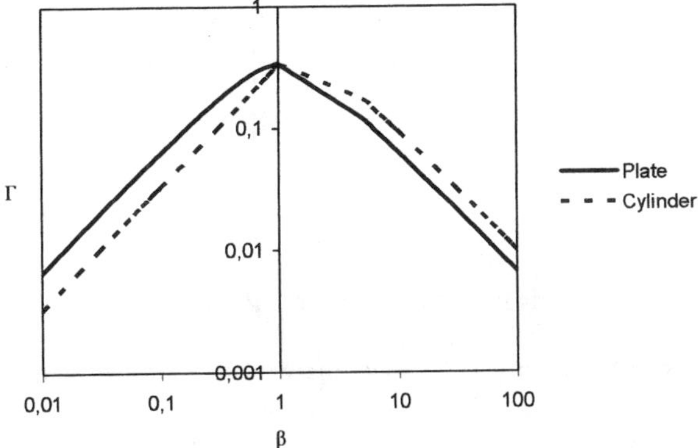

Figure 8: power factor Γ versus β

We observe that the factor of losses has the same evolution according to β for both studied cases. However, values of this loss factor differ slightly. When β is lower than unit, therefore in the case of an incomplete penetration, losses are more important for a plate.. However, when β is superior to the unit the phenomenon is reversed. It is because the physical phenomenon and the model remain the same, but the geometric structures are different.

2.3.4 Losses in a superconducting cylinder submitted to its magnetic self-field.

The magnetic self-field of a wire is the magnetic field due to the current supplied in this cable. We shall present, the calculation of losses in a cylinder fed by an alternating current, therefore it is only submitted to its magnetic self-field.

We take as example a cylinder of radius R, which is supplied by a periodic alternating current i(t) of frequency f and varying between $-I_{max}$ and I_{max} (figure 9).

The axis of the cylinder is oriented according to $\vec{u}_z$ and so $\vec{J} = J.\vec{u}_z$.

2.3.4.1 First rise of the current from 0 to I_{max}.

When the current rises, it cannot appear first in the center because there would be a magnetic electric field and a current in the entire superconductor which is impossible. Firstly, the current exists in the external part of the cylinder. Then, it progresses toward the inside until a radius r. This radius varies between r_S and R as indicated in figure 10.

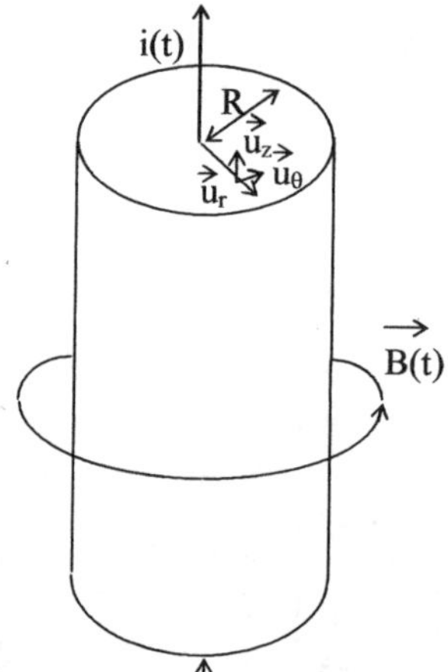

Figure 9: Superconducting cylinder

The presence of the current i(t) leads the existence of a magnetic induction B(r,t) according θ:

$$\vec{B}(r,t) = B(r,t).\vec{u}_\theta \qquad (\text{II}.39)$$

The time variation of the induction creates an electric field E(r,t) according Oz and so create some losses. According to the Bean's model, the density of current, J is equal to J_C.

The critical current according to the complete penetration of the current in the material is worth:

$$I_C = \pi.R^2.J_C \qquad (\text{II}.40)$$

If the maximum current I_{max} is lower than the critical current, we obtain:

$$I_{max} = \pi.(R^2 - r_s^2)J_C \qquad (\text{II}.41)$$

So, we deduce an expression of r_s:

$$r_S = R\sqrt{1 - \frac{I_{max}}{I_C}} \qquad (II.42)$$

We calculate the magnetic field with the help of ampere's law and the theorem of Stockes:

$$\oint \vec{B}.d\vec{l} = \mu_0 . \iint_{\text{à l'intérieur}} \vec{J}.d\vec{s} \qquad (II.43)$$

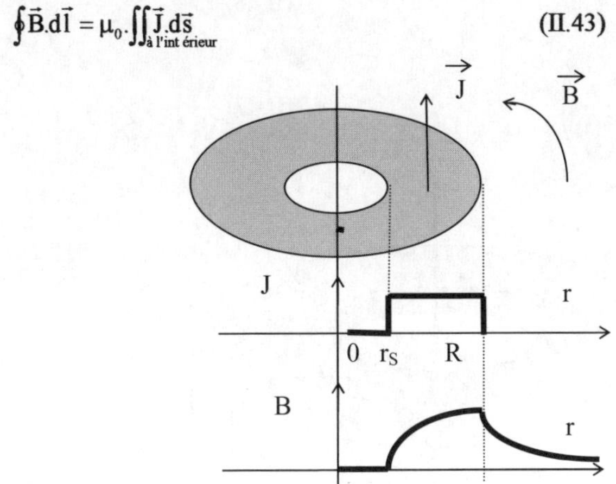

Figure 10: Distribution of J and B at the first rise of the current in a cylinder.

Therefore, we obtain the following expressions of B:

$$\begin{cases} \text{If } r < r_S & : & B = 0 \\[2mm] \text{If } r_S < r < R & : & B = \dfrac{\mu_0.J_C}{2}\left(r - \dfrac{r_S^2}{r}\right) \\[2mm] \text{If } r > R & : & B = \dfrac{\mu_0.J_C.(R^2 - r_S^2)}{2.r} \end{cases}$$

Variation of the current from I_{max} to $-I_{max}$

When the current decreases from its maximum value, it appears a negative current density in z-axis, in the external part as represented in figure 11:

For $c(t) < r < R$, the current density is worth: $\vec{J} = -J_C.\vec{u}_z$

For $r_S < r < c(t)$ the magnetic induction is trapped and don't vary. The current remains equal to J_C.

The radius $c(t)$ varies with the current $i(t)$ as follows:

$$c(t) = R\sqrt{1 - \frac{I_{max}}{2.I_C} + \frac{i(t)}{2.I_C}} \qquad (II.44)$$

For the calculation of losses, we consider the part where the induction varies, that is to say for $c(t) < r < R$. Using the Ampere's law we obtain:

$$B = \frac{\mu_0 . i(t)}{2.\pi.r} - \frac{\mu_0 J_C}{2}\left(r - \frac{R^2}{r}\right) \qquad (\text{II}.45)$$

We calculate the electric field using the Faraday's law:

$$\frac{\partial E}{\partial r} = \frac{\partial B}{\partial t} \qquad (\text{II}.46)$$

so

$$E(r,t) = \frac{\mu_0}{2.\pi} . \frac{\partial i}{\partial t} . \ln\left(\frac{r}{c(t)}\right) \qquad (\text{II}.47)$$

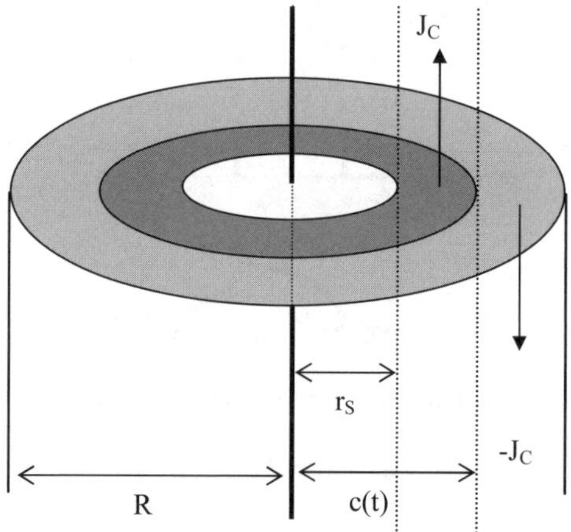

Figure 11: Distribution of J in a superconducting tube.

In figure 12, we have represented curves of distributions of J, B and E in the cylinder, versus the radius r, calculated using the previous equations.

Thanks to E(r,t), we calculate the instantaneous dissipated electromagnetic power per unit of volume:

$$p_{VC}(t) = \frac{2}{R^2} . \int_{c(t)}^{R} E.J_C.r.dr$$

$$= \frac{\mu_0 . J_C}{\pi.R^2} \frac{\partial i}{\partial t} \int_{c(t)}^{R} r.\ln\left(\frac{r}{c(t)}\right) dr$$

$$p_V(t) = \frac{\mu_0 . J_C}{4.\pi} \frac{\partial i}{\partial t}\left(\frac{i(t)}{2I_C} - \frac{I_{max}}{2I_C} - \ln\left(1 - \frac{I_{max}}{2I_C} + \frac{i(t)}{2I_C}\right)\right) \qquad (\text{II}.49)$$

Therefore, losses per unit of volume are equal to:

$$P_V = 2f . \int_{T/2} p_V(t).dt = \mu_0 . \frac{J_C}{2.\pi} . f . \int_{\frac{T}{2}}\left[\frac{1}{4I_C} . \frac{\partial i^2}{\partial t} - \frac{I_{max}}{2I_C} . \frac{\partial i}{\partial t} - 2.I_C . \frac{\partial f_i(t)}{\partial t}\right].dt$$

with $\quad f_i(t) = \left(1 - \dfrac{I_{max}}{2.I_C} + \dfrac{i(t)}{2.I_C}\right).\ln\left(1 - \dfrac{I_{max}}{2.I_C} + \dfrac{i(t)}{2.I_C}\right) - \left(1 - \dfrac{I_{max}}{2.I_C} + \dfrac{i(t)}{2.I_C}\right)$

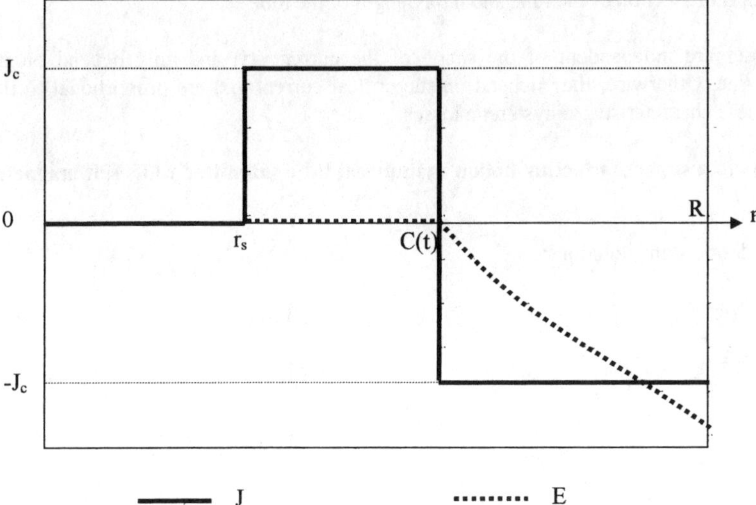

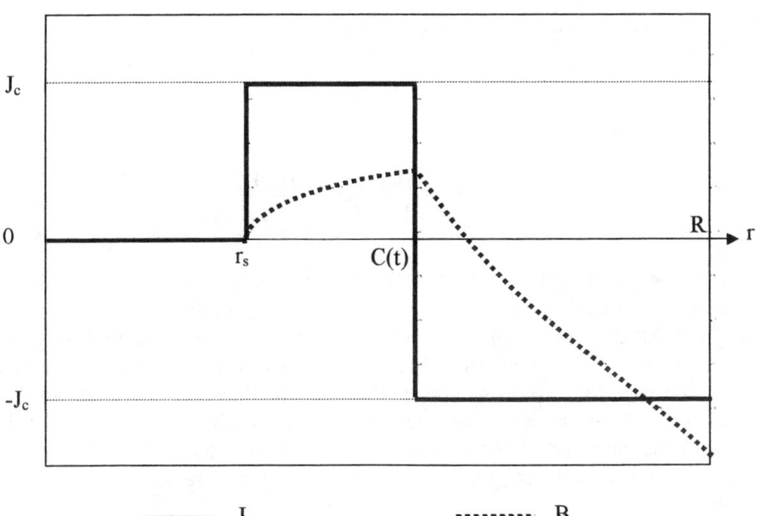

Figure 12: Distribution of J, E and B calculated versus radius r

We finally obtain losses P_{BC}:

$$P_{BC} = S.h.\mu_0.J_C^2.R^2.f.\left[\frac{I_{max}}{I_C} - \frac{I_{max}^2}{2.I_C^2} + \left(1 - \frac{I_{max}}{I_C}\right)\ln\left(1 - \frac{I_{max}}{I_C}\right)\right] \qquad (II.50)$$

With S the section of the tube and h the length of the tube.

Losses are independent of the shape of the current i(t) and only depend on its maximal value. Otherwise, they depend on the critical current and are proportional to the frequency. It is characteristic of hysteretic losses.

2.3.5 Losses in a superconducting hollow cylindrical tube submitted to its self magnetic field

We define some notations:

$$I_C = \pi R_e^2 \alpha_R J_c, \qquad (II.51)$$

$$r_S = R_e\sqrt{1 - \alpha_R \frac{I_{max}}{I_C}}, \qquad (II.52)$$

$$c(t) = R_e\sqrt{1 - \alpha_R \frac{I_{max}}{2I_C} + \alpha_R \frac{i(t)}{2I_C}} \qquad (II.53)$$

with $\alpha_R = 1 - \dfrac{R_i^2}{R_e^2}$

In the same way that the one adopted classically for the full cylinder [1], we get the instantaneous power in the superconductor and losses P_{BT}:

$$P_{BT} = S h \mu_0 J_C^2 \frac{R_e^2}{\alpha_R} f \, g\left(\frac{\alpha_R I_{max}}{I_C}\right) \qquad (II.54)$$

with

$$S = \pi(R_e^2 - R_i^2) \text{ And}$$

$$g\left(\frac{\alpha_R I_{max}}{I_C}\right) = \left(\frac{\alpha_R I_{max}}{I_C} - \frac{\alpha_R^2 I_{max}^2}{2.I_C^2} + \left(1 - \frac{\alpha_R I_{max}}{I_C}\right)\ln\left(1 - \frac{\alpha_R.I_{max}}{I_C}\right)\right)$$

It is interesting to draw curves of P_{BT} versus I_{max}, for various values of the factor of shape α_R, for a constant critical current I_C and for a constant section S (figure 13).

For a given critical current and, a given section of tube, one notices that losses decrease when the ratio between inner radius and outer radius of the tube decreases. In fact, when this ratio decreases, for a same section of the tube, the thickness decreases and the outer radius increases. Therefore, to decrease losses we have to reduce the thickness of the tube and to enlarge the outer radius of the tube. Thereafter it will be interesting to have an approached formula of P_{BT} and so of $g(\alpha_R I_{max}/I_c)$. We can write the function g under a more compact expression:

$$g(x) = x - x^2/2 + (1-x)*\ln(1-x) \qquad \text{with} \qquad x = \alpha_R \frac{I_{max}}{I_c}$$

Supposing x sufficiently small, we develop the function g in a series expansion, we obtain $g(x) \approx x^3/6$. Therefore, we will take the following approximation:

$$g\left(\frac{\alpha_R I_{max}}{I_C}\right) \approx \frac{1}{6}\left(\frac{\alpha_R I_{max}}{I_C}\right)^3 .$$

Finally, we obtain an approached formula giving losses in self-field P_{BTa}:

$$P_{BTa} = h \frac{\alpha_R}{6\pi} \frac{\mu_0 . f . I_{max}^3}{I_C} \qquad \text{for} \quad \alpha_R \frac{I_{max}}{I_c} \ll 1 \qquad (\text{II}.55)$$

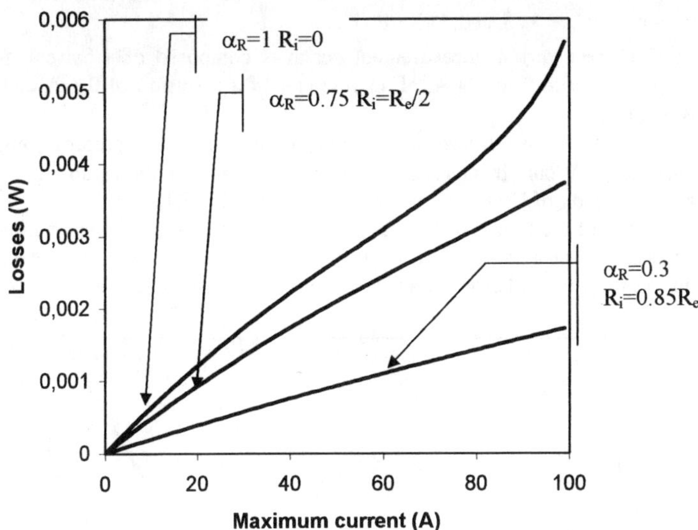

Figure 13: Losses in self-field of a tube versus I_{max}, for three values of the factor of shape and Ic = 100A.

3 STUDIED SAMPLES

The studied samples are a high critical temperature superconducting current lead constituted of a compacted composite of Bi-Pb-Sr-Ca-CuO 2223. Its critical temperature is 108K [5].

These superconducting current leads are tubes. Their geometric characteristics are the following ones:
- length h 80mm,
- section S 22mm² roughly (±10%),
- outer radius R_e de 4,8mm,
- inner radius R_i de 4mm.

At each extremity, there is a ribbon made of silver of 10mm large for electrical connexion. Therefore, the distance between potential pick off, l_{pp}, is worth 60 mm.

The minimal critical current (at 77K with the criterion 1µV/cm) given by the manufacturer is 100A, according to a minimal current density of 4,6A/mm².

To use the analytic formulas of losses and to permit the simulation of distributions of electrical field E and current density J in the material, it is necessary to know the characteristic of the electric field versus the density of current.

When the sample is supplied by direct current, one considers that the current is distributed almost uniformly in all the material. On the one hand, we have some simple relations, between the current I and the current density J, and on the other hand between the electric field E and voltage U:

$$I = J.S \qquad \text{and} \qquad U = E.l_{pp}$$
$$\text{So} \qquad : I_c = J_c.S \text{ and } U_c = E_c.l_{pp}$$

The direct current measurement bench is composed of a current generator whose value can vary from 0 to 200A, of an ammeter of a precision of 0,1 A and voltmeter of a precision of 1µV.

We chose to use in accordance with the manufacturer's prescriptions, the criteria of E_c equal to 1µV/cm. It gives us a critical voltage, which is about 6 µV. From the experimental study of U(I), we deduce relations E(I) and E(J).

Curves of the figure 14 have been obtained with three different samples. We observe that for samples of a same manufacturer, that have the same shape and the same critical current, the electric field characteristics versus current are relatively different.

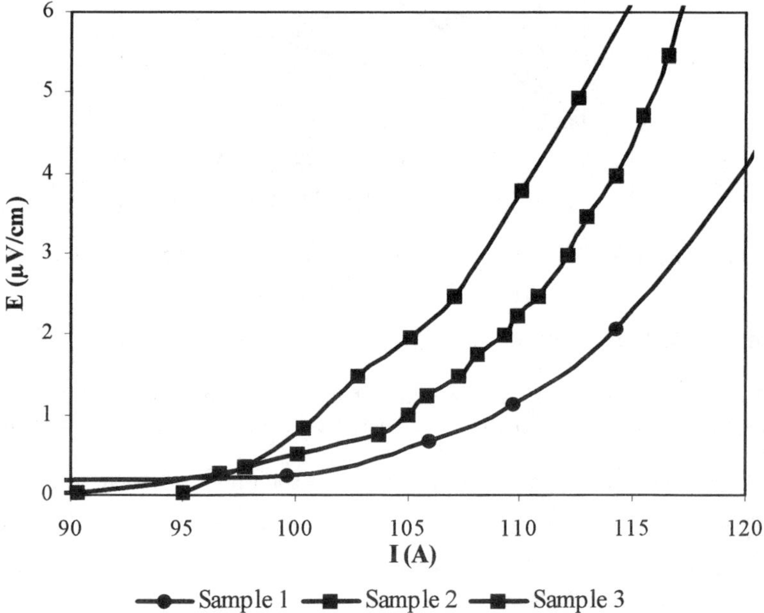

Figure 14: Electrical characteristics the different samples.

For the three samples, curves $E(I^n)$ can be fitted with the following parametres: critical current I_C and n:

The variations of values of critical current and of " n " according to the samples are about 6%. Even if this scattering is weak, we estimate that it exists different values of I_C and n for the different samples. Nevertheless, measured critical current density and the manufacturer values (100A) are close to 10%.

E_C=1μV/cm	Sample n°1	Sample n°2	Sample n°3
I_C	108	102	105
n	15	13	15

Table 1: Comparison between the different samples.

4 CALCULATION OF LOSSES IN INCOMPLETE PENETRATION

4.1 Problematic

This part is devoted to losses in incomplete penetration of the current in a high critical temperature current lead.

In a first place, we will present the numeric method used to calculate the distribution of the current density and the electric field in the material without analytic solution. Then, we are going to use the numerical results and a dimensional analysis to construct an analytic relation that gives losses.

To calculate losses in self-field inside a superconductor, it is necessary to know the distribution of the current density and electric field inside. The evolution versus time and space of this distribution also permits to analyse phenomenon accurately when a superconductor sample is supplied by sinusoidal alternating current [6-7].

The current density and the electrical field only depend on r and t, and are noted $J(r, t)$ and $E(r, t)$, $r \in [R_i, R_e]$. In cylindrical co-ordinates, we must solve the following problem:

$$\frac{1}{r}\cdot\frac{\partial E}{\partial r}+\frac{\partial^2 E}{\partial r^2}=\mu_0.k.\frac{\partial E^{\frac{1}{n}}}{\partial t} \tag{IV.1}$$

$$\text{Boundary conditions: } \frac{\partial E}{\partial r}(r = R_e, t) = \frac{\mu_0.\dfrac{\partial i(t)}{\partial t}}{2.\pi.R_e} \tag{IV.2}$$

$$\frac{\partial E}{\partial r}(r = R_i, t) = 0 \tag{IV.3}$$

$$\text{Initial condition} \quad E(r, t=0) = 0. \tag{IV.4}$$

To solve this problem we developed a numerical calculation program using a finite

452

difference method computed with Mathematica. The instantaneous power and losses, when the current is alternating, are got by numerical integration.

4.2 Method of resolution

The numerical resolution of the distribution of E(r,t) is only possible by doing a change of variable:

$$G(r,t) = E^{\frac{1}{n}}(r,t) \tag{IV.5}$$

Therefore, we must solve the following problem:

$$\frac{1}{r}\cdot\frac{\partial G^n}{\partial r} + \frac{\partial^2 G^n}{\partial^2 r} = \mu_0.k.\frac{\partial G}{\partial t} \tag{IV.6}$$

$$\text{Boundary conditions: } \frac{\partial G^n(r=Re,t)}{\partial r} = \frac{\mu_0\dfrac{\partial i(t)}{\partial t}}{2.\pi.R_e} = f(t) \tag{IV.7}$$

$$\frac{\partial G^n(r=Ri,t)}{\partial r} = 0 \tag{IV.8}$$

$$\text{Initial condition} \qquad E(r=R_{in}, t) = 0 \text{ and } J(r=R_{in}, t) = 0 \tag{IV.9}$$

4.2.1 Discretisation

Thereafter we will use the following conventions:
The space step is noted Δ_r and the time step Δ_t.
Space index is i and time index is t
i varies from 1 to N_r
t varies from 0 to N_t
N_r is the total number of points in the space
N_t is the total number of instants of calculation

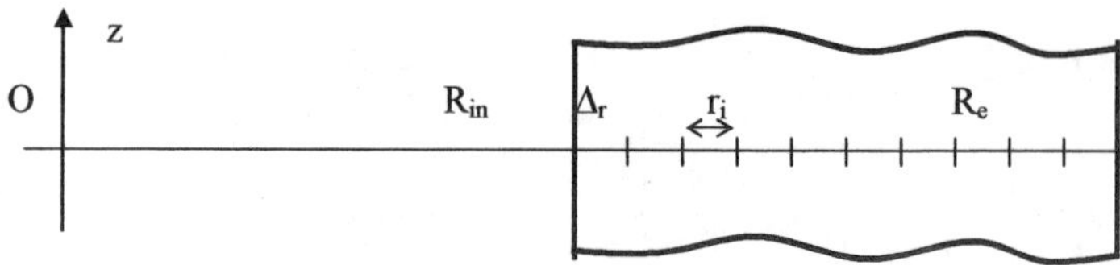

Figure 15: Discretisation in space and on boundary

We calculate the value of the function G at a radius r_i.
$$r_i = R_{in} + (i-1).\Delta_r$$

The partial derivatives equation is replaced by the equations:

$$\frac{\partial G^n}{\partial r} = \frac{G^n_{i+1,t+1} - G^n_{i-1,t+1}}{2.\Delta_r}$$

$$\frac{\partial^2 G^n}{\partial r^2} = \frac{G^n_{i+1,t+1} - 2.G^n_{i,t+1} + G^n_{i-1,t+1}}{\Delta_r^2}$$

$$\frac{\partial G}{\partial t} = \frac{G_{i,t+1} - G_{i,t}}{\Delta_t}$$

Equation (IV.6) allows us to write a set of numeric equations of $G_{i,t}$:

$$G^n_{i+1,t+1}\left(\frac{1}{\Delta_r^2} + \frac{1}{2.r_i.\Delta_r}\right) + G^n_{i,t+1}\left(-\frac{2}{\Delta_r^2}\right) + G^n_{i-1,t+1}\left(\frac{1}{\Delta_r^2} - \frac{1}{2.r_i.\Delta_r}\right) = \mu_0.k\left(\frac{G_{i,t+1}}{\Delta_t} - \frac{G_{i,t}}{\Delta_t}\right)$$

4.2.2 Boundaries conditions

The condition on the external boundary (IV.7) give, for $i = N_r$:

$$f(t) = \left(\frac{\partial G^n}{\partial r}\right)_{r=R_e} = \frac{G^n_{N_r+1,t}}{2.\Delta_r} - \frac{G^n_{N_r-1,t}}{2.\Delta_r}$$

so,

$$G^n_{N_r,t+1}\left(-\frac{2}{\Delta_r^2}\right) + G^n_{N_r-1,t+1}\left(\frac{2}{\Delta_r^2}\right) + \left(\frac{1}{\Delta_r^2} + \frac{1}{2.r_{N_r}.\Delta_r}\right).2.\Delta_r.f(t) = \mu_0.k\left(\frac{G_{N_r,t+1}}{\Delta_t} - \frac{G_{N_r,t}}{\Delta_t}\right)$$

Boundary (IV.8) for the internal boundary give for $i=1$:

$$G^n_{2,t+1}\left(\frac{2}{\Delta_r^2}\right) + G^n_{1,t+1}\left(-\frac{2}{\Delta_r^2}\right) = \mu_0.k\left(\frac{G_{1,t+1}}{\Delta_t} - \frac{G_{1,t}}{\Delta_t}\right)$$

4.2.3 Numerical resolution

We obtain N_r equations for each instant:

$$\begin{bmatrix} F_1\left[G_{i-1,t+1}, G_{i,t+1}, G_{i+1,t+1}, G_{i,t}\right] = 0 \\ \vdots \\ F_k\left[G_{i-1,t+1}, G_{i,t+1}, G_{i+1,t+1}, G_{i,t}\right] = 0 \\ \vdots \\ F_{N_r}\left[G_{i-1,t+1}, G_{i,t+1}, G_{i+1,t+1}, G_{i,t}\right] = 0 \end{bmatrix}$$

It is equivalent to solve:

$$(F_1^2 + .. + F_k^2 + .. + F_{N_r}^2) = 0$$

In fact, we have to numerically minimise the sum of F_1^2. This minimisation is directly done by Mathematica. Using values of G(r,t), we calculate E(r, t) and J(r, t).

4.2.4 Choice of the discretisation and time of calculation.

The discretisation concerns spatial and temporal dimension. They cannot be chosen in an independent way. Indeed, so that the resolution has a sense, it is necessary that the front of propagation doesn't move more than one spatial step in one time step.

Therefore, we obtain a minimal ratio between discretisation steps defined from the speed of propagation:

$$V_p \le \frac{\Delta_r}{\Delta_t}$$

4.2.5 Calculus of losses

The instantaneous power and losses in with periodic current are obtained by numerical integration

$$p(t_k) = 2.\pi.h.\Delta_r.\sum_{i=1}^{N_r} E_i.J_i.r_i$$

$$P = \frac{\sum_{k=1}^{N_t} p(t_k)}{N_t}$$

4.3 Distribution of the current density and of the electric field

In this part, we present some results coming from our calculation. The current supply is defined by $i(t) = I_{max}.\sin(\omega t)$.

We are going to studies:
- n constant, for different instant,
- at a given instant, for different n,
- a study of losses versus n and critical current

4.3.1 Temporal study

It exists a transient state of a quarter of period corresponding to the first penetration of the current. We didn't represent it because its contribution to losses is negligible.

We study a superconducting tube having the following characteristics: a critical current of 100A and a value of n equals 13. These values represent the average of tested samples. Calculations are done with a supplied current of the shape $i(t)=I_{max}\sin(\omega t)$, with a maximum current of 100A and a frequency of 50 Hz.

In figures 16 and 17, we have drawn the distributions of current and of the electric field inside the conductor. On these curves, the time t varies between 20ms and 26ms. During this interval of time the current is positive. We can observe the whole phenomenon of distribution of the current, for more than a quarter of period. The current is equal to 0 at t=20ms, increasing from 20ms to 25ms and decreasing from 25ms to 26ms.

The general behaviour of the current distribution is very close to the one that we would have found thanks to the Bean's model.

It exists a small difference after the maximum's current passage, where we have distributions of current represented in figure 18. The distribution of current using the model J^n is not constant and equal $\pm J_C$ but takes some variable values according to the radius. Besides when i(t) decreases from I_{max} this distribution of current remains positive for some values of time ($t=t_2$) (figure 18b).

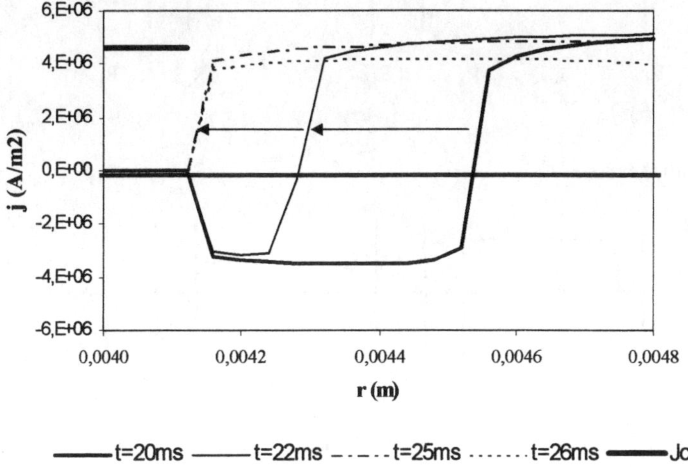

Figure 16: Distribution of the current density in the thickness of the superconductor at different instants, for n =13.

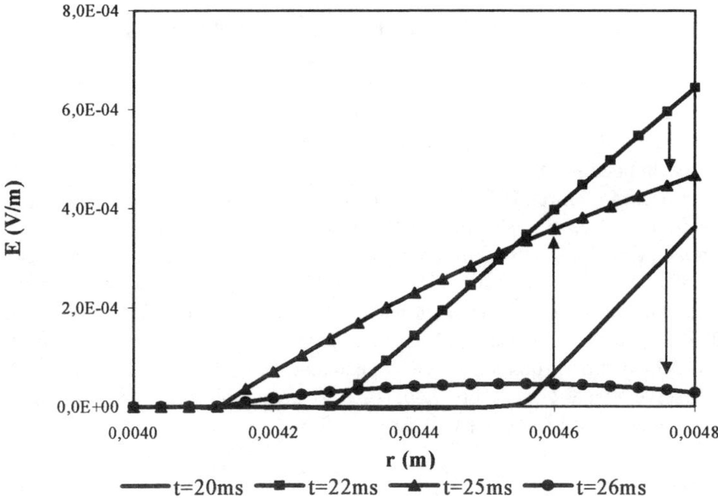

Figure 17: Distribution of the electric field in the thickness of the superconductor at different instants, for n=13.

with the Bean's model

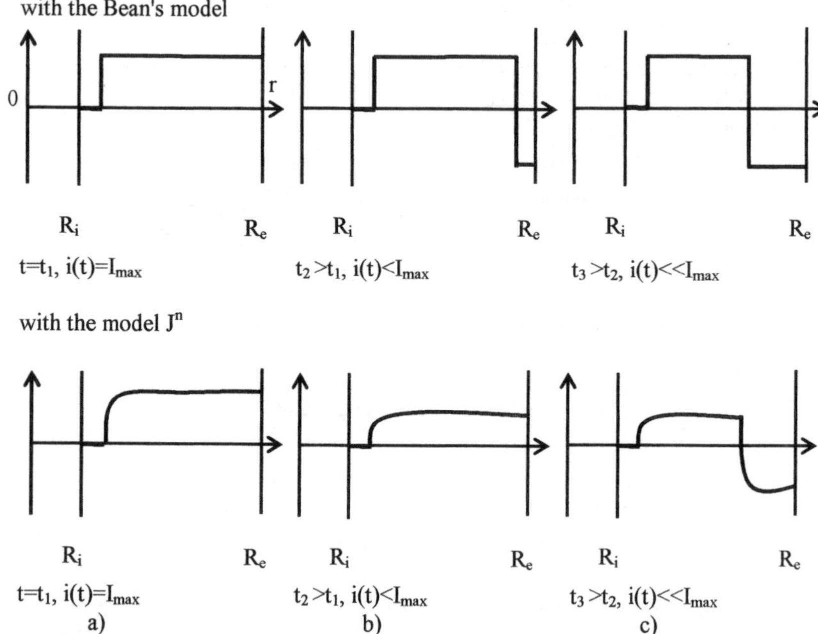

Figure 18: Distributions of the current obtain with the Bean's model and with the model J^n,

4.3.2 Influence of the factor n

It is interesting to compare distributions, of electric field and current density, given by the Bean's model with our model for different values of n.

In this studied case of we chose to simulate a sample of critical current Ic equal 100 A. Calculations are done with a supplied current of the shape $i(t)=I_{max}\sin(\omega t)$, with a maximum current of 90A and a frequency of 50 Hz. At a given instant, 24 ms, $i(t)$ is worth 80A, it is increasing and positive.

We represent in figures 19 and 20 distributions of E and J simulated according to the radius, for different values of n and with the Bean's model.

We notice that the penetration of the electric field is as much more important as n is important. It affects slightly the value of the current for which there is complete penetration of the material. However for variable n from 7 to 30, the variation of this value is relatively weak, about 2%. Therefore, we can say, that there is complete penetration of the field in the material for a critical current practically wich is equal to the current given by the constructor.

We observe that the more the value of n is great, the more the maximal current density is lower than the critical current density.

4.3.3 Parametric study

We studied the influence of characteristics of the material as the value of n and the critical current I_c on losses.

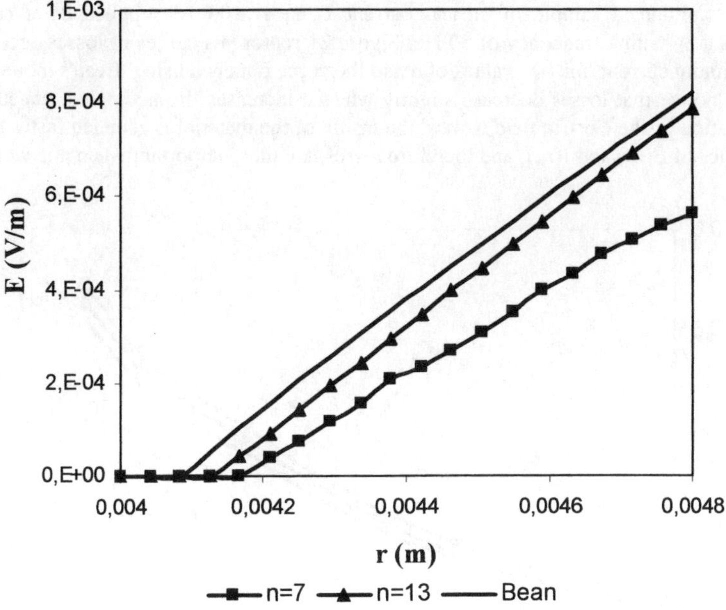

Figure 19: Electric field versus radius at t = 24ms.

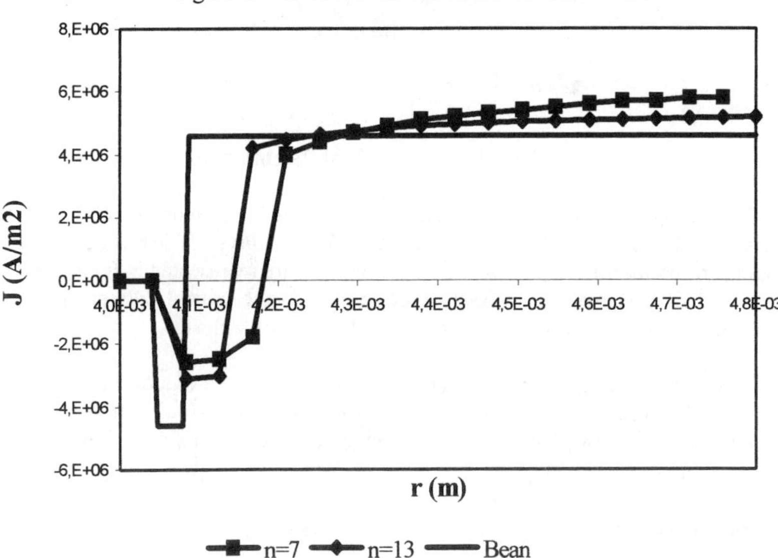

Figure 20: Current density versus radius to t = 24ms.

We simulate a sample of critical current I_c equal 100 A, supplied by a current $i(t)=I_{max}\sin(\omega t)$, with a frequency of 50 Hz. Figure 21 represents curves of losses according to the maximum current, for two values of n and the curve obtained using Bean's model.

We notice that losses decrease slightly when n increases. In incomplete penetration the propagation of the electric field toward the inside of the material is as much faster than n is big. Values of $E(r,t)$ and $J(r,t)$, and therefore losses, are more important when n is weak.

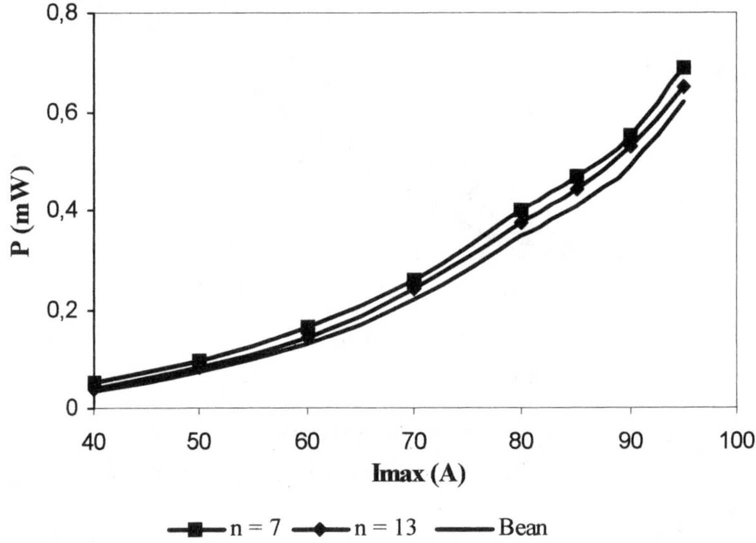

Figure 21: Calculated losses for different values of n.

Concerning the study of the variation of the critical current, we choose a sample possessing a value n equal 13 and a frequency of 50Hz. Curves of losses according to the maximum current, for different values of critical current are represented in figure 22.

In this case losses are inversely proportional to the critical current. These losses decrease when the critical current increases but values are very close.

As a conclusion, comparing results obtain by numeric resolution to those based on the Bean's model, we have some visible differences in instantaneous or local behaviour term. With regard to simulations, losses become identical as n tend to infinity. It proves that our method is valid.

4.4 Dimensional analysis

L. Dresner proposed a relation [8-10], constructed from a dimensional analysis, giving losses. This work has been done with a ribbon of thickness a and of section S, supplied by a sinusoidal current of frequency f and of amplitude I_{max}, submitted to its self-field.

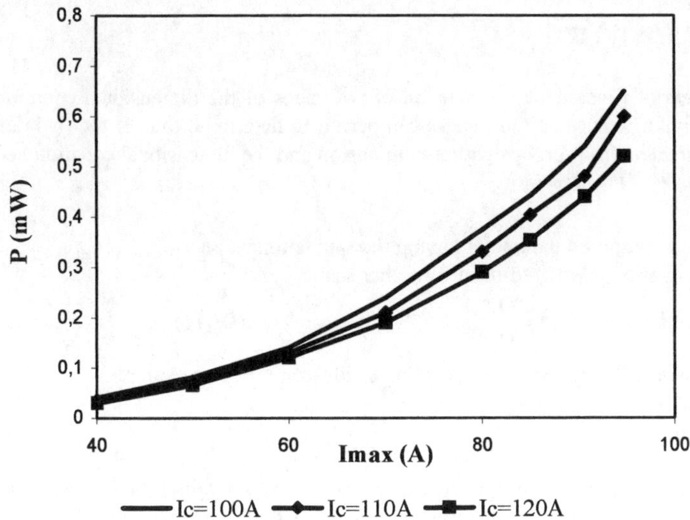

Figure 22: Calculated losses for different critical currents and n = 13.

In the same way, using the dimensional analysis [11] and previous conclusions of the previous paragraph, we searched a new analytic relation giving losses according to features of the material.

4.4.1 Principle

The dimensional analysis is based on the fact that a product of independent fundamental dimension can express the dimension of all physical quantity. These are dimensions of mass noted [M], of length [L], of time [T] and of current [I].

For example if a quantity has a dimension $[Q_i]$, this one is defined according to fundamental dimension, each one raised to a determined power.

It gives the general dimensional equation:

$$[Q_i] = [M^x, L^y, T^z, I^w]$$

If a quantity Q_1 is function of parameters:

$$Q_1 = f(Q_2, Q_3, ..., Q_n) \qquad (IV.10)$$

It exists a dimensional equation wich is the following:

$$[Q_1] = [f(Q_2, Q_3, ..., Q_n)]$$

Any continuous function can be expressed as a product, which each term is a parameter raised to a certain power. The previous equation becomes:

$$[Q_1] = [Q_2^{a_2} Q_3^{a_3} ... Q_n^{a_n}]$$

And now, if one replaces each parameter by its fundamental dimension one gets for the parameter Q_i:

$$\left[M^{a_i x}, L^{a_i y}, T^{a_i z}, I^{a_i w} \right]$$

Powers of fundamental dimension of two sides of the dimensional equation can be equalised, giving four equations. These can permit to determine four of the (n-1) of powers a_i. So we can make (n-5) groups without dimension and the dimensional equation becomes:

$$[Q_1] = \left[Q_1' \Pi_2^{a_2} ... \Pi_{n-4}^{a_{n-4}} \right]$$

Q_1' is a composed parameter having the same dimension as Q_1.
The equation (IV.10) turns into another shape:

$$\Pi_1 = g\left(\Pi_2, ..., \Pi_{n-4}\right) = \frac{Q_1}{Q_1'} \qquad\qquad (IV.11)$$

To understand our way, we are going to illustrate it by an example.

4.4.2 Example

We take, as example, a resistance of resistivity ρ, of length l_g, and of constant section S, which is fed by a direct current I. We want to know the dissipated power by the resistance according to these different parameters. The dimensional equation is the following:

$$[P] = \left[\rho^a, l_g^b, S^c, I^d \right]$$

We can establish a table in which are listed powers of the fundamental dimensions of each parameters and of the power.

Parameters	Power	Fundamental dimensions of power			
		M	L	T	I
ρ	a	1	3	-3	-2
l_g	b	0	1	0	0
S	c	0	2	0	0
I	d	0	0	0	1
Power P		1	2	-3	0

Table 2: Dimensional analysis of the electric power

It gives us the following system of four equations:

$$\begin{cases} a = 1 \\ 3a + b + 2c = 2 \\ -3a = 3 \\ -2a + d = 0 \end{cases}$$

After simplification, we get a system of three equations with four unknown. It is necessary to fix one of unknowns to calculate the other. In an arbitrary manner, we decided to fix c:

$$\begin{cases} a = 1 \\ d = 2 \\ b = -1 - 2c \end{cases}$$

So:

$$[P] = \left[\rho, l_g^{-1-2c}, S^c, I^2 \right] = \left[\rho, l_g^{-1}, \left(\frac{S}{l_g^2} \right)^c, I^2 \right]$$

and

$$P = \rho . \frac{I^2}{l_g} . F_{onct} \left(\frac{S}{l_g^2} \right)$$

where

$$\frac{P}{\rho \dfrac{I^2}{l_g}} = F_{onct} \left(\frac{S}{l_g^2} \right)$$

In this last equation the two members of the equality have no dimension. So, it sufficient to use values of simulation or measures of the power according to the different parameters to find the final equation. Hence, we find:

$$F_{onct} \left(\frac{S}{l_g^2} \right) = \left(\frac{S}{l_g^2} \right)^{-1}$$

and so:

$$P = \rho . \frac{l_g}{S} . I^2$$

4.4.3 Application to the calculation of losses

As in any case, losses are proportional to the height of the tube, we are going to study losses per unit of length P_h.

The important physical parameters in hysteretic losses are:
- permeability μ_0,
- frequency f,
- maximal current I_{max},
- critical current I_C.

Let us define k_i by: $E = k_i . I^n$ with $k_i = \dfrac{E_C}{I_C^n}$

We take k_i as a characteristic parameter of the material instead of the critical current. Therefore, the power of losses per unit of length will be a function of μ_0, f, k_i and I_{max}.

It gives us the following dimensional equation:

$$[P_h] = \left[\mu_0^a . f^b . k_i^c . I_{max}^d \right]$$

We summarize in table 3 all the useful physical quantities, we need to calculate losses.

Parameters	M	L	T	I	
μ_0	1	1	-2	-2	a
f	0	0	-1	0	b
k_i	1	1	-3	-1-n	c
I_{max}	0	0	0	1	d
Power by unit of length P_{li}	1	1	-3	0	

Table 3: Dimensional analysis of losses

It gives the following system of equations:

$$\begin{cases} a + c = 1 \\ -2a - b - 3c = -3 \\ -2a - (n+1)c + d = 0 \end{cases}$$

We choose to fix the parameter "a". In this case the previous equation resolution gives us the following system.

$$\begin{cases} b = a \\ c = 1 - a \\ d = a(1-n) + (n+1) \end{cases}$$

Therefore, the dimensional equation is the following:

$$[P_h] = \left[\mu_0^a, f^a, k_i^{1-a}, I_{max}^{a(1-n)+(n+1)} \right] \tag{IV.12}$$

where

$$[P_h] = \left[\left(\frac{\mu_0 f}{k_i} \right)^a, I_{max}^{(n+1)+a(1-n)}, k_i \right] \tag{IV.13}$$

We choose arbitrarily the shape of P_{li}:

$$P_h = \alpha_{li}(n) \cdot \left(\frac{\mu_0 f}{k_i} \right)^a \cdot I_{max}^{a(1-n)+(n+1)} \cdot k_i \tag{IV.14}$$

$\alpha_{li}(n)$ is a coefficient depending on "n" and that we shall calculate afterwards.

We need to know the power a. For that, we use the approached formula deduced from the Bean's model P_{BTa}. If n tends to infinity, P_h tends to P_{Bta}/h:

$$P_h \xrightarrow[n \to \infty]{} \frac{P_{BTa}}{h} = \frac{\alpha_R}{6\pi} \cdot \frac{\mu_0 \cdot f \cdot I_{max}^3}{I_c} \tag{IV.15}$$

By identification of power of I_{max}, we deduce that $a(1-n)+(n+1)$ tends to 3 as n tends to infinity. It exists for "a" an infinity possible values permitting to satisfy the criterion that has just been defined. We chose an analogous solution to the one proposed by L. Dresner [8-9]:

$$a(1-n)+(n+1)=\frac{3n+1}{n+1}, \text{ what gives: } a=\frac{n}{n+1}$$

We put, from (III.14) and (III.15), $\alpha_{li}(n)$ under the shape: $\alpha_{li}(n)=F(n).\alpha_R$

with $F(n)$ a function of "n" with no dimensions, $\alpha_R=1-\dfrac{R_i^2}{R_e^2}$ independent of "n" and

with $F(\infty)=\dfrac{1}{6\pi}$.

So, we have a new relation for P_h:

$$P_h=F(n).\alpha_R.\left(\frac{\mu_0 f}{k_i}\right)^{\frac{n}{n+1}} I_{max}^{\frac{3n+1}{n+1}}.k_i \qquad (IV.16)$$

There are values of $F(n)$ left to determinate. To do so, we use results of simulation of losses presented in the paragraph 2.3.

Figure 23 represents losses calculated for a sample whose critical current is worth 100A, and supplied by the following current: $i(t)=I_{max}\sin(\omega t)$, with I_{max} equal to 80A and a frequency equal to 50Hz.

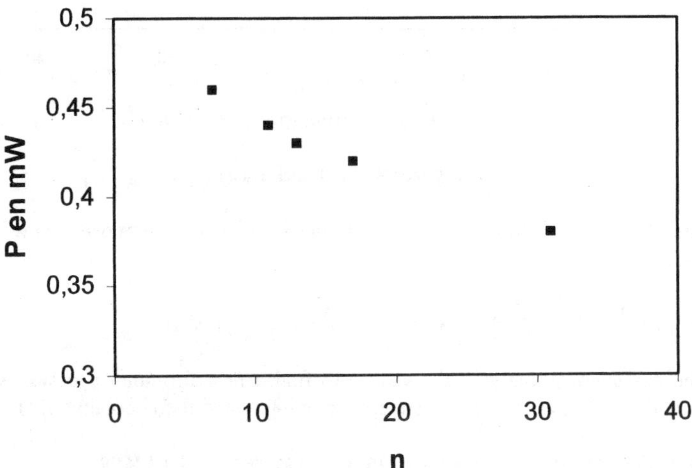

Figure 23: Losses versus n.

We have the expression of $F(n)$ from the equation (IV.16):

$$F(n)=\frac{P_{simul}}{\alpha_R.\left(\dfrac{\mu_0 f}{k_i}\right)^{\frac{n}{n+1}} I_{max}^{\frac{3n+1}{n+1}}.k_i h} \qquad (IV.18)$$

From the asymptote $F(\infty)$, from the results presented on the curve 24 and from the expression (IV.18) we found a simple formulation of the function F(n), noted Fa(n), fitting the points of calculations.

$$Fa(n) = \frac{1}{6\pi} + \frac{0,3}{n} \qquad (IV.19)$$

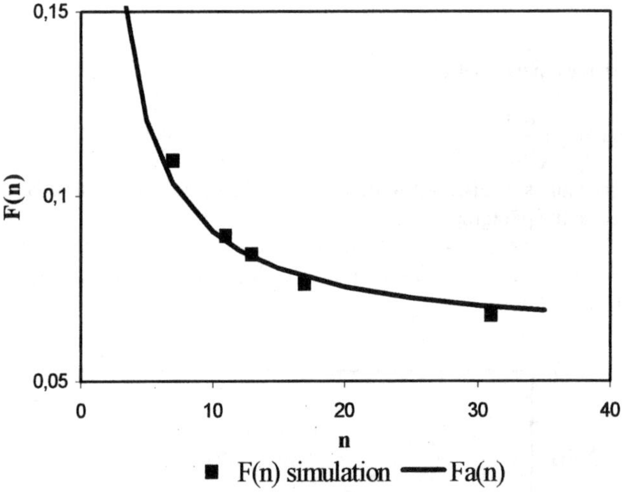

Figure 24: F(n) and $F_a(n)$.

The characteristic function of our sample is we use to propose a new formula of losses in self-field and in incomplete penetration:

$$P_I = h \left(\frac{\mu_0.f}{I_C} \right)^{\frac{n}{n+1}} . E_C^{\frac{1}{n+1}} . I_{max}^{\frac{3n+1}{n+1}} . \alpha_R \left(\frac{1}{6\pi} + \frac{0,3}{n} \right) \qquad (IV.20)$$

The dimensional analysis allowed us to find a new formula of losses, which takes into account the characteristics of the superconducting material and notably of "n".

5 CALCULATION OF LOSSES IN COMPLETE PENETRATION

In the previous part, we establish a relation giving losses in self-field in a superconducting tube in incomplete penetration of the current. Now, we shall calculate losses in the case of a complete penetration of the current.

5.1 Non linear diffusion of the current density

Because of the relation between E and J in a high critical temperature superconductor, the current density J(x,t) can take any value and not only its critical value J_C as in the Bean's model. This value varies according to time and the position in the material. I. Mayergoyz described the diffusion of J in a plate submitted to an external applied magnetic field $H_0(t)$ represented on the figure 25 [11]. We shall recall here the main results developed by this author. These results will be useful in the continuation of our study.

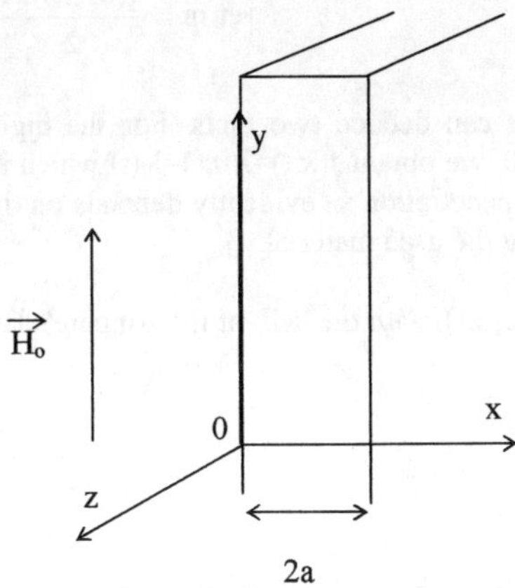

Figure 25: Superconducting plate submitted to a magnetic field.

The electric field E and the current density J are oriented according to z-axis and vary according to x.

We have to solve the same equation as previously,.

$$\Delta E = \mu_0 . k . \frac{\partial E^{\frac{1}{n}}}{\partial t} \qquad (V.1)$$

or

$$\Delta J^n = \mu_0 . k^n . \frac{\partial J}{\partial t} \qquad (V.2)$$

with

$$k = \frac{J_c}{E_c^{\frac{1}{n}}}, J = k.E^{\frac{1}{n}} \qquad (V.3)$$

In the case of a plate submitted to a field **B** I. Mayergoyz prescribe the following boundary and initial conditions:

$$J(0,t) = c.t^q \quad (t \geq 0, q > 0) \qquad (V.4)$$
$$J(x,0) = 0 \quad (x > 0) \qquad (V.5)$$

The boundaries conditions correspond to a large number of monotonous increasing functions.

Once the problem is solved, the solution is the following:

$$J(x,t) = \begin{cases} c.t^q \left(1 - \dfrac{x}{d.t^m}\right)^{\frac{1}{n-1}} & \text{si } x < x_0 \\[2ex] 0 \text{ si } x \geq x_0 \end{cases} \quad \text{with} \quad \begin{cases} x_0 = d.t^m \\[2ex] d = \sqrt{\dfrac{n.c^{n-1}}{\mu_0.k^n.m.(n-1)}} \\[2ex] \text{et } m = \dfrac{q(n-1)+1}{2} \end{cases} \qquad (V.6)$$

From these relations, we can deduce two facts. For the high critical temperature superconductor where $5 < n < 20$, we obtain $J(x,t) \approx J(0,t) = J_0(t)$ which is constant in space at one instant t. Then, the depth of penetration x_0 evidently depends on time, but also on n and on the coefficient k characterising the used material.

H(x,t) is calculated using J(x,t), with the help of the Ampere's law:

$$\frac{\partial H(x,t)}{\partial x} = J_0(t) \qquad (V.7)$$

So:

$$H(x_0,t) = 0$$
$$H(0,t) = H_0(t) = J_0(t).x_0 \qquad (V.8)$$

The author has also calculated the evolution of the position $x_0(t)$ where H is equal to zero:

$$x_0(t) = \frac{1}{H_0(t)} \cdot \left[\frac{2(n+1)}{\mu_0.k^n} \int_0^t H_0^{2n}(\tau).d\tau \right]^{\frac{1}{n+1}} \qquad (V.9)$$

In figure 26, we have represented the possible evolutions of $x_0(t)$, and $J(x,t)$ and of H(x,t).

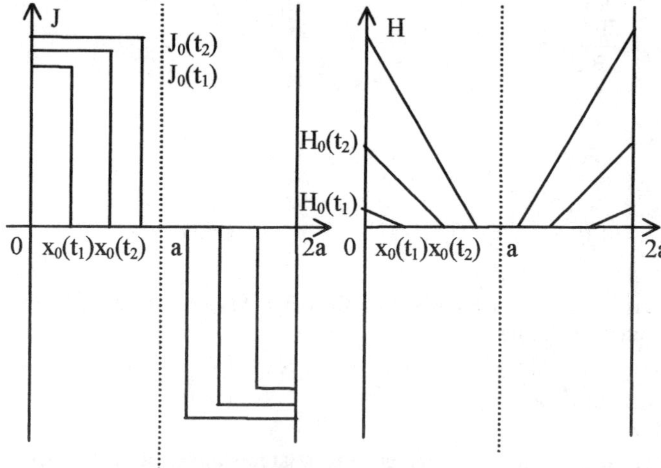

Figure 26: Diffusion of the current density in a to high critical temperature superconductor in incomplete penetration

In fact, J(x,t) quickly reaches a value close to J_C and remains there. We are close to what happened with the Bean's model. Once the current density has completly penetrated of, $J_0(t)$ continues to increase whereas x_0 remains equal to half thickness of "a".

Now, we are going to use these theoretical results to study the case of a superconducting tube.

5.2 Distribution of the current density and of the electric field

5.2.1 Principle of the modelling

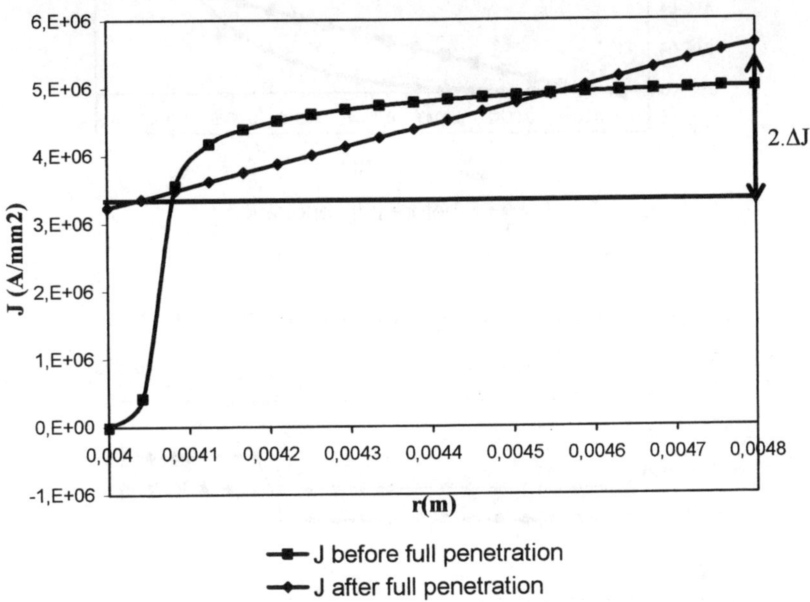

Figure 27: Current Density calculated before and after complete penetration.

Using the established results by I. Mayergoyz, we are looking for a new distribution of J(r,t) in complete penetration permitting to obtain a distribution of E(r,t) which is more closer to the one calculated just before the complete penetration. The simplest solution consists in taking a distribution of J(r,t) where this one increases linearly with the radius of tube. It gives a distribution E(r, t) represented in figure 28.

However, this distribution of current doesn't permit to calculate losses easily. Therefore, we have chosen a new model of distribution permitting a more comfortable calculation. To do so, we separated in two equal surfaces the section of the tube as we can see on the figure 29.

In the internal section we impose a current density: $J=(i(t)-\Delta I)/S$ and in the external section: $J=(i(t)+\Delta I)/S$.

The parameter ΔI is very important because it is the one that permits to connect the calculation in incomplete penetration and in complete penetration of the current.

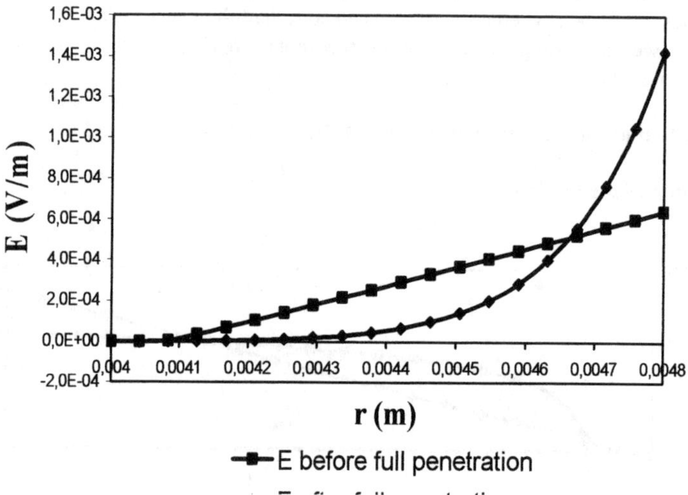

Figure 28: Electric field calculated before and after penetration

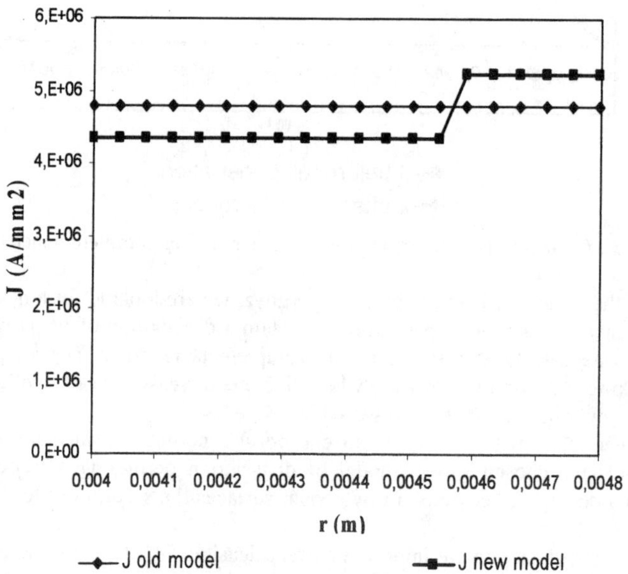

Figure 29: Current density of in complete penetration.

5.2.2 Calculation of losses

If the maximum of the current is sufficiently high, it exists two phenomenons separated in time. First, for the absolute values of i(t) relatively weak (lower than I_C), the penetration is incomplete, the density of current is only present in a part of the material. We call P_{vi} these losses. Then, when the absolute value of i(t) is high enough (higher than I_C) the penetration is complete, the current is present in all the section of the tube; we call P_{vc} these losses. The total losses are noted P_v and $P_v = P_{vi} + P_{vc}$.

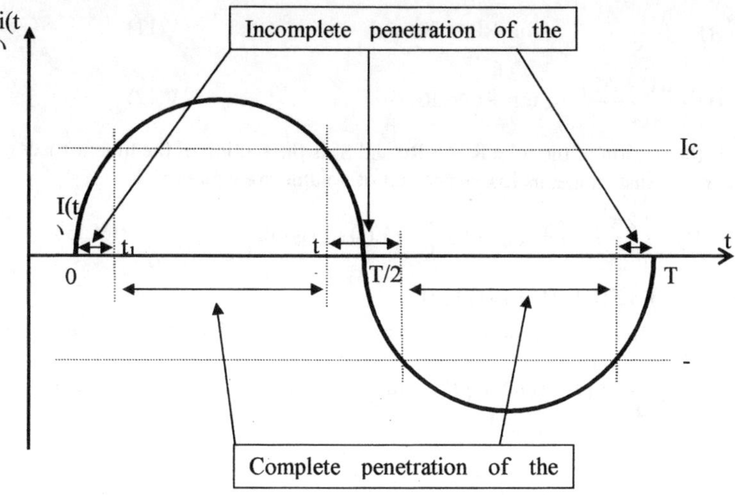

Figure 30: Representation the current.

5.2.2.1 Incomplete penetration

We shall consider that, while the current is increasing and decreasing, it is a monotonous function, as in the case of a sinusoidal signal.

To simplify, we consider that the penetration is complete when i(t) reaches and pass the critical current I_C given by the criteria $1\,\mu V/cm$. Therefore, there is incomplete penetration for:

$|i(t)| < I_C$ (figure 30).

As in the paragraph 5.1, J is according to z-axis and varies according to r. According to what has just been said, one considers that the current density $J(r,t) \approx J(R_e,t) = J_0(t)$ for $r_S < r < R$ and that it is equal to zero elsewhere.

We have already given the relation giving the losses in incomplete penetration.

$$P_{VI} = h.\left(\frac{\mu_0.f}{I_C}\right)^{\frac{n}{n+1}}.E_C^{\frac{1}{n+1}}.I_{max}^{\frac{3n+1}{n+1}}.\alpha_R.\left(\frac{1}{6\pi}+\frac{0,3}{n}\right) \qquad (V.10)$$

As we have seen previously, it is hysteretic losses that do not depend on the shape of the current i(t).

5.2.2.2 Complete penetration

There is complete penetration of the current in the material for $|i(t)| > I_C$, that is to say for $t_1 < t < t_2$ where t_1 and t_2 are instants where the current i(t) is equal to critical current.

In complete penetration we chose to take $J(r,t)=J_0(t)$ constant by part in the thickness of the tube as we presented it in the part 5.2.1. Therefore, for a current i(t) and a section of the tube S we obtain:

$$J_0(t) = \frac{i(t) - \Delta I}{S_i} \qquad \text{for } R_i < r < R_c \text{ and} \qquad (V.11)$$

$$J_0(t) = \frac{i(t) + \Delta I}{S_e} \qquad \text{for } R_c < r < R_e \qquad (V.12)$$

S_i is the section of the tube $R_i < r < R_c$ and S_e is the section of the tube for $R_c < r < R_e$. Then, the instantaneous losses per unit of volume are equal to:

$$p_{VCp}(t) = \left\{ \frac{2\pi}{S_i} \int_{R_i}^{R_c} E(t)J_{0i}(t).r.dr + \frac{2\pi}{S_e} \int_{R_c}^{R_e} E(t).J_{0e}(t).r.dr \right\} \qquad (V.13)$$

so $p_{VCp}(t) = E(t)J_{0i}(t) + E(t).J_{0e}(t)$ (V.14)

We get, with the expression (II.5), losses per unit of volume in complete penetration:

$$P_{VCp} = \frac{2}{T} . \frac{E_C}{J_C^n} . \left\{ \int_{t_1}^{t_2} J_{0i}^{n+1}(t).dt + \int_{t_1}^{t_2} J_{0e}^{n+1}(t).dt \right\} \qquad (V.15)$$

so

$$P_{VCp} = 2.f. \frac{E_C}{J_C^n} . \left\{ \int_{t_1}^{t_2} J_{0i}^{n+1}(t).dt + \int_{t_1}^{t_2} J_{0e}^{n+1}(t).dt \right\} \qquad (V.16)$$

Taking into account i(t), we obtain:

$$P_{VCp} = 2.f. \frac{E_C}{I_C^n} . \left\{ \int_{t_1}^{t_2} \frac{(i(t) - \Delta I)^{n+1}}{S_i} .dt + \int_{t_1}^{t_2} \frac{(i(t) + \Delta I)^{n+1}}{S_e} .dt \right\} \qquad (V.17)$$

Knowing that the internal volume is worth Si*h and the external volume Se*h, the total losses are worth:

$$P_{Cp} = 2.f.h. \frac{E_C}{I_C^n} . \left\{ \int_{t_1}^{t_2} (i(t) - \Delta I)^{n+1} .dt + \int_{t_1}^{t_2} (i(t) + \Delta I)^{n+1} .dt \right\} \qquad (V.18)$$

We observe that in complete penetration, losses depend on the shape of the current i(t), in opposition to the case of the incomplete penetration.

To have losses in complete penetration P_C, if $I_{max} > I_C$, it is necessary to add to P_{VCp}, losses in incomplete penetration P_{VI} for I_{max} equal to I_C:

$$P_{VI}(I_{max} = I_c) = h.(\mu_0.f)^{\frac{n}{n+1}} . E_C^{\frac{1}{n+1}} . I_c^{\frac{2n+1}{n+1}} .\alpha_R . \left(\frac{1}{6\pi} + \frac{0,3}{n} \right) \qquad (V.19)$$

$P_C = P_{VI}(I_{max} = I_C) + P_{VCp}$

To summarize losses are equal to:

$$P_I = h.\left(\frac{\mu_0.f}{I_C}\right)^{\frac{n}{n+1}}.E_C^{\frac{1}{n+1}}.I_{max}^{\frac{3n+1}{n+1}}.\alpha_R.\left(\frac{1}{6\pi}+\frac{0,3}{n}\right) \qquad \text{Si } I_{max}<I_C$$

$$P_C = h.(\mu_0.f)^{\frac{n}{n+1}}.E_C^{\frac{1}{n+1}}.I_{cx}^{\frac{2n+1}{n+1}}.\alpha_R.\left(\frac{1}{6\pi}+\frac{0,3}{n}\right)+$$

$$2.f.h.\frac{E_C}{I_C^n}.\left\{\int_{t_1}^{t_2}\left(i(t)-\Delta I\right)^{n+1}dt+\int_{t_1}^{t_2}\left(i(t)+\Delta I\right)^{n+1}dt\right\}dt \qquad \text{Si } I_{max}>I_C$$

5.2.3 Calculation of losses for a sinusoidal current

We shall calculate losses in the case of a sample supplied by a sinusoidal current $i(t) = I_{max}.\sin\omega t$.

5.2.3.1 Incomplete penetration

As we have just seen it, if $I_{max} < I_C$ we obtain the formula of losses per unit of volume (IV.20).

5.2.3.2 Complete penetration

If $I_{max} > I_C$ there is complete penetration, that corresponds to the first half period of the following interval:

$$\theta_1 < \omega.t < \pi-\theta_1$$

with

$$\theta_1 = \text{Arcsin}\frac{I_C}{I_{max}}$$

So, we obtain:

$$P_{VCp} = 2.f.\frac{E_C}{I_C^n}.\left\{\frac{(Imax-\Delta I)^{n+1}}{Si}\int_{\theta_1}^{\pi/2}\sin^{n+1}\theta.d\theta + \frac{(Imax+\Delta I)^{n+1}}{Se}\int_{\theta_1}^{\pi/2}\sin^{n+1}\theta.d\theta\right\} \quad (V.20)$$

Therefore

$$P_{VCp} = 2.f.\frac{E_C}{I_C^n}.\left\{\frac{(Imax-\Delta I)^{n+1}}{Si}+\frac{(Imax+\Delta I)^{n+1}}{Se}\right\}\int_{\theta_1}^{\pi/2}\sin^{n+1}\theta.d\theta \qquad (V.21)$$

Integral is equal to a function hypergeometric $F(a,b,c,z)$:

$$\int_{\theta_1}^{\frac{\pi}{2}}\sin^{n+1}\theta.d\theta = \cos\theta_1.F\left(\frac{1}{2},\frac{-n}{2},\frac{3}{2},\cos^2\theta_1\right) \qquad (V.22)$$

The result of the calculation of the integral depends on n, on θ_1 and on I_{max}/I_C. We introduce the function $F_1(n, I_{max}/I_C)$ as follows:

472

$$F_1\left(n, \frac{I_{max}}{I_C}\right) = \cos\theta_1 . F\left(\frac{1}{2}, \frac{1-n}{2}, \frac{3}{2}, \cos^2\theta_1\right) \qquad (V.23)$$

If the maximum current I_{max} is greater than critical current, θ_1 tends to 0. In this case we get closer to the loss formula in extreme full penetration proposed by Dresner. This one uses the gamma function $\Gamma(n)$:

$$\int_0^{\frac{\pi}{2}} \sin^{n+1}\theta . d\theta = \frac{\sqrt{\pi}.\Gamma\left(\dfrac{n+2}{2}\right)}{2.\Gamma\left(\dfrac{n+3}{2}\right)} = F_2(n) \qquad (V.24)$$

We represente $F_1(n, I_{max}/I_C)$ versus I_{max}/I_C for different values of n in figure 31. The value n=13 represents an average of values that we obtain with our samples. The other values are only used for a comparison.

We observe that the more "n" is weak, the more the function F_1 is great. Otherwise, the value of this function increases quickly as the ratio I_{max}/I_C increases. Then, as soon as I_{max}/I_C is great enough, the function F_1 reaches its final value that corresponds to the function F_2.

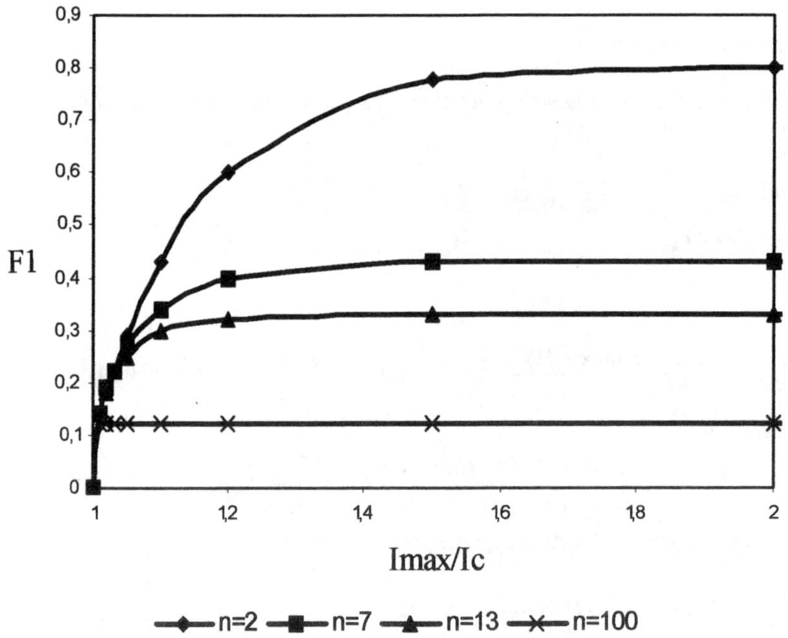

Figure 31: F_1 versus I_{max}/I_C for n=2, 7, 13 and 100

Therefore, we get losses per unit of volume for $I_{max} > I_C$:

$$P_C = h.\left(\frac{\mu_0.f}{I_C}\right)^{\frac{n}{n+1}} .E_C^{\frac{1}{n+1}} .I_{max}^{\frac{3n+1}{n+1}} .\alpha_R.\left(\frac{1}{6\pi} + \frac{0{,}3}{n}\right) +$$

$$\frac{2}{\pi}.f.h.E_C.\frac{1}{I_C^n}\left(\left(I_{max} - \Delta I\right)^{n+11} + \left(I_{max} + \Delta I\right)^{n+1}\right)F_1\left(n, \frac{I_{max}}{I_C}\right)$$

(IV.25)

We represent in figure 32 curves giving losses in complete penetration according to the maximum current, for different values of n, and with a critical current equal 100A. We have chosen here values of n corresponding to the superconducting samples that we have tested.

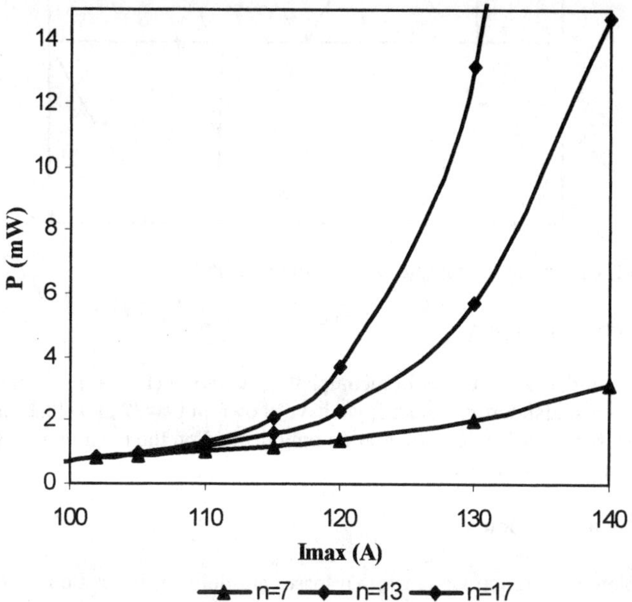

Figure 32: Losses in complete penetration versus the maximum current.

Unlike the case of the incomplete penetration, values of losses vary a lot versus "n". The more "n" is great, the more losses are important and their increase is high. The increase of losses versus I_{max}^{n+1} are little linked to the reduction of $F_1(n)$, as n increase.

5.2.4 Calculation of losses for a non sinusoidal current

Henceforth, we supply our superconducting tube by a non-sinusoidal current. Then, we insert this tube in a circuit supplied by sinusoidal voltage, a resistance R and a dimmer switch. The shape of the current in the superconductor tube is given in figure 33:

474

$i(\theta) = \hat{I}.\sin\theta$ for $\psi + k\pi < \theta < \pi - \psi + k\pi$

$i(\theta) = 0$ for $k\pi < \theta < \psi + k\pi$

k is an integer

ψ is the firing angle of the thyristor.

If $\psi > \pi/2$, in this case, the maximum of the current I_{max} is not equal to $\hat{I}$. That is why we wrote them with different symbols.

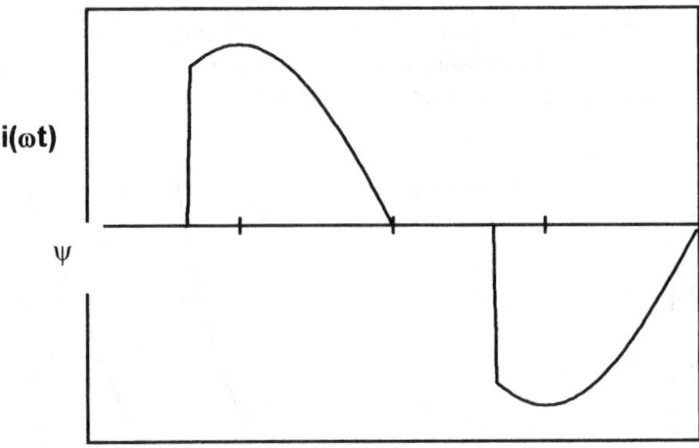

Figure 33: Current in the sample with a dimmer switch.

5.2.4.1 Incomplete penetration

The formula that gives losses in incomplete penetration ($I_{max}<I_C$), is the same as the one given for a sinusoidal current. Indeed, we have shown in part (2..3.4) that for any shape of the current, losses in incomplete penetration only depend on the maximum of the current, and not on the shape of the current.

5.2.4.2 Complete penetration

In complete penetration we recall that losses are defined, in relation to θ_1 and θ_2, as follows:

$$P_C = P_I + 2.f.h.\frac{E_C}{I_C^n}.\left\{\int_{t_1}^{t_2}\left(i(t) - \Delta I\right)^{n+1}.dt + \int_{t_1}^{t_2}\left(i(t) + \Delta I\right)^{n+1}.dt\right\} \qquad (V.26)$$

Now, we are going to calculate losses per unit of volume for $I_{max} > I_C$. According to the value of ψ we get two different cases.

For $\psi < \dfrac{\pi}{2}$:

In this case: $I_{max} = \hat{I}$ and $\theta_2 = \pi - \arcsin\dfrac{I_C}{\hat{I}}$

According to the value of I_{max} and ψ, we have two different values for θ_1.

If $I_{max} \sin \psi < I_C$

In this case $\theta_1 = \arcsin(I_C / I_{max})$ like with sinusoidal current, it is represented in figure 34. Then, we obtain the same relation giving losses as for sinusoidal current.

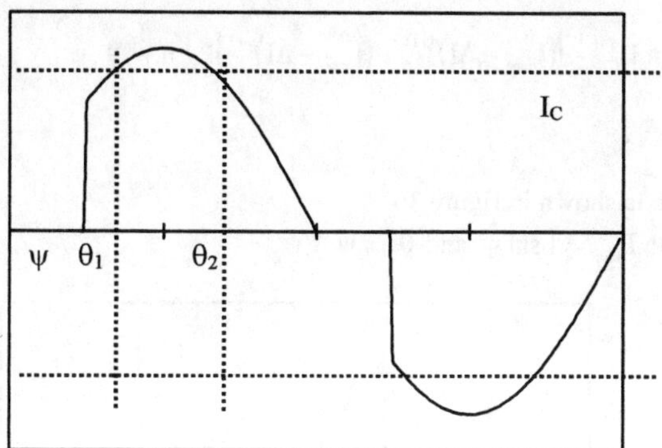

Figure 34: Case where $I_{max} \sin \psi < I_C$ for $\psi < \pi/2$

If $I_{max} \sin \psi > I_C$

So, in this case $\theta_1 = \psi$ as it is shown in figure 35. Then, we get for the integral i^{n+1}:

$$\int_{\psi}^{\theta_2} I_{max}^{n+1} \cdot \sin^{n+1} \theta . d\theta = I_{max}^{n+1}\left(\cos\psi . F(\frac{1}{2},-\frac{n}{2},\frac{3}{2},\cos^2 \psi) - \cos\theta_2 . F(\frac{1}{2},-\frac{n}{2},\frac{3}{2},\cos^2 \theta_2)\right) \quad (V.27)$$

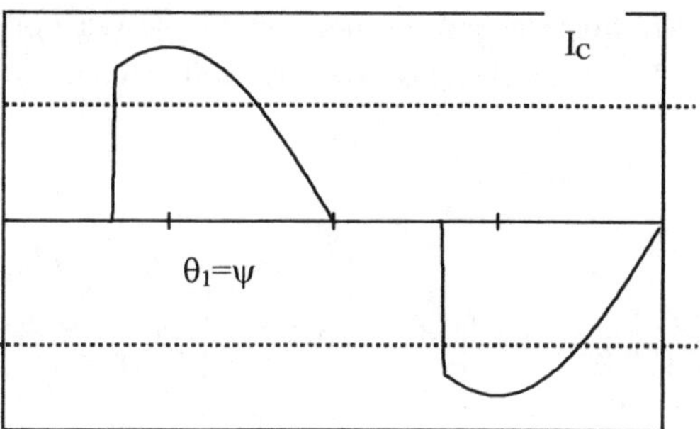

Figure 35: Case where $I_{max} \sin \psi > I_C$ for $\psi < \pi/2$

We notice $F_3(n, \dfrac{I_{max}}{I_C})$ the following function:

$$F_3\left(n, \frac{I_{max}}{I_C}, \psi \right) = \frac{1}{2}\left[\cos\psi . F(\frac{1}{2},-\frac{n}{2},\frac{3}{2},\cos^2 \psi) - \cos\theta_2 . F(\frac{1}{2},-\frac{n}{2},\frac{3}{2},\cos^2 \theta_2)\right] \quad (V.28)$$

Then, we have a relation giving losses per unit of volume, analogous to the one of case sinusoidal:

$$P_C = h.\left(\frac{\mu_0.f}{I_C}\right)^{\frac{n}{n+1}}.E_C^{\frac{1}{n+1}}.I_{max}^{\frac{3n+1}{n+1}}.\alpha_R.\left(\frac{1}{6\pi}+\frac{0,3}{n}\right)+$$

$$\frac{2}{\pi}.f.h.E_C.\frac{1}{I_C^n}\left((I_{max}-\Delta I)^{n+11}+(I_{max}+\Delta I)^{n+1}\right)F_3\left(n,\frac{I_{max}}{I_C},\psi\right)$$

(V.29)

for $\psi > \dfrac{\pi}{2}$:

This case is shown in figure 36:

We have: $I_{max} = \hat{I}.\sin\psi$ and $\theta_1 = \psi \ \forall\psi$.

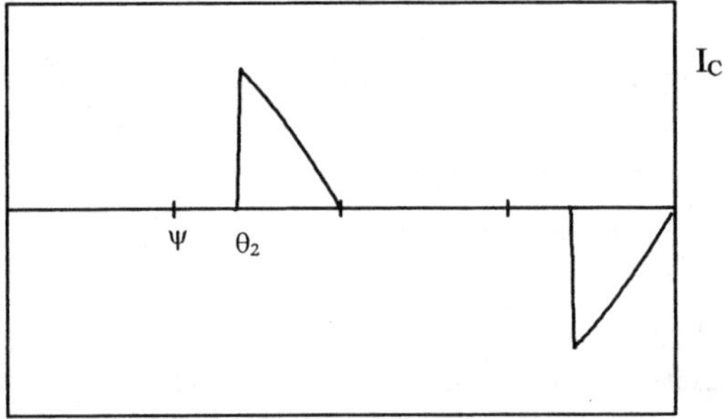

Figure 36: **Case where** $\psi > \pi/2$

    This case differs from the previous one only by the value of the maximum of the current, which is now $\hat{I}^{n+1}$. We deduce losses of this result according to I_{max}:

$$P_C = h.\left(\frac{\mu_0.f}{I_C}\right)^{\frac{n}{n+1}}.E_C^{\frac{1}{n+1}}.I_{max}^{\frac{3n+1}{n+1}}.\alpha_R.\left(\frac{1}{6\pi}+\frac{0,3}{n}\right)+$$

$$\frac{2}{\pi}.f.h.E_C.\frac{1}{I_C^n}\left((I_{max}-\Delta I)^{n+11}+(I_{max}+\Delta I)^{n+1}\right)\frac{F_3\left(n,\frac{I_{max}}{I_C},\psi\right)}{(\sin\psi)^{n+1}}$$

(V.30)

    We have calculated losses using relations previously established in the case of a sample of which critical current Ic was worth 100 A and whose "n" factor was equal to 13.

    These losses are represented in the figure 37, in complete penetration, according to I_{max} for three values of firing angle ψ (60°, 90° and 120°).

    Curves obtain with sinusoidal current and the ones obtain with ψ lower or equal to 60°, are very close. The difference would be more appreciable if the maximum current was really superior to 140 A. For ψ superior to 60°, we observe that losses decrease as the firing angle increases. It is because the interval of time $[t_1, t_2]$ decreases as ψ increases and so, the function $F_3\left(n, I_{max}/I_C\right)$ decreases too.

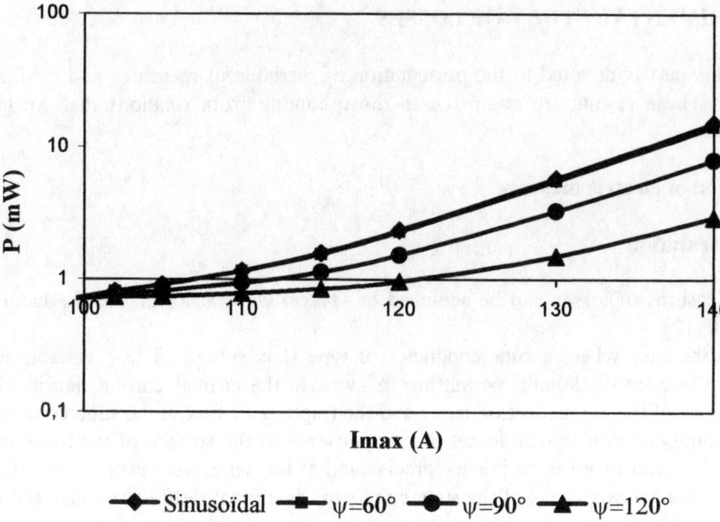

Figure 37: Losses versus maximal current for different values of firing angle, with I_C=100A and n =13.

Therefore, we keep the following relations for losses in self-field for a superconducting tube.

$$P = h.\left(\frac{\mu_0.f}{I_C}\right)^{\frac{n}{n+1}}.E_C^{\frac{1}{n+1}}.I_{max}^{\frac{3n+1}{n+1}}.\alpha_R.\left(\frac{0,2}{\pi}+\frac{0,3}{n}\right)$$

If $I_{max}<I_C$

For any current

$$P = h.(\mu_0.f)^{\frac{n}{n+1}}.E_C^{\frac{1}{n+1}}.I_C^{\frac{2n+1}{n+1}}.\alpha_R.\left(\frac{0,2}{\pi}+\frac{0,3}{n}\right)+$$

$$2.f.h.\frac{E_C}{I_C^n}.\left[\int_{t_1}^{t_2}(i^{n+1}(t)-\Delta I)dt + \int_{t_1}^{t_2}(i^{n+1}(t)-\Delta I)dt\right]$$

If $I_{max}>I_C$

For sinusoidal current

$$P = h.(\mu_0.f)^{\frac{n}{n+1}}.E_C^{\frac{1}{n+1}}.I_C^{\frac{2n+1}{n+1}}.\alpha_R.\left(\frac{0,2}{\pi}+\frac{0,3}{n}\right)+$$

$$\frac{2}{\pi}.f.h.E_C.\frac{1}{I_C^n}\left((I_{max}-\Delta I)^{n+11}+(I_{max}+\Delta I)^{n+1}\right)\cos\theta_1.F\left(\frac{1}{2},\frac{-n}{2},\frac{3}{2},\cos^2\theta_1\right)$$

The function F is a hypergeometric function F(a,b,c,z) and θ_1 is the crossing angle from passage of the incomplete penetration to the complete penetration.

It remains us to verify the validity of our results for the samples that we have used.

6 EXPERIMENTAL STUDY OF LOSSES

This part is devoted to the presentation of methods of measures and results of loss measures. These results are compared to those coming from relations that we have just established.

6.1 Method of electric measure

6.1.1 Presentation

Measures of losses can be achieved of several ways according to the nature of the supply [10, 12]

In the case where a superconductor of type II is submitted to a variable magnetic field, the presence of default, permitting to increase the critical current density, lead the irreversibility of the magnetisation curve and the trapping of flux in the superconductor. The corresponding electromagnetic losses are proportional to the surface of the hysteretic cycle [13-15]. This kind of measure is very precise and at the same time it provides information about the physical properties of the superconductor as critical field values H_{C1} and H_{C2}. On the other hand, it does not permit to measure losses due to transport current.

Calorimetric method consists in measuring the gas quantity given off by the refrigerating liquid because of losses. This measure is global, and one must take into account of powers provided by all sources of warming-up, notably current leads. It is necessary to reach the thermal stability for every measure, which takes times. Besides, the precision is only acceptable for important loss values [10].

Mostly we use the electric method in the case where losses are due to the passage of current in the superconductor and in self-field [16-20]. It is instantaneous and precise, if we bring many cares to the measurement of tension corresponding to losses. Indeed, the voltage is about the microvolt for our sample.

We want to measure losses for any the shape of the current. [21]. Indeed, in the future, these materials would be able to constitute some electric transportation cables. Static converters are likely to be used, the question of the supplementary loss importance linked to the harmonic is important. To generate the harmonic in a very simple way with high current, but meaningful, we opted for a resistive load supplied by a dimmer switch.

6.1.2 Experimental bench

We have measured losses in a current lead, as described in part 3. This current lead, immersed in nitrogen liquid at 77K, is supplied by an alternating current of frequency equal 50Hz. To adjust the voltage, we used an autotransformer. This one supplies a constant resistor. The maximal value of the current is about 110Arms.

We have two possibilities, either the current is sinusoidal, or it is supplied by dimmer switch, whose firing angle of thyristors ψ can be adjusted. The switch K makes the choice between a sinusoidal current and a non-sinusoidal current.

We collect the rms value and the shape of the current with a Hall probe, which provides an associated voltage $u_i(t)$ of current $i(t)$. A numeric integration of $u(t)$ and $i(t)$ leads to the value of the average dissipated power.

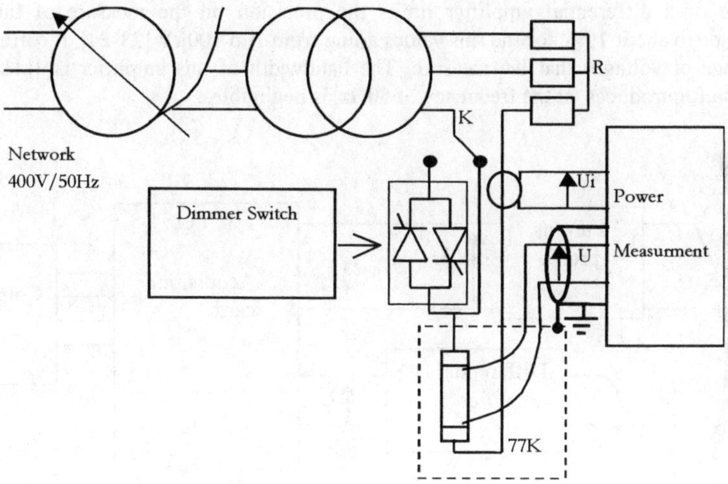

Figure 38: experimental bench

6.1.3 Methods of measures

Two electric measure types are used to determine losses. The most reliable method, when one has a cable transporting sinusoidal current, requires the implementation of a lock-in amplifier. However, when the current is not sinusoidal, this method does not fit.. Therefore, we use another method, efficient for any shape of the current. Now, we are going to expose its main points.

6.1.3.1 Utilisation of differential amplifier

The mean power is calculated by numerical integration of the instantaneous power:

$$P = \frac{1}{T}\int_0^T u(t).i(t).dt$$

The system presented in figure 39 is composed of a differential amplifier of gain G equal 1000, of a current probe, an acquisition card linked to a computer achieving the multiplication and the calculation of the mean value.

Tensions collected at the entry of the acquisition card are $u_{ad}= G.u(t)$ and $u_i(t)=\alpha.i(t)$. These tensions are sampled at a frequency f_e that is worth 10kHz, then, one has the numeric values $i_n(t)$ and $u_n(t)$. We achieve with the help of the computer the multiplication and the numeric integration of $p(t)$ to obtain losses:

$$P = \frac{1}{N}.\sum_{k=1}^{N} i_k.u_k \text{ with } NT_e=T \text{ period of the alternative signal and } T_e \text{ period of sampling.}$$

For our measures we have T=20ms and N=200. Here, the main problem is the precision of the measure of u(t) when its rms value reaches only some microvolts. Indeed, the use of a differential amplifier limits the precision on the measure of the voltage amplitude to about 10%, for the rms values going from 1 to 100µV [23-24]. It corresponds to the range of voltages that we measure. The bandwidth of this amplifier is 1kHz and the phase shift introduces, at the frequency of 50Hz, is negligible.

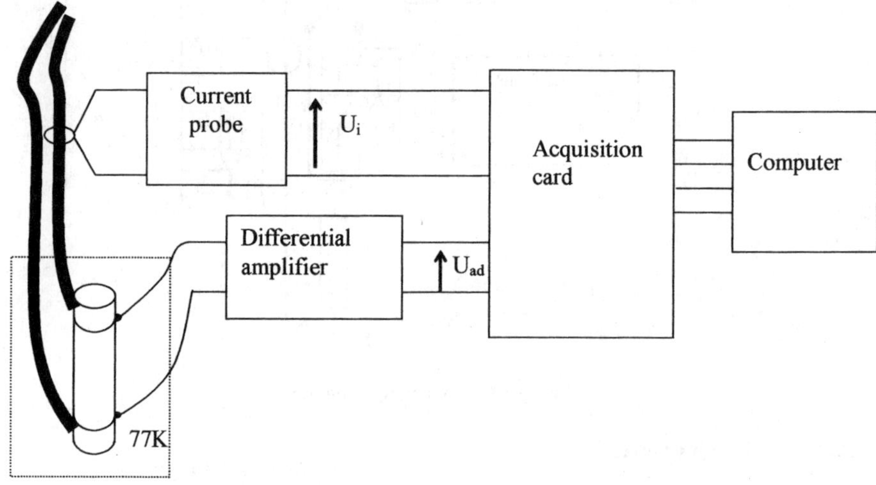

Figure 39: Measure of losses with differential amplifier

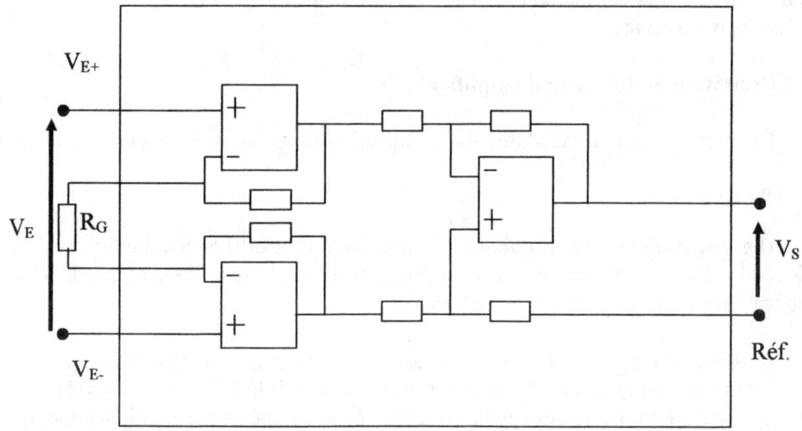

Figure 40: Differential amplifier

The used amplifier is the circuit INA114 drawn in figure 40. Its gain is defined as follows:

$$G = \frac{V_S}{V_E} = 1 + \frac{50000}{R_G}$$

With R_G equal $50,0\Omega$ we obtain G equal to 1001 ± 2. From now on, one takes G=1000. When we measure voltage of a few tens microvolts, we must pay a lot of attention to two points. The first one concerns the place where is located the voltage pick off. We choose to fix them on the internal side of silver ribbons to limit the influence of their losses the measurement. The second point consists in limiting the noise in all the system of measure. For this, it is necessary to screen and to limit the length of cables between the voltage pick off and the amplifier. The cryostat of nitrogen and the screen are joined to the ground by a link made in star.

The phase shift between the current and tension, at the exit of the acquisition card, must be as close as possible to the real phase shift. The situation becomes critical for the calculation of the mean value when this phase shift is close to $\pi/2$. The phase shift between the real current and tension at the exit of the probe $u_{sc}(t)$ is not important for a frequency of 50 Hz.

On the following figures, we have represented curves of currents, tensions and instantaneous powers.

For I_{max} equals 60A we notice that the amplitude of tension is only worth some tens microvolts and therefore the precision is weak. After the multiplication we obtain an instantaneous power, which amplitude of the alternative part is big compared to the mean value. That is implies an imprecision on the mean value.
For I_{max} equals 120A, it is the contrary, the amplitude of tension approaches the millivolt and the mean value of the power is about 10mW. In this case the precision is quite satisfactory.

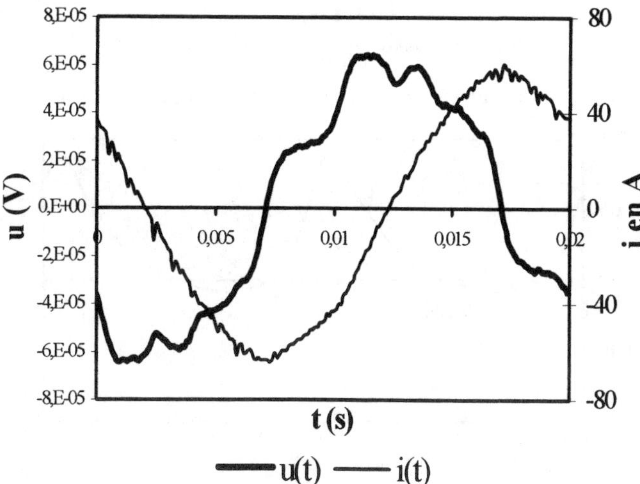

Figure 41: Tension and current of the sample with a maximum current of 60A

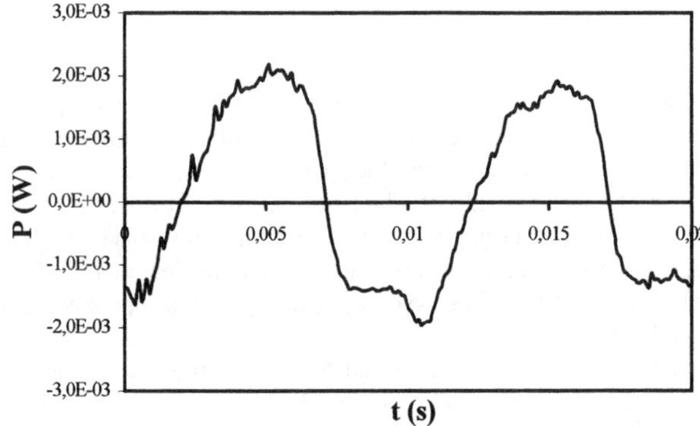

Figure 42: Instantaneous power of the sample with a maximum current of 60A

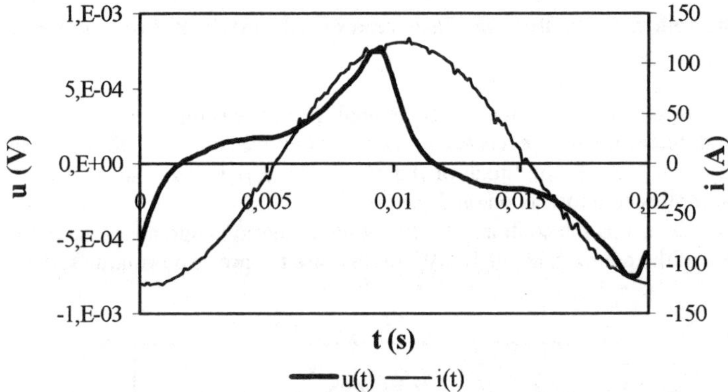

Figure 43: Tension and current of the sample with a maximum current of 120A

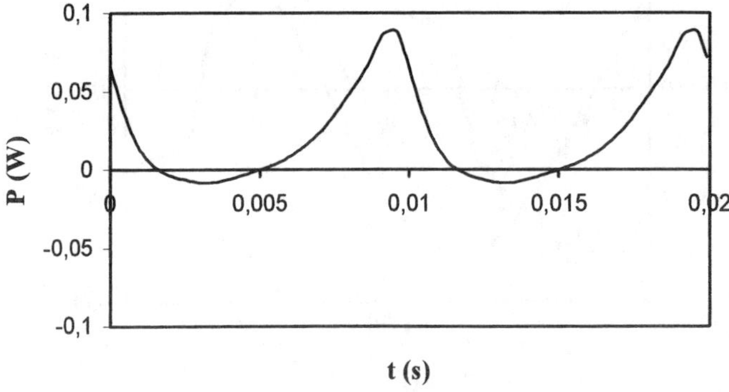

Figure 44: Instantaneous power of the sample with a maximum current of 120A

6.1.3.2 Compensation of the reactive voltage

It is often recommended to compensate the reactive tension to increase the precision of the measure.

The principle of the compensation is the following one. One considers that voltage u(t) to voltage pick off of the coil or the superconducting sample is the sum of an ohmic voltage drop $u_p(t)$, which corresponds to losses and to an inductive voltage drop:

$$u(t) = u_p(t) + L.\frac{di}{dt}$$ With the inductance L of the sample.

One compensates the reactive part Ldi/dt by one induced tension Mdi/dt to current lead of a coil placed near the cable transporting the current i(t). The position of the coil must be as follows:

$$L.\frac{di}{dt} = -M.\frac{di}{dt}$$

Therefore, the voltage is theoretically equal to tension due to losses. One uses the compensation because the reactive part of tension is in general very high compared to tension representing losses. One increases the precision of the loss measure.

In the case of a coil, whose the inductance L can be known roughly, and whose inductive voltage is high, the method of compensation is justified. In the case of a simple superconducting conductor, the theoretical justification of the method of the compensation of the reactive tension is more difficult. Indeed, the inductance of a superconducting wire used with a periodic current, and submitted to its self-field is very weak. Nevertheless, this method of compensation is also used in the case of a simple wire. To implement the compensation we have achieved the following installation (figure 45).

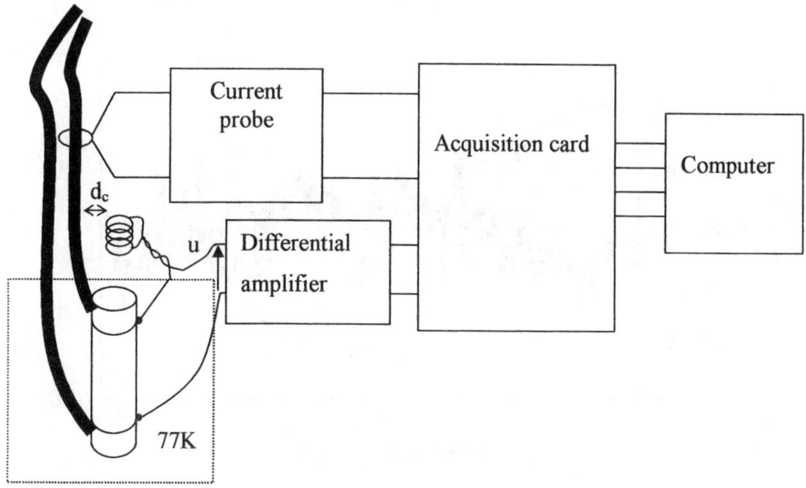

Figure 45: Measure of losses with compensation of the reactive part of tension

The coil of compensation, constituted of some turns, is put in series with the sample. We approach these turns of the cable of the supply of the sample to a distance d_c. When this distance varies, the rms voltage of the amplifier's output changes. The compensation is obtained when we have a minimum of the rms voltage. It corresponds to an instantaneous power (u(t).i(t)) solely positive that must be equal to the instantaneous power of losses.

In figures 46 to 53 we have represented voltages and currents without compensation for different maximum current values.

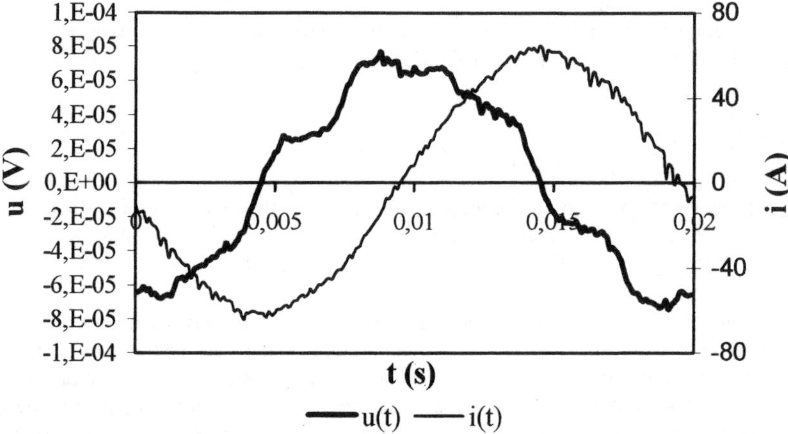

Figure 46: Voltage and current without compensation for a maximum current of 60 A.

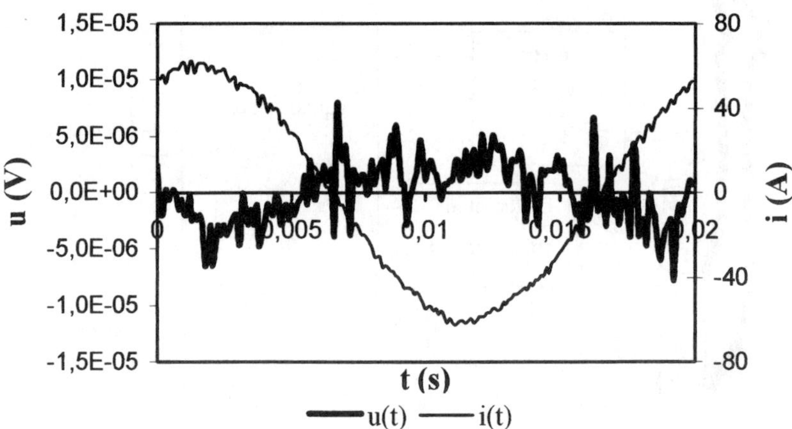

Figure 47: Voltage and current without compensation for a maximum current of 60 A.

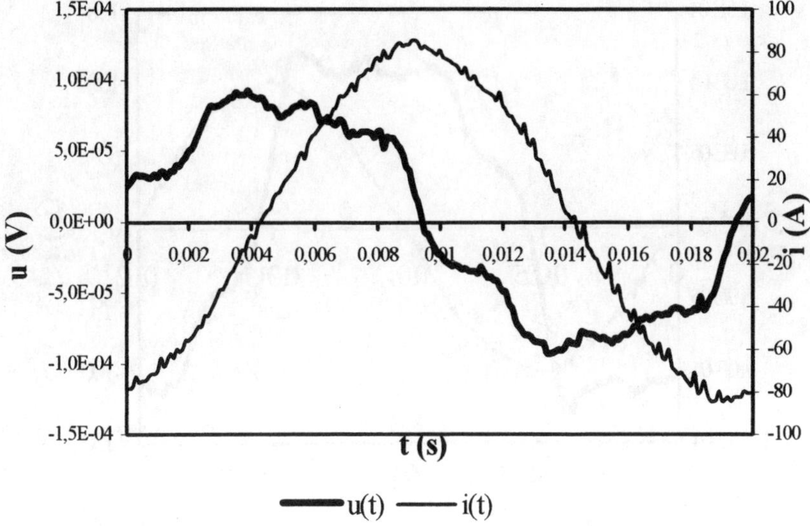

Figure 48: Voltage and current without compensation for a maximum current of 80 A.

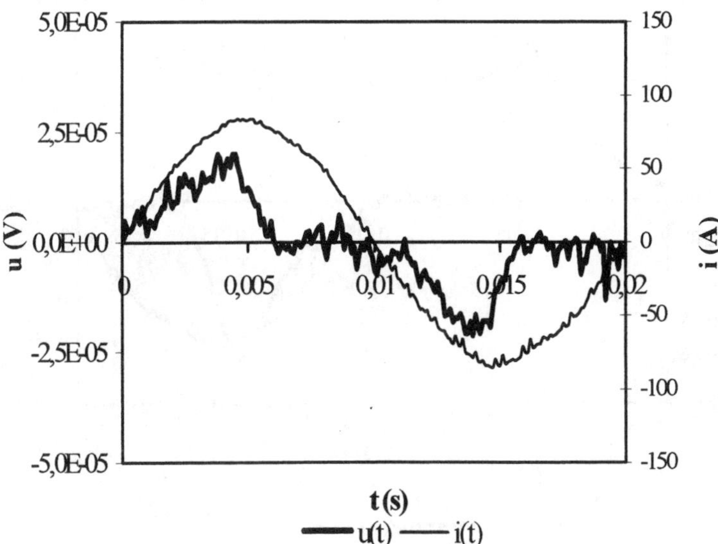

Figure 49: Voltage and current without compensation for a maximum current of 80 A.

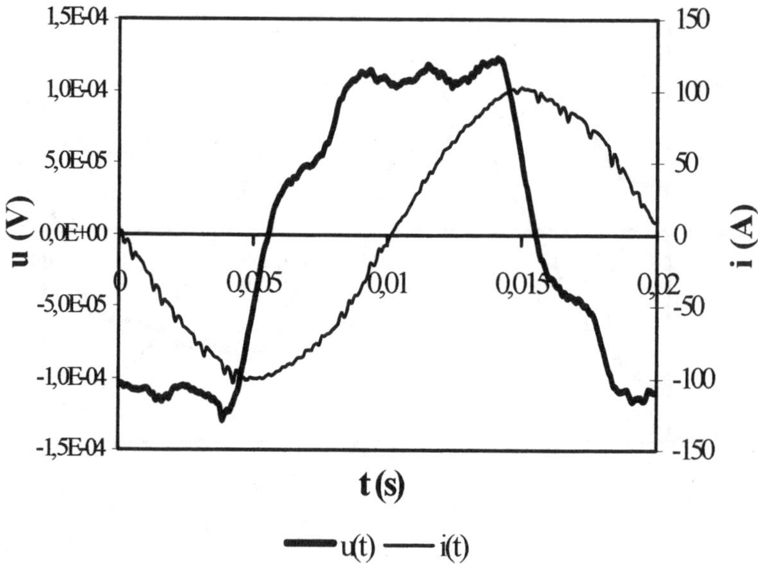

Figure 50: Voltage and current without compensation for a maximum current of 100 A.

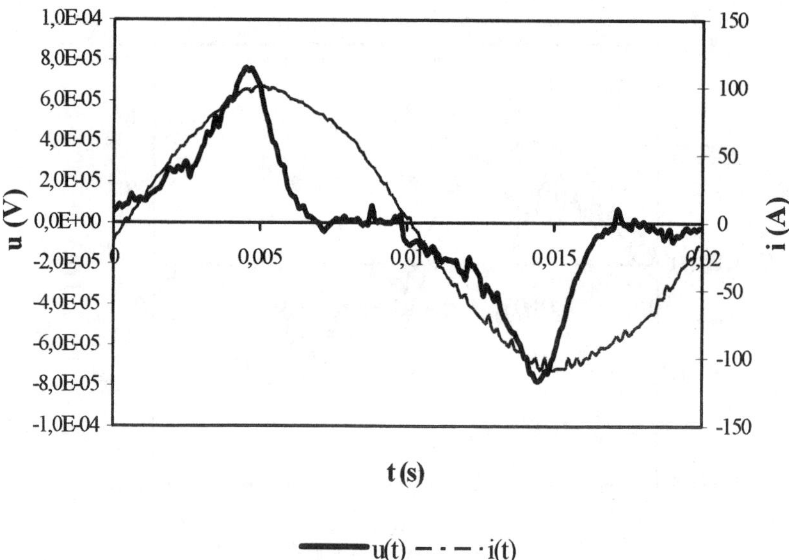

Figure 51: Voltage and current without compensation for a maximum current of 100 A.

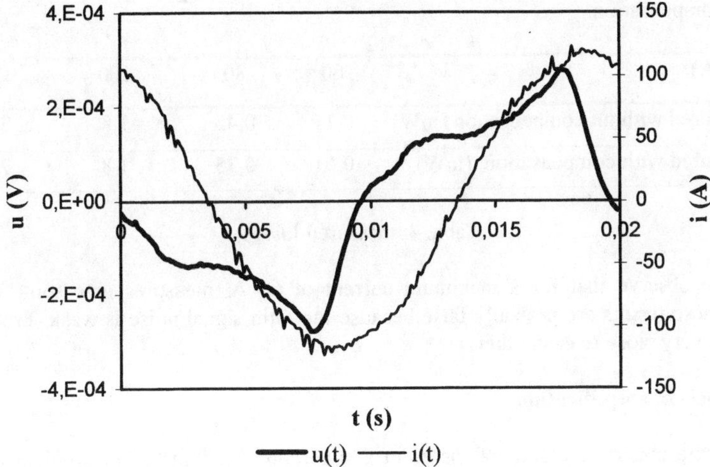

Figure 52: Voltage and current without compensation for a maximum current of 120 A.

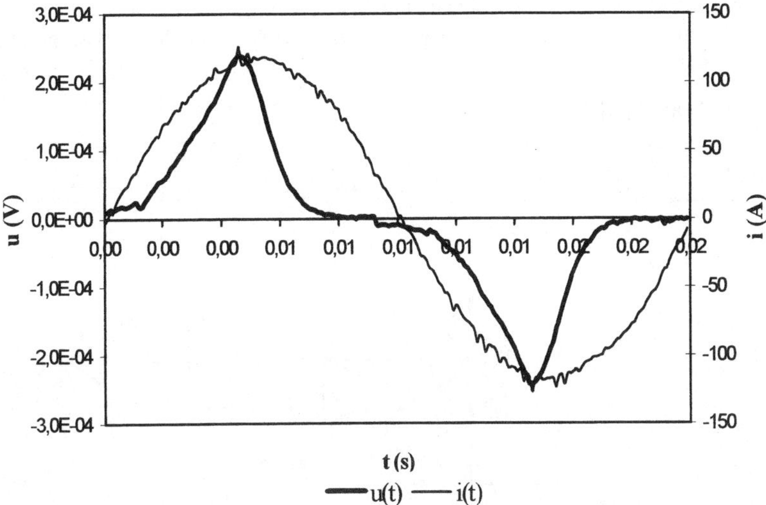

Figure 53: Voltage and current without compensation for a maximum current of 120 A.

When the maximum current is worth 60 A, the curve of voltage obtained, with compensation, is visibly in the noise. In this case, the measure of losses, got with the method of measure exposed previously, does not give conclusive results. In case where the maximum current passes 80 A, the part of tension corresponding to losses becomes important compared to the noise. We compare results losses measurement got with and without compensation.

Imax (A)	60	80	100	120
P measured without compensation (mW)	0,15	0,45	1,8	7,2
P measured with compensation (mW)	0,01	0,35	1,8	7,3

Table 4: measured losses

We observe that for a maximum current of 60 A, measures give some different results. These results are probably false because the ratio signal/noise is weak. From 80 A, results are very close to each other.

6.1.3.3 Lock-in amplification

In the electric measure of losses in a superconductor, the main problem is in the precision of the voltage measure, notably after compensation. Indeed, it can have amplitude comparable or inferior to the present noise, especially when the losses values are weak. It is the case if the maximum current is much lower than critical current. With the previous method, one amplifies the useful signal, but also the noise. Ratio signal/noise remains constant. To increase this ration, we need to use a lock-in amplifier. The principle is presented in figure 54.

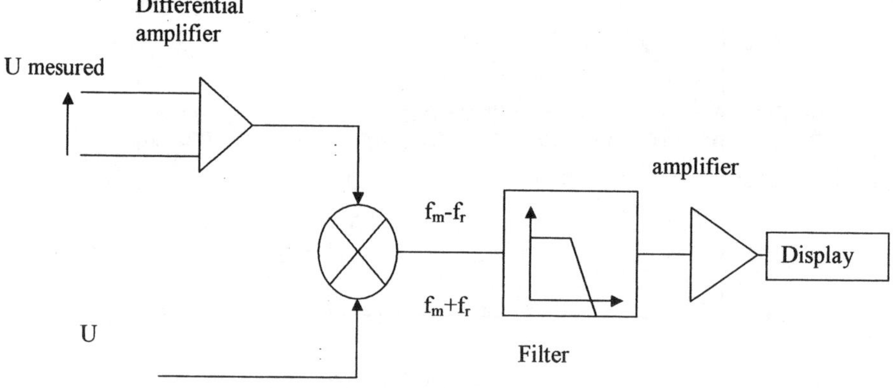

Figure 54: lock-in amplifier

We have the measured signal:
$$u_m(t) = U_m . \sqrt{2} . \sin(\omega_m . t - \phi)$$

We multiply this signal by a reference signal $u_{réf}(t) = U_{réf} . \sqrt{2} . \sin \omega_r t$, so, we obtain:

$$U_m.U_{réf}\left[\cos\left[(\omega_m - \omega_r).t - \phi\right] - \cos\left[(\omega_m + \omega_r).t - \phi\right]\right]$$

This signal is filtered through a low-pass filter of very small cutoff frequency, and we get:

$$U_m.U_{réf}.\cos\phi$$

The lock-in amplifier gives the voltage $U_m.\cos\Phi$.

This value is proportional to the rms value U_m of the signal, and to the cosine of the phase shift of the signal according to the signal of reference. If the measured signal does not have the same frequency as the signal of reference, the mean value in output becomes negligible. In most cases, the measured tension includes harmonics. Then, only the harmonic in the same frequency as the reference has a non-zero output voltage value.

Therefore if:

$$u_m(t) = \sum_{n=1}^{\infty} U_n.\sqrt{2}.\sin(n.\omega.t - \phi_n)$$

and

$$u_{réf}(t) = U_{réf}.\sqrt{2}.\sin \omega.t$$

we obtain:

$$U_1.U_{réf}.\cos \phi_1$$

If the signal measured with the voltmeter is noisy, only the noise of equal frequency to the frequency of the useful signal is going to subsist. The other noises will be strongly attenuated; it is as much truer than the cutoff frequency of the filter is low. Therefore, one has a very important increase of the ratio signal/noise.

Therefore, measures of losses are possible for the very weak values of voltage ($<10\mu V$). If the signal of reference is in phase with the current in the sample and if the current is merely sinusoidal, then the active power P dissipated by the sample is equal to:

$$P = I.U_1.\cos\phi_1$$

To increase the precision of measures we compensate the reactive part of tension as indicated in figure 55.

The only flaw of this method is that it only works if the current, therefore the signal of reference, is sinusoidal. For non sinusoidal current, it is impossible to use it. In this case, we shall use the differential amplifier associated with an acquisition card.

6.2 Results of measurement

6.2.1 Measures of losses with sinusoidal current

We present here results of measured losses for different samples. We compare results given by the two methods of measurement for a sinusoidal current. Then, we measure losses in the case of a non-sinusoidal current.

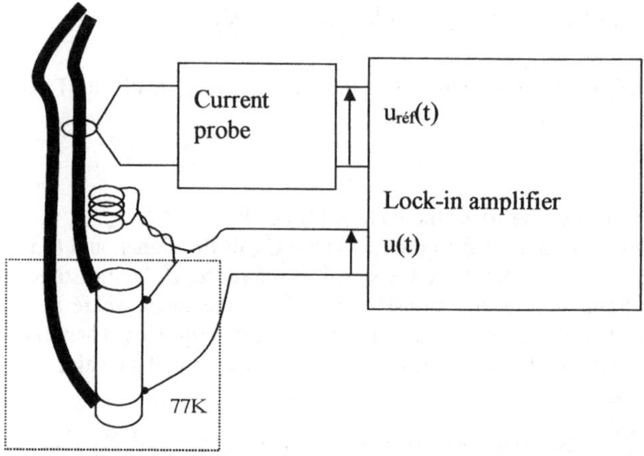

Figure 55: Measure of losses with lock-in amplifier

We have seen that the different samples present differents characteristics E(I). We compare measured losses according to the maximum of the current applied for every sample to know if there is a concordance between the characteristic I_C and "n". We present here the previously measured characteristics:

$E_C=1\mu V/Cm$	Sample n°1	Sample n°2	Sample n°3
I_C	108	102	105
n	15	13	15

Table 5: Feature of samples

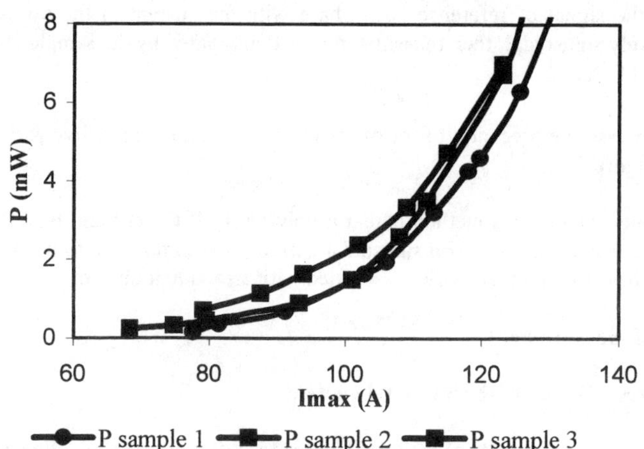

Figure 56: Comparisons of losses for three samples

We note that losses are more important when the critical current is weak, which is compliant to the theoretical forecast.

These curves show us that the influence of the value of "n" on losses seems weak, notably because of the low difference between values of "n" according to samples.

6.2.2 Measures of losses with non sinusoidal current

We were interested in losses in a high critical temperature superconductor supplied by a non-sinusoidal current. The shape of the current had to correspond to a real case, but also had to be easily feasible. The use of a dimmer switch provides an interesting solution. Therefore, we measure losses according to the maximum current for several values of firing angle ψ.

We notice in figure 57 that curves of losses for a sample are very near, when the maximum current is inferior or slightly superior to the critical current. We observe a small difference when the maximum current reaches values greater than 120% of critical current which is indicated by the manufacturer. For these values of maximal current and for all samples, losses seem to decrease when the firing angle increases.

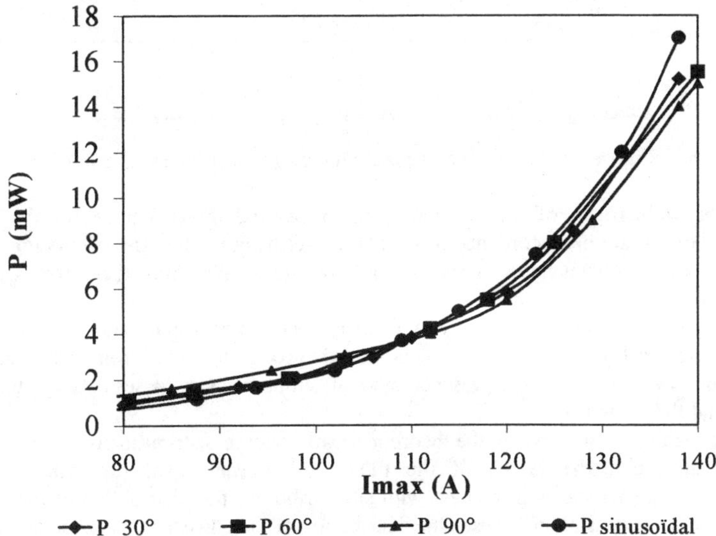

Figure 57: Measured losses of the sample n°2

6.3 Comparison between calculations and measures

We have compared measured losses of the three samples to the calculated losses. We can observe the influence of the features of every sample and we can validate our calculations of losses [22].

Losses increase very quickly when the maximum of the current passes the critical current. We have represented, in figure 58, for a sample, the measured losses curves, calculated with the help of a numeric resolution, and valued from the previously established analytic expressions.

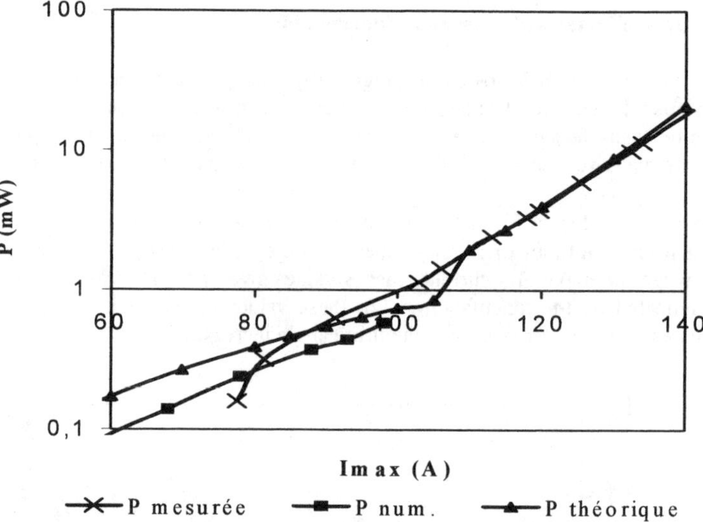

Figure 58: Losses versus maximum current for the sample n°1

The comparison of the calculated and measured losses curves for the different samples shows a good concordance of results. Nevertheless, when the maximum current is weak, curves have some quite important relative gaps. We think that there are several explanations.

Firstly, the measures are not very precise when the maximum current is weak that is to say for losses lower than 2 mW. This lack of precision comes from the measurement system and likely also from the presence of weak supplementary losses between the voltage pick up and the silver ribbon.

Secondly, we suppose, in the theoretical part, that the distribution of the current is the same as all along the length of the tube. At extremities of the tube, where the superconducting material is in contact with silver ribbons, the distribution of the current is different. It is not possible to foresee the distribution of the current in this zone and therefore it can explain the differences between theory and experiment.

Formulas that we have established give a quite precise value of losses for I greater than 80A. Nevertheless, for strong currents formulas of losses that we have established describe the evolution of losses correctly.

7 CONCLUSION

We chose a "$E=E_c(J/J_c)^n$" model instead of Bean's model because this model suits better to the experiments. Moreover, It enables to impose a current greater than the critical current.

A previous characterisation of samples, with direct current, is indispensable to the knowledge of characteristic E=f(J), that is to say for the determination of the value of "n" and of the critical current.

We have established new analytic relations permitting to calculate losses in self field in complete penetration from the knowledge of the characteristic E=f(J) of the superconductors material.

We have measured losses of a high critical temperature superconducting sample fed by sinusoidal or non-sinusoidal current. We have shown that harmonic components of current do not have much influence on the value of losses. This last study is important for the industrial applications of these conductors.

8 REFERENCES

[1]   C. P. Bean, "Magnetization of high-field superconductors", *Reviews of modern physics*, pp.31-39, January 1964.

[2]   M. N. Wilson "Superconducting Magnets", *Oxford Science Publications*, 1983.

[3]   J. Carr, "AC loss and macroscopic theory of superconductors", *Gordon and breach science publishers*, 1983.

[4]   W. T. Norris, "Calculation of hysteresis losses in hard superconductors carrying ac: conductors and edges of thin sheets", *Journal of Physics D (Applied Physics)*, Volume 3, Issue 4, Pages 489-507, 1970.

[5]   CAN Superconductors, Ltd, Praha, Czech Republic

[6]   N. Amemiya, K. Miyamoto, N. Banno, O. Tsukamoto, "Numerical analysis of AC losses in High T_C superconductors based on E-j characteristics represented with n-value", *IEEE transactions on applied superconductivity*, vol. 7, n° 2, pp.2110-2113, 1997.

[7]   J. Lehtonen, A. Korpela, J. Paasi, J Pitel, P. Kovac "Computational comparison of magnetization losses in HTS solenoids wound of tape conductors having different aspect ratios", -*Supercond. Sci. Tech.*, 12, pp.450-455, 1999.

[7]   R.L. Stoll, "The analysis of eddy currents", *Clarendon Press*, Oxford, 1974.

[8]   L. Dresner, "Hysteresis losses in power-law cryoconductors. Part II", *Applied superconductivity*, vol. 4, n° 7/8, pp. 337-341, 1996.

[9]   L. Dresner, "Hysteresis losses in power-law cryoconductors.", *Applied superconductivity*, vol. 4, n° 4, pp. 167-172, 1996.

[10]   J.W. Lue, J.A. Demko, L. Dresner., R.L. Hugey, U. Sinha , JC Tolbert., S.K.

Olsen,. "AC losses of prototype HTS transmission cables", *I.E.E.E. transactions on applied superconductivity*, vol. 9, n° 2, pp. 416-419, 1999.

[11]   I. Mayergoyz, "Non linear diffusion of electromagnetic fields", *Academic press*, 1998.

[12]   E. Beghin, J. Bock, G. Duperray, D. Legat, P.F. Herrmann, "AC loss measurements in high T_C superconductors", *Applied superconductivity*, vol. 3, n°6, pp. 339-349, 1995.

[13]   P. Vanderbemden, "Determination of critical current in bulk high temperature superconductors by magnetic flux profile measuring methods", Thèse, Université de Liège, Faculté des sciences appliquées, 1999.

[14]   H. Isshi, S. Hirano, T. Hara, T. Yamamoto, "Estimation of AC losses of polycrystalline YBCO by two different methods", *I.E.E.E. Transactions on applied superconductivity*, vol. 3, n°1, mars 1993, pp. 1417-1420.

[15]   J. Paasi, A. Tuohimaa, "Hysteresis losses in Bi-2223 superconductors", *I.E.E.E. trans. on magnetics*, vol.32, n°4, pp. 2796-2800, juillet 1996.

[16]   J. Paasi, M. Polak, P. Kottman, D. Suchon, M. Lahtinen, J. Kokavec, "Electric field and losses in BSCCO-2223/Ag tapes carrying Ac transport current", *I.E.E.E. Transactions on applied superconductivity*, vol. 5, n°2,pp. 713-716, June 1995.

[17]   H. Isshi, S. Hirano, T Hara, T. Yamamoto, "Estimation of AC losses of polycrystalline YBCO by two different methods", *I.E.E.E. Transactions on applied superconductivity*, vol. 3, n°1, pp. 1417-1420, mars 1993.

[18]   H. Daffix, P. Tixador, "Electrical ac losses measurements in superconducting coils", *IEEE Transactions on applied superconductivity*, vol.7, n°2, pp. 286-289, June 1997.

[19] J. Kokavec, I Hlasnik, S. Fukui, "Very sensitive electric method for AC loss measurement in SC coils", *I.E.E.E. Transactions on applied superconducting*, vol. 3, n°1, pp.153-155., mars 1993

[20]   J.P. Ozelis, "AC loss measurement of model and full size 50mm SSC collider dipole magnets at Fermilab", *I.E.E.E. Transactions on applied superconducting*, vol.3, n°1, pp.678-682, mars 1993.

[21]   B. Douine, J. Lévêque, A. Rezzoug, "AC losses measurements of a high critical superconductor transporting sinusoidal or non sinusoidal current", *I.E.E.E. transactions on applied superconductivity*, vol. 10, n°1, pp.1489-1492, mars 2000.

[22]   B. Douine, D. Netter, J Lévêque, A. Rezzoug, "AC losses in a BSCCO Current lead: comparison between calculation and measurement", *IEEE Transaction on AS*, Vol 12, N°1, March 2002.

9 NOMENCLATURE

a	[m]	half-thickness of a superconducting plate
B	[T]	magnetic induction
B_0	[T]	magnetic induction on the surface
B_{max}	[T]	maximal value of the magnetic induction
B_p	[T]	magnetic induction in complete penetration
$c(t)$	[m]	instantaneous penetration of electrical field in a cylinder or a hollow tube
D	[m]	diameter of the current lead
$d(t)$	[m]	instantaneous penetration of electrical field in a superconducting plate
d_c	[m]	distance between compensation coil and the alimentation wire
E	[V/m]	electrical field
E_C	[V/m]	critical electrical field
f	[Hz]	frequency
f_m	[Hz]	frequency of the measured voltage
f_r	[Hz]	frequency of the reference voltage
$F(a,b,c,z)$		hypergeometric function
H	[A/m]	magnetic field
H_a	[A/m]	applied magnetic field
H_0	[A/m]	magnetic field of a face of the superconductor
H_{max}	[A/m]	maximal value of magnetic field
$i(t)$	[A]	instantaneous current
I	[A]	RMS current
I_C	[A]	critical current
I_{max}	[A]	maximal value of the current
J	[A/m^2]	current density
J_C	[A/m^2]	critical current density
L	[H]	inductance of the sample
l_{pp}	[m]	distance between the voltage pick off
h	[m]	length of the current lead
M	[H]	mutual inductance
n		power of J in relation E(J) for superconductor
p	[m]	depth of penetration of magnetic induction in a superconducting plate
P	[W]	losses
P_{BC}	[W]	losses in a cylinder calculate thanks to Bean's model
P_{BT}	[W]	losses in a hollow cylinder calculate thanks to Bean's model
P_{BTa}	[W]	losses in a tube, calculate thanks to an approximate formula
P_C	[W/m^3]	losses in complete penetration in a hollow cylinder
P_{Cp}	[W/m^3]	partial losses in complete penetration in a hollow cylinder
P_h	[W/m]	losses by unit length
P_I	[W/m^3]	losses in incomplete penetration in a hollow cylinder
$p_V(t)$	[W/m^3]	instantaneous electromagnetic power by unit length
P_V	[W/m^3]	losses by unit volume

$P_{VC}(t)$ [W/m^3]   instantaneous electromagnetic power by unit volume in complete penetration

P_{VCp} [W/m3]   partial electromagnetic power by unit volume in complete penetration

$P_{VCp}(t)$ [W/m3]   instantaneous partial electromagnetic power by unit volume in complete penetration

R [m]   radius of a cylinder

R_{in} [m]   inner radius of a hollow cylinder

R_e [m]   outer radius of a hollow cylinder

r_S [m]   radius of penetration of current density

S [m^2]   section of a cylinder or a tube

U_m [V]   RMS measured voltage

$U_{réf}$ [V]   RMS reference voltage

$x_0(t)$ [m]   depth of penetration of magnetic induction

α_R shape of a tube $= 1 - R_i^2 / R_e^2$

β ratio B_{max}/B_p

$\Gamma(\beta)$ factor of losses

Φ [rad]   shift phase between measured voltage and reference voltage

μ_0 [H/m]   magnetic permeability of vacuum

ω [rad/s]   pulsation

ω_m [rad/s]   pulsation of the measured voltage

ω_r [rad/s]   pulsation of the reference voltage

SUBJECT INDEX

Printing: Mercedes-Druck, Berlin
Binding: Stein + Lehmann, Berlin